J. Fernandes · J.-M. Saudubray · G. Van den Berghe (Eds.): **Inborn Metabolic Diseases**

Dear MyCopy Customer,

This Springer book is a monochrome print version of the eBook to which your library gives you access via SpringerLink. It is available to you at a subsidized price since your library subscribes to at least one Springer eBook subject collection.

Please note that MyCopy books are only offered to library patrons with access to at least one Springer eBook subject collection. MyCopy books are strictly for individual use only.

You may cite this book by referencing the bibliographic data and/or the DOI (Digital Object Identifier) found in the front matter. This book is an exact but monochrome copy of the print version of the eBook on SpringerLink.

Springer-Verlag Berlin Heidelberg GmbH

J. Fernandes  J.-M. Saudubray  G. Van den Berghe (Eds.)

# Inborn Metabolic Diseases

**Diagnosis and Treatment**

3rd, Revised Edition

With 66 Figures and 57 Tables

 Springer

Dr. JOHN FERNANDES
Professor Emeritus of Pediatrics
Veldweg 87
8051 NP Hattem, The Netherlands

Professor JEAN-MARIE SAUDUBRAY
Hospital Necker-Enfants Malades
149 Rue de Sevres
75743 Paris Cedex 15, France

Professor GEORGES VAN DEN BERGHE
Christian de Duve
Institute of Cellular Pathology
Université Catholique de Louvain
Avenue Hippocrate 75
1200 Brussels, Belgium

---

Cover Illustration: Putti by della Robbia at Spedale degli Innocenti, Firenze, Italy

---

Library of Congress Cataloging in Publication Data applied for.
Die Deutsche Bibliothek – CIP-Einheitsaufnahme
**Inborn metabolic diseases:** diagnosis and treatment/J. Fernandes... – 3., rev. ed.

DOI 10.1007/978-3-662-04285-4

This work is subject to copyright. All rights are reserved, whether the whole or part of the material is concerned, specifically the rights of translation, reprinting, reuse of illustrations, recitation, broadcasting, reproduction on microfilm or in any other way, and storage in data banks. Duplication of this publication or parts thereof is permitted only under the provisions of the German Copyright Law of September 9, 1965, in its current version, and permission for use must always be obtained from Springer-Verlag Berlin Heidelberg GmbH. Violations are liable for prosecution under the German Copyright Law.

© Springer-Verlag Berlin Heidelberg 2000
Originally published by Springer-Verlag Berlin Heidelberg New York in 2000
MyCopy version of the original edition 2000

The use of general descriptive names, registered names, trademarks, etc. in this publication does not imply, even in the absence of a specific statement, that such names are exempt from the relevant protective laws and regulations and therefore free for general use.

*Product liability*: The publishers cannot guarantee the accuracy of any information about dosage and application contained in this book. In every individual case the user must check such information by consulting other relevant literature.

Cover Design: Springer-Verlag, E. Kirchner
Production: Pro Edit GmbH, 69126 Heidelberg, Germany
Typesetting: Scientific Publishing Services (P) Ltd, Madras

Printed on acid-free paper    SPIN: 10708260       22/3134/göh - 5 4 3 2 1 0
www.springer.com/mycopy

# Preface

Five years have passed since the second edition of *Inborn Metabolic Diseases; Diagnosis and Treatment* was published. The third edition, now being presented, has been thoroughly updated and revised. Again, the clinical presentation, the methods to arrive at the diagnosis and the treatment of the patient have remained the focus of the book, that for a large part has been written by clinicians for clinicians. The scope of the readership has enlarged: from the original teams of pediatricians, biochemists and dieticians it now also encompasses neurologists, internists, geneticists and psychosocial workers. This reflects the fact that for many inborn metabolic diseases the survival of the patients and their quality of life have improved. A combination of facts has contributed to this: above all refined methods for biochemical monitoring of the condition of the patients, and treatments tailored to various stages of the disease and ages of the patients. Furthermore, refined molecular genetic investigation has uncovered not only a great heterogeneity of most metabolic disorders, but also contributed to genetic counselling.

The book can be used in two main situations. When the diagnosis is already suspected, the reader can go directly to the relevant chapter, which is always presented in the design common to all chapters. However, when the diagnosis is unknown, the physician can first refer to Chapter 1, which presents symptoms according to four main entries: symptoms in the newborn infant, acute symptoms of later onset and recurrent attacks, chronic and general progressive symptoms, or symptoms related to an organ (system). Each symptom can be found in the index list of the chapter, which refers either to the text, a table, a figure, an algorithm or a list of disorders, or a combination of these. From this introductory information it is easy to proceed to a relevant chapter in order to obtain more specific information.

With respect to the contents of the book, most chapters have been rewritten or extensively revised. A few chapters from the second edition have been deleted, since their contents are discussed in other chapters of the present edition. A few new chapters have been introduced, such as a general chapter on treatment, which gives a comprehensive list of present treatments and new trends, a chapter on persistent hyperinsulinemic hypoglycemia, a combined chapter on disorders of ketogenesis and ketolysis, a chapter on disorders of proline and serine metabolism, a chapter on disorders of cholesterol synthesis, and a chapter on defective leukotriene synthesis. For more detailed information, particularly with respect to pathophysiology and genetics, we recommend the eighth edition of the *Molecular and Metabolic Bases of Inherited Disease*, edited by Charles R. Scriver et al (Mc Graw-Hill, 2000).

In view of the rapidly expanding knowledge of clinical, metabolic and genetic variability and their influence on prognosis and treatment, it is important to be aware of the existence of networks of experts and laboratories specialized in the diagnosis and treatment of inherited metabolic disorders. For countries of the European Union such a list is compiled by the Society for the Study of Inborn Errors of Metabolism (SSIEM), for the United States and Canada, Japan and Australia by the American, Japanese and Australian Societies of Inherited Metabolic Diseases (SIMD, JIMD, and AIMD, respectively).

The editors welcome new authors of old and new chapters and pay tribute to the authors who, though not participating this time, laid the framework for this book.

Spring 2000

John Fernandes
Jean-Marie Saudubray
Georges van den Berghe

# Contents

**Part I**
**Diagnosis and Treatment:**
**General Principles**

1. Clinical Approach
to Inherited Metabolic Diseases
J.-M. Saudubray, H. Ogier de Baulny,
and C. Charpentier ..................... 3

2. Diagnostic Procedures:
Function Tests and Postmortem Protocol
J. Fernandes, J.-M. Saudubray, and J. Huber ..... 43

3. Emergency Treatments
H. Ogier de Baulny and J.-M. Saudubray ....... 53

4. Psychosocial Care of the Child and Family
J.C. Harris ............................ 63

5. Treatment: Present Status and New Trends
J.H. Walter and J.E. Wraith ................ 75

**Part II**
**Disorders of Carbohydrate Metabolism**

6. The Glycogen-Storage Diseases
J. Fernandes and G.P.A. Smit ............... 87

7. Disorders of Galactose Metabolism
R. Gitzelmann ......................... 103

8. Disorders of Fructose Metabolism
G. van den Berghe ...................... 111

9. Persistent Hyperinsulinemic Hypoglycemia
P. de Lonlay and J.-M. Saudubray ............ 117

**Part III**
**Disorders of Mitochondrial Energy**
**Metabolism**

10. Disorders of Pyruvate Metabolism
and the Tricarboxylic Acid Cycle
D.S. Kerr, I.D. Wexler, and A.B. Zinn ......... 127

11. Disorders of Fatty Acid Oxidation
C.A. Stanley .......................... 139

12. Disorders of Ketogenesis and Ketolysis
A.A.M. Morris ........................ 151

13. Defects of the Respiratory Chain
A. Munnich ........................... 157

**Part IV**
**Disorders of Amino Acid Metabolism**
**and Transport**

14. The Hyperphenylalaninaemias
I. Smith and P. Lee ..................... 171

15. Disorders of Tyrosine Metabolism
E.A. Kvittingen and E. Holme .............. 185

16. Branched-Chain Organic Acidurias
H. Ogier de Baulny and J.-M. Saudubray ....... 195

17. Disorders of the Urea Cycle
J.V. Leonard .......................... 213

18. Disorders of Sulfur Amino Acid Metabolism
G. Andria, B. Fowler, and G. Sebastio ........ 223

19. Disorders of Ornithine
and Creatine Metabolism
V.E. Shih and S. Stöckler-Ipsiroglu ........... 233

20. Disorders of Lysine Catabolism
and Related Cerebral Organic-Acid Disorders
G.F. Hoffmann ........................ 241

21. Nonketotic Hyperglycinemia
K. Tada .............................. 255

22. Disorders of Proline and Serine Metabolism
J. Jaeken ............................. 259

23. Transport Defects of Amino Acids
at the Cell Membrane: Cystinuria, Hartnup
Disease, and Lysinuric Protein Intolerance
O. Simell, K. Parto, and K. Näntö-Salonen ...... 265

**Part V**
**Vitamin-Responsive Disorders**

24. Biotin-Responsive Multiple Carboxylase
Deficiency
E.R. Baumgartner and T. Suormala ........... 277

**25. Disorders of Cobalamin and Folate Transport and Metabolism**
D.S. Rosenblatt .......................... 283

## Part VI
## Neurotransmitter and Small Peptide Disorders

**26. Disorders of Neurotransmission**
J. Jaeken, C. Jakobs, and R. Wevers .......... 301

**27. Disorders in the Metabolism of Glutathione and Imidazole Dipeptides**
A. Larsson and J. Jaeken .................... 313

## Part VII
## Disorders of Lipid and Bile Acid Metabolism

**28. Dyslipidemias**
A. Rodriguez-Oquendo and
P.O. Kwiterovich Jr. ........................ 321

**29. Disorders of Cholesterol Synthesis**
G.F. Hoffmann and D. Haas .................. 337

**30. Disorders of Bile-Acid Synthesis**
P.T. Clayton .............................. 343

## Part VIII
## Disorders of Nucleic Acid and Heme Metabolism

**31. Disorders of Purine and Pyrimidine Metabolism**
G. Van den Berghe, M.-F. Vincent, and S. Marie ............................. 355

**32. The Porphyrias**
N.G. Egger, D.E. Goeger, and K.E. Anderson .... 369

## Part IX
## Disorders of Metal Transport

**33. Copper Transport Disorders: Wilson Disease and Menkes Disease**
D.W. Cox and Z. Tümer .................... 385

**34. Genetic Defects Related to Metals Other Than Copper**
F. Jochum and I. Lombeck ................... 393

## Part X
## Organelle-Related Disorders: Lysosomes, Peroxisomes, and Golgi and Pre-Golgi Systems

**35. Disorders of Sphingolipid Metabolism**
P.G. Barth ............................... 401

**36. Mucopolysaccharidoses and Oligosaccharidoses**
M. Beck ................................. 413

**37. Peroxisomal Disorders**
B.T. Poll-The and J.-M. Saudubray ............ 423

**38. Congenital Defects of Glycosylation: Disorders of *N*-Glycan Synthesis**
J. Jaeken ................................ 433

**39. Cystinosis**
M. Broyer ............................... 439

**40. Primary Hyperoxalurias**
P. Cochat and M.-O. Rolland ................ 445

**41. Leukotriene-$C_4$-Synthesis Deficiency**
E. Mayatepek ............................. 453

# List of Contributors

K.E. ANDERSON
Department of Preventive Medicine
and Community Health,
The University of Texas Medical Branch,
Galveston, TX 77555-1109, USA

G. ANDRIA
Department of Pediatrics,
Federico II University,
Via Sergio Pansini 5,
80131 Naples, Italy

P.G. BARTH
Pediatric Neurology Unit,
Emma Children's Hospital,
University of Amsterdam,
P.O.B. 22700, 1100 DE Amsterdam,
The Netherlands

E.R. BAUMGARTNER
University Children's Hospital Basel,
Metabolic Unit,
Römergasse 8,
4005 Basel, Switzerland

M. BECK
Universitätskinderklinik,
Langenbeckstrasse 1,
55101 Mainz, Germany

M. BROYER
Nephrologie Pédiatrique
Hôpital Necker Enfants Malades,
149 Rue de Sèvres,
75743 Paris Cedex 15, France

C. CHARPENTIER
Hôpital Necker Enfants Malades,
149 Rue de Sèvres,
75743 Paris Cedex 15, France

P.T. CLAYTON
Institute of Child Health,
30 Guilford Street,
London WC1N 1EH, UK

P. COCHAT
Unité de Néphrologie Pédiatrique,
Hôpital Edouard Herriot
and Université Claude Bernard,
69437 Lyon Cedex 03, France

D.W. COX
Department of Medical Genetics,
University of Alberta,
8-39 Medical Sciences Building,
Edmonton, AB T6G 2H7, Canada

N.G. EGGER
Department of Preventive Medicine
and Community Health,
The University of Texas Medical Branch,
Galveston, TX 77555-1109, USA

J. FERNANDES
Department of Pediatrics,
University Hospital, Groningen
*Private address*:
Veldweg 87, 8051NP Hattem,
The Netherlands

B. FOWLER
Basler Kinderspital, Basel University,
Römergasse 8, 4005 Basel,
Switzerland

R. GITZELMANN
Abteilung für Stoffwechsel
und Molekularkrankheiten
Universitätskinderklinik,
Steinwiesstrasse 75,
8032 Zürich, Switzerland

D.E. GOEGER
Department of Preventive Medicine
and Community Health,
The University of Texas Medical Branch,
Galveston, TX 77555-1109, USA

D. HAAS
Universitätskinderklinik,
Im Neuenheimer Feld 150,
69120 Heidelberg, Germany

J.C. HARRIS
Department of Psychiatry
and Behavioral Sciences,
Johns Hopkins Medical Institutions,
600 North Wolfe Street, CMSC 341,
Baltimore, MD 21287-3325, USA

G.F. HOFFMANN
Universitätskinderklinik,
Im Neuenheimer Feld 150,
69120 Heidelberg, Germany

E. Holme
Department of Clinical Chemistry,
Goteborg University,
Sahlgrenska University Hospital,
41345 Gothenburg, Sweden

J. Huber
Pathology Department,
Wilhelmina Children's Hospital,
University of Utrecht,
P.O. Box 85090,
3508 AB Utrecht, The Netherlands

J. Jaeken
Centre for Metabolic Disease,
Department of Paediatrics,
University Hospital Gasthuisberg,
Herestraat 49, 3000 Leuven, Belgium

C. Jakobs
Metabolic Unit 949B,
Department of Clinical Chemistry,
Free University Hospital Amsterdam,
De Boelelaan 1117,
1081 HV Amsterdam, The Netherlands

F. Jochum
Charité Universitätsklinikum
der Humboldt-Universität zu Berlin,
Campus Virchow-Klinikum,
Klinik für Neonatologie,
Augustenburger Platz 1,
13353 Berlin, Germany

D.S. Kerr
Center for Human Genetics,
Rainbow Babies' and Children's Hospital,
Case Western Reserve University,
11100 Euclid Ave.,
Cleveland, OH 44106-6004, USA

E.A. Kvittingen
Institute of Clinical Biochemistry,
University of Oslo,
Rikshospitalet,
0027 Oslo, Norway

P.O. Kwiterovich Jr.
Lipid Research-Atherosclerosis Division,
Department of Pediatrics,
The Johns Hopkins Hospital,
600 N. Wolfe Street,
Baltimore, MD 21287-3654, USA

A. Larsson
Dept. of Pediatrics, Karolinska Institutet,
Huddinge University Hospital,
141-86 Huddinge, Sweden

Ph. Lee
Metabolic Unit, The Middlesex Hospital,
The University College Hospitals,
NHS Trust, Mortimer Street,
London, W1N 8AA, UK

J.V. Leonard
Biochemistry, Endocrinology
and Metabolic Unit,
Institute of Child Health,
30 Guilford Street,
London WC1N 1EH, UK

I. Lombeck
Medizinische Einrichtungen
der Heinrich Heine Universität,
Zentrum für Kinderheilkunde,
Postfach 101007,
40001 Düsseldorf, Germany

P. de Lonlay
Hôpital Necker-Enfants Malades,
149 Rue de Sèvres,
75743 Paris Cedex 15, France

S. Marie
Laboratory of Physiological Chemistry,
Christian de Duve Institute
of Cellular Pathology,
University of Louvain Medical School,
Avenue Hippocrate 75.39,
1200 Brussels, Belgium

E. Mayatepek
Department of General Pediatrics,
Universitätskinderklinik,
Im Neuenheimer Feld 150,
69120 Heidelberg, Germany

A.A.M. Morris
Department of Child Health,
University of Newcastle upon Tyne,
Royal Victoria Infirmary,
Newcastle upon Tyne, NE1 4LP, UK

A. Munnich
Department of Genetics,
Hôpital Necker-Enfants Malades,
149 rue de Sèvres,
75743 Paris Cedex 15, France

K. Näntö-Salonen
Department of Pediatrics,
University of Turku,
Kiinamyllynkatu 4-8,
20520 Turku, Finland

H. Ogier de Baulny
Service de Neuropédiatrie,
Maladies Métaboliques,
Hôpital Robert Debré,
48 Boulevard Sérurier,
75019 Paris, France

K. Parto
Department of Pediatrics,
University of Turku,
Kiinamyllynkatu 4-8,
20520 Turku, Finland

B.T. Poll-The
Wilhelmina Children's Hospital,
University of Utrecht,
P.O. Box 85090,
3508 AB Utrecht, The Netherlands

E.A. Roberts
Division of Gastroentrology
and Nutrition,
University of Toronto,
555 University Avenue,
Toronto, Ontario M5G1X8,
Canada

A. Rodriguez-Oquendo
Department of Medicine,
Sinai Hospital of Baltimore,
Baltimore, MD 21287-3654, USA

M.-O. Rolland
Unité de Néphrologie Pédiatrique,
Hôpital Edouard Herriot
and Université Claude Bernard,
69437 Lyon Cedex 03, France

D.S. Rosenblatt
Division of Medical Genetics,
Department of Medicine,
McGill University Health Centre,
687 Pine Avenue West,
Montreal, Quebec H3A 1A1, Canada

J.M. Saudubray
Service de Maladies Métaboliques,
Département de Pédiatrie,
Hôpital Necker Enfants Malades,
149 Rue de Sèvres,
75043 Paris Cedex 15, France

G. Sebastio
Department of Pediatrics,
Federico II University,
Via Sergio Pansini 5,
80131 Naples, Italy

V.E. Shih
Amino Acid Disorder Laboratory,
Pediatrics and Neurology Services,
Massachusetts General Hospital,
Building 149, 13th Street,
Boston, MA 02129, USA

O. Simell
Department of Pediatrics,
University of Turku,
Kiinamyllynkatu 4-8,
20520 Turku, Finland

G.P.A. Smit
Pediatrics Department,
University Hospital,
P.O. Box 30001,
9700 RB Groningen,
The Netherlands

I. Smith
Metabolic Unit,
The Hospital for Sick Children,
NHS Trust,
Great Ormond Street,
London, WC1N 3JH, UK

C.A. Stanley
Endocrine Division,
The Children's Hospital of Philadelphia,
34th Street & Civic Center Boulevard,
Philadelphia, PA 19104, USA

S. Stöckler-Ipsiroglu
Department of Pediatrics,
National Newborn-Screening Laboratory,
University of Vienna,
Währinger Gürtel 18-20, 1090 Vienna
Austria

T. Suormala
University Children's Hospital Basel,
Metabolic Unit, Römergasse 8,
4005 Basel, Switzerland

K. Tada
NTT Tohoku Hospital,
2-29-1, Yamatomachi,
Wakabayashi, Sendai 983, Japan

Z. Tümer
Department of Medical Genetics,
University of Copenhagen,
Panum Institute,
Blegdamsvej 3,
2200 KBH N Copenhagen,
Denmark

G. Van den Berghe
Laboratory of Physiological Chemistry,
Christian de Duve Institute
of Cellular Pathology,
University of Louvain Medical School,
Avenue Hippocrate 75,
1200 Brussels, Belgium

M.-F. Vincent
Laboratory of Physiological Chemistry,
Christian de Duve Institute
of Cellular Pathology,
University of Louvain Medical School,
Avenue Hippocrate 75,
1200 Brussels, Belgium

J.H. Walter
Willink Biochemical Genetics Unit,
Royal Manchester Children's Hospital,
Manchester, M27 4HA, UK

R. Wevers
Institute of Neurology,
University Hospital Nijmegen,
Reinier Postlaan 4,
PO Box 9101, 6500 HB Nijmegen,
The Netherlands

I.D. Wexler
Department of Pediatrics,
Case Western Reserve University,
11100 Euclid Ave.,
Cleveland, OH 44106-6004, USA

J.E. Wraith
Willink Biochemical Genetics Unit,
Royal Manchester Children's Hospital,
Manchester, M27 4HA, UK

A.B. Zinn
Center for Human Genetics,
University Hospital of Cleveland,
Case Western Reserve University,
11100 Euclid Ave.,
Cleveland, OH 44106-6004, USA

# PART I
# DIAGNOSIS AND TREATMENT: GENERAL PRINCIPLES

CHAPTER 1

# Clinical Approach to Inherited Metabolic Diseases

J.M. Saudubray, H. Ogier de Baulny, and C. Charpentier

CONTENTS

I General Principles of Classification .......... 5
  1 Pathophysiology ............................ 5
    Group 1: Disorders Involving Complex
    Molecules .................................. 5
    Group 2: Disorders that Give Rise to
    Intoxication ............................... 5
    Group 3: Disorders Involving Energy Metabolism .... 5
  2 Clinical Presentation ....................... 5

II Acute Symptoms in the Neonatal Period
    and Early Infancy (<1 Year) ................ 5
  1 Clinical Presentation ....................... 5
    Neurologic Deterioration (Coma, Lethargy) ........ 6
    Seizures ................................... 6
    Hypotonia .................................. 6
    Hepatic Presentation ....................... 7
    Cardiac Presentation ....................... 7
  2 Metabolic Derangements and Diagnostic Tests ... 7
    Initial Approach, Protocol of Investigation ........ 7
    Identification of Five Major Types
    of Metabolic Distress ...................... 8

III Later-Onset Acute and Recurrent Attacks
    (Childhood and Beyond) .................... 11
  1 Clinical Presentations ...................... 11
    Coma, Strokes, and Attacks of Vomiting
    with Lethargy .............................. 11
    Recurrent Attacks of Ataxia
    and for Psychiatric Symptoms .............. 14
    Abdominal Pain ............................ 14
    Arrhythmias, Conduction Defects
    (Heartbeat Disorders) ...................... 15
    Bleeding Tendency, Hemorrhagic Syndromes ....... 15
    Bone Crisis ................................ 15
    Cardiac Failure, Collapse .................. 15
    Dehydration (Attacks) ..................... 15
    Exercise Intolerance (Myoglobinuria, Cramps,
    Muscle Pain) ............................... 15
    Hematological Symptoms (Anemia,
    Leukopenia, Thrombopenia, Pancytopenia) ........ 15
    Hyperventilation Attacks .................. 15
    Liver Failure (Ascites, Edema) ............ 16
    Psychiatric Symptoms (Hallucinations, Delirium,
    Anxiety, Schizophrenic-Like Episodes) ..... 16
    Reye Syndrome ............................. 16
    Skin Rashes (Photosensitivity) ............ 16
    Sudden Infant Death (and Near-Miss) ....... 16
    Vomiting .................................. 16
  2 Metabolic Derangements and Diagnostic Tests ... 16
    Initial Approach, Protocol of Investigation ........ 16
    Main Metabolic Presentations .............. 16

IV Chronic and Progressive General Symptoms ... 20
  1 Digestive Symptoms ........................ 21
  2 Neurologic Symptoms ....................... 22
    Progressive Neurologic and Mental
    Deterioration Related to Age (Overview) ... 22
    Ataxia (Chronic) ........................... 30
    Calcifications (Intracranial) ............... 31
    Cerebellar Hypoplasia
    (and Olivopontocerebellar Atrophy) ........ 31
    Cherry-Red Spot ............................ 31
    Corpus-Callosum Agenesis .................. 31
    Deafness (Sensorineural) .................. 31
    Dementia, Psychosis, Schizophrenia,
    Behavior Disturbances ..................... 31
    Extrapyramidal Signs (Choreoathetosis,
    Dyskinesia, Dystonia, Parkinsonism) ....... 31
    Hypotonia in the Neonatal Period .......... 32
    Leigh Syndrome ............................ 32
    Macrocephaly .............................. 32
    Mental Regression ......................... 32
    Microcephaly .............................. 32
    Myoclonic Epilepsy (Polymyoclonia) ........ 32
    Nystagmus ................................. 33
    Ophthalmoplegia, Abnormal Eye Movements ... 33
    Peripheral Neuropathy ..................... 33
    Retinitis Pigmentosa ...................... 33
    Self Mutilation, Auto-Aggressiveness ...... 33
    Spastic Paraplegia, Paraparesia ........... 33
  3 Muscular Symptoms ......................... 33

V Specific Organ Symptoms .................... 33
  1 Cardiology ................................ 34
    Arrhythmias, Conduction Defects ........... 34
    Cardiac Failure ........................... 34
    Cardiomyopathy ............................ 34
    SIDS ...................................... 34
  2 Dermatology ............................... 34
    Acrocyanosis (Orthostatic) ................ 34
    Alopecia .................................. 34
    Angiokeratosis ............................ 34
    Brittle Hair .............................. 34
    Hemangiomas .............................. 34
    Hyperkeratosis ............................ 34
    Ichthyosis (with Congenital Erythrodermia) ...... 34
    Laxity (Dysmorphic Scarring, Easy Bruising) ..... 35
    Nodules ................................... 35
    Photosensitivity and Skin Rashes .......... 35
    Pili Torti ................................ 35
    Telangiectasias, Purpuras, Petechiae ...... 35
    Trichorrhexis Nodosa ...................... 35
    Ulcerations (Skin Ulcers) ................. 35
    Vesiculo Bullous Skin Lesions ............. 35
  3 Dysmorphology ............................. 35
    Coarse Facies ............................. 35

Congenital Malformations and Dysmorphia......... 35
Intrauterine Growth Retardation ................ 36
**4 Endocrinology** ................................... 36
Diabetes (and Pseudodiabetes).................. 36
Hyperthyroidism ............................... 36
Hypogonadism, Sterility......................... 36
Hypoparathyroidism............................ 36
Salt-Losing Syndrome........................... 36
Sexual Ambiguity ............................... 36
Short Stature, Growth-Hormone Deficiency......... 36
**5 Gastroenterology** ................................. 36
Abdominal Pain (Recurrent) .................... 36
Acute Pancreatitis.............................. 36
Chronic Diarrhea, Failure to Thrive,
Osteoporosis.................................. 36
Hypocholesterolemia ........................... 36
Babies Born to Mothers with Hemolysis, Elevated
Liver Enzyme, and Low Platelet Count (HELLP)
Syndrome.................................... 36
**6 Hematology** ..................................... 36
Acanthocytosis ................................ 36
Anemias (Megaloblastic) ....................... 36
Anemias (Non Macrocytic, Hemolytic
or Due to Combined Mechanisms)................ 37
Bleeding Tendency ............................. 37
Pancytopenia, Thrombopenia, and Leukopenia ...... 37
Vacuolated Lymphocytes ....................... 37
**7 Hepatology** ..................................... 37
Cholestatic Jaundice............................ 37
Cirrhosis...................................... 37
Liver Failure (Ascites, Edema) .................. 37
Reye Syndrome................................ 37
**8 Myology** ....................................... 37
Exercise Intolerance and Myoglobinuria,
Muscle Pain .................................. 37
Myopathy (Progressive) ........................ 38
**9 Nephrology**..................................... 38
Hemolytic Uremic Syndrome..................... 38
Nephrolithiasis, Nephrocalcinosis................. 38
Nephrotic Syndrome............................ 38
Nephropathy (Tubulointerstitial) ................ 38
Renal Polycystosis ............................. 38
Tubulopathy................................... 38
Urine (Abnormal Color)......................... 38
Urine (Abnormal Odor) ........................ 38
**10 Neurology** ..................................... 38
**11 Ophthalmology** ................................. 38
Cataracts..................................... 38
Cherry-Red Spot............................... 39
Corneal Opacities (Clouding).................... 39
Ectopia Lentis (Dislocation of the Lens) .......... 39
Keratitis ..................................... 39
Microcornea .................................. 39
Ptosis, External Ophthalmoplegia,
Abnormal Eye Movements ..................... 39
Retinitis Pigmentosa............................ 39
**12 Osteology**...................................... 40
Osteoporosis.................................. 40
Punctate Epiphyseal Calcifications ............... 40
**13 Pneumology** ................................... 40
Pneumopathy (Interstitial) ..................... 40
Stridor ...................................... 40
**14 Psychiatry** .................................... 40
**15 Rheumatology** ................................. 40
Arthritis, Joint Contractures, Bone Necrosis ........ 40
Bone Crisis................................... 40
**16 Vascular Symptoms** ............................. 40
Raynaud Syndrome ............................ 40
Thromboembolic Accidents (Stroke-Like Episodes) .. 40
**VI References** ................................... 41

## Introduction

Inborn errors of metabolism are individually rare but collectively numerous. As a whole, they cannot be recognized through systematic neonatal screening tests, which are too slow, too expensive or unreliable. This makes it an absolute necessity to use a simple method of clinical screening before deciding to initiate sophisticated biochemical investigations. Clinical diagnosis of inborn errors of metabolism may, at times, be difficult for many reasons.

Many physicians think that, because individual inborn errors are rare, they should be considered only after more common conditions (such as sepsis) have been excluded. In view of the large number of inborn errors, it might appear that their diagnosis requires precise knowledge of a large number of biochemical pathways and their interrelationships. As a matter of fact, an adequate diagnostic approach can be based on the proper use of only a few screening tests.

The neonate has an apparently limited repertoire of responses to severe overwhelming illness, and the predominant clinical signs and symptoms are nonspecific: poor feeding, lethargy, failure to thrive, etc. It is certain that many patients with such defects succumb in the newborn period without having received a specific diagnosis, death often having been attributed to sepsis or other common causes. Classical autopsy findings in such cases are often nonspecific. Infection is often suspected as the cause of the death, since sepsis is a common accompaniment of metabolic disorders.

Many general practitioners and pediatricians only think of inborn errors of metabolism in very nonspecific clinical circumstances, such as psychomotor retardation or seizures. Conversely, they ignore most of the highly specific symptoms, which are excellent keys to the diagnosis. Another common mistake is to confuse a "syndrome" (such as Leigh syndrome or Reye syndrome), which is a set of symptoms possibly due to different causes, with the etiology itself. Although most genetic metabolic errors are hereditary and are transmitted as recessive disorders, the majority of cases appear to be sporadic because of the small size of sibships in developed countries.

Finally, "hereditary" does not mean "congenital", and many patients can present a late-onset form of disease in childhood, adolescence, or even adulthood. Based mostly upon personal experience over 30 years, this chapter gives an overview of clinical keys to the diagnosis of inborn errors of metabolism.

# I General Principles of Classification

## I.1 Pathophysiology

From a pathophysiologic perspective, the metabolic disorders can be divided into the following three diagnostically useful groups.

### Group 1: Disorders Involving Complex Molecules

This group includes diseases that disturb the synthesis or the catabolism of complex molecules. Symptoms are permanent, progressive, independent of intercurrent events, and are not related to food intake. All lysosomal disorders, peroxisomal disorders, disorders of intracellular trafficking and processing [such as α-1-antitrypsin deficiency and congenital defects of glycosylation (CDG) syndrome], and inborn errors of cholesterol synthesis belong to this group.

### Group 2: Disorders that Give Rise to Intoxication

This group includes inborn errors of intermediary metabolism that lead to an acute or progressive intoxication from accumulation of toxic compounds proximal to the metabolic block. In this group are the aminoacidopathies [phenylketonuria, maple-syrup-urine disease (MSUD), homocystinuria, tyrosinemia, etc.], most organic acidemias (methylmalonic, propionic, isovaleric, etc.), congenital urea-cycle defects, and sugar intolerances (galactosemia, hereditary fructose intolerance). All the conditions in this group present clinical similarities, including a symptom-free interval followed by clinical signs of "intoxication", which may be acute (vomiting, lethargy, coma, liver failure, thromboembolic complications, etc.) or chronic (progressive developmental delay, ectopia lentis, cardiomyopathy, etc.). Clinical expression is often both late in onset and intermittent. Biologic diagnosis is easy and relies mostly on plasma and urine amino-acid or organic-acid chromatography. Treatment of these disorders requires removal of the toxin by special diets, exchange transfusion, peritoneal dialysis, or hemodialysis.

### Group 3: Disorders Involving Energy Metabolism

This group consists of inborn errors of intermediary metabolism, with symptoms due at least partly to a deficiency in energy production or utilization resulting from a defect in the liver, myocardium, muscle, or brain. Included in this group are glycogenosis, gluconeogenesis defects, congenital lactic acidemias [deficiencies of pyruvate carboxylase (PC) and pyruvate dehydrogenase (PDH)], fatty-acid-oxidation (FAO) defects, and Kreb's-cycle and mitochondrial respiratory-chain disorders. Symptoms common to this group include: failure to thrive, hypoglycemia, hyperlactacidemia, severe generalized hypotonia, myopathy, cardiomyopathy, cardiac failure, circulatory collapse, sudden infant death syndrome (SIDS), and malformations [1].

## I.2 Clinical Presentation

A great diversity of signs and symptoms can lead to the diagnosis of inborn errors of metabolism. Besides systematic screening in the general population (as for phenylketonuria) or in at-risk families (in which diagnosis depends on specific biologic or genetic tests), there are four groups of clinical circumstances in which physicians are faced with the possibility of a metabolic disorder:

1. Acute symptoms in the neonatal period
2. Later-onset acute and recurrent attacks of disorders, such as coma, ataxia, vomiting, or acidosis
3. Chronic and progressive general symptoms, which can be mainly digestive, neurological, or muscular
4. Specific and permanent organ symptoms suggestive of an inborn error of metabolism, such as cardiomyopathy, hepatomegaly, lens dislocation, etc.

These four categories of clinical abnormalities are presented in the following sections. For the two first categories, which often present as emergencies, clinical presentations, metabolic derangements, and laboratory tests required for a tentative diagnostic approach are described in detail. Chronic, progressive, general symptoms that should raise suspicion of an inborn error of metabolism are listed in tables, which take into account system and organ involvement, leading symptoms and other signs, and age of onset. Specific organ symptoms are listed in alphabetical order. For each symptom, the diagnostic possibilities mentioned encompass not only inborn errors of metabolism but also diverse inherited syndromes that mimic (and are possibly related to) inborn errors and a number of non-inherited disorders, which should be considered in the differential diagnosis.

# II Acute Symptoms in the Neonatal Period and Early Infancy (<1 Year)

## II.1 Clinical Presentation

The neonate has a limited repertoire of responses to severe illness and, at first, presents with nonspecific symptoms, such as respiratory distress, hypotonia, poor sucking reflex, vomiting, diarrhea, dehydration, lethargy, or seizures, all symptoms that could easily be attributed to infection or some other common cause

[2-5]. If these symptoms are present, the deaths of affected siblings may have been falsely attributed to sepsis, heart failure, or intraventricular hemorrhage, and it is important to review clinical records and autopsy reports critically when they are available.

In the *intoxication type of metabolic distress*, an extremely evocative clinical setting is the course of a full-term baby born after normal pregnancy and delivery who, after an initial symptom-free period during which the baby is completely normal, deteriorates relentlessly for no apparent reason and does not respond to symptomatic therapy. The interval between birth and clinical symptoms may range from hours to weeks, depending on the nature of the metabolic block and the environment.

Investigations routinely performed in all sick neonates, including chest X-ray, cerebrospinal-fluid (CSF) examination, bacteriologic studies, and cerebral ultrasound yield normal results. This unexpected and "mysterious" deterioration of a child after a normal initial period is the most important signal of the presence of an inherited disease of the intoxication type. If present, careful re-evaluation of the child's condition is warranted. Signs previously interpreted as nonspecific manifestations of neonatal hypoxia, infection, or other common diagnoses take on a new significance in this context. In energy deficiencies, clinical presentation is often less evocative and displays variable severity. A careful reappraisal of the child is warranted for the following aspects.

### Neurologic Deterioration (Coma, Lethargy)

Most inborn errors of both the intoxication or energy-deficiency type are brought to a doctor's attention because of neurologic deterioration. In the intoxication type, the initial symptom-free interval varies in duration depending on the condition (Chaps. 16, 17). Typically, the first reported sign is poor sucking and feeding, after which the child sinks into an unexplained coma despite supportive measures. At a more advanced state, neurovegetative problems with respiratory abnormalities, hiccups, apneas, bradycardia, and hypothermia can appear. In the comatose state, characteristic changes in muscle tone and involuntary movements appear. Generalized hypertonic episodes with opisthotonus are frequent, and boxing or pedaling movements and slow limb elevations, spontaneously or upon stimulation, are observed. Conversely, most nonmetabolic causes of coma are associated with hypotonia, so the presence of "normal" peripheral muscle tone in a comatose child reflects a relative hypertonia. Another neurologic pattern suggesting metabolic disease is axial hypotonia and limb hypertonia with large-amplitude tremors and myoclonic jerks, which are often mistaken for convulsions. An abnormal urine and body odor is present in some diseases in which volatile metabolites accumulate. If one of the preceding symptoms is present, metabolic disorders should be given a high diagnostic priority.

In energy deficiencies, the clinical presentation is less evocative and displays a more variable severity. In many conditions, there is no symptom-free interval. The most frequent symptoms are a severe generalized hypotonia, rapidly progressive neurologic deterioration, possible dysmorphia, and malformations. However, in contrast to the intoxication group, lethargy and coma are rarely inaugural signs. Hyperlactacidemia with or without metabolic acidosis is a very frequent symptom. Cardiac symptoms and hepatic symptoms are frequently associated (below).

Only a few lysosomal disorders with storage symptoms are expressed in the neonatal period. By contrast, most peroxisomal disorders present immediately after birth, with dysmorphia and severe neurological dysfunction.

### Seizures

True convulsions occur late and inconsistently in inborn errors of intermediary metabolism, with the exception of pyridoxine-dependent seizures and some cases of non-ketotic hyperglycinemia (NKH), sulfite-oxidase (SO) deficiency [4], and peroxisomal disorders, where convulsions may be important inaugural elements in the clinical presentation. Convulsions are the unique symptom in pyridoxine-dependent convulsions. This rare disorder should be considered with all refractory seizures in children under 1 year of age. In contrast, newborns with MSUD, organic acidurias, and urea-cycle defects rarely experience seizures in the absence of pre-existing stupor, coma, or hypoglycemia. The electroencephalogram (EEG) often shows a periodic pattern in which bursts of intense activity alternate with nearly flat segments.

### Hypotonia

Hypotonia is a very common symptom in sick neonates. Whereas many nonmetabolic inherited diseases can give rise to severe generalized neonatal hypotonia (mainly all severe fetal neuromuscular disorders), only a few inborn errors of metabolism present with isolated hypotonia in the neonatal period. Discounting disorders in which hypotonia is included in a very evocative clinical context (such as a well-known polymalformative syndrome) or visceral symptoms, the most severe metabolic hypotonias are observed in hereditary hyperlactacidemias, respiratory-chain disorders, urea-cycle defects, NKH, SO deficiency, peroxisomal disorders, and trifunctional-enzyme deficiency. In all these circumstances, the diagnosis is mostly based upon the association of the central hypotonia with lethargy, coma, seizures, and neurologic symptoms. At first, severe forms of Pompe disease (glycogenosis type II) can mimic respiratory-chain disorders or trifunctional-enzyme deficiency when generalized hypotonia is associated with cardiomyopathy. However, Pompe disease does not strictly start in the neonatal period. Finally, one of the most frequent diagnoses is Willi-Prader syndrome, where hypotonia is central and is apparently an isolated symptom at birth.

### Hepatic Presentation

Three main clinical groups of hepatic symptoms can be identified:

1. Hepatomegaly with hypoglycemia and seizures suggest glycogenosis types I and III, gluconeogenesis defects, or severe hyperinsulinism.
2. Liver-failure syndrome (jaundice, hemorrhagic syndrome, hepatocellular necrosis with elevated transaminases, and hypoglycemia with ascites and edema) suggests hereditary fructose intolerance (in the case of a fructose-containing diet), galactosemia, tyrosinosis type I (after 3 weeks), neonatal hemochromatosis, and respiratory-chain disorders. A new disorder has been recently described in 15 newborns from Finland, who all presented with severe fetal growth retardation, lactic acidosis, failure to thrive, hyperaminoaciduria, very high serum ferritin, hemosiderosis of the liver, and early death. The etiology of this syndrome is unknown [6].
3. Predominantly cholestatic jaundice with failure to thrive is observed in α-1-antitrypsin deficiency, Byler disease, inborn errors of bile-acid metabolism, peroxisomal disorders, Niemann-Pick type-C disease, CDG syndrome and cholesterol biosynthesis defects.

Hepatic presentations of inherited FAO disorders and urea-cycle defects consist of acute steatosis or Reye syndrome with normal bilirubin, slightly prolonged prothrombin time, and moderate elevation of transaminases rather than true liver failure, with the exception of long-chain 3-hydroxyacyl-coenzyme-A (CoA)-dehydrogenase (LCHAD) deficiency, which can present early in infancy (but not strictly in the neonatal period) as cholestatic jaundice, liver failure, and hepatic fibrosis. There are frequently difficulties in investigating patients with severe hepatic failure. At an advanced state, many nonspecific symptoms secondary to liver damage can be present. Mellituria (galactosuria, glucosuria, fructosuria), hyperammonemia, hyperlactacidemia, short-fast hypoglycemia, hypertyrosinemia (>200 μmol/l), and hypermethioninemia (sometimes higher than 500 μmol/l) are encountered in all advanced hepatocellular insufficiencies.

### Cardiac Presentation

Sometimes metabolic distress can strike with predominant cardiac symptoms. Cardiac failure revealing or accompanying a cardiomyopathy (dilated hypertrophic) and most often associated with hypotonia, muscle weakness, and failure to thrive, suggests respiratory-chain disorders, Pompe disease, or FAO disorders. Recent observations suggest that some respiratory-chain disorders are tissue specific and are only expressed in the myocardium. The new multisystemic CDG syndrome can sometimes present in infancy with cardiac failure due to pericardial effusions, cardiac tamponade, and cardiomyopathy. Many defects of long-chain FAO can be revealed by cardiomyopathy and/or arrhythmias and conduction defects (auriculoventricular block, bundle-branch blocks, ventricular tachycardia) responsible for cardiac arrest [7, 8].

## II.2 Metabolic Derangements and Diagnostic Tests

### Initial Approach, Protocol of Investigation

Once clinical suspicion of an inborn metabolic error is aroused, general supportive measures and laboratory investigations must be undertaken simultaneously (Table 1.1). Abnormal urine odors can be detected from a drying filter paper or by opening a container of urine that has been closed at room temperature for a few minutes. The most important examples are the maple syrup odor of MSUD and the sweaty-feet odor of isovaleric acidemia (IVA) and type-II glutaric acidemia (GA). Although serum ketone bodies reach 0.5–1 mmol/l in early neonatal life, acetonuria, if observed in a newborn, is always abnormal and an important sign of metabolic disease. The dinitrophenylhydrazine (DNPH) test screens for the presence of α-keto acids, such as seen in MSUD. The test can be considered significant only in the absence of glucosuria and acetonuria, which also react with DNPH. Hypocalcemia and elevated or reduced blood glucose are frequently present in metabolic diseases. The physician should be cautious of attributing marked neurologic dysfunction merely to these findings.

The metabolic acidosis of organic acidurias is usually accompanied by an elevated anion gap. Urine pH should be below 5; otherwise, renal acidosis is a consideration. A normal serum pH does not exclude hyperlactacidemia, as neutrality is usually maintained until serum levels exceed 5 mmol/l. Ammonia and lactic acid should be determined systematically in newborns at risk. An elevated ammonia level in itself can induce respiratory alkalosis; hyperammonemia with ketoacidosis suggests an underlying organic acidemia. Elevated lactic-acid levels in the absence of

**Table 1.1.** Emergency protocol for investigations

| | Immediate investigations | Storage of samples |
|---|---|---|
| Urine | Smell (special odor) | Urine collection: collect separately each fresh sample and put it in the refrigerator |
| | Look (special color) | Freezing: freeze samples collected before and after treatment at −20°C, and collect an aliquot 24 h after treatment. Do not use them without having expert metabolic advice |
| | Acetone (Acetest, Ames)<br>Reducing substances (Clinitest, Ames)<br>Keto acids (DNPH)<br>pH (pHstix Merck)<br>Sulfitest (Merck)<br>Brand reaction<br>Electrolytes (Na, K), urea, creatinine<br>Uric acid | |
| Blood | Blood cell count<br>Electrolytes (search for anion gap)<br>Glucose, calcium | Plasma (5 ml) heparinized at −20°C<br>Blood on filter paper: 2 spots (as "Guthrie" test)<br>Whole blood (10–15 ml) collected on EDTA and frozen (for molecular-biology studies) |
| | Blood gases (pH, $pCO_2$, $HCO_3H$, $pO_2$)<br>Uric acid<br>Prothrombin time<br>Transaminases (and other liver tests)<br>Ammonemia<br>Lactic, pyruvic acids<br>3-Hydroxybutyrate, acetoacetate<br>Free fatty acids | |
| Miscellaneous | Lumbar puncture<br>Chest X-ray<br>Cardiac echography, ECG<br>Cerebral ultrasound, EEG<br>Autopsy | Skin biopsy (fibroblast culture)<br>CSF (1 ml), frozen<br>Postmortem: liver, muscle biopsies (Chap. 2) |

*CSF*, cerebrospinal fluid; *DNPH*, dinitrophenylhydrazine; *ECG*, electrocardiogram; *EDTA*, ethylenediaminetetra-acetic acid; *EEG*, electroencephalogram

infection or tissue hypoxia are a significant finding. Moderate elevations (3–6 mmol/l) are often observed in organic acidemias and in the hyperammonemias; levels greater than 10 mmol/l are frequent in hypoxia. It is important to measure lactate (L), pyruvate (P), 3-hydroxybutyrate (3OHB), and acetoacetate (AA) on a plasma sample immediately deproteinized at the bedside in order to appreciate cytoplasmic and mitochondrial redox states through the measurement of L/P and 3OHB/AA ratios, respectively. Some organic acidurias induce granulocytopenia and thrombocytopenia, which may be mistaken for sepsis.

The storage of adequate amounts of plasma, urine, blood (on filter paper), and CSF is an important element in diagnosis. The utilization of these precious samples should be carefully planned after taking advice from specialists in inborn errors of metabolism.

Once the above clinical and laboratory data have been assembled, specific therapeutic recommendations can be made (Chap. 3). This process is completed within 2–4 h and often precludes long waiting periods for sophisticated diagnostic results. On the basis of this evaluation, most patients can be classified into one of five groups (Table 1.2). The experienced clinician will, of course, have to carefully interpret the metabolic data, especially in relation to time of collection and ongoing treatment. It is important to insist on the need to collect at the same time all the biologic data listed in Table 1.1. Some very significant symptoms (such as metabolic acidosis and especially ketosis) can be moderate and transient, largely depending on the symptomatic therapy. Conversely, at an advanced state, many nonspecific abnormalities (such as respiratory acidosis, severe hyperlactacidemia, or secondary hyperammonemia) can disturb the original metabolic profile. This applies particularly to disorders with a rapid fatal course, such as urea-cycle disorders, in which the initial characteristic presentation of hyperammonemia with respiratory alkalosis and without ketosis shifts rapidly to a rather nonspecific picture of acidosis and hyperlactacidemia.

### Identification of Five Major Types of Metabolic Distress

According to the major clinical presentations and the proper use of the laboratory data described above, most patients can be schematically assigned to one of five syndromes. In our experience, type I (MSUD), type II (organic acidurias), type IVa (urea-cycle defects), type IVb (NKH, respiratory-chain and FAO disorders) encompass more than 80% of the newborn infants with inborn errors of intermediary metabolism.

### Type I: Predominant Ketosis

*Neurologic deterioration of the intoxication type with ketosis* is represented by MSUD. It is one of the most common aminoacidopathies (Table 1.2).

### Type II: Predominant Ketoacidosis (with Hyperammonemia)

*Neurologic deterioration of the intoxication type with metabolic acidosis and hyperammonemia* encompasses many of the organic acidurias. The presence of ketosis is also a very important key symptom. In addition to methylmalonic (MMA), propionic (PA) and isovaleric (IVA) acidemias, which mostly present with ketoacidosis, a large number of rare organic acidurias have been uncovered in recent years as sophisticated organic-acid-analysis techniques have become more available. Among them, GA type II or multiple acyl-CoA-dehydrogenase deficiency, long-chain FAO defects, and 3-hydroxy-3-methylglutaryl (HMG)-CoA-lyase deficiency have many similarities with MMA, PA, and IVA, except that ketosis is absent and hypoglycemia is frequent. FAO disorders frequently present with cardiac symptoms and moderate liver dysfunction and can be responsible for sudden death [8]. Very rare conditions in this group are succinyl-CoA-transferase deficiency, biotin-dependent multiple-carboxylase deficiency (MCD) due to holoenzyme-synthase deficiency, short-chain acyl-CoA-dehydrogenase (SCAD) deficiency, 3-methylglutaconicuria, and glycerol-kinase deficiency, which all display ketoacidosis. Pyroglutamic aciduria is a rare condition that can start in the first few days of life, with a severe metabolic acidosis but without ketosis or abnormalities of blood glucose, lactate, or ammonia. The final diagnosis of all these organic acidurias is made by identifying specific abnormal metabolites by gas chromatography-mass spectrometry (MS) of plasma and urine, or by investigating the acylcarnitine profile of a blood spot collected on filter paper by fast-atom-bombardment MS or ordinary MS.

### Type III: Predominant Lactic Acidosis with Neurological Distress of the Energy-Deciency Type

The clinical presentation of these children varies. Unlike the previous disease category, in which moderate acidosis is noted during the evaluation of an acutely ill, comatose child, the main medical preoccupation in group-III patients is the acidosis itself, which clinically may be surprisingly well tolerated. The acidosis can sometimes be mild. An elevated anion gap exists, which can be explained in part by the presence of equimolar amounts of lactic acid in the blood. Often, the acidosis recurs soon after bicarbonate therapy, in the absence of adequate treatment.

If a high lactic-acid concentration is found, it is urgent that readily treatable causes, especially hypoxia, be ruled out. Ketosis is present in most of the primary lactic acidemias but is absent in acidosis secondary to tissue hypoxia. Biotin-responsive MCD may present as lactic acidosis, and biotin therapy is indicated in all patients with lactic acidosis of unknown cause after baseline blood and urine samples have been taken. Primary lactic acidoses form a complex group. A definite diagnosis is often elusive and is attained with specific enzyme assays after considering metabolite levels, redox-potential states, and fluxes under fasting and fed conditions. However, many cases remain unexplained.

### Type IVa: Neurologic Distress of the Intoxication Type with Hyperammonemia and Without Ketoacidosis

Type IV is divided into two groups: IVa and IVb. Type IVa encompasses urea-cycle defects and triple H. As mentioned above, this group of patients is one of the most important among those with neonatal inborn errors of metabolism. A diagnostic clue to separate urea-cycle defects from organic acidurias with hyperammonemia is the universal absence of ketonuria in the former group. As already stated, at an advanced state, neurovegetative disorders rapidly give rise to nonspecific findings including acidosis and hyperlactacidemia. An elusive diagnostic consideration is transient hyperammonemia of the neonate. Long-chain FAO disorders [mainly carnitine-palmitoyltransferase-II (CPT-II), and carnitine-translocase deficiencies] can also (though rarely) present in the neonatal period with hyperammonemia and mimic urea-cycle disorders. They are usually associated with hypoglycemia, hepatic dysfunction, muscular and cardiac symptoms, or sudden infant death.

### Type IVb: Neurologic Deterioration without Ketoacidosis and without Hyperammonemia

The most frequent diseases of type IVb are NKH, SO, and inborn errors of peroxisomal metabolism [9, 10]. SO is probably underdiagnosed, as its clinical pattern shares many similarities with common acute fetal distress. In addition, some patients with respiratory-chain disorders can present in the neonatal period without evidence of lactic acidosis. Beside these four disorders, an increasing number of other rare conditions has been described in recent years, including neurotransmitter disorders, CDG syndrome, and cholesterol-biosynthesis defects, and we can assume that

**Table 1.2.** Classification of inborn errors revealed in the neonatal period and early in infancy

| Type | Clinical type | Acidosis/ketosis | Other signs | Usual diagnosis | Elective methods of investigation |
|---|---|---|---|---|---|
| I | Neurologic deterioration, "intoxication" type, abnormal movements, hypertonia | Acidosis 0/+; DNPH +++; Acetest 0/+ | $NH_3$ N or ↑ +; lactate N; blood count N; glucose N; calcium N | MSUD (special odor) | AAC (plasma, urine) |
| II | Neurologic deterioration, "intoxication" type, dehydration | Acidosis ++; Acetest ++; DNPH 0/+ | $NH_3$ ↑ +/++; lactate N or ↑ +; blood count: leukopenia, thrombopenia; glucose N or ↑ +; calcium N or ↓ + | Organic acidurias (MMA, PA, IVA, MCD), ketolysis defects | Organic-acid chromatography by GLCMS (urine, plasma), carnitine (plasma); carnitine esters (urine, plasma, blood on filter paper) |
|  | Neurologic deterioration, "energy-deficiency" type, with liver or cardiac symptoms | Acidosis ++/+; Acetest 0; DNPH 0 | $NH_3$ ↑ +/++; lactate ↑ +/++; blood count N; glucose ↓ +/++; calcium N or ↓ + | Fatty-acid oxidation and ketogenesis defects [GA-II, CPT-II, CAT, VLCAD, MCKAT, HMGCoA lyase] | As above (acylcarnitine on filter paper ++); also function test, fatty-acid-oxidation studies on lymphocytes or fibroblasts |
| III | Neurologic deterioration, "energy deficiency" type, polypnea, hypotonia | Acidosis +++/+; Acetest ++/0; lactate +++/+ | $NH_3$ N or ↑ +; blood count: anemia or N; glucose N or ↓ +; calcium N | Congenital lactic acidoses (PC, PDH, Krebs-cycle and respiratory-chain disorders), MCD | Plasma redox states ratios (L/P, 3OHB/AA), organic-acid chromatography (urine), polarographic studies, enzyme assays (muscle, lymphocytes, fibroblasts) |
| IVa | Neurologic deterioration, moderate hepatocellular disturbances, hypotonia, seizures, coma | Acidosis 0 (alkalosis); Acetest 0; DNPH 0 | $NH_3$ ↑ +/+++; lactate N or ↑ +; glucose N; calcium N; blood count N | Urea-cycle defects; triple-H deficiency, fatty-acid-oxidation defects (GAII, CPTII, VLCAD, LCHAD, CAT) | AAC (plasma, urine), orotic acid (urine), liver or intestine enzyme studies (CPS, OTC) |
| IVb | Neurologic deterioration, seizures, myoclonic jerks, severe hypotonia | Acidosis 0; Acetest 0; DNPH 0 | $NH_3$ N; lactate N or ↑ +; blood count N; glucose N | NKH, SO plus XO, pyridoxine dependency, peroxisomal disorders, trifunctional-enzyme deficiency, respiratory-chain disorders, neurotransmitter disorders, CDG syndrome, cholesterol-disorders | AAC (NKH, SO), VLCFA, phytanic acid in plasma (PZO), acylcarnitine profile (Guthrie card), lactate (plasma), dopa, HVA, 5HIAA (CSF), glycosylated transferrin (plasma), cholesterol (plasma) |
| V | Hepatomegaly, hypoglycemia | Acidosis ++/+; Acetest + | $NH_3$ N; lactate ↑ +/++; blood count N; glucose ↓ ++ | Glycogenosis type I (acetest –), glycogenosis type III (acetest ++), fructose-bisphosphatase | Fasting test, loading test, enzyme studies (liver, lymphocytes, fibroblasts) |
|  | Hepatomegaly, jaundice, liver failure, hepatocellular necrosis | Acidosis +/0; Acetest +/0 | $NH_3$ N or ↑ +; lactate ↑ +/++; glucose N or ↓ ++ | HFI, galactosemia, tyrosinosis type I, hemochromatosis, respiratory-chain | Enzyme studies (HFI, galactosemia), organic-acid and enzyme studies (tyrosinosis type I) |
|  | Hepatomegaly, cholestatic jaundice with failure to thrive and chronic diarrhea | Acidosis 0; ketosis 0 | $NH_3$ N; lactate N; glucose N | α-1-Antitrypsin, inborn errors of bile acid metabolism, peroxisomal defects, CDG, LCHAD, Niemann-Pick type C | Protein electrophoresis, organic-acid chromatography (plasma, urine, duodenal juice), acylcarnitine profile, VLCFA, phytanic acid, pipecolic acid, glycosylated transferrin, fibroblast studies |

**Table 1.2.** (*Contd.*)

| | | | | |
|---|---|---|---|---|
| Hepatosplenomegaly, "storage" signs bone changes, cherry-red spot, vacuolated lymphocytes failure to thrive, chronic diarrhea | Acidosis 0; Acetest 0; ketosis 0; DNPH 0 | NH₃ N; lactate N or ↑; blood count N; glucose N; hepatic signs + | GM1 gangliosidosis, sialidosis type II, I-cell disease, Niemann-Pick IA, MPS-VII, galactosialidosis, CDG | Oligosaccharides, sialic acid (urine), mucopolysacchardies (urine), enzyme studies (lymphocytes, fibroblasts) |

↑, elevated; ↓, decreased; –, slight; +, moderate; ++, marked; +++, significant/massive; 0, absent (acidosis) or negative (acetest, DNPH); *3OHB*, 3-hydroxybutyrate; *5HIAA*, 5-hydroxyindoleacetic acid; *AA*, acetoacetate; *AAC*, amino-acid chromatography; *CAT*, carnitine acylcarnitine translocase; *CDG*, Congenital disorders of glycosylation; *CPS*, carbamyl phosphate synthase; *CPT-II*, carnitine palmitoyltransferase II; *CSF*, cerebrospinal fluid; *DNPH*, dinitrophenylhydrazine; *GA-II*, glutaric acidemia type II; *GLCMS*, gas–liquid chromatography–mass spectrometry; *HFI*, hereditary fructose intolerance; *HMGCoA*, 3-hydroxy-3-methylglutaryl coenzyme A; *HVA*, homovanillic acid; *IVA*, isovaleric acidemia; *L*, lactate; *LCHAD*, 3-hydroxy long-chain acyl-coenzyme-A dehydrogenase; *MCD*, multiple carboxylase; *MCKAT*, medium-chain 3-ketoacyl-coenzyme-A thiolase; *MMA*, methylmalonic acidemia; *MPS-VII*, mucopolysaccharidosis type VII; *MSUD*, maple syrup urine disease; *N*, normal (normal values: NH₃ < 80 µM, lactate < 1.5 mM, glucose 3.5–5.5 mM); *NKH*, nonketotic hyperglycinemia; *OTC*, ornithine transcarbamylase; *P*, pyruvate; *PA*, propionic acidemia; *PC*, pyruvate carboxylase; *PDH*, pyruvate dehydrogenase; *PZO*, peroxisomal disorders; *SO*, sulfite oxidase; *VLCAD*, very-long-chain acyl-coenzyme-A dehydrogenase; *VLCFA*, very-long-chain fatty acids; *XO*, xanthine oxidase

the list of disorders of this group will expand substantially in the near future.

FAO disorders can also be observed in the neonatal period without acidosis or hyperammonemia. They present with hypoglycemia, hepatic dysfunction possibly associated with muscular and cardiac symptoms, or sudden infant death. Trifunctional-enzyme deficiency can be revealed by severe hypotonia and neurologic distress without obvious metabolic disturbances.

### Type V: Hepatomegaly and Liver Dysfunction

In this type, four main clinical groups of hepatic symptoms lead to the diagnosis of more than 20 inborn errors of metabolism (Table 1.2).

## III Later-Onset Acute and Recurrent Attacks (Childhood and Beyond)

### III.1 Clinical Presentations

In approximately 50% of the patients with inborn errors of intermediary metabolism, disease onset is late. The symptom-free period is often longer than 1 year and may extend into late childhood, adolescence, or even adulthood. Each attack can present a rapid course toward either spontaneous improvement or unexplained death despite supportive measures in the intensive care unit. Between attacks, the child may appear normal. Onset of acute disease may be precipitated by a minor viral infection, fever, or even severe constipation or may occur without overt cause. Excessive protein intake, prolonged fast, prolonged exercise, and all conditions that enhance protein catabolism may exacerbate such decompensations. In the following pages are listed the 17 most common acute and recurrent clinical symptoms that can inaugurate an inborn error of metabolism beyond infancy. Only the attacks of coma, recurrent vomiting, ataxia, and psychiatric disturbances are described in detail. All other symptoms are listed in alphabetical order with their corresponding diagnostic possibilities.

### Coma, Strokes, and Attacks of Vomiting with Lethargy

All types of comas in pediatrics can signal inborn errors of metabolism, including those presenting with focal neurologic signs (Table 1.3). Neither the age at onset, the accompanying clinical signs (hepatic, digestive, neurologic, psychiatric, etc.), the mode of evolution (improvement, sequelae, death), nor the routine laboratory data allow an inborn error of metabolism to be ruled out a priori. Three categories can be distinguished.

#### Metabolic Coma without Neurologic Signs

The main varieties of metabolic comas may all be observed in the following late-onset, acute diseases: coma with predominant metabolic acidosis, coma with predominant hyperammonemia, coma with predominant hypoglycemia, and combinations of these three major abnormalities. A rather confusing finding in some organic acidurias and ketolytic defects is ketoacidosis with hyperglycemia and glycosuria that mimics diabetic coma. The diagnostic approach to the metabolic derangements is developed below (in "Metabolic Derangements and Diagnostic Tests").

#### Neurologic Coma with Focal Signs, Seizures, or Severe Intracranial Hypertension

Although most recurrent metabolic comas are not accompanied by neurologic signs other than encepha-

**Table 1.3.** Diagnostic approach for recurrent attacks of coma and vomiting with lethargy

| Clinical presentation | Metabolic derangements or other important signs | Additional symptoms | Most frequent diagnosis | Differential diagnosis |
|---|---|---|---|---|
| Metabolic coma (without focal neurological signs) | Acidosis (metabolic); pH < 7.20; $HCO_3^-$ < 10 mmol/l; $pCO_2$ < 25 mmHg | Ketosis + (acetest ++) | Respiratory-chain disorders MCD, PC Organic acidurias (MMA, PA, IVA, GA-I, MSUD) Ketolysis defects, Gluconeogenesis defects | Diabetes Intoxication Encephalitis |
| | | Ketosis – | PDH, ketogenesis defects, fatty-acid-oxidation defects, FDP, EPEMA syndrome | |
| | Hyperammonemia; $NH_3$ > 100 µmol/l; gaseous alcalosis; pH > 7.45 $pCO_2$ < 25 mmHg | Normal glucose | Urea-cycle disorders Triple-H deficiency LPI | Reye syndrome Encephalitis Intoxication |
| | | Hypoglycemia | Fatty-acid-oxidation defects, HMGCoA-lyase deficiency | |
| | Hypoglycemia (<2 mmol/l) | Acidosis + | Gluconeogenesis defects MSUD HMGCoA-lyase deficiency Fatty-acid-oxidation defects | Drugs and toxins Ketotic hypoglycemia Adrenal insufficiency GH deficiency, hypopituitary coma |
| | Hyperlactacidemia (>4 mmol/l) | Normal glucose | PC, MCD, Krebs cycle defects, respiratory-chain disorders, PDH (without ketosis), EPEMA syndrome | |
| | | Hypoglycemia | Gluconeogenesis defects (ketosis variable), fatty-acid-oxidation defects (moderate hyperlactacidemia, no ketosis) | |
| Neurologic coma (with focal signs, seizures, or intracranial hypertension) | Biological signs are very variable, can be absent or moderate; see "Metabolic coma" | Cerebral edema Hemiplegia (hemianopsia) Extrapyramidal signs | MSUD, OTC MSUD, OTC, MMA, PA, PGK MMA, GA-I, Wilson disease, homocystinuria | Cerebral tumor, migraine, encephalitis |
| | | Stroke-like | Urea-cycle defects, MMA, PA, IVA Respiratory chain disorders (MELAS) Homocystinurias CDG Thiamine-responsive megaloblastic anemia, Fabry disease (rarely revealing) | Moya-Moya syndrome Vascular hemiplegia Cerebral thrombophlebitis Cerebral tumor |
| | Abnormal coagulation, hemolytic anemia | Thromboembolic accidents | Antithrombin-III deficiency, protein-C or -S deficiency, homocystinurias, sickle-cell anemia, CDG, PGK | |
| Hepatic coma (hepatomegaly, cytolysis or liver failure) | Normal bilirubin, slight elevation of transaminases | Steatosis and fibrosis | Fatty-acid-oxidation defects, urea-cycle disorders | |

**Table 1.3.** (*Contd.*)

| Hepatic coma | Hyperlactacidemia | Liver failure | Respiratory-chain disorders | Reye syndrome, hepatitis, intoxication |
|---|---|---|---|---|
| | Hemolytic jaundice | Cirrhosis, chronic hepatic dysfunction | Wilson disease | |
| | Hypoglycemia | | Hepatic fibrosis with enteropathy | |
| | Exsudative enteropathy | | CDG | |

−, slight; +, moderate; ++, marked; *CDG*, congenital disorders of glycosylation; *EPEMA*, encephalopathy, petechiae, ethylmalonic aciduria syndrome; *FBP*, fructose 1,6-bisphosphatase; *GA*, glutaric acidemia; *GH*, growth hormone; *HMG-CoA*, 3-hydroxy-3-methylglutaryl coenzyme A; *IVA*, isovaleric acidemia; *LPI*, lysinuric protein intolerance; *MCD*, multiple carboxylase deficiency; *MELAS*, mitochondrial encephalopathy lactic-acidosis stroke-like episodes; *MMA*, methylmalonic acidemia; *MSUD*, maple-syrup-urine disease; *OTC*, ornithine transcarbamoylase; *PA*, propionic acidemia; *PC*, pyruvate carboxylase; *PDH*, pyruvate dehydrogenase; *PGK*, phosphoglycerate kinase

lopathy, some patients with organic acidemias and urea-cycle defects present with focal neurologic signs or cerebral edema. These patients can be mistakenly diagnosed as having a cerebrovascular accident or cerebral tumor. Classic homocystinuria (vitamin-B6 responsive) can strike in late childhood with an acute cerebrovascular accident. A few patients with MMA may have had acute extrapyramidal disease and corticospinal-tract involvement after metabolic decompensation. The neurologic findings result from bilateral destruction of the globus pallidus, with variable involvement of the internal capsule. Cerebellar hemorrhage has also been observed in IVA, PA, and MMA.

EPEMA syndrome (encephalopathy, petechiae and ethylmalonic aciduria) is a recently described entity which starts early in infancy and is characterized by the association of progressive encephalopathy with mental retardation, pyramidal signs, bilateral lesions in striatum resembling Leigh syndrome, relapsing petechiae, orthostatic acrocyanosis, and recurrent attacks of metabolic decompensation with lactic acidosis and without ketosis, during which there is an excess of ethylmalonic and methylsuccinic acid. The etiology of this syndrome remains unknown [11, 12]. In one patient, a profound defect in cytochrome-C-oxidase activity was found in muscle [12].

Two patients with 3-hydroxyisobutyric aciduria presenting with recurrent episodes of vomiting and ketoacidotic coma have recently been described [13]. Cyclic vomiting syndrome associated with intermittent lactic acidosis has been described as a revealing sign in mitochondrial DNA mutations [14, 15].

GA type I is frequently revealed by encephalopathic episodes mimicking encephalitis with acute metabolic derangements that occur in connection with gastrointestinal and viral infections. Mitochondrial-encephalopathy lactic-acidosis stroke-like-episode (MELAS) syndrome is another important diagnostic consideration in such late-onset and recurrent comas. Early episodic central-nervous-system problems possibly associated with liver insufficiency or cardiac failure have been inaugural symptoms in some cases of CDG syndrome.

In summary, all these disorders should be considered in the expanding differential-diagnosis list of strokes or stroke-like episodes. Certain vaguely defined and/or undocumented diagnoses, such as encephalitis, basilar migraine, intoxication, or cerebral thrombophlebitis should, therefore, be questioned, especially when even moderate ketoacidosis, hyperlactacidemia, or hyperammonemia is present. In fact, these apparent initial acute manifestations are frequently preceded by other premonitory symptoms, which may be unrecognized or misinterpreted. Such symptoms include acute ataxia, persistent anorexia, chronic vomiting, failure to thrive, hypotonia, and progressive developmental delay – all symptoms that are often observed in urea-cycle disorders, respiratory-chain defects, and organic acidurias.

Certain features or symptoms are characteristic of particular disorders. For example, macrocephaly is a frequent finding in GA type I, unexplained episodes of dehydration may occur in organic acidurias, and hepatomegaly at the time of coma is an important (though inconsistent) finding in fructose-bisphosphatase deficiency. Severe hematologic manifestations and recurrent infections are common in IVA, PA, and MMA.

### Hepatic Coma with Liver Dysfunction or Hepatomegaly

When coma is associated with hepatic dysfunction, Reye syndrome secondary to disorders of FAO and of the urea cycle should be considered. Hepatic coma with liver failure and hyperlactacidemia can be a revealing sign of respiratory-chain disorders. Finally, hepatic coma with cirrhosis, chronic hepatic dysfunction, hemolytic jaundice, and various neurologic signs (psychiatric, extrapyramidal) is a classic but underdiagnosed manifestation of Wilson disease.

## Recurrent Attacks of Ataxia and for Psychiatric Symptoms

Intermittent acute ataxia and abnormal behavior (Table 1.4) can be the presenting signs of late-onset MSUD and organic aciduria, in which they are associated with ketoacidosis and (sometimes) hyperglycemia, which can mimic ketoacidotic diabetes. Late-onset forms of congenital hyperammonemia, especially partial ornithine transcarbamoylase (OTC) deficiency, can strike late in childhood or in adolescence with psychiatric symptoms. Because hyperammonemia and liver dysfunction can be mild even at the time of acute attacks, these intermittent late-onset forms of urea-cycle disorder can be easily misdiagnosed as hysteria, schizophrenia, or intoxication. Acute ataxia associated with peripheral neuropathy is a frequent presenting sign of PDH deficiency; moderate hyperlactacidemia with a normal L/P ratio supports this diagnosis. Acute intermittent porphyria and hereditary coproporphyria present classically with recurrent attacks of vomiting, abdominal pain, neuropathy, and psychiatric symptoms. Finally, patients affected with homocysteine-remethylation defects may present with schizophrenia-like episodes that are folate responsive.

## Abdominal Pain

- With meteorism, diarrhea, loose stools:
  - Lactose malabsorption
  - Congenital sucrase-isomaltase deficiency
- With vomiting, lethargy, ketoacidosis:
  - Urea-cycle defects (OTC deficiency, argininosuccinic aciduria)
  - Organic acidurias (MA, PA, IVA)
  - Ketolysis defects
  - Respiratory-chain disorders
  - Diabetes
- With neuropathy, psychiatric symptoms:
  - Porphyrias
  - Tyrosinemia type I
  - OTC deficiency (late onset)
- With hepatomegaly (and splenomegaly):
  - Cholesteryl-ester-storage disease
  - Lipoprotein-lipase deficiency
  - Lysinuric protein intolerance
  - Hemochromatosis
- With pain in extremities:
  - Fabry disease
  - Aminolevulinic-acid-dehydratase deficiency
  - Sickle-cell anemia

**Table 1.4.** Diagnostic approach for RECURRENT ATTACKS OF ATAXIA

| Clinical presentation | Metabolic derangements or other important signs | Additional symptoms | Most frequent diagnosis | Differential diagnosis |
|---|---|---|---|---|
| Acute ataxia | Ketoacidosis | Special odor<br>Neutropenia, thrombopenia, hyperglycemia | Late-onset MSUD, MMA, PA, IVA | Diabetes |
| | Hyperammonemia | Respiratory alkalosis, hepatomegaly, | Urea-cycle defects (OTC, ASA) | Intoxication<br>Encephalitis |
| | Hyperlactacidemia | Normal L/P ratio<br>No ketosis<br>Peripheral neuropathy | PDH | Migraine<br>Cerebellitis (varicella), polymyoclonia |
| | | High L/P ratio<br>Ketosis, cutaneous signs | MCD<br>Respiratory-chain | Acetazolamide responsive ataxia<br>Acute exacerbation in chronic ataxias |
| | No metabolic disturbance | Skin rashes, pellagra, sun intolerance | Hartnup disease | |
| Psychiatric symptoms (hallucinations, delirium, dizziness, aggressivity, anxiety, schizophrenic-like behavior, agitation) | Hyperammonemia | Slight liver dysfunction, vomiting, failure to thrive | Urea-cycle disorders (OTC, ASA, arginase deficiency, LPI) | |
| | Ketoacidosis | Ataxia, neutropenia | Organic acid disorders, MSUD | |
| | Port-wine urine | Abdominal pain<br>All kinds of neuropathy, vomiting | Acute intermittent porphyria<br>Hereditary coproporphyria | Hysteria |
| | Positive Brand reaction | Stroke, seizures, myelopathy | Methylene-tetrahydrofolate-reductase deficiency | Schizophrenia |

*ASA*, argininosuccinic aciduria; *IVA*, isovaleric acidemia; *L*, lactate; *LPI*, lysinuric protein intolerance; *MMA*, methymalonic acidemia; *MSUD*, maple syrup urine disease; *OTC*, ornithine transcarbamoylase deficiency; *P*, pyruvate; *PA*, propionic acidemia; *PDH*, pyruvate dehydrogenase deficiency; *MCD*, multiple carboxylase deficiencies

- With hemolytic anemia:
  - Coproporphyria
  - Hereditary spherocytosis
  - Sickle-cell anemia
  - Nocturnal paroxystic hemoglobinuria

### Arrhythmias, Conduction Defects (Heartbeat Disorders)

- Primitive heartbeat disorders:
  - FAO disorders [carnitine-palmitoyl transferase II, carnitine translocase, long-chain acyl-CoA dehydrogenases (LCHAD), very-long-chain acyl-CoA dehydrogenase (VLCAD), trifunctional enzyme]
  - Kearn-Sayre syndrome (respiratory-chain disorders)
- With cardiac/multiorgan failure: see "Cardiac Failure, Collapse"
- With cardiomyopathy (see "Cardiomyopathy"):
- Triose-phosphate-isomerase deficiency
- Adrenal dysfunction (hyperkalemia)
- Hypoparathyroidism (hypocalcemia)
- Thiamine-deficiency-dependent states

### Bleeding Tendency, Hemorrhagic Syndromes

- Glycogenosis types Ia and Ib
- Gaucher disease
- Inborn errors with severe liver failure
- Inherited disorders of coagulation
- Severe thrombopenias

### Bone Crisis

See "Rheumatology".

### Cardiac Failure, Collapse

- With tamponade, multiorgan failure
  - CDG
- With apparently primitive heartbeat disorders: see "Arrhythmias, Conduction Defects (Heartbeat Disorders)"
  - With cardiomyopathy: see "Cardiomyopathy"

### Dehydration (Attacks)

- With severe diarrhea (digestive causes):
  - Glucose–galactose malabsorption
  - Congenital lactase deficiency
  - Congenital chloride diarrhea
  - Sucrase-isomaltase deficiency
  - Acrodermatitis enteropathica
- With ketoacidosis (organic acidurias):
  - Diabetic coma
  - MA, PA, IVA
  - 3-Ketothiolase deficiency
  - Hydroxyisobutyric aciduria
- With renal tubular dysfunction (tubulopathies; see "Tubulopathy"):
  - Cystinosis
  - Nephrogenic diabetes insipidus
  - Renal tubular acidosis (RTA) types I, II and IV
  - Inborn errors of energy metabolism
- With salty sweat:
  - Cystic fibrosis
- With salt-losing syndrome:
  - Adrenal dysfunctions
  - Tubulopathies (see "Tubulopathy")

### Exercise Intolerance (Myoglobinuria, Cramps, Muscle Pain)

- Glycolytic defects (muscle "glycogenosis"):
  - Phosphorylase deficiency (MacArdle)
  - Phosphofructokinase deficiency
  - Phosphoglycerate-kinase deficiency
  - Phosphoglycerate-mutase deficiency
  - Lactate dehydrogenase deficiency
  - Glucose-6-phosphate-dehydrogenase deficiency
  - Phosphorylase-B-kinase deficiency
- FAO defects:
  - Carnitine-palmitoyl transferase II
  - VLCAD, LCHAD, carnitine translocase, trifunctional enzyme
  - Short-chain 3-hydroxy-acyl-CoA dehydrogenase (SCHAD; restricted to muscles), medium-chain 3-ketoacyl-CoA-thiolase
  - Others undescribed (CPT-I, SCAD?)
- Miscellaneous:
  - Myoadenylate-deaminase deficiency
  - Respiratory chain disorders
  - Lipoamide dehydrogenase deficiency
  - Duchenne and Becker muscular dystrophies
  - Idiopathic, familial, recurrent myoglobinuria

### Hematological Symptoms (Anemia, Leukopenia, Thrombopenia, Pancytopenia)

See "Hematology".

### Hyperventilation Attacks

  - Hyperammonemias
  - Metabolic acidosis
  - Joubert syndrome

- Leigh syndrome (idiopathic or due to various inborn errors)
- Rett syndrome (only girls)

### Liver Failure (Ascites, Edema)

- Age at onset: congenital (hydrops fetalis)
  - GM-1 gangliosidosis (Landing)
  - Niemann-Pick A and C
  - Galactosialidosis
  - Sialidosis type II
  - Mucopolysaccharidosis type VII
  - Barth hemoglobin syndrome
- Age at onset: neonatal (<1 month)
  - Galactosemia
  - Hereditary fructose intolerance
  - Fructose-bisphosphatase deficiency
  - Neonatal hemochromatosis
  - Respiratory chain disorders
  - Tyrosinemia type I (after 3 weeks)
  - FAO disorders
  - Mevalonic aciduria
- Age at onset: infancy
  - Same defects as in neonatal period
  - Ketogenesis defects
  - Phosphoenolpyruvate carboxykinase deficiency
  - PC deficiency
  - α1-Antitrypsin deficiency
  - Urea cycle defects
  - Wolman disease
  - Cholesteryl-ester storage disease
  - S-adenosylhomocysteine hydrolase deficiency
  - Familial hepatic fibrosis with exsudative enteropathy
  - CDG
  - Cystic fibrosis
- Age at onset: childhood to adolescence
  - Wilson disease

### Psychiatric Symptoms (Hallucinations, Delirium, Anxiety, Schizophrenic-Like Episodes)

See "Recurrent Attacks of Ataxia and Psychiatric Symptoms" and "Dementia, Psychosis, Schizophrenia, Behavior Disturbances".

### Reye Syndrome

- FAO disorders
- Ketogenesis defects
- Urea cycle defects
- Gluconeogenesis defects
- Respiratory chain disorders
- Organic acidurias
- Fructose intolerance
- GA type I

### Skin Rashes (Photosensitivity)

- Age at onset: neonatal to childhood
  - Congenital erythropoietic porphyria
  - Erythrohepatic porphyria
  - Erythropoietic protoporphyria
  - Hartnup disease
  - Respiratory chain disorders
  - Mevalonic aciduria (with fever and arthralgia)
  - Xeroderma pigmentosum (nine varieties)
- Age at onset: adulthood
  - Porphyria variegata
  - Hereditary coproporphyria
  - Porphyria cutanea tarda

### Sudden Infant Death (and Near-Miss)

See "Reye syndrome".

### Vomiting

See "Coma, Strokes, and Attacks of Vomiting with Lethargy".

## III.2 Metabolic Derangements and Diagnostic Tests

### Initial Approach, Protocol of Investigation

The initial approach to the late-onset, acute forms of inherited metabolic disorders, like the approach to acute neonatal distress, is based on the proper use of a few screening tests. As with neonates, the laboratory data listed in Table 1.1 must be collected during the acute attack, all at the same time, both before and after treatment.

### Main Metabolic Presentations

### Metabolic Acidosis

Metabolic acidosis (Fig. 1.1) is a very common symptom in pediatrics. It can be observed in a large variety of acquired circumstances, including infections, severe catabolic states, tissue anoxia, severe dehydration, and intoxication, all of which should be ruled out. However, these circumstances can also trigger an acute decompensation of an unrecognized inborn error of metabolism. The presence or absence of ketonuria associated with metabolic acidosis is the major clinical key to the diagnosis.

When metabolic acidosis is not associated with ketosis, PDH deficiency, FAO disorders, and some disorders of gluconeogenesis should be considered, particularly when there is moderate to severe hyper-

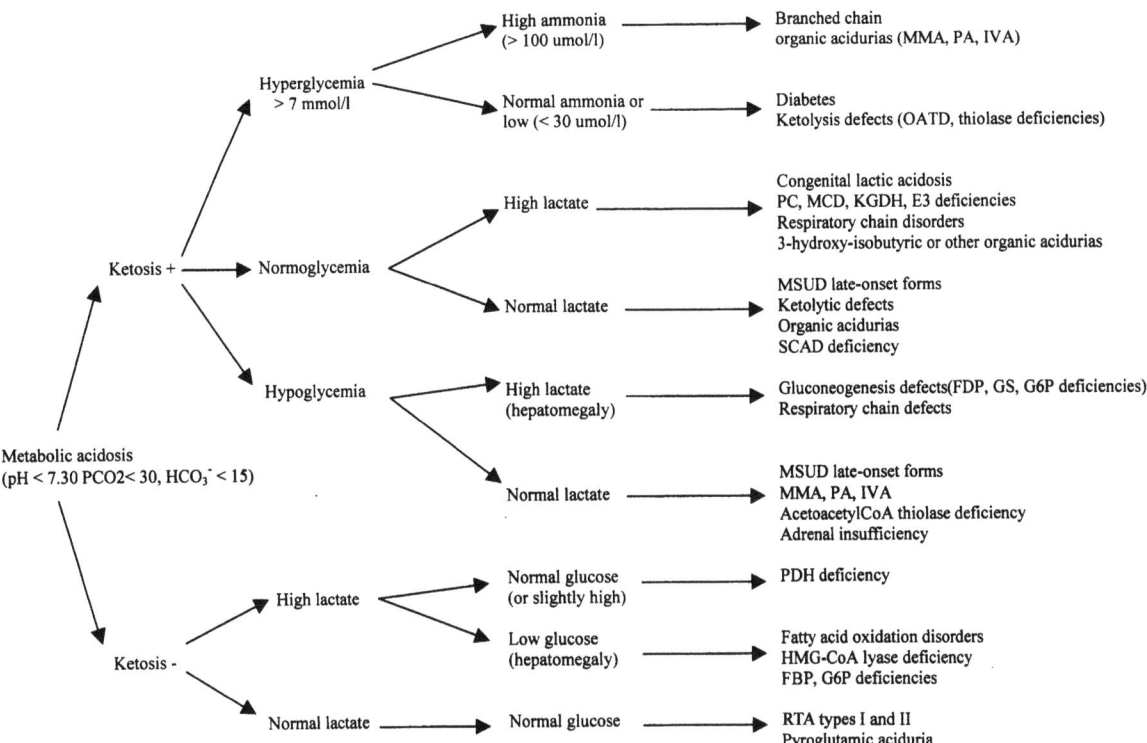

**Fig. 1.1. Metabolic acidosis.** *E3*, lipoamido oxido reductase; *FBP*, fructose bisphosphatase; *G6P*, glucose-6-phosphatase; *GS*, glycogen synthase; *HMG-coenzyme A*, 3-hydroxy-3-methylglutaryl coenzyme A; *IVA*, isovaleric acidemia; *KGDH*, α-ketoglutarate dehydrogenase; *MCD*, multiple carboxylase deficiency; *MMA*, methylmalonic acidemia; *MSUD*, maple syrup urine disease; *OATD*, oxoacid coenzyme-A transferase; *PA*, propionic acidemia; *PC*, pyruvate carboxylase; *PDH*, pyruvate dehydrogenase; *RTA*, renal tubular acidosis; *SCAD*, short-chain acyl-coenzyme-A dehydrogenase

lactacidemia. All these disorders (except PDH deficiency) have concomitant fasting hypoglycemia. Although fructose bisphosphatase deficiency is classically considered to give rise to ketoacidosis, some patients were referred with the tentative diagnosis of FAO disorders because of low concentrations of ketone bodies during hypoglycemia.

When metabolic acidosis occurs with a normal anion gap and without hyperlactacidemia or hypoglycemia, the most frequent diagnosis is RTA types I and II. Pyroglutamic aciduria also can present early in life, with permanent, isolated metabolic acidosis, which can be mistaken for RTA type II.

Ketoacidotic states compose the second large group of inherited metabolic disorders. The range of serum ketone body concentration varies with age and nutritional state (Chap. 2).

Many metabolic disorders of childhood may lead to ketoacidosis, including insulin-dependent diabetes, inborn errors of branched-chain-amino-acid metabolism, congenital lactic acidoses (such as MCD and PC deficiencies), inherited defects in enzymes of gluconeogenesis and glycogen synthesis (glycogen synthase, GS), and ketolytic defects.

When metabolic acidosis is associated with ketosis, the first parameter to be considered is the glucose level, which can be elevated, normal, or low. When ketoacidosis is associated with hyperglycemia, the classic diagnostic is diabetic ketoacidosis. However, organic acidurias (such as PA, MA, and IVA) and ketolytic defects can also be associated with hyperglycemia and glycosuria, mimicking diabetes. The distinction between the different disorders is also based on ammonia and lactate levels, which are generally increased in organic acidemias and normal or low in ketolytic defects. When ketoacidosis is associated with hypoglycemia, the first classical group of diseases to be considered is the gluconeogenesis and glycogenosis defects. The main suggestive symptoms are hepatomegaly and hyperlactacidemia, though they are not constant findings. When there is no significant hepatomegaly, late-onset forms of MSUD and organic aciduria must also be considered. A classic differential diagnosis is adrenal insufficiency, which can strike as a ketoacidotic attack with hypoglycemia.

When blood glucose levels are normal, congenital lactic acidosis must be considered in addition to the disorders discussed above. According to this schematic approach to inherited ketoacidotic states, a simplistic diagnosis of fasting ketoacidosis or ketotic hypoglycemia should be questioned when there is concomitant severe metabolic acidosis.

## Ketosis

While ketonuria should always be considered abnormal in neonates, it is a physiological finding in many circumstances of late infancy, childhood, and even adolescence. However, as a general rule, one can assume that hyperketosis at a level that produces metabolic acidosis is not physiologic.

Ketosis (Fig. 1.2), which is not associated with acidosis, hyperlactacidemia, or hypoglycemia, is likely to be a normal physiological reflection of the nutritional state (fasting, catabolism, vomiting, medium-chain triglyceride-enriched or other ketogenic diets). Of interest are ketolytic defects (succinyl-CoA-transferase and 3-ketothiolase deficiencies) that can present as permanent moderate ketonuria occurring mainly in the fed state at the end of the day.

Significant fasting ketonuria without acidosis is often observed in glycogenosis type III (in childhood) and in the very rare GS defect (in infancy). In both disorders, there is hepatomegaly (inconstant in GS), fasting hypoglycemia, and postprandial hyperlactacidemia.

Finally, ketosis without acidosis is observed in ketotic hypoglycemias of childhood and in association with hypoglycemias due to adrenal insufficiency. SCHAD, SCAD, and MCAD deficiencies can present as recurrent attacks of ketotic hypoglycemia, as these enzymes are both sufficiently downstream the β-oxidation pathway to allow the synthesis of some ketones from long-chain fatty acids [16]. Absence of ketonuria in hypoglycemic states, fasting, and catabolic circumstances induced by vomiting, anorexia, or intercurrent infections, is an important observation, suggesting an inherited disorder of FAO or ketogenesis. It can also be observed in hyperinsulinemic states at any age and in growth-hormone deficiency in infancy.

## Hyperlactacidemia

Lactate and pyruvate are normal metabolites. Their plasma levels reflect the equilibrium between their cytoplasmic production from glycolysis and their consumption by different tissues. The blood levels of lactate and pyruvate and the L/P ratio reflect the cytosolic redox states of the cells.

Blood lactate accumulates in circulatory collapse, hypoxic insult, and other conditions involving failure of cellular respiration. These conditions must be ruled out before an inborn error of lactate–pyruvate oxidation is sought. Persistent hyperlactacidemias (Table 1.5) can result from many acquired conditions, such as diarrhea, persistent infections (mainly of the urinary tract), hyperventilation, and hepatic failure. Ketosis is absent in most hyperlactacidemias secondary to tissue hypoxia, while it is a nearly constant finding in most inborn errors of metabolism (except in PDH deficiency, glycogenosis type I, and FAO disorders). However, the level of lactic acidemia is not discriminating; some acquired disorders are associated with very high levels, whereas some inborn errors of lactate–pyruvate metabolism cause only moderate hyperlactacidemia. Nutritional state also influences the levels of lactate and pyruvate.

Once the organic acidurias, urea-cycle defects (mainly citrullinemia), and FAO defects that cause secondary hyperlactacidemia have been excluded as possible diagnoses, four types of inherited disorders remain to be considered:

1. Disorders of liver glycogen metabolism
2. Disorders of liver gluconeogenesis
3. Abnormalities of lactate–pyruvate oxidation, the PDH complex, or Krebs cycle defects

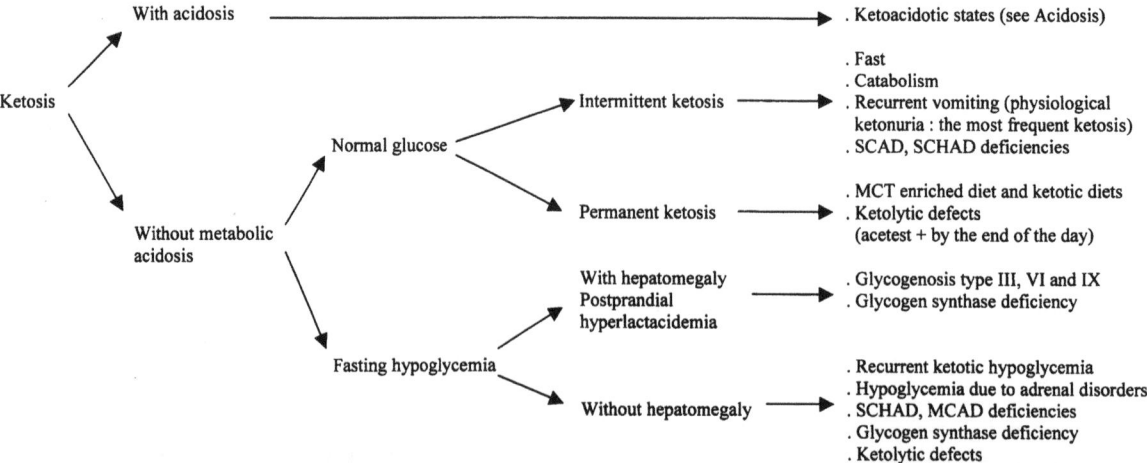

**Fig. 1.2.** Ketosis (also Fig. 1.1). *MCAD*, medium-chain acyl-coenzyme-A dehydrogenase; *MCT*, medium-chain triglycerides; *SCAD*, short-chain acyl coenzyme A dehydrogenase; *SCHAD*, hydroxy short-chain acyl-coenzyme-A dehydrogenase

**Table 1.5.** HYPERLACTACIDEMIA: Diagnostic approach

| Time of occurrence | Main clinical signs | Redox potential states | Diagnosis |
| --- | --- | --- | --- |
| Only after feeding (or exacerbated after feeding) | Hepatomegaly Fasting ketotic hypoglycemia Neurologic signs Encephalomyopathy | Not diagnostic | Glycogenosis type III Glycogen-synthase deficiency |
| | | Normal L/P ratio No ketosis | PDH deficiency |
| | | L/P high, 3OHB/AA low, postprandial hyperketosis | PC (citrullinemia), MCD, α-KGDH (isolated or multiketodecarboxylase) deficiency |
| | | L/P high, 3OHB/AA high, postprandial ketosis | Respiratory-chain disorders (3-methylglutaconic aciduria, Krebs-cycle intermediates) |
| | | L/P high, no ketosis | Respiratory-chain disorders |
| Only after fasting (or exacerbated after fasting) | Prominent hepatomegaly Hypoglycemia | Not diagnostic | Glycogenosis type I FBP (ketosis inconstant) |
| | Moderate hepatomegaly Hypoketotic hypoglycemia | Not diagnostic | Fatty-acid-oxidation disorders (cardiac, muscle symptoms) FBP |
| Permanent | Moderate hyperlactacidemia, recurrent attacks of ketoacidosis | Not diagnostic | Organic acidurias (MMA, PA, IVA) |
| | Predominant hyperammonemia | Not diagnostic | Urea-cycle defects (in neonates) |
| | Predominant hypoglycemia Hepatomegaly | Not diagnostic | Glycogenosis type I Fructose-bisphosphatase deficiency |
| | Neurological signs, encephalomyopathy, important hyperlactacidemia (>10 mM) | Highly diagnostic (see "After feeding") | Congenital lactic acidemias (see "After feeding") |

*3OHB*, 3-hydroxybutyrate; *AA*, acetoacetate; *IVA*, isovaleric acidemia; *KGDH*, ketoglutarate dehydrogenase; *L*, lactate; *MCD*, multiple carboxylase deficiency; *MMA*, methylmalonic acidemia; *P*, pyruvate; *PA*, propionic acidemia; *PC*, pyruvate carboxylase; *PDH*, pyruvate dehydrogenase; *FBP*, Fructose bisphosphatase deficiency

4. Deficient activity in one of the components of the respiratory chain

The diagnosis of hyperlactacidemia is largely based on two metabolic criteria: time of occurrence of lactic acidosis relative to feeding and determinations of L/P and ketone body ratios.

In disorders of gluconeogenesis (fructose-bisphosphatase and glucose-6-phosphatase deficiencies), hyperlactacidemia reaches its maximum level when the patient is fasting and hypoglycemic (up to 15 mM). By contrast, in glycogenosis types III and VI and in GS deficiency, hyperlactacidemia is observed only in the postprandial period in patients on a carbohydrate-rich diet. Here, hyperlactacidemia never exceeds 7 mM. In PC deficiency, hyperlactacidemia is present in both the fed and fasted states but tends to decrease with a short fast. In disorders of PDH, α-ketoglutarate-dehydrogenase deficiencies, and respiratory-chain disorders, maximum lactate levels are observed in the fed state (although all hyperlactacidemias exceeding 7 mM appear more or less permanent). In these disorders, there is a real risk of missing a moderate (though significant) hyperlactacidemia when the level is checked only before breakfast after an overnight fast (as is usual for laboratory determinations).

The second clue to diagnosis is the simultaneous determination of plasma L/P and 3OHB/AA ratios before and after meals. These ratios indirectly reflect cytoplasmic (L/P) and mitochondrial (3OHB/AA) redox-potential states. They must be measured in carefully collected blood samples (Chap. 2).

Three hyperlactacidemia/pyruvicacidemia profiles are nearly pathognomonic of an inborn error of lactate–pyruvate metabolism:

1. When hyperpyruvicemia is associated with a normal or low (<10) L/P ratio without hyperketonemia, PDH deficiency is highly probable, regardless of the lactate level
2. When the L/P ratio is very high (>30) and is associated with postprandial hyperketonemia and a normal or low 3OHB/AA ratio (<1.5), a diagnosis of PC deficiency (isolated or secondary to biotinidase or holocarboxylase-synthase deficiency) or α-ketoglutarate-dehydrogenase deficiency is highly probable

3. When both L/P and 3OHB/AA ratios are elevated and associated with a significant postprandial hyperketonemia, respiratory-chain disorders should be suspected

All other situations, especially when the L/P ratio is high without hyperketonemia, are compatible with respiratory-chain disorders, but all acquired anoxic conditions should also be ruled out (see above).

## Hypoglycemia

Our approach to hypoglycemia (Table 1.6) is based on the following clinical criteria:

1. Liver size
2. Characteristic time schedule of hypoglycemia (unpredictable, postprandial, or after fasting)
3. Association with lactic acidosis
4. Association with hyperketosis or hypoketosis

Other clinical signs of interest are hepatic failure, vascular hypotension, dehydration, short stature, neonatal body size (head circumference, weight, height), and evidence of encephalopathy, myopathy, or cardiomyopathy. Liver size can be used to separate the hypoglycemias into two large groups, as discussed below.

*Hypoglycemia with Permanent Hepatomegaly*
Most hypoglycemias associated with permanent hepatomegaly are due to inborn errors of metabolism. All conditions, acquired or inherited, that are associated with severe liver failure can give rise to severe hypoglycemia, which appears after 2–3 h of fasting and involves moderate lactic acidosis and no ketosis. When hepatomegaly is the most prominent feature without liver insufficiency, gluconeogenesis defects (involving glucose-6-phosphatase or fructose bisphosphatase), glycogenosis type III, and GS deficiency are the most probable diagnoses. Disorders presenting with hepatic fibrosis and cirrhosis, such as hereditary tyrosinemia type I, can also give rise to hypoglycemia. The late-onset form of hereditary fructose intolerance is rarely, if ever, revealed by isolated postprandial hypoglycemic attacks. S-adenosyl-homocysteine-hydrolase deficiency presents with fasting hypoglycemia and hepatocellular insufficiency, often triggered by high protein or methionine ingestion, and is associated with hepatic fibrosis, mental retardation, and marked hypermethioninemia. Respiratory-chain disorders can present with hepatic failure and hypoglycemia. The amazing familial association of hepatic fibrosis and exsudative enteropathy can strike by hypoglycemia early in infancy [17] and is probably related to CDG syndrome [18, 19].

*Hypoglycemia without Permanent Hepatomegaly*
It is important to discover the timing of hypoglycemia and to search for metabolic acidosis and ketosis when the patient is hypoglycemic. As a general rule, most (if not all) hypoglycemias that are not accompanied by hepatomegaly and are caused by inborn errors of metabolism appear after at least 8 h of fasting. This is especially true of hypoglycemias due to inherited FAO disorders (except in the neonatal period). Conversely, unpredictable postprandial or very short-fasting hypoglycemias (2–6 h) are mostly due to hyperinsulinism and growth-hormone deficiency or related disorders. When ketoacidosis is present at the time of hypoglycemia, organic acidurias, ketolytic defects, late-onset MSUD, and glycerol-kinase deficiencies should be considered. Here, hypoglycemia is very rarely the initial metabolic abnormality. Adrenal insufficiencies should be systematically considered in the differential diagnosis, especially when vascular hypotension, dehydration, and hyponatremia are present. Severe hypoglycemia with metabolic acidosis and absence of ketosis, in the context of Reye syndrome, suggests HMG-CoA lyase deficiency or FAO disorders. Fasting hypoglycemia with ketosis occurring mostly in the morning and in the absence of metabolic acidosis suggests recurrent functional ketotic hypoglycemia, which presents mostly in late infancy or childhood in children who were small for their gestational age or in children with macrocephaly. All types of adrenal insufficiencies (peripheral or central) can share this presentation. SCHAD and MCAD can also present as recurrent attacks of ketotic hypoglycemia [16]. Conversely, in our experience, this pattern is rarely associated with inborn errors of metabolism. Hypoketotic hypoglycemias encompass several groups of disorders, including hyperinsulinemic states, growth-hormone deficiency, inborn errors of FAO, and ketogenesis defects (see "Ketosis" above).

## Hyperammonemia

The diagnostic approach to hyperammonemia is developed in Chap. 17.

## IV Chronic and Progressive General Symptoms

As already stated, many apparently delayed-onset acute presentations of inherited disorders are preceded by insidious premonitory symptoms that have been ignored or misinterpreted. These symptoms are grouped schematically into three categories: digestive, neurologic, and muscular symptoms.

**Table 1.6.** HYPOGLYCEMIA: general approach

| Leading symptoms | Other signs | Age at onset | Diagnosis |
|---|---|---|---|
| *With permanent hepatomegaly* | | | |
| Permanent short-fast hypoglycemia | Severe liver failure, hepatic necrosis | Neonatal to early in infancy | Galactosemia, HF1, tyrosinosis type I, neonatal hemochromatosis, respiratory-chain disorders, other severe hepatic failure |
| Fibrosis, cirrhosis | Postprandial hypoglycemia (triggered by fructose), vomiting | Neonatal to early infancy | HF1, glycerol intolerance |
| | Mental retardation Hypermethioninemia Hepatic failure induced by methionine | Early in infancy | Glycogenosis type IV SAH hydrolase deficiency Respiratory-chain disorders |
| | Exsudative enteropathy, cholangitis attacks | Early in infancy | Familial hepatic fibrosis with exsudative enteropathy |
| | Short-fast hypoglycemia | | CDG syndrome |
| Isolated hepatomegaly | Fasting hypoglycemia and lactic acidosis | Infancy | G6P deficiency, FDP deficiency |
| | Ketosis | | PEPCK deficiency |
| | Protuberant abdomen Fasting hypoglycemia and ketosis, postprandial hyperlactacidemia | Infancy | Glycogenosis type III Glycogen-synthase deficiency |
| Hypotonia | Abnormal glycosylated transferrin | Infancy | CDG syndrome |
| Failure to thrive, chronic diarrhea | Fanconi-like tubulopathy, postprandial hyperglycemia | Infancy | Fanconi-Bickel syndrome (GLUT-II mutations) |
| *Without permanent hepatomegaly* | | | |
| With ketoacidosis | Recurrent attacks, hyperlactacidemia | Infancy to childhood | Organic acidurias, late-onset MSUD, ketolysis defects, glycerol-kinase deficiency, FDP deficiency, SCHAD, MCAD, respiratory-chain disorders |
| | Dehydration, collapse, hyponatremia | Neonatal to childhood | Adrenal insufficiency (central or peripheral) |
| Acidosis without ketosis | Moderate hyperlactacidemia, Reye syndrome (with muscle/cardiac symptoms) | Neonatal to infancy | HMG-CoA lyase (frequent), HMG-CoA synthase (rare), FAO disorders (frequent), idiopathic Reye syndrome |
| Ketosis without acidosis | Fasting hypoglycemia, low lactate levels, small size for age, macrocephaly | 1–6 years | Recurrent ketotic hypoglycemia, adrenal insufficiency (central or peripheral), SCHAD, MCAD, glycogen synthase, ketolysis defects |
| Without acidosis or ketosis | Unpredictable and postprandial hypoglycemia reactive to glucagon | Neonatal to childhood | Hyperinsulinisms, cortisol deficiency, CDG syndrome, Munchausen by proxy |
| | Short stature, short-fast hypoglycemia | Infancy | Growth-hormone deficiency and related disorders |
| | Long-fast hypoglycemia, Reye syndrome, moderate hepatomegaly, transient cytolysis | Neonatal to infancy | FAO disorders (frequent), HMG-CoA lyase (rare), FDP (rare), HMG-CoA synthase (rare) |

*CDG*, congenital defects of glycosylation; *FAO*, fatty-acid oxidation; *FDP*, fructose bisphosphatase; *G6P*, glucose-6-phosphatase; *GLUT*, glucose transporter; *HFI*, hereditary fructose intolerance; *HMG-CoA*, 3-hydroxy-3-methylglutaryl-coenzyme-A; *MCAD*, medium-chain acyl-coenzyme-A dehydrogenase; *MSUD*, maple-syrup-urine disease; *PEPCK*, phosphoenolpyruvate carboxykinase; *SAH*, S-adenosyl homocysteine hydrolase; *SCHAD*, short-chain 3-hydroxy-acyl-coenzyme-A dehydrogenase

## IV.1 Digestive Symptoms

Gastrointestinal symptoms (anorexia, failure to thrive, osteopenia, chronic vomiting) occur in a wide variety of inborn errors of metabolism. Unfortunately, their cause often remains unrecognized, thus delaying the correct diagnosis. Persistent anorexia, feeding difficulties, chronic vomiting, failure to thrive, frequent infections, osteopenia, and generalized hypotonia in a context of chronic diarrhea are the presenting symptoms in a number of constitutional and acquired diseases in pediatrics. They are easily misdiagnosed as cow's-milk-protein intolerance, celiac disease, chronic ear, nose, or throat infections, late-onset chronic pyloric stenosis, etc. Congenital immunodeficiencies are also frequently considered, although only a few present early in infancy with such a clinical picture.

From a pathophysiologic viewpoint, it is possible to define two groups of inborn errors of metabolism presenting with chronic diarrhea and failure to thrive:

1. Disorders of the intestinal mucosa or the exocrine function of the pancreas – for example, congenital chloride diarrhea, glucose–galactose malabsorption, lactase- and sucrase-isomaltase deficiencies, aβ-lipoproteinemia type II (Anderson disease), enterokinase deficiency, acrodermatitis enteropathica, and selective intestinal malabsorption of folate and vitamin B12
2. Systemic disorders that are also accompanied by digestive abnormalities

In clinical practice, these groups are sometimes very difficult to distinguish, because a number of specific intestinal disorders can elicit various systemic clinical abnormalities and vice versa. This is summarized in Table 1.7.

## IV.2 Neurologic Symptoms

Neurologic symptoms are very frequent in inborn errors and encompass progressive psychomotor retardation, seizures, a number of neurologic abnormalities in both the central and peripheral nervous system, and sensorineural defects. It must be stressed that a large number of inborn errors of intermediary metabolism present with nonspecific early progressive developmental delay, poor feeding, hypotonia, some degree of ataxia, and frequent autistic features. The list has lengthened rapidly as new laboratory techniques have been applied. The relationship between clinical symptoms and biochemical abnormalities is not always firmly established. Some aminoacidopathies that were first described in the late 1950s and 1960s, when plasma and urine amino-acid chromatography was systematically used in studying mentally retarded children, must now be questioned as definite causes of neurologic disturbance. This is the case for histidinemia, hyperlysinemia, hyperprolinemia, α-amino-adipic aciduria, and Hartnup "disease".

The same story may now be developing with organic acidurias, so it is more and more important to try to define pathophysiologic links between clinical symptoms and metabolic disturbances. Conversely, it is more and more difficult to screen patients on clinical grounds when the clinical symptoms consist only of developmental delay, microcephaly, hypotonia, and convulsions. Among the new categories of inborn errors of intermediary metabolism that can present with uninformative clinical manifestations are, for example, adenylosuccinase deficiency, dihydropyrimidine-dehydrogenase deficiency, 4-hydroxybutyric aciduria, L-2- and D-2-hydroxyglutaric acidurias, and many other kinds of inborn errors (bottom of Table 1.8). These disorders rarely, if ever, cause true development arrest; rather, they cause progressive subacute developmental delay. Conversely, there is still an important gap between neurologic descriptions and biologic investigations. Many well-known heritable neurologic or polymalformative syndromes have not been considered from a pathophysiologic perspective and should be submitted to a comprehensive biochemical evaluation. This is illustrated, for example, by the story of Canavan disease, in which N-acetylaspartic aciduria was only found in 1988, even though the clinical phenotype had been identified in 1949 and the procedure for identifying N-acetylaspartate in urine was available in 1972.

In the following pages, neurological symptoms are presented according to two different viewpoints. The first is a synthetic view according to the age at onset and the association of neurologic and extraneurologic signs ["Progressive Neurologic and Mental Deterioration Related to Age (Overview)" and Tables 1.8–1.11]. The second is an analytical view giving an extensive alphabetical list of disorders for each neurologic symptom (remainder of the section). Of course, these two views are complementary and involve inevitable redundancies. It is always recommended that one look at both.

### Progressive Neurologic and Mental Deterioration Related to Age (Overview)

Tables 1.8–1.11 present a general approach to inborn errors of metabolism involving neurologic and/or mental deterioration. Diseases are classified according to their age at onset, the presence or absence of associated extraneurologic signs, and the neurologic presentation itself; the last is based largely on the clinical classification of Lyon and Adams [20]. Inborn errors of metabolism with neurologic signs presenting in the neonate (birth to 1 month; Table 1.2) and those presenting intermittently as acute attacks of coma, lethargy, ataxia, or acute psychiatric symptoms, were presented earlier (Tables 1.3, 1.4).

### Early Infancy (1–12 Months)

Three general categories can be identified.

*Category 1: Disorders Associated with Extraneurologic Symptoms*
Visceral signs appear in lysosomal disorders. A cardiomyopathy (associated with early neurologic dysfunction, failure to thrive, and hypotonia), sometimes responsible for cardiac failure, is suggestive of respiratory-chain disorders, D-2-hydroxyglutaric aciduria (with atrioventricular block), or CDG syndrome.

Table 1.7. Chronic diarrhea, poor feeding, vomiting, failure to thrive

| Leading symptoms | Other signs | Age at onset | Diagnosis |
| --- | --- | --- | --- |
| Severe watery diarrhea, attacks of dehydration | No meconium, non-acidic diarrhea, metabolic alkalosis, hypochloremia, and hypochloruria | Congenital to infancy | Congenital chloride diarrhea |
| | Acidic diarrhea, reducing substances in stools | Neonatal | Glucose–galactose malabsorption, lactase deficiency |
| | Acidic diarrhea, reducing substances in stools; after weaning, starch or dextrins are added to the diet | Neonatal to infancy | Sucrase isomaltase deficiency |
| | Skin lesions, alopecia (late onset), failure to thrive | Neonatal or after weaning | Acrodermatitis enteropathica |
| Fat-soluble vitamin malabsorption Hypocholesterolemia, osteopenia, steatorrhea | Unexplained cholestatic jaundice | Neonatal to infancy | Bile acid synthesis defects, infantile Refsum disease |
| | Hepatomegaly, hypotonia, slight mental retardation, retinitis pigmentosa, deafness (after at least 1 year) | Infancy | Infantile Refsum disease, CDG syndrome type I |
| | Abdominal distension, failure to thrive, anorexia, acanthocytosis, peripheral neuropathy, ataxia, retinitis pigmentosa[a] | Infancy | Aβ-lipoproteinemia types I and II, (no acanthocytes, no neurological sign in type II) |
| | Pancreatic insufficiency, neutropenia, pancytopenia | Early in infancy | Pearson syndrome, Schwachman syndrome |
| Severe failure to thrive, anorexia, poor feeding, with predominant hepatosplenomegaly | Severe hypoglycemia, inflammatory bowel disease, neutropenia, recurrent infections, hepatomegaly without splenomegaly | Neonatal to early infancy | Glycogenosis type Ib (glucose-6-phosphate translocase deficiency) |
| | Hypotonia, vacuolated lymphocytes, adrenal-gland calcifications | Neonatal | Wolman disease |
| | Recurrent infections, inflammatory bowel disease, dermatitis, stomatitis | Infancy | Chronic granulomatosis (X-linked) |
| | Megaloblastic anemia, stomatitis, muscle weakness, peripheral neuropathy, homocystinuria, MMA | 1–5 years | Intrinsic factor deficiency |
| | Neutropenia, osteopenia, recurrent attacks of hyperammonemia, interstitial pneumonia, orotic aciduria | Infancy | Lysinuric protein intolerance |
| | Recurrent fever, inflammatory bowel syndrome, hyper IGD, skin rashes, arthralgias | Infancy | Mevalonate kinase deficiency |
| Severe failure to thrive, anorexia, poor feeding, with megaloblastic anemia | Oral lesions, stomatitis, neuropathy, infections, pancytopenia, peripheral homocystinuria, MMA | 1–2 years | Transcobalamin-II deficiency, intrinsic factor deficiency |
| | Stomatitis, infections, peripheral neuropathy, intracranial calcifications | Infancy | Congenital folate malabsorption |
| | Severe pancytopenia, vacuolization of marrow precursors, exocrine pancreas insufficiency, lactic acidosis | Neonatal | Pearson syndrome |

**Table 1.7.** (*Contd.*)

| | | | |
|---|---|---|---|
| Severe failure to thrive, anorexia, poor feeding, hypotonia, no significant hepatosplenomegaly, no megaloblastic anemia | Severe hypoproteinemia, putrefaction diarrhea | Infancy | Enterokinase deficiency |
| | Diarrhea after weaning, cutaneous lesions (periorificial), alopecia, modest diarrhea, low alkaline phosphatase, low plasma zinc | Infancy | Acrodermatitis enteropathica |
| | Ketoacidosis, metabolic attacks, frequent infections, vomiting | Infancy | Organic acidurias (MMA, PA), mitochondrial-DNA deletions (lactic acidosis) |
| | Vomiting, lethargy, hypotonia, metabolic attacks, hyperammonemia | Infancy | Urea-cycle defects (mainly OTC) |
| | Frequent infections, lymphopenia, bone changes, severe combined immune deficiency | Infancy | Adenosine deaminase deficiency |
| | Developmental delay, relapsing petechiae, orthostatic acrocyanosis, recurrent attacks | | Ethylmalonic aciduria |

*CDG*, congenital defects of glycosylation; *MMA*, methylmalonic acidemia; *OTC*, ornithine transcarbamoylase; *PA*, propionic acidemia
[a] While abdominal distension, failure to thrive, anorexia, and acanthocytosis start in infancy, peripheral neuropathy, ataxia, and retinitis pigmentosa only appear after 5 years of evolution

**Table 1.8.** Progressive neurologic and mental deterioration with obvious extraneurological symptoms (1–12 months; see also Table 1.2)

| Leading symptoms | Other signs | Diagnosis |
|---|---|---|
| Visceral signs | Hepatosplenomegaly | Landing, I-cell disease |
| | Storage signs | Sialidosis type II, Niemann-Pick A |
| | | Lactosyl ceramidosis |
| | Hepatosplenomegaly, opisthotonos, spasticity, vegetative state | Gaucher type II |
| | Hepatomegaly | Peroxisomal disorders |
| | Retinitis pigmentosa | CDG syndrome |
| Hair and cutaneous symptoms | Steely, brittle hair | Menkes disease (X-linked), trichothiodystrophy, argininosuccinic aciduria |
| | Ichtyosis, spastic paraplegia | Sjögren-Larsson syndrome |
| | Alopecia, Cutaneous rashes | Biotinidase deficiency |
| | | Respiratory-chain disorders |
| | Peculiar fat pads on buttocks | CDG syndrome |
| | Cyanosis (generalized), hypertonicity | Cytochrome-b5-reductase deficiency |
| | Kernicterus, athetosis | Crigler-Najjar disease |
| | Acrocyanosis, petechiae | Ethylmalonic aciduria, EPEMA syndrome |
| Megabloblastic anemia | Failure to thrive, feeding difficulties, pigmentary retinopathy | Inborn errors of folate and cobalamin metabolism (see text), orotic aciduria |
| Cardiac symptoms | Cardiomyopathy | D-2-Hydroxyglutaric acidemia |
| | Heart failure | Respiratory-chain disorders |
| | Heartbeat disorders | CDG syndrome |
| Ocular symptoms | Cherry-red spot | Landing disease |
| | Hydrops fetalis | Galactosialidosis, sialidosis type I |
| | Myoclonic jerks | Tay-Sachs disease |
| | Macrocephaly | Sandhoff disease |
| | Optic atrophy | Canavan disease |
| | Nystagmus, dystonia, stridor | Pelizaeus-Merzbacher (X-linked) |
| | Retinitis pigmentosa | See text |
| | Abnormal eye movements | Aromatic-amino-acid-decarboxylase deficiency |
| | Strabism | CDG syndrome |
| | Supranuclear paralysis | Gaucher disease, Niemann-Pick type C |

*CDG*, congenital defects of glycosylation; *EPEMA*, encephalopathy, petechiae, and ethylmalonic aciduria

Abnormal hair and cutaneous signs appear in Menkes disease, Sjögren-Larsson syndrome, biotinidase deficiency, and respiratory-chain disorders. Peculiar fat pads of the buttocks and thick and sticky skin (like "tallow, *peau d'orange*"), and inverted nipples are highly suggestive of CDG syndrome. A generalized cyanosis, unresponsive to oxygen, suggests methemoglobinemia, which is associated with severe hypertonicity in cytochrome-b5-reductase deficiency. Kernicterus and athetosis are complications of Criggler-Najjar syndrome. The recently described EPEMA syndrome is characterized by an orthostatic acrocyanosis, relapsing petechiae, pyramidal signs, mental retardation, and recurrent attacks of lactic acidosis. The presence of megaloblastic anemia suggests an inborn error of folate and cobalamin (Cbl) metabolism. Ocular symptoms can be highly diagnostic signs, like cherry-red spot, optic atrophy, nystagmus, abnormal eye movements, and retinitis pigmentosa.

*Category 2: Disorders with Specific or Suggestive Neurological Signs*
Predominant extrapyramidal symptoms are associated with inborn errors of biopterin and aromatic-amino-acid metabolism, Lesch-Nyhan syndrome, cytochrome-b5-reductase deficiency, Criggler-Najjar syndrome, the early-onset form of GA type I, and cerebral creatine deficiency. Dystonia can also be observed as a subtle but revealing sign in X-linked Pelizaeus-Merzbacher syndrome.

Macrocephaly with startled response to sound, incessant crying, and irritability are frequent early signs in GM-2 gangliosidosis, Canavan disease, Alexander leukodystrophy, infantile Krabbe disease, and GA type I. Macrocephaly can be also a revealing sign in L-2-hydroxyglutaric aciduria and in respiratory-chain disorders due to complex-I deficiency (association with hypertrophic cardiomyopathy).

Recurrent attacks of neurologic crisis associated with progressive neurologic and mental deterioration suggest Leigh syndrome, which can present at any age from early in infancy to late childhood. Leigh syndrome is not a specific phenotype but, rather, is the clinical phenotype of any of several inborn errors of metabolism, some of which still remain to be identified. Recurrent stroke-like episodes often associated with anorexia, failure to thrive, and hypotonia can be revealing symptoms in urea-cycle defects (mostly OTC), late-onset MSUD, organic acidurias, GA type I, CDG syndrome, and respiratory-chain disorders. Thromboembolic accidents can be revealing signs of classical homocystinuria and CDG syndrome. Angelman syndrome sometimes displays a very suggestive picture, with early-onset encephalopathy, happy-puppet appearance, and epilepsy with a highly suggestive EEG pattern.

*Category 3: Disorders with Nonspecific Developmental Delay*
A large number of inborn errors present with nonspecific early progressive developmental delay, poor feeding, hypotonia, some degree of ataxia, frequent autistic features, and seizures. The list has lengthened rapidly as new laboratory techniques have been applied. The relationship between clinical symptoms and biochemical abnormalities is not always firmly established. It is more and more difficult to screen patients on clinical grounds when the clinical symptoms consist only of developmental delay, hypotonia, and convulsions. An increasing number of inborn errors of metabolism can masquerade as a cerebral palsy by presenting as a permanent impairment of movement or posture (Table 1.9).

Late-onset subacute forms of hyperammonemia (usually OTC deficiency in girls) can also be revealed by an apparently nonspecific early encephalopathy. Inborn errors of neurotransmitter synthesis, especially dopa-responsive dystonia due to cyclohydrolase deficiency, tyrosine-hydroxylase deficiency, and aromatic-L-amino-acid-decarboxylase deficiency, can masquerade as cerebral palsy. These disorders rarely if ever give rise to true development arrest; rather, they give rise to progressive subacute developmental delay. Recurrent attacks of seizures unresponsive to anticonvulsant drugs occurring in the first year of life is the revealing symptom of blood–brain-barrier glucose-transporter (GLUT-1) defect. The diagnosis relies on the finding of a low glucose level in the CSF while the simultaneous blood glucose level is normal.

## Late Infancy to Early Childhood (1–5 Years)

In this period, diagnosis becomes easier. Five general categories can be defined (Table 1.10).

*Category 1: with Visceral, Craniovertebral, Ocular, or Other Somatic Abnormalities*
When these symptoms are present and associated with a slowing or regression of development, diagnosis is usually easy. Mucopolysaccharidosis types I and II, mucolipidosis type III, oligosaccharidosis, Austin disease, Niemann-Pick disease type C, Gaucher disease type III, and lactosyl ceramidosis are usually easy to recognize. Mucolipidosis type IV, which causes major visual impairment by the end of the first year of life, sometimes associated with dystonia, presents with characteristic cytoplasmic membranous bodies in cells. In SanFilippo syndrome, coarse facies and bone changes may be very subtle or absent. Peroxisomal disorders may present at this age, with progressive mental deterioration, retinitis pigmentosa, and deafness, very similar to Usher syndrome type II. Pyrro-

**Table 1.9.** Progressive neurologic and mental deterioration with and without specific or suggestive neurologic signs (1–12 months)

| Leading symptoms | Other signs | Diagnosis |
|---|---|---|
| *With specific or suggestive neurologic signs* | | |
| Extrapyramidal signs | Major parkinsonism | Inborn errors of biopterin metabolism |
| | Abnormal neurotransmitters | Aromatic-amino-acid-decarboxylase deficiency, tyrosine-hydroxylase deficiency |
| | Choreoathetoid movements, self-mutilation | Lesch-Nyhan (X-linked) |
| | Bilateral athetosis, hypertonicity | Cytochrome-b5-reductase deficiency |
| | Dystonia | Pelizaeus Merzbacher (X-linked) |
| | Kernicterus syndrome | Criggler-Najjar syndrome |
| | Acute-onset pseudoencephalitis | GA type I |
| | Low creatinine | Creatine deficiency (GAMT deficiency) |
| Macrocephaly | Cherry red spot | Tay Sachs disease, Sandhoff disease |
| Startled response to sound | Myoclonic jerks | Canavan, Van Bogaert, Bertrand (aspartoacylase deficiency), Alexander disease |
| Ocular symptoms | Optic atrophy, incessant crying, irritability | Krabbe (infantile) |
| | Dystonia, choreoathetosis | GA-I, L-2-hydroxyglutaric aciduria |
| | Developmental delay, progressive irritability | Respiratory-chain disorders (complex I) |
| Recurrent attacks of neurologic crisis (Table 1.3) | Mental regression, failure to thrive, hyperventilation attacks | Leigh syndrome (PC and PDH deficiencies, respiratory chain disorders) |
| | Stroke-like episodes | Urea-cycle defects, MSUD, organic acidemias, GA-I, CDG syndrome, respiratory-chain disorders |
| | Thromboembolic accidents | Classical homocystinuria, CDG syndrome |
| *Without suggestive neurologic signs* | | |
| Evidence of developmental arrest | Infantile spasms, hypsarrhythmia, autistic features | Classical untreated phenylketonuria, inborn errors of biopterin metabolism, peroxisomal disorders, Rett syndrome |
| Nonspecific symptoms, apparently non-progressive disorder | Frequent autistic feature, poor feeding, failure to thrive, hypotonia, seizures | Hyperammonemia (late-onset, subacute), 4-hydroxybutyric aciduria, L-2-hydroxyglutaric aciduria, D-2-hydroxyglutaric aciduria, mevalonic aciduria |
| | Diverse neurologic findings simulating cerebral palsy | Adenylosuccinase deficiency, dihydropyrimidine-dehydrogenase deficiency, 3-methylglutaconic aciduria, fumarase deficiency, other organic acidurias, creatine deficiency, 3-phosphoglycerate-dehydrogenase deficiency, 3-phosphoserine-phosphatase deficiency, homocystinuria, Salla disease, neurotransmittors disorders, Angelman syndrome, blood–brain-barrier glucose-carrier deficiency |

*CDG*, congenital defects of glycosylation; *GA*, glutaric acidemia; *GAMT*, guanidino-acetate methyltransferase; *MSUD*, maple-syrup-urine disease; *PC*, pyruvate carboxylase; *PDH*, pyruvate dehydrogenase

line-5-carboxylate-synthase deficiency presents with slow progressive neurologic and mental deterioration, severe hypotonia, joint laxity, and congenital cataracts.

*Category 2: with Progressive Paraplegia and Spasticity*
Progressive paraplegia and spasticity reveal three disorders. Metachromatic leukodystrophy and neuroaxonal dystrophy strike between 12 months and 24 months of age and present with flaccid paraparesis, hypotonia, and weakness. CSF protein content and nerve conduction velocity are disturbed in the former but normal in the latter. Schindler disease is roughly similar to neuroaxonal dystrophy, though it is often associated with myoclonic jerks. Argininase deficiency is a rare disorder that presents early in infancy to childhood (2 months–5 years) with progressive spastic

**Table 1.10.** Progressive neurologic and mental deterioration (1–5 years)

| Symptoms | Diagnosis |
|---|---|
| *With visceral, craniovertebral, or other somatic abnormalities* | |
| Coarse facies, skeletal changes, hirsutism, corneal opacities | Hurler (MPS-I), Hunter (X-linked; MPS-II), SanFilippo (MPS-III), pseudo Hurler polydystrophy (MLP-III) |
| Coarse facies, subtle bone changes, hepatosplenomegaly, vacuolated lymphocytes, and lens/corneal opacities | Mannosidosis (gingival hyperplasia), fucosidosis (angiokeratoma), aspartylglucosaminuria (macroglossia, joint laxity), Austin disease (ichthyosis) |
| Hepatosplenomegaly, progressive dementia, myoclonic jerks vertical supranuclear ophthalmoplegia | Niemann-Pick type C and related disorders (late infantile form) |
| Splenomegaly with hepatomegaly, osseous lesions, ataxia, myoclonus, ophthalmoplegia | Gaucher type III (subacute neuronopathy) |
| Major visual impairment, blindness | Mucolipidosis type IV (corneal clouding) |
| Retinitis pigmentosa, deafness | Peroxisomal disorders, Usher type II |
| Cataract, joint laxity, hypotonia | Pyrroline-5-carboxylate-synthase deficiency |
| *With progressive paraplegia, weakness, hypotonia, or spasticity due to corticospinal-tract involvement or peripheral neuropathy* | |
| Flaccid paraparesis with pyramidal signs, high protein content in CSF | Metachromatic leukodystrophy (abnormal nerve conduction velocity) |
| Flaccid paraparesis, no change in CSF, optic atrophy, early mental regression | Neuroaxonal dystrophy, Schindler disease (normal nerve conduction velocity) |
| Progressive spastic diplegia, scissoring or "tiptoe" gait | Arginase deficiency (hyperargininemia, high orotic-acid excretion) |
| *With unsteady gait, uncoordinated movements due to cerebellar syndrome, sensory defects or involuntary movements* | |
| • Without disturbances of organic-acid excretion | |
| Ataxia with choreoathetosis, oculocephalic asynergia | Ataxia telangiectasia |
| Ataxia, difficulty in walking, mental deterioration (speech) | GM-1 gangliosidosis (Landing, late infantile form), spastic quadriparesis, pseudobulbar signs |
| Ataxia, spinocerebellar degeneration, psychotic behavior | GM-2 gangliosidosis (Tay-Sachs, Sandhoff, late infantile form) |
| Ataxia, pyramidal signs, (hemiplegia, paraplegia), vision loss | Krabbe disease (late infantile form, peripheral neuropathy) |
| Ataxia, muscular atrophy in lower extremities (peripheral neuropathy) | CDG<br>Peroxisomal biogenesis disorder |
| Seizures and myoclonic jerks, postictal coma, transient hemiplegia | Alpers syndrome (respiratory-chain disorders), hepatic syndrome, hyperlactacidemia |
| • With disturbances of organic- and amino-acid excretion | |
| Progressive ataxia, intention tremor, cerebellar atrophy | L-2-Hydroxyglutaric aciduria (spongiform encephalopathy) |
| Combined degeneration of the spinal cord | Cobalamin deficiencies (homocystinuria, low methionine) |
| Ataxia, peripheral neuropathy | PDH deficiency (hyperlactactemia) |
| Ataxia, muscular weakness, retinitis pigmentosa, myoclonic epilepsy | Respiratory-chain disorders, MERRF syndrome (hyperlactacidemia, Krebs cycle intermediate) |
| Extrapyramidal signs | GAMT deficiency (guanidino acetic acid, low creatinine) |
| Ataxia, peripheral neuropathy, retinitis pigmentosa | 3-Hydroxy-acyl-CoA-dehydrogenase deficiency (3-hydroxydicarboxylic aciduria) |
| Acute attacks resembling encephalitis, temporal-lobe atrophy | Glutaric acidemia type I (glutaryl-CoA-dehydrogenase deficiency, glutaric acidemia) |
| Dystonia, athetosis | Methylmalonic and proprionic acidemias, Homocystinuria |
| *With convulsions, seizures and myoclonus, ataxia, frequent falling due to intention myoclonus or the cerebellar ataxia* | |
| Rapidly advancing psychomotor degeneration, myoclonic jerks, blindness | Santavuori-Hagberg (infantile ceroid lipofuscinosis, early-flattening EEG) |
| Akinetic myoclonic petit mal, retinitis pigmentosa, typical EEG on slow rate, photic stimulation | Jansky Bielchowski (late infantile ceroid lipofuscinosis); do not misdiagnose with Lennox-Gastaut syndrome (vacuolated lymphocytes) |
| Rapid regression, myoclonic seizures, spasticity | Schindler disease (α-N-acetyl galactosaminidase deficiency), optic atrophy, severe osteoporosis |
| Myoclonic epilepsy, volitional and intentional myoclonias, muscular weakness | MERRF syndrome (respiratory chain disorders; hyperlactacidemia)<br>Niemann-Pick type C, Gaucher type III (supranuclear ophthalmoplegia, hepatosplenomegaly) |
| Seizures and myoclonic jerks, uncoordinated movements | Alpers syndrome (respiratory-chain disorders), hepatic symptoms, hyperlactatemia |
| *Disorders with arrest or regression of psychic and perceptual functions as the preponderant or unique revealing symptom* | |
| Autistic behavior, regression of high-level achievements, stereotyped movements of fingers | Rett syndrome (only girls), sporadic cases of unknown etiology (acquired microcephaly, secondary epilepsy) |
| Regression of high-level achievements, loss of speech, agitation | San Filippo (hirsutism) |

*CoA*, coenzyme A; *CDG*, congenital defects of glycosylation; *CSF*, cerebrospinal fluid; *EEG*, electroencephalogram; *GAMT*, guanidinoacetate methyltransferase; *MERRF*, myoclonic-epilepsy ragged red fibers; *MLP*, mucolipidosis; *MPS*, mucopolysaccharidosis; *PDH*, pyruvate dehydrogenase

**Table 1.11. Progressive neurologic and mental deterioration (5–15 years)**

| Symptoms | Diagnosis |
|---|---|
| *With predominant extrapyramidal signs, Parkinson syndrome, dystonia, choreoathetosis [see also "Extrapyramidal Signs"]* | |
| Torsion, dystonia, no mental retardation | Dystonia musculorum deformans |
| Dystonia on lower extremities, gait difficulties, normal intellect | Segawa disease, Tyrosine hydroxylase deficiency |
| Lens dislocation, marfanoid morphology | Classic homocystinuria |
| Progressive disorder of locomotion, dystonic posture, severe mental regression | Hallervorden-Spatz (retinitis pigmentosa, acanthocytosis) |
| Generalized Parkinsonian rigidity, scholastic failure | Wilson disease |
| Rigidity, fine tremor abolished by movements, dementia, seizures | Huntington chorea (adult) |
| Parkinsonism, difficulties in reading and writing, alacrima, dysphagia due to achalasia | Familial glucocorticoid deficiency (hypoglycemia due to selective cortisol deficiency) |
| *With severe neurologic and mental deterioration, diffuse central nervous system disorders, seizures, visual failure, dementia* | |
| With visceral signs (hepatosplenomegaly) | Niemann-Pick type C, Gaucher type III |
| Without visceral signs | Metachromatic leukodystrophy (juvenile form), adrenal leukodystrophy (X-linked, many variants), peroxisomal-biogenesis defects, Krabbe disease (infantile), GM-1 and GM-2 (juvenile), Leigh syndrome, respiratory-chain disorders |
| *With polymyoclonia* | |
| Generalized epilepsy, dementia | Lafora disease (adult) |
| Intellectual deterioration, loss of sight, retinitis | Spielmeyer-Vogt (juvenile neuronal ceroid lipofuscinosis; vacuolated lymphocytes) |
| Proeminent seizures, myoclonic epilepsy | Gaucher type III |
| Cerebellar ataxia, cherry-red spot | Late GM-2 gangliosidosis (Sandhoff, Tay-Sachs) |
| Hepatomegaly, splenomegaly | Niemann-Pick type C |
| Myoclonic epilepsy, lactic acidosis | Respiratory-chain disorders (MERRF, etc.) |
| *With predominant cerebellar ataxia* | |
| • Without significant mental deterioration | |
| Dysarthria | Friedreich ataxia (pes cavus, cardiomyopathy) |
| Spinocerebellar degeneration | Other hereditary ataxias, peroxisomal disorders |
| Chronic diarrhea | Aβ-lipoproteinemia (low cholesterol, acanthocytosis) |
| Retinitis pigmentosa, peripheral neuropathy | Peroxisomal disorders, Refsum disease, CDG |
| Oculocephalic asynergia | Ataxia telangiectasia (Table 1.10), conjunctival telangiectasias |
| • With deterioration and dementia | Lafora disease, cerebrotendinous xanthomatosis, GM-1, GM-2, Gaucher, Niemann-Pick type C, Krabbe (juvenile), metachromatic leukodystrophy, respiratory chain disorders |
| *With predominant polyneuropathy* | |
| Acute attacks | Porphyrias, tyrosinemia type I |
| Progressive | Metachromatic leukodystrophy, Krabbe disease, Refsum disease, peroxisomal-biogenesis defects, Aβ-lipoproteinemia, Leigh syndrome, respiratory-chain disorders, PDH deficiency, trifunctional enzyme, LCHAD deficiency, CDG |
| *With psychiatric symptoms as the only presenting sign* | |
| Behavior disturbances, personality and character changes, mental regression, dementia, schizophrenia before any significant neurologic or extraneurologic sign | San Filippo, metachromatic leukodystrophy, Krabbe disease, Niemann-Pick type C, X-linked adrenoleukodystrophy, Leigh syndrome, Spielmeyer-Vogt disease (lipofuscinosis), Hallervorden-Spatz disease, Wilson disease, cerebrotendinous xanthomatosis, Huntington chorea (juvenile form), OTC deficiency, classic homocystinuria, methylene-tetrahydrofolate-reductase deficiency, cobalamin defects (Cbl-C) |

*CDG*, congenital defects of glycosylation; *E3*, lipoamido oxido reductase; *KGDH*, ketoglutarate dehydrogenase; *LCHAD*, 3-hydroxy long-chain acyl-coenzyme-A dehydrogenase; *MCD*, multiple carboxylase deficiency; *MERRF*, myoclonic epilepsy ragged red fibers; *OTC*, ornithine transcarbamoylase; *PDH*, pyruvate dehydrogenase

diplegia, scissoring or "tiptoe" gait, and developmental arrest. A rapidly progressive flaccid paraparesis resembling subacute degeneration of the cord can be the presenting sign of inherited Cbl-synthesis defects.

*Category 3: with Unsteady Gait and Uncoordinated Movements (When Standing, Walking, Sitting, Reaching for Objects, Speaking, and Swallowing)*
Several groups of disorders must be considered. A careful investigation of organic-acid and amino-acid metabolism is always mandatory, especially in stressing circumstances.

Disorders without disturbances of urinary organic-acid excretion and lactic-acid metabolism are the late-onset forms of GM-1 and GM-2 gangliosidosis, late infantile Krabbe disease, ataxia telangiectasia, and CDG syndrome; each presents with signs suggestive enough to warrant specific investigation. A severe early-onset encephalopathy with seizures and myoclonic jerks associated with hepatic symptoms is highly suggestive of Alpers syndrome due to respiratory-chain disorders. Creatine deficiency due to guanidinoacetate-methyltransferase deficiency can strike in infancy, with an extrapyramidal disorder associated with epilepsy, neurologic regression, and failure to thrive.

Disorders with disturbances of organic- and amino-acid metabolism are numerous. PDH deficiency presents frequently with peripheral neuropathy, intermittent ataxia, and slight or moderate hyperlactacidemia (see "Hyperlactacidemias" above). Several respiratory-chain disorders first cause ataxia, intention tremor, dysarthria, epilepsy, myopathy, and (eventually) various multivisceral failures. 3-Hydroxyacyl-CoA-dehydrogenase deficiency, L-2-hydroxyglutaric aciduria, 3-methylglutaconic aciduria, MMA, and PA disturb organic-acid excretion significantly, though sometimes only slightly and intermittently. In these disorders, the acylcarnitine profile determined (by a tandem MS–MS technique) from blood spots collected on dry filter paper can be very helpful in identifying characteristic acylcarnitine compounds. GA type I due to glutaryl-CoA-dehydrogenase deficiency can also present with permanent unsteady gait due to choreoathetosis and with dystonia developing abruptly after an acute episode resembling encephalitis.

*Category 4: with Predominant Epilepsy and Myoclonus*
Predominant epilepsy and myoclonus result in ataxia and frequent falling and include two ceroid lipofuscinoses: Santavuori-Hagberg disease and Jansky-Bielchowski disease, which roughly resembles Lennox-Gastaut syndrome (akinetic myoclonic petit mal). Late-onset forms of Niemann-Pick type C and Gaucher disease are easily suspected because of hepatosplenomegaly and supranuclear paralysis. Two other disorders must also be considered: myoclonic-epilepsy ragged red fiber (MERRF) syndrome and Schindler disease, which roughly resembles neuroaxonal dystrophy.

*Category 5: Isolated Arrest or Psychic Regression*
Only a few disorders present between 1 year and 5 years of age with an isolated arrest or regression of psychic and perceptual functions and without significant neurologic or extraneurologic signs. SanFilippo disease is one, although regression of high-level achievements, loss of speech, and agitation usually begin at later than 5 years of age. Although nonmetabolic, Rett syndrome is another such disease; it should be considered when a girl without familial precedent presents with autistic behavior, mental regression, special stereotyped movements of fingers, and microcephaly at between 1 year and 2 years of age.

## Late Childhood to Adolescence (5–15 Years)

It is important to distinguish between conditions in which mental functions are primarily affected and "pure" neurologic disorders with normal or subnormal intellectual functioning. According to Lyon and Adams [20], there are six clinical categories.

*Category 1: with Predominant Extrapyramidal Signs (Parkinsonian Syndrome, Dystonia, Choreoathetotis)*
See "Extrapyramidal Signs (Dyskinesia, Dystonia, Choreoathetosis, Parkinsonism)" below.

*Category 2: with Severe Neurologic and Mental Deterioration and "Diffuse" Central Nervous System Disorders*
Category-2 patients have in common severe neurologic dysfunction with bipyramidal paralysis, incoordination, seizures, visual failure, impaired scholastic performance, and dementia. In association with splenomegaly or hepatomegaly, these signs suggest Niemann-Pick disease type C or Gaucher disease type III. When visceral signs are absent, they may indicate juvenile metachromatic leukodystrophy, X-linked adrenoleukodystrophy, Krabbe disease, juvenile GM-1 and GM-2 gangliosidoses, or respiratory-chain disorders. It has been recently described in a patient with a peroxisomal biogenesis defect who presented with peripheral neuropathy mimicking (at first) Charcot-Marie-Tooth type II disease in the second decade of life and who developed a pyramidal syndrome and intellectual deterioration with dementia and a neurovegetative state shortly thereafter.

*Category 3: with Polymyoclonia and Epilepsy*
The juvenile form of Spielmeyer-Vogt disease (or Batten disease due to CLN3 gene mutations), which

presents with loss of sight, retinitis, ataxia, and (at an advanced stage) extrapyramidal signs, should be suspected on presentation of polymyoclonia and epilepsy. After puberty, Lafora disease is another important diagnostic consideration. Gaucher disease type III, late GM-2 gangliosidosis, Niemann-Pick disease type C, and respiratory-chain disorders can also begin with polymyoclonia as an early important sign.

*Category 4: with Predominant Cerebellar Ataxia*
Friedreich ataxia and other hereditary ataxias must be considered. They are recognized on clinical and genetic grounds. Aβ-lipoproteinemia and ataxia telangiectasia are usually easily suspected because of the associated extraneurologic signs. Peroxisomal disorders, CDG syndrome, and Refsum disease (which can all present like a peripheral neuropathy and retinitis pigmentosa) can be easily demonstrated by the investigation of plasma very-long-chain fatty acids, glycosylated transferrin profile, and plasma phytanic acid, respectively. Cerebellar ataxia in a context of progressive mental deterioration, dementia, and epilepsy suggests Lafora disease, cerebrotendinous xanthomatosis, late-onset forms of gangliosidosis, Krabbe disease, Gaucher disease, Niemann-Pick disease type C, and metachromatic leukodystrophy. Respiratory-chain disorders also can be revealed by predominant ataxia.

*Category 5: with Predominant Polyneuropathy*
Porphyrias and tyrosinemia type I can present in acute decompensations as polyneuropathy mimicking Guillain-Barré syndrome. Many other disorders can present with a late-onset progressive polyneuropathy that can mimic heredoataxia, such as Charcot-Marie-Tooth disease or other hereditary ataxias. These disorders include lysosomal diseases (Krabbe disease and metachromatic leukodystrophy), defects of energy metabolism (Leigh syndrome, respiratory-chain disorders, PDH deficiency, LCHAD and trifunctional-enzyme deficiencies), aβ-lipoproteinemia, CDG syndrome, and peroxisomal disorders, including Refsum disease.

*Category 6: with Behavioral Disturbances as Revealing Signs*
Some inborn errors of metabolism can appear between 5 years and 15 years of age as psychiatric disorders. Behavioral disturbances (personality and character changes), loss of speech, scholastic failure, mental regression, dementia, psychosis, and schizophrenia-like syndrome are the most frequent symptoms. Among the disorders listed in Table 1.11, we have personal experience of such pure psychiatric presentations in SanFilippo disease, X-linked adrenoleukodystrophy, Wilson disease, and classic homocystinuria. In addition to this list of inborn errors that classically present as psychiatric disorders, OTC deficiency can present with intermittent abnormal behavior and character change until hyperammonemia and coma reveal the true situation (see "Recurrent Attacks of Coma" above). Homocystinuria due to methylene-tetrahydrofolate-reductase deficiency has presented as isolated schizophrenia.

**Onset in Adulthood (15–70 Years)**

In adults, neurodegenerative diseases, such as psychosis, dementia, ataxia, epilepsy, dystonia, and peripheral neuropathy, can be revealing signs of variant forms of inborn errors. It is highly probable that, in the near future, many other disorders will be discovered in adults when neurologists extend metabolic investigations to late-onset progressive neurologic deterioration, which is now considered to be "degenerative", inflammatory, or of vascular origin.

*Ataxia (Chronic)*

- Leigh syndrome
- Joubert syndrome
- Gaucher type III
- Inborn errors of Cbls (Cbl-C, Cbl-D)
- Folic acid, transport defect
- Phosphoribosylpyrophosphate (PRPP)-synthetase superactivity
- Arginase deficiency
- Ataxia telangiectasia
- Landing
- Late-onset GM-2 gangliosidosis (Tay Sachs, Sandhoff) (A)
- Niemann-Pick type C (A)
- Krabbe disease, metachromatic leukodystrophy (A)
- CDG syndrome
- Alpers syndrome
- LCHAD deficiency
- Galactosialidosis (A)
- Cerebrotendinous xanthomatosis (A)
- L-2-Hydroxyglutaric aciduria
- PDH deficiency
- Respiratory-chain disorders
- 3-Methylglutaconic aciduria
- MERRF syndrome
- Batten disease (Santavuori-Jansky-Bielchowski)
- Peroxisomal biogenesis defects
- Refsum disease
- Lafora disease
- Spielmeyer-Vogt syndrome
- Friedreich ataxia (and other hereditary ataxias)
- Aβ-lipoproteinemia
- Mevalonic aciduria
- Hartnup disease
- Pyroglutamic aciduria

## Calcifications (Intracranial)

- Inborn errors of folic-acid metabolism
- Inborn errors of biopterin metabolism
- Congenital lactic acidemias
- Respiratory-chain disorders
- 3-Hydroxyisobutyric aciduria
- Leigh syndrome
- Cockayne syndrome
- MELAS syndrome
- Kearn-Sayre syndrome
- Krabbe disease
- GM-2 gangliosidosis
- Aicardi-Goutieres syndrome

## Cerebellar Hypoplasia (and Olivopontocerebellar Atrophy)

- CDG syndrome
- 3-Hydroxy-isobutyric aciduria
- Peroxisomal disorders
- L-2-Hydroxyglutaric aciduria
- 3-Methylglutaconic aciduria
- Mevalonic aciduria
- Joubert syndrome (vermis atrophy)

## Cherry-Red Spot

See "Ophthalmology".

## Corpus-Callosum Agenesis

- PDH deficiency
- Peroxisomal disorders
- Respiratory-chain disorders
- 3-Hydroxy-isobutyric aciduria
- NKH
- Adrenocorticotrophic-hormone (ACTH) deficiency
- Aicardi syndrome

## Deafness (Sensorineural)

- Detectable in neonatal to early infancy:
  - Zellweger and variants
  - Rhizomelic chondrodysplasia punctata
  - Acyl-CoA-oxidase deficiency
  - Cockayne syndrome
  - Alport syndrome
  - Encephalopathy with hyperkinurininuria
- Detectable in late infancy to childhood:
  - Infantile Refsum disease (pseudo Usher syndrome)
  - PRPP-synthetase superactivity
  - Mucopolysaccharidosis types I, II, and IV
  - Mannosidosis (α)
  - Mucolipidosis type II (I-cell disease)
  - Biotinidase deficiency (biotin responsive)
  - Megaloblastic anemia and diabetes (thiamine responsive)
  - Wolframm syndrome
  - Neutral-lipid-storage disorder
  - Mitochondrial encephalomyopathy
  - MELAS
  - MERRF
  - Kearn-Sayre syndrome
- Detectable in late childhood to adolescence:
  - β-Mannosidosis
  - Refsum disease (adult form)
  - Usher syndrome type II
  - MERRF
  - Kearn-Sayre syndromes

## Dementia, Psychosis, Schizophrenia, Behavior Disturbances

- Acute attacks of delirium, hallucinations, mental confusion, hysteria, psychosis:
  - OTC deficiency
  - Homocystinurias
  - Porphyrias
  - Acute and inaugural attacks along the course of a progressive disorder
- Progressive disorders with intellectual disintegration, mental regression, psychosis:
  - Metachromatic leukodystrophy
  - GM-1, GM-2 gangliosidosis (late-onset form)
  - Hexosaminidase deficiencies
  - Krabbe disease (juvenile form)
  - Niemann-Pick type C
  - SanFilippo disease
  - Ceroid lipofuscinosis (Kufs, Spielmeyer-Vogt)
  - Cerebrotendinous xanthomatosis (adult)
  - Huntington chorea (adult)
  - Wilson disease
  - Lafora disease (adult)
  - Defects of Cbl metabolism (Cbl-C)
  - Peroxisomal disorders
  - Respiratory-chain disorders
  - Hallervorden-Spatz syndrome
  - Untreated phenylketonuria
  - Rett syndrome (only girls)
  - Gaucher type III
  - X-linked adrenoleukodystrophy

## Extrapyramidal Signs (Choreoathetosis, Dyskinesia, Dystonia, Parkinsonism)

- Aromatic-amino-acid-decarboxylase deficiency
- 4-Hydroxy-butyric aciduria
- Crigler-Najjar (glucuronyl-transferase deficiency)

- GA type I
- 3-Methylglutaconic aciduria
- Homocystinuria
- MMA
- Tyrosinemia type I
- Oligosaccharidosis
- Segawa disease (dopa-responsive dystonia)
- Gaucher type II (infantile type) and type III
- Cytochrome-b5-reductase deficiency
- 2-Hydroxyglutaric aciduria
- GM-1 gangliosidosis (adult form)
- Niemann-Pick type C (juvenile dystonic lipidosis)
- Krabbe disease
- Familial hypoparathyroidism (isolated)
- Pseudo hypoparathyroidism
- Hallervorden-Spatz disease
- Lesch-Nyhan syndrome [X-linked; hypoxanthine-guanine-phosphoribosyltransferase (HGPRT) deficiency]
- Neuroaxonal dystrophy
- Pelizaeus-Merzbacher disease
- Wilson disease
- Leigh syndrome
- Familial glucocorticoid deficiency
- Huntington chorea

### Hypotonia in the Neonatal Period

See also "Acute Symptoms in the Neonatal Period and Early Infancy (<1 year)".

- Evocative clinical context (dysmorphia, bone changes, visceral symptoms, malformations):
  - Hypophosphatasia
  - Calciferol-metabolism defects
  - Osteogenesis imperfecta
  - Peroxisomal disorders
  - Lowe syndrome (X-linked)
  - Chromosomal abnormalities
  - Other multiple-malformation syndromes (such as Walker-Warburg, Fukuyama, muscular dystrophy)
  - Lysosomal disorders (Table 1.2)
  - Smith-Lemli-Opitz syndrome
  - Organic acidurias
  - Pompe disease
  - Respiratory-chain disorders
- Neurological neonatal distress: see "Identification of Five Major Types of Metabolic Distress" and Table 1.2.
- Apparently isolated at birth:
  - Severe fetal neuromuscular disorders
  - Steinert
  - Myasthenia
  - Congenital myopathy
  - Hereditary sensorimotor neuropathy
  - Familial dysautonomy
  - Congenital dystrophy
  - Werdnig-Hoffmann (spinal muscular atrophy type I)
  - Prader-Willi syndrome
  - Pelizaeus-Merzbacher
  - CDG syndrome
  - Peroxisomal disorders

### Leigh Syndrome

- Respiratory-chain disorders
- PC deficiency
- PDH deficiency
- Biotinidase deficiency
- Fumarase deficiency
- Sulfite oxidase deficiency
- 3-Methylglutaconic aciduria
- EPEMA syndrome

### Macrocephaly

- Tay-Sachs
- Sandhoff
- GA type I
- L-2-Hydroglutaric aciduria
- Canavan (acetylaspartaturia)
- Respiratory-chain disorders
- Krabbe (infantile form)
- Bannayan-Riley-Ruvalcaba

### Mental Regression

See "Progressive Neurologic and Mental Deterioration Related to Age (Overview)".

### Microcephaly

- Congenital
  - Baby born to an untreated phenylketonuric mother
  - Sulfite oxidase deficiency
- Acquired
  - Rett syndrome
- Many untreated disorders in which microcephaly is a symptom of a nonspecific cerebral atrophy that complicates the outcome of the disorder

### Myoclonic Epilepsy (Polymyoclonia)

- Age at onset: neonatal period [see also "Acute Symptoms in the Neonatal Period and Early Infancy (<1 year)"]
  - NKH
  - D-Glyceric aciduria

- Peroxisomal disorders
- Sulfide oxidase deficiency
- Familial idiopathic neonatal myoclonic epilepsy
- Age at onset: infancy to childhood
  - Ceroid lipofuscinosis (Santavuori-Hagberg, Jansky-Bielchowski, Spielmeger- Vogt, Kufs)
  - Schindler disease (α-N-acetyl-galactosaminidase deficiency)
  - Neuroaxonal dystrophy
  - Sialidosis type I
  - Respiratory-chain disorders (MERRF, Alpers syndrome, etc.)
  - 3-Methylglutaconic aciduria
  - Niemann-Pick type C
  - Gaucher type III
  - Tay-Sachs, Sandhoff (late-onset forms)
  - Lafora disease (adult onset)
  - Galactosialidosis

### Nystagmus

- Pelizaeus-Merzbacher (revealing sign)
- Neuroaxonal dystrophy, Canavan disease
- Schindler disease
- All causes of optic atrophy
- All causes of retinitis pigmentosa

### Ophthalmoplegia, Abnormal Eye Movements

See "Ophthalmology".

### Peripheral Neuropathy

- Acute attacks:
  - Porphyrias
  - Tyrosinemia type I
- Progressive:
  - Krabbe
  - Leigh syndrome
  - Respiratory-chain disorders
  - PDH deficiency
  - Peroxisomal disorders
  - Metachromatic leukodystrophy
  - Neuroaxonal dystrophy
  - Schindler disease
  - CDG
  - LCHAD and trifunctional-enzyme deficiencies
  - Aβ-lipoproteinemia
  - Vitamin-E deficiency
  - Refsum disease
  - Farber lipogranulomatosis
  - Austin disease
  - Triose-phosphate-isomerase deficiency
  - Pyroglutamic aciduria

### Retinitis Pigmentosa

See "Ophthalmology".

### Self mutilation, Auto-Aggressiveness

- Lesch-Nyhan syndrome
- Tyrosinemia type I (crisis)
- Phenylketonuria (untreated)
- 3-Methylglutaconic aciduria

### Spastic Paraplegia, Paraparesia

- Hyperargininemia
- Triple-H syndrome
- Metachromatic leukodystrophy
- Pyroglutamic aciduria
- Sjögren-Larsson syndrome
- L-2-Hydroxyglutaric aciduria

## IV.3 Muscular Symptoms

Many inborn errors of metabolism can present with severe hypotonia, muscular weakness, and poor muscle mass. These include most of the late-onset forms of urea-cycle defects and many organic acidurias. Severe neonatal generalized hypotonia and progressive myopathy associated (or not associated) with a nonobstructive idiopathic cardiomyopathy can be specific revealing symptoms of a number of inherited energy deficiencies. The most frequent conditions actually observed are mitochondrial respiratory-chain disorders and other congenital hyperlactacidemias, FAO defects, peroxisomal disorders, muscular glycogenolysis defects, glycogen-storage disease type II, and some other lysosomal disorders. See also "Hypotonia" and "Myology".

## V Specific Organ Symptoms

A number of clinical or biologic symptoms can reveal or accompany inherited inborn errors of metabolism. Some of these phenotypes are rare and very distinctive (lens dislocation and thromboembolic accidents in homocystinuria), whereas others are common and rather nonspecific (hepatomegaly, seizures, mental retardation). The most important ones are listed in the Appendix. The diagnostic checklist presented in the Appendix is mostly based upon the author's personal experience and, of course, is not exhaustive. It should be progressively extended by the personal experiences of all readers.

In searching for the diagnosis, we must re-emphasize the importance of not confusing a syndrome due to different causes with the etiology itself. Hence, Leigh syndrome and Reye syndrome have been incorporated in the list of symptoms and must also be considered as actual diagnoses. Some other well-known recessive syndromes (such as Joubert, Usher, Cockayne, etc.) have been listed under inborn errors of metabolism in order to highlight the necessity of performing extensive metabolic and genetic investigations before attributing a label of false security to a patient. The recent demonstration of a cholesterol-synthesis defect in Smith-Lemli-Opitz syndrome is an illustration of this statement [21].

## V.1 Cardiology

### Arrhythmias, Conduction Defects

See "Later-Onset Acute and Recurrent Attacks (Childhood and Beyond)".

### Cardiac Failure

See "Later-Onset Acute and Recurrent Attacks (Childhood and Beyond)".

### Cardiomyopathy

- Respiratory-chain disorders (revealing sign)
- FAO disorders
- Pompe disease (revealing sign)
- Glycogenosis types III and IV
- Phosphorylase-B kinase (revealing sign)
- 3-Methylglutaconic aciduria
- PA
- MMA (Cbl-C)
- D-2-Hydroxyglutaric aciduria
- CDG syndrome
- Mucopolysaccharidosis
- Friedreich ataxia
- Steinert disease
- Myotonic dystrophy
- Congenital muscle dystrophies

### Sudden Infant Death Syndrome

See "Reye Syndrome".

## V.2 Dermatology

### Acrocyanosis (Orthostatic)

- EPEMA syndrome

### Alopecia

- Age at onset: neonatal to infancy
  - Menkes disease (X-linked)
  - Biotin-responsive MCDs
  - MMA and PA
  - Acrodermatitis enteropathica
  - Essential fatty-acid deficiency
  - Zinc deficiency
  - Hepatoerythropoietic porphyria
  - Congenital erythropoietic porphyria
  - Calciferol-metabolism defects (vitamin-D-dependent rickets)
  - Ehlers-Danlos type IV
  - Netherton syndrome
  - Conradi-Hunermann syndrome
- Age at onset: adulthood
  - Steinert
  - Porphyria cutanea tarda

### Angiokeratosis

- Fabry disease
- Fucosidosis
- Galactosialidosis
- Aspartylglucosaminuria
- β-Mannosidosis
- Schindler disease (adult form)

### Brittle Hair

- Trichothiodystrophy
- Menkes syndrome
- Arginosuccinic aciduria
- Citrullinemia
- Pollitt's syndrome

### Hemangiomas

- Bannayan-Riley-Ruvalcaba

### Hyperkeratosis

- Ichthyosis (see below)
- Tyrosinemia type II (keratosis on palms and soles)

### Ichthyosis (with Congenital Erythrodermia)

- Conradi-Hunermann syndrome (chondrodysplasia punctata, X-linked)
- Multisystemic triglyceride-storage disease
- Sjögren-Larsson syndrome

- Austin disease
- Steroid-sulfatase deficiency (X-linked)
- Netherton syndrome
- Refsum disease (adult form)

### Laxity (Dysmorphic Scarring, Easy Bruising)

- Inborn errors of collagen:
  - Ehlers-Danlos syndrome (nine types)
  - Occipital-horn syndrome
  - Cutis laxa

### Nodules

- Farber lipogranulomatosis
- CDG syndrome

### Photosensitivity and Skin Rashes

See "Later-Onset Acute and Recurrent Attacks (Childhood and Beyond)".

### Pili Torti

- Menkes disease
- Netherton syndrome

### Telangiectasias, Purpuras, Petechiae

- Prolidase deficiency
- Ethylmalonic aciduria (EPEMA syndrome)

### Trichorrhexis Nodosa

- Argininosuccinic aciduria
- Argininemia
- Lysinuric protein intolerance
- Menkes disease
- Netherton syndrome

### Ulceration (Skin Ulcers)

- Prolidase deficiency

### Vesiculo Bullous Skin Lesions

- Acrodermatitis enteropathica
- Holocarboxylase-synthase deficiency (biotin responsive)
- Biotinidase deficiency (biotin responsive)

- MMA, PA
- Zinc deficiency

## V.3 Dysmorphology

### Coarse Facies

- Age at onset: present at birth
  - Landing
  - Sialidosis type II
  - Galactosialidosis (early infancy)
  - Sly [mucopolysaccharidosis (MPS) type VII] (rare)
  - I-cell disease
- Age at onset: early infancy
  - Fucosidosis type I
  - Sialidosis type II
  - Salla disease
  - Hurler (MPS type Is)
  - Sly (MPS type VII)
  - Mannosidosis
  - Austin disease
  - Maroteaux-Lamy (MPS type V)
- Age at onset: childhood
  - Hunter (MPS type II)
  - Aspartylglucosaminuria
  - Pseudo Hurler polydystrophy
  - SanFilippo (MPS type III)

### Congenital Malformations and Dysmorphia

- Inborn errors affecting the fetus:
  - 3-Hydroxy-isobutyryl-CoA-deacylase deficiency
  - Mevalonic aciduria (mevalonate kinase deficiency)
  - GA type II (multiple acyl-CoA-dehydrogenase deficiency)
  - Carnitine palmitoyl-transferase-II deficiency
  - Peroxisomal disorders (Zellweger and variants, chondrodysplasia punctata)
  - PDH deficiency
  - Respiratory-chain defects
  - Inborn errors of collagen
  - Hypoparathyroidism
  - Hypophosphatasia
  - Leprechaunism
  - Lysosomal storage disorders
  - Smith-Lemli-Opitz syndrome (inborn error of cholesterol synthesis)
  - NKH
- Metabolic disturbances of the mother:
  - Phenylketonuria
  - Alcohol, drugs
  - Diabetes
  - Vitamin deficiencies (riboflavin)

*Intrauterine Growth Retardation*

- Lysosomal storage disorders
- Peroxisomal disorders
- Cholesterol-biosynthesis defects (Smith-Lemli-Opitz)
- Respiratory-chain disorders
- Alcoholic fetal syndrome
- Babies born to untreated phenylketonuric mothers
- Many non-metabolic polymalformative syndromes

## V.4 Endocrinology

*Diabetes (and Pseudodiabetes)*

- Respiratory-chain disorders
- Wolfram syndrome
- Diabetes, deafness, and thiamine-responsive megaloblastic anemia
- Organic acidurias (MA, PA, IVA, ketolytic defects)
- Abnormal pro-insulin cleavage

*Hyperthyroidism*

- GA (glutaryl-CoA oxidase?)

*Hypogonadism, Sterility*

- Galactosemia

*Hypoparathyroidism*

- Respiratory-chain disorders
- LCHAD deficiency
- Trifunctional-enzyme deficiency

*Salt-Losing Syndrome*

- Disorders of adrenal steroid metabolism
- FAO disorders (carnitine palmitoyl transferase II)
- Respiratory-chain disorders (mitochondrial-DNA deletions)
- CDG syndrome

*Sexual Ambiguity*

- Disorders of adrenal-steroid metabolism
- Congenital adrenal hyper- and hypoplasias
- Smith Lemli Opitz syndrome

*Short Stature, Growth-Hormone Deficiency*

- Respiratory-chain disorders

## V.5 Gastroenterology

*Abdominal Pain (Recurrent)*

See "Later-Onset Acute and Recurrent Attacks (Childhood and Beyond)".

*Acute Pancreatitis*

- Organic acidurias (MA, PA, IVA, MSUD)
- Hyperlipoproteinemia types I and IV
- Lysinuric protein intolerance
- Respiratory-chain disorders (Pearson, MELAS)

*Chronic Diarrhea, Failure to Thrive, Osteoporosis*

See "Digestive Symptoms" and Table 1.7.

*Hypocholesterolemia*

- Peroxisomal disorders
- Infantile Refsum disease
- Mevalonic aciduria
- Smith-Lemli-Opitz syndrome
- CDG syndrome
- Aβ-lipoproteinemia types I and II
- Tangier disease (α-lipoprotein deficiency)

*Babies Born to Mothers with Hemolysis, Elevated Liver Enzyme, and Low Platelet Count (HELLP) Syndrome*

- LCHAD deficiency
- Carnitine palmitoyl-transferase-I deficiency
- Respiratory-chain defects

## V.6 Hematology

*Acanthocytosis*

- Wolman disease
- Aβ-lipoproteinemia
- Inborn errors of Cbl (Cbl-C)
- Hallervorden-Spatz syndrome

*Anemias (Megaloblastic)*

- Inborn errors of folate metabolism:
  - Dihydrofolate-reductase deficiency
  - Glutamate-formimino-transferase deficiency
  - Congenital folate malabsorption
- Inborn errors of Cbl metabolism:
  - Imerslund disease
  - Intrinsic-factor deficiency
  - Transcobalamin-II deficiency

- Cbl-C, Cbl-E, and Cbl-G deficiencies
- Methionine-synthase deficiency
- Thiamine-responsive megaloblastic anemia
- Respiratory-chain disorders
- Pearson syndrome (due to mitochondrial-DNA deletion)
- Hereditary orotic aciduria
- Mevalonic aciduria
- Dyserythropoiesis type II (respiratory-chain disorder)

### Anemias (Non Macrocytic, Hemolytic or Due to Combined Mechanisms)

- Red blood cell glycolysis defects
- Pyroglutamic aciduria
- Galactosemia
- Wolman disease
- Wilson disease
- Aβ-lipoproteinemia
- Hemochromatosis
- Erythropoietic protoporphyria
- Congenital erythropoietic porphyria
- Erythropoietic porphyria
- Lecithin-cholesterol-acyltransferase deficiency
- Carnitine-transport defect
- Severe liver failure

### Bleeding Tendency

See "Bleeding Tendency, Hemorrhagic Syndromes".

### Pancytopenia, Thrombopenia, and Leukopenia

- Inborn errors of Cbl metabolism
- Inborn errors of folate metabolism
- Organic acidurias (MA, PA, IVA)
- Respiratory-chain disorders
- Pearson syndrome
- Johansson-Blizzard syndrome
- Schwachman syndrome
- Lysinuric protein intolerance
- Glycogenosis type Ib (neutropenia)
- Gaucher types I and III
- Other conditions with marked splenomegaly
- Aspartylglucosaminuria

### Vacuolated Lymphocytes

- Aspartylglucosaminuria
- I-cell disease (mucolipidosis type II)
- Landing disease (GM1)
- Niemann-Pick type Ia
- Wolman disease
- Ceroid lipofuscinosis
- Mucopolysaccharidosis
- Austin disease
- Sialidosis
- Pompe disease

## V.7 Hepatology

### Cholestatic Jaundice

- α1-Antitrypsin deficiency, cystic fibrosis
- Byler disease
- Inborn errors of bile-acid metabolism
- Peroxisomal disorders
- Niemann-Pick type C
- LCHAD deficiency
- CDG syndrome
- Mevalonic aciduria
- Cholesterol synthesis defects (Smith, Lemli, Opitz)

### Cirrhosis

- Hereditary fructose intolerance
- Galactosemia
- Glycogenosis type IV
- S-Adenosylhomocysteine hydrolase deficiency
- Phosphoenol-pyruvate-carboxykinase deficiency
- Tyrosinemia type I
- α1-Antitrypsin deficiency
- Alpers progressive infantile polydystrophy
- Cystic fibrosis
- Familial hepatic fibrosis with exsudative enteropathy
- Gaucher disease
- Wolman disease
- Cholesteryl-ester-storage disease
- Hemochromatosis
- Niemann-Pick disease
- Wilson disease
- CDG syndrome
- Peroxisomal disorders
- LCHAD deficiency

### Liver Failure (Ascites, Edema)

See "Later-Onset Acute and Recurrent Attacks (Childhood and Beyond)".

### Reye Syndrome

See "Later-Onset Acute and Recurrent Attacks (Childhood and Beyond)".

## V.8 Myology

### Exercise Intolerance, Myoglobinuria, Muscle Pain

See "Later-Onset Acute and Recurrent Attacks (Childhood and Beyond)".

### Myopathy (progressive)

- Adenylate-deaminase deficiency
- Glycogenosis type II (acid-maltase deficiency)
- Glycogenosis type III
- FAO disorders
- Respiratory-chain disorders (Kearns-Sayre and others)
- Multisystemic triglyceride-storage disease
- Steinert disease

## V.9 Nephrology

### Hemolytic Uremic Syndrome

- Inborn errors of Cbl metabolism (Cbl-C, Cbl-G)

### Nephrolithiasis, Nephrocalcinosis

- Cystinuria (cystine)
- Hyperoxaluria types I and II (oxalic)
- Xanthine-oxidase deficiency (xanthine)
- Molybdenum-cofactor deficiency (xanthine)
- Lesch-Nyhan (uric acid)
- PPRP-synthetase superactivity (uric acid)
- Hereditary renal hypouricemia (uric acid)
- Adenine-phosphoribosyl-transferase deficiency (2,8-dihydroxy adenine)
- Hereditary hyperparathyroidism (calcium)
- RTA type I

### Nephrotic Syndrome

- Respiratory-chain disorders
- CDG syndrome

### Nephropathy (Tubulointerstitial)

- Glycogenosis type I
- MMA
- Respiratory-chain disorders (pseudo Senior-Loken syndrome)

### Renal Polycystosis

- Zellweger syndrome
- GA type II
- Carnitine palmitoyl-transferase II deficiency
- CDG syndrome

### Tubulopathy

- Fanconi syndrome:
  - Hereditary fructose intolerance
  - Galactosemia
  - Respiratory-chain disorders (complex IV or mitochondrial DNA deletion)
  - Tyrosinemia type I
  - Glycogenosis with tubulopathy (Bickel-Fanconi syndrome: Glut-II mutations)
  - Lowe syndrome (X-linked)
  - Cystinosis
- RTA:
  - RTA type I (distal)
  - RTA type II (proximal)
  - PC deficiency
  - MMA
  - Glycogenosis type I
  - Carnitine palmitoyl-transferase-I deficiency

### Urine (Abnormal Color)

- Alkaptonuria (black)
- Myoglobinuria (red)
- Porphyria (red)
- Indicanuria (blue)

### Urine (Abnormal Odor)

- 3-Methyl-crotonylglycinuria (cat)
- GA type II (sweaty feet)
- IVA (sweaty feet)
- Trimethylaminuria (fish)
- MSUD (maple syrup)
- Tyrosinemia type I (boiled cabbage)
- Phenylketonuria (musty odor)
- Dimethylglycine dehydrogenase (fish)

## V.10 Neurology

See "Neurologic Symptoms".

## V.11 Ophthalmology

### Cataracts

- Detectable at birth (congenital):
  - Lowe syndrome (X-linked)
  - Peroxisomal biogenesis defects (Zellweger and variants)
  - Rhizomelic chondrodysplasia punctata
  - Cockayne syndrome
  - Sorbitol-dehydrogenase deficiency
  - Phosphoglycerate-dehydrogenase deficiency
- Detectable in the newborn period (first week to first month):
  - Galactosemias
  - Peripheral epimerase deficiency (homozygotes and heterozygotes)

- Marginal maternal galactokinase deficiency
- Detectable in infancy (first month to first year):
  - Galactokinase deficiency
  - Galactitol or sorbitol accumulation of unknown origin
  - Sialidosis
  - α-Mannosidosis
  - Pyrroline-5-carboxylic-acid-synthase deficiency
  - Hypoglycemia (various origins)
  - Respiratory-chain disorders
- Detectable in childhood (1–15 years):
  - Hypoparathyroidism
  - Pseudo-hypoparathyroidism
  - Diabetes mellitus
  - Wilson disease
  - Sjögren-Larsson syndrome
  - Lysinuric protein intolerance
  - Neutral-lipid-storage disorders (unknown cause)
  - Mevalonic aciduria
  - Dominant cataract with high serum ferritin
- Detectable in adulthood (>15 years):
  - Heterozygotes for galactose-uridyltransferase and galactokinase deficiencies
  - Carriers for Lowe syndrome
  - Lactose malabsorption
  - Ornithine-aminotransferase (OAT) deficiency
  - Cerebrotendinous xanthomatosis
  - Glucose-6-phosphate-dehydrogenase deficiency
  - Steinert dystrophy (cataract can be revealing sign)

### Cherry-Red Spot

- Gangliosidosis GM1 (Landing)
- Galactosialidosis (neuraminidase deficiency)
- Cytochrome-C-oxidase deficiency
- Sialidosis type I
- Niemann-Pick types A, C, and D
- Nephrosialidosis
- Sandhoff disease
- Tay-Sachs disease

### Corneal Opacities (Clouding)

- Visible in early infancy (3–12 months):
  - Tyrosinosis type II (revealing sign)
  - Cystinosis (revealing sign)
  - I-cell disease (mucolipidosis type II)
  - Hurler, Scheie (MPS-I)
  - Maroteaux-Lamy (MPS-VI)
  - Steroid-sulfatase deficiency
- Visible in late infancy to early childhood (1–6 years):
  - Morquio (MPS-IV)
  - Mucolipidosis type IV (revealing sign)
  - α-Mannosidosis (late-onset form)
  - Tangier disease
  - Lecithin-cholesterol-acyltransferase deficiency
  - Pyroglutamic aciduria (revealing sign)
- Visible in late childhood, adolescence to adulthood:
  - Fabry disease (X-linked)
  - Galactosialidosis (juvenile form)
  - Wilson disease (green Kaiser-Fleischer ring)

### Ectopia Lentis (Dislocation of the Lens)

- Classical homocystinuria
- SO deficiency
- Marfan syndrome
- Marchesani syndrome

### Keratitis

See also "Corneal Opacities (Clouding)".

- Tyrosinemia type II
- Fabry disease (X-linked)

### Microcornea

- Ehlers-Danlos type IV

### Ptosis, External Ophthalmoplegia, Abnormal Eye Movements

- Respiratory-chain disorders (Kearns-Sayre)
- Niemann-Pick types C and D (supranuclear paralysis)
- Gaucher type III (supranuclear paralysis)
- Ataxia telangiectasia (ocular contraversion)
- Cogan syndrome (ocular contraversion)
- Steinert disease
- Aromatic amino-acid decarboxylase deficiency (oculogyric crisis)
- CDG syndrome (congenital strabism and oculogyric crisis)

### Retinitis Pigmentosa

- Inborn errors of lipid metabolism:
  - Aβ-lipoproteinemia
  - Vitamin-E malabsorption
  - 3-Hydroxy-acyl-CoA-dehydrogenase deficiency (LCHAD)
  - Sjögren-Larsson syndrome
- Ceroid lipofuscinosis:
  - Infantile forms of Santavuori-Hagberg, Jansky-Bielchovsky, Spielmeyer-Vogt (Batten)
- Peroxisomal disorders:
  - Zellweger and variant forms

- Neonatal adrenoleukodystrophy
- Infantile Refsum disease
- Isolated FAO defects
- Classical Refsum disease (adult form)
- CDG syndrome
- Respiratory-chain disorders:
  - Kearn-Sayre syndrome
  - Other mitochondrial DNA deletions
- Cbl-metabolism defects (Cbl-C)
- Recessive autosomal syndromes (Cockayne, Hallervorden-Spatz, Laurence-Moon-Biedl, Usher type II, Joubert, Senior-Loken etc.)
- Gyrate atrophy with OAT deficiency
- Primary retinitis pigmentosa (X-linked, autosomal recessive or dominant)

## V.12 Osteology

### Osteopenia

- Lysinuric protein intolerance
- Infantile Refsum disease
- CDG syndrome
- Homocystinuria
- I-cell disease (mucolipidosis type II)
- Cerebrotendinous xanthomatosis
- Glycogenosis type I

### Punctate Epiphyseal Calcifications

- Peroxisomal disorders:
  - Zelwegger and variants
  - Chondrodysplasia punctata, rhizomelic type
- Conradi-Hunermann syndrome (cholesterol biosynthesis defect)
- Familial resistance to thyroid hormone
- Warfarin embryopathy
- β-Glucuronidase deficiency

## V.13 Pneumology

### Pneumopathy (Interstitial)

- Dibasic aminoaciduria
- Niemann-Pick type B
- Gaucher disease

### Stridor

- Hypocalcemia
- Hypomagnesemia
- Pelizaeus-Merzbacher
- Biotinidase deficiency

- Multiple acyl-CoA-dehydrogenase deficiency (riboflavin responsive)

## V.14 Psychiatry

See "Dementia, Psychosis, Schizophrenia, Behavior Disturbances".

## V.15 Rheumatology

### Arthritis, Joint Contractures, Bone Necrosis

- Alkaptonuria
- Gaucher type I
- Lesch-Nyhan syndrome
- Farber disease
- Familial gout
- PRPP-synthease deficiency
- HGPRT deficiency
- I-cell disease
- Mucolipidosis type III
- Homocystinuria
- Mucopolysaccharidosis type IS
- Mevalonic aciduria (recurrent crisis of arthralgia)

### Bone Crisis

- With bone changes (rickets):
  - Calciferol-metabolism deficiency
  - Hereditary hypophosphatemic rickets
- With hemolytic crises (and abdominal pain):
  - Porphyrias
  - Tyrosinemia type I
  - Sickle-cell anemia
- With progressive neurologic signs:
  - Krabbe disease
  - Metachromatic leukodystrophy
  - Gaucher type III
- Apparently isolated (revealing symptom):
  - Fabry disease
  - Gaucher type I

## V.16 Vascular Symptoms

### Raynaud Syndrome

- Fabry disease

### Thromboembolic Accidents (Stroke-Like Episodes)

- Homocystinuria (all types)
- Ehlers-Danlos type IV
- Respiratory-chain disorders (MELAS and others)
- Urea-cycle disorders (OTC deficiency)

- Organic acidurias (MMA, PA)
- CDG
- Fabry disease
- Menkes disease

## VI References

1. McKusick VA (1992) Mendelian inheritance in man, 10th edn. Johns Hopkins University Press, Baltimore
2. Burton BK (1987) Inborn errors of metabolism: the clinical diagnosis in early infancy. Pediatrics 79:359-369
3. Poggi F, Billette de Villeneuve T, Munnich A, Saudubray JM (1994) The newborn with suspected metabolic disease: an overview. In: Clayton BE, Round JM (eds) Clinical biochemistry and the sick child, 2nd edn. Blackwell Scientific, Oxford, pp 60-86
4. Saudubray JM, Ogier H, Bonnefont JP et al. (1989) Clinical approach to inherited metabolic diseases in the neonatal period: a 20 year survey. J Inher Metab Dis 12 [Suppl]:25-41
5. Nyhan WL (1977) An approach to the diagnosis of overwhelming metabolic diseases in early infancy. Curr Probl Pediatr 7:1-15
6. Fellman V, Rapola J, Pihko H, Varilo T, Raivio KO (1998) Iron-overload disease in infants involving fetal growth retardation, lactic acidosis, liver haemosiderosis, and aminoaciduria. Lancet 351:490-493
7. Saudubray JM, Martin D, De Lonlay P et al. (1999) Recognition and management of fatty acid oxidation defects: a series of 107 patients. J Inher Metab Dis 22:488-502
8. Bonnet D, Martin D, De Lonlay P et al. (1999) Arrhythmias and conduction defects as a presenting symptom of fatty-acid oxidation disorders in children. Circulation 100:2248-2253
9. Baumgartner MR, Poll-The BT, Verhoeven N et al. (1998) Clinical approach to inherited peroxisomal disorders: a series of 27 patients. Ann Neurol 44:720-730
10. Baumgartner MR, Verhoeven N, Jakobs C et al. (1998) Defective peroxisome biogenesis with a neuromuscular disorder resembling Werdnig-Hoffmann disease. Neurology 51:1427-1432
11. Burlina AB, Dionisi-Vici C, Bennett MJ et al. (1994) A new syndrome with ethylmalonic aciduria and normal fatty acid oxidation in fibroblasts. J Pediatr 124:79-86
12. Garcia-Silva MT, Ribes A, Campos Y, Garavaglia B, Arenas J (1997) Syndrome of encephalopathy, petechiae, and ethylmalonic aciduria. Pediatr Neurol 17:165-170
13. Sasaki M, Kimura M, Sugai K, Hashimoto T, Yamaguchi S (1998) 3-Hydroxyisobutyric aciduria in two brothers. Pediatr Neurol 18:253-255
14. Boles RG, Chun N, Senadheera D, Wong LJC (1997) Cyclic vomiting syndrome and mitochondrial DNA mutations. Lancet 350:1299-1300
15. Chinnery PF, Turnbull DM (1997) Vomiting, anorexia, and mitochondrial DNA disease. Lancet 6351:448
16. Bennett MJ, Weinberger MJ, Kobori JA, Rinaldo P, Burlina AB (1996) Mitochondrial short-chain L-3-hydroxyacyl-coenzyme A dehydrogenase deficiency: a new defect of fatty acid oxidation. Pediatr Res 39:185-188
17. Pelletier VA, Gal'ano N, Brochu P et al. (1986) Secretory diarrhea with protein-losing enteropathy, enterocolitis cystica superficialis, intestinal lymphangiectasia, and congenital hepatic fibrosis: a new syndrome. J Pediatr 108:61-65
18. Niehues R, Hasilik M, Alton G et al. (1998) Carbohydrate-deficient glycoprotein syndrome type Ib: phosphomannose isomerase deficiency and mannose therapy. J Clin Invest 101:1414-1420
19. Jaeken J, Matthijs G, Saudubray JM et al. (1998) Phosphomannomutase isomerase deficiency: a carbohydrate-deficient glycoprotein syndrome with hepatic-intestinal presentation. Am J Hum Genet 62:1535-1539
20. Lyon G, Adams RD, Kolodny EH (1996) Neurology of hereditary metabolic diseases of children, 2nd edn. McGraw-Hill, New-York
21. Tint GS (1993) Cholesterol defect in Smith-Lemli-Opitz Syndrome. Am J Med Genet 47:5

# CHAPTER 2

# Diagnostic Procedures: Function Tests and Postmortem Protocol

J. Fernandes, J.-M. Saudubray, and J. Huber

## CONTENTS

Introduction ............................... 43
Function Tests ............................ 43
   Metabolic Profile over the Course of a Day ......... 43
      Indications ............................. 43
      Procedure .............................. 43
      Interpretation .......................... 44
   Fasting Test ............................. 45
      Indications ............................. 45
      Preparation ............................ 45
      Procedure .............................. 45
      Interpretation .......................... 45
      Complications .......................... 46
   Glucose Test ............................ 46
      Indications ............................. 46
      Procedure .............................. 46
      Interpretation .......................... 46
      Note of Caution ........................ 46
   Galactose Test .......................... 46
      Indication and Contraindication ............. 46
      Procedure .............................. 46
      Interpretation .......................... 47
   Fructose Test ........................... 47
      Indications ............................. 47
      Procedure .............................. 47
      Interpretation .......................... 47
      Note of Caution ........................ 47
   Fat-Loading Test ........................ 47
      Indications ............................. 47
      Procedure .............................. 47
      Interpretation .......................... 47
      Side Effects ............................ 47
   Tetrahydrobiopterin Test ................. 47
      Indications ............................. 47
      Procedure .............................. 47
      Interpretation .......................... 48
   Exercise Test ........................... 48

Postmortem Protocol ..................... 49
   Cells and Tissues for Enzyme Assays ........... 49
   Cells and Tissues for Chromosome and DNA
   Investigations .......................... 49
   Skin Fibroblasts ......................... 49
   Body Fluids for Chemical Investigations ........ 49
   Imaging ................................ 50
   Autopsy ............................... 50
   Notes .................................. 50

References ............................... 51

## Introduction

The best function test is elicited by nature itself during acute metabolic stresses, such as those caused by an acute infection, inadvertent fasting, or consumption of a nutrient for which a metabolic intolerance exists. As discussed in Chap. 1, if symptoms lead one to suspect the existence of an inborn metabolic disease, blood, urine, and cerebrospinal fluid should be investigated and/or stored in the correct way to perform the emergency protocol (Table 1.1). If no material is available or if the results are incomplete or ambiguous, a function test that challenges a metabolic route may provide a tentative diagnosis.

When performing a function test, it is very important to adhere to a strictly defined protocol in order to attain a maximum of diagnostic information and to minimize the risk of metabolic complications. Protocols applicable to a variety of inborn errors are discussed here. It should be kept in mind that some function tests are used less often, since more direct assays of metabolites and DNA have diminished their diagnostic value. This applies to the galactose test and the fat-loading test. Other tests have fallen into disuse for comparable reasons and have been deleted here. This applies to the glucagon test for the differentiation of glycogen-storage diseases (GSDs) and the phenylpropionic-acid test for uncovering medium-chain acyl-coenzyme-A (CoA)-dehydrogenase deficiency.

## Function Tests

### Metabolic Profile over the Course of a Day

#### Indications

Previous acute or recurrent clinical incident of unknown etiology.

#### Procedure

Blood samples from an indwelling venous catheter (kept open with a saline infusion) are taken before and after meals and once during the night, as outlined in Table 2.1. For a reliable interpretation of the results, the correct method of sampling and processing of blood and urine is specified in Table 2.5.

### Interpretation

Exploration over the course of the day may reveal differences in the metabolic and endocrinologic profiles during the fasting and fed states. It may lead to a tentative diagnosis. For instance:

1. Repeated assays are required for glucose, insulin and free fatty acids (FFA), as hyperinsulinemia is sometimes erratic and difficult to (dis)prove. A plasma insulin concentration greater than 10 mU/l combined with a glucose concentration less than 2.8 mmol/l and a FFA concentration less than 0.6 mmol/l may suggest the existence of hyperinsulinemia (Table 2.2; Chap. 9).
2. The level of some metabolites may characteristically differ in the fasting and the fed states or may remain continuously increased. This applies to lactate, which (during fasting) decreases in cases of pyruvate-dehydrogenase (PDH) deficiency but increases in cases of GSD type I. However, lactate may stay at elevated levels before and after meals in cases of mitochondriopathies involving the citric-acid cycle and the respiratory chain [1].
3. Ketone bodies may increase paradoxically after meals in disorders of the citric-acid cycle and the respiratory chain. For comparison with normal fasting values, see Table 2.2 and, for a further discussion and diagnostic work-up, see Chap. 1 and Figs. 1.1 and 1.2, or proceed to the fasting test.
4. The lactate/pyruvate ratio (L/P) and the 3-hydroxybutyrate/acetoacetate ratio (3OHB/AcAc), reflecting the redox states of the cytoplasm and the mitochondrion, respectively, may provide additional information [2] as follows.

   - L/P increased, 3OHB/AcAc normal or decreased: Pyruvate-carboxylase deficiency or 3-ketoglutarate-dehydrogenase deficiency (Chap. 10)
   - L/P and 3OHB/AcAc both increased with persistent hyperlactacidemia: Respiratory-chain disorder (Chap. 13)
   - L/P decreased or normal and 3OB/AA normal, combined with a varying hyperlactacidemia: PDH deficiency (Chap. 10)

**Table 2.1.** Exploration of intermediary metabolism in the fed and the fasting states. Urine collected overnight and during the daytime during 12-h periods is assayed for amino acids, organic acids, ketone bodies, and carnitine

| Parameters in blood or plasma | Breakfast | | Lunch | | Dinner | | Night |
|---|---|---|---|---|---|---|---|
| | 0 h | After 1 h | 0 h | After 1 h | 0 h | After 1 h | 4 h |
| Glucose[a] | x | x | x | x | x | x | x |
| Acid–base | x | x | | | | | |
| Amino acids | x | x | | | | | |
| Ammonia | x | x | x | x | x | x | x |
| Carnitine and acylcarnitines | x | | | | | | |
| Profile[b] | x | x | x | x | x | x | x |

[a] Glucose is determined immediately
[b] Lactate, pyruvate, 3-hydroxybutyrate, acetoacetate, free fatty acids, and insulin

**Table 2.2.** Metabolic profile during fasting tests in children of different ages (from [3]). Normal blood values at the end of the fast or when the patient is hypoglycemic, irrespective of his age, are: insulin less than 10 mU at a glucose level of less than 2.8 mmol/l; cortisol greater than 120 ng/ml; adrenocorticotrophic hormone (ACTH) less than 80 pg/ml; growth hormone greater than 10 ng/ml

| Age | Less than 12 months | 1–7 years | | 7–15 years | |
|---|---|---|---|---|---|
| | 20 h | 20 h | 24 h | 20 h | 24 h |
| Glucose (mM) | 3.5–4.6 | 2.8–4.3 | 2.8–3.8 | 3.8–4.9 | 3.0–4.3 |
| Lactate (mM) | 0.9–1.8 | 0.5–1.7 | 0.7–1.6 | 0.6–0.9 | 0.4–0.9 |
| FFA (mM) | 0.6–1.3 | 0.9–2.6 | 1.1–2.8 | 0.6–1.3 | 1.0–1.8 |
| KB (mM) | 0.6–3.2 | 1.2–3.7 | 2.2–5.8 | 0.1–1.3 | 0.7–3.7 |
| 3-OH-B (mM) | 0.5–2.3 | 0.8–2.6 | 1.7–3.2 | <0.1–0.8 | 0.5–1.3 |
| 3-OH-B/AcAc | 1.9–3.1 | 2.7–3.3 | 2.7–3.5 | 1.3–2.8 | 1.6–3.1 |
| FFA/KB | 0.3–1.4 | 0.4–1.5 | 0.4–0.9 | 0.7–4.6 | 0.5–2.0 |
| Carnitine (free; μM) | 15–26 | 16–27 | 11.5–18 | 24–46 | 18–30 |

*AcAc*, acetoacetate; *FFA*, free fatty acids; *KB*, ketone bodies; *3-OH-B*, 3-hydroxybutyrate

## Fasting Test

### Indications

The test is used for the clarification of hypoglycemia observed in disorders of gluconeogenesis, fatty-acid oxidation and ketolysis and in organic acidemias, some endocrinopathies and myopathies.

### Preparation

The fasting test is not without risk and should always be performed under close medical supervision. Preferably, if a metabolic abnormality has been noted before the test, the results of the following investigations should be known: L/P, 3OHB/AcAc, ammonia, blood gases, pH, organic acids, amino acids, and carnitine status with differentiation of acylcarnitines. If permanent abnormalities exist, the diagnostic work-up should be changed accordingly. The patient should be adequately fed and the energy intake appropriate for his age during the last three days before the test. No intercurrent infection or metabolic incident should have occurred during the last week.

### Procedure

Tolerance for fasting differs considerably depending on the age of the patient and on the disorder. The recommended period of fasting is as follows: 12 h for age less than 6 months, 20 h for age 6–12 months and 24 h from age 1 year onwards (Table 2.2). The planning of the time schedule of the test should ensure that its last and most important part (during which complications may arise) takes place during the daytime, when the best facilities for close supervision are available.

An indwelling venous catheter with a saline drip is inserted at zero time. The patient is encouraged to drink tea or mineral water (without sugar) during fasting. Table 2.3 gives the time schedule for the laboratory investigations.

The main metabolic "monitors" for the safe continuation of the test are glucose and $HCO_3^-$ concentrations of the blood. Blood for a complete metabolic and endocrinologic profile is sampled at the start and twice at the end of the test. If glucose drops below 3.3 mmol/l, glucose and $HCO_3^-$ should be determined subsequently at 60-min or 30-min intervals. If glucose drops below 2.6 mmol/l and/or $HCO_3^-$ drops below 15 mmol/l, or if neurologic symptoms develop, the test should be terminated. At that time, blood is sampled for the complete metabolic and endocrinologic profile. The urine is initially sampled on ice during 8-h periods and is later sampled during 4-h periods. From each 4-h or 8-h urine collection, a sample of 10 ml has to be frozen at –70°C for the determination of lactate, ketone bodies, amino acids and organic acids.

### Interpretation

The results are compared with the normal values for the particular age (Table 2.2).

**Blood**

The tentative diagnoses are as follows:

- Hyperinsulinemia Glucose < 2.8 mmol/l, insulin > 10 mU/l and FFA < 0.6 mmol/l (simultaneously)
- Fatty-acid-oxidation defect Glucose < 2.8 mmol/l, FFA increase and FFA/total ketone body ratio (mmol/mmol) > 2 (normal < 1)
- Gluconeogenesis defect-Lactate > 3.0 mmol/l

**Table 2.3.** Fasting-test flow sheet. The duration of the test is adapted to the age of the patient (see text)

| Time (h) | 0 | 8 | 12 | 16 | 20 | 24 |
|---|---|---|---|---|---|---|
| Blood | | | | | | |
|   Glucose | + | + | + | + | + | + |
|   $HCO_3^-$ | + | + | + | + | + | + |
|   Lactate | + | + | + | + | + | + |
|   Pyruvate | + | + | + | + | + | + |
|   3-Hydroxybutyrate | + | + | + | + | + | + |
|   Acetoacetate | + | + | + | + | + | + |
|   FFA | + | + | + | + | + | + |
|   Carnitines[a] | + | + | + | + | + | + |
|   Amino acids | + | + | + | + | + | + |
|   Insulin | + | + | | | | + |
|   Cortisol | + | + | | | | + |
|   ACTH | + | + | | | | + |
|   HGH | + | + | | | | + |
| Urine | 0 | 8 | | 16 | 20 | 24 |
|   Amino acids | ←——————→ | | | ←——————→ | | |
|   Organic acids | ←——————→ | | | ←——————→ | | |

*ACTH*, adrenocorticotrophic hormone; *FFA*, free fatty acids; *HGH* human growth hormone
[a] Total, free, and esters

- PDH defect: Lactate and pyruvate decrease during the test
- Defects of the citric-acid cycle and the respiratory chain: Variable levels of lactate and ketone bodies
- Adrenal-cortex insufficiency: Cortisol < 250 nmol/l and glucose < 2.8 mmol/l
- Human growth-hormone (HGH) deficiency: HGH < 10 ng/ml
- Adrenocorticotrophic-hormone (ACTH) deficiency: ACTH < 80 pg/l

### Urine

The safest approach is to compare the results of the last period with those of the first period.

### Complications

Hypoglycemia, metabolic acidosis, heart dysrhythmia, cardiomyopathy, organ failure; see Emergency Treatment (Chap 3).

## Glucose Test

### Indications

Hypoglycemia and/or hyperlactacidemia of unknown etiology.

### Procedure

Previous fasting for 3–8 h, depending on the duration of the intervals between meals. In the case of previously recorded hypoglycemia, the test is started at a plasma glucose concentration between 3.3 mmol/l and 2.8 mmol/l. Insert an indwelling venous catheter 30 min before the expected start of the test and flush with saline drip. Glucose (2 g/kg, with a maximum of 50 g) as a 10% solution in water is administered orally or through a gastric tube in 5–10 min. The blood is sampled from the indwelling venous catheter twice at zero time (just before glucose administration) and every 30 min thereafter until 3–4 h after the completion of the glucose ingestion.

All blood samples are assayed for glucose, lactate, pyruvate, 3OHB and AcAc. A urine sample collected just before the test and a second sample from urine collected during 8 h after glucose administration are tested for lactate, ketone bodies and organic acids.

### Interpretation

### Glucose

A short-lived increase followed by a precipitous decrease is observed in hyperinsulinemia.

### Lactate

A marked decrease from an elevated fasting level occurs in disorders of gluconeogenesis and GSD caused by glucose-6-phosphatase deficiency [4]. An exaggerated increase from a normal fasting level occurs in other GSDs and glycogen-synthase deficiency. Lactate remains increased or increases even further after glucose administration in PDH deficiency and other mitochondrial diseases [5]. The L/P ratio, normally 10:1, is usually increased in pyruvate-carboxylase deficiency and in mitochondrial disorders and is normal or low in PDH deficiency.

### Ketone Bodies

Ketone bodies may increase paradoxically in pyruvate-carboxylase deficiency (with a low 3OHB/AcAc ratio) and in respiratory-chain disorders (with a high 3OHB/AcAc ratio). Fasting ketone bodies are very low in hyperinsulinemia, defects of fatty-acid oxidation and defects of ketogenesis. In contrast, they are very high in disorders of ketolysis.

### Note of Caution

1. In patients with PDH deficiency, a glucose test might precipitate lactic acidosis.
2. The test should be stopped if plasma glucose drops below 2.6 mmol/l. The complete metabolic profile should be taken at that time.

## Galactose Test

### Indication and Contraindication

The galactose test, an old method, is still a valid way to screen for GSD (except in patients presumed to have GSD-I, for whom the test implies a continuation of the fasting state [4]). It should never be applied in patients with galactose in the urine, because galactose administration is very toxic for patients with galactose intolerance.

### Procedure

The preparation is similar to that for the glucose test. Galactose (2 g/kg, with a maximum of 50 g) as a 10% solution in water is administered orally in 5–10 min. Blood is sampled from an indwelling venous catheter, twice at zero time and then every 30 min until 3–4 h after the completion of galactose ingestion. All blood samples are assayed for glucose and lactate.

## Interpretation

Serum lactate rises above 3.5 mmol/l and up to 10 mmol/l in GSD-III, GSD-VI and GSD-IX [4].

## Fructose Test

### Indications

The fructose test is used for the differentiation of disorders of fructose metabolism, fructose-1,6-bisphosphatase deficiency included.

### Procedure

A diet devoid of fructose and sucrose is prescribed for 2 weeks before the test. An indwelling venous catheter is inserted for the collection of blood samples at zero time (twice) and 5, 10, 15, 30, 45, 60 and 90 min after fructose administration. Fructose 0.2 g/kg as a 10% solution in water is injected intravenously during a 2-min period. The intravenous route is chosen instead of the oral route, since the oral administration of fructose is accompanied by severe gastrointestinal side effects in cases of hereditary fructose intolerance. Blood samples are collected from the catheter and assayed for glucose, fructose, phosphate, magnesium and urate.

### Interpretation

Glucose and phosphate decrease within 10–20 min and magnesium and urate increase in parallel in cases of hereditary fructose intolerance. In fructose-1,6-bisphosphatase deficiency, these changes tend to be less pronounced. A marked fructose rise is the only abnormality in fructokinase deficiency; for a review, see [6].

### Note of Caution

When starting the fructose test, the plasma glucose level should not be less than 3.3 mmol/l in view of a potential hypoglycemic effect of fructose. The test is toxic in cases of impaired liver function.

## Fat-Loading Test

### Indications

The fat-loading test is used to investigate whether the in vivo production of ketone bodies from fatty-acid β-oxidation is intact. The additional investigation of some metabolic markers in blood plasma and urine allows one to distinguish the enzyme defects at various levels of the chain of fatty-acid β-oxidation (Chap. 11, Table 11.2). These metabolic markers encompass several acylcarnitines and organic acids [7]. The test is not useful and is even contraindicated if a basal disturbance of organic-acid or acylcarnitine profile has been found.

### Procedure

After a normal diet lasting at least 3 days, the patient is kept fasting during 8–10 h, depending on his tolerance for fasting. Sunflower oil or corn oil (2 g/kg; 2 ml/kg) is administered orally or via a gastric tube, which is afterwards flushed with water. At zero time and after 1, 2, 3, 4, 5 and 6 h, blood from an indwelling venous catheter is sampled for assays of the following substances in plasma: triglycerides, 3OHB, AcAc, FFA, total blood gases (anion gap) and acylcarnitines. The test is terminated earlier at glucose levels less than or equal to 2.6 mmol/l. Urine is collected for assays of dicarboxylic acids in urine 12 h before and after fat loading.

### Interpretation

A dissociation between high levels of FFA and steady, very low levels of ketone bodies while blood glucose drops to hypoglycemic levels is highly suggestive of enzyme defects of fatty-acid oxidation. The test is not reliable in medium-chain and short-chain acyl-CoA-dehydrogenase deficiencies.

### Side Effects

Muscle cramps and gastrointestinal symptoms like abdominal pain, vomiting and diarrhea may occur.

## Tetrahydrobiopterin Test

### Indications

This test is performed in newborns with hyperphenylalaninemia in order to explore whether tetrahydrobiopterin ($BH_4$), the cofactor of phenylalanine hydroxylase, is deficient (1–2% of all patients).

### Procedure

The test is carried out immediately after the detection of hyperphenylalaninemia and before a phenylalanine-restricted diet is initiated. A *simple* $BH_4$-loading test is applied if the plasma phenylalanine level is more than

400 µmol/l. A *combined* BH$_4$ and phenylalanine loading is applied in newborns with an intermediately elevated plasma phenylalanine between 120 µmol/l and 400 µmol/l. The additional phenylalanine is a prerequisite for a elevated phenylalanine concentration in blood sufficient for measurement of the BH$_4$ effect. Before the test, blood and urine should be sampled for the following assays:

- Plasma or serum – Phenylalanine and tyrosine (blood sample 1)
- Test blood spots on filter paper for dihydropteridine-reductase activity
- Urine Pterins – (urine sample 1)

### Simple BH$_4$ Test

BH$_4$ (20 mg/kg) dissolved in 20–30 ml water is administered in a bottle approximately 30 min before a normal bottle feeding. Blood samples 2 and 3 are collected at 4 h and 8 h, respectively, after BH$_4$ loading; urine sample 2 is collected between 4 h and 8 h after BH$_4$ loading. The plasma and urine samples should be protected against light and heat and should immediately be frozen.

### Combined Test

L-Phenylalanine (100 mg/kg) dissolved in fruit juice is administered 3 h before the BH$_4$ load. Blood and urine are sampled at time intervals after BH$_4$ loading that are the same as those used for the simple test. After the completion of the test, a phenylalanine-restricted diet is instituted pending the availability of the results.

### *Interpretation*

In patients with defects of pterin metabolism, a precipitous drop of the blood phenylalanine concentration occurs within 4–8 h after BH$_4$ challenge. In patients with classical phenylketonuria, phenylalanine and tyrosine levels are not influenced by BH$_4$ [8].

### Exercise Test

The exercise test is a means to disclose patients suspected of having a metabolic myopathy. Several methods exist:

1. Semi-ischemic forearm-exercise test [9]
2. Bicycle ergometer test [10]
3. Treadmill test

The forearm test and the bicycle test are only applicable in adults and older children who can squeeze the sphygmomanometer balloon or ride a bicycle. The treadmill test offers the advantage that it can be used from the age at which the child is able to walk. All exercise tests should be carried out at a *sub*maximal workload. This is a safeguard to prevent severe complications, such as rhabdomyolysis, myoglobinuric anuria and metabolic acidosis.

The *forearm test*, first described by McArdle in 1951 to define a "myopathy due to a defect in muscle glycogen breakdown", was an ischemic exercise test. Its successor, the semi-ischemic modification, is much safer. In that version, a sphygmomanometer cuff is inflated around the upper arm to the mean arterial pressure, and not above arterial pressure. A hand manometer is squeezed by the patient for 2 min (or less) with submaximal exertion (to prevent muscle cramps).

In the *bicycle ergometer test*, the duration of the exercise and a submaximal workload associated with a pulse rate below 150 beats/min for adults or between 150 beats/min and 180 beats/min for children are adapted to the condition of the patient. In the *treadmill test*, the speed of the belt and its angle of inclination can be manipulated to a walking velocity of 3–5 km/h and a pulse rate of 150–180 beats/min. Exhaustion arises rapidly in myopathies due to defects of glycolysis, since glycolysis is the first source of energy. It also arises rapidly in patients with defects of the citric-acid cycle and the respiratory chain, since mitochondria are the most essential providers of energy. It comes slower in defects of fatty-acid oxidation (after the exhaustion of energy from glycogen via aerobic and anaerobic glycolysis). The interpretation of the results of each exercise test should take this time sequence into consideration. In plasma and urine, the parameters to be compared before, during and after exercise are the following:

- Plasma – Compare lactate, pyruvate, 3OHB and AcAc regularly; compare carnitine, acylcarnitines, ammonia, creatine kinase and potassium at the start and the end of the test
- Urine – Lactate, ketone bodies, organic acids, carnitine
- Lactate excess – Reflects a disturbed equilibrium between its production from glycolysis and its expenditure in the citric-acid cycle
- Ammonia excess – Arises if deamination of adenosine monophosphate (AMP) – from adenosine triphosphate (ATP) – into inosine monophosphate is increased and the regeneration of ATP from AMP is reduced during a fuel deficit in muscles

In conclusion, the exercise test is a suitable means for screening some metabolic myopathies and for comparing the results of treating a metabolic myopathy. The best exercise test for the widest span of age is the treadmill test. However, the apparatus is mainly used by cardiologists and pulmonologists to assess cardiac and lung function during exertion, and, to a lesser extent, by neurologists and pediatricians for the differentiation of myopathies. The development of standardized metabolic nomograms is, therefore, an urgent issue. For a tentative interpretation of the exercise test, see Table 2.4.

## Postmortem Protocol

Since the first description of a postmortem protocol for suspected genetic disease by Kronick [11], some refinements have become available to enhance the diagnostic value of the original recommendations [12, 13]. In the protocol given below, the time schedule for proper conservation of specimens determines the sequence of the diagnostic procedures. It comprises the following elements of investigation:

1. Cells and tissues for enzyme assays
2. Cells and tissues for chromosome and DNA investigations
3. Body fluids for chemical investigation
4. Imaging
5. Autopsy

### Cells and Tissues for Enzyme Assays

Liver (minimum 10–20 mg wet weight) and muscle (minimum 20–50 mg wet weight) biopsies are taken by needle puncture or, even better, via open incision. The advantage of obtaining samples of several muscles and heart muscle is that some enzyme defects are expressed in a tissue-specific pattern. If possible, a brain biopsy should be taken within 2–4 h postmortem. The tissues are immediately frozen in small plastic cups in liquid nitrogen, followed by storage at -70°C. If the muscle biopsy is taken from muscle above the fascia lata, it can be combined with a fascia-lata biopsy (diameter 3 mm) for a fibroblast culture. As most (but not all) enzymes are stable at that temperature, the biochemist should be consulted regarding whether the processing of fresh material before freezing is also indicated for additional biochemical assays. Part of the liver biopsy should be fixed for histological and electron-microscopic investigation prior to freezing (see "Autopsy" below). Blood cells must be conserved for enzyme assays and for chromosome and DNA investigation. A total of 20 ml of blood is collected by peripheral or intracardiac puncture in a heparin-coated syringe; 10 ml is transferred to the laboratory for isolation of erythrocytes or white blood cells, and the biochemist is notified. At least 10 ml is conserved for chromosome analysis and DNA extraction (see below).

### Cells and Tissues for Chromosome and DNA Investigations

Of the above-mentioned 10 ml of fresh heparinized blood, 1–2 ml is reserved for chromosome analysis; the remaining 8–9 ml can be used for DNA extraction. Alternatively, blood spots dried on filter paper (as in the Guthrie test) are useful for many investigations and should always be collected. These samples can also be used for DNA analysis after polymerase-chain-reaction (PCR) amplification.

### Skin Fibroblasts

At least two biopsies (diameter 3 mm) are taken under sterile conditions as early as possible: one from the forearm, one from the upper leg (fascia lata, see above), and one from the pericardium (in cases of delayed autopsy). These samples are conserved in culture medium or, alternatively, on sterile gauze wetted in sterile saline and sealed in a sterile tube for one night at room temperature. Cultured fibroblasts provide an unlimited source of cells for chromosome and DNA investigations and enzyme assays. If none of the above-mentioned measures have been taken and, on hindsight, one needs DNA from the patient, one should bear in mind that, with PCR amplification, one needs only small amounts of DNA to perform certain analyses. Sources for DNA in these situations are manifold, e.g., blood spots on filter paper, fresh frozen tissue samples provided by the pathologist, or even paraffin-embedded tissue (although DNA extraction of the latter is more complicated).

### Body Fluids for Chemical Investigations

Plasma from the centrifuged blood sample, urine (~10 ml), and cerebrospinal fluid (~4 ml) are immediately frozen at -20°C (Table 2.5). If no urine can be obtained by suprapubic puncture or catheterization, the bladder may be filled with 20 ml of saline solution, and diluted urine may be harvested. Alternatively, vitreous humor can also be collected (by intraocular

**Table 2.4.** Exercise-induced changes of lactate, ammonia, and potassium in metabolic myopathies at 50% of estimated maximum exertion

| Disorder | Lactate | Ammonia | Potassium |
|---|---|---|---|
| Glycolysis defects | – | ± | + |
| Mitochondriopathies | + | + | + |
| MAD deficiency | + | – | + |
| Controls | – | ± | – |

– no change; + abnormal increase; ± slight increase; *MAD*, myoadenylate deaminase

puncture) and frozen. This liquid is comparable to blood plasma with respect to its solubility for organic acids. Moreover, it is more stable than other body fluids after death, and there is little exchange of substances with blood constituents. Vitreous humor may be easily collected by piercing a small-bore needle, attached on a 2 ml syringe, laterally through the sclera into the center of the eyeball, and aspirating 0.5–1.5 ml of fluid. If the eyeball would shrink, the aspirated fluid can be replaced by saline. Recently, bile, readily available at autopsy, has been found to be useful material for the postmortem assay of acylcarnitines in disorders of fatty-acid oxidation [14].

It must be realized that many biochemical parameters are impossible to interpret postmortem due to rapid tissue lysis. Among them are carnitine (total and free), ammonia, lactate, and amino acids, which rapidly increase without any specific significance. In contrast, the acylcarnitine-ester profile (determined from dried blood spots) is highly diagnostic for many disorders of fatty-acid oxidation and for organic acidemias.

## Imaging

Photographs are made of the whole body and of specific dysmorphic anomalies, if present. Total-body radiographs in anteroposterior and lateral views are performed, as is ultrasound of the skull, thorax and abdomen. In those cases where autopsy is refused, much additional information might be gathered from postmortem magnetic resonance imaging (MRI) [15].

## Autopsy

The autopsy should be complete, including the cranium, provided that the parents give their permission. The pathologist freezes fresh samples of liver, spleen, muscle, heart, kidney and brain and conserves important tissues for histology and electron microscopy in buffered formaldehyde (4%) and Karnofski fixative, respectively.

Mentioning the autopsy as a last item does not mean that it is the least important. On the contrary, the first three items are complementary to it, and only the rapid decay of enzymes and vital cells gives their conservation a higher priority.

## Notes

An explanation of the importance of a complete work-up is generally well received by modern parents. However, even if a complete autopsy is refused, permission to take photographs, X-rays and MRI scans, blood, urine, and cerebrospinal fluid samples and to do needle biopsies of liver and muscle is usually given. An assembly kit containing all the material for collecting and conserving the specimens is a highly recommended means of enhancing the speed and completeness of the postmortem protocol.

**Table 2.5.** Collection, processing and storage of blood, urine, and cerebrospinal fluid for metabolic and endocrinologic investigation. The volumes of blood, urine, and cerebrospinal fluid are subject to local customs, which must be taken into account

**Blood**
    Hematology: 0.5 ml in EDTA tube (no capillary)
    Blood gases: 0.5 ml heparin-coated syringe (eject air bubble, cap syringe immediately)
    Electrolytes, urea, creatinine, urate, total protein, liver function tests: 1–2 ml (centrifuge after clotting)
    Glucose: 0.3 ml fluoride-heparin cup (dry heparin and fluoride salts, no solution); lactate/pyruvate, 3-hydroxybutyrate/acetoacetate: 1 ml blood (no forcing), mix immediately with 0.5 ml perchloric acid (18% v/v, keep on ice, centrifuge under refrigeration, store supernatant at $-20°C$
    Ammonia: 0.5 ml in heparin-coated syringe on ice (eject air bubble, cap syringe immediately)
    Amino acids: 1–2 ml in EDTA or heparin tube
    Carnitine (free and total): 2 ml in EDTA tube on ice, centrifuge immediately under refrigeration, store at $-20°C$
    Free fatty acids: 0.3 ml in fluoride-heparin cup (dry heparin and fluoride salts, no solution)
    Insulin: 1 ml in EDTA tube, centrifuge under refrigeration, store at $-20°C$
    ACTH: 1 ml in plastic, heparin-coated syringe (keep on ice, centrifuge under refrigeration in plastic tube, store in plastic tube at $-20°C$)
    Growth hormone: 1 ml (centrifuge under refrigeration after clotting, store at $-20°C$)
    Glucagon: 3 ml heparin tube (centrifuge under refrigeration, store in plastic vial at $-20°C$)
**Urine**
    *pH*, amino acids, organic acids, ketone bodies, lactate, reducing substances: 5 ml (at least), freeze at $-20°C$
**Cerebrospinal fluid**
    Cells, protein, glucose: 0.5–1.0 ml in plastic tube
    Lactate/pyruvate: 1 ml, add to 0.5 ml perchloric acid (18% v/v, keep on ice, centrifuge under refrigeration, store supernatant at $-20°C$)
    Amino acids: 1 ml in plastic tube
    Culture: 0.5–1 ml in sterile tube

*ACTH*, adrenocorticotrophic hormone; *EDTA*, ethylenediaminetetraacetic acid

The results of the postmortem protocol provide an essential source of information for parents and for genetic counseling, which would otherwise be impossible. Cells cultured for chromosome and DNA studies should be transferred to a repository for future reference.

## References

1. Touati G, Rigal O, Lombes A et al. (1997) In vivo functional investigations of lactic acid in patients with respiratory chain disorders. Arch Dis Child 76:16-21
2. Poggi-Travert F, Martin D, Billette de Villemeur T et al. (1996) Metabolic intermediates in lactic acidosis: compounds, samples, and interpretation. J Inher Metab Dis 19:478-488
3. Bonnefont JP, Specola NB, Vassault A et al. (1990) The fasting test in paediatrics: application to the diagnosis of pathological hypo- and hyperketotic states. Eur J Pediatr 150:80-85
4. Fernandes J, Huying F, Van de Kamer JH (1969) A screening method for liver glycogen diseases. Arch Dis Childh 44:311-317
5. Ching-Shiang Chi, Suk-Chun Mak, Wen-Jye Shian, Chao-Huei Chen (1992) Oral glucose lactate stimulation in mitochondrial diseases. Pediatr Neurol 8:445-449
6. Steinmann B, Gitzelmann R (1981) The diagnosis of hereditary fructose intolerance. Helv Paediatr Acta 36:297-316
7. Costa CCG, Tavares de Almeida I, Jakobs C, Poll-The BT, Duran M (1999) Dynamic changes of plasma acylcarnitine levels induced by fasting and sunflower oil challenge test in normal children. Pediatr Res 46:440-444
8. Ponzone A, Guardamagna O, Spada M et al. (1993) Differential diagnosis of hyperphenylalaninaemia by a combined phenylalanine-tetrahydrobiopterin loading test. Eur J Pediatr 152:655-661
9. Kono N, Tarui S (1990) The exercise test. In: Fernandes J, Saudubray J-M, Tada K (eds) Inborn metabolic diseases. Diagnosis and treatment. Springer, Berlin Heidelberg New York
10. Kono N, Mineo I, Shimizu T et al. (1986) Increased plasma uric acid after exercise in muscle phosphofructokinase deficiency. Neurology 36:106-108
11. Kronick JB, Scriver CR, Goodyer PR, Kaplan PB (1983) A perimortem protocol for suspected genetic disease. Pediatr 71:960-963
12. Helweg-Larsen K (1993) Postmortem protocol. Acta Paediatr [Suppl] 389:77-79
13. Poggi F, Rabier D, Vassault A et al. (1994) Protocole d'investigations métaboliques dans les maladies héréditaires du métabolisme. Arch Pediatr 1:667-673
14. Rashed MS, Ozand PT, Bennett J et al. (1995) Inborn errors of metabolism diagnosed in sudden death cases by acylcarnitine analysis of postmortem bile. Clin Chem 41:1109-1114
15. Brookes JAS, Hall-Craggs MA, Sams VR, Lees WR (1996) Non-invasive perinatal necropsy by magnetic resonance imaging. Lancet 348:1139-1141

# CHAPTER 3

# Emergency Treatments

H. Ogier de Baulny and J.M. Saudubray

CONTENTS

General Principles .......................... 53
   Supportive Care ........................... 53
   Nutrition ................................. 53
   Toxin-Removal Procedures .................. 54
   Additional Therapies ...................... 54
Neonatal Emergencies ....................... 54
   Neurological Distress ..................... 54
      Supportive Care ........................ 55
      Nutrition .............................. 55
      Toxin-Removal Procedures ............... 57
      Additional Therapies ................... 58
      Specific Approaches in Primary Hyperlactacidemia ... 58
   Liver Failure ............................ 59
   Neonatal Hypoglycemia .................... 59
      GSD and Fructose-1,6-Biphosphatase Deficiency ... 59
      Neonatal Hyperinsulinism ............... 59
      Fatty-Acid-Oxidation Defects ........... 59
   Cardiac Failure .......................... 60
Late-Onset Coma ............................ 60
   Supportive Care .......................... 60
      Cerebral Edema ......................... 60
      Hydration .............................. 60
   Nutrition ................................ 60
   Toxin-Removal Procedures ................. 60
   Additional Therapies ..................... 60
References ................................. 61

As soon as the diagnosis of a metabolic disorder is suspected, an emergency management has to be scheduled. As already stated in Chap. 1, both the signs and management mainly depend on the physiopathology involved. However, this chapter focuses on the main age-related clinical presentations in neonates and children for which emergency treatment can be life saving. In neonates, four main clinical presentations can be identified.

1. The most frequent situation is neurological distress, in which branched-chain organic acidurias (BCOAs) and urea-cycle defects (UCDs) are the main disorders concerned.
2. Some patients may present with liver failure, in which only galactosemia, hereditary fructose intolerance and tyrosinemia type I are amenable to emergency treatment and must be considered at once.
3. In patients presenting with hypoglycemia, blood glucose levels must be corrected. Progress following glucose provision is useful in recognizing the three disorders usually implicated, which are hyperinsulism, glycogen-storage disease (GSD) and mitochondrial β-oxidation defects.
4. In neonates presenting with cardiac failure, the sole treatable disorder is a fatty-acid β-oxidation defect, which requires high-glucose infusion and a fat-free diet.

In addition, in patients presenting with intractable convulsions, vitamin responsiveness (biotin, pyridoxine, folate) must be systematically considered. All these clinical situations can also arise in older children, and any type of coma can be an especially revealing sign of metabolic disorders. In addition, children may present with recurrent attacks of unexplained dehydration, abdominal pain, muscle pain and myolysis. Such situations require careful and urgent biochemical investigation before starting potentially risky diagnostic or therapeutic procedures.

## General Principles

### Supportive Care

Many of these very ill patients (especially newborns) require ventilatory and circulatory support. Most of them require rehydration and correction of electrolyte, calcium and phosphate imbalance, since specific therapeutic approaches will fail unless these derangements are corrected. Conversely (and despite their importance), these treatments must not postpone the initiation of the specific therapeutic means.

Patients affected with a metabolic crisis frequently suffer from concomitant septicemia. This, in turn, results in persistent catabolism and, therefore, in therapeutic failure. Therefore, infections must be thoroughly searched for and prevented.

### Nutrition

Whatever the disease, nutrition is a pivotal therapeutic approach, and both its composition and mode of administration must be rapidly considered. Briefly, four types of composition can be considered: a normal diet, a low-protein diet, a carbohydrate-restricted diet and a high-glucose diet with or without lipid restriction (Table 3.1).

The mode of administration selected depends on the disorder and the clinical status. When the clinical status allows oral nutrition, it is the first choice. However, continuous enteral tube feeding can be temporarily useful in many patients with poor initial condition. In some cases (such as intestinal intolerance, high-energy or high-glucose requirements or invasive techniques of toxin removal), effective enteral nutrition is precluded. In those cases, total parenteral nutrition (TPN) is the method of choice.

**Toxin-Removal Procedures**

Toxin-removal procedures are considered for those patients affected with disorders of the intoxication types, such as BCOA and UCD, during which the neurologic outcome is compromised. The procedures are also used if symptomatic care associated with specific nutrition is insufficient to rapidly correct the metabolic imbalance. Exchange transfusion, peritoneal dialysis (PD), hemofiltration (HF), and hemodialysis (HD) are the main techniques that can be used, keeping in mind that the choice of the technique may be influenced by the experience of the medical staff.

**Additional Therapies**

Depending on the diseases involved, some specific therapies can be used. They mainly comprise vitamin supplementation, carnitine and specific substrates that allow the outlet by alternate pathways (Table 3.2) (see also Table 5.1).

**Table 3.2.** Cofactors used in various metabolic disorders

| Cofactor (doses) mg/day | Disorders |
| --- | --- |
| Biotin (10–20) | Propionic acidemia |
|  | Multiple carboxylase deficiency |
|  | Hyperlactacidemia (PC) |
| Carnitine (50–100 po, 400 i.v.) | Branched-chain organic aciduria |
|  | Dicarboxylic aciduria |
|  | Primary hyperammonemia |
|  | Hyperlactacidemia |
| Cobalamin, B12 (1–2) | Methylmalonic aciduria |
| Folic acid (10–40) | Folate-responsive seizures |
| Pyridoxine, vitamin B6 (50–100) | Pyridoxine-responsive seizures |
| Riboflavin, vitamin B2 (20–40) | Glutaric aciduria |
|  | β-oxidation defects |
| Thiamin, vitamin B1 (10–50) | Maple-syrup-urine disease |
|  | Hyperlactacidemia (PDH) |

*PC*, pyruvate-carboxylase deficiency; *PDH*, pyruvate-dehydrogenase deficiency

**Neonatal Emergencies**

After the diagnosis of an inborn error of metabolism is suspected, an emergency treatment initially aimed at treatable disorders has to be scheduled. As already stated in Chap. 1, the management depends on the physiopathology involved (neonatal distress of intoxication type or of energy-deficiency type). Four main clinical presentations can be identified.

**Neurological Distress**

Treatable disorders are mostly BCOAs and UCDs. Some patients with inborn errors of energy metabolism can present with predominant hyperlactacidemias. After the diagnosis of an intoxication type is suspected, the treatment is directed towards the suppression of the production of toxic metabolites from catabolism of endogenous protein and the stimulation of their elimination by extrarenal procedures and specific alternate pathways, if available.

**Table 3.1.** Types of diet with regard to the type of metabolic disease

| Type of disease | Type of diet |
| --- | --- |
| Intoxication | Suppression of the harmful metabolite |
|   BCOA and UCD | Low protein/protein free |
|   Galactosemia/HFI | Galactose/fructose exclusion |
| Hypoglycemia | Control of glucose and fat supplies |
|   GSD/gluconeogenesis defects | Normal |
|   Mitochondrial β-oxidation defects | High glucose + low fat (±MCT) |
|   Hyperinsulinism | High glucose |
| Primary hyperlactacidemia | Control of glucose and fat supplies |
|   PC | Normal |
|   PDH | Normal/low glucose ± high fat |
|   RCD | Normal/low glucose ± high fat |

*BCOA*, branched-chain organic aciduria; *GSD*, glycogen-storage diseases; *HFI*, hereditary fructose intolerance; *MCT*, medium-chain triglycerides; *PC*, pyruvate-carboxylase deficiency; *PDH*, pyruvate-dehydrogenase deficiency, *RCD*, respiratory-chain disorders; *UCD*, urea-cycle defects

## Supportive Care

### Hydration and Acid–Base Equilibrium

Poor feeding, polypnea, increased renal fluid losses and consequent hypovolemia and prerenal failure are frequently observed during metabolic crisis. Thus, rehydration is often necessary.

Many patients with ketoacidosis due to organic-acid accumulation present with intracellular dehydration, which is often underestimated. In this situation, aggressive rehydration with hypotonic fluids and alkalinization may cause or exacerbate pre-existing cerebral edema. Therefore, overhydration should be prevented, and a rehydration schedule should be planned over a 48-h period, with fluid infusion less than 3 l/m²/day. The repair fluid should contain an average concentration of 70–85 mmol/l of $Na^+$ (4–5 g/l of NaCl), 30–40 mmol/l of $K^+$ (2–3 g/l of KCl) and 5% glucose.

Acidosis, if severe (pH < 7.10, $HCO_3^-$ < 10 mEq/l), can be partially corrected with i.v. bicarbonate, especially if it does not improve with the first measures of toxin removal. However, it should be stressed that aggressive therapy with repeated boluses of i.v. bicarbonate may induce cerebral edema, hypernatremia and even cerebral hemorrhage. In order to compensate for bicarbonate consumption, sodium bicarbonate may be substituted for one quarter to one half of the sodium requirements during the first 6–12 h of rehydration. To prevent precipitation with calcium, the bicarbonate solution should be connected to the infusion line with a Y-connector. Some patients affected with UCD may present with mild acidemia, a situation which should not be corrected, as acidosis protects against $NH_4$ dissociation and toxicity.

### Venous Catheterization

Insertion of a central venous catheter should be considered at once to rapidly meet a high energy requirement (see below).

### Nutrition

Whatever the disease, suppression of toxic metabolite production from breakdown of endogenous protein is essential. Therefore, anabolism should be induced using a high-energy and low-protein diet. However, we must be aware that hypercaloric nutrition by itself is seldom sufficient to correct a metabolic imbalance rapidly, and its use as a sole therapeutic means might compromise the neurologic outcome [1, 2]. Conversely, any toxin-removal procedure will fail without concomitant anabolism.

### Enteral Nutrition

When the clinical status allows continuous enteral nutrition, it must be considered as a first choice. The composition of the enteral formula is based on a glucose–lipid mixture (Table 3.3 for examples). To prevent acute protein malnutrition, a protein-free diet must not be used for more than 2 days. Depending on the defect involved, an appropriate amino acid mixture can be added to cover the protein requirement. In order to titrate gastric tolerance, nutrition is given at a low rate – for instance, 10 ml/3 h and increased every 3–6 h until the full fluid requirement is met. Simultaneously, the parenteral infusion rate is decreased reciprocally. Finally, the diet should provide 130–150 kcal/kg/day. Micronutrients, osmolarity, and renal solute load must be checked in order to provide the recommended dietary allowance (RDA) and prevent diarrhea and dehydration.

Once the toxic metabolites have normalized, natural proteins are introduced using quantified amounts of infant formula. At this step, attention must be paid to both the protein and essential amino acid requirements. For patients with an inborn error blocking an amino acid catabolic pathway, intake of natural protein and essential amino acids must cover the minimal requirements (protein accretion + non-urinary losses), which are 50–60% below the normal requirements (protein accretion + non-urinary losses + urinary losses) and should not follow the RDA [3]. These minimal requirements represent the basis for initiation of a protein-controlled diet. Next, natural protein and amino acid intakes are adjusted to growth and to specific biochemical controls.

### Parenteral Nutrition

In most cases, digestive intolerance or application of invasive toxin-removal techniques preclude effective enteral feeding. In that case, TPN is the method of choice to provide hypercaloric nutrition during the emergency treatment. As an example, parenteral solutions providing 100 kcal/kg/day to a 3.5-kg baby are shown in Table 3.4. If a suitable amino acid solution is not immediately available, start with a mixture of glucose (15–20%) and lipid (2–3 g/kg/day) solutions. To prevent acute protein malnutrition, this protein-free solution should not be used for more than 48 h. After that, protein must be added using either the oral route (with measured amounts of milk) or, in case of prolonged digestive intolerance, the i.v. route (using a commercially available amino acid solution in the TPN) is used. Initially, the amino acids are introduced in amounts sufficient to meet minimal daily requirements and are then titrated according to biochemical

**Table 3.3.** Examples of protein-free diets for enteral nutrition in a 3.5-kg neonate. These diets must be checked for all micronutrients and vitamins and supplemented in order to cover normal requirements for a newborn

| Nutrient | Amount | Energy (kcal) | Prot (g) | Fat (g) | Glc (g) | Na (mEq) | K (mEq) | Ca (mg) | Pi (mg) |
|---|---|---|---|---|---|---|---|---|---|
| **Diet 1** | | | | | | | | | |
| 80056 (Mead Johnson) | 70 g | 343 | | 15.4 | 50.4 | 2.2 | 6 | 378 | 210 |
| Maltodextrin | 20 g | 80 | | | 20 | | | | |
| Oil | 2 g | 18 | | 2 | | | | | |
| Powder for oral rehydration | 1 unit | 35 | | | | 8 | 10 | 5 | |
| Water | Sufficient for a total volume of 500 ml | | | | | | | | |
| Total | | 476 | 0 | 17.4 | 78.4 | 12.2 | 11 | 378 | 210 |
| Total energy provision (%) | | | 0 | 33 | 66 | | | | |
| **Diet 2** | | | | | | | | | |
| Maltodextrin | 70 g | 280 | | | 70 | | | | |
| Oil | 17 g | 153 | | 17 | | | | | |
| Powder for oral rehydration | 1 unit | 35 | | | | 8 | 10 | 5 | |
| Water | Sufficient for a total volume of 500 ml | | | | | | | | |
| Total | | 468 | 0 | 17 | 78 | 10 | 5 | | |
| Total energy provision (%) | | | 0 | 33 | 67 | | | | |

*Glc*, glucose and glucose polymers; *Pi*, inorganic phosphate; *Prot*, proteins

**Table 3.4.** Examples of the composition of parenteral nutrition for a 3.5-kg newborn (minerals, vitamins and nutrients should be added at the recommended allowances for age)

| | Without adequate i.v. amino acid solution | | | | | With i.v. amino acid solution free of BCAA | | | | |
|---|---|---|---|---|---|---|---|---|---|---|
| | Vol. (ml) | N (g) | Fat (g) | Glc (g) | Energy (kcal) | Vol. (ml) | N (g) | Fat (g) | Glc (g) | Energy (kcal) |
| **Diet 1** | | | | | | | | | | |
| Amino acids (9%)[a] | | | | | | 20 | 0.27 | | 4 | 23 |
| Intralipid (20%) | 50 | | 10 | | 90 | 50 | | 10 | | 90 |
| Glc (15%) | 400 | | | 60 | 240 | 375 | | | 56 | 225 |
| Total | 450 | | 10 | 60 | 330 | 445 | 2.6 | 10 | 60 | 338 |
| Water (ml/kg/day) | 130 | | | | | 127 | | | | |
| Total energy provision (%) | | | 27 | 73 | | | 2 | 27 | 71 | |
| **Diet 2** | | | | | | | | | | |
| Amino acids (6.53%)[b] | 28 | 0.26 | | | 13 | | | | | |
| Intralipid (20%) | 50 | | 10 | | 90 | | | | | |
| Glc (15%) | 400 | | | 60 | 240 | | | | | |
| Total | 475 | 2.6 | 10 | 60 | 337 | | | | | |
| Water (ml/kg/day) | 135 | | | | | | | | | |
| Total energy provision (%) | | 2 | 27 | 71 | | | | | | |

*BCAA*, branched-chain amino-acids; *Glc*, glucose
[a] Calculation based on the i.v. BCAA-free amino acids solution used by Berry et al. [1] in patients affected with maple-syrup-urine disease. It contains 9 g/dl of amino acids mixed with 20–25% Glc
[b] Calculation based on a commercially available i.v. amino acid solution (Vaminolact; Pharmacia) containing 9.3 g/l of total nitrogen and 65.3 g/l amino acids; 28 ml provides 200 mg leucine, 103 mg valine, 89 mg isoleucine, 37 mg methionine and 103 mg threonine

checks. The method is safe if the amino acid solution is evenly distributed over the whole day [4].

During this process, one must be aware of the fact that the minimal isoleucine requirement in neonates is at least equal to the minimal valine requirement and that many i.v. amino acid solutions provide less isoleucine than valine. Thus, additional oral supplementation with L-isoleucine (25–100 mg/day) is often necessary when nutrition only provides the minimal requirement for L-valine.

As soon as the digestive route is available, the switch from parenteral to enteral nutrition is scheduled over a 4-day to 5-day period (Table 3.5). The first step is to progressively give the desired amount of protein (or controlled amino acid) using human milk or infant formula. Next, calories are slowly added using either glucose polymer and lipids or 80056 (Mead Johnson). Minerals, vitamins and micronutrients are also added. Addition of an amino acid mixture, if necessary, is the last step, because it increases the osmolarity of the solution and can induce diarrhea. During this process, water supply is increased to cover the requirement for age and weight (500–600 ml/day), and calories are adapted to reach 130–150 kcal/kg/day.

## Toxin-Removal Procedures

### Exchange Transfusion

Theoretically, exchange transfusion (ET) is an inadequate removal procedure for metabolites distributed throughout total body water. However, ET with large volumes of fresh blood has long been recognized as an effective means in numerous inborn errors of metabolism, such as maple syrup urine disease (MSUD), methylmalonic, propionic, and isovaleric acidemias (MMA, PA, IVA) and even UCD [5]. However, its transient effect limits its use, and ET should only be applied in association with other methods (such as PD [6]) or as a result of long-standing patterns, such as multiple or continuous exchanges. Multiple exchanges use 1.5–4 volume exchanges repeated four to six times within 24 h. Continuous exchange using 600 ml/kg body weight within 15 h has been successfully performed in patients affected with MSUD [7].

### Peritoneal Dialysis

PD for the emergency treatment of newborns with various metabolic disorders was long ago demonstrated to be efficacious [8]. Manual PD requires minimal technical expertise and can be rapidly initiated in any pediatric intensive care unit (PICU). Warmed dialysate solutions buffered with bicarbonate with volumes of 40–50 ml/kg body weight are delivered by gravity. One-hour cycles (15-min fill-up, 30-min dwell time, 15-min drainage) are repeated over 24–36 h, during which most of the toxin removal occurs. Prolonged PD is usually not necessary, except for UCD. Continuous-flow PD with inflow and outflow catheters could be a way to circumvent severe technical problems, such as poor drainage and leakage of dialysate [9]. Careful records of the inflow and outflow must be kept, and net water exchange must be checked frequently by weighing the patient. Dehydration must be prevented by i.v. infusion or TPN when available. When overhydration exists, the dialysis fluid should be made hypertonic by addition of glucose (3 g/100 ml = 100 mOsm/l). Due to glucose absorption from the dialysis fluid (200–300 mg/kg/h), hyperglycemia may develop, and insulin therapy may be required.

In terms of clearance, PD is far less efficient than HD and HF. It has, however, the advantage of simplicity. The main cause of failure is poor splanchnic flow secondary to shock and septicemia. Clearances average 6–12 ml/min/m$^2$ for ammonia [5, 10, 11], leucine [6, 12, 13], PA, and MMA [6]. Only methylmalonate has a spontaneous renal clearance twice as high as that of PD; MMA patients, therefore, do not require PD [6].

### Continuous Hemofiltration

HF appears to be an effective means of treatment in newborns and infants suffering acute decompensation of various metabolic disorders. The procedure consists of a low-resistance extracorporeal circuit connected to a small-fiber hemofilter that is permeable to water and non-protein-bound small solutes [14]. Continuous veno-venous HF is the most suitable technique, as venous dual-luminal catheters and extracorporeal material adapted for low-weight children are now available. The ultrafiltrate of plasma is concurrently replaced by an electrolyte and TPN solution. The advantages are: simplicity of logistics, high tolerance in neonates or infants who present with hemodynamic

**Table 3.5.** Example of switching from a total parenteral diet to an enteral diet in a neonate (weight: 3.5 kg). This diet must be checked for other micronutrients and vitamins and supplemented in order to cover normal requirements for a newborn

|  | Vol. (ml) | Prot (g) | Fat (g) | CHO (g) | Energy (kcal) |
|---|---|---|---|---|---|
| **First step** | | | | | |
| *Parenteral nutrition* | | | | | |
| Glucose (25%) | 200 | | | 50 | 200 |
| Intralipid (20%) | 10 | | 2 | | 18 |
| *Enteral nutrition* | | | | | |
| Human milk | 300 | 3.9 | 10.5 | 20.4 | 192 |
| *Total* | 510 | 3.9 | 12.5 | 70.4 | 410 |
| Water (ml/kg/day) | 146 | | | | |
| Total energy provision (%) | | 3.8 | 27.5 | 68 | |
| **Second step** | | | | | |
| *Parenteral nutrition* | | | | | |
| Glucose (25%) | 120 | | | 30 | 120 |
| *Enteral nutrition* | | | | | |
| Human milk | 300 | 3.9 | 10.5 | 20.4 | 192 |
| Maltodextrin (25%) | 100 | | | 25 | 100 |
| Oil | 4 | | 4 | | 36 |
| *Total* | 532 | 3.9 | 14.5 | 75.4 | 448 |
| Water (ml/kg/day) | 152 | | | | |
| Total energy provision (%) | | 3.5 | 29.1 | 67.3 | |

*Cal*, calories; *CHO*, carbohydrates; *Prot*, protein

instability, multiorgan failure, and hypercatabolic state, and the ability to use a large volume of TPN without the risk of overhydration. Nevertheless, application of such procedures requires a PICU trained in the techniques of extracorporeal circulation. Hemodiafiltration increases solute removal by the addition of diffusive transport exerted by a dialysis solution flowing upstream through the ultrafiltrate compartment of the hemofilter [15]. However, most of the basic simplicity of the HF is then lost [16].

The ultrafiltrate formed during HF has essentially the same small-solute composition as plasma water. Therefore, the clearance of these solutes approximates the ultrafiltration rate. Clearances of leucine and ammonia have been reported to vary from 8 ml/min/m$^2$ to 50 ml/min/m$^2$ and from 8 ml/min/m$^2$ to 21 ml/min/m$^2$, respectively [17–20]. Even though the procedure is pursued for 18–48 h, toxin removal is achieved much earlier (8–10 h), without any rebound in the circulation of toxic metabolites. This allows resumption of effective anabolism through TPN, a major prerequisite for final success [21].

### Hemodialysis

HD is the most effective and rapid method of removing small solutes [5, 11]. However, the logistics are such that it is difficult to mobilize this procedure for acute management of a newborn infant, and it cannot be performed without the assistance of a dialysis staff. With this method, clearances of PA, branched-chain amino acids and keto acids attain around 60 ml/min/m$^2$ [22, 23], and that of ammonia reaches 80–100 ml/min/m$^2$, results which are undoubtedly better than those obtained by any other procedure [11, 23]. HD cycles of 2–4 h appear to be sufficient to sustain improvement in MSUD and PA patients. By contrast, the procedure has not allowed hyperammonemic neonates to survive, despite large ammonia removal [5, 23, 24]. These poor results could be due to the difficulty in obtaining prompt anabolism in severely hyperammonemic neonates, who often develop hemodynamic instability and multiorgan failure.

### Assessment of Biochemical Progress

In order to evaluate the efficiencies of toxin-removal procedures, the general rule is to schedule regular monitoring of both specific and non-specific biologic values in blood, urine and dialysate or ultrafiltrate within timed periods. Attention must be paid to blood glucose, plasma electrolytes, and calcium, which should be appropriately corrected. Regular blood cell counts are also important since, in organic aciduria, neutropenia and thrombocytopenia may be present or may develop after the initiation of treatment and may require specific transfusions. Repeated searches for septicemia must be systematically done and treatment initiated as soon as suspicion arises.

## Additional Therapies

### Enhancing Anabolism

Owing to its well known anabolic effect, insulin is used to suppress severe catabolism. However, to attain this goal, dehydration and acidosis must be corrected. High infusion doses (0.2–0.3 IU/kg/h) used in association with high glucose infusion provided by TPN may be useful [25, 26]. During this process, insulin doses are frequently adapted to control glycemia. Sustained normalization of blood glucose levels allows insulin withdrawal. This situation is, in fact, an indirect marker of effective anabolism. Human growth hormone should not be used, as it is unlikely to have a beneficial role in sustaining protein anabolism in a short-time situation.

### Alternate Pathways

The stimulation of alternate pathways depends on the catabolic pathway involved. It is detailed in Chaps. 11, 14 and 20, which deal with those emergency situations. As a rule, L-carnitine supplementation is never contraindicated in these disorders.

### Vitamin Therapy

Megadoses of specific vitamins must be systematically tested in each case of potentially vitamin-dependent disorders (Table 3.5). As the response to the vitamin may be masked by simultaneous use of other therapies, this trial should be repeated in a later stable metabolic period and compared with in vitro studies. When seizures are the preponderant or revealing sign, pyridoxine (50 mg), biotin (10 mg) and folic acid (10–40 mg/day) must be systematically tested.

### Specific Approaches in Primary Hyperlactacidemia

Whatever the enzymatic defect, most newborns affected with primary hyperlactacidemia may present with acute ketoacidosis and dehydration requiring supporting care similar to that described for BCOA. Usually, this treatment is sufficient to reduce hyperlactacidemia to levels that do not lead to severe metabolic acidosis. In some cases, sustained hyperlactacidemia is due to high glucose infusion and can be reduced by using 5% or even 2.5% glucose i.v. solutions. Thus, none of these patients require any toxin-removal procedures. Dichloroacetate (50 mg/kg/day in one or two divided doses), an inhibitor of PDH kinase, can be an effective means to lower lactate accumulation in both PDH and respiratory-chain disorders [27]. As soon as clinical and metabolic status allow nutrition, continuous enteral feeding can be instituted, with a normal diet for the age of the patient.

Because some patients affected with multiple carboxylase deficiency may present with signs similar to primary hyperlactacidemia, biotin (10–20 mg/day) must be systematically tested. Thiamine (vitamin B1) supplementation (100–250 mg/day) must also be tried, as some PDH deficiencies are (theoretically) thiamine responsive. Also, L-carnitine supplementation (100 mg/kg/day) may be an additive therapeutic means, since many patients have a secondary carnitine deficiency. The acute and long-term management of congenital hyperlactacidemias is presented in Chaps. 10, 13 and 24.

## Liver Failure

Three disorders require urgent and specific treatment: galactosemia, hereditary fructose intolerance and tyrosinemia type I. In addition to the general, supportive measures common to all sick neonates, many acquired disorders, such as sepsis and severe neonatal hepatitis, may elicit liver failure. At an advanced state, many unspecific symptoms secondary to liver damage, such as mellituria, hyperammonemia, hyperlactacidemia, short fast hypoglycemia, hypertyrosinemia and hypermethioninemia, can be present.

In neonates, fructose is largely excluded from the diet. Thus, except for a few newborns, the main carbohydrate intolerance is due to galactosemia. Later in childhood, both disorders may be involved. Neonatal-onset forms and late-onset forms may present with acute deterioration, vomiting, seizures, dehydration, hypoglycemia, liver failure and tubulopathy. Tyrosinemia type I rarely (if ever) starts before the third week of life. As soon as these disorders are considered, galactose, fructose and protein must be excluded from the diet [with normal (according to patient age) intake of all other nutrients] while waiting for the diagnosis. When galactosemia is confirmed, protein can be reintroduced. When tyrosinemia is confirmed, a low-protein diet and 2-(2-nitro-4-trifluoromethylbenzoyl)-1,3-cyclohexanedione (NTBC) treatment must be initiated (Chap. 15). The emergency treatment of acute metabolic derangement caused by intolerance for galactose and fructose is discussed in Chaps. 7 and 8, respectively.

## Neonatal Hypoglycemia

Whatever the cause of hypoglycemia, blood glucose levels must be corrected with an acute glucose administration (0.5–1 g/kg) followed by a permanent glucose supply. The route of administration could be oral or i.v., depending on the clinical status. Thereafter, observation of patient progress under glucose provision is useful for both diagnostic and therapeutic approach.

### Glycogen Storage Disease (GSD) and Fructose-1,6-Bisphosphatase Deficiency

In GSD type I and fructose-1,6-bisphosphatase deficiency, fasting hypoglycemia is associated with hyperlactacidemia and metabolic acidosis. In GSD type III, a moderate hyperlactacidemia is observed after glucose administration. In these disorders, alkalinization with i.v. bicarbonate is not necessary.

As soon as the abnormal blood values have returned to normal, continuous enteral feeding is substituted for glucose infusion. At first, a milk-based, lactose-free, sucrose-free formula enriched with maltodextrin is used. The total amount of glucose should allow an average glucose provision of 10–12 mg/kg/min. This is easily reached using a normal energy intake, in which 50–60% of the energy is supplied by glucose. This diet is later adapted to the correct diagnosis (Chaps. 6, 8).

### Neonatal Hyperinsulinism

Neonatal hyperinsulinism presents with recurrent intractable hypoglycemia without ketoacidosis. The newborn requires a continuous high-glucose provision that exceeds the capacities of the peripheral i.v. route and continuous enteral feeding. Thus, central venous catheterization is quite unavoidable in meeting the excessive glucose requirement.

In cases of persistent hypoglycemia, continuous i.v. or subcutaneous glucagon administration (0.1–0.2 mg/kg/day) can be instituted. The emergency treatment in neonatal hyperinsulinism is presented in Chap. 9.

### Fatty-Acid-Oxidation Defects

Fatty-acid-oxidation defects can be suspected in both newborns and children who present with fasting hypoglycemia and/or an acute deterioration associated with lethargy, hepatomegaly and liver failure, cardiac dysrhythmia and high blood creatine-kinase levels. This severe condition, which may require resuscitation, is due to severe energy deprivation and must be treated with high-glucose provision.

At first, an i.v. solution providing 10–12 mg/kg/min of glucose is necessary (120–150 ml/kg/day of a 12–15% dextrose solution). If hyperammonemia due to N-acetylglutamate-synthase inhibition by acyl-coenzyme A does not spontaneously resolve with i.v. glucose, treatment with carbamoyl glutamate (50 mg/kg/day in two to four divided doses) can be tried. Hypocarnitinemia is most often present. The usefulness and safe character of carnitine supplementation is still controversial.

Once acute problems are resolved, continuous enteral feeding is progressively instituted. Acute and long-term management of fatty-acid-oxidation disorders are presented in Chap. 11.

### Cardiac Failure

The only treatable disorders starting with cardiac failure in the neonatal period are mitochondrial fatty-acid-oxidation defects. They present with myocardiopathy or conduction defects. In addition to the usual cardiac drugs and symptomatic treatment of cardiac failure, specific emergency treatment encompasses both glucose infusion to suppress lipolysis and a fat-free diet. Carnitine supplementation is still debated (Chap. 11). Medium-chain-triglyceride supplementation must be postponed until the exact site of the defect is known (Chap. 11).

### Late-Onset Coma

All types of coma in pediatrics, including those presenting with focal neurological signs, can signal inborn errors of metabolism. In this context, it is always important to check for ammonia, amino acids and organic acids before starting possibly risky procedures under general anesthesia. Compared with the approach for neonates, the management of these late acute forms is more supportive; exogenous removal is not necessary as often as with neonates. Specific therapies are essentially similar to those for neonates.

### Supportive Care

#### Cerebral Edema

Patients with metabolic disorders are at particular risk of cerebral edema and, during emergency treatment, care should be taken to reduce this risk [28]. Cerebral edema due to hypotonic fluid overload is probably an underestimated cause of death in these late decompensations [29]. In an early phase, restriction of fluids to maintenance, intubation and hyperventilation may be sufficient. If insufficient, mannitol (0.25–0.50 g/kg), furosemide (1 mg/kg) and even phenobarbital must be considered. For any seriously ill patient, monitoring of intracranial pressure or of cerebral perfusion should be considered.

#### Hydration

As in the neonatal period, many patients with metabolic ketoacidosis present with intracellular dehydration. In this situation, where the patient is at risk of cerebral edema, the first priority is to use colloids (10–20 ml/kg within 30 min) to replenish the intravascular space if shock is present. Thereafter, rules of rehydration and alkalinization similar to those described for the newborn must be applied.

#### Nutrition

Early high-energy nutrition is essential to prevent further protein and fat catabolism. Continuous enteral feeding must be considered first, as it might be tolerated even if recurrent vomiting has previously occurred. There are important advantages: it allows the provision of more energy than is provided by peripheral venous infusion and it is often sufficient to obtain rapid clinical and biological recovery. Initially, in parallel with intravenous infusion, energy and water requirements can be provided through 80056 powder (Mead Johnson) at 16% dilution. Depending on the disorder, specific amino acid mixtures with additional water are added. Once toxic metabolites have been suppressed, natural proteins are introduced, and appropriate long-term dietary treatment is initiated. When clinical status prevents effective oral nutrition, TPN following a general pattern similar to that already described for neonates [1] must be considered. In many cases, its application prevents the use of exogenous toxin-removal procedures.

### Toxin-Removal Procedures

In some cases, the situation deteriorates so rapidly that toxin-removal procedures become necessary. The choice of the technique is quite limited. Due to a lower peritoneum area relative to body weight, PD appears to be far less effective in children than in newborns. Emergency blood ET using a single main vein cannulated with a dual luminal catheter could be rapidly (though only transiently) effective in BCOA. It fails in primary hyperammonemia; HD and HF are probably better choices. However, the selection of the procedure is also influenced by local facilities and experience.

### Additional Therapies

Each therapy that has already been discussed for neonates should be considered in treating late-onset coma. Vitamin responsiveness is more likely in late-onset forms than in neonatal diseases.

# References

1. Berry GT, Heidenreich R, Kaplan P et al. (1991) Branched-chain amino acid-free parenteral nutrition in the treatment of acute metabolic decompensation in patients with maple syrup urine disease. N Engl J Med 324:175-179
2. Parini R, Sereni LP, Bagozzi DC et al (1993) Nasogastric drip feeding as the only treatment of neonatal maple syrup urine disease. Pediatrics 92:280-283
3. Ruch T, Kerr D (1982) Decreased essential aminoacid requirements without catabolism in phenylketonuria and maple syrup urine disease. Am J Clin Nutr 35:217-228
4. Khaler SG, Millington DS, Cederbaum SD et al (1989) Parenteral nutrition in propionic and methylmalonic acidemia. J Pediatr 115:235-241
5. Donn SM, Swartz RD, Thoene JG (1979) Comparison of exchange transfusion, peritoneal dialysis, and hemodialysis for the treatment of hyperammonemia in an anuric newborn infant. J Pediatr 95:67-70
6. Saudubray JM, Ogier H, Charpentier C et al. (1984) Neonatal management of organic acidurias - clinical update. J Inherit Metab Dis 7:2-9
7. Wendel U, Langenbeck U, Lombeck I, Bremer HJ (1982) Exchange transfusion in acute episodes of maple syrup urine disease: studies on branched-chain amino and keto acids. Eur J Pediatr 138:293-296
8. Goertner L, Leupold D, Pohlandt F, Bartmann P (1989) Peritoneal dialysis in the treatment of metabolic crises caused by inherited disorders of organic and amino acid metabolism. Acta Pediatr Scand 78:706-711
9. Brusilow SW, Batshaw ML, Waber L (1982) Neonatal hyperammonemic coma. Adv Pediatr 29:69-103
10. Herrin JT, McCredie DA (1969) Peritoneal dialysis in the reduction of blood ammonia levels in a case of hyperammonaemia. Arch Dis Childhood 44:149-151
11. Wiegang C, Thompson T, Bock GH, Mathis RK (1980) The management of life -threatening hyperammonemia: a comparison of several therapeutic modalities. J Pediatr 96:142-144
12. Harris RJ (1971) Treatment in maple syrup urine disease. Lancet 2:813-814
13. Wendel U, Becker K, Przyrembel H et al. (1980) Peritoneal dialysis in maple syrup urine disease: studies on branched-chain amino and keto acids. Eur J Pediatr 134:57-63
14. Alexander SR (1990) Continuous arteriovenous hemofiltration. In: Levin DL, Morris FC (eds) Essential of pediatric intensive care. Quality Medical, St. Louis, Missouri, pp 1022-1048
15. Gouyon JB, Desgres J, Mousson C (1994) Removal of branched-chain amino acids by peritoneal dialysis, continuous arteriovenous hemofiltration, and continuous arteriovenous hemodialysis in rabbits: implications for maple syrup urine disease treatment. Pediatr Res 35:357-361
16. Jouvet P, Poggi F, Rabier D et al. (1997) Continuous venovenous haemodiafiltration in the acute phase of neonatal maple syrup urine disease. J Inherit Metab Dis 20:463-472
17. Ring E, Zobel G, Stöckler S (1990) Clearance of toxic metabolites during therapy for inborn errors of metabolism J Pediatr 117:349-350
18. Casadevall I, Ogier H, Germain JF et al. (1992) Hemofiltration arterioveineuse continue: prise en charge d'un cas de leucinose néonatale. Arch Fr Pediatr 49:803-805
19. Thompson GN, Butt WW, Shann FA et al (1991) Continuous venovenous hemofiltration in the management of acute decompensation in inborn errors of metabolism. J Pediatr 118:879-884
20. Falk MC, Knight JF, Roy LP et al. (1994) Continuous venovenous haemofiltration in the acute treatment of inborn errors of metabolism. Pediatr Nephrol 8:330-333
21. Sperl W, Geiger R, Maurer H et al. (1992) Continuous arteriovenous haemofiltration in a neonate with hyperammonaemic coma due to citrullinemia. J Inherit Metab Dis 15:158-159
22. Roth B, Younossi-Hartenstein A, Skopnik H, Leonard JV (1987) hemodialysis for metabolic decompensation in propionic acidaemia. J Inherit Metab Dis 10:147-151
23. Rutledge SL, Havens PL, Haymond MW et al. (1990) Neonatal hemodialysis: effective therapy for the encephalopathy of inborn errors of metabolism. J Pediatr 116:125-128
24. Maestri NE, Hauser ER, Bartholomew D, Brusilow SW (1991) Prospective treatment of urea cycle disorders. J Pediatr 119:923-928
25. Biggemann B, Zass R, Wendel U (1993) Postoperative metabolic decompensation in maple syrup urine disease is completely prevented by insulin. J Inherit Metab Dis 16:912-913
26. Leonard JV, Umpleby AM, Naughten EM, Boroujerdy MA, Sonksen PH (1983) Leucine turnover in maple syrup urine disease. J Inherit Metab Dis 6 [Suppl 2]:117-118
27. Stacpoole PW, Barnes CL, Hurbanis MD, Cannon SL, Kerr DS (1997) Treatment of congenital lactic acidosis with dichloroacetate. Curr Top Arch Dis Child 77:535-541
28. Surtees R, Leonard JV (1989) Acute metabolic encephalopathy: a review of causes, mechanisms and treatment. J Inherit Metab Dis 12 [Suppl 1]:42-54
29. Riviello JJ, Rezvani I, DiGeorge AM, Foley CM (1991) Cerebral edema causing death in children with maple syrup urine disease. J Pediatr 119:42-45

# CHAPTER 4

# Psychosocial Care of the Child and Family

J.C. Harris

CONTENTS

General Issues.............................. 63
   Counseling................................. 63
      Orientation to the Family Interview............... 63
      Ongoing Counseling............................ 63
      Genetic Counseling............................. 64
   Crisis Intervention............................ 64
      Coping at Times of Predictable Crisis .............. 64
   Interviewing.................................. 64
      Coping ....................................... 65
   Life-Long Treatment with Potentially
   Favorable Outcome: Phenylketonuria ............. 65
   Gradual Deterioration and Unfavorable
   Outcome: Lysosomal Storage Diseases ............ 67
   Onset in Infancy and Chronic Course:
   Lesch-Nyhan Syndrome ......................... 69
   Period of Normal Development with
   Deterioration and Death: Leukodystrophies ........ 69

Issues in Regard to Specific Treatments........ 70
   Bone-Marrow Transplantation................... 70
   Organ Transplantation .......................... 72
   Gene Therapy ................................. 73

Conclusion.................................. 73
References .................................. 73

Since Garrod's initial description of inborn errors of metabolism in 1923 [1], a large number and a great variety of inherited metabolic disorders have been identified. Increasingly, physicians have come to recognize the complexities of psychosocial care for children with inborn errors of metabolism. Although the special features of each condition dictate an individualized treatment approach, there are general issues of adaptation to the illness, styles of coping, and mechanisms of defense against anxiety that are pertinent to all families with affected children.

The types of problems that occur in representative disorders illustrate the complexity and range of problems in behavior that may result from inborn errors. In each of these conditions, acknowledging the diagnosis, finding school programs, and participating in treatment of associated physical and behavioral disorders are all tasks the family must confront. The physician must be available in times of crisis and, when there is the need, must support the family through the child's chronic deterioration or terminal illness.

## General Issues

### Counseling

#### Orientation to the Family Interview

When the parents and child come to the physician for treatment, they are concerned. For this reason, it is advisable to consider that the interview begins with a statement of concern rather than the traditional chief complaint. An emphasis on the family's concerns rather than on complaints subtly reframes the interview to one that may facilitate the doctor–patient relationship in the treatment of illnesses that are long lasting and require the establishment of considerable trust from the time of first contact with the patient and family. The family's disquietude must be appreciated as symptoms are elicited and signs understood so that confidence in the physician and in the treatment program can be established. The physician has an important role: to help family members to become confident as they cope with the genetic [2–4], medical, and psychosocial [5, 6] aspects of these conditions. Instilling confidence through a confiding relationship helps the family member and the child develop the capacity to confront, actively struggle and persevere despite emotional frustration and confusion. The physician interview is an opportunity to establish understanding, develop confidence, and encourage interpersonal rapport so that recommendations can be carried out effectively and appropriately. This approach to the patient is biopsychosocial [7] and not exclusively biomedical in nature. It addresses the biological presentation of the disease, its social and interpersonal antecedents, the consequences of the illness, and facilitates the child's and parents' capacity for psychological adjustment.

#### Ongoing Counseling

Following the initial interviews, supportive counseling is a continuous process for the family and the child, who must deal with major psychosocial issues in care, such as confusion and delay in diagnosis, uncertainties in clinical course, questions about the screening of siblings, and day-to-day stresses inherent in management. Considerable emphasis has been placed on the generic family's psychological adjustment. This has often led to the consideration of families as homogeneous in their responses. However, each family is unique and must be given individual consideration. It is a particular family's response to the stress of the illness that is at issue. It remains essential to (1) identify characteristic vulnerabilities to stress in parents and family members, (2) provide sensitive support, and (3) appreciate the psychological phases involved in the adaptation to illness. It is also imperative to appreciate the reality of the stresses of unmet service needs and to recognize that repeated questions by family members do not necessarily reflect a failure in emotional adaptation. Respite or day care during holidays and weekends, childcare arrangements, and help with transportation are critical to facilitate family adaptation [8].

### Genetic Counseling

Genetic counseling is individualized according to the type of case. Specific issues with regard to counseling adults about recurrence risk for future pregnancies and sibling screening must be addressed according to the particular disorder. The majority of these conditions are associated with mental retardation or problems in learning and behavior that require support for parents.

## Crisis Intervention

Crisis intervention [9] includes both anticipatory guidance and preventive intervention. Anticipatory guidance is the approach taken when a crisis can be predicted. Preventive intervention is a method of guidance for parents and children during the crisis itself. These approaches recognize the family's increasing dependency and need for continuing support during the crisis. Help is provided by facilitating communication among family members and by discussion of their concerns and their plans for coping. During these discussions, the physician should point out that negative feelings are normal, empathize with the family's frustration, and encourage the sharing of tasks among family members in recognition of the fatigue that may develop with ongoing family care of chronic conditions. Interview sessions should focus on present problems and not emphasize discussion of past failures. Identification of psychological needs and the development of confiding relationships [9] among family members is essential in preventive intervention. If this natural support does not occur spontaneously, it is appropriate for the professionals involved to arrange active support with professional counselors.

### Coping at Times of Predictable Crisis

Times when family crises may develop include:

1. During establishment of a diagnosis and discussion of its implications
2. While living with the child at home and participating in the specific management program
3. The time of school entry
4. During entry into adolescence
5. When dealing with loss of function and deterioration
6. During family planning

The way a family copes depends on the support that family members can offer one another, the ability of each member to adapt to loss, both potential and actual, and the availability of community care. It is the physician's first responsibility to convene a support group at the time of diagnosis. The first step in this is to counsel both parents together rather than either parent alone, and then to assist the family in finding local resources through the extended family, recognized parent-support groups for the given condition, community agencies, and religious organizations. Meeting with other parents who have faced similar crises can be particularly helpful. Kazak [10] studied three samples of families with disabled children with regard to stress and social networks. Mothers and fathers of 125 handicapped/chronically-ill children were compared with parents of 127 matched, nondisabled children from three separate samples with respect to personal stress, marital satisfaction, and social-network size and density. Only mothers of disabled children experienced higher levels of stress than comparison parents. No differences were found in marital satisfaction. Although few group differences were found for social-network variables, mothers of handicapped children had higher-density networks than comparison mothers, illustrating the importance of extensive psychosocial support.

### Interviewing

Successful interviews with the parents should result in an acknowledgment of the nature of the child's illness and an awareness of what they can do for their child. During the interview, the parent is reassured not only by what is said, but also by how it is said. Self-awareness by the physician is critical to empathetic listening to the parents' perceptions of the child's difficulties. To understand the parents' adaptation to illness, it is essential that the physician understand the psychological mechanisms that are normally present in a time of stress.

When experiencing stress, individuals may use a variety of defense mechanisms to minimize experienced anxiety. If the individual is not able to cope with the disorder, the use of these defense mechanisms against anxiety may be heightened and may interfere with the ability to understand and utilize appropriate recommendations for care. The most commonly used defenses are denial, guilt or self-blame, projection or blaming others, and excessive dependency on the physician, family, or community members by the parent [11]. Unexpected and unwanted information is normally responded to with denial and disbelief followed by mixed feelings of sadness and anger before eventual acknowledgment occurs. To determine the degree of the parents' adaptation, the following questions are suggested for the interviewer:

- Whom do you talk to when you are concerned about your child? This question is needed to establish the degree to which the parent has become isolated and whether there is a confiding relationship with another person. It also helps to clarify if the parent is denying the seriousness of the child's illness, thereby putting the child at risk. If the parent expresses excessive shame or denial and feels isolated from others, additional counseling is needed.
- Do you blame yourself for your child's illness? This question clarifies whether the parent is experiencing guilt. Evidence of self-blame should be pursued, since the self-blaming parent may be demonstrating symptoms of depression.
- Do you have doubts about the staff's ability to provide care for your child? This question deals with projection and excessive suspiciousness. Not uncommonly, parents criticize caregivers as an expression of their own projected fears and anxiety.
- Do you feel adequate to take care of the child yourself, or do you frequently seek directions from others and feel dependent on them? This question deals with dependency and passivity, which may present in the overwhelmed parent. When this occurs, the physician often sees signs of helplessness and hopelessness in the parent, receives frequent telephone calls, and is asked to make more and more decisions for family members.

## Coping

All these processes, i.e., denial, guilt, projection, and dependency, occur normally [11] and are problematic only if one of these means of coping becomes predominant and persists. Throughout the coping process, parental adaptation and efforts at maintaining confidence take place as the child's condition changes. Several phases in adaptation are identified: acknowledgment, the progressive realization of the seriousness of the condition; grieving, the experience and expression of impending loss; and reconciliation, the process of developing the perspective that restores the parents' confidence in the worth of the child's life [12].

In order to cope with day-to-day events, the parents must gain and maintain confidence. This is accomplished by mastery operations and an affirmation of life. Mastery operations include efforts to obtain information about everything involved in the child's care, searching out the best care available, and coming to terms with feeling responsible for their child's illness. Participation in care is of vital importance for many families. They need to help with procedures and "do everything possible" for their child. These efforts should, of course, be balanced with care of other family members and continuation of daily tasks in order to maintain psychological equilibrium. Mastery operations are most prominent in the phases of acknowledgment and grieving and lead to reconciliation. Affirmation of life is a response to the fact that the child's illness often threatens the parents' optimism about the value of life. The majority of families gradually come to terms with their feelings of resentment and hopelessness through devotion to the child and make full use of treatment facilities.

There are a variety of inborn errors that illustrate the complexity and range of psychosocial issues that may result in the care of children with intracellular enzymatic defects. The following examples represent types of problems that may be encountered.

## Life-Long Treatment with Potentially Favorable Outcome: Phenylketonuria

Phenylketonuria (PKU), is the most common of the amino acid disorders and, if untreated, leads to mental subnormality (PKU is the most common biochemical cause of mental retardation). Early dietary control, an environmental intervention, requires that the parent act as a co-therapist to help prevent mental retardation and subsequent behavioral and emotional problems. Moreover, current practice suggests that the diet must be carefully monitored during the school years, since early dietary discontinuation may result in learning and behavioral problems [13, 14]. Maintenance of diet at recommended levels seems to be protective with regard to executive function and psychosocial adjustment [15-17] of children tested up to age 10-13 years. Additional follow-up data is needed for adolescence and adulthood. However, behavioral disturbance and neuropsychological disturbance [15, 16] are present in many patients with early-treated PKU (ETPKU), and these conditions are more prevalent in those with higher phenylalanine concentrations. With early discontinuation of diet, the patients or their families may complain of lack of concentration and emotional instability [13]. However, after returning to a phenylalanine-restricted, tyrosine-enriched diet, the impaired neuropsychological and behavioral functions appear to be reversible. Moreover, when the diet is introduced, benefits in behavioral management are described even in elderly, untreated patients [18] and in affected mentally retarded patients with destructive behavior [19].

A re-evaluation of dietary treatment [20, 21] was undertaken, because it was found that even children with PKU receiving dietary treatment demonstrated impaired attention control and, despite normal intelligence quotient (IQ), exhibited diminished academic achievement. The main influence on intelligence seems to occur in the first decade of life, but the neurocognitive problems may persist, so additional information is needed about their course. Neurocognitive problems seem linked to dopamine depletion and dorsolateral prefrontal-cortical dysfunction, which is linked to the biochemical deficits (failure to convert phenylalanine to tyrosine, the precursor of dopamine; low levels of tyrosine) resulting from this genetic disorder [22, 23]. Diamond et al., in a 4-year longitudinal study, found that children (age: 3.5–7 years) and infants (6–12 months) with PKU whose plasma phenylalanine levels were three to five times higher than normal (360–600 µM) performed significantly worse on tasks of working memory and inhibitory control than their siblings, those with PKU and lower blood levels, matched controls, and children in the general population. The higher the plasma phenylalanine level, the worse the performance; girls were more affected than boys. The deficit seems selective, because those with phenylalanine levels three to five times higher than normal scored in the normal range on 13 control tasks. The authors also have found dopamine reductions in an animal model of PKU [24].

Behavioral problems may compound the intellectual problems that are often present; emotional stress and neurologic dysfunction are their likely causal factors. Thus, an increased frequency of deviant behavior may be a result of an interaction of psychological stress and neurologic impairment. Behavioral disorders, including attention-deficit disorder and pervasive developmental disorder may be associated with PKU [25]. Careful monitoring and family counseling is critical to prevent these complications. When they occur, appropriate referral is needed for treatment of the disorder and to counsel parents on their role in treatment.

Adolescence may be a difficult time for individuals with PKU. In one study [16] of early-treated, normally intelligent adolescents, patients and their mothers were contacted regarding their psychosocial situation and knowledge of disease and diet. The adolescents described their emotional lives as restricted, and their knowledge of disease and diet were poor. The majority had problems with dietary management without parental assistance. Thus, adolescence is an important time for ongoing intervention.

Parents of children with PKU face difficulties that may disrupt family life. In an early study in Glasgow, Scotland, McBean [26] evaluated and followed-up 59 families having a total of 204 children, 79 with PKU.

The families were interviewed to gather pilot information. Parents were asked about the following specific topics:

1. Their reaction to the PKU-screening test
2. Their reaction to the diagnosis of PKU in the infant and their appreciation of what the diagnosis meant
3. Their understanding of the diet and the problems involved in following through with the diet
4. Their thoughts about the future for the infant
5. The effect of the child with PKU on the family and on their marriage
6. Their attitude toward having additional children

All the parents, when told of the diagnosis of PKU, acknowledged feelings of anxiety and disbelief and found it difficult to assimilate the facts about the disorder given at the first interview. Most appreciated that both parents are involved in the transmission of the disorder, but grandparents tended to blame one or the other side of the family. The parents frequently had not understood what was thought by the professional staff to have been an adequate and full explanation.

In regard to understanding the administration of the diet, all parents expressed considerable difficulty. When they could not understand or accept the consequences of dietary neglect, their belief that the diet was essential to the well-being of the child was often modified. When there was no previous experience of a retarded child with PKU, it was difficult for the parents to appreciate the form the mental retardation might take or how quickly it would occur. When some mothers found that dietary indiscretion does not lead to immediate mental retardation, they were inclined to question the diagnosis.

When asked about the future of the child with PKU, few parents looked beyond their child's entry into school and feared that special schooling might be necessary. Others were more concerned with the supervision of the diet while the child was away from home

Based on this pilot study, McBean [26] suggests that the parents be given information and simple and repeated explanations of the cause of PKU, the course of the disease, and the reasons underlying treatment. Although pamphlets may be helpful, they are inadequate if they are the sole source of information. For all of the families, the moment of initial diagnosis was traumatic. Genetic counseling can be very difficult and is particularly hard if the family already has two or more affected children or knows of other families where this situation has occurred. There is a need for constant emotional and practical support, which must be available for as long as the diet continues and may be necessary for an even longer period to help in dealing with the problems of adolescence in patients who began treatment for PKU in childhood.

In the 20 years since McBean's study was carried out, there has been considerable experience in working with families. Pueschel et al. [27] used questionnaires, informal group meetings, and individual interviews with parents and their children with PKU to understand their attitudes and experiences surrounding discontinuance of the phenylalanine-restricted diet. These authors stressed the importance of understanding changing social interactions as termination of the restricted diet progresses. Preparatory discussions with parents and children prior to the change in diet should be held to avoid undue stress and conflict in such families.

Reber et al. [28] have provided a recent systematic study of family factors. They studied a population of 41 young children with ETPKU and included family investigation to determine relationships between dietary phenylalanine control and patient and family functioning. Children received neuropsychological tests, and parents completed behavior checklists on their child. They also completed four self-report measures aimed at evaluating family adjustment, stress, and social interaction. Significant correlations were found between concurrent phenylalanine control and patients' intelligence-test scores and between lifetime phenylalanine control and patients' social competence. Children with PKU had lower social-competence scores than a comparison control group. Parent-report measures of family psychological adjustment, stress, interaction, and socioeconomic status showed no significant association with children's dietary phenylalanine control. Family cohesion and adaptability correlated positively with the patients' cognitive performance. Mothers of children with PKU perceived their families to be significantly less cohesive (more separated) and less adaptable (more rigid) than matched mothers of non-PKU children. Fathers of children with PKU perceived their families to be less adaptable. The reported reduced cohesion and rigidity may have negative implications on test performance by children with PKU. These findings suggest that both metabolic and family factors be considered in evaluating the outcome of ETPKU.

A longitudinal study of families whose children have ETPKU is needed to better understand the effects of the illness on family cohesion and adaptability. Koch et al. [29] followed-up 43 adults; 19 remained on the diet and 24 stopped the diet at an average of 7.8 years of age. Follow-up at an average age of 22 years showed that the cohort remaining on dietary treatment had achieved substantially better social and academic achievement, based on the Wechsler adult-intelligence scale, college attendance, employment and marital status. Ris et al. [30] evaluated the adult psychosocial outcome (18 years of age and older) in ETPKU because of data showing increased behavioral risk in children with ETPKU, evidence of impairment in adults with ETPKU, evidence of neuroimaging abnormalities in adults with ETPKU, and possible increased rates of psychiatric disorder. Although similar to sibling controls on most issues, on a self report of psychiatric symptoms, 20% reported increased morbidity. There was a strong relationship between neurocognitive measures (intelligence and executive functioning) and psychosocial morbidity.

Although most adults were found to cope with the challenges of young adulthood, neuropsychological monitoring in childhood and adolescence is needed to identify those at risk. Another study [31] assessed psychiatric disorders in adults with ETPKU and considered whether biochemical control, intellectual functioning, white-matter abnormalities on magnetic resonance imaging (MRI), and/or style of parenting influence psychopathology. Although the overall rate of psychiatric disorder (25.7% vs 16.1% in controls) was not statistically significant, one quarter of the adults studied had psychiatric symptoms. This tallies with the 20% rate in the Ris et al. study. Moreover, the pattern of disorder differed from that in the controls; externalizing disorders (disruptive behavior) were actually reduced (0% vs 7.8%), while internalizing disorders (anxiety, depression) were increased (25.7% vs 8.3%) and were largely accounted for by depression in females. The authors found a correlation between both IQ and parental education up to age 12 years. It was found that a restricting, controlling style of parenting is a risk factor for the emergence of psychiatric symptoms. The authors concluded that optimal medical treatment and psychiatric monitoring, psychological support, and parent training are all critical to successful adult outcome.

## Gradual Deterioration and Unfavorable Outcome: Lysosomal Storage Diseases

The lysosomal storage diseases lead to particular problems in psychosocial management. These conditions are normally identifiable by the characteristic facial appearance along with skeletal deformities and physical features associated with involvement of other organ systems. The descriptive term "gargoylism" graphically illustrates the altered physical appearance, which can be of particular psychosocial importance. However, in SanFillipo's syndrome [mucopolysaccharidosis (MPS) type III], it is not physical appearance but disruptive behavior and developmental delay after early normal development that is most distinctive [32].

Mental deterioration with mental retardation is characteristic of the severe forms, but milder variants with normal intelligence exist [33]. A period of early apparent normality with a later decline in function is important to address with families. The severe form is a particular challenge to family adaptation. Behavioral problems related to central-nervous-system involvement may be present in the severe forms and complicated by interpersonal management problems. Those severely affected may die in childhood. In the milder variants, survival into adulthood requires ongoing specific support for the young person who is affected. A specific metabolic diagnosis is important with regard to prognosis and genetic counseling [34].

In the UK [35], psychosocial problems were investigated in a national study of MPS-II (Hunter's syndrome), a sex-linked recessive condition. The sample consisted of families who volunteered to give interviews and hospital records; there was no control group, and specific rating instruments were not used. Visits were made to 33 sets of parents with a total of 44 affected sons, 27 with the severe form and 17 with the mild form of the disease. Information about the behavioral patterns of another 22 boys was obtained from hospital records. Serious behavioral disturbance was reported in 36 of the 38 severely affected boys. The mildly affected boys generally adapted to the condition but often suffered from stigma related to their physical appearance. Adaptation to adult life after leaving special schooling is problematic for the mildly involved, highlighting the need for long-term support for the families and the boys themselves.

In the early-onset, severe group, the initial behavioral complaint was overactivity (29 of 38 boys), commonly beginning in the second year and continuing until 8–9 years of age, when the disease process progressed to the point that the boys were more inactive and lethargic. Aggression towards others was reported in 16 out of 38 cases and, in some instances, was related to rapid growth in the early years. However, ten were described as particularly affectionate and playful. The prevalence of aggression and oppositional behavior is comparable to the prevalence of behavioral disorder in other reports of severe mental retardation. The rate of hyperactivity is quite high and, according to the authors, is not responsive to pharmacotherapy. In the mild form of MPS-II, overactivity, sleep disturbance, and violence were not reported. In regard to family adjustment, most families indicated that they had received considerable psychosocial support from professionals, though particular concern was expressed with regard to adult-support services.

Crocker and Cullinane [36] addressed both clinical and educational issues for families and personality development in children with both MPS-IH and -II. They emphasized specific problems faced by families, including orthopedic, cardiac, and ear, nose, and throat management, and described the work of an interdisciplinary team involved in the management of three cases. They studied: (1) response to the diagnosis; (2) the family's view of long-term needs; (3) reaction to genetic consequences; (4) continuing parental adjustment; (5) household emotional tone; and (6) response to guidance. The first family denied the disorder, did not plan for long-term care, avoided the mental retardation initially then showed painful acknowledgment, and avoided the genetic issues (which led to three interrupted pregnancies). The mother became clinically depressed and the father left home; the home atmosphere was one of mourning, and the diagnosis was not accepted. The second family showed prolonged bereavement regarding the diagnosis, was cautious about the future, showed partial acceptance of the mental retardation, showed disappointment and avoided thinking about the hereditary aspect, maintained a stressful and precarious marriage, and showed partial understanding regarding efforts at support. The third family accepted the diagnosis, made plans for the future, accommodated to the diagnosis but had some difficulty regarding siblings' anxieties, coped together with problems, oriented their attention to the child, and effectively used medical and psychosocial guidance. The first family remained disorganized, the second was in a process of continuous reintegration, and the third was maturely adapted.

In SanFillipo syndrome, three phases are described in the emergence of the neuropsychiatry disorder [32]. There is an early period of normal development with subsequent developmental delay in language development, upper airway and bowel disturbances. By age 3–4 years, severe temper tantrums are noted, and the child becomes overactive and inattentive. Panic may be noted with environmental change, and sleep disturbances are noted, with reversal of the day–night pattern. By age 10 years, the aggressive behavior lessens, but coordination and swallowing problems become more apparent, and seizures may occur. A variant form, type B, shows a longer period of apparent normality, with onset of mental retardation in elementary school and with sleep disturbance but no seizures. In early adulthood after age 20 years, behavioral disturbance and dementia are characteristic. Families need considerable help in treatment. Structured daily routines and the creation of a safe home environment to prevent injury are essential, and behavioral techniques may be helpful for sleep problems [37, 38].

Parent associations have been formed to provide group support to family members to help them cope with difficulties associated with these various disorders. In a multiethnic culture, understanding of the parents' belief systems is critical to gain their cooperation in treatment and participation in parent groups [39].

Neuropsychologic testing is essential in providing documentation of cognitive function. To facilitate mature adaptation, Crocker and Cullinane [36] suggest:

1. Establishment of a relationship with the family early on, preferably before the diagnosis is reached
2. Identification of parental attitudes relevant to positive adaptation, and initiation of contact with parent organizations and community support groups
3. Formulation of specifics of the patient-care program and a clear outline of how they can be accomplished
4. Orientation of the program toward the family's rights and needs
5. Attention to the needs of siblings
6. Continuous regular support for parents
7. Provision of genetic counseling and ongoing provision of information about new research related to the disorder

## Onset in Infancy and Chronic Course: Lesch-Nyhan Syndrome

Lesch-Nyhan syndrome is a disorder of purine metabolism associated with gross uric-acid overproduction, dystonias, problems in speech articulation, mental retardation (usually mild), and chronic self-injury [40]. The self-biting has been described by Nyhan as a "behavioral phenotype", suggesting that a specific behavioral pattern may be a characteristic feature of this disorder. It is of psychosocial and psychiatric importance, because of the association with mental retardation and self-injury. In type, the behavior is different from that seen in other mental-retardation syndromes of self-injury, where self-hitting and head banging are the most common presentations. The self-injury occurs even though all sensory modalities, including the pain sense, are intact. Because of the self-injurious behavior, the patient may be restrained. Despite their dystonias, when restraints are removed, the patient may appear terrified and may quickly and accurately place his hand in his mouth. The child may ask for restraints to prevent elbow movement. When restraints are placed, the child may appear relaxed and more good-humored. Their dysarthric speech may result in interpersonal communication problems; however, the higher-functioning children can express themselves and participate in their treatment. Hemiballismic arm movements can also create difficulty, since the raised arm is sometimes interpreted by others as a threatening gesture rather than as a neurological symptom [41]. Psychosocially, throughout the child's life, parents must cope with multiple hospitalizations and continuous surveillance of their child.

The self-mutilation is conceptualized as an obsessive behavior that the child tries to control but generally is unable to resist. As the patient becomes older, he becomes more adept at finding ways to control his behavior, including enlisting the help of others to protect him against these impulses. Some older children show aggression towards others by pinching, grabbing, or using verbal forms of aggression.

In regard to treatment, the motivation for self-injury [42] must be considered, as must the biological basis of self-injury [41]. Behavioral techniques using operant conditioning approaches alone may have limited effectiveness in Lesch-Nyhan disease. Pharmacological approaches that aim to use medication to reduce anxiety and spasticity have also met with mixed results. However, combined psychological, behavioral, and pharmacological approaches have been more successful. Parents report that attending to physical comfort, adjusting restraints, discussing concerns, and other stress-management procedures are most helpful [43]. Of the psychopharmacological agents used, the drug reported by parents to be most effective is diazepam [43]. An emphasis on parent training is of particular importance for drug compliance and the generalization of treatment effects. Continuous education and family support is essential. Behavioral treatment must focus on reducing self-injury and treating the phobic anxiety associated with being unrestrained. Stress management is needed to help families develop more effective coping skills. Use of restraints, dental management, anxiety intervention, parental intervention and pharmacotherapy are all of importance in treatment [37]. Combining drug and behavioral treatments is being explored.

## Period of Normal Development with Deterioration and Death: Leukodystrophies

Adrenoleukodystrophy is a disorder of peroxisomal fatty-acid metabolism, with secondary manifestations of neuroinflammatory disease in about 50% of cases; there are infantile, juvenile, and adult onset forms. Early juvenile symptoms include learning and attention problems, with progression to deterioration in motor and cognitive abilities and, finally, dementia and death.

The leukodystrophies are inherited progressive, nonselective disorders of the central and peripheral nervous system [44, 45]. The child comes to medical attention during infancy or early childhood. Subtle changes in affect, behavior, and attention are among the early symptoms of the leukodystrophies. Frequently, nonmedical management is attempted prior to the discovery of the specific diagnosis. In some instances, psychiatric disorders are diagnosed, and children are placed in school programs for the emotionally disturbed or, in other instances, in programs for learning disabilities. A common behavioral disturbance is an alteration of the sleep–wake cycle. The child may have difficulty getting to sleep and staying asleep. Muscle spasms may awaken the child at night and may be part of the disease.

As the leukodystrophic process develops, a major therapeutic focus is to maintain muscle tone and support bulbar-muscle function. Bulbar-muscle control is needed for normal respiratory toilet, eating, and normal gastrointestinal activity. With deterioration in muscle control, handling of oral secretions and the child's ability to feed himself are frequently impaired. This requires changes in feeding patterns, with the use of pureed and soft foods. In some instances, as the ability to handle oral liquids is lost, invasive measures, including nasogastric tubes, nasoesophageal tubes, or a gastrostomy, may be required to administer medication, food, and water.

Since children with leukodystrophies have problems in learning and attention and may later have cognitive disorders, the coordination of school services becomes particularly important. The purpose of the education program is to enhance the quality of life and to provide as much normalization as possible.

When the child can no longer attend school, home and hospital teaching services that focus on helping parents with positioning, handling, and transporting techniques and relaxation activities are needed. These latter services may involve the use of physical and occupational therapy and a teacher certified to work with the multiply handicapped.

Medication may be needed for a variety of problems that present during the course of the illness, including the sleep disorder, attention deficits, and treatment of muscle spasms. The management of these symptoms is often problematic for parents, since the child frequently requires one-on-one supervision. As the child experiences difficulties in understanding and processing auditory experience and interpreting what is seen, the child requires increasing support. The family must fill in the missing information to help compensate for the child's loss in understanding and interpreting his experiences.

From a psychosocial perspective, grief, frustration and anger about the lack of a specific therapy and the experience of dealing with progressive deterioration affect the patient, the family, the physician, and other professionals who are involved in care. There are experimental dietary treatments that are currently being tested but, in conducting them, one must work carefully with the family to minimize discomfort, inconvenience, and additional time involved in care. There is interest in dietary treatments that might have an effect on disease progression if initiated during the presymptomatic period. When one child in the family is affected, the recognition that there are genetically positive, asymptomatic siblings is very stressful for parents. The use of neuroimaging techniques [46] and neuropsychological testing data is essential to evaluate disease progress in such cases. New drug treatments being investigated include lovastatin and phenylbutyrate [47] and pharmacologic gene therapy [48]. Criteria may vary among medical centers regarding the introduction of dietary treatments. Moreover, the role of bone-marrow transplantation (BMT) is being investigated for use in asymptomatic children being monitored with neuropschological testing and MRI/magnetic resonance spectroscopy [45]. Close family surveillance with psychosocial support is needed for dietary treatment and particularly for BMT because of its associated morbidity and mortality.

The severity of medical problems is compounded by the necessity of coping with a fatal illness that threatens family functioning. Parent organizations are available, and families should be encouraged to contact them. Family organizations provide newsletters and monthly mailings regarding current research, provide opportunities to talk to other parents about patient management, and offer personal experiences in how to cope with the disorder. Through helping one another, the families can enhance their own ability to cope with the situation. As the family begins to understand the severity of the condition, they may begin to withdraw investment in the child. The physician, then, has a major role in helping to maintain confidence and hope. To the degree that the final phase of the illness is adequately supported, the parents may develop a realistic and meaningful perspective on the individual child's life as part of the family.

## Issues in Regard to Specific Treatments

### Bone-Marrow Transplantation

Various forms of treatment have both risks and benefits; for example, BMT [49–59] is an approach that has been applied to more than 50 inborn errors of metabolism, and gene therapy may become available in the future [60–62]. BMT represents a major medical advance and provides hope to children and families. Increasingly, it is being made available to children with a variety of disorders. Pfefferbaum, Lindamood and Wiley [53] have reported that, although psychosocial factors do not influence survival for BMT, they may be critical in the management of many cases.

The first psychosocial issues emerge at the time of referral for treatment. Because of limited bed availability and the location of BMT programs in tertiary-care facilities, children are commonly referred to facilities that are distant from their homes for BMT. Such social disruption may place a considerable burden on family members. When proposing BMT, in addition to providing knowledge of the procedure itself, an important psychosocial issue is the child's motivation and willingness to have this procedure done. Moreover, feelings of guilt and misgivings about the procedure by the parents should be discussed prior to the transplantation. Considerable psychosocial support may be needed to help the child and parent work through their feelings about an impending transplantation.

In preparing parents and children for transplantation, it is important to elicit a history of the child's previous emotional responses to stress and hospitalization. Assessment procedures may include pretreatment psychological evaluation regarding the child's level of intelligence, history of past emotional problems, and typical coping strategies used by the child. Common problems include anxiety related to the procedure itself, the feeling of being a burden to the family, low self-esteem, and a sense of vulnerability. During the hospital stay, depression, anxiety, excessive dependence, aggressive behavior, and anger may be noted. This may be demonstrated by less tolerance for repeated procedures and periodic refusal to cooperate with treatment. For a family, the emotional events linked to BMT ultimately relate to the recipient's medical course, the type and extent of the family members' psychological strengths and weaknesses, and the disruptions in family organization that are required for the transplantation.

Psychological stress may occur when adapting to disruptions brought about by the move. Moreover, during BMT procedures, stress can be severe and prolonged. Patients may be isolated in sterile environments and subjected to high doses of chemotherapy. The child and family must then cope with the secondary effects of treatment, which include the risk of infection and graft-versus-host disease. The possibility of infection, rejection of the graft, relapse, and graft-versus-host responses may lead to chronic high anxiety. Commonly, patients stay in the hospital for several months and require close follow-up at discharge, requiring that family members remain with them, away from the rest of the family, for considerable lengths of time. Following discharge, the readjustment phase can also be stressful.

During the hospitalization, the child's responses to other patients must also be considered. It is not uncommon for children to be placed in adult units, where BMTs are carried out for both children and adults with potentially terminal disorders, such as aplastic anemia and leukemia. Although the child with an inborn error does not have these conditions, the possible negative effect of this exposure to other patient groups must be anticipated, and measures must be taken to provide appropriate support. For example, the parents of other hospitalized children or relatives of other patients may become emotionally close to a child on a BMT unit. Should another patient on the unit die, then the child not only loses that person but also their visitors and family, with whom he may have become familiar.

Stresses within the marriage and problems with siblings must also be considered. The experience of participating in a life-saving or life-enhancing procedure may produce a strong and intimate bond between parent and child. In a BMT, the donor may be a parent or sibling. If it is a sibling, competition among siblings may occur for the role of donor. In cases of successful transplantation, a special relationship may develop between the child and the donor. However, in some instances, the donors may feel that they have not received adequate recognition for their contribution. Moreover, if medical complications develop, the donor may worry about the part played by his marrow in the illness. If death occurs, the donor may experience a sense of guilt about failing a sibling or child. Siblings may experience a sense of loss if a parent is the donor and goes to a distant medical center for the transplantation. Parent expectations about the transplantation may also be excessive and can lead to unrealistic promises about how things might turn out after the transplantation.

Differences may also be noted in coping with transplantation if one parent is primarily involved with the child and the other parent is not. There is always a dilemma when parents are differentially involved in the day-to-day events concerning the illness and its treatment. Due to differences in life experiences regarding the child's illness, there may be misunderstandings about what is happening, the meaning that the parents assign to the event, and the timing of their emotional responses.

The medical professional should try to ensure that the patient and family enter discharge planning with realistic expectations. This involves not only helping them to develop a positive attitude toward recovery but also the recognition of possible complications. Following the successful completion of a BMT, patients and their families may experience ambivalence about leaving the protected environment of the BMT unit, and there may be unrealistic expectations about outcome. Children may become quite dependent during the time of the procedure and, when isolation has been necessary, there may be concerns about the home being a sufficiently safe environment. Parents sometimes may establish excessive precautions at home in planning for discharge. Therefore, a discharge meeting where the parents and siblings are encouraged to articulate expectations and to review the course of the hospital stay can be very helpful. Issues, such as the time to return to school and participation in peer group activities, should be included. Sensitization of parents to the possible presence of siblings' unresolved feelings about the transplantation procedure may act as a form of psychological immunization if difficulties do emerge [54].

After discharge, parents require reassurance and recognition that a new equilibrium in their relationship may have to be established following prolonged separation and that stresses may be involved with the procedure [55]. Follow-up visits are particularly important and should be scheduled (1) immediately after discharge, when parents may be apprehensive about the expected course, and (2) subsequently, when lingering unresolved feelings may re-emerge.

Children receiving a bone-marrow transplant are at risk for neuropsychological sequelae due to potentially neurotoxic chemotherapy and total-body irradiation [56]. Kramer et al. studied 67 children and found reductions in IQ and adaptive function at 1 year but no further deterioration at 3 years when total-body irradiation was used for cancer.

In BMT, the psychosocial effects on the siblings must also be considered. Packman et al. [59] studied 44 siblings (21 donors, 23 nondonors) with regard to psychological distress and behavior. Although more anxious and with lower self-esteem on their own self report, their teachers rated them as having effective adaptive skills in school, suggesting that these children were coping, despite their self evaluations. However, one-third of both donors and nondonors reported symptoms of post-traumatic stress. Successful factors in management in these studies include attention to parent's coping styles, encouragement of activity, anxiety-reduction procedures for the children, and the availability of a consistent staff member over time.

The stresses on families who participate in these BMT studies are considerable. Although survival may be superior, particularly for those who have sibling matches, the risks remain high compared with risks for non-transplanted cases, and the timing of the procedure is important. Some suggest that the procedure be carried out at a younger age, when the child is not symptomatic. Nevertheless, advances are needed in bone-marrow technology to reduce the risk inherent in the transplantation procedure itself. Considerable work with family members of children who are presymptomatic will be needed regarding the decision to proceed with BMT.

In summary, factors that affect families coping with BMT include the length of the patient's previous illness, the degree and duration of family geographical dislocation from home, and pre-existing and intercurrent stresses within the family, such as marital conflict, separation/divorce, and changes in family role relationships. High levels of physical and emotional stress, intense relationships that develop between the BMT team members and recipients during the procedure, and psychosocial phases in the transplantation procedure must also be considered. Other issues include the psychological impact of BMT on donors, the long-term family consequences of BMT, and the long-term cognitive, neuroendocrine, sexual, reproductive, and psychosocial status of BMT survivors.

**Organ Transplantation**

In regard to pediatric organ transplantation, Gold et al. [63] have reviewed the parents' perspectives. They report that there are three specific stages that the family must cope with, which they call the preoperative, perioperative, and long-term postoperative stages. Preoperative psychosocial issues include the initial hospital experience, which involves preparation for the hospital stay and an initial building of trust in staff members. It also involves making plans for both family members who will be waiting at home and financial issues. During the perioperative period, the first 24 h are associated with anxiety about outcome, often followed by exhilaration during the first 2 weeks if the transplantation has been successful. They describe a "roller-coaster phase" following this, where there may be fear of rejection or infectious disease, continued guilt and fear of loss, and ongoing isolation and marital stress. They emphasize the need for preparation for hospital discharge. In doing so, there is the need to build confidence and work through the dependency that has developed during the hospital stay. Finally, in the postoperative period, issues regarding returning home involve re-adaptation to parental roles and readjustments in family structure. Uncertainty must be endured until the final results of the procedure become apparent.

The stress, coping resources, and family functioning of parents awaiting transplantation for a child was evaluated by Rodriguez et al. [64, 65]. Of 36 mothers who were evaluated, 20% reported considerable increased stress. The issue is how to deal with reducing stress and identify those most at risk. Stress was higher for solid organs than for BMT. Finally, there are multicultural issues [62] that must be considered in assessing coping strategies.

## Gene Therapy

Gene therapy is a new approach to treatment of inborn errors that is currently under development [60]. A variety of techniques are now available for inserting specific genes into either living individuals or into cells maintained in culture. Many of these investigations are at the stage of gene-transfer studies demonstrating the possibility of correction of metabolic derangements in fibroblasts [61, 66]. Clear data on psychosocial adjustment to gene therapy is not available; however, one would expect that the psychological adjustment may be similar to that for other potentially corrective procedures. One psychosocial study dealing with gene transfer was carried out as a phase-I safety trial in cystic fibrosis to determine the attitudes, expectations, knowledge, and psychological functioning of participants [67]. The majority of subjects were in good psychological health; however, a significant minority had anxiety disorders and expressed greater concern about the safety and efficacy of gene therapy than other participants did. These authors' approach might be applied to other disorders as gene-transfer procedures become more common. They utilized a family-assessment scale, a quality-of-life scale, and anxiety and depression measures.

## Conclusion

The challenge for society in dealing with these families is to provide care through offering comprehensive services, including genetic counseling, modern treatment, and follow-up supportive counseling. These approaches must ensure confidentiality and freedom of choice to avoid misunderstanding and stigmatization. The primary objective of screening programs should be to maximize the options available to families at risk. Whenever possible, the ongoing treatment should be coordinated with and include the involvement of the primary health-care physician working with consultation from the metabolic specialist.

## References

1. Garrod AE (1923) Inborn errors of metabolism. Oxford University Press, London
2. Childs B (1982) Genetic decision making and pastoral care: the dimensions of the problem. Hosp Pract 17 (12):96D-96E, 96I-96L, 96n passim
3. Jarvinen L, Autio S (1983) Psychological obstacles to genetic education. Scand J Soc Med 11:7-10
4. Polani PE, Alberman E, Alexander BJ et al. (1979) Sixteen years' experience of counseling, diagnosis, and prenatal detection in one genetic centre: progress, results, and problems. J Med Genet 16:166-175
5. Schneiderman G, Lowden JA (1986) Fatal metabolic disease and family breakdown. Psychiatr J Univ Ottawa 11:35-37
6. Thelin T, McNeil TF, Aspegren-Jansson E, Sveger T (1985) Identifying children at high somatic risk: parents' long-term emotional adjustment to their children's α-1 antitrypsin deficiency. Acta Psychiatr Scand 72:323-330
7. Engel GL (1977) The need for a new medical model: a challenge for biomedicine. Science 196:127-136
8. Byrne EA, Cunningham CC (1985) The effects of mentally handicapped children on families - a conceptual review. J Child Psychol Psychiatry 26:847-864
9. Caplan G (1980) An approach to preventive intervention in child psychiatry. Can J Psychiatry 25:671-682
10. Kazak AE (1987) Families with disabled children: stress and social networks in three samples. J Abnormal Child Psychol 15:137-146
11. Richmond JB (1972) The family and the handicapped child. Clin Proc Child Hosp Nat Med Cent 8:156-164
12. Futterman EH (1975) Studies of family adaptational responses to a specific threat. In: Anthony EJ (ed) Explorations in child psychiatry. Plenum, New York, pp 287-301
13. Seashore MR, Freidman E, Novelly RA, Bapat V (1985) Loss of intellectual function in children with phenylketonuria after relaxation of dietary phenylalanine restriction. Pediatr 75:226-232
14. Azen C, Koch R, Friedman E, Wenz E, Fishler K (1996) Summary of findings from the United States Collaborative Study of children treated for phenylketonuria. Eur J Pediatr 155 [Suppl 1]:S29-S32
15. Griffiths P, Tarrini M, Robinson P (1997) Executive function and psychosocial adjustment in children with early treated phenylketonuria: correlation with historical and concurrent phenylalanine levels. J Intellect Disabil Res 41:317-323
16. Weglage J, Pietsch M, Funders B, Koch HG, Ullrich K (1996) Deficits in selective and sustained attention processes in early treated children with phenylketonuria: result of impaired frontal lobe functions? Eur J Pediatr 155:200-204
17. Arnold GL, Kramer BM, Kirby et al. (1998) Factors affecting cognitive, motor, behavioral, and executive functioning in children with phenylketonuria. Acta Paediatr 87:565-570
18. Williams K (1998) Benefits of normalizing plasma phenylalanine: impact on behaviour and health. J Inherit Metab Dis 21:785-790
19. Baumeister AA, Baumeister AA (1998) Dietary treatment of destructive behavior associated with hyperphenylalaninemia. Clin Neuropharmacol 21:18-27
20. Weglage J, Funders B, Ullrich K, Rupp A, Schmidt E (1996) Psychosocial aspects in phenylketonuria. Eur J Pediatr 155 [Suppl]:S101-S104
21. Koch R, Fishler K, Azen C, Guldberg P, Guttler F (1997) The relationship of genotype to phenotype in phenylalanine hydroxylase deficiency. Biochem Mol Med 60:92-101
22. Welsh MC (1996) A prefrontal dysfunction model of early-treated phenylketonuria. Eur J Pediatr 155 [Suppl]:S87-S89
23. Diamond A (1996) Evidence for the importance of dopamine for prefrontal cortex functions early in life. Philos Trans R Soc Lond [B] Biol Sci 351:1483-1493
24. Diamond A, Prevor MB, Callender G, Druin DP (1997) Prefrontal cortex cognitive deficits in children treated early and continuously for PKU. Monogr Soc Res Child Dev 62:i-v
25. Realmuto G, Garfinkel BD, Tuchman M et al. (1986) Psychiatric diagnosis and behavioral characteristics of phenylketonuric children. J Nerv Ment Dis 174:536-540
26. McBean MS (1971) The problems of parents of children with phenylketonuria. In: Bickel B, Hudson FP, Woolf LI (eds) Phenylketonuria. Thieme, Stuttgart, pp 180-282
27. Pueschel SM, Yeatman S, Hum C (1977) Discontinuing the phenylalanine-restricted diet in young children with PKU. Psychosocial aspects. J Am Diet Assoc 70:506-509
28. Reber M, Kazak AE, Himmelberg P (1987) Phenylalanine control and family functioning in early treated phenylketonuria. J Dev Behav Pediatr 8:311-317
29. Koch R, Azen C, Friedman EG et al. (1996) Care of the adult with phenylketonuria. Eur J Pediatr 155 [Suppl]:S90-S92

30. Ris MD, Weber AM, Hunt MM et al. (1997) Adult psychosocial outcome in early-treated phenylketonuria. J Inherit Metab Dis 20:499–508
31. Pietz J, Fatkenheuer B, Burgard P et al. (1997) Psychiatric disorders in adult patients with early-treated phenylketonuria. Pediatr 99:345–350
32. Cleary MA, Wraith JE (1993) Management of the mucopolysaccharidosis, Type III. Arch Dis Childhood 69:403–406
33. Colville GA, Bax MA (1996) Early presentation in the mucopolysaccharide disorders. Child Care Health Dev 22:31–36
34. Epstein CJ, Yatziv S, Neufeld E, Liebaers I (1976) Genetic counseling for Hunter syndrome [letter]. Lancet 2:73–78
35. Young ID, Harper PS (1981) Psychosocial problems in Hunter's syndrome. Child Care Health Dev 7:201–209
36. Crocker AC, Cullinane MM (1972) Families under stress: the diagnosis of Hurler's syndrome. Postgrad Med 51:22–39
37. Harris J (1998) Developmental neuropsychiatry: assessment, diagnosis, and treatment of developmental disorders, vol 2. Oxford University Press, New York, chap 11.1: Lesch-Nyhan Disease, pp 306–319; chap 11.6: mucopolysaccharidoses, pp 355–361
38. Colville GA, Watters JPW, Bax M (1996) Sleep problems in children with Sanfilippo syndrome. Dev Med Child Neurol 38:538–544
39. Handelman L, Menahem S, Eisenbruch IM (1989) Transcultural understanding of a hereditary disorder: mucopolysaccharidosis VI in a Vietnamese family. Clin Pediatr 28:470–473
40. Lesch M, Nyhan WL (1964) A familial disorder of uric acid metabolism and central nervous system function. Am J Med 36:561–570
41. Cataldo MF, Harris J (1982) The biological basis for self-injury in the mentally retarded. Anal Intervent Dev Disabil 2:21–39
42. Carr EG (1977) The motivation of self-injurious behavior: a review of some hypotheses. Psychol Bull 84:800–816
43. Anderson LT, Ernst M (1994) Self-injury in Lesch-Nyhan Disease. J Autism Dev Dis 24:67–81
44. Brown FR III, Stowens DW, Harris JC, Moser HG (1985) Leukodystrophies. In: Johnson RT (ed) Current therapy in neurologic disease. Decker, Philadelphia, pp 313–317
45. Moser HW (1997) Adrenoleukodystrophy: phenotype, genetics, pathogenesis, and therapy. Brain 120 (8):1485–1508
46. Rajanayagam V, Balthazor M, Shapiro EG et al. (1997) Proton MR spectroscopy and neuropsychological testing in adrenoleukodystrophy. Am J Neurorad 18:1909–1914
47. Singh I, Pahan K, Khan M (1998) Lovastatin and sodium phenylacetate normalize the levels of very long chain fatty acids in skin fibroblasts of X-adrenoleukodystrophy. FEBS Lett 426:342–346
48. Kemp S, Wei HM, Lu JF et al. (1998) Gene redundancy and pharmacological gene therapy: implications for X-linked adrenoleukodystrophy. Nat Med 4:1261–1268
49. Peters C, Balthazor M, Shapiro EG et al. (1996) Outcome of unrelated donor bone marrow transplantation in 40 children with Hurler syndrome. Blood 87:4894–4902
50. Shapiro EG, Lockman LA, Balthazor M, Krivit W (1995) Neuropsychological outcomes of several storage diseases with and without bone marrow transplantation. J Inherit Metab Dis 18:413–429
51. Malm G, Ringden O, Anvret M et al. (1997) Treatment of adrenoleukodsytrophy with bone marrow transplantation. Acta Paed 86:484–492
52. Wolcott DL, Fawzy FI, Wellisch DK (1986) Psychiatric aspects of bone marrow transplantation: a review and current issues. Psychiatr Me 4:299–317
53. Pfefferbaum B, Lindamood MM, Wiley FM (1977) Pediatric bone marrow transplantation: psychosocial aspects. Am J Psychiatry 134:1299–1301
54. Atkins DM, Patenaude AF (1987) Psychosocial preparation and follow-up for pediatric bone marrow transplant patients. Am J Orthopsychiatry 57:246–252
55. Freund BL, Siegel K (1986) Problems in transition following bone marrow transplantation: psychosocial aspects. Am J Orthopsychiatry 56:244–252
56. Kramer JH, Crittenden MR, DeSantes K, Cowan MJ (1997) Cognitive and adaptive behavior 1 and 3 years following bone marrow transplant. Bone Marrow Transplant 19:606–613
57. Peters C, Balthazor M, Shapiro EG et al. (1997) Outcome of unrelated donor bone marrow transplantation in 40 children with Hurlers' syndrome. Blood 87:4894–4902
58. Guffon N, Souillet G, Maire I, Straczek J, Guibaud P (1998) Follow-up of nine patients with Hurler syndrome after bone marrow transplantation. J Pediatr 133:119–125
59. Packman WL, Crittenden MR, Schaeffer E et al. (1997) Psychosocial consequences of bone marrow transplantation in donor and nondonor siblings. J Dev Behav Pediatr 18:244–253
60. Eisensmith RC, Woo SL (1996) Somatic gene therapy for phenylketonuria and other hepatic deficiencies. J Inherit Metab Dis 19:412–423
61. Huang MM, Wong A, Yu X, Kakkis E, Kohn DB (1997) Retrovirus-mediated transfer of the human alpha-L-iduronidase cDNA into human hematopoietic progenitor cells leads to correction in trans of Hurler fibroblasts. Gene Ther 4:1150–1159
62. Prieto LR, Miller DS, Gaoyowski T, Marino IR (1997) Multicultural issues in organ transplantation: the influence of patients' cultural perspectives on compliance with treatment. Clin Transplant 11:529–535
63. Gold LM, Kirkpatrick BS, Fricker FJ, Zitelli BJ (1986) Psychosocial issues in pediatric organ transplantation: the parents' perspective. Pediatr 77:738–744
64. Rodriguez JR, Hoffman RG, MacNaughton K et al. (1996) Mothers of children evaluated for transplantation: stress, coping resources, and perceptions of family functioning. Clin Transplant 10:447–450
65. Rodriguez JR, MacNaughton K, Hoffman RG et al. (1996) Perceptions of parenting stress and family relations by fathers of children evaluated for organ transplantation. Psychol Rep 79 (3/1):723–727
66. Bielicki J, Hopwood JJ, Anson DS (1996) Correction of Sanfilippo A skin fibroblasts by retroviral vector-mediated gene transfer. Hum Gene Ther 7:1965–1970
67. Blair C, Kacser E, Porteous D (1998) Gene therapy for cystic fibrosis: a psychosocial study of trial participants. Gene Ther 5:218–222

# CHAPTER 5

# Treatment: Present Status and New Trends

J.H. Walter and J.E. Wraith

CONTENTS

| | |
|---|---|
| Introduction | 75 |
| General Principles for Treatment | 75 |
| Reducing the Load on the Affected Pathway | 75 |
| Substrate Deprivation by Diet | 75 |
| Limiting the Availability of Ingested Substrate | 76 |
| Removing Toxic Metabolites | 76 |
| Replenishing Depleted Products | 76 |
| Giving Increased Substrate | 77 |
| Blocking Production of Toxic Metabolites | 77 |
| Blocking the Effects of Toxic Metabolites | 77 |
| Stimulating Any Residual Enzyme | 77 |
| New Trends | 78 |
| Enzyme Replacement | 78 |
| Direct Enzyme-Replacement Therapy | 78 |
| Bone-Marrow Transplantation | 78 |
| Other Organ Transplantation | 79 |
| Gene Transfer | 79 |
| Pharmacological Gene Therapy | 79 |
| Symptomatic Treatment | 79 |
| Conclusions | 79 |
| Appendix, medications used in the treatment of inborn errors | 80 |
| References | 83 |

## Introduction

Improvements in the understanding of the biochemical and molecular basis of inborn errors have led to significant improvements in our ability to treat many of these disorders. Such improvements, coupled with an ability to make more rapid diagnoses and advances in general medical care, particularly intensive care, are resulting in better long-term prognosis for many patients. However, the rarity of individual disorders has often made it difficult or impossible to obtain sufficient data for evidence-based assessment of treatments. This should be kept in mind when considering the efficacy of particular therapies. Anecdotal reports of improvements should be reviewed critically, but it is equally important to remain open to new advances.

This chapter discusses recent progress in the development of treatments. We have also included a list of medications (with recommended dosages) that may be used in the treatment of inborn errors (Table 5.1). Readers should refer to the relevant chapters for detailed information about the management of specific disorders and to Chap. 4, "*Psychosocial Care of the Child and Family*", for discussion of the psychological consequences of treatment.

## General Principles for Treatment

### Reducing the Load on the Affected Pathway

#### Substrate Deprivation by Diet

Restrictive diets are the treatment of choice for a number of inborn errors. Such diets are highly efficacious in phenylketonuria (PKU), maple-syrup-urine disease (MSUD) and homocystinuria, disorders in which the defective enzyme's substrate can be effectively limited in the diet and for which substrate levels in the body can be monitored. Dietary therapy is less successful in disorders in which the defect is further downstream in a metabolic pathway – for example, propionic and methylmalonic acidaemias and disorders of the urea cycle. Improvements in the understanding of basic human nutritional requirements, food technology and the biochemical abnormalities in specific disorders have led to continued development. This is exemplified by PKU. The relative frequency of this disorder in the developed world and the recommendation that dietary therapy should be continued into adulthood have made it commercially viable for specialist food manufactures to invest in the necessary research. There have been some improvements in the palatability of amino acid supplements and in the range of available products. The need for dietary flexibility, particularly important for adolescents and adults, has led to the development of products, such as the Phlexy-10 system (SHS, Liverpool). Products have been reformulated to increase the content of various minerals and trace elements (such as selenium) that have been recognised to be low in individuals on semi-synthetic

**Table 5.1.** Co-factor responsive disorders: treatment and dosage

| Disorder | Co-factor | Therapeutic dose | Frequency of responsive variants |
| --- | --- | --- | --- |
| Biotinidase deficiency | Biotin | 5–10 mg/day | All cases |
| Folinic-acid-responsive seizures | Folinic acid | 5–15 mg/day | All cases |
| Glutaric aciduria type I | Riboflavin | 20–40 mg/day | Rare |
| Homocystinuria | Pyridoxine | 50–500 mg/day | ~50% |
| Hyperphenylalaninaemia due to disorders of biopterin metabolism | Tetrahydrobiopterin | 5–20 mg/day | All, but no improvement in CNS neurotransmitter levels |
| Methylmalonic acidaemia | Vitamin B12 | 1 mg IM/day | Some |
| MSUD | Thiamine | 10–50 mg/day | Rare |
| Multiple carboxylase deficiency | Biotin | 10–40 mg/day | Most |
| OAT | Pyridoxine | 20–600 mg/day | ~30% |
| Propionic acidaemia | Biotin | 5–10 mg/day | Possibly never |
| Pyridoxine-responsive seizures | Pyridoxine | 50–100 mg/day | All cases |
| Respiratory-chain disorders | Ubiquinone | 100–300 mg/day | Anecdotal evidence |

*CNS*, central nervous system; *MSUD*, maple-syrup urine disease; *OAT*, ornithine aminotransferase deficiency

diets [1]. Clearly, though, there is more work needed before these diets can be considered attractive.

### Limiting the Availability of Ingested Substrate

Limiting the availability of ingested substrate for absorption by the gut is a further method for reducing substrate. Examples include the treatment of Wilson disease with zinc and the use of resins in hyperlipidaemias. A more novel approach is under investigation for treatment of PKU using microencapsulated phenylalanine–ammonia lyase [2].

As the biochemical bases of various disorders have been determined, a theoretical reason for dietary therapy has become evident in some. The efficacy of such treatments varies; for example, the use of medium-chain triglycerides as a fat source in patients with very-long-chain acyl-coenzyme-A (CoA)-dehydrogenase deficiency and long-chain hydroxyacyl-CoA-dehydrogenase deficiency is generally accepted [3–5], whereas the use of a low-fat, high-carbohydrate diet in medium-chain acyl-CoA-dehydrogenase deficiency now appears unnecessary. Other disorders in which dietary therapy has recently been implemented include a high-cholesterol diet in Smith-Lemli-Opitz syndrome [6] and Lorenzo's oil (glycerol trioleate and glycerol trierucate) in X-linked adrenoleucodystrophy [7], although the efficacy of such treatments remains unproven.

### Removing Toxic Metabolites

A number of medications are used to expedite the removal of metabolites that accumulate due to particular inborn errors. These include well-established treatments, such as sodium benzoate and sodium phenylbutyrate in disorders associated with hyperammonaemia, glycine in isovaleric acidaemia and L-carnitine in organic acidaemias. Although there have been no rigorous clinical trials to demonstrate its efficacy, L-carnitine has become an established treatment for organic acidaemias, where there is a reduction in plasma carnitine and an increase in the acyl:free L-carnitine ratio [8–10]. The role of L-carnitine in the treatment of medium-chain acyl-CoA-dehydrogenase deficiency is not established, and it is our experience that, even without this treatment, children do well following diagnosis [11]. However, it remains possible that L-carnitine may improve exercise tolerance and have a protective effect on metabolic decompensation. Further studies will be necessary to clarify its role in this disorder. In carnitine-transport defect, however, the effect of supplementation is dramatic, with a resolution of cardiomyopathy and prevention of further episodes of hypoketotic hypoglycaemia [12]. Animal studies suggest that carnitine may also have a protective role in hyperammonaemia [13].

### Replenishing Depleted Products

Where the deficiency of an enzyme's product is important in the aetiology of clinical illness, its replacement may form the basis of treatment; for example, the administration of carbohydrate in glycogen-storage disease (GSD), arginine or citrulline in urea-cycle disorders and tyrosine in PKU. More recent developments are the use of serine and glycine in 3-phosphoglycerate-dehydrogenase deficiency [14, 15], creatine in guanidinoacetate-methyltransferase deficiency [16, 17] and neurotransmitters in defects of biopterin synthesis and primary disorders of neurotransmitter metabolism [18].

## Giving Increased Substrate

Giving pharmacological amounts of substrate may be effective, particularly in inborn errors of membrane transport proteins - for example, L-carnitine in carnitine transporter deficiency and ornithine in triple-H syndrome. Such therapy, of course, requires that the substrate itself has low toxicity. Similar treatment has been attempted with copper histidine in Menkes disease (with variable effect) [19, 20] and in disorders with single-enzyme deficiencies, such as mannose in CDG type 1b (phosphomannose-isomerase deficiency) [21].

## Blocking Production of Toxic Metabolites

The use of 2-(2-nitro-4-trifluoro-methylbenzoyl)-1,3-cyclohexanedione (NTBC) in tyrosinaemia type 1 demonstrates a novel approach to inherited metabolic disease (Chap. 15). Inhibition of 4-hydroxyphenylpyruvate dioxygenase (an enzyme proximal to fumarylacetoacetase) by NTBC prevents the production of maleylacetoacetate, fumarylacetoacetate and succinylacetone, compounds that are the major toxic agents in this disease [22]. In lysosomal disorders, this approach is also under assessment, and successful prevention of glycosphingolipid accumulation in the Tay-Sachs mouse (using N-butyldeoxynojirimycin) would suggest that further clinical studies are indicated [23].

## Blocking the Effects of Toxic Metabolites

Disorders in which the phenotype is related to metabolites binding to receptors may be amenable to treatments that block this effect. For example, the use of N-methyl-D-aspartate (NMDA)-channel agonists, such as dextromethorphan and ketamine in nonketotic hyperglycinaemia (NKH), limit the neuroexcitatory effect of glycine on the NMDA receptor (Chap. 21) [24-26]. Although there have been reports of successful treatments, the variable phenotypes have made the efficacy difficult to access. Our experience with a number of infants presenting with the severe neonatal form of this disorder has not been encouraging.

## Stimulating Any Residual Enzyme

Some metabolic disorders are caused by mutations that affect the metabolism or binding of a coenzyme or cofactor necessary for normal enzyme activity. Treatment with the coenzyme may lead to a complete return of the clinical phenotype to normal, as occurs, for example, with biotin in biotinidase deficiency. Most disorders with coenzyme-responsive variants show a more limited improvement; for example, a majority of patients with vitamin-B12-responsive methylmalonic acidaemia continue to produce abnormal, albeit smaller, quantities of methylmalonic acid. Other disorders may not be fully correctable because of difficulties in getting the coenzyme to the appropriate location. This is the case in disorders of biopterin synthesis (guanosine-triphosphate-cyclohydrolase deficiency, 6-pyruvoyltetrahydrobiopterin deficiency), where oral tetrahydrobiopterin ($BH_4$) rapidly corrects the hyperphenylalaninaemia by its effect on liver phenylalanine hydroxylase. However, $BH_4$ does not easily cross the blood-brain barrier and is consequently not readily available to tyrosine hydroxylase or tryptophan hydroxylase within the central nervous system (CNS). The profound neurotransmitter deficiency associated with these conditions is, therefore, not significantly improved by $BH_4$ therapy.

Many single-enzyme disorders are due to mutations that prevent any enzyme production and cannot, therefore, improve as a result of coenzyme therapy. For example, this is probably the case with all patients with propionic acidaemia and nearly all those with MSUD, even though both propionyl-CoA carboxylase and α-ketoacid dehydrogenase require a coenzyme (biotin and thiamine, respectively). Disorders with known coenzyme-responsive variants are listed in Table 5.1.

On occasion, it may be difficult to determine whether a particular patient is responsive to coenzyme therapy. This may arise due to confounding factors that cause a concomitant decrease in a particular biochemical marker (such as the concentration of a particular metabolite) unrelated to the administration of the coenzyme. For example, the patient may have entered a recovery phase following a period of metabolic decompensation or may have recently started a dietary therapy. Standard protocols may be helpful in such situations, but further assessment may be required when the patient is more stable.

Most coenzymes are safe even in large doses, and it is appropriate to treat patients who have disorders that are known to have responsive variants with the relevant coenzyme. The arguments for giving a "cocktail" of various vitamins to patients before a diagnosis has been made are less strong. Disorders presenting in the newborn period or soon after are likely to be due to severe enzyme deficiencies that are not coenzyme responsive. Furthermore, the majority of clinicians have access to rapid metabolic investigations by specialist laboratories. However, if there is likely to be a delay in diagnostic investigation, the intelligent use of a number of vitamins or cofactors may be indicated (Table 5.2).

**Table 5.2.** The vitamin cocktail

| | | |
|---|---|---|
| Biotin | 10 mg/day | Oral or IV |
| Thiamine | 200 mg/day | Oral |
| Lipoic acid | 100 mg/day | Oral |
| L-Carnitine | 25 mg/kg every 6 h | Oral or IV |
| Co-$Q_{10}$ | 5 mg/kg/day | Oral |
| Vitamin B12 | 1 mg/day | IM |
| Vitamin C | 100 mg/kg/day | Oral |
| Riboflavin | 100–300 mg/day | Oral |

*Co-Q*, coenzyme Q

Certain enzymes may also be stimulated by specific medication. For example, dichloroacetate (DCA) increases pyruvate-dehydrogenase (PDH) activity by its inhibitory effect on PDH-kinase. There are anecdotal reports of biochemical and clinical improvement following its use in congenital lactic acidosis [27–29]. However, the (unpublished) experience of many units is that DCA has not improved the outcomes of most patients. A randomised, controlled trial is reported to be underway to investigate the efficacy of DCA in congenital lactic acidosis [30].

## New Trends

### Enzyme Replacement

#### Direct Enzyme-Replacement Therapy

Initial enthusiasm for enzyme replacement therapy (ERT) in the 1970s waned as a consequence of difficulties in enzyme production, purification and targeting. However, the spectacular success of β-glucosidase replacement in non-neuronopathic Gaucher disease has stimulated renewed interest in this area.

Although there now appears to be general agreement about the efficacy of ERT in Gaucher disease, there are no consistent dosage recommendations. The recombinant enzyme imiglucerase (Cerezyme, Genzyme) has been approved for use after two clinical trials using different enzyme schedules. In the first trial, a high-dose, low-frequency regimen (60 units/kg every 2 weeks) was used [31] and, in the other, a low-dose, variable-frequency regimen (30 units/kg every 4 weeks) was used either once every other week or three times per week [32]. Clinical efficacy is seen with both dosage regimens although, in paediatric practice, we favour the higher-dose schedule. In contrast to the rapid visceral and haematological improvement, the response of the skeletal system to ERT is very much slower and incomplete. This mirrors the response of the skeleton in other lysosomal disorders, e.g. mucopolysaccharidoses (MPS), following a successful bone-marrow transplant. For this reason, it has been suggested that anti-osteoclastic therapy with bisphosphonates may be a useful adjunct to therapy in those patients with severe bone disease [33].

In paediatric practice, there are a significant number of patients with type-III (subacute neuropathic) Gaucher disease. The outcome of enzyme therapy in these patients remains unclear, and concerns remain that, even with high-dose therapy, progression of the CNS disease may not be prevented. For this reason it is our practice to use a higher dosage (120 units/kg every 2 weeks) and to not reduce the dosage despite haematological and visceral improvement. In type-II (neuronopathic) Gaucher disease, enzyme therapy has not prevented the progressive neurological deterioration seen in these patients. Despite the good results in type-I (non-neuronopathic) Gaucher disease, concerns remain about antibody production [34] and cost of therapy [35]. Other disorders, including Fabry disease, MPS type I and GSD-II, have also become the targets of potential enzyme-replacement strategies.

### Bone-Marrow Transplantation

Bone-marrow transplantation (BMT) readily corrects the enzyme deficiencies associated with lysosomal and peroxisomal disorders, at least in cells of haematopoietic origin. An attempt has been made to treat almost all of the lysosomal disorders by this method. Unfortunately, almost all of the reported cases are either anecdotal or consist of very few cases from the same centre. The first disorder treated by this method was MPS type I (Hurler syndrome) [36], and most clinical experience applies to this disorder. There can be no doubt that a successful BMT in MPS-I alters the natural history of the disease, but there are considerable residual problems, especially with the spinal deformity and joint disease. In addition, BMT must be performed early (ideally at less than 18 months at age) if neurological progression is to be prevented. Whilst the risk of graft-versus-host disease has lessened with improvements in tissue typing and drug therapy, primary graft rejection remains a problem in this group of patients [37].

With other disorders, the role of transplantation is less clear-cut and is probably contraindicated in disorders with very aggressive neurodegeneration, e.g. MPS-III (SanFilippo syndrome). It is possible that the use of bone-marrow stromal cells as stem cells for non-haematopoietic tissues will lead to a better result from BMT [38].

## Other Organ Transplantation

Liver transplantation has been used as a successful therapy for a number of inborn errors of metabolism, although infants with organic acid disorders or urea-cycle defects are usually given a very low priority by most transplant services. Whilst the indications for transplant in disorders (such as Crigler-Najjar syndrome, GSD-IV or fulminant hepatic failure secondary to Wilson's disease) may be clear-cut, the indications in other inborn errors are not, and the decision-making process is often very difficult. In addition, the mortality associated with liver transplantation is likely to be high in patients with severe disorders of intermediary metabolism, e.g. propionic acidaemia. Nevertheless, because of the grave prognosis associated with conventional medical therapy, increasing numbers of infants are likely to be referred to transplant centres. Combined liver–kidney transplantation has been performed in both methylmalonic acidaemia [39] and primary hyperoxaluria type I [40]. However, a pre-emptive liver transplant from a live, related donor may be the best approach in such disorders, as this will avoid the problems associated with the low priority of cadaveric organs and will protect the kidneys from the harmful effects of the metabolic lesion [41].

## Gene Transfer

Initial enthusiasm for "gene therapy" as a potential treatment for single-gene disorders has been dampened by a number of important challenges, which include the targeting and efficiency of gene transfer and the magnitude and duration of subsequent gene expression. However, a great deal of work has been achieved, and there has been steady progress within the field, even though there have been no successful treatment protocols. Successful therapy for metabolic disorders must combine appropriate disease targeting with an efficient delivery system that ensures long-term expression with no toxicity. This has proved to be an elusive goal, but early results using adenovirus-mediated transfer in GSD-II cells [42] and the use of a lentivirus vector to target the nervous system [43] have been promising.

## Pharmacological Gene Therapy

Stimulating the expression of endogenous "redundant" genes by pharmacological agents may provide treatment for some inborn errors. In haemoglobinopathies, foetal haemoglobin may be stimulated by various agents, such as hydroxyurea, 5-azacytidine and sodium phenylbutyrate [44–46]. Similarly, in cystic fibrosis, sodium phenylbutyrate increases cystic-fibrosis trans-membrane-conductance-regulator gene expression [47]. Recently, Kemp et al. demonstrated that, in cells from patients with X-linked adrenoleukodystrophy (X-ALD) and from X-ALD knockout mice, sodium phenylbutyrate increased β-oxidation and decreased very-long-chain fatty acids (VLCFAs) by enhancing the expression of a peroxisomal membrane adenosine-triphosphate-binding-cassette transporter-protein gene coding for adrenoleukodystrophy (ALD)-related protein (ALDRP) [48]. Furthermore, dietary treatment of the X-ALD mice with sodium phenylbutyrate led to a substantial decrease in VLCFAs in both the brain and adrenal glands. ALDRP is closely related to ALD protein, the product of the X-ALD gene. The ALDRP gene appears to be redundant in normal individuals. Studies of sodium phenylbutyrate in the treatment of humans with X-ALD are likely to be undertaken in the near future.

## Symptomatic Treatment

Despite recent advances, therapies that directly effect pathology remain unavailable for many inborn errors. Improving the quality of life for patients and their families often relies on symptomatic treatment. For example, many disorders are associated with seizures which, if difficult to control, may benefit from the use of new anticonvulsant medication. Some other disorders have particular complications that require treatment. These include middle-ear and upper-airway disease in the MPSs and pain in Fabry disease. In addition to medical and surgical treatments, providing a nutritionally adequate diet is of fundamental importance for all patients. Anorexia is common in organic acidaemias and some other disorders. In other conditions, disturbed swallowing mechanisms may decrease dietary intake. For these patients, early intervention with nasogastric feeding or feeding via gastrostomy is often essential to provide the best possible outcome.

## Conclusions

Although, the outcome for many inborn errors remains poor at present, over the next decade, there will be significant advances in our ability to offer more effective treatment. Formidable obstacles remain, particularly for those disorders where there is significant in utero damage or where the CNS is primarily affected.

*Acknowledgement.* We are grateful to Annette Adams, pharmacist at Manchester Children's Hospital, for information included in the Appendix.

**Appendix.** Medications used in the treatment of inborn errors. The evidence for the efficacy of many of these drugs is anecdotal. The majority are unlicensed. Refer to the relevant chapters or published literature for further discussion about their use

| Medication | Mode of action | Disorders | Recommended dose | Route | Remarks |
|---|---|---|---|---|---|
| 5-Hydroxytryptophan | Neurotransmitter replacement | Disorders of neurotransmitter synthesis | 1–2 mg/kg increasing every 4–5 days to 8–10 mg/kg in four divided doses | Oral | Monitor CSF 5HIAA levels; adjust dose accordingly; use with L-dopa/carbidopa |
| Ammonium tetrathiomolybdate | Chelating agent | Wilson's disease | 160 mg/day in six divided doses | Oral | May be preferred to penicillamine if severe neurological disease |
| Arginine | Replenishes arginine | CPS deficiency; OTC deficiency; AS deficiency (citrullinaemia); AL deficiency | 200 mg/kg (OTC and CPS deficiency) up to 700 mg/kg in AL and AS deficiency) | Oral, IV | IV loading dose: 200 mg/kg (OTC or CPS) or 600 mg/kg (AL or AS) over 90 min (Chap. 17); dilute IV solutions in 10% glucose (30 ml/kg) |
| Betaine | Remethylates homocysteine to methionine | Classical homocystinuria; remethylation deficiencies | 100 mg/kg/day in two divided doses, max dose 20 g/day | Oral | In classical homocystinuria associated with an increase in plasma methionine levels |
| $BH_4$ | Replacement of $BH_4$ | Disorders of $BH_4$ synthesis or recycling | 1–3 mg/kg/day | Oral | |
| Biotin | Co-factor for carboxylases | Biotinidase deficiency; multiple carboxylase deficiency | 5–20 mg/day | Oral, IV | |
| Carbamoylglutamate | Stimulates N-acetylglutamate synthase | N-acetylglutamate-synthase deficiency; CPS deficiency | 50–100 mg/kg/day in four divided doses | Oral | IV |
| Carnitine | Replenishes body stores; removes toxic acyl-CoA intermediates from within the mitochondria | Primary and secondary carnitine deficiency | 100–200 mg/kg/day | Oral or IV | Use L-carnitine, not the racemic mixture; diarrhoea and fishy body odour with high dosage |
| Chenodeoxycholic acid | Inhibits cholesterol 7α-hydroxylase (rate-limiting enzyme in bile-acid biosynthesis) | 3β-Dehydrogenase deficiency; as an additive therapy in SLO syndrome | 12–18 mg/kg/day for first 2 months, then 9–12 mg/kg/day; 7 mg/kg/day in SLO syndrome | Oral | |
| Cholesterol | Replenishes cholesterol | SLO syndrome | 20–40 mg/kg/day in three or four divided doses | Oral | |
| Cholestyramine | Anion-exchange resin; prevents bile reabsorption | Familial hypercholesterolaemia | Adults: 12–24 g/day<br><br>Children: (weight in kilograms/70 × adult dose) in four divided doses | Oral | Possible vitamin A, D, and K deficiencies with prolonged treatment |
| Citrulline | Replenishes citrulline and arginine | Used as an alternative to arginine in CPS deficiency and OTC deficiency | 170 mg/kg/day or 3.8 gm/m²/day in divided doses | Oral | |
| Copper histidine | Increases intracellular copper | Menkes disease | 100–200 μg Cu/day (newborn), increasing to 450 μg by 9 weeks; 1 mg Cu/day after 12 months | IM | |

**Appendix.** (*Contd.*)

| Medication | Mode of action | Disorders | Recommended dose | Route | Remarks |
|---|---|---|---|---|---|
| Creatine monohydrate | Replenishes creatine | GAMT deficiency | 400–670 mg/kg/day | Oral | |
| Cysteamine/phospho-cysteamine | Depletes lysosomal cystine | Cystinosis | 1.3 g/m$^2$/day of free base, given every 6 h | Oral (and eye drops) | Phosphocysteamine more palatable |
| Dextromethorphan | NMDA-channel antagonist | NKH | 5–7 mg/kg/day in four divided doses | Oral | Doses up to 35 mg/day have been used |
| DHA and arachidonic acid | Replenishes deficient product | Peroxisomal disorders with low plasma DHA levels | | Oral | Still under investigation [49] |
| Dichloroacetate | Stimulates PDH activity | Primary lactic acidosis | 50 mg/kg/day in three or four divided doses | Oral | May cause polyneuropathy with prolonged use |
| Diazoxide | Opens K$^+$ channel | Hyperinsulinism | 10–15 mg/kg/day | Oral | May cause oedema and hypertension |
| Folic acid | Provides accessible source of folate for CNS | DHPR deficiency; UMP-synthase deficiency (hereditary orotic aciduria); methylenetetrahydrofolate reductase deficiency; methionine-synthase deficiency; hereditary folate malabsorption; some disorders of cobalamin metabolism; folic acid responsive seizures | 5–15 mg/day | Oral, IV | |
| G-CSF | Stimulates granulocyte production | Neutropenia in GSD Ib | 5 µg/kg/day | SC | |
| Glycine | Forms isovalerylglycine with high renal clearance | Isovaleric acidaemia | 150 mg/kg/day in three divided doses | Oral | Up to 600 mg/kg/day during decompensation |
| Heme arginate | Replenishing depleted product | Porphyrias | 3–4 mg/kg/day once daily for 4 days | IV | |
| Hydroxycobalamin (vitamin B12) | Co-factor for methylmalonyl-CoA mutase | Vitamin-B12-responsive forms of MMA | 1 mg IM/day; oral dose 10 mg once or twice daily | IM, Oral | IM dose may be reduced to once or twice weekly, according to response |
| Imiglucerase | Genetically engineered β-glucerebrosidase | Gaucher disease | 60 units/kg every 2 weeks | IV | Dose may be deceased after favourable response |
| Ketamine | NMDA Channel antagonist | NKH | 1–30 mg/kg/day in four divided doses | Oral, IV | |
| L-Dopa | Replacement of neurotransmitters | Disorders of L-dopa synthesis, including disorders of tetrahydrobiopterin synthesis | 1–2 mg/kg increasing every 4–5 days to 10–12 mg/kg in four divided doses | Oral | Give as L-dopa/carbidopa (1:10 or 1:5); monitor CSF HVA levels; adjust dose accordingly. Use with 5-hydroxytryptophan |
| L-Tryptophan | Increases kynurenic acid which is an endogenous antagonist of the NMDA receptor | NKH | 100 mg/kg/day in three divided doses | Oral | |
| Mannose | Increases glycosylation | CDG-Ib (PMI deficiency) | 100 mg/kg every 3 h | Oral | Not of proven benefit in CDG Ia [50, 51] |

## Appendix. (Contd.)

| Medication | Mode of action | Disorders | Recommended dose | Route | Remarks |
|---|---|---|---|---|---|
| Metronidazole | Reduces propionate production by gut bacteria | Propionic and methylmalonic acidaemia | 7.5 mg three times per day | Oral | |
| NTBC (2-[2-nitro-4-trifluoro-methylbenzoyl]-1,3-cyclohexanedione) | Inhibits 4-hydroxyphenyl pyruvate dioxygenase | Tyrosinaemia type I | 1 mg/kg in one or two divided doses | Oral | Combine with low-tyrosine, low-phenylalanine diet to maintain plasma tyrosine < 600 µmol/l |
| Ornithine | Increased transport of ornithine in mitochondria | Triple H | 75–150 mg/kg/day | Oral | |
| Pantothenic acid | Source of CoA | Type-II 3-methylglutaconic aciduria | 15–150 mg/day in three divided doses | | [52] |
| Penicillamine | Chelating agent | Wilson disease | Up to 20 mg/kg/day in divided doses (minimum of 500 mg/day) | Oral | Side effects include anorexia, fever, skin reactions, thrombo cytopenia, neutropenia |
| Proline | Replenishes proline | Gyrate atrophy due to ornithine aminotransferase deficiency | 2–10 g/day | Oral | |
| Pyridoxine | Co-factor | Pyridoxine responsive γ-cystathionase deficiency; pyridoxine-responsive CBS deficiency; pyridoxine dependency with seizures; ornithine aminotransferase deficiency | 50–500 mg/day; 20–600 mg/day<br><br>For pyridoxine dependency with seizures: 100 mg IV with EEG monitoring or 30 mg/kg/day for 7 days | Oral | Peripheral neuropathy can occur with doses >1000 mg daily |
| Riboflavin | Co-enzyme | Glutaric acidurias I and II; congenital lactic acidosis | 100 mg/day in two or three divided doses | Oral | Large doses discolour urine |
| Serine | Replenishes serine | 3-Phosphoglycerate-dehydrogense deficiency | Up to 500 mg/day | Oral | |
| Simvastatin | HMGCoA-reductase inhibitor | Familial hypercholesterolaemia | Age < 10 years: 5 mg/day, increasing slowly to 20 mg/day, according to response<br>Age > 10 years: 10 mg/day, increasing slowly to 20–40 mg/day | Oral | Side effects may include reversible elevation of liver transaminases and myopathy |
| Sodium benzoate | Combines with glycine to form hippuric acid which has high renal clearance | Hyperammonaemia; NKH | Oral 250 mg/day in divided doses; IV 250 mg/day by continuous infusion. Dose may be doubled if severe hyperammonaemia or with NKH | Oral, IV | IV loading dose: 250 mg/kg over 90 min; high sodium load: 3.5 mmol in 500 mg of sodium benzoate |

**Appendix.** (*Contd.*)

| Medication | Mode of action | Disorders | Recommended dose | Route | Remarks |
|---|---|---|---|---|---|
| Sodium phenylbutyrate | Converted to phenylacetate, which combines with glutamine to form phenylglutamine, which has high renal clearance | Hyperammonaemia | 250–600 mg/kg/day in three or four divided doses IV by continuous infusion | Oral, IV | Maximum oral dose 20 g/day; high sodium load |
| Thiamine | Co-factor | Thiamine responsive variants of MSUD; CLA | 10–15 mg/day; doses of up to 300 mg in CLA | Oral | |
| Triethylene tetramine (trientine) | Chelating agent | Wilson's disease | 600 mg/day in divided doses, increasing to a maximum of 2.4 g/day if necessary | Oral | May reduce serum iron; iron supplements may be necessary |
| Ubiquinone (coenzyme Q10) | Improves oxidative phosphorylation (?) | Mitochodrial cytopathies | 100–300 mg/day | Oral | |
| Uridine | Replenishes pyrimidine stores | UMP synthase deficiency | 150 mg/kg | Oral | |
| | | 5′ nucleotidase superactivity | 1 g/kg | Oral | |
| Ursodeoxycholic acid | Suppression of endogenous bile-acid production | Disorders of bile-acid synthesis and anecdotal use as additive therapy for cholesterol in SLO syndrome | 15 mg/kg/day | Oral | |
| Vitamin C | Cofactor; antioxidant | Hawkinsinuria; tyrosinaemia III (4-hydroxyphenylpyruvate-dioxygenase deficiency); transient tyrosinaemia of the newborn; glutathione-synthase deficiency; CLA | 200–4000 mg/day | Oral | |
| Vitamin E (α-tocopherol) | Replenishes vitamin-E stores | Glutathione-synthetase deficiency | 10 mg/kg | Oral | |
| | | Aβ-lipoproteinaemia | 100 mg/kg | | |
| Zinc sulphate | Increases zinc | Acrodermatitis enteropathica | 30–100 mg Zn/day | Oral | Higher doses may be needed during periods of rapid growth |
| | Impairs copper absorption | Wilson's disease | 100–150 mg in three or four divided doses | | |

5HIAA, 5-hydroxyindoleacetic acid; AL, arginosuccinate lyase; AS, argininosuccinate synthetase; $BH_4$, tetrahydrobiopterin; CBS, cystathionine-β synthase; CDG, congenital defects of glycosylation; CLA, congenital lactic acidosis; CNS, central nervous system; CoA, coenzyme A; CPS, carbamoyl phosphate synthetase; CSF, colony-stimulating factor; DHA, docosahexaenoic acid; DHPR, dihydropteridine reductase; GAMT, guanidinoacetate methyltransferase; GSD, glycogen-storage disease; HMG, 3-hydroxy-3-methylglutaryl; HVA, homovanillic acid; MMA, methylmalonic acidaemia; NKH, non-ketotic hyperglycinaemia; NMDA, N-methyl-D-aspartate; NTBC, 2-[2-nitro-4-trifluoro-methylbenzoyl]-1,3-cyclohexanedione; OTC, ornithine transcarbamoylase; PDH, pyruvate dehydrogenase; PMI, phosphomannose isomerase; SLO, Smith-Lemli-Opitz; UMP, uridine monophosphate

# References

1. Acosta PB, Stepnick Gropper S, Clarke Sheehan N et al. (1987) Trace element status of PKU children ingesting an elemental diet. JPEN J Parenter Enteral Nutr 11:287–292
2. Safos S, Chang TM (1995) Enzyme replacement therapy in ENU2 phenylketonuric mice using oral microencapsulated phenylalanine ammonia-lyase: a preliminary report. Artif Cells Blood Substit Immobil Biotechnol 23:681–692
3. Brown-Harrison MC, Nada MA, Sprecher H et al. (1996) Very long chain acyl-CoA dehydrogenase deficiency: successful treatment of acute cardiomyopathy. Biochem Mol Med 58:59–65
4. Pollitt RJ (1995) Disorders of mitochondrial long-chain fatty acid oxidation. J Inherit Metab Dis 18:473–490
5. Morris AA, Clayton PT, Surtees RA et al. (1997) Clinical outcomes in long-chain 3-hydroxyacyl-coenzyme A dehydrogenase deficiency. J Pediatr 131:938
6. Irons M, Elias ER, Abuelo D et al. (1997) Treatment of Smith-Lemli-Opitz syndrome: results of a multicenter trial. Am J Med Genet 68:311–314

7. Moser HW (1995) Komrower lecture. Adrenoleukodystrophy: natural history, treatment and outcome. J Inherit Metab Dis 18:435-447
8. Davies SE, Iles RA, Stacey TE et al. (1991) Carnitine therapy and metabolism in the disorders of propionyl-CoA metabolism studied using 1H-NMR spectroscopy. Clin Chim Acta 204:263-277
9. De Sousa C, Chalmers RA, Stacey TE et al. (1986) The response to L-carnitine and glycine therapy in isovaleric acidaemia. Eur J Pediatr 144:451-456
10. Rutledge SL, Berry GT, Stanley CA et al. (1995) Glycine and L-carnitine therapy in 3-methylcrotonyl-CoA-carboxylase deficiency. J Inherit Metab Dis 18:299-305
11. Walter JH (1996) L-Carnitine. Arch Dis Child 74:475-478
12. Waber LJ, Valle D, Neill C et al. (1982) Carnitine deficiency presenting as familial cardiomyopathy: a treatable defect in carnitine transport. J Pediatr 101:700-705
13. Igisu H, Matsuoka M, Iryo Y (1995) Protection of the brain by carnitine. Sangyo Eiseigaku Zasshi 37:75-82
14. Jaeken J, Detheux M, Van Maldergem L et al. (1996) 3-Phosphoglycerate dehydrogenase deficiency: an inborn error of serine biosynthesis. Arch Dis Child 74:542-545
15. Vasiliauskas E, Rosenthal P (1994) Is acute fatty liver of pregnancy a metabolic defect? Am J Gastroenterol 89:1908-1910
16. Stockler S, Isbrandt D, Hanefeld F et al. (1996) Guanidinoacetate methyltransferase deficiency: the first inborn error of creatinine metabolism in man. Am J Hum Genet 58:914-922
17. Sullivan CA, Magann EF, Perry KGJ et al. (1994) The recurrence risk of the syndrome of hemolysis, elevated liver enzymes, and low platelets (HELLP) in subsequent gestations. Am J Obstet Gynecol 171:940-943
18. Bennett MJ, Weinberger MJ, Kobori JA et al. (1996) Mitochondrial short-chain L-3-hydroxyacyl-coenzyme A dehydrogenase deficiency: a new defect of fatty acid oxidation. Pediatr Res 39:185-188
19. Kreuder J, Otten A, Fuder H et al. (1993) Clinical and biochemical consequences of copper-histidine therapy in Menkes disease. Eur J Pediatr 152:828-832
20. Kaler SG, Buist NR, Holmes CS et al. (1995) Early copper therapy in classic Menkes disease patients with a novel splicing mutation. Ann Neurol 38:921-928
21. Niehues R, Hasilik M, Alton G et al. (1998) Carbohydrate-deficient glycoprotein syndrome type Ib. Phosphomannose isomerase deficiency and mannose therapy. J Clin Invest 101:1414-1420
22. Holme E, Lindstedt S (1998) Tyrosinaemia type I and NTBC (2-(2-nitro-4-trifluoromethylbenzoyl)-1,3- cyclohexanedione). J Inherit Metab Dis 21:507-517
23. Platt FM, Butters TD (1998) New therapeutic prospects for the glycosphingolipid lysosomal storage diseases. Biochem Pharmacol 56:421-430
24. Hamosh A, McDonald JW, Valle D et al. (1992) Dextromethorphan and high-dose benzoate therapy for nonketotic hyperglycinemia in an infant. J Pediatr 121:131-135
25. Alemzadeh R, Gammeltoft K, Matteson K (1996) Efficacy of low-dose dextromethorphan in the treatment of nonketotic hyperglycinemia. Pediatrics 97:924-926
26. Matsuo S, Inoue F, Takeuchi Y et al. (1995) Efficacy of tryptophan for the treatment of nonketotic hyperglycinemia: a new therapeutic approach for modulating the N- methyl-D-aspartate receptor. Pediatrics 95:142-146
27. Saitoh S, Momoi MY, Yamagata T et al. (1998) Effects of dichloroacetate in three patients with MELAS. Neurology 50:531-534
28. Takanashi J, Sugita K, Tanabe Y et al. (1997) Dichloroacetate treatment in Leigh syndrome caused by mitochondrial DNA mutation. J Neurol Sci 145:83-86
29. Elpeleg ON, Ruitenbeek W, Jakobs C et al. (1995) Congenital lacticacidemia caused by lipoamide dehydrogenase deficiency with favorable outcome. J Pediatr 126:72-74
30. Stacpoole PW, Barnes CL, Hurbanis MD et al. (1997) Treatment of congenital lactic acidosis with dichloroacetate. Arch Dis Child 77:535-541
31. Grabowski GA, Barton NW, Pastores G et al. (1995) Enzyme therapy in type 1 Gaucher disease: comparative efficacy of mannose-terminated glucocerebrosidase from natural and recombinant sources. Ann Intern Med 122:33-39
32. Zimran A, Elstein D, Levy-Lahad E et al. (1995) Replacement therapy with imiglucerase for type 1 Gaucher's disease. Lancet 345:1479-1480
33. Allgrove J (1997) Biphosphonates. Arch Dis Child 76, 73-75
34. Brady RO, Murray GJ, Oliver KL et al. (1997) Management of neutralizing antibody to ceredase in a patient with type 3 Gaucher disease. Pediatrics 100:E11
35. Thomson D (1993) Miracle drug: only $350,000 a year. (Ceredase). Time 141:54-55
36. Hobbs JR, Hugh-Jones K, Barrett AJ et al. (1981) Reversal of clinical features of Hurler's disease and biochemical improvement after treatment by bone-marrow transplantation. Lancet 2:709-712
37. Peters C, Shapiro EG, Anderson J et al. (1998) Hurler syndrome: II. Outcome of HLA-genotypically identical sibling and HLA-haploidentical related donor bone marrow transplantation in fifty-four children. The Storage Disease Collaborative Study Group. Blood 91:2601-2608
38. Prockop DJ (1997) Marrow stromal cells as stem cells for nonhematopoietic tissues. Science 276:71-74
39. Van't Hoff WG, Dixon M, Taylor J et al. (1998) Combined liver-kidney transplantation in methylmalonic acidemia. J Pediatr 132:1043-1044
40. Jamieson NV, Watts RW, Evans DB et al. (1991) Liver and kidney transplantation in the treatment of primary hyperoxaluria. Transplant Proc 23:1557-1558
41. Gruessner RW (1998) Preemptive liver transplantation from a living related donor for primary hyperoxaluria type I. N Engl J Med 338:1924
42. Nicolino MP, Puech JP, Kremer EJ et al. (1998) Adenovirus-mediated transfer of the acid alpha-glucosidase gene into fibroblasts, myoblasts and myotubes from patients with glycogen-storage disease type II leads to high level expression of enzyme and corrects glycogen accumulation. Hum Mol Genet 7:1695-1702
43. Blomer U, Naldini L, Kafri T et al. (1997) Highly efficient and sustained gene transfer in adult neurons with a lentivirus vector. J Virol 71:6641-6649
44. Charache S (1993) Pharmacological modification of hemoglobin F expression in sickle cell anemia: an update on hydroxyurea studies. Experientia 49:126-132
45. Dover GJ, Humphries RK, Moore JG et al. (1986) Hydroxyurea induction of hemoglobin F production in sickle cell disease: relationship between cytotoxicity and F cell production. Blood 67:735-738
46. Charache S, Dover G, Smith K et al. (1983) Treatment of sickle cell anemia with 5-azacytidine results in increased fetal hemoglobin production and is associated with nonrandom hypomethylation of DNA around the γδβ-globin gene complex. Proc Natl Acad Sci USA 80:4842-4846
47. Rubenstein RC, Egan ME, Zeitlin PL (1997) In vitro pharmacologic restoration of CFTR-mediated chloride transport with sodium 4-phenylbutyrate in cystic fibrosis epithelial cells containing delta F508-CFTR. J Clin Invest 100:2457-2465
48. Kemp S, Wei HM, Lu JF et al. (1998) Gene redundancy and pharmacological gene therapy: implications for X-linked adrenoleukodystrophy. Nat Med 4:1261-1268
49. Martinez M, Vazquez E (1998) MRI evidence that docosahexaenoic acid ethyl ester improves myelination in generalized peroxisomal disorders. Neurology 51:26-32
50. Kjaergaard S, Kristiansson B, Stibler H et al. (1998) Failure of short-term mannose therapy of patients with carbohydrate-deficient glycoprotein syndrome type 1A. Acta Paediatr 87:884-888
51. Mayatepek E, Schroder M, Kohlmuller D et al. (1997) Continuous mannose infusion in carbohydrate-deficient glycoprotein syndrome type I. Acta Paediatr 86:1138-1140
52. Ostman-Smith I, Brown G, Johnson A et al. (1994) Dilated cardiomyopathy due to type-II X-linked 3-methylglutaconic aciduria: successful treatment with pantothenic acid. Br Heart J 72:349-353

# PART II
# DISORDERS OF CARBOHYDRATE METABOLISM

## Glycogen Metabolism

Glycogen is the giant molecule in which glucose is stored in all tissues, most abundantly in liver and muscle. Synthesis and degradation of glycogen are catalyzed by enzymes that are activated (or inactivated) by hormones (Fig. 6.1).

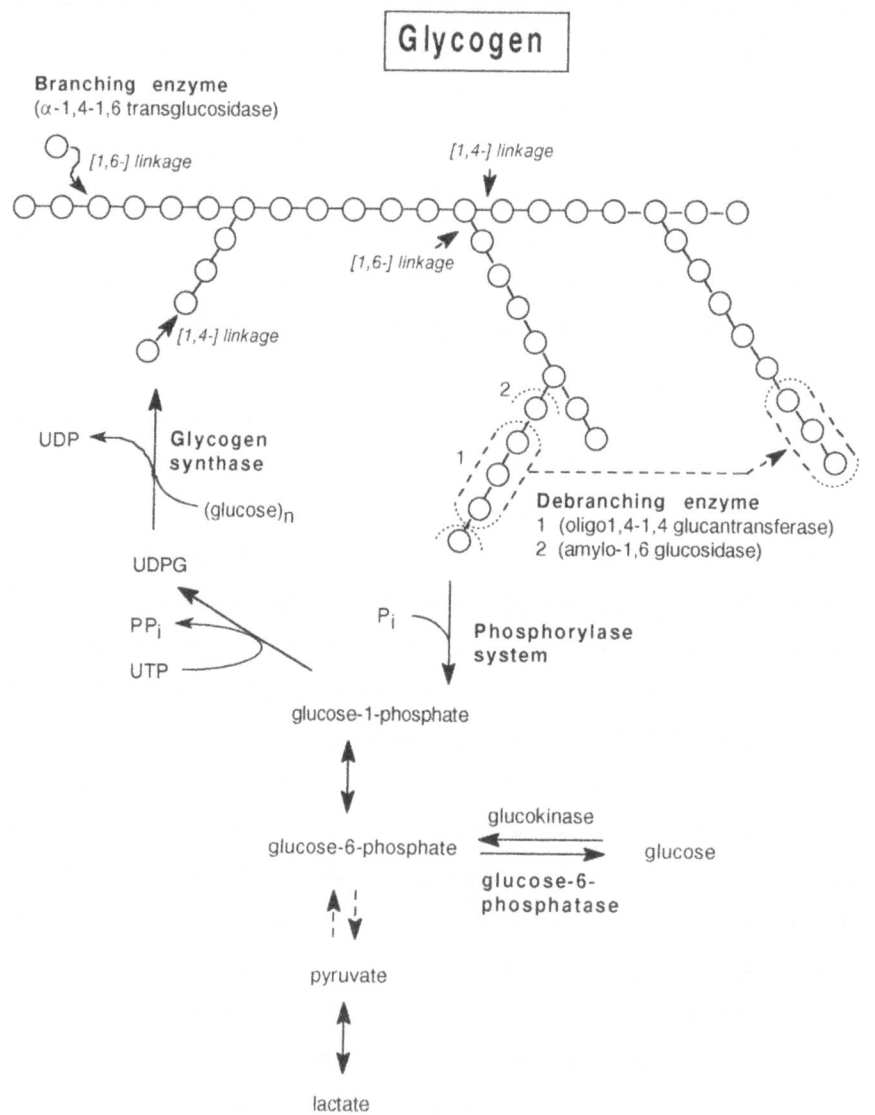

**Fig. 6.1.** The enzyme defects in glycogen metabolism. Glycogen synthase and branching enzyme are the enzymes for glycogen synthesis. The phosphorylase system, debranching enzyme and glucose-6-phosphatase are the enzymes for glycogen degradation and glucose liberation. Glycogen degradation by lysosomal α-glucosidase is not depicted in the figure. *Pi*, phosphate, *PPi*, pyrophosphate, *UDP*, uridine diphosphate, *UDPG*, uridine diphosphate glucose, *UTP*, uridine triphosphate

CHAPTER 6

# The Glycogen-Storage Diseases

J. Fernandes and G.P.A. Smit

CONTENTS

Glycogen Storage Disease-Ia:
Glucose-6-Phosphatase Deficiency,
Von Gierke's Disease ....................... 87
   Clinical Presentation......................... 87
   Metabolic Derangements .................... 88
   Diagnostic Tests ............................ 88
   Complications .............................. 89
      Liver Adenoma........................... 89
      Progressive Renal Failure.................. 89
      Gout..................................... 89
      Xanthomas............................... 89
      Pancreatitis.............................. 90
      Anemia.................................. 90
      Osteopenia .............................. 90
      Atherosclerosis........................... 90
      Polycystic Ovaries........................ 90
      Vascular Abnormalities ................... 90
   Treatment and Prognosis .................... 90
      Dietary Treatment ........................ 90
      Treatment with Drugs .................... 92
      Treatment of Complications................ 92
   Genetics ................................... 93

Glycogen Storage Disease-Ib:
Glucose-6-Phosphate Translocase Deficiency.... 93

Glycogen Storage Disease-II:
Lysosomal α-1,4-Glucosidase Deficiency,
Pompe's Disease............................. 94

Glycogen Storage Disease-III:
Debranching-Enzyme Deficiency............... 95

Glycogen Storage Disease-IV:
Branching-Enzyme Deficiency, Amylopectinosis 96

Deficiencies of the Phosphorylase System:
Phosphorylase Deficiency of the Liver (Glycogen
Storage Disease-VI) and Phosphorylase-B-
Kinase Deficiency of the Liver
(Glycogen Storage Disease-IX)................ 97

Muscle Glycogen Storage Disease ............. 98
   Glycogenolytic Defects Located in Muscles......... 99
   Glycolytic Defects Located in Muscles............. 99

Other Rare Glycogen Storage Disease Types.... 99
   Glycogen Storage Disease-0:
   Glycogen Synthase Deficiency................... 99
   Hepatorenal Glycogen Storage Disease with the
   Fanconi-Bickel Syndrome...................... 99

References ................................. 99

The glycogen-storage diseases (GSDs) are caused by enzyme defects of glycogen degradation. Some enzyme defects are mainly localized in the liver. Hepatomegaly and hypoglycemia are the main abnormalities. Some enzyme defects are localized in the muscles. Muscle cramps during exertion, progressive weakness and other features of myopathy are predominant. Other enzyme defects are generalized, and they present with variable symptoms related to the organs that are most seriously involved. GSDs are denoted either by the deficient enzyme or by a type number that reflects the historical sequence of elucidation. The latter presentation is used in this chapter. The main characteristics of each GSD type are summarized in Table 6.1.

## Glycogen Storage Disease-Ia: Glucose-6-Phosphatase Deficiency, Von Gierke's Disease

### Clinical Presentation

"Hepatonephromegalia glykogenia" was the title of the first description by von Gierke [1]. A protruded abdomen (because of marked hepatomegaly), truncal obesity, a rounded "doll face", hypotrophic muscles and short stature are conspicuous clinical findings. The liver may already be enlarged at birth. Its size increases gradually, and the lower border may extend to well below the umbilicus.

Initially, the liver has a normal consistency and a smooth surface but, after the age of ~15 years, the surface may become uneven and the consistency much firmer because of the development of adenomas. Cirrhosis does not develop. The kidneys are moderately enlarged, whereas the spleen remains normal sized. Usually, the patient's growth lags behind unless intensive dietary treatment is started early. The patient bruises easily, and nosebleeds may be troublesome due to impaired platelet function [2]. Episodes of diarrhea or loose stools are presumably due to impaired active absorption of glucose [3]. Profound hypoglycemia

**Table 6.1.** Classification of glycogen-storage diseases (GSDs)

| Type | Defective enzyme or transporter | Tissue involved | Main clinical symptoms |
|---|---|---|---|
| Ia | Glucose-6-phosphatase | Liver, kidney | Hepatomegaly, hypoglycemia, lactic acidosis, hyperlipidemia |
| Ib | Glucose-6-phosphate translocases | Liver, leukocytes | In addition to GSD-Ia symptoms, neutropenia, infections, IBD |
| II | Acid α glucosidase | Generalized | Infant form: cardiomyopathy, hypotonia<br>Later forms: myopathy |
| III | Debranching enzyme | Liver, muscle | Hepatomegaly, hypoglycemia, myopathy |
| IV | Branching enzyme | Liver | Hepatosplenomegaly, cirrhosis |
| VI | Phosphorylase | Liver | Hepatomegaly, hypoglycemia |
| IX | Phosphorylase-B kinase | Liver | Hepatomegaly, hypoglycemia |
|  | Glycogenolytic and glycolytic defects | Muscle (erythrocytes) | Myopathy, exercise intolerance (hemolytic anemia) |
| 0 | Glycogen synthase | Liver | Hypoglycemia |
| Fanconi-Bickel | GLUT-2 | Liver, kidney | Hepatomegaly, rickets, hypoglycemia, tubulopathy |

*GLUT*, glucose transporter; *IBD*, inflammatory bowel disease

occurs frequently and can be elicited by trivial events, such as a short delay of a meal or a lower food intake induced by an intercurrent illness. Hypoglycemic symptoms are usually accompanied by hyperventilation, a symptom of lactic acidosis. Exceptionally, hypoglycemic symptoms do not occur even during prolonged fasting, though other clinical features and metabolic abnormalities persist. This resistance to fasting appears to be due to residual glucose-6-phosphatase activity and probably not to glycogen cycling via the concerted action of glycogen synthase and debranching enzyme (Fig. 6.1) [4]. Long-term cerebral function is normal if hypoglycemic damage is prevented. Complications are numerous; as they are usually related to the metabolic derangements, they are discussed below.

## Metabolic Derangements

Hypoglycemia, hyperlactacidemia, hyperlipidemia and hyperuricemia are the most characteristic metabolic derangements. Hypoglycemia occurs as soon as exogenous sources of glucose are exhausted, because the enzyme defect between glucose-6-phosphate and glucose blocks glucose release from both glycogenolysis and gluconeogenesis (Fig. 6.2).

However, the degradation of glycogen to pyruvate is intact and is intensified under hormonal stimulation as soon as the provision of glucose fails. The resulting increased pyruvate and lactate production is a useful mechanism as long as pathological lactic acidosis does not develop, because lactate may serve as a fuel for the brain [5]. This alternate brain fuel may protect some patients against cerebral symptoms, even when the blood glucose concentration is very low.

More important is the increased synthesis of fatty acids and cholesterol from the conversion of excess lactate/pyruvate into acetyl-coenzyme A (CoA). Excess acetyl-CoA is converted to malonyl-CoA, a potent activator of liponeogenesis and inhibitor of fatty-acid oxidation. Thus, hyperlipidemia is observed in most patients, and the serum may even show a milky appearance. Serum triglycerides predominate; cholesterol (esters) and phospholipids are less elevated. The hyperpre-β-lipoproteinemia is characterized by high levels of apolipoproteins (Apos) B, C (particularly C-III) and E, and low levels of Apo A and D [6]. Ketosis does not occur, in contrast with the situation in the other types of GSD [7]. This is due to the elevation of malonyl-CoA. Elevated malonyl-CoA, by inhibiting carnitine palmitoyl transferase 1, prevents entry of long-chain fatty acyl-CoA into the mitochondria, and hence its oxidation and conversion into ketone bodies.

Hyperuricemia is another metabolic derangement. It is caused by both a decreased renal clearance of urate by inhibition from lactic acid and increased production of uric acid. The overproduction of uric acid is elicited by a mechanism similar to that elicited by fructose administration in a patient with hereditary fructose intolerance (Chap. 8).

## Diagnostic Tests

The clinical approach is to perform an oral glucose-tolerance test and to determine the blood glucose and lactate concentrations. It is not the glucose but the lactate curve that provides the clue for the diagnosis. The initially increased lactate concentration (resulting from increased gluconeogenesis during fasting) drops when glucose increases in a reciprocal way. This is the opposite of the normal situation, in which blood lactate is low at fasting and increases slightly. For details of the test, see Chap. 2. Some authors prefer a glucagon test, which shows a flat or descending blood glucose curve and a markedly rising lactate curve. However, in view of the risk of hypoglycemia due to the continuation of fasting, this test is not recommended as a *first* test. It might be used as a second test to differentiate other GSDs. A biochemical assay of glucose-6-phosphatase activity in a liver biopsy is indicated to confirm the tentative diagnosis.

Differentiation between deficiency of glucose-6-phosphatase itself (GSD-Ia) and that of other proteins of the multi-enzyme system is very important (see "GSD-Ib"). At histological examination, fat accumulation often predominates over that of glycogen. Confirmation of the diagnosis is obtained by DNA analysis.

## Complications

Many complications have been observed (particularly at adult age [8]), such as liver adenoma, proteinuria and progressive renal failure, kidney stones, gout, xanthomas, pancreatitis, anemia, osteopenia, ovarian cysts and vascular abnormalities.

### Liver Adenoma

Liver adenoma, single or multiple, develops in the majority of patients in their second decade [9]. Its presence can be suspected because of palpation of one or more liver nodules and can be confirmed by ultrasound. Liver adenomas are mostly benign, but malignant transformation (carcinoma) may occur. Acute hemorrhage in a liver adenoma has also been reported. It should be treated with continuous intravenous glucose administration during a few days in order to restore the presumed abnormal thrombocyte aggregation. In view of the risk of malignancy, serum α-fetoprotein, a tumor marker, and ultrasound of the liver should be determined regularly (at least once per year). A nodule that increases in size or changes from circumscribed to having poorly defined margins, or the appearance of a "nodule in a nodule" should be checked by other investigations, such as computed tomography (CT) scans or magnetic resonance imaging. Enucleation of a carcinoma or orthotopic liver transplantation is the final option. The cause of adenoma development is not known. It is speculated that hyperglucagonemia or long-term toxicity of excess of substrates are contributing factors, and this speculation is supported by regression of adenomas following intensive dietary treatment. This regression is transient, however, as adenomas usually remain constant during many years of optimal treatment.

### Progressive Renal Failure

Focal glomerulosclerosis is a complication at adult age. It starts with a "silent" glomerular hyperperfusion and hyperfiltration, followed by microalbuminuria. This slowly evolves to overt proteinuria and a decreased glomerular filtration rate, due to focal segmental glomerulosclerosis and interstitial fibrosis [10]. Deteriorating kidney function and hypertension are the final results, and hemodialysis and kidney transplantation are the ultimate therapeutic options. Glomerulosclerosis, with its long latency, is observed in almost all older patients. Its cause is still unknown. Other renal abnormalities include both proximal and distal tubular dysfunction [11], of which the latter is frequently associated with hypercalciuria. Hypercalciuria may contribute to kidney stones and osteopenia [12].

### Gout

Hyperuricemia occurs in almost all patients and may lead to gout and urate kidney stones.

### Xanthomas

Xanthomas may be observed on the buttocks, elbows and knees, similar to those observed in some primary lipid disorders (Chap. 11). Its occurrence has become

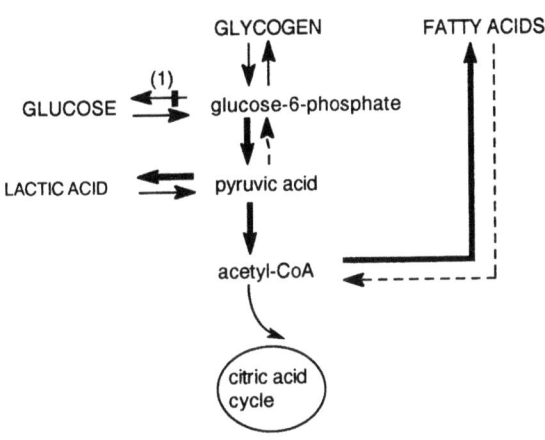

**Fig. 6.2.** Simplified presentation of the main metabolic derangements in glucose-6-phosphatase deficiency. *Bold arrows* denote increased flow

rare since the introduction of intensive dietary treatment.

### Pancreatitis

Pancreatitis, reflected by bouts of abdominal pain and diarrhea, can be elicited by severe hyperlipidemia. Its presence can be "screened" by looking for elevated levels of serum amylase, lipase and trypsin and can be proved by CT and endoscopic retrograde cholangiopancreatography [13].

### Anemia

Older patients are prone to develop normochromic anemia. Its cause is not known [14].

### Osteopenia

Osteopenia is frequently observed. Two factors may contribute: hypercalciuria and increased desorption of calcium from bones, elicited by the chronic lactic-acid load. Bone densitometry should be performed at least once in every older patient.

### Atherosclerosis

Atherosclerosis is remarkably rare despite the chronic hyperlipidemia and the atherogenic profile of serum lipoproteins [6]. The conservation of normal arterial endothelial function might be due to diminished platelet interaction with the vascular endothelium [15].

### Polycystic Ovaries

Polycystic ovaries have been observed in adolescent female patients. Ultrasound might be performed in that age group to search for this complication [16].

### Vascular Abnormalities

*Pulmonary hypertension* followed by progressive heart failure in the second decade of life is a rare fatal complication ([17] and our own observations). *Moyamoya disease*, manifested by acute hemiplegia, has been reported thrice [18]. As both disorders are caused by vasoconstrictive processes, it would be justified to look for vascular abnormalities in each GSD-Ia patient.

## Treatment and Prognosis

### Dietary Treatment

The aim of dietary treatment is to prevent hypoglycemia and suppress secondary metabolic derangements as much as possible. The methods employed to achieve this have changed considerably during the last 20 years. The first breakthrough was the introduction of nocturnal drip feeding via a nasogastric tube [19]. It allowed the patient and the parents to sleep during the night instead of waking up every 2–3 h for obligatory feeding. The second improvement was the introduction of uncooked starch, from which glucose is much more protractedly released than from cooked starch [20].

Both methods have been compared during follow-up studies, and the results are equally favorable [21]. Their application is shown in Table 6.2.

The glucose requirements for formula, gastric drip feeding, meals containing precooked starch (cereals, etc.) and in uncooked starch are calculated from the theoretical glucose production rate, which decreases with age. Three elements of the diet need special consideration.

### Formula Feeding

If breast milk is not available, a sucrose-free, milk-based formula with a low lactose content is enriched with maltodextrin until the calculated maltodextrin requirement is reached. This sugar is gradually replaced by precooked starch (rice, corn) up to a maximum of 6% in order to prolong gastric emptying time. This might allow a wider spacing of the feeding frequency from 2-h to 3-h intervals to 3-h to 4-h intervals.

### Gastric Drip Feeding

Gastric drip feeding at night may be introduced in young infants at the time of the diagnosis. The feeding may be identical to the above-mentioned formula, or it may contain only glucose or a glucose polymer in water. Theoretically, the former complete formula is preferable.

Gastric drip feeding should not start later than 1 h after the last meal, and breakfast should be within 30 min of removal of the drip tube. The technical outfit of the pump and the connection of the tube and its fixation to the patient should be meticulously explained to the parents, as fatal outcome due to disconnection of the system has been described. Some parents cannot cope with the technical and emotional implications of infusion pumps and tube feeding and prefer to switch to uncooked cornstarch.

### Uncooked Starch

Initially, uncooked cornstarch was used to replace nocturnal drip feeding in order to allow adolescents more freedom from their tight dietary schedules. Its effect is based on the fact that glucose is slowly

Table 6.2. Feeding schedule of patients with glycogen-storage disease I

| Age | Schedule | Nocturnal gastric drip[a] | Starch processed in meals | Uncooked starch (g/kg body weight) | Glucose[b] (mg/kg/min) |
|---|---|---|---|---|---|
| 0–12 months | Formula, 2- to 3-h intervals | Necessary | Rice, corn 1–6% in low-lactose and sucrose-free formula | | 7–9 |
| 1–3 years | 3 meals, 2 snacks | 35% energy in 12 h | Cereals, bread, rice, macaroni, legumes | Cornstarch 2 × 1.0–1.5 | 7 |
| 3–6 years | 3 meals, 2 snacks | See above | See above | Slow: 2 × 1.75–2; semislow[c]: 2 × 1–1.5 | 6–7 |
| 6–14 years | 3 meals, 2 snacks | 30% energy in 10 h | See above | See above | 5–6 |
| Adolescents | 3 meals, 1–2 snacks | Possibly | See above | Slow: 2 × 1.5; semislow: 2 × 1.0 | 4–5 |
| Adults | See above | | At night | Slow 2 × 1.5 | 3–4 |

[a] Gastric drip feeding and uncooked starch are mutually exchangeable or may be complementary
[b] The total requirement of glucose expressed in this way is used for the calculation of the amount of glucose in gastric drip feeding, starch in meals or uncooked starch
[c] One or two doses of slow starch can be interchanged for semislow starches (see text)

released and absorbed from uncooked cornstarch, so normoglycemia may be maintained for 6–8 h, instead of 3 h after an equivalent intake of glucose in water. Cornstarch can be introduced directly at age 1 year, the approximate age that pancreatic-amylase activity has sufficiently matured [22]. Side-effects of bowel distension, flatulence and loose stools are usually transient and can be mitigated by slowly increasing the dose. For infants under 2 years of age, the cornstarch can be introduced at a dose of 1.0–1.5 g/kg every 4 h. The response is variable. As the child grows older, the regimen can be changed to every 6 h, and the dose can be increased to 1.75–2.0 g/kg. The starch is mixed with water in a starch:water ratio of 1:2. Adding glucose is contraindicated, as its insulin stimulation offsets the advantage of the starch. However, mixing the uncooked starch with yogurt, diet drink or milk increases its palatability for some children without significantly affecting the period of euglycemia. The efficacy of cornstarch in keeping the blood glucose concentration at 3.5 mM (63 mg/dl) or higher is investigated by means of a starch-tolerance test. Its procedure is similar to that of a glucose-tolerance test (Chap. 2). Depending on the results of the test, the 4- or 6-h dose of the diet is adapted (Table 6.2). With a large amount of cornstarch consumed, the dietary plan to provide other essential nutrients for growth and development needs to be carefully designed and followed. Cornstarch can be interchanged with other "lente" carbohydrates with similar slow-releasing properties, such as rice, wheat or tapioca.

Another approach is to prescribe lente carbohydrates twice per night and "semilente" carbohydrates twice during the daytime to cover the periods between meals. Semilente carbohydrates, which release glucose during a 4-h period, are: cooked rice, macaroni, rolled oats, barley groats, millet, couscous, legumes and lentils. Variation between lente and semilente carbohydrates may prevent the "cornstarch monotony" that would occur if only that starch were administered four times per day.

With respect to "rapid" carbohydrates, the consumption of lactose is limited to the amount present in 0.5 l of milk per day, as milk and dairy products are important sources of protein and calcium. Sucrose and fructose are prohibited except from fruits, as these sugars enhance the production of lactate [23]. The total amount of carbohydrates should provide 60–65% of the total energy intake; protein should provide 10–15%, and fat should provide the remainder.

The laboratory parameters used to control the adequacy of the dietary treatment are summarized in Table 6.3. In this approach, slightly elevated blood lactate levels between 2.0 mM and 5.0 mM (normal < 2.0 mM) are acceptable. Higher levels may reflect insufficient carbohydrate intake or excessive spacing of the meals. Lower levels, though preferred theoretically, may reflect overtreatment with carbohydrate, which increases susceptibility to preprandial hypoglycemia. The lactate concentrations in urine, collected at home in 12-h portions or as freshly voided samples and delivered to the lab in the frozen state, provide valuable information about the adequacy of the diet in the home situation. The optimal lactate concentration is no greater than 0.6 mM, but slightly higher concentrations of 0.6–1.0 mM are acceptable.

The dietary refinements of gastric drip feeding and uncooked cornstarch are very effective at inducing catch-up growth or maintaining growth close to the expected target curve in most patients. However, a small group of "hyporesponders" exists, in whom growth lags behind despite all efforts to attain optimal

**Table 6.3.** Laboratory parameters for dietary adjustment of glycogen-storage-disease-I patients

| Parameter | Value |
|---|---|
| **Blood** | |
| Glucose profile during the daytime before meals and during gastric drip feeding | ≥3.5 mM (63 mg/dl) |
| Lactate before each meal or 2 h after the start of drip feeding | 2.0–5.0 mM (18–45 mg/dl) |
| **Urine** lactate concentration | |
| Optimal | ≤0.6 mM |
| Acceptable | 0.6–1.0 mM |

metabolic adjustment [24]. In these patients, urinary lactate is higher than in responders although, even in the latter group (which has a more favorable course), serum lactate, lipids and urate usually do not normalize.

For adults, two doses of 1.5 g uncooked starch/kg body weight both before and halfway through the night is the safest approach [25]. However, many adults prefer to take only one dose before night and do not experience untoward symptoms. Each of the two possibilities has to be tried to determine the safest choice.

### Treatment with Drugs

Diazoxide, an insulin inhibitor that increases blood glucose levels in patients with hyperinsulinemic hypoglycemia, has been used in two prepubertal girls; it had favorable effects on some metabolic parameters and growth [26]. However, spontaneous pubertal growth could have coincided with the growth attributed to the drug. Therefore, this observation does not justify use of a drug known for its side-effects. Growth hormone for the treatment of patients with retarded growth has not been proven to be of any benefit.

### Treatment of Complications

### Progressive Renal Failure

Preventive treatments for this late complication are not yet known. However, intensive dietary treatment might delay and abate the symptoms [27]. In analogy to diabetic nephropathy, patients with proteinuria are treated with a moderate restriction of protein. More promising is the treatment with an inhibitor of angiotensin-converting enzyme (ACE), for instance Enalapril. It induces intrarenal efferent vasodilatation, thereby reducing intraglomerular capillary pressure. The decreased filtration pressure presumably contributes to the antiproteinuric effect and to long-term renoprotection. The dosage of the ACE inhibitor should be low, and renal function should be monitored at regular intervals.

### Hyperuricemia

Hyperuricemia, defined as a serum urate concentration greater than 0.36 mM (6 mg/dl), should be treated with Allopurinol, a xanthine-oxidase inhibitor, at a dosage between 10–15 mg/kg/day orally. Alkalinization of urine with 1–2 mmol (85–170 mg)/kg/day $NaHCO_3$ is optional.

### Hyperlipidemia

Premature atherosclerosis does not appear to be a feature of GSD-I. Nevertheless, a low-fat, low-cholesterol diet and the use of vegetable oils with high linoleic-acid contents is emphasized, although this regime in itself does not improve hyperlipidemia. In short-term experiments, the administration of 10 g of fish oil per day lowered serum triglycerides and cholesterol markedly and improved the lipoprotein profile, presumably by enhancing fat catabolism [28].

### Osteopenia

The risk of osteopenia is probably increased, because chronic lactacidemia and the gradual development of renal insufficiency are expected to promote calcium desorption and to limit peak bone mass at adult age. Therefore, preventive measures are indicated: 10–20 μg (400–800 IU)/day vitamin $D_3$ in countries with a high latitude, and 0.5–1.0 g/day calcium in cases of limited milk intake.

### Intercurrent Infections

During infections, dietary treatment is endangered because of anorexia and vomiting. This may lead to hypoglycemia and lactic acidosis, which both increase nausea. The parents should administer a small amount of a polycose syrup (low osmolarity) while bringing their child to the hospital. For further treatment of hypoglycemia, see Chap. 3.

### The Preparation for Major Elective Surgery

Bleeding time and platelet adhesiveness must be investigated before an operation. If abnormal, continuous gastric drip feeding for one week or intensive intravenous glucose infusion for 24–48 h should be instituted to normalize the bleeding tendency prior to surgery.

## Liver Transplantation

Orthotopic liver transplantation followed by complete resolution of the metabolic derangements is a final option for patients with malignancy of a liver adenoma and for patients not responding to dietary treatment [29]. It is not yet known whether the kidney abnormalities regress. It is too early to judge the effect of combined liver and kidney transplantation.

## Preganancies

Pregnancies under close supervision and with intensive dietary treatment have terminated with the premature birth of low-birthweight children who developed normally ([30] and our own observations). The liver adenomas of the mothers did not change.

In conclusion, dietary treatment and a few drugs have considerably improved the prognosis of this GSD-I, and a few patients have passed the age of 50 years. However, the late kidney abnormalities are of great concern.

## Genetics

The inheritance is autosomal recessive. The gene that encodes for the catalytic unit of human glucose-6-phosphatase is mapped to chromosome 17q21. Several mutations have been identified [31], indicating allelic heterogeneity. Antenatal diagnosis is feasible by DNA assay of chorionic villi.

## Glycogen Storage Disease-Ib: Glucose-6-Phosphate Translocase Deficiency

Glucose-6-phosphatase is localized at the inner luminal wall of the endoplasmic reticulum of the cell, whereas the other glycogenolytic enzymes are in the cytoplasm. This means that the products of the latter enzymes must cross the membrane of the endoplasmic reticulum to allow glucose-6-phosphatase to act. A few proteins for this transport have been isolated: T1 for the entry of glucose-6-phosphate, T2 for the exchange of pyrophosphate and phosphate and T3 for the export of glucose. The latter protein belongs to a family of plasma-membrane facilitative glucose-transport proteins, termed GLUTs. Deficiencies have been described in all three translocases, known as GSD Ib, GSD Ic, and GSD Id, respectively. However, new evidence contradicts the existence of GSD Ic and GSD Id as separate entities, since they retrospectively do not differ from GSD Ib clinically, enzymatically, or genetically [32]. Therefore, GSD-Ib, GSD-Ic and GSD-Id are at present categorized under one heading: GSD-Ib.

The relative incidence of GSD-Ia:GSD-Ib is in the order of 5–10:1. In a large Collaborative European Study of GSD-I, 56 patients with GSD-Ib were found among a total of 256 GSD-I patients.

Patients with GSD-Ib are not clinically or metabolically discernible from those with GSD-Ia, except for an additional susceptibility to infections and immunologic abnormalities. The description that follows is limited to those abnormalities that differ from those in GSD-Ia.

## Clinical Presentation and Complications

Bacterial infections (cutaneous, pulmonary) occur frequently. Even brain abscesses have been reported [8]. A decreased number of neutrophils and defective neutrophil and monocyte functions usually underlie the infections. The number of neutrophils is usually below $1500/\mu l$. Bone-marrow examination shows hypercellularity instead of the expected hypocellularity. The main complication is inflammatory bowel disease, which closely resembles Crohn's disease. It is usually preceded or accompanied by oral, perioral and perianal infections with fistulae and abscesses. Splenomegaly occurs in one third of the patients. Other complications are systemic fungemia [8], acute myelogenous leukemia [33] and generalized amyloidosis [34].

## Metabolic Derangements and Diagnosis

As the metabolic derangements do not permit one to differentiate GSD-Ia from GSD Ib, biochemical assay of a liver biopsy and/or DNA analysis is mandatory for obtaining the diagnosis. A fresh liver biopsy in which the hepatocytes and their microsomes are intact shows deficient glucose-6-phosphatase activity, as the mutated translocase does not deliver the required substrate into the microsomal lumen. However, microsomes disrupted by solubilization show normal glucose-6-phosphatase activity, because all substrates have free entry to the enzyme. Thus, it is emphasized that consultation of the biochemist before performing the biopsy is crucial for a differentiation between GSD Ia and GSD Ib.

## Treatment and Prognosis

Treatment with antibiotics (cotrimoxazole might be used simultaneously) is indicated for patients with infections but without inflammatory bowel disease. In the latter serious condition, the use of recombinant human granulocyte-colony-stimulating factor (GCSF) reverses the gradually deteriorating clinical course of the disorder, causing rapid abatement of infections and the healing of abscesses, ulcers and inflammatory bowel disease [35]. Impressive weight gain and improved vitality occur simultaneously. GCSF is a cytokine that induces proliferation and differentiation of bone-marrow precursor cells into mature neutrophils. It prolongs the survival of neutrophils and enhances several functions, but not all. We recommend GCSF in the non-glycosylated form in doses of 3–10 μg/kg body weight subcutaneously two to four times per week. This frequency can gradually be tapered off to one or two times per week, depending on its clinical effects and the occurrence of side effects like splenomegaly, arthralgia and bone pain, which gradually disappear at lower doses. The duration of treatment of our patients is up to 6 years. The choice between treatment with GCSF and/or antibiotics depends on the severity of the disease and the side-effects of GCSF treatment.

### Genetics

The inheritance of GSD-Ib is autosomal recessive. The gene that encodes for glucose-6-phosphate translocase is mapped to chromosome 11q23 [36].

## Glycogen Storage Disease-II: Lysosomal α-1,4-Glucosidase Deficiency, Pompe's Disease

### Clinical Presentation and Complications

Three entities exist: infantile, juvenile and adult forms, with transitional forms in between. The *infantile form* is the most severe. It presents with profound hypotonia, muscle weakness, hyporeflexia and an enlarged tongue. The heart is extremely enlarged, and the electrocardiogram is characterized by huge QRS complexes and shortened PR intervals. There are usually no cardiac murmurs. The liver has a normal size unless enlarged by cardiac decompensation. The cerebral development is normal. The clinical course is rapidly downward, and the child dies from cardiopulmonary failure or aspiration pneumonia in the first year of life.

The *juvenile form* shows retarded motor milestones, hypotonia and weakness of limb girdle and truncal muscles, but shows no overt cardiac disease. Myopathy deteriorates gradually, and the patient dies from respiratory failure before adult age.

The *adult form* mimics other myopathies with a long latency. Decreased muscle strength and weakness develop in the third or fourth decade of life. Cardiac involvement is minimal or absent. The slow, progressive weakness of the pelvic girdle, paraspinal muscles and diaphragm results in walking difficulty and respiratory insufficiency, but old age can be attained. Experience with the adult form has increased during the last years, leading to the detection of hitherto unknown complications. Rupture of aneurysms of cerebral arteries (due to accumulation of glycogen in vascular smooth muscle) with fatal outcome has been observed [37].

The mental development of all three phenotypes is normal. For a review of the three phenotypes and genotypes, see Kroos et al. [38]. Metabolic abnormalities are not present in any form, because the lysosomal enzyme defect is outside intermediary metabolism.

### Diagnosis

In the infantile form, a tentative diagnosis can be based on the typical electrocardiogram abnormalities. For confirmation, α-1,4-glucosidase should be determined in tissues containing lysosomes. The preferred tissues are fibroblasts or muscle. The enzyme has its pH optimum between 4.0 and 4.5; the acid environment of lysosomes and the activity of this "acid maltase" must be differentiated from contamination with a nonspecific cytosolic neutral maltase. Although the enzyme defect is generalized in all three GSD-II types, the site of glycogen accumulation is different in each, and the amount varies greatly in different organs and even in different muscles. Residual enzyme activity is found in the adult form, whereas the enzyme is absent in the infantile form. At histological examination of a muscle biopsy, large glycogen-laden vacuoles surrounded by a membrane are found next to freely dispersed glycogen outside the lysosomes. Gradually, the huge vacuoles lead to impairment of cell function. Prenatal diagnosis is possible by enzyme assay or DNA analysis of chorionic villi.

### Treatment

So far, there is no effective treatment. Improvement of muscle function has been obtained by a high-protein diet, particularly a high-protein diet fortified with branched-chain amino acids [39]. This treatment presumably diminishes the catabolism of muscle protein. Imminent respiratory insufficiency in the adult type can be successfully treated with assisted ventilation. Enzyme-replacement therapy might become available in the near future.

### Genetics

The mode of inheritance is autosomal recessive. The incidence is estimated at ~1:50,000. The gene for acid α-glucosidase is mapped to 17q25.2-q25 [40]. A large heterogeneity has been found in adults [41] and even within one family [42].

# Glycogen Storage Disease-III: Debranching-Enzyme Deficiency

## Clinical Presentation and Complications

The clinical features vary in relation to the localization of the enzyme defect. Two clinical entities exist: a combined hepatic–myogenic form (GSD-IIIa) and a purely hepatic form of the disease (GSD-IIIb). Although GSD-IIIb is the less frequently observed form, it will be discussed first for clarity's sake.

GSD-IIIb occurs in ~15% of the patients. An enormously enlarged liver and protruding abdomen, which can be observed shortly after birth, are the most striking symptoms. Truncal obesity and a doll face develop, and the child closely resembles patients with GSD-I. Profound hypoglycemia may occur – for instance, if a meal is inadvertently delayed or if food intake is hampered by anorexia during an infection. The liver size decreases gradually in the course of childhood and is barely palpable at puberty. Serum transaminases also decrease from markedly elevated to the normal range. The reason for this "normalization" is not known. However, complications may occur. A smooth surface of the liver may become nodular, and liver adenoma (identified by ultrasonography) was detected in 14 of 27 GSD-III patients [9]. Cirrhosis may develop from periportal fibrosis, and hepatocellular carcinoma has been (rarely) observed [43]. Hypoglycemia, most frequent and menacing at a young age, becomes milder at an increased age, and this compares favorably with GSD-I patients.

GSD-IIIa occurs in ~78% of the patients [44]. Children who are very weak in childhood may improve during adolescence. Other patients experience their first symptoms of weakness after their third decade, many years after the liver abnormalities have abated. Weakness starts proximally and extends to distal muscles. Muscle wasting often has both myopathic and neurogenic origins. Increased levels of serum creatine kinase are a reliable indicator of muscle involvement. The generalized myopathy includes cardiomyopathy in many patients [45] and is usually preceded by electrocardiographic abnormalities. The hepatic abnormalities closely resemble those of GSD-IIIb. Mental development is usually normal. Complications other than those stemming from the liver are polycystic ovaries [16], obstructive hypertrophic cardiomyopathy [46], renal tubular acidosis [47], kidney stones, proteinuria and hyperuricemia [8].

## Metabolic Derangements

Glycogen degradation, initiated by the phosphorylase system, stops at a distance of four glucose molecules from the outer branches of the glycogen molecule (Fig. 6.1). Next, debranching enzyme (having two catalytic sites) acts. First, the transferase transfers the three outer glucose molecules to another chain. Second, the glucosidase detaches the last glucose molecule from the branch. Both catalytic sites are simultaneously affected in patients with GSD-IIIa or GSD-IIIb. Exceptionally, only the transferase is deficient in liver and muscles. The clinical presentation of this rare GSD-IIId is identical to that of GSD-IIIa. The site of the enzyme defect implies limited liberation of glucose from the outer branches of glycogen in cases of fasting. This results in hypoglycemia unless the fuel deficit is replenished by glucose from increased gluconeogenesis and ketones from increased fatty-acid oxidation (Fig. 6.3). Indeed, ketosis in GSD-III and lactic acidemia in GSD-I are important distinctions in the metabolic fasting profiles of both disorders [7].

Limit dextrin, produced by partial degradation of glycogen, is supposed to behave like a foreign body and to trigger the development of liver fibrosis, cirrhosis and adenoma. Another metabolic derangement is hyper-β-lipoproteinemia with high cholesterol levels but without excessive triglyceride levels.

## Diagnostic Tests

An oral glucose test is performed to differentiate GSD-III from GSD-I and other enzyme defects of gluconeogenesis. A moderate increase of serum lactate is observed in GSD-III, whereas serum lactate drops precipitously in GSD-I. An oral galactose test shows a pronounced increase of serum lactate, whereas the

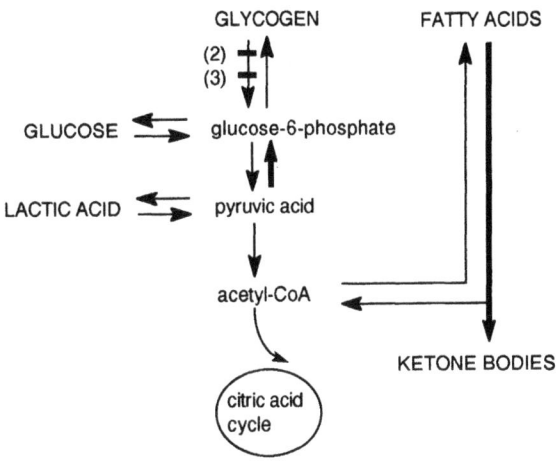

(2) phosphorylase and phosphorylase-b-kinase
(3) debranching enzyme

**Fig. 6.3.** Simplified presentation of the main metabolic derangements in deficiency of phosphorylase or phosphorylase-B kinase (2) and deficiency of debranching enzyme (3). *Bold arrows* denote increased flow

ensuing glucose curve is unremarkable. For a definite diagnosis, one should proceed to enzyme and/or DNA assays of erythrocytes, lymphocytes or fibroblasts [48]. A muscle biopsy should be performed if serum creatine kinase is abnormally increased.

### Treatment and Prognosis

Dietary treatment is less demanding than in GSD-I. Carbohydrates should be given frequently when the patient is young. Milk products and fruits can be allowed without restriction, as galactose and fructose (constituents of the disaccharides lactose and sucrose, respectively) can normally be converted into glucose. Gastric drip feeding at night and uncooked starch induce catch-up growth, reduce liver size and decrease serum transaminases [49]. Drip feeding and protein enrichment of meals may counteract the drain from muscle protein. The results on muscle function are, however, ambiguous [50, 51]. The composition of the diet should be approximately 55–60% carbohydrates and 15–20% protein, and the remainder should be fat (predominantly polyunsaturated). The prognosis, favorable for the purely hepatic GSD-IIIb, is less favorable for the other GSD-III subtypes, as severe myopathy and cardiomyopathy may develop even after a long latency.

### Genetics

The inheritance is autosomal recessive. The gene for debranching enzyme is mapped to chromosome 1p21. Heterogeneity for GSD-IIIb has been found, and compound heterozygotes exist [52]. Antenatal diagnosis is possible by assay of amniotic fluid cells and chorionic villi.

## Glycogen Storage Disease-IV: Branching-Enzyme Deficiency, Amylopectinosis

### Clinical Presentation

The most common clinical presentation is the hepatic presentation. Hepatomegaly, which is observed in the young infant, progresses to liver cirrhosis and liver failure with symptoms of hepatosplenomegaly, esophageal varices and ascites within the first 3–5 years. A non-progressive hepatic form, in which the abnormal liver function normalizes over time, even until the age of 20 years, is also observed [53]. Intermediate hepatic variants with juvenile onset exist. Hypoglycemia is not a prominent feature. If it occurs, it is due to impeded degradation of the abnormally structured glycogen and the gradually decreasing number of normal hepatocytes. A fatal neuromuscular form is the third presentation [54], and this form may also encompass cardiomyopathy [55]. The neuromuscular form presents with muscle hypotonia and weakness, hypo- or areflexia and retarded motor milestones. A late onset variant at adult age also exists. Furthermore, all kinds of combinations of hepatic and myopathic forms occur [53].

### Metabolic Derangement and Diagnosis

The generalized enzyme defect causes insufficient branching of the glycogen molecule, and its prolonged inner and outer chains give it the appearance and properties of amylopectin. Apparently, this abnormal glycogen acts as a foreign body towards the development of liver cirrhosis. Furthermore, glucose release from it is hampered. Because of the unspecific liver symptoms, the diagnosis is usually only suspected at the histological examination of a liver biopsy. The hepatocytes contain large deposits that are periodic-acid–Schiff-staining but are partially resistant to diastase digestion. Electron microscopy shows accumulation of fibrillar aggregations that are typical for amylopectin. The diagnosis is confirmed by enzyme assay and/or DNA assays of a liver biopsy or fibroblasts.

### Treatment and Prognosis

Nutritional management with cornstarch and/or gastric drip feeding during the night can improve the clinical condition. These conservative measures may suffice in exceptional variants with a mild course. However, in severe cases, liver transplantation is the only ultimate option and, of seven transplanted patients, five survived during observation until 73 months after transplantation [56]. Remarkably, sequential muscle biopsies over prolonged periods after transplantation showed marked reduction of amylopectin deposits in some patients. This "chimerism" is supposed to be caused by the migration of cells of the implanted normal allograft, which carry enzyme into the deficient tissues of the recipient [29, 57].

### Genetics

The inheritance is autosomal recessive. Many allelic variants of the gene (which is localized to chromosome 3p12) exist. They are associated with the large variability of clinical expression of the enzyme defect [54]. Antenatal diagnosis seems feasible by enzyme or DNA assay of cultured amniotic-fluid cells or chorionic villi.

## Deficiencies of the Phosphorylase System: Phosphorylase Deficiency of the Liver (Glycogen Storage Disease-VI) and Phosphorylase-B-Kinase Deficiency of the Liver (Glycogen Storage Disease-IX)

Enzyme defects of the phosphorylase system show a wide range of clinical expression due to different mutations of different isoenzymes localized in different organs [58]. The discussion that follows is focussed on diseases that are expressed only in the liver or in both liver and muscles, irrespective of the type of enzyme defect within the phosphorylase system. Diseases that are only expressed in the muscles are assembled in the group of myopathies caused by enzyme defects of glycogenolysis or glycolysis.

### Clinical Presentation and Complications

Pronounced hepatomegaly, a protruded abdomen and a mild tendency to fasting hypoglycemia are the most striking features in early childhood. The liver enlargement decreases slowly and usually disappears at puberty. The spleen and kidneys are normal in size. Muscle hypotonia and weakness are mild, except in cases in which phosphorylase-B kinase is deficient in both liver and muscles [59]. Slightly retarded motor development and retarded growth normalize gradually. Puberty is delayed in some patients. Mental development is normal. In some cases, the course is so mild that hepatomegaly is accidentally found at routine examination. Complications of GSD-VI are not known, except for brain involvement in a case in which the enzyme defect could not be sufficiently defined. Other rare complications are associated with GSD-IX: cirrhosis of the liver [60] and renal tubular dysfunction [61].

### Metabolic Derangements

The phosphorylase system is a complicated cascade of enzymes that is sequentially activated by neuronal or hormonal stimulation as follows: glucagon (adrenaline) → activation of adenylate cyclase → production of cyclic adenosine monophosphate (cAMP) from adenosine triphosphate (ATP) → activation of cAMP-dependent protein kinase → activation of phosphorylase-B kinase → activation of phosphorylase. Only deficiencies of the last two enzymes are clinically relevant. Both impair the cleavage of glycosyl-molecules from the straight chains of glycogen to the same degree. Since the enzyme defects are usually partial, glycogen degradation is not totally blocked. Also, gluconeogenesis remains intact. Mild fasting hypoglycemia and fasting ketosis, moderate hyperlipidemia (serum cholesterol more elevated than serum triglycerides) and elevated serum transaminases are evident at a young age but normalize completely before or during puberty [62].

### Diagnostic Tests

An oral glucose test is performed, as is the usual procedure in all patients with hepatomegaly and a tendency to hypoglycemia. The normal or slightly exaggerated rise of serum lactate is a valid differentiation from all cases with impaired gluconeogenesis (see GSD-I). An additional glucagon test is of no help for the diagnosis. Instead, it is better to proceed directly to enzyme assays of both phosphorylase and phosphorylase-B kinase. Enzyme assays of a liver biopsy have been abandoned in favor of assays in peripheral tissues, such as white or red blood cells [58]. However, the differentiation between the two enzymes is technically difficult. Also, normal phosphorylase-B kinase activities have been measured in some rare GSD-IX variants [63]. These problems can be solved by DNA analysis for both enzyme defects.

### Treatment and Prognosis

Dietary treatment, if necessary at all, should be limited to young children. For this age group, a late supper should be inserted if the child has an infectious disease. Prolonged fasting must be prevented. The inclusion of polyunsaturated fat in the diet is a very effective means of suppressing hypercholesterolemia. Retarded growth, which is a worry for many patients, usually normalizes without special treatment. In such a patient, however, a remarkable catch-up growth and improvement of the metabolic derangements has been elicited by the introduction of uncooked cornstarch before night [64]. The prognosis is good for most patients with GSD-VI and for hepatic GSD-IX, but insufficient follow-up is known for GSD-IX with muscle involvement [65].

### Genetics

The relative frequencies of GSD-VI and GSD-IX in a large retrospective study was about 24 and 61 patients, respectively [58]. For a review of genetic deficiencies of the phosphorylase system, see [66].

#### Phosphorylase

Three isoforms exist in liver, muscle and brain and are encoded by three different genes. The gene for the liver isoform is located at 14q21-q22, and the first mutations in three patients have recently been described [67]. The inheritance is presumably autosomal recessive.

### *Phosphorylase-B Kinase*

Four subunits exist: α, β, γ and δ. The γ subunit harbors the catalytic activity, which is regulated by the α, β and δ subunits. Various isoforms have been identified for all four subunits, and some isoforms are encoded by different genes. Thus, it can be easily understood that many mutations have been found, leading to different clinical presentations (liver, muscle) and to different modes of inheritance (X-linked [68] or autosomal recessive [69]).

## Muscle Glycogen Storage Diseases

Muscle enzyme defects of the glycogenolytic pathway (from glycogen to glucose-6-phosphate) and the glycolytic pathway (from glucose-6-phosphate to pyruvate and lactate) share several clinical and metabolic abnormalities despite the fact that they are located at very different sites of the glycogenolytic–glycolytic pathway. Most forms are very rare, myophosphorylase deficiency (GSD-V) and phosphofructokinase deficiency (GSD-VII) excepted. All defects discovered to date are discussed together. A list of their names, and references to the recent literature, are given under "Genetics". For a review, see [70].

### Clinical Presentation

The clinical features vary with age and the biochemical heterogeneity of a particular enzyme. Usually, the patient starts to complain of exercise intolerance at adult age. The complaints are tiredness, diminished muscle strength, stiffness and myalgia after exertion. Brief exertions of great intensity and less intensive but sustained activities, such as climbing stairs, may cause severe muscle cramps, which can be accompanied by muscle tenderness and swelling, myoglobinuria and even anuria. The complaints subside after resting. Gradually, the patient learns to avoid symptoms by adjusting his activities. Other patients have very mild symptoms consisting of tiredness, which may be dismissed as psychogenic. Progressive weakness at adult age without a previous history of cramps has been reported [71], as has early-onset myopathy with a fatal outcome [72].

### Metabolic Derangements

The metabolic derangements are caused by insufficient fuel supply for muscle function. The first provision of energy from glycogen or glucose degradation fails and is not promptly followed and replenished by energy from fatty-acid oxidation, the activation of which proceeds slower. An acute shortage of ATP (the main source of chemical energy) ensues, as it is insufficiently regenerated from adenosine diphosphate (ADP). Instead, ADP is increasingly converted to AMP, and this compound is further catabolized to inosine and other purines, which are ultimately degraded to uric acid. Thus, hyperuricemia is a marker for cell energy-crisis [73]. In some patients, the hyperuricemia leads to gout or renal calculi.

Serum creatine-kinase levels are usually elevated at rest, with further increases after exercise as a sign of increased muscle catabolism and leakage. Occasionally, a slight and chronic hemolytic jaundice is observed because the erythrocytes share an isoenzyme of the defective muscle enzyme. This applies to phosphofructokinase deficiency.

### Diagnostic Tests

An exercise test (semi-ischemic forearm test, bicycle ergometer test or treadmill test) is used to demonstrate the failure of venous lactate and pyruvate to rise and the excessive increase in production of uric acid, inosine, hypoxanthine and ammonia (Chap. 2) [74]. Ammonia is produced by the increased deamination of AMP to inosine monophosphate which, in turn, is degraded to inosine, hypoxanthine and uric acid. The exercise test, which is immediately stopped at the first sign of myalgia or cramps, can be resumed without untoward symptoms at a lower level of exertion after a short rest. This remarkable "second-wind" phenomenon can be explained by the fact that (1) fatty acids become gradually available as an alternative fuel and/or (2) muscle oxygenation improves by vasodilatation. If the results of the exercise test are abnormal, a metabolic myopathy is probable and should be verified both by assaying all enzymes of the glycogenolytic–glycolytic pathway in a muscle biopsy and by DNA assays. Histological examination of the muscle biopsy is often non-specific. Glycogen accumulation is observed between myofibrils (GSD-V), or the glycogen is composed of fine fibrillar materials like amylopectin (GSD-VII).

### Treatment

Muscle function may be influenced favorably or unfavorably by the diet. The effect of protein may be favorable, as it compensates for increased muscle catabolism [75]. The effect of glucose is ambiguous; it is favorable for phosphorylase deficiency, as the localization of the enzyme defect leaves glycolysis intact. However, it is unfavorable for enzyme defects distal to glucose-6-phosphate, because it not only overloads the blocked glycolytic pathway, it also inhibits lipolysis and, thus, deprives muscle of free fatty acids and ketone bodies [76]. Regular physical exercise at a submaximal level is useful, whereas intensive exertion is to be avoided.

## Genetics

All enzyme defects presumably have an autosomal recessive inheritance, except phosphoglycerate kinase deficiency, which is an X-linked disorder. Different mutations exist in many diseases, and even compound heterozygotes have been found. The Appendix provides some references.

## Glycogenolytic Defects Located in Muscles

- Debranching-enzyme deficiency, GSD-III [77]
- Myophosphorylase deficiency, GSD-V, McArdle disease [78, 79]
- Phosphorylase-B-kinase deficiency, GSD-IX [80, 81]
- Phosphoglucomutase deficiency [82, 83]

## Glycolytic Defects Located in Muscles

- Phosphofructokinase deficiency, GSD-VII, Tarui disease [76, 84–86]
- Phosphoglycerate-kinase deficiency [87]
- Phosphoglycerate-mutase deficiency [88]
- Lactate-dehydrogenase deficiency [89]

## Other Rare Glycogen Storage Disease Types

### GSD-0: Glycogen-Synthase Deficiency

This rare enzyme defect leads to hypoglycogenosis instead of glycogen storage, because glycogen synthesis is reduced. Nevertheless, it is grouped under the GSDs, and this is due to the fact that it shares some metabolic derangements with GSD-III, such as fasting hypoglycemia and ketosis and postprandial hyperlactacidemia. However, there is more resemblance to the so-called ketotic hypoglycemia.

Clinically, the patient presents with drowsiness, uncoordinated eye movements and occasional hypoglycemic convulsions before breakfast or at inadvertent fasting. The liver is not enlarged or is only slightly enlarged, and its glycogen content is low. Growth lags gradually behind. Oral tolerance tests with glucose, galactose and fructose are characterized by a marked increase of serum lactate. The glucagon test is of no help. The enzyme defect can only be demonstrated in the liver, not in other tissues [90].

Treatment with frequent protein-rich meals benefits the clinical course. A normal pregnancy was observed in the first patient described in the literature [91]. Inheritance is probably autosomal recessive.

### Hepatorenal GSD with the Fanconi-Bickel Syndrome

A protruded abdomen due to enlargement of the liver and the kidneys (caused by glycogen storage), stunted growth, osteopenia, hyperuricemia, hyperlipidemia, renal glomerular hyperfiltration, and (micro)albuminuria closely resemble the abnormalities of GSD-Ia. Other symptoms differ, such as rickets, glucosuria, generalized aminoaciduria, phosphaturia and loss of bicarbonate, which reflect proximal tubular dysfunction. This syndrome was first described by Fanconi and Bickel. Mild hypoglycemia and ketosis may occur at fasting. It is not an enzyme defect of glycogenolysis that underlies the disorder; it is a combined transport defect of glucose and galactose. This is caused by mutations in GLUT2, a member of a family of transport proteins that transport monosaccharides across membranes [92]. GLUT2 is expressed in liver, pancreas, intestine and kidneys, where it facilitates the transport of glucose and galactose but not fructose. It is the finding of hypergalactosemia at neonatal screening for disorders of galactose metabolism that may initiate further exploration towards the existence of the Fanconi-Bickel syndrome (Chap. 7).

The treatment is symptomatic and is mainly directed at the correction of renal spill-over. Adequate supplementation of calcium and vitamin D, a ketogenic diet and fructose-containing carbohydrates might be of use. The GLUT2 gene is localized to chromosome 3q26.1-q26.3. For an excellent review of 82 cases from the literature, see Santer [93].

## References

1. Von Gierke E (1929) Hepato-nephro-megalia glykogenika (Glykogenspeicherkrankheit der Leber und Nieren). Beitr Pathol Anat 82:497–513
2. Corby DC, Putnam CW, Greene HL (1974) Impaired platelet function in glucose-6-phosphatase deficiency. J Pediatr 85: 71–76
3. Milla PJ, Atherton DA, Leonard JV, Wolff OH, Lake BD (1978) Disordered intestinal function in glycogen storage disease. J Inherit Metab Dis 1:155–157
4. Rother KI, Schwenk WF (1995) Glucose production in glycogen storage disease I is not associated with increased cycling through hepatic glycogen. Am J Physiol 269:E774–778
5. Fernandes J, Berger R, Smit GPA (1984) Lactate as a cerebral metabolic fuel for glucose-6-phosphatase deficient children. Pediatr Res 18:335–339
6. Alaupovic P, Fernandes J (1985) The serum apolipoprotein profile of patients with glucose-6-phosphatase deficiency. Pediatr Res 19:380–384
7. Fernandes J, Pikaar NA (1972) Ketosis in hepatic glycogenosis. Arch Dis Child 47:41–46
8. Talente GM, Coleman RA, Craig C (1994) Glycogen storage disease in adults. Review. Ann Intern Med 120:218–226
9. Labrune P, Trioche P, Duvaltier I et al. (1997) Hepatocellular adenomas in glycogen storage disease type I and III: a series of 43 patients and review of the literature. J Pediatr Gastroenterol Nutrit 24:276–279
10. Chen Y-T, Coleman RA, Scheinman JI et al. (1990) Amelioration of proximal renal tubular dysfunction in type I

glycogen storage disease with dietary therapy. N Engl J Med 323:590–593
11. Restaino I, Kaplan BS, Stanley C, Baker L (1993) Nephrolithiasis, hypercitraturia, and a distal renal tubular acidification defect in type I glycogen storage disease. J Pediatr 122:392–396
12. Lee PJ, Patel JS, Fewtrill M et al. (1995) Bone mineralisation in type I glycogen storage disease. Eur J Pediatr 154:483–487
13. Kikuchi M, Hasegawa K, Handa I et al. (1991) Chronic pancreatitis in a child with glycogen storage disease type I. Eur J Pediatr 150:852–853
14. Smit GPA (1993) The long-term outcome of patients with glycogen storage disease type Ia. Eur J Pediatr 152:S52–55
15. Lee PJ, Celermajer DS, Robinson J et al. (1994) Hyperlipidaemia does not impair vascular endothelial junction in glycogen storage disease type I a. Atherosclerosis 110:95–100
16. Lee PJ, Patel A, Hindmarsch PC et al. (1995) The prevalence of polycystic ovaries in the hepatic glycogen storage diseases: its association with hyperinsulinism. Clin Endocrinol 42:601–606
17. Kishani P, Bengun AR, Chen YT (1996) Pulmonary hypertension in glycogen storage disease Type I a. J Inherit Metab Dis 19:213–216
18. Goutieres F, Bourgois M, Trioche P et al. (1997) Moyamoya disease in a child with glycogen storage disease type I a. Neuropediatrics 28:133–134
19. Greene HL, Slonim AE, Burr IM, Moran JR (1980) Type I glycogen storage disease: five years of management with nocturnal intragastric feeding. J Pediatr 96:590–595
20. Chen Y-T, Cornblath M, Sidbury JB (1984) Cornstarch therapy in type I glycogen storage disease. N Engl J Med 310:171–175
21. Chen Y-T, Bazzarre CH, Lee MM, Sidbury JB, Coleman RA (1993) Type I glycogen storage disease: nine years of management with cornstarch. Eur J Pediatr 152:S56–59
22. Hayde M, Widhalm K (1990) Effects of cornstarch treatment in very young children with type I glycogen storage disease. Eur J Pediatr 149:630–633
23. Fernandes J (1974) The effect of disaccharides on the hyperlactacidaemia of glucose-6-phosphatase-deficient children. Acta Paediatr Scand 63:695–698
24. Fernandes J, Alaupovic P, Wit JM (1989) Gastric drip feeding in patients with glycogen storage disease type I: its effects on growth and plasma lipids and apolipoproteins. Pediatr Res 25:327–331
25. Wolsdorf JI, Crigler JF (1997) Cornstarch regimens for nocturnal treatment of young adults with type I glycogen storage disease. Am J Clin Nutr 65:1507–1511
26. Nuoffer JM, Mullis PE, Wiesmann UN (1997) Treatment with low-dose diasoxide in two growth-retarded prepubertal girsl with glycogen storage disease type I a resulted in catch-up growth. J Inherit Metab Dis 20:790–798
27. Wolsdorf JI, Laffel LM, Crigler JF (1997) Metabolic control and renal dysfunction in type I glycogen storage disease. J Inherit Metab Dis 20:559–568
28. Levy E, Thibault L, Turgeon J et al. (1993) Beneficial effects of fish-oil supplements on lipids, lipoproteins, and lipoprotein lipase in patients with glycogen storage disease type I. Am J Clin Nutr 57:922–929
29. Matern D, Starzl TE, Amaout W et al. (1999) Liver transplantation for glycogen storage diseases types I, III and IV. Eur J Pediatr 158[Suppl. 2]:S43–48
30. Ryan IP, Havel RJ, Laros RK (1994) Three consecutive pregnancies in a patient with glycogen storage disease type I A (von Gierke's disease). Am J Obstet Gynecol 170:1687–1691
31. Lei KJ, Shelly LL, Pan CJ, Sidbury JB, Chou JY (1993) Mutations in the glucose-phosphatase gene that cause glycogen storage disease type Ia. Science 262:580–583
32. Veiga-da-Cunha M, Gerin I, van Schaftingen E (2000) How many forms of glycogen storage disease type I? Eur J Pediatr (in press)
33. Simmons PS, Smithson WA, Gronert GA, Haymond MW (1984) Acute myelogenous leukemia and malignant hyperthermia in a patient with type Ib glycogen storage disease. J Pediatr 105:428–431
34. Kikuchi M, Haginoya K, Miyabayashi S et al. (1990) Secondary amyloidosis in glycogen storage disease type I b. Eur J Pediatr 149:344–345
35. Hoover EG, Du Bois JJ, Samples TL et al. (1996) Treatment of chronic enteritis in glycogen storage disease type 1 B with granulocyte colony-stimulating factor. J Pediatr Gastroenterol Nutr 22:346–350
36. Veiga-da-Cunha M, Gerin I, Chen Y-T et al. (1998) A gene on chromosome 11q23 coding for a putative glucose-6-phosphatase translocase is mutated in glycogen storage disease types 1b and 1c. Am J Hum Genet 63:976–983
37. Makos MM, Mc Comb RD, Hart MN, Bennett DR (1987) Alpha-glucosidase deficiency and basilar artery aneurism: report of a sibship. Ann Neurol 22:629–633
38. Kroos MA, van der Kraan M, van Diggelen OP et al. (1995) Glycogen storage disease type II: frequency of three common mutant alleles and their associated clinical phenotypes studied in 121 patients. J Med Genet 32:836–837
39. Mobarhan S, Pintozzi RL, Damle P, Friedman H (1990) Treatment of acid maltase deficiency with a diet high in branched-chain aminoacids. J Parent Enteral Nutr 14:210–212
40. Kuo WL, Hirschhorn R, Huie ML, Hirschhorn K (1996) Localisation and ordering of acid alpha-glucosidase (GAA) and thymidine kinase (TKI) by fluorescence in situ hybridization. Hum Genet 97:404–406
41. Wokke JHJ, Ausems MGEM, van den Boogaard MJH et al. (1995) Genotype-Phenotype correlation in adult-onset acid maltase deficiency. Ann Neurol 38:450–454
42. Hoefsloot LH, van der Ploeg AT, Kroes MA et al. (1990) Adult and infantile glycogenosis type II in one family, explained by allelic diversity. Am J Hum Genet 46:45–52
43. Haagsma EB, Smit GPA, Niezen-Koning KE et al. (1997) Type III b glycogen storage disease associated with end-stage cirrhosis and hepatocellular carcinoma. Hepatology 25:537–540
44. Coleman RA, Winter HS, Wolf B, Gilchrist JM, Chen Y-T (1992) Glycogen storage disease type III (glycogen debranching enzyme deficiency): correlation of biochemical defects with myopathy and cardiomyopathy. Ann Intern Med 116:896–900
45. Moses SW, Wandermann KL, Myroz A, Frydman M (1989) Cardiac involvement in glycogen storage disease III. Eur J Pediatr 148:764–766
46. Cuspidi C, Sampieri L, Peliazoli S et al. (1997) Obstructive hypertrophic cardiomyopathy in Type III glycogen-storage disease. Acta Cardiol 52:117–123
47. Cohen J, Friedman M (1979) Renal tubular acidosis associated with type III. glycogenosis. Acta Paediatr Scand 68:779–782
48. Ding JH, de Barsy T, Brown BI et al. (1990) Immunoblot analysis of glycogen debranching enzyme in different subtypes of glycogen storage disease type III. J Pediatr 116:95–100
49. Gremse DA, Bucavalas JC, Balistreri WF (1990) Efficacy of cornstarch therapy in type III glycogen storage disease. Am J Clin Nutr 52:671–674
50. Slonim AE, Coleman RA, Moses SW (1984) Myopathy and growth failure in debrancher enzyme deficiency: improvement with high-protein nocturnal enteral therapy. J Pediatr 105:906–911
51. De Parscau L, Gibaud P, Hermier M, Francois R (1989) l'Alimentation intragastrique nocturne continue dans les glycogenoses de type I et III. Pediatrie 41:197–203
52. Shen J, Bao Y, Liu HM et al. (1996) Mutations in exon 3 of the glycogen debranching enzyme gene are assiociated with glycogen storage disease type III that is differentially expressed in liver and muscle. J Clin Invest 98:352–357
53. Mc Conkie-Rosell A, Wilson C, Piccoli DA et al. (1996) Clinical and laboratory findings in four patients with the non-progressive hepatic form of type IV glycogen storage disease. J Inherit Metab Dis 19:51–58
54. Bao Y, Kishumi P, Wu JY, Chen YT (1996) Hepatic and neuromuscular forms of glycogen storage disease type IV caused by mutations in the same glycogen-branching enzyme gene. J Clin Invest 97:941–948

55. Servidei S, Riepe RE, Langston C et al. (1987) Severe cardiopathy in branching enzyme deficiency. J Pediatr 111:51-56
56. Selby R, Starzl TE, Yunis E et al. (1993) Liver transplantation for type I and type IV glycogen storage disease. Eur J Pediatr 152:S71-S76
57. Starzl TE, Demetris AJ, Trucoo M et al. (1993) Chimerism after liver transplantation for type IV glycogen storage disease and type I Gaucher disease. N Engl J Med 328:745-749
58. Maire I, Baussan C, Moatti N, Mathieu M, Lemonnier A (1991) Biochemical diagnosis of hepatic glycogen storage diseases: 20 years French experience. Clin Biochem 24:169-178
59. Madiom M, Besley GTN, Cohen PTW, Marrian VJ (1989) Phosphorylase b kinase deficiency in a boy with glycogenosis affecting both liver and muscle. Eur J Pediatr 14(9):52-53
60. Kagalwalla AF, Yasmeen A, Sulaiman Al Ajaji et al. (1995) Phosphorylase b kinase deficiency glycogenosis with cirrhosis of the liver. J Pediatr 127:602-605
61. Sanjad SA, Kaddoura RE, Nazir HM et al. (1993) Fanconi's syndrome with hepatorenal glycogenosis associated with phosphorylase b kinase deficiency. Am J Dis Child 147: 957-959
62. Willems PJ, Gerver WJM, Berger R, Fernandes J (1990) The natural history of liver glycogenosis due to phosphorylase kinase deficiency: a longitudinal study of 41 patients. Eur J Pediatr 149:268-271
63. Hendrickx J, Dams E, Couche P et al. (1996a) X-linked liver glycogenosis type II (XLG II) is caused by mutations in PHKA2, the gene encoding the liver α subunit of phosphorylase kinase. Hum Mol Genet 5:649-652
64. Nakai A, Shigematsu Y, Takano T et al. (1994) Uncooked starch treatment for hepatic phosphorylase kinase deficiency. Eur J Pediatr 153:581-583
65. Smit GPA, Fernandes J, Leonard JV et al. (1990) The long-term outcome of patients with glycogen storage diseases. J Inherit Metab Dis 13:411-418
66. Hendrickx J, Willems PJ (1996b) Genetic deficiencies of the glycogen phosphorylase system. Hum Genet 97:551-556
67. Burwinkel B, Bakker HD, Herschkovitz E et al. (1998) Mutations in the liver glycogen phosphorylase gene (PYGL) underlying glycogenosis type VI (Hers disease). Am J Hum Genet 62:785-791
68. Van de Berg IET, van Beurden EACM, Malingre HEM et al. (1995) X-linked liver phosphorylase deficiency associated with mutations in the human liver phosphorylase kinase α subunit. Am J Hum Genet 56:381-387
69. Burwinkel B, Maichelle AJ, Aagenaes Ø et al. (1997) Autosomal glycogenosis of liver and muscle due to phosphorylase kinase deficiency is caused by mutations in the phosphorylase kinase β subunit (PHKB). Hum Mol Genet 6:1109-1115
70. Servidei S, DiMauro S (1989) Disorders of glycogen metabolism of muscle. Neurol Clin 7:159-178
71. Abarbanel JM, Bashan N, Potashnik R et al. (1986) Adult muscle phosphorylase "b" kinase deficiency. Neurology 86:560-562
72. Servidei S, Metlay LA, Chodosh J, Di Mauro S (1988) Fatal infantile cardiopathy caused by phosphorylase b kinase deficiency. J Pediatr 113:82-85
73. Fox IH, Palella TD, Kelley WN (1987) Hyperuricemia: a marker for cell energy crisis. N Engl J Med 317:111-112
74. Mineo I, Kono N, Hara N et al. (1987) Myogenic hyperuricemia. A common pathophysiologic feature of glycogenosis types III, V and VII. N Engl J Med 317:75-80
75. Slonim AE, Goans PJ (1985) Myopathy in Mc Ardle's syndrome. Improvement with a high-protein diet. N Engl J Med 312:355-359
76. Haller RG, Lewis SF (1991) Glucose-induced exertional fatigue in muscle phosphofructokinase deficiency. N Engl J Med 324:364-369
77. Yang BZ, Ding JH, Enghild JJ, Bao Y, Chen YT (1992) Molecular cloning and nucleotide sequence of cDNA encoding human muscle glycogen debranching enzyme. J Biol Chem 267:9294-9299
78. El-Schahawi M, Tsujino S, Shanske S, di Mauro S (1996) Diagnosis of Mc Ardle's disease by molecular genetic analysis of blood. Neurology 47:579-580
79. Vosgerd M, Kubisch C, Burwinkel B et al. (1998) Mutation analysis in myophosphorylase deficiency (Mc Ardle' disease). Ann Neurol 43:326-331
80. Wilkinson DA, Tonin P, Shanske S et al. (1994) Clinical and biochemical features of 10 adult patients with muscle phosphorylase kinase deficiency. Neurol 44:461-466
81. Wehner M, Clemens PR, Engel AG, Kilimann M (1994) Human muscle glycogenosis due to phosphorylase kinase deficiency associated with a nonsense mutation in the muscle isoform of the α subunit. Hum Mol Genet 3:1983-1987
82. Sugie H, Kobayashi J, Sugie Y et al. (1988) Infantile muscle glycogen storage disease: phosphoglucomutase deficiency with decreased muscle and serum carnitine levels. Neurology 38:602-605
83. Whitehouse DB, Putt W, Lovegrove JU et al. (1992) Phosphoglucomutase 1: complete human and rabbit mRNA sequences and direct mapping of this highly polymorphic marker on human chromosome 1. Proc Natl Acad Sci USA 89:411-415
84. Vissing J, Galbo H, Haller RG (1996) Paradoxically enhanced glucose production during exercise in humans with blocked glycolysis caused by muscle phosphofructokinase deficiency. Neurology 47:766-771
85. Servidei S, Bonilla E, Diedrich RG et al. (1986) Fatal infantile form of muscle phosphofructokinase deficiency. Neurology 36:1465-1470
86. Howard TD, Akots G, Bowden DW (1996) Physical and genetic mapping of the muscle phosphofructokinase gene (PFKM): reassignment to human chromosome 12q. Genomics 34:122-127
87. Tonin P, Shanske S, Miranda AF et al. (1993) Phosphoglycerate kinase deficiency: biochemical and molecular genetic studies in a new myopathic variant. (PGK Alberta). Neurology 43:387-391
88. Tsujino S, Shanske S, Sakoda S et al. (1993) The molecular genetic basis of muscle phosphoglycerate mutase (PGAM) deficiency. Am J Hum Genet 52:472-477
89. Kanno T, Sudo K, Takuchi I et al. (1980) Hereditary defcency of lactate dehydrogenase M-subunit. Clin Chim Acta 108:267
90. Gitzelmann R, Spycher MA, Fiel G et al. (1996) Liver glycogen synthase deficiency: a rarely diagnosed entity. Eur J Pediatr 155:561-567
91. Byrne BM, Killmer MD, Turner RC, Aynsley-Green A (1995) Glucose homoiostasis in adulthood and in pregnancy in a patient with hepatic glycogen synthase deficiency. Br J Obstet Gynaecol 102:931-933
92. Santer R, Schneppenheim R, Dombrowski A et al. (1997) Mutations in GLUT 2, the gene for the liver-type glucose transporter in patients with Fanconi-Bickel syndrome. Nature Genet 17: 324-326 (published erratum in Nature Genet 18:298, 1998)
93. Santer R, Schneppenheim R, Suter D, Schaub J, Steinmann B (1998) Fanconi-Bickel syndrome - the original patient and his natural history, historical steps leading to the primary defect, and a review of the literature. Review. Eur J Pediatr 157: 783-797

## Galactose Metabolism

Together with its 4′-epimer glucose, galactose forms the disaccharide lactose, which is the principal carbohydrate in milk, providing 40% of its total energy. Ingested lactose is hydrolyzed to galactose and glucose in the small intestine. After absorption, galactose is converted to galactose-1-phosphate, uridine diphosphate(UDP)–galactose, and UDP–glucose (Fig. 7.1). UDP–glucose is the glycosyl carrier in several reactions, including the synthesis of glycogen and the metabolism of galactose-1-phosphate. In the latter, glucose-1-phosphate is produced and metabolized to glucose in the liver.

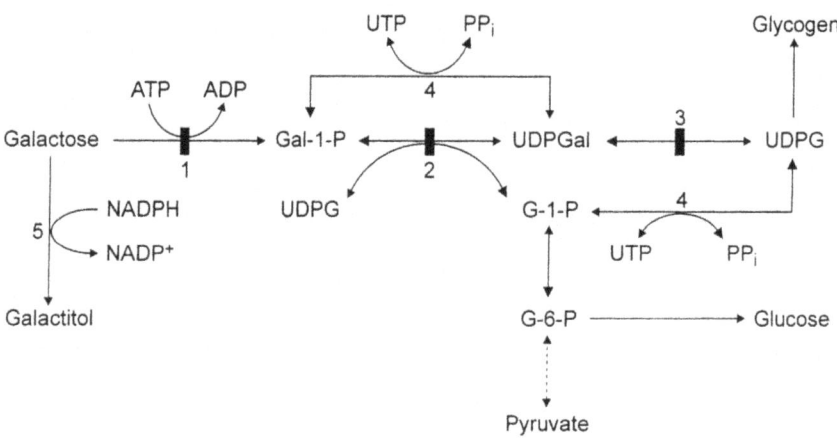

**Fig. 7.1.** Galactose metabolism (simplified). *1* Galactokinase, *2* galactose-1-phosphate uridyltransferase, *3* uridine diphosphate (UDP)-galactose 4′-epimerase, *4* UDP-glucose (UDP-galactose) pyrophosphorylase, *5* aldose reductase; *G-1-P*, glucose-1-phosphate; *G-6-P*, glucose-6-phosphate; *Gal-1-P*, galactose-1-phosphate; *NADP*, nicotinamide adenine dinucleotide phosphate; *NADPH*, reduced NADP; *UDPG*, uridine diphosphoglucose; *UDPGal*, uridine diphosphogalactose. The three enzyme defects in galactose metabolism are depicted by *solid bars* across the *arrows*

# CHAPTER 7

# Disorders of Galactose Metabolism

R. Gitzelmann

> "Whenever you consider a galactose disorder, stop milk feeding first and only then seek a diagnosis!"

CONTENTS

| | |
|---|---|
| Galactokinase Deficiency | 103 |
| Deficiency of Galactose-1-Phosphate Uridyltransferase | 104 |
|   Clinical Presentation | 104 |
|   Metabolic Derangement | 104 |
|   Diagnostic Tests | 105 |
|   Treatment and Prognosis | 105 |
|     Treatment of the Newborn Infant | 106 |
|     Spoon-Feeding | 106 |
|     Vegetables and Fruits | 106 |
|     Cheese | 106 |
|     Breaks of Discipline | 106 |
|     Complications of Treated Galactosemia | 106 |
|     Long-Term Results | 106 |
|     Treatment of Partial Transferase Deficiency | 107 |
|     Dietary Treatment in Pregnant Woman at Risk | 107 |
|   Genetics | 107 |
| Uridine Diphosphate-Galactose 4'-Epimerase Deficiency | 107 |
| Fanconi-Bickel Syndrome | 108 |
| Portosystemic Venous Shunting | 108 |
| References | 108 |

Three inborn errors of galactose metabolism are known. Galactokinase deficiency is the most insidious, since it results in the formation of nuclear cataracts without provoking symptoms of intolerance. Galactose-1-phosphate uridyltransferase deficiency exists in two forms. The complete or near-complete deficiency is life-threatening and affects not only the eye lens but also liver, kidney and brain. Partial deficiency is usually, if not always, benign. Uridine diphosphate (UDP)–galactose 4'-epimerase deficiency also exists in two forms. The very rare profound deficiency clinically resembles classical galactosemia. The more frequent partial deficiency is benign. Review articles are recommended for detailed information [1–3]. Fanconi-Bickel syndrome (Chap. 6) and portosystemic venous shunting are congenital errors of galactose transport leading to hypergalactosemia.

## Galactokinase Deficiency

### Clinical Presentation

Cataracts are the only consistent manifestation of the untreated disorder, though pseudotumor cerebri has been described. Liver, kidney and brain damage, as seen in transferase deficiency (below), are not features of untreated galactokinase deficiency, and hypergalactosemia and galactose–galactitol–glucose diabetes are the only chemical signs.

### Metabolic Derangement

Persons with galactokinase deficiency lack the ability to phosphorylate galactose (Fig. 7.1). Consequently, nearly all of the ingested galactose is excreted, either as such or as its reduced metabolite, galactitol, formed by aldose reductase. Cataracts result from the accumulation of galactitol in the lens, causing osmotic swelling of lens fibers and denaturation of proteins.

### Diagnostic Tests

Provided they have been fed mother's milk or a lactose-containing formula prior to the test, newborns with the defect are discovered by mass screening methods for detecting elevated blood galactose [4]. If they have been fed glucose-containing fluid, the screening test could be false-negative. Any chance finding of a *reducing substance* in urine, especially in children or adults with nuclear cataracts, calls for the identification of the excreted substance. In addition to galactose, galactitol and glucose may be found. Every person with nuclear cataracts ought to be examined for galactokinase deficiency.

Final diagnosis is made by assaying galactokinase activity in heparinized whole blood, red cell lysates, liver or fibroblasts [4, 5]. Heterozygotes have intermediate activity in erythrocytes [6, 7]. Reports of galactokinase variants have appeared [1, 2].

### Treatment and Prognosis

Treatment may be limited to the elimination of milk from the diet. Minor sources of galactose, such as milk products, green vegetables, legumes, drugs in tablet form, etc., can probably be disregarded, since it can be assumed that the small amounts of ingested galactose are either metabolized or excreted before significant amounts of galactitol can be formed. When diagnosis is made rapidly and treatment begun promptly, i.e., during the first 2–3 weeks of life, cataracts can clear. When treatment is late, and cataracts too dense, they will not clear completely (or at all) and must be removed surgically. In patients who have had their lenses removed, recurring cataracts may appear, originating from remnants of the posterior lens capsule. This can be avoided by continuing the diet.

The speculation [7] that heterozygosity predisposes to the formation of presenile cataracts remains unproven [1, 2]. Still, it may be reasonable for heterozygotes to restrict their milk intake, though scientific proof of the merits of this measure is lacking.

### Genetics

The mode of inheritance is autosomal recessive. In most parts of Europe, in the USA and in Japan, birth incidence is in the order of one in 150,000 to one million. It is higher in the Balkan countries [8], the former Yugoslavia, Rumania and Bulgaria, where it favors Gypsies (below). In Gypsies, birth incidence was calculated as one in 2500.

Two genes have been reported to encode galactokinase: GK1 on chromosome 17 [9] and GK2 on chromosome 15 [10]. Until recently, only two disease-causing mutations were known, both in GK1 [9]. Now, a third mutation, also in GK1, is identified as the founder mutation responsible for galactokinase deficiency in Gypsies [11].

## Deficiency of Galactose-1-Phosphate Uridyltransferase

### Clinical Presentation

Two forms of the deficiency exist. Infants with complete or near-complete deficiency of the enzyme (*classical galactosemia*) have normal weight at birth but, as they start drinking milk, lose more weight than their healthy peers and fail to regain birth weight. Symptoms appear in the second half of the first week and include refusal to feed, vomiting, jaundice, lethargy, hepatomegaly, edema and ascites. Death from liver and kidney failure and sepsis may follow within days. Mental disability may take years to become apparent. Symptoms are milder and the course is less precipitous when milk is temporarily withdrawn and replaced by intravenous nutrition. Nuclear cataracts appear within days or weeks and become irreversible within weeks of their appearance.

In many countries, galactosemic newborns are discovered through mass screening for blood galactose, the transferase enzyme or both; this screening is performed using dried blood spots usually collected between the third and fifth days [4]. At the time of discovery, the first symptoms may already have appeared, and the infant may already have been admitted to a hospital, usually for jaundice.

Where newborns are not screened for galactosemia or when the results of screening are not yet available, diagnosis rests on clinical awareness. It is crucial that milk feeding be stopped as soon as galactosemia is considered, and resumed only when a galactose disorder has been excluded. The presence of a reducing substance in a routine urine specimen may be the first diagnostic lead. Galactosuria is present provided the last milk feed does not date back more than a few hours and vomiting has not been excessive. However, owing to the early development of a proximal renal tubular syndrome, the acutely ill galactosemic infant may also excrete some glucose, together with an excess of amino acids. While hyperaminoaciduria may aid in the diagnosis, glucosuria often complicates it. When both reducing sugars (galactose and glucose) are present and reduction and glucose tests are done, and when the former test is strongly positive and the latter is weakly positive, the discrepancy is easily overlooked. Glucosuria is recognized, and galactosuria is missed. On withholding milk, galactosuria ceases, but galactitol and amino acids in excess continue to be excreted for a few days.

*Partial transferase deficiency* is, as a rule, asymptomatic. It is more frequent than classical galactosemia and is most often discovered in mass newborn screening because of moderately elevated blood galactose and/or low transferase activity [12].

### Metabolic Derangement

Individuals with a profound deficiency of galactose-1-phosphate uridyltransferase can phosphorylate ingested galactose but fail to metabolize galactose-1-phosphate. As a consequence, galactose-1-phosphate and galactose accumulate, and galactitol is formed. As in galactokinase deficiency, cataract formation can be explained by galactitol accumulation. The pathogenesis of the hepatic, renal and cerebral disturbances is less clear but is probably related to the accumulation of galactose-1-phosphate and (perhaps) of galactitol.

## Diagnostic Tests

Diagnosis is made by assaying transferase in heparinized whole blood or erythrocyte lysates, and/or by measuring abnormally high levels of galactose-1-phosphate in red cells. Where rapid shipment of whole blood is difficult, blood dried on filter paper can also be used for a semiquantitative assay. In patients with classical galactosemia, deficiency of galactose-1-phosphate uridyltransferase is complete or nearly complete. It should be noted that, when an infant has received an exchange transfusion, as is often the case, assays in blood must be postponed for 3-4 months. In some hospitals, a blood specimen, liquid or dried on filter paper, is collected prior to every exchange transfusion. In this situation, the finding of reduced transferase activity in parental blood may provide welcome preliminary information since, in heterozygotes, the enzyme activity in red cells is ~50% of normal. Cultured skin fibroblasts can also be used for the enzyme assay, as can liver or kidney cortex. If taken post mortem, the latter specimens should be adequately collected and frozen, since in vivo cell damage and/or autolysis may result in decreased enzyme activity. *Antenatal diagnosis* is possible by measuring transferase activity in cultured amniotic fluid cells, biopsied chorionic villi, or amniotic fluid galactitol [13]. Restricting maternal lactose intake does not interfere with a diagnosis based on galactitol measurements in amniotic fluid.

In *partial transferase deficiency*, activities of 10-50% of the normal activity are measured. As a rule, red cell galactose-1-phosphate is also elevated. Several variants of the partial enzyme defect have been reported, of which the best known is the *Duarte* variant. Some variants can be distinguished by enzyme electrophoresis and, more recently, DNA analysis. Our own experiences and published studies [1, 12] suggest that the Duarte variant is usually, if not always, benign. Each newborn with partial transferase deficiency must nevertheless be observed closely, because allelic variants other than Duarte may be operative [1, 2]. Assessment involves examination of blood galactose, erythrocyte galactose-1-phosphate and transferase activity levels, testing for aminoaciduria, enzyme electrophoresis and investigation of the parents [12]. Galactose-tolerance tests are notoriously noxious to the child with classical galactosemia and have no place in evaluating the need for treatment of partial deficiencies [1].

## Treatment and Prognosis

Treatment of the newborn with classical galactosemia consists of the exclusion of all galactose from the diet and must be started at once, even before the results of diagnostic tests are available. When a galactose-free diet is instituted early enough, symptoms disappear promptly, jaundice resolves within days, cataracts may clear, liver and kidney functions return to normal and liver cirrhosis may be prevented.

For dietary treatment, the following facts are worthy of consideration:

- From early embryonic life on, man is capable of synthesizing UDP-galactose from glucose through the epimerase reaction [14-16]. Therefore, man does not depend on exogenous galactose. Raising a galactosemic child on a diet completely devoid of galactose would cause it no harm, yet such a diet does not exist!
- Milligram amounts of galactose cause an appreciable rise of galactose-1-phosphate in red blood cells, and one must assume that the same happens in sensitive tissues, such as brain, liver and kidney. It is impossible to define toxic tissue levels of galactose-1-phosphate and, therefore, "safe" amounts of dietary galactose – if they exist at all – cannot be defined. For these reasons, it is advisable to be alert for traces of the sugar and to eliminate it as much as possible.
- Galactosemics certainly synthesize galactose from glucose [1, 15-19]. In galactosemic newborns first exposed to milk, then diagnosed and treated properly, erythrocyte galactose-1-phosphate stays high for several weeks. This fact and other observations [16-19] are evidence for continuous "self-intoxication" [17, 18] by the galactosemic, a matter of concern because of some late complications, such as premature ovarian failure [21] and central nervous system dysfunction. In adults on a strict lactose-exclusion diet, galactose intake was estimated at 20-40 mg/day (which almost certainly is an underestimate); at the same time, they produced a gram or more of galactose endogenously [19, 22]. Minimal amounts of galactose from food and hidden sources probably contribute to erythrocyte galactose-1-phosphate, but only real breaks in the diet cause a rise above 6 mg/dl. Such breaks do not cause any discomfort to the patient who, therefore, never develops aversion to galactose-containing food. The measurement of urinary galactitol for monitoring treatment has not been successful, as only large amounts of ingested galactose are reflected, and those are only detected after some delay [23].

### Treatment of the Newborn Infant

Treating newborns is comparatively easy, as adequate lactose-free formulae with meat or soy bases are available. Soya formulae from which raffinose and stachyose have been removed are preferable over others. Elimination of milk and milk products is the mainstay of lifelong treatment [24].

### Spoon-Feeding

When spoon-feeding is started, parents must learn to know all other sources of galactose and need assistance from the pediatrician and dietitian, who must have recourse to published recommendations [25–29]. A critical workshop report is especially helpful [27]. Parents are advised to do the following:

- Prepare meals from basic foodstuffs
- Avoid canned food, byproducts and preserves unless they are certified not to contain lactose or galactose
- Read and reread labels and declarations of ingredients which may change without notification
- Look out for hidden sources of galactose and lactose from milk powder, milk solids, "hydrolyzed whey" (a sweetener labeled as such), drugs in tablet form, toothpaste, baking additives, fillers, sausages, etc.
- Support campaigns for complete food and drug labeling

### Vegetables and Fruits

Parents must be trained to understand that *eliminating all galactose from the diet must remain the goal*, although it can never be reached. The reason for this is that galactose is present in a great number of vegetables and fruits [29], as a component of galactolipids and glycoproteins, in the disaccharide melibiose and in the oligosaccharides raffinose and stachyose [30, 31]. The latter two contain galactose in an α-galactosidic linkage not hydrolyzable by human small intestinal mucosa in vitro or in vivo [31]. They are often considered safe for consumption by galactosemics. However, this may not be the case when the small intestine is colonized by bacteria capable of releasing galactose. Raffinose- and stachyose-rich vegetables (beans, peas, lentils, etc.) should not be eaten by a galactosemic who has diarrhea. While beans, peas and lentils should never make up a full meal or a large dish, a few small seeds, e.g., in a dish of young string beans, should not cause concern. Nevertheless, gastroenterologists insist that the small intestine may be colonized in the absence of diarrhea; obviously, the issue is not closed.

### Cheese

It is not generally known that Swiss cheeses of the Emmentaler, Gruyères, and Tilsiter types are galactose- and lactose-free, as these sugars are cleared by the fermenting microorganisms [32]. Other hardened cheeses may prove equally safe for galactosemics. Calcium supplements should be prescribed before cheese is introduced to the child's diet; supplements may also be needed by older children and young adults [33]. Calcium prescriptions containing lactobionate [29] must be avoided, because the β-galactosidase of human intestinal mucosa hydrolyses lactobionate, freeing galactose [34].

### Breaks of Discipline

Whether single or repeated breaks of discipline (such as an occasional ice cream by a school-age child or adult galactosemic) will cause any damage is unknown. Uridine supplements for galactosemics have been proposed [35]; the concept was born from frustration. No therapeutic value has been demonstrated and, therefore, they have absolutely no place in the management of galactosemia [36]. Dietary treatment of female patients is continued during pregnancy [37].

### Complications of Treated Galactosemia

Mild growth retardation, delayed speech development, verbal dyspraxia, difficulties in spatial orientation and visual perception, and mild intellectual deficit have been variably described as complications of treated galactosemia. The complete set of sequelae is not necessarily present in every patient, and the degree of handicap appears to vary widely. Puberty is often delayed and may be induced by hormonal therapy. *Ovarian dysfunction*, an almost inescapable consequence of galactosemia is not prevented even by strict diet and is often signaled early in infancy or childhood by hypergonadotropinism [21]. Since, in galactosemic women, the number of expected ovulatory cycles is limited, it may be wise to temporarily suppress cycles by birth-control medication, which is lifted when the young woman wishes to become pregnant. Prescription is hampered by the fact that seemingly all drug tablets contain lactose, providing 100 mg or more of the noxious sugar per treatment day.

### Long-Term Results

Several multicenter, long-term studies have cast a shadow of doubt on the effectiveness of dietary treatment. It must be stressed here that said studies were retrospective, not prospective, and were probably marred by negative selection of patients. They have spread unfounded pessimism among physicians and dieticians who must overcome it for the benefit of their patients.

### Treatment of Partial Transferase Deficiency

Because it is impossible to decide whether partial transferase deficiency needs to be treated (above), some centers have adopted a pragmatic approach, prescribing lactose-free formula (limited to 4 months) to all such infants discovered by newborn screening [12]. The formula must not be completely free of galactose. At the end of a 1-week trial with a daily supplement of 2–3 dl of cows milk, if the erythrocyte galactose-1-phosphate level is below 2 mg/dl and aminoaciduria is normal, the infant is returned to normal nutrition and declared healthy.

### Dietary Treatment in Pregnant Woman at Risk

Based on the presumption that toxic metabolites deriving from galactose ingested by the heterozygous mother accumulate in the galactosemic fetus, mothers are often counseled to refrain from drinking milk for the duration of pregnancy. However, despite dietary restriction by the mother, galactose-1-phosphate and galactitol accumulate in the fetus [1, 16, 26] and in the amniotic fluid [13]. It is hypothesized [15–18] that the affected fetus produces galactose-1-phosphate endogenously from glucose-1-phosphate via the pyrophosphorylase–epimerase pathway (Fig. 7.1), which also provides UDP-galactose and, thus, secures the biosynthesis of galactolipids and galactoproteins indispensable for cell differentiation and growth. Since the affected fetus does not depend on (but may suffer from) the galactose he receives from his mother via the placenta, galactose restriction is the prudent stance for pregnant mothers. Affected newborns of treated mothers appear healthy at birth.

## Genetics

The mode of inheritance is autosomal recessive. In a large European screening series, birth incidence of classical galactosemia was one in ~55,000. The gene is situated on chromosome 9. The gene structure has been characterized, and mutation analysis is in full progress [38]. Nevertheless, some genotype–phenotype matching is already possible. For instance, homozygosity for the Q188R mutation, unfortunately prevalent, has been associated with unfavorable clinical outcome [38]. Because transferase polymorphism abounds [1, 2, 38], partial transferase deficiency is more frequent than classical galactosemia. Owing to the high gene frequency of the allelic variant Duarte, compound heterozygosity for galactosemia/Duarte is relatively common, occurring once in approximately 3000–4000 newborns.

## Uridine Diphosphate-Galactose 4'-Epimerase Deficiency

### Clinical Presentation

This disorder [1, 2] exists in two forms, both of which are discovered through newborn screening using suitable tests sensitive to both galactose and galactose-1-phosphate in dried blood [4]. In the very few children with the severe form of the disorder, the enzyme defect was subtotal. These newborns presented with vomiting, jaundice and hepatomegaly reminiscent of untreated classical galactosemia; one was found to have elevated blood methionine on newborn screening. All had galactosuria and hyperaminoaciduria; one had cataracts, and one had sepsis.

Infants with the mild form (sometimes falsely labeled "localized" or "peripheral") appear healthy [14]. The enzyme defect is incomplete; reduced stability and greater than normal requirement for the coenzyme nicotinamide adenine dinucleotide have been described [39]. Milk-fed newborns with the mild form detected in newborn screening are healthy and have neither hypergalactosemia, galactosuria nor hyperaminoaciduria.

### Metabolic Derangement

The enzyme deficiency provokes an accumulation of UDP-galactose after milk feeding. This build-up also results in the accumulation of galactose-1-phosphate (Fig. 7.1).

### Diagnostic Tests

The deficiency should be suspected when red cell galactose-1-phosphate is measurable while galactose-1-phosphate uridyltransferase is normal. Diagnosis is confirmed by the assay of epimerase in erythrocytes. Heterozygous parents have reduced epimerase activity, a finding that usually helps in the evaluation. Diagnosis of the severe from is based on the clinical symptoms, chemical signs and more marked deficiency of epimerase in red cells.

### Treatment and Prognosis

The child with the severe form of epimerase deficiency is unable to synthesize galactose from glucose and is, therefore, galactose-dependent. Dietary galactose in excess of actual biosynthetic needs will cause accumulation of UDP-galactose and galactose-1-phosphate, the latter being one presumptive toxic metabolite. When the amount of ingested galactose does not meet biosynthetic needs, synthesis of galactosylated compounds, such as galactoproteins and galactolipids, is impaired. As there is no easily available chemical parameter on which to base the daily galactose allowance (such as, e.g., blood phenylalanine in phenylketonuria) treatment is extremely difficult. Children known to suffer from the disorder have impaired psychomotor development.

Infants with the mild form of epimerase deficiency described thus far have not required treatment, but it is advisable that the family physician or pediatrician examine one or two urine specimens for reducing substances and exclude aminoaciduria within a couple of weeks after diagnosis, while the infant is still being fed milk. He should also watch the infant's psychomotor progress without, however, causing concern to the parents.

## Genetics

Epimerase deficiency is inherited as an autosomal-recessive trait. The epimerase gene resides on chromosome 1; it has been cloned and characterized.

## Fanconi-Bickel Syndrome

This is a recessively inherited disorder of glucose and galactose transport and is thought to be rare. A few cases have been discovered during newborn screening for galactose in blood. For further details, see Chap. 6.

## Portosystemic Venous Shunting

Portosystemic bypass of splanchnic blood via ductus venosus Arantii [40] or an intrahepatic venous malformation [41] causes alimentary hypergalactosemia, which is discovered during metabolic newborn screening.

## References

1. Gitzelmann R, Hansen RG (1980) Galactose metabolism, hereditary defects and their clinical significance. In: Burman D, Holton JB, Pennock CA (eds) Inherited disorders of carbohydrate metabolism. MTP, Lancaster, pp 61–101
2. Segal S, Berry GT (1995) Disorders of galactose metabolism. In: Scriver CR, Beaudet AL, Sly WS, Valle D (eds) The metabolic and molecular basis of inherited disease, 7th edn. McGraw-Hill, New York, pp 967–1000
3. Gitzelmann R, Steinmann B (1984) Galactosemia: how does long-term treatment change the outcome? Enzyme 32:37–46
4. Gitzelmann R (1980) Newborn screening for inherited disorders of galactose metabolism. In: Bickel H, Guthrie R, Hammersen G (eds) Neonatal screening for inborn errors of metabolism. Springer, Berlin Heidelberg New York, pp 67–79
5. Beutler E, Paniker NV, Trinidad F (1971) The assay of red cell galactokinase. Biochem Med 5:325–332
6. Mayes JS, Guthrie R (1986) Detection of heterozygotes for galactokinase deficiency in a human population. Biochem Genet 2:219–230
7. Gitzelmann R (1967) Hereditary galactokinase deficiency a newly recognized cause of juvenile cataracts. Pediatr Res 1:14–23
8. Gitzelmann R (1987) This week's citation classic. Curr Contents Life Sci 30:14
9. Stambolian D, Ai Y, Sidjanin D, Nesburn K, Sathe G, Rosenberg M, Bergsma DJ (1995) Cloning the galactokinase cDNA and identification of mutations in two families with cataracts. Nature Genet 10:307–317
10. Lee RT, Peterson GL, Calman AF, Herskowitz I, O'Donell J (1992) Cloning of a human galactokinase gene (GK2) on chromosome 15 by complementation in yeast. Proc Natl Acad Sci USA 89:10887–10891
11. Kalaydjieva L, Perez-Lezaun A, Angelicheva D, Onengut S, Dye D, Bosshard NU, Tournev I, Hallmayer J, Jordanova A, Yanakiev P, Savov A, Kremensky I, Aneva L, Gitzelmann R (1999) A founder mutation in the GK1 gene is responsible for galactokinase deficiency in Gypsies. Am J Human Genet 65:1299–1307
12. Gitzelmann R, Bosshard NU (1995) Partial deficiency of galactose-1-phosphate uridyltransferase. Eur J Pediatr 154 [Suppl 2]:S40–44
13. Jakobs C, Kleijer WJ, Allen J, Holton JB (1995) Prenatal diagnosis of galactosemia. Eur J Pediatr 154 [Suppl 2]:S33–36
14. Gitzelmann R, Steinmann B (1973) Uridine diphosphate galactose 4-epimerase deficiency II. Clinical follow-up, biochemical studies and family investigation. Helv Paediatr Acta 28:497–510
15. Gitzelmann R (1969) Formation of galactose-1-phosphate from uridine diphosphate galactose in erythrocytes from patients with galactosemia. Pediatr Res 3:279–286
16. Gitzelmann R (1995) Galactose-1-phosphate in the pathophysiology of galactosemia. Eur J Pediatr 154 [Suppl 2]:S45–49
17. Gitzelmann R, Hansen RG (1974) Galactose biogenesis and disposal in galactosemics. Biochim Biophys Acta 372:374–378
18. Gitzelmann R, Hansen RG, Steinmann B (1975) Biogenesis of galactose, a possible mechanism of self-intoxication in galactosemia. In: Hommes FA, Van den Berg CJ (eds) Normal and pathological development of energy metabolism. Academic, London, pp 25–37
19. Berry GT, Nissin I, Lin Z, Masur AT, Gibson JB, Segal S (1995) Endogenous synthesis of galactose in normal men and patients with hereditary galactosaemia. Lancet 346:1073–1074
20. Donnell GN, Koch R, Fishler K, Ng WG (1980) Clinical aspects of galactosaemia. In: Burman D, Holton JB, Pennock CA (eds) Inherited disorders of carbohydrate metabolism. MTP, Lancaster, pp 103–115
21. Gibson JB (1995) Gonadal function in galactosemics and in galactose-intoxicated animals. Eur J Pediatr 154 [Suppl 2]:S14–20
22. Segal S (1998) Galactosaemia today: the enigma and the challenge. J Inherit Metab Dis 21:455–471
23. Donnell GN, Bergren WR (1975) The galactosaemias. In: Raine DN (ed) The treatment of inherited metabolic disease. MTP, Lancaster, pp 91–114
24. Brandt NJ (1980) How long should galactosaemia be treated? In: Burman D, Holton JB, Pennock CA (eds) Inherited disorders of carbohydrate metabolism. MTP, Lancaster, pp 117–124
25. Koch R, Acosta P, Ragsdale N, Donnell GN (1963) Nutrition in the treatment of galactosemia. J Am Diet Assoc 43:216–222
26. Koch R, Acosta P, Donnell G, Lieberman E (1965) Nutritional therapy of galactosemia. Management success depends on

rigid exclusion of all galactose-containing foods. Clin Pediatr 4:571-576
27. Clothier CM, Davidson DC (1983) Galactosaemia workshop. Hum Nutr Appl Nutr 37A:483-490
28. Gross KC, Acosta PB (1991) Fruits and vegetables are a source of galactose: implications in planning the diets of patients with galactosaemia. J Inherit Metab Dis 14:253-258
29. Acosta PB, Gross KC (1995) Hidden sources of galactose in the environment. Eur J Pediatr 154 [Suppl 2]:S87-92
30. Wiesmann U, Rosé-Beutler B, Schlüchter R (1995) Leguminosae in the diet: The raffinose-stachyose-question. Eur J Pediatr 154 [Suppl 2]:S93-96
31. Gitzelmann R, Auricchio S (1965) The handling of soya alpha-galactosides by a normal and a galactosemic child. Pediatrics 36:231-235
32. Steffen C (1975) Enzymatische Bestimmungsmethoden zur Erfassung der Gärungsvorgänge in der milchwirtschaftlichen Technologie. Lebensm Wiss Technol 8:1-6
33. Kaufman FR, Loro ML, Azen C, Wenz E, Gilsanz V (1993) Effect of hypogonadism and deficient calcium intake on bone density in patients with galactosemia. J Pediatr 123:365-370
34. Harju M (1990) Lactobionic acid as a substrate of β-galactosidases. Milchwissenschaft 45:411-415
35. Kaufman FR, Ng WG, Xu YK, Guidici T, Kaleita TA, Donnell GN (1989) Treatment of patients (PTS) with classical galactosemia (G) with oral uridine. Pediatr Res 25:142A
36. Manis FR, Cohn LB, McBride Chang C, Wolff JA, Kaufmann FR (1997) A longitudinal study of cognitive functioning in patients with classical galactosaemia, including a cohort treated with oral uridine. J Inherit Metab Dis 20:549-555
37. Sardharwalla IB, Komrower GM, Schwarz V (1980) Pregnancy in classical galactosemia. In: Burman D, Holton JB, Pennock CA (eds) Inherited disorders of carbohydrate metabolism. MTP, Lancaster, pp 125-132
38. Elsas LJ II, Fridovich-Keil JL, Leslie ND (1993) Galactosemia: a molecular approach to the enigma. Int Pediatr 8:101-109
39. Gitzelmann R, Steinmann B, Mitchell B, Haigis E (1976) Uridine diphosphate galactose 4'-epimerase deficiency. IV. Report of eight cases in three families. Helv Paediatr Acta 31:441-445
40. Gitzelmann R, Arbenz UV, Willi UV (1992) Hypergalactosaemia and portosystemic encephalopathy due to persistence of ductus venosus Arantii. Eur J Pediatr 151:564-568
41. Gitzelmann R, Forster I, Willi UV (1997) Hypergalactosaemia in a newborn: self-limiting intrahepatic portosystemic venous shunt. Eur J Pediatr 156:719-722

## Fructose Metabolism

Fructose is one of the main sweetening agents in the human diet. It is found in its free form in honey, fruits and many vegetables, and is associated with glucose in the form of the disaccharide sucrose in numerous foods and beverages. Sorbitol, also widely distributed in fruit and vegetables, is converted into fructose in the liver by sorbitol dehydrogenase (Fig. 8.1).

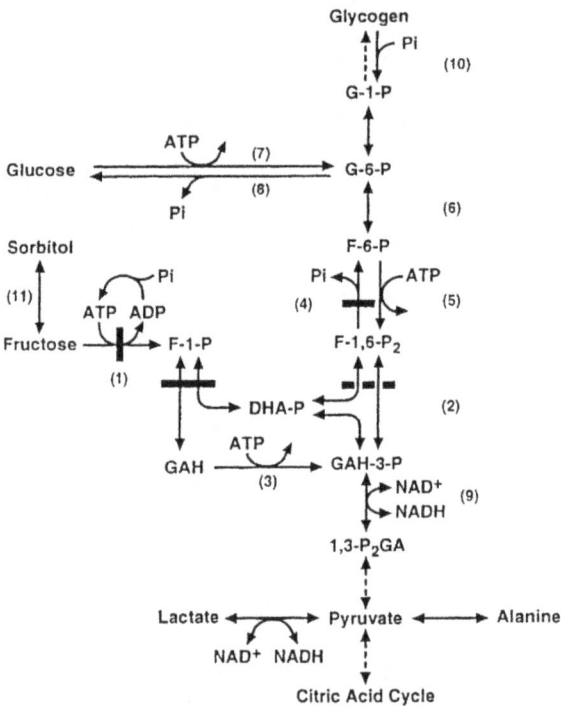

**Fig. 8.1.** Fructose is mainly metabolized by a specialized pathway found in the liver, kidney cortex and small-intestinal mucosa and composed of fructokinase (1), aldolase B (2) and triokinase (3). Aldolase B also intervenes in the glycolytic–gluconeogenic pathway, which also includes the following enzymes: fructose-1,6-bisphosphatase (4), phosphofructokinase (5), glucose-6-phosphate isomerase (6), glucokinase and hexokinase (7), glucose-6-phosphatase (8) and glyceraldehyde-3-phosphate dehydrogenase (9). Also shown are glycogen phosphorylase (10) and sorbitol dehydrogenase (11). $1,3-P_2GA$, 1,3-bisphosphoglycerate; *DHA-P*, dihydroxyacetone phosphate; *F*, fructose; *G*, glucose; *GAH*, glyceraldehyde; *P*, phosphate; *Pi*, inorganic phosphate. The three enzyme defects in fructose metabolism are depicted by *solid bars* across the *arrows*; the diminished activity of aldolase B toward fructose-1,6-bisphosphate in hereditary fructose intolerance is depicted by a *broken bar*

# CHAPTER 8

# Disorders of Fructose Metabolism

Georges Van den Berghe

## CONTENTS

| | |
|---|---|
| Essential Fructosuria | 111 |
| Clinical Presentation | 111 |
| Metabolic Derangement | 111 |
| Diagnostic Tests | 111 |
| Treatment and Prognosis | 111 |
| Genetics | 111 |
| Hereditary Fructose Intolerance | 112 |
| Clinical Presentation | 112 |
| Metabolic Derangement | 112 |
| Diagnostic Tests | 113 |
| Treatment and Prognosis | 113 |
| Genetics | 113 |
| Fructose-1,6-bisphosphatase Deficiency | 113 |
| Clinical Presentation | 113 |
| Metabolic Derangement | 114 |
| Diagnostic Tests | 115 |
| Treatment and Prognosis | 115 |
| Genetics | 115 |
| References | 115 |

Two inborn errors are known in the specialized pathway of fructose metabolism depicted in Fig. 8.1. *Essential fructosuria* is a completely harmless anomaly characterized by the appearance of fructose in the urine after the intake of fructose-containing foods. In *hereditary fructose intolerance* (HFI), fructose provokes prompt gastrointestinal discomfort and hypoglycemia upon ingestion, although sensitivity varies from patient to patient; it may cause liver and kidney failure when taken persistently and becomes life threatening when given intravenously. *Fructose-1,6-bisphosphatase* (FBPase) *deficiency* is usually also considered an inborn error of fructose metabolism although, strictly speaking, it is not a defect of the specialized fructose pathway. It is manifested by the appearance of hypoglycemia and lactic acidosis (neonatally or during fasting) and may also be life-threatening.

## Essential Fructosuria

### Clinical Presentation

Essential fructosuria is a rare "non-disease"; it is detected by the routine screening of urine for reducing sugars [1]. It is caused by a deficiency of fructokinase [2], the first enzyme of the specialized fructose pathway (Fig. 8.1). Fructokinase is normally only found in the liver, kidney and small-intestinal mucosa.

### Metabolic Derangement

In cases of deficiency, ingested fructose is partly excreted as such in the urine and is partly slowly metabolized by an alternate pathway, namely conversion into fructose-6-phosphate by hexokinase in adipose tissue and muscle.

### Diagnostic Tests

Fructose gives a positive test for reducing sugars and a negative reaction with glucose oxidase. It can be identified by various techniques, such as thin-layer chromatography [3]. Fructose-tolerance tests (see "Hereditary Fructose Intolerance") provoke neither an increase in blood glucose, as in normal subjects, nor hypoglycemia, as in HFI and FBPase deficiency.

### Treatment and Prognosis

Dietary treatment is not indicated, and the prognosis is excellent.

### Genetics

The mode of inheritance is autosomal recessive, and the frequency of the homozygotes has been estimated at 1:130,000 [4]. Two mutations, G40R and A43T, altering the same region of fructokinase, have been found in a family with three compound heterozygotes [5].

## Hereditary Fructose Intolerance

### Clinical Presentation

HFI provokes no symptoms as long as the affected subjects do not ingest fructose. Typically, babies do well during breast feeding. Symptoms appear upon introduction of cow's-milk formulas sweetened with sucrose or at weaning, when fruits and vegetables are given [6, 7]. Certain patients are very sensitive to fructose, whereas others can tolerate moderate intakes of fructose (up to 250 mg/kg/day, as compared with an average intake of 1–2 g/kg/day in Western societies). Generally, the first signs of HFI are those of gastrointestinal discomfort and hypoglycemia following meals containing fructose. Nausea, vomiting, pallor, sweating, trembling, lethargy and, eventually, jerks and convulsions may be observed. If the condition is not recognized and fructose is not excluded from the diet, failure to thrive, liver disease manifested by hepatomegaly, jaundice, bleeding tendency and, eventually, edema, ascites, and proximal renal tubular dysfunction appear. The younger the child and the higher the intake of fructose, the more severe the clinical picture which, when its cause is not recognized and treated appropriately, may lead to liver and kidney failure and, eventually, death.

In young infants, HFI can be suspected if the mother, being aware that her baby does not tolerate certain foods, has suppressed them from the diet so that the infant develops normally. In older children, a distinct aversion towards foods containing fructose develops; this protects them but is sometimes considered psychotic behavior. At school age, HFI is occasionally recognized when hepatomegaly or growth delay is found [8]. At adult age, some cases are only diagnosed after life-threatening perfusions with fructose [9] or sorbitol. Because approximately half the adults with HFI are completely free of caries, the diagnosis has also been made by dentists. Although several hundred patients with HFI have been identified, since its recognition as an inborn error of metabolism in 1957 [6], these observations indicate that affected subjects may remain undiagnosed in the general population.

### Metabolic Derangement

HFI is caused by the inability of the second enzyme of the fructose pathway, aldolase B (Fig. 8.1), to split fructose-1-phosphate into dihydroxyacetone phosphate and glyceraldehyde [10]. In tissues that possess the specialized fructose pathway (namely liver, kidney cortex and small-intestinal mucosa), fructose cannot be converted into glucose and lactate. Moreover, as a consequence of the high activity of fructokinase, ingestion and intravenous infusion of fructose results in accumulation of fructose-1-phosphate. This accumulation has two major effects [11]: it inhibits the production of glucose, hence inducing hypoglycemia, and it provokes depletion of ATP, an essential component for all cellular functions. Recently, the accumulation of fructose-1-phosphate has also been shown to result in deficient glycosylation of serum transferrin by inhibiting phosphomannomutase [12], the deficiency of which is the cause of a congenital defect of glycosylation, CDG syndrome type Ia (Chap. 38).

### *Inhibition of Glucose Production*

Inhibition of glucose production results from a block of both hepatic glycogenolysis, which maintains blood glucose in the early postprandial phase, and gluconeogenesis, which provides glucose during more prolonged fasting. The block of glycogenolysis results from inhibition (by fructose-1-phosphate) of phosphorylase, the enzyme that catalyzes the liberation of glucose-1-phosphate from glycogen. The impairment of gluconeogenesis results from inhibition (by fructose-1-phosphate) of the condensation of glyceraldehyde-3-phosphate and dihydroxyacetone phosphate into fructose-1,6-bisphosphate (which is also catalyzed by aldolase B) and of the conversion of fructose-6-phosphate into glucose-6-phosphate, which is catalyzed by glucose-6-phosphate isomerase.

### *Depletion of Adenosine Triphosphate*

Depletion of adenosine triphosphate (ATP) (and of the related nucleotide guanosine triphosphate, GTP) in fructose-metabolizing tissues is a consequence of their utilization and sequestration in the formation of high amounts of fructose-1-phosphate. It is accompanied by depletion of inorganic phosphate, which is required to regenerate ATP from adenosine diphosphate in the mitochondria (Fig. 8.1). Because both GTP and inorganic phosphate are inhibitors of liver AMP deaminase, a rate-limiting enzyme of the catabolism of the adenine nucleotides, their depletion provokes a degradation of the hepatic adenine nucleotide pool, leading to increased production of uric acid. The depletion of ATP, the energy currency of the cell, induces a series of disturbances, including inhibition of protein synthesis and ultrastructural lesions, which are responsible for the gastrointestinal discomfort and hepatic and renal dysfunction.

It should be noted that the intravenous administration of fructose to normal subjects also induces the metabolic derangements described in the previous paragraph, although higher doses are required than in patients with HFI, as demonstrated by $^{31}P$ nuclear magnetic resonance spectroscopy [13]. In normal subjects, intravenous fructose raises glycemia because of its rapid conversion into glucose. However, the equally rapid conversion of fructose into lactate may provoke metabolic acidosis. For these reasons, in parenteral nutrition, the use of fructose, sorbitol and mixtures of glucose and fructose known as invert sugars has been strongly discouraged [14].

## Diagnostic Tests

Whenever HFI is suspected, fructose should immediately be completely withdrawn from the diet. The beneficial effect of withdrawal, usually seen within days, provides a first diagnostic clue. Only after some weeks should an *intravenous fructose tolerance test* be performed, provided liver function tests have become normal. Oral tests are not recommended, because they provoke more ill effects and are less reliable [15]. Due to its potential hypoglycemic effect, the test should be started at a slightly elevated plasma glucose concentration (between 4 mmol/l and 5 mmol/l). Fructose (200 mg/kg body weight) is injected as a 10% solution intravenously in 2 min. Blood samples are taken at 0 (×2), 5, 10, 15, 30, 45, 60 and 90 min for determination of glucose and phosphate. In normal children, blood glucose increases by 0–40%, with no or minimal changes in phosphate [15]. In HFI patients, glucose and phosphate decrease within 10–20 min. As a rule, the decrease of phosphate precedes and occurs more rapidly than that of glucose.

Laboratory findings in patients with a sustained fructose intake are similar to those of liver disease (elevations of serum transaminases and bilirubin, depletion of blood clotting factors) and of proximal tubular dysfunction (proteinuria, mellituria, generalized hyperaminoaciduria, metabolic acidosis).

To confirm the diagnosis, the activity of aldolase B should be measured in a biopsy of liver, kidney cortex or intestinal mucosa. In HFI patients, the capacity of aldolase B to split fructose-1-phosphate is reduced, usually to a few percent of normal [10, 15], although residual activities as high as 30% of normal have been reported [9]. There is also a distinct (but less marked) reduction of the activity of aldolase B toward fructose-1,6-bisphosphate. As a consequence, the ratio of $V_{max}$ towards fructose-1,6-bisphosphate versus the $V_{max}$ towards fructose-1-phosphate, which is approximately 1 in control liver, is increased to 2–∞ in HFI patients. The activity of aldolase is normal in blood cells, muscle and skin fibroblasts, which contain a different isozyme, aldolase A.

## Treatment and Prognosis

Treatment of HFI consists of the elimination of all sources of fructose in the diet. This involves the suppression of all foods in which fructose and/or sucrose or sorbitol occur naturally or have been added during processing [16]. That fructose may be present in medications and in infant formulas should also be verified. A list of foods to use and to avoid is given to the parents (Table 8.1). Sucrose should be replaced by glucose, maltose and/or starch to prevent the fructose-free diet from containing too much fat. After institution of the diet, most abnormalities disappear rapidly – except hepatomegaly, which may persist for months and even years [17]. The reason for this is unclear. An insufficient restriction of fructose has been reported to cause isolated growth retardation, as evidenced by catch-up growth on a stricter diet [8]. Thus, at least in childhood, the intake of fructose should not be determined by subjective tolerance. Needless to say, patients (and their parents) should be made aware of the fact that infusions containing fructose, sorbitol or invert sugar are life-threatening, and they should report fructose intolerance on any hospital admission.

## Genetics

HFI is inherited as an autosomal-recessive trait. Studies of the aldolase-B gene in 50 predominantly European patients have shown that the most frequent mutation, accounting for 67% of alleles, is A149P [18]. This mutation also creates a new recognition site for the restriction enzyme AhaII, which renders it easily detectable. Next in frequency, accounting for 16% of alleles, is A174D, which is found in Switzerland and in Southern Europe. In patients from the United States and Canada, the A149P and A174D mutations are found in the same order of prevalence, although at a slightly lower frequency than in Europe [19]. In other patients, a variety of point mutations and deletions are found, the latter ranging in size from a single base pair to 1.65 kb [20]. Thus, the heterogeneity of HFI, evidenced by the variability of both the sensitivity to fructose and the residual activity of aldolase B toward fructose-1-phosphate, is also apparent at the gene level. The true incidence of HFI is unknown but may be estimated at 1:20,000 in Switzerland and Great Britain [21].

## Fructose-1,6-bisphosphatase Deficiency

### Clinical Presentation

In about half of the cases, the deficiency of FBPase presents in the neonatal period (day 1–4) with episodes

**Table 8.1.** Sucrose- and fructose-free diet [16]. Labels of medications and of all canned, packaged or processed foods should be checked to be sure that sugar and fruit are not used

| Food group | Foods to use | Foods to avoid |
| --- | --- | --- |
| Bread | White and brown breads (ask for composition), soda crackers, saltines | All other breads, crackers, biscuits, cookies |
| Cereals | Cooked or ready-to-eat cereals (except sugar-coated cereals) | Sugar-coated cereals |
| Cheese | Any kind | None |
| Desserts | Natural yogurt, pudding without sugar, homemade ice cream | All deserts containing sugar (cake, pie, cookies, candy, puddings, Jello, ice cream, sherbet), honey, fruit or fruit juice, most products for diabetics |
| Eggs | Any kind | None |
| Fat | Butter, margarine, oil, homemade mayonnaise and salad dressings made without sugar | Mayonnaise, salad dressings made with sugar |
| Fruits | None | All fruits and fruit juices, dates |
| Meat, fish | Beef, chicken, fish, lamb, pork, turkey, veal | Ham, bacon, lunch meats, and any other meats in which sugar is used in processing |
| Milk | Any kind | Milk preparations with added sugar |
| Miscellaneous | Coffee, tea, vegetable juices and soups from allowed vegetables, dietetic beverages with sugar substitutes, cocoa, salt, pepper | Catsup, chili sauces and other sauces containing sugar, carbonated beverages, maple syrup, preserves, honey, jam, jellies, other condiments |
| Nuts | Any kind | Sugar coated |
| Potatoes and substitutes | White potatoes, macaroni, noodles, spaghetti, rice | Sweet potatoes |
| Sweeteners | Glucose, dextrin, maltose, calorie-free sweeteners | Sucrose, fructose, sorbitol |
| Vegetables | Asparagus, cabbage, cauliflower, celery, green beans, green peppers, lettuce, peas, spinach, wax beans, root vegetables except carrots | Carrots, leeks, onions, sweetmaize, canned vegetables with added sucrose |

of severe lactic acidosis associated with marked hypoglycemia. Hyperventilation, dyspnea, tachycardia and apneic spells are accompanied by irritability or somnolence and lethargy that may lead to coma and convulsions. Muscular weakness and moderate hepatomegaly are often present. In the other half of the patients, attacks may not occur for months or even years and are usually triggered by a febrile episode accompanied by refusal to feed and vomiting, as in the first reported patient [22]. Attacks may also occur following ingestion of fructose or sucrose. FBPase deficiency is life threatening and, as in HFI, administration of intravenous fructose is particularly contraindicated and may lead to death. In between attacks, patients are usually well, although mild, intermittent or chronic acidosis may exist. The frequency of the attacks lessens with age, and the majority of the survivors display normal somatic and psychomotor development [23].

**Metabolic Derangement**

FBPase deficiency impairs the formation of glucose from all gluconeogenic precursors, including dietary fructose (Fig. 8.1). Consequently, maintenance of normoglycemia in patients with the defect is exclusively dependent on glucose (and galactose) intake and on degradation of hepatic glycogen. Thus, hypoglycemia is likely to occur when glycogen reserves are limited (as in newborns) or exhausted (as when fasting). The defect also provokes accumulation of the gluconeogenic substrates, lactate/pyruvate, glycerol and alanine. The lactate/pyruvate ratio is usually increased, owing to secondary impairment of the conversion of 1,3-bisphosphoglycerate into glyceraldehyde-3-phosphate, resulting in accumulation of reduced nicotinamide adenine dinucleotide, the other substrate of glyceraldehyde-3-phosphate dehydrogenase. Recently, attention has been drawn to the fact that hyperketonemia and ketonuria, which usually accompany hypoglycemia (owing to release of fatty acids from adipose tissue), may be absent in some patients with FBPase deficiency [24]. This may be explained by accumulation of pyruvate, resulting in build-up of oxaloacetate and, hence, in diversion of acetyl-coenzyme A (CoA) away from ketone-body formation and into citrate synthesis. This, in turn, results in increased synthesis of malonyl-CoA in the cytosol. Elevated malonyl-CoA, by inhibiting carnitine-palmitoyl transferase I, prevents entry of long-chain fatty-acyl-CoA into the mitochondria and, thereby, further reduces ketogenesis. It also promotes accumulation of fatty acids in liver and plasma, as documented in some patients.

In contrast to subjects with HFI, children with FBPase deficiency generally tolerate sweet foods. Nevertheless, loading tests with fructose induce hypoglycemia as in HFI, although higher doses are required. The hypoglycemia is caused by the inhibitory effect of rapidly formed but (owing to the FBPase defect) slowly metabolized fructose-1-phosphate on liver glycogen phosphorylase. That higher doses are required is explained by the fact that, in contrast to the aldolase-B defect, the FBPase deficiency still allows fructose-1-phosphate to be converted into lactate, with the consequent deleterious effects expressed as a lactic acidosis-prone condition. $^{31}$P Magnetic resonance spectroscopy of the liver following the intravenous administration of fructose (200 mg/kg) has documented a slower decrease of the fructose-induced accumulation of fructose-1-phosphate and a delayed recovery of the ensuing depletion of inorganic phosphate and ATP (both of which are signs of fructose toxicity) in patients with FBPase deficiency compared with healthy controls [25].

## Diagnostic Tests

Analysis of plasma during acute episodes reveals lactate accumulation (up to 15–25 mM) accompanied by decreased pH and increased lactate/pyruvate ratio (up to 30), hypoglycemia and hyperalaninemia. Hyperketonemia can be found but, as already discussed, in several patients, ketosis has been reported to be moderate or absent [24]. Increased levels of free fatty acids and uric acid may also be measured. Urinary analysis reveals increased lactate, alanine, ketones (sometimes) and the presence of glycerol and glycerol-3-phosphate.

Loading tests with fructose (or with glycerol or alanine) should not be carried out during acute episodes and provide only a tentative diagnosis. Final diagnosis requires measurement of the activity of FBPase, preferably in the liver, or in kidney cortex or jejunum. The residual activity varies from zero to 30% of normal, indicating genetic heterogeneity of the disorder. Although normal enzyme activity is low in leukocytes, diagnosis can be attempted in these cells, taking into account that a deficient activity is diagnostic but that a normal activity does not rule out FBPase deficiency in the liver, particularly in girls, as repeatedly reported [26–28]. Muscle contains a different FBPase isozyme, with normal activity in patients with liver FBPase deficiency. Cultured skin fibroblasts, amniotic fluid cells and chorionic villi do not exhibit FBPase activity.

## Treatment and Prognosis

The acute, life-threatening episodes should be treated with continuous infusion at high rates of glucose (10–12 mg/kg/min for newborns) and bicarbonate (up to 200 mEq/24 h) to control hypoglycemia and acidosis. Maintenance therapy should be aimed at avoiding fasting periods, particularly during febrile episodes. This involves frequent feeding, use of slowly absorbed carbohydrates (such as uncooked starch) and gastric drip, if necessary. In small children, restriction of fructose, sucrose and sorbitol is also recommended, as is restriction of fat to 20–25% and restriction of protein to 10% of energy requirements.

Once FBPase deficiency has been diagnosed and adequate management introduced, its course is usually benign. Growth and development are normal, and tolerance to fasting improves with age up to the point that the disorder does not present a problem in later life [23]. This might be explained by an increasing capacity to store glycogen in the liver, resulting in a lower dependency on gluconeogenesis for the maintenance of blood glucose.

## Genetics

The deficiency of FBPase is inherited as an autosomal-recessive trait. Although there is evidence for at least two FBPase genes in humans, only a single gene, FBP1, has been cloned and characterized hitherto. It is localized on chromosome 9 and encodes the liver enzyme [29]. Recent studies show that the enzyme deficiency results from a variety of mutations [30], among them point mutations (E30X, G164S, A177D), an extensive deletion and a guanine insertion at base 961 in exon 7. The latter seems to be the predominant mutation in the Japanese. In two Caucasian patients [29] and three other Japanese patients [30], no mutation was found. This suggests the existence of unidentified mutations in the promoter of FBP1 or of mutations of the bifunctional enzyme that controls the concentration of fructose-2,6-bisphosphate, the main physiological regulator of FBPase [31].

## References

1. Steinitz H, Mizrahy O (1969) Essential fructosuria and hereditary fructose intolerance. N Engl J Med 280:222
2. Shapira F, Shapira G, Dreyfus JC (1961/1962) La lésion enzymatique de la fructosurie bénigne. Enzymol Biol Clin 1:170–175
3. Kraffczyk F, Helger R, Bremer HJ (1972) Thin-layer chromatographic screening tests for carbohydrate anomalies in plasma, urine and faeces. Clin Chim Acta 42:303–308
4. Lasker M (1941) Essential fructosuria. Hum Biol 13:51–63
5. Bonthron DT, Brady N, Donaldson IA, Steinmann B (1994) Molecular basis of essential fructosuria: molecular cloning and mutational analysis of human ketohexokinase (fructokinase). Hum Mol Genet 3:1627–1631
6. Froesch ER, Prader A, Labhart A, Stuber HW, Wolf HP (1957) Die hereditäre Fructoseintoleranz, eine bisher nicht bekannte kongenitale Stoffwechselstörung. Schweiz Med Wochenschr 87:1168–1171
7. Baerlocher K, Gitzelmann R, Steinmann B, Gitzelmann-Cumarasamy N (1978) Hereditary fructose intolerance in early childhood: a major diagnostic challenge. Survey of 20 symptomatic cases. Helv Paediatr Acta 33:465–487

8. Mock DM, Perman JA, Thaler MM, Morris RC Jr (1983) Chronic fructose intoxication after infancy in children with hereditary fructose intolerance. A cause of growth retardation. N Engl J Med 309:764-770
9. Lameire N, Mussche M, Baele G, Kint J, Ringoir S (1978) Hereditary fructose intolerance: a difficult diagnosis in the adult. Am J Med 65:416-423
10. Hers HG, Joassin G (1961) Anomalie de l'aldolase hépatique dans l'intolérance au fructose. Enzymol Biol Clin 1:4-14
11. Van den Berghe G (1978) Metabolic effects of fructose in the liver. Curr Top Cell Regul 13:97-135
12. Jaeken J, Pirard M, Adamowicz M, Pronicka E, Van Schaftingen E (1996) Inhibition of phosphomannose isomerase by fructose-1-phosphate: an explanation for defective N-glycosylation in hereditary fructose intolerance. Pediatr Res 40:764-766
13. Oberhaensli RD, Rajagopalan B, Taylor DJ et al. (1987) Study of hereditary fructose intolerance by use of $^{31}$P magnetic resonance spectroscopy. Lancet ii:931-934
14. Woods HF, Alberti KGMM (1972) Dangers of intravenous fructose. Lancet ii:1354-1357
15. Steinmann B, Gitzelmann R (1981) The diagnosis of hereditary fructose intolerance. Helv Paediatr Acta 36:297-316
16. Cornblath M, Schwartz R (1991) Hereditary fructose intolerance. In: Cornblath M, Schwartz R (eds) Disorders of carbohydrate metabolism in infancy, 3rd edn. Blackwell, London, pp 391-403
17. Odièvre M, Gentil C, Gautier M, Alagille D (1978) Hereditary fructose intolerance in childhood. Diagnosis, management and course in 55 patients. Am J Dis Child 132:605-608
18. Cross NCP, DeFranchis R, Sebastio G et al. (1990) Molecular analysis of aldolase B genes in hereditary fructose intolerance. Lancet i:306-309
19. Tolan DR, Brooks CC (1992) Molecular analysis of common aldolase B alleles for hereditary fructose intolerance in North Americans. Biochem Med Metabol Biol 48:19-25
20. Tolan DR (1995) Molecular basis of hereditary fructose intolerance: mutations and polymorphisms in the human aldolase B gene. Hum Mutat 6:210-218
21. James CL, Rellos P, Ali M, Heeley AF, Cox TM (1996) Neonatal screening for hereditary fructose intolerance: frequency of the most common mutant aldolase B allele (A149P) in the British population. J Med Genet 33:837-841
22. Baker L, Winegrad AI (1970) Fasting hypoglycaemia and metabolic acidosis associated with deficiency of hepatic fructose-1,6-bisphosphatase activity. Lancet ii:13-16
23. Moses SW, Bashan N, Flasterstein BF, Rachmel A, Gutman A (1991) Fructose-1,6-bisphosphatase deficiency in Israel. Israel J Med Sci 27:1-4
24. Morris AAM, Deshpande S, Ward-Platt MP et al. (1995) Impaired ketogenesis in fructose-1,6-bisphosphatase deficiency: a pitfall in the investigation of hypoglycaemia. J Inherit Metab Dis 18:28-32
25. Boesiger P, Buchli R, Meier D, Steinmann B, Gitzelmann R (1994) Changes of liver metabolite concentrations in adults with disorders of fructose metabolism after intravenous fructose by $^{31}$P magnetic resonance spectroscopy. Pediatr Res 36:436-440
26. Bührdel P, Böhme H-J, Didt L (1990) Biochemical and clinical observations in four patients with fructose-1,6-diphosphatase deficiency. Eur J Pediatr 149:574-576
27. Shin Y (1993) Diagnosis of fructose-1,6-bisphosphatase deficiency using leukocytes: normal leukocyte enzyme activity in three female patients. Clin Invest 71:115-118
28. Besley GTN, Walter JH, Lewis MA, Chard CR, Addison GM (1994) Fructose-1,6-bisphosphatase deficiency: severe phenotype with normal leukocyte enzyme activity. J Inherit Metab Dis 17:333-335
29. El-Maghrabi MR, Lange AJ, Jiang W et al. (1995) Human fructose-1,6-bisphosphatase gene (FBP1): exon-intron organization, localization to chromosome bands 9q22.2-q22.3, and mutation screening in subjects with fructose-1,6-bisphosphatase deficiency. Genomics 27:520-525
30. Kikawa Y, Inuzuka M, Jin BY et al. (1997) Identification of genetic mutations in Japanese patients with fructose-1,6-bisphosphatase deficiency. Am J Hum Genet 61:852-861
31. Hers HG, Van Schaftingen E (1982) Fructose 2,6-bisphosphate two years after its discovery. Biochem J 206:1-12

# CHAPTER 9

## Glucose-Induced Insulin Secretion and its Modulation

Glucose entry into the pancreatic β-cell is followed by its phosphorylation to glucose-6-phosphate by glucokinase. This enzyme functions as the glucose sensor, because its $K_m$ for glucose is close to the concentration of blood glucose. A small change in the latter will, therefore, increase the rate of metabolism of glucose, the generation of ATP by the glycolytic pathway, and the concentration of ATP relative to adenosine diphosphate. Elevation of ATP results in the closure of $K^+$ channels composed of two subunits: a $K^+$-ATP channel (KIR) and the sulfonylurea receptor (SUR). Closure of the $K^+$ channels results in membrane depolarization, which opens voltage-sensitive $Ca^{++}$ channels. Influx of extracellular $Ca^{++}$ stimulates insulin secretion by exocytosis from storage granules. Leucine, a potent enhancer of insulin secretion, acts by allosteric stimulation of glutamate dehydrogenase. This results in increased formation of α-ketoglutarate (an intermediate of the Krebs cycle) and, hence, in elevation of ATP. Insulin secretion is stimulated by sulfonylureas and inhibited by diazoxide, which bind to SUR. Somatostatin and $Ca^{++}$ antagonists inhibit insulin secretion by decreasing $Ca^{++}$ influx.

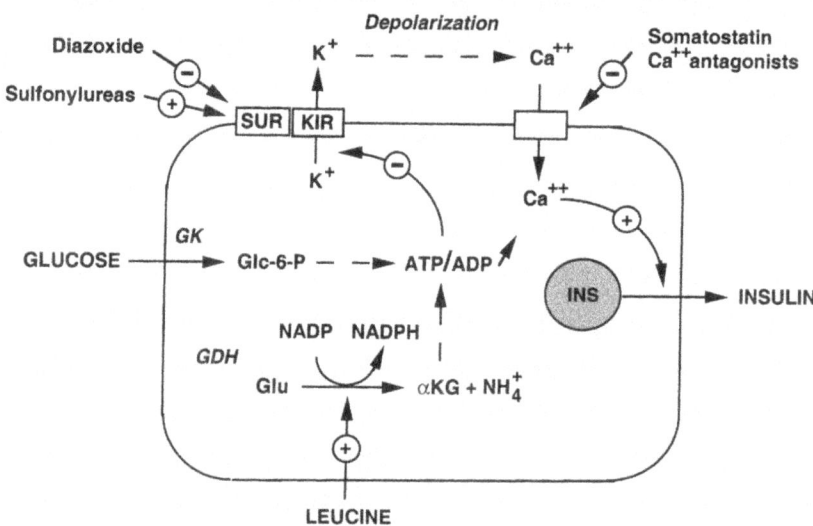

**Fig. 9.1.** Mechanisms of insulin secretion by the pancreatic β cell. +, stimulation; –, inhibition; *ADP*, adenosine diphosphate; *ATP*, adenosine triphosphate; α-*KG*, α-ketoglutarate; *G-6-P*, glucose-6-phosphate; *GDH*, glutamate dehydrogenase; *GK*, glucokinase; *Glc*, glucose; *Glu*, glutamate; *Ins*, insulin; *NADP*, nicotinamide adenine dinucleotide phosphate

# Persistent Hyperinsulinemic Hypoglycemia

Pascale de Lonlay and Jean-Marie Saudubray

CONTENTS

| | |
|---|---|
| Clinical Presentation | 119 |
| Metabolic Derangement | 119 |
| Diagnostic Tests | 120 |
| Treatment and Prognosis | 120 |
| Glucose Administration | 120 |
| Drugs | 120 |
| Surgery | 120 |
| Prognosis | 121 |
| Genetics | 121 |
| References | 122 |

Persistent hyperinsulinemic hypoglycemia of infancy (PHHI) is the most important cause of hypoglycemia in early infancy. The inappropriate oversecretion of insulin is responsible for profound hypoglycemias, which require aggressive treatment to prevent severe and irreversible brain damage. A combination of glucose and glucagon is started as an emergency treatment as soon as a tentative diagnosis of PHHI is made. It is followed by treatment with diazoxide and other drugs and, finally, by pancreatectomy if the patient is drug resistant. PHHI is a heterogeneous disorder with two histopathological lesions, *diffuse* (DiPHHI) and *focal* (FoPHHI), which are clinically indistinguishable. FoPHHI is characterized by a somatic islet cell hyperplasia. DiPHHI corresponds to a functional abnormality of insulin secretion in the whole pancreas and involves several genes with different transmissions. The therapeutic outcome differs for both histological entities, as does genetic counseling.

## Clinical Presentation

The presentation varies according to the age of onset of hypoglycemia.

In the *neonatal period*, the majority of the newborns are macrosomic at birth, with mean birth weights of 3.7 kg, and approximately 30% are delivered by cesarean section [1]. Severe hypoglycemia, with its high risk of seizures and brain damage, is the major feature of hyperinsulinism [2-4]. Hypoglycemia occurs within 72 h after birth and is revealed in half by seizures [1]. Other symptoms are abnormal movements (such as tremulousness), hypotonia, cyanosis, hypothermia or a life-threatening event. In some cases, hypoglycemia is detected by routine assay of blood glucose. The plasma glucose concentration at the time of the first symptoms is often extremely low (<1 mmol/l). Hypoglycemia is permanent (both fasting and postprandial) and requires continuous glucose administration in amounts far above the normal age-related requirements. A mild hepatomegaly is frequent and does not exclude the diagnosis of hyperinsulinism.

In *infancy*, hypoglycemia occurs between 1 month and 12 months of life and is also revealed by seizures in half of the cases. Macrosomy at birth is observed frequently (mean birth weights of 3.6 kg in our series).

In *childhood* less extreme rates of IV glucose are required; hypoglycemia occurs later, between 4 years and 8 years of age in our series. Macrosomy at birth is usual. This late-onset presentation is mostly due to pancreatic adenoma. The clinical presentation of the different histological and genetic forms is similar and depends on the age of manifestation of hypoglycemia [5].

## Metabolic Derangement

Hyperinsulinemic hypoglycemia is due to insulin hypersecretion (either focal or general) of the pancreas [2]. Insulin is the only hormone that decreases plasma glucose both by inhibiting glucose release from the liver and by increasing glucose uptake in muscle. Simultaneously, it inhibits both the release of fatty acids from adipose tissue and fatty-acid oxidation. This mode of action explains the three characteristic findings of neonatal PHHI: the high glucose requirement to correct hypoglycemia, the glucagon responsiveness and the inappropriately low levels of free fatty acids and ketone bodies during hypoglycemia. Several pathways are involved in the regulation of insulin secretion by the pancreatic β cells, explaining the modes of action of medical treatments with diazoxide, somatostatin and diet restricted in protein

(Fig. 9.1). Note that the sulfonylureas, such as tolbutamide, act by binding to sulfonylurea receptor (SUR) to stimulate insulin secretion. Conversely, diazoxide blocks insulin secretion by inhibiting SUR (Fig. 9.1). Octreotide (a long-acting somatostatin analog) inhibits $Ca^{++}$-mediated insulin release.

## Diagnostic Tests

The diagnostic criteria for PHHI include: (1) persistent fasting and postprandial hypoglycemia (<3 mmol/l) with hyperinsulinemia (plasma insulin >10 mU/l), requiring high rates of intravenous glucose administration (>10 mg/kg/min) to maintain blood glucose levels greater than 3 mmol/l and (2) a positive response to subcutaneous or intramuscular glucagon (0.5 mg) followed by increased blood glucose levels of 2–3 mmol/l, which excludes a defect of glycogenolysis and gluconeogenesis. In infancy and childhood, normal plasma insulin and C-peptide concentrations during hypoglycemia may occur. They do not allow one to exclude the diagnosis of hyperinsulinism, and the analyses have to be repeated. In the absence of clearly abnormal insulin levels during hypoglycemia, an 8- to 12-h fasting test searching for inappropriate low plasma levels of ketone bodies, free fatty acids and branched-chain amino acids can be helpful [6, 7].

A tolbutamide test has recently been suggested to separate the focal forms (which would be tolbutamide responsive) from the diffuse forms (which would be tolbutamide insensitive; see below). However, this test has to be further assessed and could be hazardous.

Hyperammonemia might be searched for in each new PHHI patient before deciding to pursue a more aggressive treatment, because the hyperammonemia/hyperinsulinism syndrome is usually amenable to medical or dietetic treatment. Furthermore, the "secondary" causes of hyperinsulinism (namely Munchhausen syndrome [8], and a congenital defect of glycosylation [9]) should be excluded.

## Treatment and Prognosis

### Glucose Administration

Glucose must be rapidly and aggressively administered to prevent irreversible brain damage and often necessitates central venous feeding combined with continuous oral alimentation with a nasogastric drip. In the neonatal period, the rate of intravenous glucose administration required to maintain plasma glucose higher than 3 mmol/l is as high as 17 mg/kg/min in our series [1]. In infancy, less extreme amounts of 12–13 mg/kg/min are needed while, in childhood (between 4 years and 8 years of age), the maintenance doses of glucose and its polymers for oral ingestion is again lower and has to be "titrated".

### Drugs

IV continuous glucagon (1–2 mg/day) can be added to glucose infusion when blood glucose levels remain unstable. Its effect remains remarkably constant. At the same time, drug treatment must be applied. Oral diazoxide is first used at 15 mg/kg/day divided into three doses [5]. It is effective in most infantile forms (60% of cases, in our experience), while most neonatal forms are resistant to diazoxide (90% of our cases). Diazoxide efficacy is defined as the normalization of blood glucose levels (>3 mmol/l) measured before and after each meal in patients fed normally, having a physiological overnight fast and after having stopped IV glucose and any other medication for at least five consecutive days. Two confirmed hypoglycemias (<3 mmol/l) in such a 24-h glucose profile lead us to consider the patient as diazoxide-unresponsive and to restart continuous drip feeding and/or other measures to regain permanent normoglycemia. Tolerance to diazoxide is usually excellent. The most frequent adverse effect is hirsutism, which can sometimes be marked and distressing in young children. Hematologic side effects and troublesome fluid retention are very rare with the usual doses.

Octreotide might be tried before surgery in case of diazoxide-unresponsiveness. Doses vary from 3–15 µg/day to 60 µg/day and are divided into three to four subcutaneous injections [10]. High doses could lead to worsening of the hypoglycemia by suppressing both glucagon and growth hormone. After initiation of octreotide treatment, many patients show vomiting and/or diarrhea and abdominal distension that resolve spontaneously within 7–10 days. Steatorrhea is also present, partially responding to oral pancreatic enzymes, and abating after several weeks or months. Gall-bladder sludge can be found and necessitates routine abdominal ultrasound. Other drugs, such as $Ca^{++}$ inhibitors (like nifedipine), have been proposed, but their efficacy is not clear. Contrary to common opinion, corticosteroid treatment is usually ineffective. A protein-restricted diet limiting the leucine intake to 200 mg leucine per meal is mandatory in the hyperammonemia/hyperinsulinism syndrome, on which it is often effective [11].

### Surgery

When medical or dietetic therapies are ineffective, surgical treatment is required. Hitherto, most pediatric surgeons recommended 95% subtotal pancreatectomy involving a high risk of later development of diabetes mellitus [5, 10, 12, 13]. However, there is strong evidence that the two types of histological lesions underlying PHHI, diffuse and focal, greatly influence the outcome of surgical treatment [1, 14, 15].

Patients who are treated surgically have to be classified according to histological criteria. The focal form, which accounts for 40% of these patients, is defined as a focal adenomatous hyperplasia [16–21]. The lesion measures 2.5–7.5 mm in diameter, differing from true adult-type pancreatic adenoma, which is clearly limited and features different topographic distribution. Diffuse PHHI shows abnormal β cell nuclei in all sections of the pancreas [22]. In the absence of any distinctive clinical feature, and because classical preoperative radiology of the pancreas (including echo tomography, computed-tomography scan and magnetic-resonance imaging) is not efficient in detecting a focal form, pancreatic venous catheterization and pancreatic arteriography are presently the only preoperative procedures available for locating the site of insulin hypersecretion [23]. For pancreatic catheterization, diazoxide and all other drugs are stopped 5 days before catheterization, and a continuous intravenous glucose infusion is given to prevent hypoglycemia. Percutaneous, transhepatic catheterization is undertaken under general anesthesia, and venous blood samples are collected from the head, isthmus, body and tail of the pancreas for measurements of plasma glucose, insulin and C-peptide. The patients with a focal lesion have high plasma insulin and C-peptide concentrations in one or several contiguous samples, with low concentrations in the remaining samples. Patients with diffuse lesions have high plasma insulin and C-peptide concentrations in all samples.

Patients suspected of having a focal lesion at catheterization undergo surgery. The others are also operated on if they resist or do not tolerate medical treatment. Intraoperative histology is systematically performed to confirm and delineate the findings of catheterization and to guide the limits of resection in case of a focal form. Diffuse lesions are characterized by β cells with large nuclei and abundant cytoplasm in all samples. Focal lesions show no abnormal β cell nuclei and shrunken cytoplasm but give a pattern of crowded β cells. The localization of focal forms is crucial because of the fact that focal lesions can be located in the head of the pancreas, whereas surgeons usually resect pancreatic tissue by first removing the tail and body of the pancreas. A subtotal pancreatectomy is performed for diffuse lesions.

## Prognosis

Although most of the patients treated medically remain drug dependent, some patients, who respond well to medical management (diazoxide and/or octreotide) may have a complete clinical remission, relatively rapid (<16 months) in cases of focal lesion and later (60 months) in cases of diffuse form. This justifies cessation of medical treatment once per year under medical supervision to search for a spontaneous recovery. A conservative attitude is preferable for patients with PHHI associated with hyperammonemia, who are usually responsive to diazoxide and a low-leucine diet and who have a spontaneous favorable outcome. Focal PHHI treated by limited pancreatectomy is completely cured without clinical or biochemical hypoglycemia [1]. By contrast, in diffuse PHHI, subtotal pancreatectomy may be followed by postoperative hypoglycemia despite extensive surgery and/or diabetes mellitus or serious alteration of glucose tolerance [1]. Pancreatic exocrine insufficiency might be treated by pancreatic enzymes. An annual investigation of residual insulin secretion (based on pre- and postprandial plasma glucose and insulin levels at various intervals, measurement of glycosylated hemoglobin and an oral glucose-tolerance test) is mandatory, as diabetes or glucose intolerance can develop later.

The long-term outcome of patients with PHHI is variable, depending on the severity of hypoglycemia. Children with neonatal hypoglycemia appear to have more severe neurological diseases than those with the late-onset form [24, 25]. Neurological sequelae are psychomotor retardation (20% in our series), epilepsy (25%), microcephaly and strabismus.

In all forms (neonatal and late-onset), neurological sequelae are correlated to the frequency and severity of hypoglycemic attacks, including those with no apparent clinical symptoms. This emphasizes the need to strictly define the criteria for drug responsiveness. In this view, it must be emphasized that most neonatal forms of PHHI are actually diazoxide unresponsive and corticoid resistant.

## Genetics

The estimated incidence of PHHI is 1/50,000 live births but, in countries with substantial inbreeding, such as Saudi Arabia, the incidence may be as high as 1/2500 [26]. The two histological forms of hyperinsulinism correspond to distinct molecular entities [16–21]. Focal islet-cell hyperplasia is associated with homozygosity of a paternally inherited mutation of the SUR1 gene on chromosome 11p15 and loss of the maternal allele in the hyperplastic islets [14, 15]. The somatic molecular abnormality in the pancreas suggests a sporadic event for focal lesions [14] which is what we actually observed in our patients [1]. However, the diffuse form is a heterogeneous disorder involving the genes encoding (1) SUR or the inward-rectifying $K^+$ channel

(Kir6.2) in recessively inherited hyperinsulinism [26-30], (2) the glucokinase gene [31] or other loci [32] in dominantly inherited hyperinsulinism and (3) the glutamate dehydrogenase gene when hyperammonemia is associated with the hyperinsulinism [11]. In the latter case, transmission can be sporadic or dominant. Non-insulin-dependent diabetes should be more frequent in PHHI related to the glucokinase gene.

Adenoma radically differs from focal PHHI by the late onset of hypoglycemia and the histological lesion. Hitherto, its etiology was unknown, except for adenoma related to type-1 multiple endocrine neoplasia (MEN1) syndrome by dominant mutation of the MEN1 gene [24, 33-38].

## References

1. de Lonlay-Debeney P, Poggi-Travert F, Fournet JC et al. (1999) Clinical aspects and course of neonatal hyperinsulinism. N Engl J Med 340:1169-1175
2. Stanley CA (1997) Hyperinsulinism in infants and children. Pediatr Clin North Am 44:363-374
3. Bruining GJ (1990) Recent advances in hyperinsulinism and the pathogenesis of diabetes mellitus. Curr Opin Pediatr 2:758-765
4. Thomas CG Jr, Underwood LE, Carney CN, Dolcourt JL, Whitt JJ (1977) Neonatal and infantile hypoglycemia due to insulin excess: new aspects of diagnosis and surgical management. Ann Surg 185:505-517
5. Touati G, Poggi-Travert F, Ogier de Baulny H et al. (1998) Long-term treatment of persistent hyperinsulinaemic hypoglycaemia of infancy with diazoxide: a retrospective review of 77 cases and analysis of efficacy-predicting criteria. Eur J Pediatr 157:628-633
6. Stanley CA, Baker L (1976) Hyperinsulinism in infancy: diagnosis by demonstration of abnormal response to fasting hypoglycemia. Pediatrics 57:702-711
7. Chaussain JL, Georges P, Gendrel D, Donnadieu M, Job JC (1980) Serum branched-chain amino acids in the diagnosis of hyperinsulinism in infanncy. J Pediatr 97:923-926
8. Scarlett JA, Mako ME, Rubenstein AH et al. (1977) Factitious hypoglycemia. Diagnosis by measurement of serum C-peptide immunoreactivity and insulin-binding aantibodies. N Engl J Med 297:1029-1032
9. de Lonlay P, Cuer M, Barrot S et al. (1999) Hyperinsulinemic hypoglycemia as presenting symptom of Carbohydrate-Deficiency Glycoproteins. J Pediatr 135:379-383
10. Thornton PS, Alter CA, Levitt Katz LE, Baker L, Stanley CA (1993) Short- and long-term use of octreotide in the treatment of congenital hyperinsulinism. J Pediatr 123:637-643
11. Stanley CA, Lieu Y, Hsu B et al. (1998) Hyperinsulinemia and hyperammonemia in infants with regulatory mutations of the glutamate dehydrogenase gene. N Engl J Med 338:1352-1357
12. Shilyanski J, Fisher S, Cutz E, Perlman K, Filler RM (1997) Is 95% pancreatectomy the procedure of choice for treatment of persistent hyperinsulinemic hypoglycemia of the neonate? J Pediatr Surg 32:342-346
13. Spitz L, Bhargava RK, Grant DB, Leonard JV (1992) Surgical treatment of hyperinsulinaemic hypoglycaemia in infancy and childhood. Arch Dis Child 67:201-205
14. de Lonlay P, Fournet JC, Rahier J et al. (1997) Somatic deletion of the imprinted 11p15 region in sporadic persistent hyperinsulinemic hypoglycemia of infancy is specific of focal adenomatous hyperplasia and endorses partial pancreatectomy. J Clin Invest 100:802-807
15. Verkarre V, Fournet JC, de Lonlay P et al. (1998) Maternal allele loss with somatic reduction to homozygosity of the paternally-inherited mutation of the SUR1 gene leads to congenital hyperinsulinism in focal islet cell adenomatous hyperplasia of the pancreas. J Clin Invest 102:1286-1291
16. Sempoux C, Guiot Y, Lefevre A et al. (1998) Neonatal hyperinsulinemic hypoglycemia: heterogeneity of the syndrome and keys for differential diagnosis. J Clin Endocrinol Metab 83:1455-1461
17. Klöppel G (1997) Nesidioblastosis. In: Soleia E, Capella C, Klöppel G (eds) Tumors of the pancreas. AFIP, Washington, pp 238-243
18. Goossens A, Gepts W, Saudubray JM et al. (1989) Diffuse and focal nesidioblastosis. A clinicopathological study of 24 patients with persistent neonatal hyperinsulinemic hypoglycemia. Am J Surg Pathol 3:766-775
19. Goudswaard WB, Houthoff HJ, Koudstaal J, Zwierstra RP (1986) Nesidioblastosis and endocrine hyperplasia of the pancreas: a secondary phenomenon. Hum Pathol 17:46-53
20. Rahier J, Fält K, Müntefering H et al. (1984) The basic structural lesion of persistent neonatal hypoglycaemia with hyperinsulinism: deficiency of pancreatic D cells or hyperactivity of B cells? Diabetologia 26:282-289
21. Jaffé R, Hashida Y, Yunis EJ (1980) Pancreatic pathology in hyperinsulinemic hypoglycemia of infancy. Lab Invest 42:356-365
22. Rahier J, Sempoux C, Fournet JC et al. (1998) Partial or near-total pancreatectomy for persistent neonatal hyperinsulinaemic hypoglycaemia: the pathologist's role. Histopathology 32:15-19
23. Dubois J, Brunelle F, Touati G et al. (1995) Hyperinsulinism in children: diagnostic value of pancreatic venous sampling correlated with clinical, pathological and surgical outcome in 25 cases. Pediatr Radiol 25:512-516
24. Kim H, Kerr A, Morehouse H (1995) The association between tuberous sclerosis and insulinoma. Am J Neuroradiol 16:1543-1544
25. Cresto JC, Abdenur JP, Bergada I, Martino R (1998) Longterm follow up of persistent hyperinsulinaemic hypoglycaemia of infancy. Arch Dis Child 79:440-444
26. Thornton PS, Sumner AE, Ruchelli ED et al. (1991) Familial and sporadic hyperinsulinism: histopathological findings and segregation analysis support a single autosomal recessive disorder. J Pediatr 119:721-724
27. Thomas PM, Cote GJ, Wohllk N et al. (1995) Mutations in the sulfonylurea receptor gene in familial persistent hyperinsulinemic hypoglycemia of infancy. Science 268:426-429
28. Nestorowicz A, Wilson BA, Schoor KP et al. (1996) Mutations in the sulfonylurea receptor gene are associated with familial hyperinsulinism in Ashkenazi Jews. Hum Mol Genet 5:1813-1822
29. Thomas P, Ye Y, Lightner E (1996) Mutation of the pancreatic islet inward rectifier Kir6.2 also leads to familial persistent hyperinsulinemic hypoglycemia of infancy. Hum Mol Genet 5:1813-1822
30. Nestorowicz A, Inagaki N, Gonoi T et al. (1997) A nonsense mutation in the inward rectifier potassium channel gene, Kir6.2, is associated with familial hyperinsulinism. Diabetes 46:1743-1748
31. Glaser B, Kesavan P, Heyman M et al. (1998) Familial hyperinsulinism caused by an activating glucokinase mutation. N Engl J Med 22:226-230
32. Kukuvitis A, Deal C, Arbour L, Polychronakos C (1997) An autosomal dominant form of familial persistent hyperinsulinemic hypoglycemia of infancy, not linked to the sulfonylurea receptor. J Clin Endocrinol Metab 82:1192-1194
33. Larsson C, Skogseid B, Oberg K, Nakamura Y, Nordenskjold M (1988) Multiple endocrine neoplasia type 1 gene maps to chromosome 11 and is lost in insulinoma. Nature 332:85-87
34. Demeure MJ, Klonoff DC, Karam JH, Duh QY, Clark OH (1991) Insulinomas associated with multiple endocrine neoplasia type 1: the need for a different surgical approach. Surgery 110:998-1004

35. Bassett JH, Forbes SA, Pannett AA et al. (1998) Characterization of mutations in patients with multiple endocrine neoplasia type 1. Am J Hum Genet 62:232–244
36. Agarwal SK, Kester MB, Debelenko LV et al. (1997) Germline mutations of the MEN1 gene in familial multiple endocrine neoplasia type 1 and related states. Hum Mol Genet 6:1169–1175
37. Guru SC, Goldsmith PK, Burns AL et al. (1998) Menin, the product of the MEN1 gene, is a nuclear protein. Proc Natl Acad Sci USA 95:1630–1634
38. Patel P, O'Rahilly S, Buckle V et al. (1990) Chromosome 11 allele loss in sporadic insulinoma. J Clin Pathol 43:377–378

# PART III
# DISORDERS OF MITOCHONDRIAL ENERGY METABOLISM

## Pyruvate Metabolism and the Tricarboxylic Acid Cycle

Pyruvate is formed from glucose and other monosaccharides, from lactate, and from the gluconeogenic amino acid alanine (Fig. 10.1). After entering the mitochondrion, pyruvate can be converted into acetyl coenzyme A (CoA) by the pyruvate dehydrogenase complex, followed by further oxidation in the TCA cycle. Pyruvate can also enter the gluconeogenic pathway by sequential conversion into oxaloacetate by pyruvate carboxylase, followed by conversion into phosphoenolpyruvate by phosphoenolpyruvate carboxykinase. Acetyl CoA can also be formed by fatty acid oxidation or used for lipogenesis. Other amino acids enter the TCA cycle at several points. One of the primary functions of the TCA cycle is to generate reducing equivalents in the form of reduced nicotinamide adenine dinucleotide and reduced flavin adenine dinucleotide, which are utilized to produce energy (as ATP) in the electron transport chain.

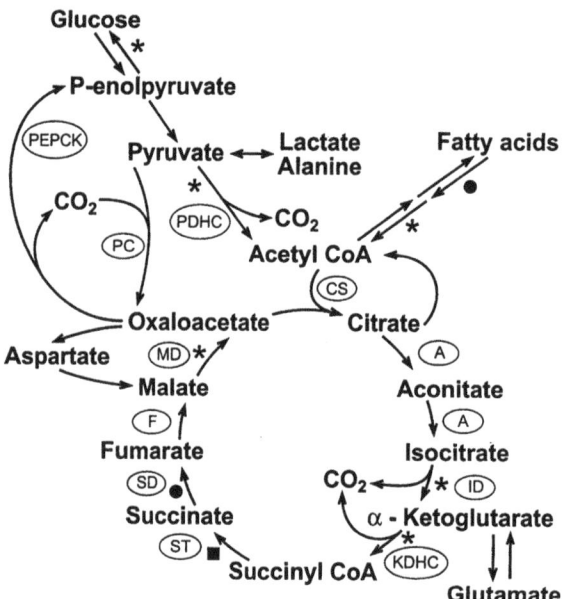

Fig. 10.1. Overview of glucose, pyruvate/lactate, fatty acid, and amino acid oxidation by the tricarboxylic acid cycle. *A*, aconitase; *CS*, citrate synthase; *F*, fumarase; *ID*, isocitrate dehydrogenase; *KDHC*, α- or 2-ketoglutarate dehydrogenase complex; *MD*, malate dehydrogenase; *PC*, pyruvate carboxylase; *PDHC*, pyruvate dehydrogenase complex; *PEPCK*, phosphoenolpyruvate carboxykinase; *SD*, succinate dehydrogenase; *ST*, succinyl coenzyme A transferase. Sites where reducing equivalents and intermediates for energy production intervene are indicated by the following symbols: ★, reduced nicotinamide adenine dinucleotide; ●, reduced flavin adenine dinucleotide; ■, guanosine triphosphate

# CHAPTER 10

# Disorders of Pyruvate Metabolism and the Tricarboxylic Acid Cycle

Douglas S. Kerr, Isaiah D. Wexler, and Arthur B. Zinn

CONTENTS

Pyruvate Carboxylase Deficiency.................... 127
Phosphoenolpyruvate Carboxykinase Deficiency........ 129
Pyruvate Dehydrogenase Complex Deficiency.......... 132
2-Ketoglutarate Dehydrogenase Complex Deficiency .... 134
Dihydrolipoamide Dehydrogenase Deficiency.......... 134
Fumarase Deficiency............................. 135
Succinate Dehydrogenase Deficiency................. 136
References....................................... 137

Owing to the role of pyruvate and the tricarboxylic acid (TCA) cycle in energy metabolism, as well as in gluconeogenesis, lipogenesis and amino acid synthesis, defects of pyruvate metabolism and of the TCA cycle almost invariably affect the central nervous system. The severity and patterns of clinical phenotypes vary tremendously among affected patients and are not specific, with the range of manifestations extending from overwhelming neonatal lactic acidosis and early death to relatively normal adult life and variable effects on systemic functions. The same clinical manifestations may be caused by other defects of energy metabolism, especially defects of the electron transport (respiratory) chain (Chap. 13). Diagnosis of these disorders depends primarily on biochemical analyses of metabolites in body fluids, followed by definitive enzymatic assays in cells or tissues, and DNA analysis if feasible. Among the three disorders of pyruvate metabolism the deficiencies of pyruvate carboxylase (PC) and phosphoenolpyruvate carboxykinase (PEPCK) constitute defects in gluconeogenesis, and therefore fasting results in hypoglycemia with worsening lactic acidosis. The deficiency of the pyruvate dehydrogenase complex (PDHC) impedes glucose oxidation and aerobic energy production, and ingestion of carbohydrate aggravates lactic acidosis. Nutritional treatment of these disorders of pyruvate metabolism comprises avoidance of fasting (PC and PEPCK) or minimizing dietary carbohydrate intake (PDHC). In some cases, vitamin or drug therapy may be helpful. The deficiencies of the TCA cycle enzymes, the 2-ketoglutarate dehydrogenase complex (KDHC) and fumarase, interrupt the cycle, resulting in accumulation of the corresponding substrates. Dihydrolipoamide dehydrogenase (E3) deficiency affects PDHC as well as KDHC and the branched-chain 2-ketoacid dehydrogenase (BCKD) complex (Chap. 16), with biochemical manifestations of all three disorders. Succinate dehydrogenase deficiency represents a unique disorder affecting both the TCA cycle and the electron transport chain. A more complex defect of iron-sulfur cluster metabolism involves aconitase, succinate dehydrogenase, and electron transport chain complexes I and III. Treatment strategies for all of these TCA cycle defects are very limited; metabolism of any dietary source of energy is impaired, and these defects are generally not vitamin responsive.

## Pyruvate Carboxylase Deficiency

### Clinical Presentation

Pyruvate carboxylase (PC) deficiency has been classified into three distinct clinical presentations [1]. The most severe form, which always presents in the neonatal period, is associated with severe lactic acidosis and neurologic dysfunction, including seizures, coma, and abnormal muscle tone. The metabolic derangements in these children are severe, and they rarely survive beyond 3 months. In the less severe form of PC deficiency, the presentation is later and neurologic characteristics of psychomotor retardation are more prominent.

Other clinical features that are often present include seizures, spasticity, failure to thrive, and renal tubular acidosis (Table 10.1). Pathologic features for both the severe and milder forms include hepatomegaly associated with lipid accumulation in hepatocytes and neuroanatomic alterations, including poor myelination, ventricular enlargement, periventricular cysts, astrocyte proliferation, and a reduction of the number of astrocytes in the cerebral cortex. A "benign" form of

PC deficiency characterized by recurrent attacks of lactic acidosis and very mild neurologic deficits has also been described.

**Metabolic Derangement**

PC is a biotinylated mitochondrial enzyme that converts pyruvate and $CO_2$ to oxaloacetate. The PC reaction is essential for replenishing oxaloacetate in the TCA cycle (anaplerosis) and providing the necessary substrate for synthetic pathways, including gluconeogenesis, glycerogenesis, lipogenesis, and formation of certain nonessential amino acids (aspartate, glutamate, glutamine, and their products; Fig. 10.1). Oxaloacetate is transaminated into aspartate, which is a required substrate for the urea cycle and the malate–aspartate shuttle; this shuttle maintains the redox potential between the mitochondrial and cytosolic compartments of cells.

The clinical presentation and the metabolic derangements of PC deficiency are related to the severity of the defect. In the severe form of PC deficiency, both the anaplerotic and synthetic functions of PC are compromised. Oxidation via the TCA cycle is compromised by lack of oxaloacetate. In affected children, this results in severe lactic acidosis, with an increased blood lactate/pyruvate (L/P) ratio contrasting with a decreased 3-hydroxybutyrate/acetoacetate (3HB/AA) ratio, consistent with a more oxidized mitochondrial compartment. Plasma ammonia, ornithine, and citrulline are elevated, presumably due to depletion of oxaloacetate, leading to reduced intracellular levels of aspartate. Alanine is high while glutamine is low. The neuroanatomic defects are related both to abnormalities of lipid synthesis (leading to decreased myelin formation) and to decreased glutamine-derived neurotransmitters. The milder form of PC deficiency is associated with a less severe lactic acidosis, which often improves as the child becomes older. The lactic acidosis is related both to the inability of the liver to convert lactate derived from other tissues into glucose and to functional inhibition of PDHC by low insulin and high acetyl coenzyme A levels. The L/P ratio is typically normal, indicating that the oxidative metabolism is intact. Quite often, attacks of severe acidosis are precipitated by fasting or the stress of an intercurrent illness and are associated with severe hypoglycemia and ketoacidosis. Ketosis is due to decreased ability of the liver to oxidize acetyl CoA because of lack of oxaloacetate (Fig. 10.1).

**Diagnostic Tests**

The possibility of a defect in PC should be considered in any child presenting with lactic acidosis and neurologic abnormalities, especially if associated with hypoglycemia, hyperammonemia, or ketosis. In neonates, a high L/P ratio associated with a low 3HB/AA ratio and postprandial hyperketonemia is nearly pathognomonic (Table 10.2) [2]. The increase of blood lactate is more pronounced in the fasting state and improves after ingestion of carbohydrate. If the lactic acidosis worsens on fasting, PC activity should be assayed in cultured skin fibroblasts. Care should be taken in preparing samples for assay, as PC protein is especially labile. In cases in which the diagnosis is uncertain, it may be appropriate to assay the activity of PC in liver, because PC enzymatic activity is much lower in fibroblasts than in liver. Immunoassay and

**Table 10.1.** Clinical features of defects of enzymes of pyruvate metabolism and the tricarboxylic acid cycle

| Clinical features | Enzyme defect | | | | | | |
|---|---|---|---|---|---|---|---|
| | PC | PEPCK[a] | PDHC | KDHC | $E_3$ | Fumarase | SDH[a] |
| Delayed growth and development | +++ | ++ | +++ | +++ | +++ | +++ | − |
| Hypotonia | ++ | + | +++ | +++ | +++ | +++ | − |
| Seizures | +++ | + | ++ | + | + | ++ | + |
| Ataxia, choreoathetosis | + | − | + | ++ | + | − | + |
| CNS degeneration (including Leigh disease) | ++ | − | ++ | +++ | + | +++ | + |
| CNS malformations | ++ | − | + | − | + | − | − |
| Apnea, hypoventilation | − | − | + | − | + | − | − |
| Sudden death | + | − | + | ++ | − | − | − |
| Skeletal myopathy | − | + | − | − | − | − | ++ |
| Cardiomyopathy | − | + | − | + | − | − | ++ |
| Hepatic dysfunction | ++ | + | − | ++ | + | + | − |
| Dysmorphic features | − | − | + | − | − | + | − |
| Renal tubular acidosis | + | + | − | − | − | − | − |
| Number of reported cases (approximate) | 20 | 5 | >200 | 10 | 15 | 20 | 10 |

+++ very common (>75%); ++ common (25–75%); + uncommon (<25%); − not noted; *CNS*, central nervous system; $E_3$, dihydrolipoamide dehydrogenase; *KDHC*, 2-ketoglutarate dehydrogenase complex; *PC*, pyruvate carboxylase; *PDHC*, pyruvate dehydrogenase complex; *PEPCK*, phosphoenolpyruvate carboxykinase; *SDH*, succinate dehydrogenase
[a] The estimated frequency of the features of these disorders is limited by few cases (PEPCK) or clinical heterogeneity (SDH)

**Table 10.2.** Metabolic abnormalities in defects of enzymes of pyruvate metabolism and the tricarboxylic acid (TCA) cycle

| Metabolic abnormality (increased) | Enzyme defect | | | | | | |
|---|---|---|---|---|---|---|---|
| | PC | PEPCK[a] | PDHC | KDHC | $E_3$ | Fumarase | SDH[a] |
| Blood lactate | +++ | +++ | +++ | +++ | +++ | + | ++ |
| Lactate/pyruvate ratio | + | + | + | ++ | + | ++ | + |
| Plasma alanine | ++ | + | ++ | + | + | ++ | - |
| Plasma branched-chain amino acids | - | - | - | - | +++ | - | - |
| Plasma glutamate, glutamine | - | - | - | ++ | + | - | - |
| Plasma ammonia | + | - | - | - | - | + | - |
| Urine lactic acid | ++ | ++ | ++ | + | ++ | + | + |
| Urine 2-ketoglutaric acid | + | - | + | +++ | +++ | + | + |
| Urine branched-chain 2-ketoacids | - | - | - | - | ++ | - | - |
| Urine fumaric acid | - | - | - | - | - | +++ | - |
| Other urine TCA acids | - | - | + | + | + | ++ | + |
| Hypoglycemia | ++ | ++ | - | + | + | + | - |
| Ketosis | + | - | - | - | - | - | - |

+++ very common (>75%); ++ common (25–75%); + uncommon (<25%); - not noted; $E_3$, dihydrolipoamide dehydrogenase; *KDHC*, 2-ketoglutarate dehydrogenase complex; *PC*, pyruvate carboxylase; *PDHC*, pyruvate dehydrogenase complex; *PEPCK*, phosphoenolpyruvate carboxykinase; *SDH*, succinate dehydrogenase
[a] The estimated frequency of the features of these disorders is limited by few cases (PEPCK) or clinical heterogeneity (SDH)

Northern blot analysis of tissue samples from patients with PC may provide prognostic information, as the severity of PC deficiency has been related to the absence of immunoreactive protein and/or mRNA [2].

PC deficiency must be differentiated from multiple carboxylase deficiency resulting from a defect in biotinidase or holocarboxylase synthase, which affect the activity of all biotinylated enzymes (Chap. 24). Multiple carboxylase deficiency has a characteristic urinary organic acid chromatogram not found in isolated PC deficiency. More definitive testing involves the measurement of both biotinidase activity and the enzymatic activity of other biotinylated enzymes, such as propionyl-CoA carboxylase.

### Treatment and Prognosis

Patients should be instructed to ingest carbohydrate before bedtime and to avoid fasting. Patients with acute attacks of lactic acidosis should be given continuous intravenous glucose (solutions containing at least 10% glucose). Some patients with persistent lactic acidosis may require bicarbonate or another source of base to correct the acidosis. Provision of intravenous aspartate or citrate as an alternative source of four-carbon intermediates may be beneficial and may ameliorate the hyperammonemia associated with PC deficiency. It has been reported that one patient with PC deficiency was clinically responsive to biotin (10 mg/day) [3].

Prognosis of individuals with PC deficiency depends on the severity of the defect. Patients with minimal residual PC activity usually do not live beyond the neonatal period, but some children with very low PC activity have survived beyond the age of 5 years [4]. Those with milder defects may survive and have neurologic deficits of varying degrees. Those patients who are susceptible to attacks of life-threatening lactic acidosis and intercurrent illnesses – especially those associated with vomiting and decreased oral intake – must be treated aggressively.

### Genetics

PC deficiency is an autosomal recessive disorder, and the human PC gene has been localized to chromosome 11 and cloned, and mutations causing the milder form of PC deficiency have been identified [5–7]. There is an increased incidence of the milder form of PC deficiency in certain native North American tribes [1]. Carrier detection is possible by measuring PC activity in fibroblasts. Prenatal diagnosis of PC deficiency has been performed, based on the enzymatic activity of PC in cultured amniocytes or chorionic villus samples [8]. Knowledge of specific mutations would facilitate carrier detection and prenatal diagnosis.

## Phosphoenolpyruvate Carboxykinase Deficiency

### Clinical Presentation

As there are so few reports of phosphoenolpyruvate carboxykinase (PEPCK) deficiency, it is difficult to derive a composite description of patients with this disorder. The age of presentation is variable, with some patients presenting in the newborn period and others presenting at several months of age. The major manifestation is hypoglycemia associated with lactic acidosis, which is similar to the symptoms of PC deficiency. There may be multisystem involvement, with neuromuscular deficits, hepatocellular damage, cardiomyopathy, and renal tubular acidosis (Table 10.1). Non-specific features include hypotonia, hepatomegaly, lethargy, and failure to thrive [2, 9].

## Metabolic Derangement

PEPCK converts oxaloacetate into phosphoenolpyruvate and plays a major role in gluconeogenesis and glycerogenesis (Fig. 10.1). PEPCK exists as two separate isoforms, mitochondrial and cytosolic, which are encoded by two distinct genes. In humans, both the cytosolic and mitochondrial isoforms are involved in gluconeogenesis [10]. Lack of PEPCK blocks the conversion of pyruvate, lactate, alanine, and TCA intermediates to glucose. As a result, there is severe fasting hypoglycemia (Table 10.2). Lactic acidosis is aggravated by physiologic inhibition of the PDHC by hypoinsulinemia, which is secondary to hypoglycemia. The hyperlipidemia and steatosis sometimes seen in PEPCK deficiency may be a consequence of the low insulin level and/or increased citrate production resulting from accumulation of oxaloacetate.

## Diagnostic Tests

The presence of separate mitochondrial and cytosolic isoforms of PEPCK complicates the diagnosis of this disorder, and measurement of PEPCK activity in whole cells may provide misleading results. Optimally, of both mitochondrial and cytosolic PEPCK activity should be measured in a fresh liver sample in which the mitochondrial and cytosolic compartments have been fractionated. In cultured fibroblasts, most of the PEPCK activity is located in the mitochondrial compartment, and low PEPCK activity in whole-cell homogenates indicates deficiency of the mitochondrial isoform. Caution must be taken in interpreting PEPCK enzymatic assays in tissues, as PEPCK activity is physiologically depressed in hyperinsulinemic states that may also cause hypoglycemia during infancy [2]. Decreased PEPCK activity might also be a secondary phenomenon, e.g. secondary to respiratory chain dysfunction.

## Treatment and Prognosis

Patients with confirmed PEPCK deficiency should be treated with intravenous glucose and sodium bicarbonate during acute episodes of hypoglycemia and lactic acidosis. Fasting should be avoided, and cornstarch or other forms of slow-release carbohydrates should be provided before bedtime (Chap. 6). There may be theoretical advantages to restricting dietary protein intake as a means of reducing production of gluconeogenic substrates that require PEPCK for further metabolism [9]. The long-term prognosis of patients with PEPCK deficiency is usually poor, with most subjects dying of intractable hypoglycemia or neurodegenerative disease.

## Genetics

The cDNA encoding the cytosolic isoform of PEPCK in humans has been sequenced and localized to human chromosome 20, but no mutations have been identified [11]. Little information is available about prenatal testing.

## Structure and Interconversion of the Pyruvate Dehydrogenase Complex

PHDC, and the two other mitochondrial α or 2-ketoacid dehydrogenases, KDHC and the BCKD complex, are similar in structure and analogous or identical in their specific mechanisms. They are composed of three components: $E_1$, an α- or 2-ketoacid dehydrogenase; $E_2$, a dihydrolipoamide acyltransferase, and $E_3$, a dihydrolipoamide dehydrogenase. $E_1$ is specific for each complex, utilizes thiamine pyrophosphate, and is composed of two different subunits, $E_1\alpha$ and $E_1\beta$. The $E_1$ reaction results in decarboxylation of the specific α- or 2-ketoacid, providing a means of assaying enzyme activity in cells and tissue samples. For PDHC, the $E_1$ component is reate-limiting and regulated by phosphorylation/dephosphorylation catalyzed by two enzymes, $E_1$ kinase and $E_1$ phosphatase, which are intrinsic to the ovall PDHC. $E_1$ phosphatase activates, and $E_1$ kinase inactivates the enzyme. $E_2$ is a transacetylase that utilizes covalently bound lipoic acid and is also unique for each substrate-specific complex. Moreover, $E_2$ serves as the structural core of the complex. $E_3$ is a flavoprotein common to all three 2-ketoacid dehydrogenases (as well as the glycine cleavage enzyme – Chapter 21). Another important structural component of PDHC, the $E_3$ binding protein (formerly called "component X"), is critical for attaching the $E_3$ subunits to the core $E_2$ subunits.

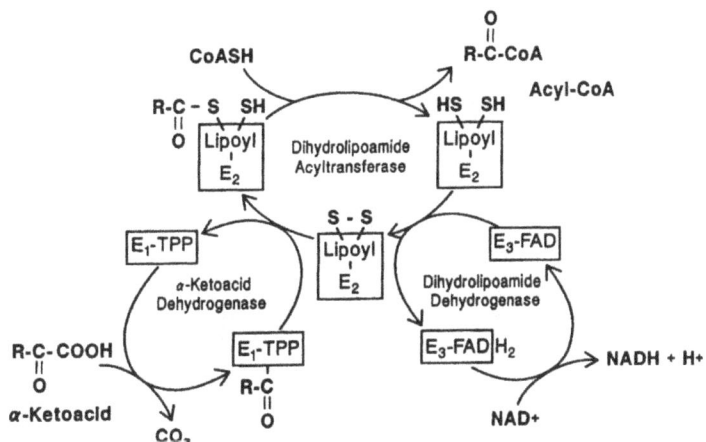

**Fig. 10.2.** Structure of the α- or 2-ketoacid dehydrogenase complexes, pyruvate dehydrogenase complex (PDHC), 2-ketoglutarate dehydrogenase complex (KDHC) and branched-chain α-ketoacid dehydrogenase complex (BCKD). *CoA*, coenzyme A; *FAD*, flavin adenine dinucleotide; *NAD*, nicotinamide adenine dinucleotide; *R*, methyl group (for pyruvate in PDHC) and the corresponding moiety for 2-ketoglutarate (in KDHC) and the branched-chain 2-ketoacids (in BCKD); *TPP*, thiamine pyrophosphate

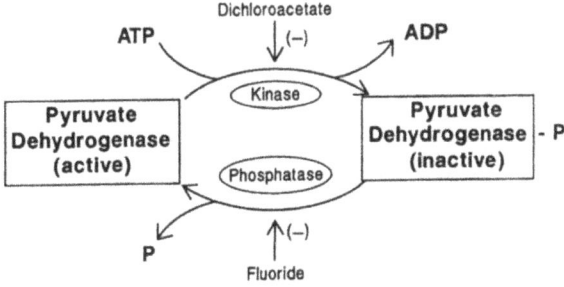

**Fig. 10.3.** Interconversion of the $E_1$ component of pyruvate-dehydrogenase complex by phosphorylation/dephosphorylation. Dichloroacetate is an inhibitor of $E_1$ kinase, and fluoride inhibits $E_1$ phosphatase. *ADP*, adenosine diphosphate; *P*, inorganic phosphate

## Pyruvate Dehydrogenase Complex Deficiency

### Clinical Presentation

PDHC deficiency is by far the most common of the group of defects described in this chapter. Over 200 cases of PDHC deficiency have been reported [2, 12, 13]. PDHC deficiency most often involves the first component of this enzyme complex, pyruvate dehydrogenase ($E_1$). The most common features associated with PDHC deficiency in infants and children are delayed development and hypotonia (Table 10.1). Seizures and ataxia are frequent features. Less common, but perhaps most characteristic, are defects of CNS respiratory control leading to apnea, dependence on assisted ventilation, or sudden unexpected death. Loss of respiratory control appears to reflect focal injury to the basal ganglia and brain stem, which may be visible by magnetic resonance imaging (MRI) of the brain. These MRI findings appear to correlate with the degenerative neuropathological findings of subacute necrotizing encephalopathy originally described by Leigh [14]. However, "Leigh disease" has been described in only a minority of cases of PDHC deficiency, and the same pathology can be associated with defects of the electron transport chain (ETC), ATP synthase, or the TCA cycle [15]. In contrast, some infants with PDHC deficiency have congenital malformations of the brain, including agenesis of the corpus callosum [16, 17]. Craniofacial dysmorphism suggestive of fetal alcohol syndrome has been described [16], but this observation needs further confirmation. At the other extreme of phenotypic variation, the oldest known adult with PDHC deficiency has intermittent ataxia with normal cognitive function, even though this male has almost no measurable PDHC activity in his cultured skin fibroblasts [18].

About five cases of $E_1$-phosphatase deficiency (Fig. 10.3) [19] and nine cases of deficiency of the second component of PDHC (dihydrolipoamide transacetylase, $E_2$) or the $E_3$-binding protein (component X) have been reported (Fig. 10.3) [20]. These other PDHC defects have clinical manifestations that are within the variable spectrum associated with PDHC deficiency due to $E_1$ deficiency (combined in Table 10.1).

### Metabolic Derangement

Without mitochondrial oxidation, pyruvate is reduced to lactate. The conversion of glucose to lactate yields less than one tenth of the total available ATP that would be derived from complete oxidation of glucose via the TCA cycle and the ETC. In the presence of oxygen and normal mitochondrial function, pyruvate can be oxidized to acetyl-CoA via PDHC. PDHC is, therefore, the "gateway" for complete oxidation of carbohydrate via the TCA cycle (Fig. 10.1). Hence, deficiency of PDHC specifically interferes with production of ATP from carbohydrate oxidation, and lactic acidemia is aggravated by consumption of carbohydrate.

PDHC deficiency, in contrast to ETC defects, impairs reduced nicotinamide adenine dinucleotide (NADH) production but not NADH oxidation. The L/P ratio is a reflection of the NADH/nicotinamide adenine dinucleotide (NAD) ratio in the cytosol of cells. *PDHC deficiency is characterized by a normal L/P ratio, whereas deficiencies of ETC complexes I, III, and IV are generally characterized by a high L/P ratio because of impaired NADH oxidation.*

### Diagnostic Tests

The most important laboratory test for initial recognition of possible PDHC deficiency is measurement of blood lactate and pyruvate. Measurement of cerebrospinal fluid (CSF) lactate and pyruvate, quantitative analysis of plasma amino acids, and urinary organic acid analysis may also be useful in recognizing PDHC defects (Table 10.2). Blood lactate and pyruvate and plasma alanine can be intermittently normal, but an increase is expected after an oral carbohydrate load. While the L/P ratio is characteristically normal, a high ratio can be found artifactually if the patient is acutely ill, if blood is very difficult to obtain without a struggle, or if the assay of pyruvate (which is unstable) is not done reliably. The practical solution to these artifacts is to obtain several samples of blood and urine, including samples collected under different dietary conditions (during an acute illness, after fasting, and postprandially after a high-carbohydrate meal). In contrast to deficiencies of PC or PEPCK, fasting hypoglycemia is not an expected feature of PDHC deficiency, and blood lactate and pyruvate usually decrease after fasting. Glucose-tolerance or carbohydrate-loading tests should be performed with caution, as acute deterioration may occur in PDHC-deficient patients after a glucose load (Chap. 2). A practical and probably safer solution for diagnostic screening is to obtain blood after overnight fasting and 1–2 h after an ordinary, carbohydrate-containing breakfast. Failure to find elevated blood lactate may be an indication that one should obtain CSF for measurement of lactate and pyruvate (and possibly organic acids), because there may be a dysequilibrium between blood and CSF metabolites in patients with primary CNS disease ("cerebral" PDHC deficiency) [17].

The most commonly utilized samples for assay of PDHC and its components are cultured skin fibroblasts. PDHC also can be assayed in fresh blood lymphocytes, which saves significant time and expense required for tissue culture but has the disadvantage of not permitting follow-up testing from the same sample.

If available, skeletal muscle and/or other tissues are very useful. When a patient with suspected but unproven PDHC deficiency dies, it is valuable to freeze samples of skeletal muscle, heart muscle, liver, and/or brain, ideally within 4 h postmortem, since PDHC deficiency can be expressed variably in different cells and tissues [21].

While PDHC can be assayed by measuring rates of formation of various products of the overall reaction (Fig. 10.3), for practical purposes, release of $^{14}CO_2$ from [1-$^{14}$C]-pyruvate has proven most useful for assays in crude homogenates of cells and tissues [18, 22]. PDHC must be activated (dephosphorylated) prior to assay, which can be done by pre-incubation of whole cells or mitochondria with dichloroacetate (DCA, an inhibitor of the kinase) or by pre-incubation of freeze-thawed cells or tissues with $E_1$ phosphatase (Fig. 10.3). $E_1$-phosphatase deficiency is implicated if PDHC cannot be activated in cells or mitochondria pre-incubated with DCA but can be activated by addition of $E_1$ phosphatase [19]. The three catalytic components of PDHC can be assayed separately, but the reactions utilized for these assays do not employ physiological substrates. The activity of the $E_1$ component, when assayed separately, is extremely low, and the results must be considered qualitative. Immunoblotting of the components of PDHC can help distinguish if a particular protein is missing. This is critical if there is no catalytic assay for the defective component, such as the $E_3$-binding protein ("component X") [20]. Mutations that result in absence of the $E_1\alpha$ subunit are almost always accompanied by loss of the $E_1\beta$ subunit [23].

## Treatment and Prognosis

The general prognosis for most individuals with PDHC deficiency is poor, and treatment is usually not very effective. Experience with early prospective treatment to prevent irreversible brain injury is lacking. Perhaps the most rational strategy for treating PDHC deficiency is the use of a ketogenic diet [24]. Oxidation of fatty acids, 3-hydroxybutyrate, and acetoacetate provides alternative sources of acetyl-CoA not derived from pyruvate. We compared the outcome of males with PDHC deficiency caused by identical $E_1\alpha$ mutations and found that the earlier the ketogenic diet was started and the more severe the restriction of carbohydrate, the better the outcome of mental development and survival [25]. However, this treatment was not adequate to assure long-term survival. Sustaining significant ketosis (4-6 mM) is an important part of the treatment strategy, because fatty acids do not cross the blood-brain barrier. Maintenance of ketosis over an extended period of time is difficult and requires restricting dietary carbohydrate to less than 5% of dietary energy and protein to less than 15%; i.e., at least 80% of energy should be derived from fat.

Thiamin has been administered in very large doses (500-2000 mg/day) to some patients with PDHC deficiency, with lowering of blood lactate and apparent clinical improvement. In vitro studies in cultured cells from some of these cases has shown decreased affinity of PDHC for thiamin pyrophosphate [26]. It is not clear that this accounts for the mechanism of clinical response or to what extent thiamin therapy affects long-term outcome. Use of other vitamins has not been effective in treating $E_1$ deficiency.

DCA offers another potential treatment for PDHC deficiency. DCA, a structural analogue of pyruvate, inhibits $E_1$ kinase, thereby keeping any residual $E_1$ activity in its active (dephosphorylated) form (Fig. 10.3). DCA can be administered in sufficient amounts to achieve a significant inhibitory concentration in vivo without apparent toxicity (about 50 mg/kg/day). Over 40 cases of congenital lactic acidosis due to various defects (including PDHC deficiency) were treated with DCA in uncontrolled studies, and most of these cases appeared to have some limited short-term benefit [27]. A controlled clinical trial of DCA in congenital lactic acidosis is currently being conducted to determine its long-term efficacy and safety.

## Genetics

All of the components of PDHC are encoded by nuclear genes and are synthesized in the cytoplasm as precursor proteins that are imported into mitochondria, where the mature proteins are assembled into a very large enzyme complex. Although most of the genes that encode the various subunits are autosomal, the $E_1\alpha$-subunit gene is located on chromosome Xp22.3 [28]. A second, intronless gene for $E_1\alpha$ is located on chromosome 4 and is expressed only in developing and mature sperm [29]. Mutations of the "somatic" gene on the X chromosome account for the vast majority of cases of PDHC deficiency in males. Therefore, most cases of PDHC deficiency are X-linked, generally with more severe consequences of a particular mutation in hemizygous males and more variable clinical manifestations in heterozygous females due to variable inactivation of the two alleles [30]. We are aware of identical twin girls with $E_1\alpha$ deficiency who have different degrees of clinical severity.

To date, over 60 different mutations of the $E_1\alpha$ subunit of PDHC have been characterized [12, 31]. About half of these are small deletions, insertions, or frame-shift mutations and the other half are missense mutations. While the consequences of most of these mutations on enzyme structure and function are not known, some affect highly conserved amino acids that are critical for mitochondrial import, subunit interaction, binding of thiamin pyrophosphate, dephosphorylation, or catalysis at the active site [12]. No "null" $E_1\alpha$ mutations have been identified in males, suggesting that such mutations are likely to be lethal. No defects of the $E_1\beta$ subunit have been identified to date. Phenotype–genotype correlations have had limited clinical usefulness; even the most common $E_1\alpha$ mutation (R234G, mature protein) is associated with variable phenotypic expression in males [13, 25]. The molecular basis of $E_3$-binding protein ("component-X") deficiency has been characterized in several cases [32].

There has been little success with prenatal diagnosis of PDHC deficiency. As most cases of PDHC deficiency appear to be the consequence of new $E_1\alpha$ mutations, the overall rate of recurrence in the same family does not appear to be greater than 10% [12, 31]. Based on measurement of PDHC activity in chorionic villus samples and/or cultured amniocytes obtained from some 30 pregnancies in families with a previously affected child, three cases of reduced activity were found [33, 34]. In two unpublished cases, prenatal testing of cultured amniocytes indicated normal PDHC activity, but testing after birth indicated PDHC deficiency. In subsequent pregnancies, prenatal testing of specific mutations determined in the proband should be the most reliable method [34], but detection of an affected fetus by this method has not yet been reported.

## 2-Ketoglutarate Dehydrogenase Complex Deficiency

### Clinical Presentation

KDHC deficiency has been reported in approximately ten children in several unrelated families [35, 36]. As in PDHC deficiency, the primary clinical manifestations are neurological impairment, including developmental delay, hypotonia, ataxia, opisthotonos and, less commonly, seizures (Table 10.1). All patients presented in early childhood, with most presenting in infancy. None of the affected children has survived past 10 years of age.

### Metabolic Derangement

KDHC is a 2-ketoacid dehydrogenase that is analogous to PDHC and BCKD (Fig. 10.2). By analogy to PDHC, the $E_1$ component, 2-ketoglutarate dehydrogenase, is a substrate-specific dehydrogenase that utilizes thiamin and is composed of two different subunits ($\alpha$ and $\beta$). In contrast to PDHC, the $E_1$ component is not regulated by phosphorylation/dephosphorylation. The $E_2$ component, dihydrolipoyl succinyltransferase, is also specific to KDHC and includes covalently bound lipoic acid. The $E_3$ component is the same as for PDHC. An $E_3$-binding protein has not been identified for KDHC. Since KDHC is integral to the TCA cycle, its deficiency has consequences similar to that of other TCA enzyme deficiencies (see "Fumarase Deficiency").

### Diagnostic Tests

The most useful test for recognizing KDHC deficiency is urine organic acid analysis, which shows increased excretion of 2-ketoglutaric acid with or without concomitantly increased excretion of other TCA cycle acids (Table 10.2). However, mildly to moderately increased urinary 2-ketoglutaric acid is a common finding and is not a specific marker of KDHC deficiency. Most patients with KDHC deficiency also have increased blood lactate with a normal (but sometimes increased) L/P ratio. Plasma glutamate and glutamine may be increased. KDHC activity can be assayed by measuring the release of $^{14}CO_2$ from [$1$-$^{14}C$]-2-ketoglutarate in crude homogenates of cultured skin fibroblasts and other cells and tissues [35, 37]. Preactivation of the enzyme is not necessary.

### Treatment and Prognosis

In contrast to deficiencies of the other two 2-ketoacid dehydrogenases, there is no known selective dietary treatment that bypasses KDHC, since this enzyme is involved in the terminal steps of virtually all oxidative energy metabolism. In theory, supplementation with thiamin might be beneficial, as has been found with deficiencies of the other 2-ketoacid dehydrogenases; however, in practice, thiamin-responsive KDHC deficiency has not been described.

### Genetics

KDHC deficiency is inherited as an autosomal recessive trait. The $E_1$ gene has been mapped to chromosome 7p13-14, cloned, and sequenced [38]. The $E_2$ gene has been mapped to chromosome 14q24.3. The molecular basis of KDHC deficiencies has not been reported. While prenatal diagnosis of KDHC should be possible by measurement of the corresponding enzyme activity in chorionic villus samples or cultured amniocytes, this has not been reported.

## Dihydrolipoamide Dehydrogenase Deficiency

### Clinical Presentation

Approximately 15 cases of $E_3$ deficiency have been reported [2, 39–42]. Since this enzyme is common to all the 2-ketoacid dehydrogenases, $E_3$ deficiency results in multiple 2-ketoacid dehydrogenase deficiency and

should be thought of as a combined PDHC and TCA cycle defect. The clinical manifestations of $E_3$ deficiency appear to be similar to those shown for PDHC and KDHC deficiencies (Table 10.1).

## Metabolic Derangement

Dihydrolipoyl dehydrogenase ($E_3$), is a flavoprotein common to all three 2-ketoacid dehydrogenases (PDHC, KDHC, and BCKD; Fig. 10.2) and the glycine cleavage enzyme. The predicted metabolic manifestations are a composite of the deficiency states for each of these enzymes: increased blood lactate and pyruvate, plasma alanine, glutamate, glutamine, and branched-chain amino acids (leucine, isoleucine, and valine), and increased urinary lactic, pyruvic, 2-ketoglutaric, and branched-chain 2-hydroxy- and 2-keto acids (Table 10.2). Glycine has not been found to be consistently elevated.

## Diagnostic Tests

The initial diagnostic screening should include all the tests used for evaluating each of the respective enzyme complex deficiencies: i.e., analyses of blood lactate and pyruvate, plasma amino acids, and urinary organic acids. However, the predicted pattern of metabolic abnormalities described above is not seen in all patients or at all times in the same patient, making the diagnosis more difficult (Table 10.2). In cultured skin fibroblasts, blood lymphocytes, or tissues, the $E_3$ component can be assayed using a straightforward spectrophotometric method based on either reduction of free D,L-lipoamide by NADH or oxidation of reduced D,L-lipoamide by NAD (preferable) [22].

## Treatment and Prognosis

As with KDHC deficiency, there is no simple dietary treatment for $E_3$ deficiency, since the affected enzymes affect carbohydrate, fat, and protein metabolism. Restriction of dietary branched-chain amino acids was reportedly helpful in one case [40]. Another case of $E_3$ deficiency improved after supplementation with D,L-lipoic acid [39] whereas, in another case, treatment with D,L-lipoic acid was not beneficial [43]. It is not clear why lipoic acid was helpful in the former case, since it is not part of the defective $E_3$ protein; if lipoic acid served as an alternate substrate for the $E_2$ component (replacing $E_3$), then it might be expected that very large (i.e., substrate-level) quantities of lipoic acid would be required. D,L-lipoic acid ($\alpha$-lipoic acid) is currently widely used as a non-prescription antioxidant for a variety of medical conditions. Another case of $E_3$ deficiency was treated with carnitine, DCA, and thiamin with a good outcome [41].

## Genetics

The gene for $E_3$ is located on chromosome 7p31-32 [44]. Deficiency of this enzyme is inherited as an autosomal-recessive trait, and the carrier status is detectable. In three separate families, the individuals affected with $E_3$ deficiency were each found to be compound heterozygotes for two different missense point mutations affecting conserved amino acids [43, 45]. A homozygous mutation of the leader sequence for $E_3$ has been identified in Ashkenazi Jews [42]. Prenatal diagnosis of $E_3$ deficiency should be possible by measurement of the corresponding enzyme activity in chorionic villus samples or cultured amniocytes; knowledge of the specific mutations involved in the proband and in parents would provide an even more reliable method for prenatal diagnosis [43].

## Fumarase Deficiency

### Clinical Presentation

Approximately 20 patients with fumarase deficiency have been reported [36, 46-49]. The clinical features of this disorder are summarized in Table 10.1. All patients have shown poor postnatal neurological function. The most severely affected patients generally develop seizures and respiratory control difficulties and die in early childhood, whereas less severely affected patients develop a static encephalopathy and survive into adolescence and adulthood. Three patients have had prenatal onset of cerebral dysgenesis (hydrocephalus or agenesis of the corpus callosum, or both), sometimes associated with polyhydramnios.

### Metabolic Derangement

Fumarase is a homotetramer that catalyzes the reversible interconversion of fumarate and malate (Fig. 10.1). No cofactors are required for this enzyme. There are two isoforms of fumarase, mitochondrial and cytosolic, which have the same primary amino acid sequence, except for the amino-terminal residue. Both proteins are encoded by the same gene and the same mRNA, but the fumarase transcript is alternately translated to generate the two isoforms.

The pathogenesis of fumarase deficiency and other TCA cycle defects includes: (1) impaired energy production caused by interrupting the flow of the TCA cycle and (2) secondary enzyme inhibition associated with accumulation of metabolites proximal to the primary enzyme deficiency. The first mechanism limits the number of enzymatic steps at which reducing equivalents can be generated and transferred to the ETC. In addition, this mechanism may lead to depletion of oxaloacetate, preventing continued influx of acetyl-CoA into the TCA cycle via citrate synthase. The second mechanism may involve other pathways of oxidative metabolism. For example, fumarase deficiency is associated with secondary inhibition of succinate dehydrogenase (SDH) and glutamate dehydrogenase in skeletal muscle mitochondria [46].

### Diagnostic Tests

The key finding in this disorder is increased urinary fumaric acid, which is associated (in some cases) with increased excretion of succinic and 2-ketoglutaric acids. Mild lactic acidosis and mild hyperammonemia are sometimes seen in infants with fumarase deficiency, but are generally not seen in older children. Fumarase is generally measured in mononuclear blood leukocytes, cultured skin fibroblasts, skeletal muscle, or liver by monitoring the formation of fumarate from malate or, more sensitively, by coupling the reaction with malate dehydrogenase and monitoring the production of NADH [46].

### Treatment and Prognosis

As yet, effective strategies to compensate for the pathogenic mechanisms of fumarase deficiency have not been developed. Protein restriction was reportedly of no benefit for one patient with fumarase deficiency. While removal of certain amino acids that are precursors of fumarate could be beneficial, removal of exogenous aspartate might deplete a potential source of oxaloacetate. Conversely, supplementation with aspartate or citrate might lead to overproduction of toxic TCA intermediates. At present, the prognosis for this disorder appears to depend more on the severity of the mutation than on the mode of medical intervention, which is largely supportive.

### Genetics

Fumarase deficiency is inherited as an autosomal recessive trait, and the carrier status is identifiable by enzyme assay. The fumarase gene has been mapped to chromosome 1q42.1 [47]. The gene that encodes for fumarase is part of a gene family that encodes for other enzymes that catalyze reactions involving fumarate, namely argininosuccinate lyase and adenylosuccinate lyase. The molecular basis of fumarase deficiency has recently been shown to be genetically heterogeneous. Different mutations have been demonstrated in several unrelated families [47, 49]. Prenatal diagnosis has been accomplished by measuring fumarase activity and/or mutational analysis in chorionic villus samples or cultured amniocytes [49].

## Succinate Dehydrogenase Deficiency

### Clinical Presentation

Numerous patients with SDH or ETC complex-II deficiency have been reported, but most of them have been relatively poorly characterized from a biochemical or genetic perspective. Of those patients who have been well characterized, not more than ten appear to have had isolated SDH deficiency, and a clear clinical picture has not yet emerged. Although the putative clinical and laboratory features of SDH deficiency are included in Tables 10.1 and 10.2, respectively, these disorders are even more heterogeneous than the other conditions described in this chapter. The clinical picture of SDH deficiency can include: Kearns-Sayre syndrome; isolated hypertrophic cardiomyopathy; combined cardiac and skeletal myopathy; Leigh syndrome; and cerebellar ataxia with optic atrophy [50-52]. Interestingly, these various phenotypes resemble the phenotypes associated with ETC defects more closely than those associated with other TCA cycle defects. There is a striking absence of impaired mental development in patients with SDH deficiency. In addition, isolated SDH deficiency has been reported more commonly in adults and is associated with less severe impairment of enzyme activity than occurs in other TCA cycle defects [36].

SDH deficiency may also present as a compound deficiency state that involves aconitase (another TCA cycle enzyme; Fig. 10.1) and complexes I and III of the ETC. This disorder has generally presented with lifelong exercise intolerance, myoglobinuria, and lactic acidosis (with a normal or increased L/P ratio at rest and a decreased L/P ratio during exercise) [53]. This disorder, thus far found only in Swedish patients, appears to be caused by a defect in the metabolism of the iron-sulfur clusters common to these various enzymes [54], and its presentation as an ETC defect is not surprising.

### Metabolic Derangement

SDH is part of a larger enzyme unit, complex II of the ETC (succinate-ubiquinone oxidoreductase). Complex II is composed of four subunits. SDH contains two of these subunits, a flavoprotein and an iron-sulfur-containing protein. The remaining subunits are intrinsic membrane proteins that attach SDH to the inner mitochondrial membrane and the other ETC complexes. None of the subunits is known to have tissue-specific isoforms, which precludes a ready explanation for the variable organ expression of SDH deficiency.

The observation that combined SDH/aconitase deficiency is associated with a reduced L/P ratio during exercise but a normal L/P ratio at rest is paradoxical. Theoretically, TCA-cycle defects should lead to a decreased L/P ratio, because of impaired production of NADH. However, too few cases of SDH deficiency (or other TCA-cycle defects) have been evaluated to determine whether this is a consistent finding. If a reduced L/P ratio was found consistently, the L/P ratio would be useful to discriminate between patients with PDHC deficiency (normal L/P ratio), patients with TCA cycle defects (reduced L/P ratio), and patients with ETC defects (increased L/P ratio).

## Diagnostic Tests

As opposed to the other TCA cycle disorders, SDH deficiency does not always lead to a characteristic organic aciduria. Many patients, especially those whose clinical phenotypes closely resemble those of patients with ETC defects, do not exhibit the expected succinic aciduria. Diagnostic confirmation of a patient with a suspected SDH deficiency requires enzyme analysis of SDH activity itself, as well as complex-II (succinate-ubiquinone oxidoreductase) activity, which reflects the integrity of SDH and the remaining two subunits of this complex. These enzyme assays can be accomplished using standard spectrophotometric procedures. Other ETC complexes and TCA enzymes (especially aconitase) should be assayed and, whenever possible, polarographic analysis of isolated mitochondria should be performed to confirm the presence of the expected functional consequences [46].

## Treatment and Prognosis

No effective treatment has been reported. As with other defects of the TCA cycle, all dietary energy sources ultimately depend on this enzyme for their metabolism. Although SDH is a flavoprotein, riboflavin-responsive defects have not been described.

## Genetics

Complex II is unique among the ETC complexes in that all four of its subunits are nuclear encoded. The flavoprotein and iron-sulfur-containing subunits of SDH have been mapped to chromosomes 5p15 and 1p35-p36, respectively, while the two integral membrane proteins have been mapped to chromosomes 1q21 and 11q23. The genes encoding for all four subunits have been cloned and sequenced. One case of SDH deficiency associated with Leigh syndrome was shown to be an autosomal recessive disorder due to compound heterozygosity of two mutations in the flavoprotein subunit of SDH [51]. The molecular basis of the other disorders associated with isolated SDH deficiency has not been established, nor has the locus for the combined aconitase/SDH deficiency associated with an apparent defect in iron-sulfur-cluster metabolism been identified.

## References

1. Robinson BH (1989) Lacticacidemia. Biochemical, clinical, and genetic considerations. Adv Hum Genet 18:151-179, 371-372
2. Robinson BH (1995) Lactic acidemia (Disorders of pyruvate carboxylase, pyruvate dehydrogenase). In: Scriver CR, Beaudet AL, Sly WS, Valle D (eds) Metabolic and molecular bases of inherited disease, 7th edn. McGraw-Hill, New York, pp 1479-1499
3. Higgins JJ, Glasgow AM, Lusk MM, Kerr DS (1994) MRI, clinical, and biochemical features of partial pyruvate carboxylase deficiency. J Child Neurol 9:436-439
4. Stern HJ, Nayar R, Depalma L, Rifai N (1995) Prolonged survival in pyruvate carboxylase deficiency: lack of correlation with enzyme activity in cultured fibroblasts. Clin Biochem 28:85-89
5. Wexler ID, Du Y, Lisgaris MV et al. (1994) Primary amino acid sequence and structure of human pyruvate carboxylase. Biochim Biophys Acta 1227:46-52
6. Carbone MA, MacKay N, Ling M et al. (1998) Amerindian pyruvate carboxylase deficiency is associated with two distinct missense mutations. Am J Hum Genet 62:1312-1319
7. Wexler ID, Kerr DS, Du Y et al. (1998) Molecular characterization of pyruvate carboxylase deficiency in two consanguineous families. Pediatr Res 43:579-584
8. Van Coster RN, Janssens S, Misson JP, Verloes A, Leroy JG (1998) Prenatal diagnosis of pyruvate carboxylase deficiency by direct measurement of catalytic activity on chorionic villi samples. Prenat Diagn 18:1041-1044
9. Van den Berghe G (1996) Disorders of gluconeogenesis. J Inherit Metab Dis 19:470-477
10. Hanson RW, Patel YM (1994) Phosphoenolpyruvate carboxykinase (GTP): the gene and the enzyme. Adv Enzymol Relat Areas Mol Biol 69:203-281
11. Stoffel M, Xiang KS, Espinosa R et al. (1993) cDNA sequence and localization of polymorphic human cytosolic phosphoenolpyruvate carboxykinase gene (PCK1) to chromosome 20, band q13.31: PCK1 is not tightly linked to maturity-onset diabetes of the young. Hum Mol Genet 2:1-4
12. Kerr DS, Wexler ID, Tripatara A, Patel MS (1996) Defects of the human pyruvate dehydrogenase complex. In: Patel MS, Roche T, Harris RA (eds) Alpha keto acid dehydrogenase complexes. Birkhauser, Basel, pp 249-270
13. Otero LJ, Brown RM, Brown GK (1998) Arginine 302 mutations in the pyruvate dehydrogenase E1alpha subunit gene: identification of further patients and in vitro demonstration of pathogenicity. Hum Mutat 12:114-121
14. Medina L, Chi TL, DeVivo DC, Hilal SK (1990) MR findings in patients with subacute necrotizing encephalomyelopathy (Leigh syndrome): correlation with biochemical defect. AJNR 11:379-384
15. De Vivo DC (1998) Leigh syndrome: historical perspective and clinical variations. Biofactors 7:269-271
16. Robinson BH, MacMillan H, Petrova-Benedict R, Sherwood WG (1987) Variable clinical presentation in patients with defective $E_1$ component of pyruvate dehydrogenase complex. J Pediatr 111:525-533
17. Shevell MI, Matthews PM, Scriver et al. (1994) Cerebral dysgenesis and lactic acidemia: an MRI/MRS phenotype associated with pyruvate dehydrogenase deficiency. Pediatr Neurol 11:224-229
18. Sheu KFR, Hu CWC, Utter MF (1981) Pyruvate dehydrogenase complex activity in normal and deficient fibroblasts. J Clin Invest 67:1463-1471
19. Ito M, Kobashi H, Naito E et al. (1992) Decrease of pyruvate dehydrogenase phosphatase activity in patients with congenital lactic acidemia. Clin Chim Acta 209:1-7

20. Robinson BH, MacKay N, Petrova-Benedict R et al. (1990) Defects in the E₂ lipoyl transacetylase and the X-lipoyl containing component of the pyruvate dehydrogenase complex in patients with lactic acidemia. J Clin Invest 85:1821–1824
21. Kerr DS, Berry SA, Lusk MM, Ho L, Patel MS (1988) A deficiency of both subunits of pyruvate dehydrogenase which is not expressed in fibroblasts. Pediatr Res 24:95–100
22. Kerr DS, Ho L, Berlin CM et al. (1987) Systemic deficiency of the first component of the pyruvate dehydrogenase complex. Pediatr Res 22:312–318
23. Huq AHMM, Ito M, Naito E et al. (1991) Demonstration of an unstable variant of pyruvate dehydrogenase protein (E₁) in cultured fibroblasts from a patient with congenital lactic acidemia. Pediatr Res 30:11–14
24. Falk RE, Cederbaum SD, Blass JP et al. (1976) Ketogenic diet in the management of pyruvate dehydrogenase deficiency. Pediatrics 58:713–721
25. Wexler ID, Hemalatha SG, McConnell J et al. (1997) Outcome of pyruvate dehydrogenase deficiency treated with ketogenic diets. Studies in patients with identical mutations. Neurology 49:1655–1661
26. Naito E, Ito M, Takeda E et al. (1994) Molecular analysis of abnormal pyruvate dehydrogenase in a patient with thiamine-responsive congenital lactic acidemia. Pediatr Res 36:340–346
27. Stacpoole PW, Barnes CL, Hurbanis MD, Cannon SL, Kerr DS (1997) Treatment of congenital lactic acidosis with dichloroacetate: a review. Arch Pediatr Adolesc Med 77:535–541
28. Brown RM, Dahl HHM, Brown GK (1989) X-chromosome localization of the functional gene for the E₁α subunit of the human pyruvate dehydrogenase complex. Genomics 4:174–181
29. Young JC, Gould JA, Kola I, Iannello RC (1998) Review: Pdha-2, past and present. J Exp Zool 282:231–238
30. Dahl HHM (1995) Pyruvate dehydrogenase E₁α deficiency: males and females differ yet again. Am J Hum Genet 56:553–557
31. Dahl HHM, Brown GK, Brown RM et al. (1992) Mutations and polymorphisms in the pyruvate dehydrogenase E₁α gene. Hum Mutat 1:97–102
32. Aral B, Benelli C, Ait-Ghezala G et al. (1997) Mutations in PDX1, the human lipoyl-containing component X of the pyruvate dehydrogenase-complex gene on chromosome 11p1, in congenital lactic acidosis. Am J Hum Genet 61:1318–1326
33. Kerr DS, Lusk MM (1992) Infrequent expression of heterozygosity or deficiency of pyruvate dehydrogenase (E₁) among parents and sibs of affected patients. Pediatr Res 31:133 A (abstract)
34. Brown RM, Brown GK (1994) Prenatal diagnosis of pyruvate dehydrogenase E₁ alpha subunit deficiency. Prenat Diagn 14:435–441
35. Bonnefont JP, Chretien D, Rustin P et al. (1992) Alpha-ketoglutarate dehydrogenase deficiency presenting as congenital lactic acidosis. J Pediatr 121:255–258
36. Rustin P, Bourgeron T, Parfait B et al. (1997) Inborn errors of the Krebs cycle: a group of unusual mitochondrial diseases in human. Biochim Biophys Acta 1361:185–197
37. Chuang DT, Hu CW, Patel MS (1983) Induction of the branched-chain 2-oxo acid dehydrogenase complex in 3T3-L1 adipocytes during differentiation. Biochem J 214:177–181
38. Koike K (1998) Cloning, structure, chromosomal localization and promoter analysis of human 2-oxoglutarate dehydrogenase gene. Biochim Biophys Acta 1385:373–384
39. Matalon R, Stumpf DA, Michals K et al. (1984) Lipoamide dehydrogenase deficiency with primary lactic acidosis: favorable response to treatment with oral lipoic acid. J Pediatr 104:65–69
40. Sakaguchi Y, Yoshino M, Aramaki S et al. (1986) Dihydrolipoyl dehydrogenase deficiency: a therapeutic trial with branched-chain amino acid restriction. Eur J Pediatr 145:271–274
41. Elpeleg ON, Ruitenbeek W, Jakobs C et al. (1995) Congenital lacticacidemia caused by lipoamide dehydrogenase deficiency with favorable outcome. J Pediatr 126:72–74
42. Elpeleg ON, Shaag A, Glustein J Z et al. (1997) Lipoamide dehydrogenase deficiency in Ashkenazi Jews: an insertion mutation in the mitochondrial leader sequence. Hum Mutat 10:256–257
43. Hong YS, Kerr DS, Liu TC et al. (1997) Deficiency of dihydrolipoamide dehydrogenase due to two mutant alleles (E340 K and G101del). Analysis of a family and prenatal testing. Biochim Biophys Acta 1362:160–168
44. Scherer SW, Otulakowski G, Robinson BH, Tsui LC (1991) Localization of the human dihydrolipoamide dehydrogenase gene (DLD) to 7q31 > q32. Cytogenet Cell Genet 56:176–177
45. Hong YS, Kerr DS, Craigen WJ et al. (1996) Identification of two mutations in a compound heterozygous child with dihydrolipoamide dehydrogenase deficiency. Hum Mol Genet 5:1925–1930
46. Zinn AB, Kerr DS, Hoppel CL (1986) Fumarase deficiency: a new cause of mitochondrial encephalomyopathy. N Engl J Med 315:469–475
47. Coughlin EM, Chalmers RA, Slaugenhaupt SA et al. (1993) Identification of a molecular defect in a fumarase deficient patient and mapping of the fumarase gene. Am J Hum Genet 53:869
48. Bourgeron T, Chretien D, Poggi-Bach J et al. (1994) Mutation of the fumarase gene in two siblings with progressive encephalopathy and fumarase deficiency. J Clin Invest 93:2514–2518
49. Coughlin EM, Christensen E, Kunz PL et al. (1998) Molecular analysis and prenatal diagnosis of human fumarase deficiency. Mol Genet Metab 63:254–262
50. Rivner MH, Shamsnia M, Swift TR et al. (1989) Kearns-Sayre syndrome and complex II deficiency. Neurology 39:693–696
51. Bourgeron T, Rustin P, Chretien D et al. (1995) Mutation of a nuclear succinate dehydrogenase gene results in mitochondrial respiratory chain deficiency. Nat Genet 11:144–149
52. Taylor RW, Birch-Machin MA, Schaefer J et al. (1996) Deficiency of complex II of the mitochondrial respiratory chain in late-onset optic atrophy and ataxia. Ann Neurol 39:224–232
53. Haller RG, Henriksson KG, Jorfeldt L et al. (1991) Deficiency of skeletal muscle succinate dehydrogenase and aconitase. Pathophysiology of exercise in a novel human muscle oxidative defect. J Clin Invest 88:1197–1206
54. Hall RE, Henriksson KG, Lewis SF, Haller RG, Kennaway NG (1993) Mitochondrial myopathy with succinate dehydrogenase and aconitase deficiency. Abnormalities of several iron-sulfur proteins. J Clin Invest 92:2660–2666

# CHAPTER 11

## Fatty Acid Oxidation

Fatty acid oxidation comprises four components: the carnitine cycle, the β-oxidation cycle, the electron-transfer path, and the synthesis of ketone bodies. Long-chain free fatty acids of exogenous and endogenous origin are activated toward their coenzyme A (CoA) esters in the cytosol. These fatty acyl-CoAs enter the mitochondria as fatty acylcarnitines via the carnitine cycle. Medium- and short-chain fatty acids enter the mitochondria directly and are activated toward their CoA derivatives in the mitochondrial matrix. Each step of the four-step β-oxidation cycle shortens the fatty acyl-CoA by two carbons until it is completely converted to acetyl-CoA. The electron-transfer path transfers some of the energy released in β-oxidation to the respiratory chain, resulting in the synthesis of ATP. In the liver, most of the acetyl-CoA from fatty acid β-oxidation is used to synthesize the ketone bodies 3-hydroxybutyrate and acetoacetate. The ketones are then exported for terminal oxidation (chiefly in the brain). In other tissues, such as muscle, the acetyl-CoA enters the Krebs' cycle of oxidation and ATP production.

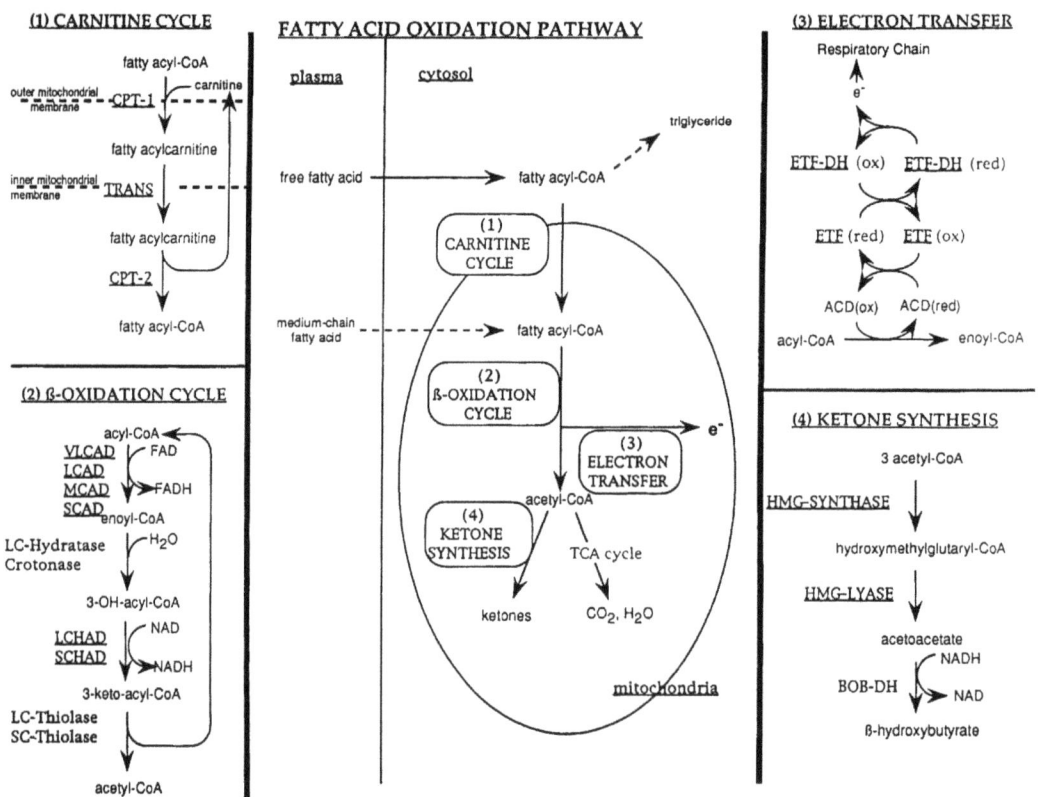

**Fig. 11.1.** Mitochondrial fatty acid-oxidation pathway. In the center panel, the pathway is subdivided into its four major components, which are shown in detail in the side panels. Sites of identified defects are *underscored*. *CoA*, coenzyme A; *CPT*, carnitine palmitoyl tranferase; *ETF*, electron-transfer flavoprotein; *ETF-DH*, ETF dehydrogenase; *FAD*, flavin adenine dinucleotide; *FADH*, reduced FAD; *HMG*, 3-hydroxy-3-methylglutaryl; *LCAD*, long-chain acyl-CoA dehydrogenase; *MCAD*, medium-chain acyl-CoA dehydrogenase; *NAD*, nicotinamide adenine dinucleotide; *NADH*, reduced NAD; *SCAD*, short-chain acyl-CoA dehydrogenase; *SCHAD*, short-chain 3-hydroxyacyl-CoA dehydrogenase; *TCA*, tricarboxylic acid; *TRANS*, carnitine/acylcarnitine translocase; *VLCAD*, very-long-chain acyl-CoA dehydrogenase

# CHAPTER 11

# Disorders of Fatty Acid Oxidation

C.A. Stanley

CONTENTS

Introduction .................................. 141
   Clinical Presentation......................... 141
      Carnitine-Cycle Defects ..................... 141
      Fatty Acid β-Oxidation Defects ................ 142
      Electron-Transfer Defects..................... 144
      Ketone-Synthesis Defects .................... 144
   Metabolic Derangement ....................... 144
   Diagnostic Tests ............................ 145
      Disease-Related Metabolites .................. 145
      Tests of Overall Pathway .................... 146
      Enzyme Assays............................ 147
   Treatment and Prognosis ..................... 147
      Management of Acute Illness .................. 147
      Long-Term Diet Therapy ..................... 147
      Carnitine Therapy ......................... 147
      Other Therapies ........................... 148
      Prognosis ............................... 148
   Genetics .................................. 148
Rare Related Disorders .......................... 148
   Transport Defect of Fatty Acids ................. 148
   Defect of Fatty-Alcohol Metabolism ............... 149
References ................................... 149

A dozen genetic defects in the fatty acid-oxidation pathway are currently known. Nearly all of these defects present in early infancy as acute, life-threatening episodes of hypoketotic, hypoglycemic coma induced by fasting [1–3]. In some of the disorders, there may also be chronic skeletal-muscle weakness or acute exercise-induced rhabdomyolysis and acute or chronic cardiomyopathy. Recognition of the fatty acid-oxidation disorders is often difficult, because patients can appear well until they undergo prolonged fasting.

## Introduction

The oxidation of fatty acids in mitochondria plays an important role in energy production. During late stages of fasting, fatty acids are used for hepatic ketone synthesis and to provide 80% of total body energy needs via oxidation in muscle. Fatty acids are the preferred fuel for the heart and serve as an essential source of energy for skeletal muscle during sustained exercise. Free fatty acids are released from adipose-tissue triglyceride stores and circulate bound to albumin. The oxidation of free fatty acids to $CO_2$ and $H_2O$ in peripheral tissues, such as muscle, spares glucose consumption and the need to convert body protein to glucose. The use of fatty acids by the liver provides energy for gluconeogenesis and ureagenesis. Equally important, the liver uses fatty acids to synthesize ketones, which serve as a fat-derived fuel for the brain and, thus, further reduce the need for glucose utilization.

## Clinical Presentation

The clinical phenotypes of most of the disorders of fatty acid oxidation are very similar [1–3]. Table 11.1 presents the three major types of presentation: a form with signs of mainly hepatic involvement, one with predominantly cardiac involvement, and one chiefly involving skeletal muscle. The individual defects are discussed below under the four components of the fatty acid-oxidation pathway outlined in Fig. 11.1 and Table 11.1.

### Carnitine-Cycle Defects

CARNITINE-TRANSPORTER DEFECT (CTD). Although most of the fatty acid-oxidation disorders affect the heart, skeletal muscle, and liver, cardiac failure is seen as the major presenting manifestation only in CTD [4]. In this disorder, sodium-dependent transport of carnitine across the plasma membrane is absent in muscle and kidney. This leads to severe reduction (<2–5% of normal) of carnitine in plasma and in heart and skeletal muscle. These levels of carnitine are low enough to impair fatty acid oxidation. Over half of the 20–30 known cases of CTD first presented with progressive heart failure and generalized muscle weakness. The age of onset of the cardiomyopathy or skeletal-muscle weakness ranged from 12 months to 7 years. The cardiomyopathy in CTD patients is most evident on echocardiography, which shows poor contractility and thickened ventricular walls. Electro-

**Table 11.1.** Inherited disorders of mitochondrial fatty acid oxidation

| Defect | Clinical manifestations of defect | | | |
|---|---|---|---|---|
| | Hepatic | Cardiac | Skeletal muscle | |
| | | | Acute | Chronic |
| **Carnitine cycle** | | | | |
| CTD | + | + | | (+) |
| CPT-1 | + | | | |
| Trans | + | + | | + |
| CPT-2 | + | + | (+) | + |
| **β-Oxidation cycle** | | | | |
| Acyl-CoA dehydrogenases | | | | |
| VLCAD | + | + | | + |
| MCAD | + | | | |
| SCAD | | | | + |
| 3-Hydroxyacyl-CoA dehydrogenases | | | | |
| LCHAD | + | + | | |
| SCHAD | | | | + |
| MCKT | | | + | + |
| DER | | | | + |
| **Electron transfer** | | | | |
| ETF | + | + | (+) | + |
| ETF-DH | + | + | (+) | + |
| **Ketone synthesis** | | | | |
| HMG-CoA synthase | + | | | |
| HMG-CoA lyase | + | | | |

*CPT*, carnitine-palmitoyl transferase; *CTD*, carnitine-transporter defect; *DER*, 2,4-dienoyl-coenzyme-A reductase; *ETF*, electron-transfer flavoprotein; *ETF-DH*, ETF dehydrogenase; *LCHAD*, long-chain 3-hydroxyacyl-coenzyme-A dehydrogenase; *MCAD*, medium-chain acyl-coenzyme-A dehydrogenase; *MCKT*, medium-chain ketoacyl-CoA thiolase; *SCAD*, short-chain acyl-coenzyme-A dehydrogenase; *SCHAD*, short-chain 3-hydroxyacyl-coenzyme-A dehydrogenase; *TRANS*, carnitine/acylcarnitine translocase; *VLCAD*, very-long-chain acyl-coenzyme-A dehydrogenase

cardiograms may be normal or show increased T-waves. Without carnitine treatment, the cardiac failure can progress rapidly to death by 2–4 years of age.

During the first years, extended fasting stress may provoke an attack of hypoketotic, hypoglycemic coma with or without evidence of cardiomyopathy. This hepatic presentation occurs less frequently than the myopathic forms, because the liver has a separate transporter for carnitine and can usually maintain levels of carnitine sufficient to support ketogenesis.

CARNITINE PALMITOYL-TRANSFERASE-1 (CPT-1) DEFICIENCY. In patients with CPT-1 deficiency, only the nonmuscle isoenzyme of CPT-1, which is expressed in liver and kidney, is affected. Patients with this defect usually present with attacks of fasting hypoketotic coma during the first 2 years of life [5–7]. They do not have cardiac or skeletal-muscle involvement. CPT-1 deficiency is the only fatty acid-oxidation disorder with elevated plasma total carnitine levels (see below) [6]. The defect is also noteworthy for unusually severe abnormalities in liver-function tests (including massive increases in serum transaminases and hyperbilirubinemia) during and for several weeks after acute episodes of illness. Transient renal tubular acidosis has also been described in a patient with CPT-1 deficiency, probably reflecting the importance of fatty acids as fuel for the kidney [8].

CARNITINE/ACYLCARNITINE-TRANSLOCASE (TRANS) DEFICIENCY. Eight cases of this recently described defect have been reported [9–12]. Most were severely affected, with onset in the neonatal period and death occurring before 3 months of age. Presentations included fasting hypoketotic hypoglycemia, coma, cardiopulmonary arrest, and ventricular arrhythmias. One of the children with neonatal onset survived until 3 years of age; he succumbed to progressive skeletal-muscle weakness and liver failure that were unresponsive to intensive feeding. Two milder cases, with attacks of fasting hypoketotic coma similar to those seen in cases of medium-chain acyl-coenzyme A (CoA) dehydrogenase (MCAD) deficiency, have been reported. The severe form of translocase deficiency, and the severe forms of CPT-2 and electron-transfer flavoprotein/electron-transfer flavoprotein dehydrogenase (ETF/ETF-DH) deficiencies, appear to have a very poor prognosis.

CARNITINE PALMITOYL TRANSFERASE-2 (CPT-2) DEFICIENCY. Two forms of CPT-2 deficiency are known: a mild adult-onset form characterized by exercise-induced attacks of rhabdomyolysis and a severe neonatal-onset form that presents with life-threatening coma, cardiomyopathy, and weakness [13–16]. Neonatal-onset CPT-2 deficiency and the severe form of ETF/ETF-DH deficiency have been associated with congenital brain and renal malformations.

Patients with the milder adult form of CPT-2 deficiency begin to have attacks of rhabdomyolysis in the second and third decades of life. These attacks are triggered by catabolic stresses, such as prolonged exercise, fasting, or cold exposure. Episodes are associated with aching muscle pain, elevated plasma creatine-phosphokinase levels, and myoglobinuria, which may lead to renal shutdown [13].

## Fatty Acid β-Oxidation Defects

These comprise deficiencies of acyl-CoA dehydrogenase and 3-hydroxy-acyl-CoA dehydrogenase.

VERY-LONG-CHAIN ACYL-COA-DEHYDROGENASE DEFICIENCY. This defect was originally reported as a defect of the long-chain acyl-CoA dehydrogenase (LCAD) enzyme before the existence of two separate enzymes capable of acting on long-chain substrates was recognized [17]. Very-long-chain acyl-CoA-dehydrogenase deficiency (VLCAD) is bound to the inner mitochondrial membrane, whereas LCAD is a matrix enzyme. All of the known patients appear to have defects in VLCAD activity [18]. A separate disorder of the LCAD enzyme has yet to be identified, perhaps because this enzyme acts primarily on branched-chain rather than straight-chain fatty acids [19, 20]. Many patients with VLCAD deficiency have had severe clinical manifestations, including chronic cardiomyopathy, weakness, and episodes of fasting coma. Several have presented in the newborn period with life-threatening coma similar to patients with TRANS or severe CPT-2 deficiencies. However, milder cases of VLCAD deficiency have also been identified with a phenotype very similar to MCAD deficiency.

MEDIUM-CHAIN ACYL-COENZYME A DEHYDROGENASE (MCAD) DEFICIENCY. This is the single most common fatty acid-oxidation disorder [1, 21, 22]. It is also one of the mildest, with no evidence of chronic muscle or cardiac involvement. In addition, it is unusually homogeneous, because 90% of patients are homozygous for a single A985G missense mutation originating in northwestern Germany and the British Isles [23]. The estimated incidence in Britain is 1 in 10,000 births [24].

As shown in Table 11.1, patients with MCAD deficiency have an exclusively hepatic type of presentation. Affected individuals appear to be entirely normal until an episode of illness is provoked by an excessive period of fasting. This may occur with an infection that interferes with normal feeding or may result simply because breakfast is delayed. The first episode typically occurs between 3-24 months of age, after nocturnal feedings have ceased. A few neonatal cases in which attempted breast-feeding was sufficient fasting stress to cause illness have been reported. Attacks become less frequent after childhood, because fasting tolerance improves with increasing body mass.

The response to fasting in MCAD deficiency identifies many of the pathophysiologic features of the hepatic presentation of the fatty acid-oxidation disorders (Fig. 11.2). No abnormalities occur during the first 12-14 h, because lipolysis and fatty acid oxidation have not yet been activated. By 16 h, plasma levels of free fatty acids have risen dramatically, but ketones remain inappropriately low, reflecting the defect in hepatic fatty acid oxidation. Hypoglycemia develops shortly thereafter, probably because of excessive glucose utilization due to the inability to switch to fat as a fuel. Severe symptoms of lethargy and nausea develop in association with the marked increase in plasma fatty acids. It should be stressed that patients with fatty acid-oxidation defects can become dangerously ill before plasma glucose falls to hypoglycemic values. This may be due to several factors, including the absence of ketones, the loss of energy supply to heart and skeletal muscle, and toxic effects of elevated plasma free-fatty acid intermediates. An acute attack in

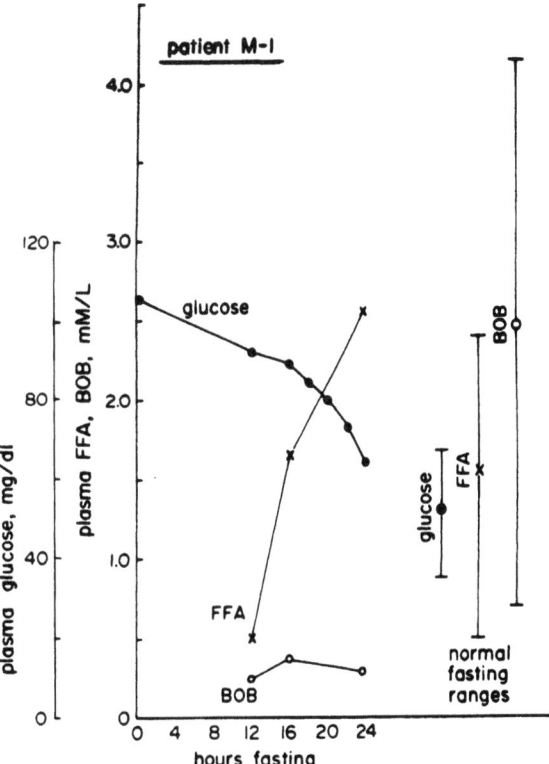

**Fig. 11.2.** Response to fasting in a patient with medium-chain acyl-CoA-dehydrogenase deficiency. Shown are plasma levels of glucose, free fatty acids (*FFA*), and β-hydroxybutyrate (*BOB*) in the patient, and the mean and range of values in normal children who fasted for 24 h. At 14-16 h of fasting, the patient became ill, with pallor, lethargy, nausea, and vomiting

MCAD deficiency usually features lethargy, nausea, and vomiting, which rapidly progresses to coma within 1-2 h. Seizures may occur, and patients may die suddenly from acute cardiorespiratory arrest. They may also die or suffer permanent brain damage from cerebral edema. Up to 25% of MCAD patients die during their first attack. Because there is no forewarning, the first episode may be misdiagnosed as Reye syndrome, sudden infant-death syndrome (SIDS), etc. MCAD deficiency can be considered as one cause of Reye syndrome, although the age of onset is younger than usual, and the liver biopsy shows steatosis but not mitochondrial swelling.

At the time of an acute attack in MCAD deficiency, the liver may be slightly enlarged, or it may become enlarged during the first 24 h of treatment. Chronic cardiac and skeletal-muscle abnormalities are not seen in MCAD deficiency, perhaps because the block in fatty acid oxidation is incomplete. However, the enzyme defect is probably expressed in cardiac and skeletal muscle, and these organs are probably responsible for the sudden death that may occur during attacks of illness in MCAD-deficient infants and children.

SHORT-CHAIN ACYL-COENZYME A DEHYDROGENASE (SCAD) DEFICIENCY. Clinical manifestations of this disorder have primarily been chronic failure to thrive, developmental regression, and acidemia rather than the acute life-threatening episodes of coma and hypoglycemia associated with most of the fatty acid-oxidation disorders [25–27]. Similar evidence of chronic toxicity occurs in other short-chain fatty acid-oxidation disorders, medium-chain 3-ketoacyl-CoA thiolase (MCKT; β-ketothiolase) deficiency, and short-chain 3-hydroxyacyl-CoA dehydrogenase deficiency. Although a significant number of cases have been identified, the molecular basis of the disease remains unclear, since the most commonly found mutation appears to be a frequent polymorphism in normal individuals.

3-HYDROXY-ACYL-COA-DEHYDROGENASE DEFICIENCIES. Recent work indicates that the activities of long-chain enoyl-CoA hydratase, 3-hydroxy-acyl-CoA dehydrogenase, and β-ketoacyl-CoA thiolase are combined in a single trifunctional protein. Some patients have isolated *long-chain 3-hydroxyacyl-CoA dehydrogenase* (LCHAD) deficiency, while others are also deficient in *long-chain enoyl-CoA hydratase* and *long-chain β-ketoacyl-CoA thiolase* activities [28–32]. The clinical phenotype ranges from a mild disorder resembling MCAD deficiency to more severe involvement of heart and skeletal muscle, similar to VLCAD deficiency. Some patients have had retinal degeneration or peripheral neuropathy, suggesting a toxicity effect. Several cases of heterozygote mothers developing acute fatty-liver-of-pregnancy syndrome when carrying affected fetuses have been reported [33, 34]. This reinforces the suggestion of peculiar toxicity effects in LCAD deficiency.

SHORT-CHAIN 3-HYDROXY-ACYL-COA-DEHYDROGENASE (SCHAD) DEFICIENCY. Two reports of patients with potential defects of SCHAD have appeared, but these featured inconsistent clinical phenotypes. The first was a child with recurrent myoglobinuria and hypoglycemic coma, who appeared to have SCHAD deficiency in muscle, but not in fibroblasts [35]. The second report was of two children with recurrent episodes of fasting ketotic hypoglycemia who had reduced SCHAD-enzyme activity in fibroblast mitochondria [36]. If the latter observation is correct, short-chain β-oxidation defects, such as SCHAD and short-chain acyl-coenzyme A dehydrogenase (SCAD) deficiencies, which affect only the last few steps of the fatty acid-oxidation cycle, may have little impact on acetyl-CoA generation and ketogenesis. It is possible that illness provoked by fasting in patients with these defects may be attributable to toxic effects of short-chain fatty acid intermediates rather than to hypoglycemia.

MEDIUM-CHAIN 3 KETOACYL-COA THIOLASE (MCKT) DEFICIENCY. One case of a defect in MCKT has been reported: a baby boy who died in the newborn period after presenting with vomiting and acidosis on day two of life [37]. At 2 weeks of age, he had terminal rhabdomyolysis and myoglobinuria. Urine showed elevated ketones, suggesting fairly good acetyl-CoA generation from partial oxidation of long-chain fatty acids, similar to what has been noted in other defects that are specific for short-chain fatty acids.

2,4-DIENOYL-COA-REDUCTASE (DER) DEFICIENCY. Only a single case of DER deficiency has been reported in the pathway required for oxidation of unsaturated fatty acids [38]. The patient was hypotonic from birth and died at age 4 months. The disorder was suspected based on low plasma total carnitine levels and urinary excretion of an unusual unsaturated fatty acylcarnitine in the urine.

## Electron-Transfer Defects

ETF/ETF-DH defects in the pathway for transferring electrons from the first step in β-oxidation to the electron-transport system are grouped together [39]. They are also known as *glutaric aciduria type 2* or *multiple acyl-CoA dehydrogenase deficiencies*. These defects block not only fatty acid oxidation, but also the oxidation of branched-chain amino acids (leucine, isoleucine, and valine) and of glutaryl-CoA on the catabolic pathway of lysine, tryptophan and hydroxylysine (Chap. 20; Fig. 20.1). Patients with severe or complete deficiencies of the enzymes present with hypoglycemia, acidosis, hypotonia, cardiomyopathy, and coma in the neonatal period. Some neonates with ETF/ETF-DH deficiencies have had congenital anomalies (polycystic kidney, midface hypoplasia). Partial deficiencies of ETF/ETF-DH are associated with milder diseases resembling MCAD or LCAD deficiency. Some patients have been reported to respond to supplementation with riboflavin, the co-factor for these enzymes. The urine organic-acid profile is usually diagnostic, especially in the severe forms of these deficiencies.

## Ketone-Synthesis Defects

Genetic defects in ketone synthesis also present with episodes of fasting hypoketotic hypoglycemia, 3-hydroxy-3-methylglutaryl-CoA synthase deficiency, and 3-hydroxy-3-methylglutaryl-CoA lyase deficiency [40, 41]. These defects are described in Chap. 12.

## Metabolic Derangements

The metabolic derangements associated with different defects in the mitochondrial fatty acid-oxidation and ketone-synthesis pathways vary significantly depending on the exact site of the defect and are, therefore, discussed in the section "Diagnostic Tests".

## Diagnostic Tests

Recently, the diagnostic investigation of disorders of fatty acid oxidation has been radically simplified by the assay of the plasma or urine acylcarnitine profile by tandem mass spectrometry or other refined methods [42, 43]. This method and other assays to detect disease-related abnormal metabolites are to be used first, since they do not burden the patient and carry no risk. If no abnormalities are detected, a test that loads the overall pathway of fatty acid oxidation is indicated. This diagnostic approach is followed in the discussion below.

### Disease-Related Metabolites

URINE/PLASMA ACYLCARNITINES. Since acyl-CoA intermediates proximal to blocks in the fatty acid oxidation pathway can be transesterified to carnitine, most of the fatty acid oxidation disorders can be detected by analysis of either urinary or (preferably) plasma acylcarnitine profiles (Table 11.2) [42, 43]. The newer tandem-mass-spectrometry methods can use small volumes of plasma or filter-paper blood spots.

In a few regions, the latter method has also been used successfully for extended screening of newborn infants for inborn errors of metabolism. It allows detection of about 20 defects, including fatty acid-oxidation disorders caused by deficiencies of MCAD, VLCAD, ETF/ETF-DH, 3-hydroxy-3-methylglutaryl (HMG)-CoA lyase, and (possibly) SCAD. In the state of Pennsylvania, a large commercial screening program using tandem mass spectrometry revealed an incidence of MCAD deficiency approaching 1 in 5000 infants [44]. In some areas, extended screening by mass-spectrometry methods is replacing more traditional, limited screening programs.

PLASMA AND TISSUE TOTAL CARNITINE CONCENTRATIONS. A peculiar feature of the fatty acid-oxidation disorders is that all but one are associated with either decreased or increased concentrations of total carnitine in plasma and tissues [9]. In CTD, plasma total carnitine levels are severely decreased (<5% of normal) [4]. In CPT-1 deficiency, total carnitine levels are increased (150%–200% of normal) [6]. In all of the other defects except HMG-CoA-synthase deficiency, total carnitine levels are reduced to 25–50% of normal (secondary carnitine deficiency). Thus, simple measurement of plasma total carnitine is often helpful in determining the presence of a fatty acid-oxidation disorder. It should be emphasized that samples must be taken during the well-fed state with normal dietary carnitine intake, because patients with disorders of fatty acid oxidation may show acute increases in plasma total carnitine during prolonged fasting or during attacks of illness.

The basis of the carnitine deficiency in CTD has been shown to be a defect in the plasma-membrane carnitine-transporter activity. The reason for the increased carnitine levels in CPT-1 deficiency and the decreased carnitine levels in other fatty acid-oxidation disorders is unclear. Both phenomena can actually be explained by the competitive inhibitory effects of long-chain and medium-chain acylcarnitines on the carnitine transporter [45]. Thus, in patients with MCAD or

**Table 11.2.** Fatty acid-oxidation disorders with distinguishing metabolic markers

| Disorder | Plasma acylcarnitines | Urinary acylglycines | Urinary organic acids |
|---|---|---|---|
| VLCAD | Tetradecenoyl- | | |
| MCAD | Otanoyl-<br>Decenoyl- | Hexanoyl-<br>Suberyl-<br>Phenylpropionyl- | |
| SCAD | Butyryl- | Butyryl- | Ethylmalonic |
| LCHAD | 3-Hydroxy-palmitoyl-<br>3-Hydroxy-oleoyl-<br>3-Hydroxy-linoleoyl- | | 3-Hydroxydicarboxylic |
| DER | Dodecadienoyl- | | |
| ETF and ETF-DH | Butyryl-<br>Isovaleryl-<br>Glutaryl- | Isovaleryl-<br>Hexanoyl- | Ethylmalonic<br>Glutaric<br>Isovaleric |
| HMG-CoA lyase | Methylglutaryl- | | 3-Hydroxy-3-methyl-glutaric |

*DER*, 2,4-dienoyl-coenzyme A reductase; *ETF*, electron-transfer flavoprotein; *ETF-DH*, ETF dehydrogenase; *HMG-CoA*, 3-hydroxy-3-methylglutaryl-coenzyme A; *LCHAD*, long-chain 3-hydroxyacyl-coenzyme A dehydrogenase; *MCAD*, medium-chain acyl-coenzyme A dehydrogenase; *SCAD*, short-chain acyl-coenzyme A dehydrogenase; *VLCAD*, very-long-chain acyl-coenzyme A dehydrogenase

TRANS deficiency, the blocks in acyl-CoA oxidation lead to accumulation of acylcarnitines, which inhibit renal and tissue transport of free carnitine and result in lowered plasma and tissue concentrations of carnitine. Conversely, the inability to form long-chain acyl-CoA in CPT-1 deficiency results in below-normal inhibition of carnitine transport from long-chain acylcarnitine and, therefore, increases renal carnitine thresholds and plasma levels of carnitine to values greater than normal.

URINARY ORGANIC ACIDS. The urinary organic-acid profile is usually normal in patients with fatty acid-oxidation disorders when they are well. During times of fasting or illness, all of the disorders are associated with an "inappropriate" dicarboxylic aciduria, i.e., urinary medium-chain dicarboxylic acids are elevated, while urinary ketones are not. This reflects the fact that dicarboxylic acids, derived from partial oxidation of fatty acids in microsomes and peroxisomes, are produced whenever plasma free-fatty acid concentrations are elevated. In MCAD deficiency, the amounts of dicarboxylic acids excreted are two- to fivefold greater than in normal fasting children. However, in other defects, only the ratio of ketones to dicarboxylic acids is abnormal. In a few of the disorders, specific abnormalities of urine organic acid-profiles may be present (Table 11.2) but are not likely to be found except during fasting stress.

URINARY ACYLGLYCINES. In MCAD deficiency, urine contains increased concentrations of the glycine conjugates of hexanoate, suberate (C-8 dicarboxylic acid), and phenylpropionate, which are derived from their coenzyme-A esters [46]. When these are quantified by isotope dilution–mass spectrometry, specific diagnosis of MCAD deficiency is possible, even when using random urine specimens. Abnormal glycine conjugates are present in urine from patients with some of the other disorders of fatty acid oxidation (Table 11.2). A *phenylpropionate loading test* (20 mg/kg per os) followed by assay of urinary phenylpropionylglycine excretion can be used as a test specific for MCAD deficiency [47]. However, this test is falling into disuse because of the availability of the plasma acylcarnitine-profile test (see above).

PLASMA FATTY ACIDS. In MCAD deficiency, specific increases in plasma concentrations of the medium-chain fatty acids octanoate and *cis*-4-decenoate have been identified, which can be useful for diagnosis. Abnormally elevated plasma concentrations of these fatty acids are most apparent during fasting. It is not known whether specific abnormalities in plasma fatty acid profiles might be found in other disorders.

## Tests of Overall Pathway

These include in vivo fasting test, in vitro fatty acid oxidation, and histology.

IN VIVO FASTING TEST. In diagnosing the fatty acid-oxidation disorders, it is frequently useful to first demonstrate an impairment in the overall pathway before attempting to identify the specific site of the defect. Blood and urine samples collected immediately prior to treatment of an acute episode of illness can be used for this purpose, e.g., by showing elevated plasma free-fatty acid levels but inappropriately low ketone levels at the time of hypoglycemia. If such important data are not available, a carefully monitored study of fasting ketogenesis can provide this information (Fig. 11.2). This provocative test can put the patient at risk and should only be done under controlled circumstances, with careful supervision. Some investigators prefer using fat loading as an alternative means of testing hepatic ketogenesis [3, 48] (Chap. 2).

IN VITRO FATTY ACID OXIDATION. Using $^{14}$C- or $^{3}$H-labeled substrates, cultured skin fibroblasts or lymphoblasts from patients can also be used to demonstrate a general defect in fatty acid oxidation. In addition, fatty acid substrates with different chain lengths can be used with these cells to localize the probable site of defect. Very low rates of labeled-fatty acid oxidation are found in CPT-1, TRANS, CPT-2, and ETF/ETF-DH deficiencies. However, high residual rates of oxidation (50–80% or more of normal) frequently make identification of the β-oxidation-enzyme defects difficult. In CTD, oxidation rates are normal unless special steps are taken to grow cells in carnitine-free media. The in vitro oxidation assays do not detect defects in ketone synthesis.

HISTOLOGY. The appearance of increased triglyceride droplets in affected tissues sometimes provides a clue to the presence of a defect in fatty acid oxidation. In the hepatic presentation of any of the fatty acid-oxidation disorders, a liver biopsy obtained during an acute episode of illness shows an increase in neutral fat deposits, which may have either a micro- or macrovesicular appearance. Between episodes, the amount of fat in the liver may be normal. More severe changes, including hepatic fibrosis, have been seen in VLCAD patients who where ill for prolonged periods [49]. This damage appears to reflect persistent efforts to metabolize fatty acids, since it may resolve as patients are adequately nourished. On electron microscopy, mitochondria do not show the severe swelling described in Reye syndrome, but may show minor changes, such as crystalloid inclusion bodies. The fatty acid-oxidation disorders that are expressed in muscle may be associated with increased fat-droplet accumulation in muscle fibers and may have the appearance of "lipoid myopathy" on biopsy.

### Enzyme Assays

Cultured skin fibroblasts and cultured lymphoblasts have become the preferred material in which to measure the in vitro activities of specific steps in the fatty acid-oxidation pathway. All of the known defects except HMG-CoA-synthase deficiency are expressed in these cells, and the results of assays in cells from both control and affected patients have been reported. Because these assays are not widely available, they are most usefully applied to confirm a site of defect that is suggested by other clinical and laboratory data.

## Treatment and Prognosis

The following sections focus on treatment of the hepatic presentation of fatty acid-oxidation disorders, since this is the most life-threatening aspect of these diseases. Although there is a high risk of mortality or long-term disability during episodes of fasting-induced coma, with early diagnosis and treatment, patients with most of the disorders have an excellent prognosis. The mainstay of therapy is prevention of recurrent attacks by adjusting the diet to minimize fasting stress.

### Management of Acute Illnesses

When patients with fatty acid-oxidation disorders become ill, treatment with intravenous glucose should be given immediately. Delay may result in sudden death or permanent brain damage. The goal is to provide sufficient glucose to stimulate insulin secretion to levels that will suppress fatty acid oxidation in liver and muscle and will block adipose-tissue lipolysis. Solutions of 10% dextrose (rather than the usual 5%) should be used at infusion rates of 10 mg/kg/min or greater to maintain high to normal levels of plasma glucose [above 100 mg/dl (5.5 mmol/l)]. Resolution of coma may not be immediate, perhaps because of the toxic effects of fatty acids noted above; it may take 2–4 h in mildly ill patients or as long as 1–2 days in severely ill patients.

### Long-Term Diet Therapy

It is essential to prevent any period of fasting that would require the use of fatty acids as a fuel. This can be done by simply ensuring that patients consume carbohydrates at bedtime and do not fast for more than 12 h overnight. During intercurrent illnesses when appetite is diminished, care should be taken to give small extra feedings of carbohydrates during the night. In a few patients with severe defects in fatty acid oxidation who had developed weakness and/or cardiomyopathy, we have gone further, completely eliminating fasting by the addition of continuous nocturnal intragastric feedings. Uncooked cornstarch at bedtime might be considered as a source of slowly released glucose (Chap. 6), although this has not been formally tested in these disorders. Some authors recommend restricting fat intake. Although this seems reasonable in patients with severe defects, we have not routinely restricted dietary fat in milder defects, such as MCAD deficiency.

### Carnitine Therapy

In patients with CTD, treatment with carnitine improves cardiac and skeletal-muscle function to nearly normal within a few months [4]. It also corrects any impairment in hepatic ketogenesis that may be present [4]. With oral carnitine at doses of 100 mg/kg/day, plasma carnitine levels can be maintained in the low to normal range, and liver carnitine levels may be normal. However, muscle carnitine concentrations remain less than 5% of normal. Since these low levels are adequate to reverse myopathy in CTD, it appears that the threshold for defining carnitine deficiency is a tissue concentration less than ~5% of normal.

A possible role for carnitine therapy in disorders of fatty acid oxidation that are associated with secondary carnitine deficiency remains controversial [42]. Since these disorders involve blocks at specific enzyme steps that do not involve carnitine, it is obvious that carnitine treatment cannot correct the defect in fatty acid oxidation. It has been proposed that carnitine might help to remove metabolites in these disorders, because the enzyme defects might be associated with accumulation of acyl-CoA intermediates. However, there has been no direct evidence that this is true, and some evidence to the contrary has been presented [50]. In addition, as noted above, the mechanism of secondary carnitine deficiency is not a direct one (via loss of acylcarnitines in urine) but appears to be indirect (via inhibition of the carnitine transporter in kidney and other tissues by medium- or long-chain acylcarnitines). It should also be noted that secondary carnitine deficiency could be a protective adaptation, since there is data showing that long-chain acylcarnitines may have toxic effects. Our current practice is not to recommend the use of carnitine except as an investigational drug in fatty acid-oxidation disorders other than CTD.

### Other Therapies

Since medium-chain fatty acids bypass the carnitine cycle (Fig. 11.1) and enter the midportion of the mitochondrial β-oxidation spiral directly, it is possible that they might be used as fuels in defects that block either the carnitine cycle or long-chain β-oxidation. For example, dietary medium-chain triglyceride (MCT) was suggested to be helpful in a patient with LCHAD deficiency. The benefits of MCT have not been thoroughly investigated, but MCT clearly must not be used in patients with MCAD, SCAD, SCHAD, ETF/ETF-DH, HMG-CoA-synthase, or HMG-CoA lyase deficiencies. MCT is probably also contraindicated in translocase deficiency, since translocase is involved in both long-chain and short-chain β-oxidation. Some patients with mild variants of ETF/ETF-DH and SCAD deficiencies have been reported to respond to supplementation with high doses of riboflavin (100 mg/day), the cofactor for these enzymes.

### Prognosis

Although acute episodes carry a high risk of mortality or permanent brain damage, many patients can be easily managed by avoidance of prolonged fasts. These patients have an excellent long-term prognosis. Patients with chronic cardiomyopathy or skeletal-muscle weakness have a more guarded prognosis, since they seem to have more severe defects in fatty acid oxidation. For example, TRANS deficiency or the severe variants of CPT-2 and ETF/ETF-DH deficiencies frequently lead to death in the newborn period. However, the mild form of CPT-2 deficiency may remain silent as long as patients avoid exercise stress.

### Genetics

All of the genetic disorders of fatty acid oxidation that have been identified are inherited in autosomal-recessive fashion. Heterozygous carriers are clinically normal (with the possible exception, noted above, of the occurrence of acute fatty liver of pregnancy in LCHAD heterozygote mothers carrying an affected fetus). Carriers of the fatty acid-oxidization disorders show no biochemical abnormalities except for carriers of CTD, who have half the normal levels of plasma total carnitine concentrations. Since some of the disorders, such as MCAD deficiency, may be present without having caused an attack of illness, siblings of patients with fatty acid-oxidation disorders should be investigated to determine whether they might be affected.

*Prenatal diagnosis* by assay of labeled-fatty acid oxidation and/or enzyme activity in amniocytes or chorionic villi should theoretically be possible for those disorders that are expressed in cultured skin fibroblasts (i.e., all of the currently known defects except HMG-CoA-synthase deficiency). This has been done in a few instances of MCAD deficiency, although new molecular methods are now available for this particular defect. Metabolite screening of amniotic fluid has not been useful for most defects. No general newborn-screening test has been developed for the fatty acid-oxidation disorders, although filter-paper blood spots can be used to diagnose MCAD deficiency by analysis of fatty acid or acylcarnitine profiles or demonstration of the common c.985A → G mutation. Analysis of acylcarnitine profiles from newborn blood spots might also prove useful in neonatal detection of other fatty acid-oxidation disorders, although only a few have been associated with specific abnormalities (Table 11.2).

Rapid progress has been made in establishing the molecular basis of several of the defects in fatty acid oxidation [1]. This has become especially useful in MCAD deficiency. About 80% of MCAD patients are homozygous for a single missense mutation, c.985A → G, resulting in a K329E amino acid substitution; 17% carry this mutation in combination with another mutation. This probably represents a "founder" effect and explains why most MCAD patients share a northwestern-European ethnic background. Simple polymerase-chain-reaction assays have been established to detect the c.985A → G mutation using DNA from many different sources, including newborn blood-spot cards. This method has been used to diagnose MCAD deficiency in a variety of circumstances, including prenatal diagnosis, postmortem diagnosis of affected siblings, and for surveys of disease incidence. It has been suggested that this would also provide a possible method for newborn screening, but it is not as reliable or as simple as tandem-mass-spectrometry analysis of blood-spot acylcarnitines.

## Rare Related Disorders

### Transport Defect of Fatty Acids

Two children with liver failure for whom a genetic defect in the transport of free fatty acids across the plasma membrane was suggested have been reported [51]. One of these children had reduced levels of long-chain free fatty acids in liver tissue; cultured fibroblasts from both showed modest reductions in both oxidation and uptake of long-chain fatty acids. Although five putative fatty acid transporters have been described, their function as carrier proteins remains speculative.

## Defect of Fatty-Alcohol Metabolism

The fatty-alcohols comprise a minor component of membrane lipids. Children with Sjogren-Larsson syndrome (congenital ichthyosis, mental retardation, spastic di- or tetraplegia) have a recessive genetic defect in the oxidation of long-chain fatty alcohols (fatty-aldehyde-dehydrogenase deficiency) [52]. Fatty-alcohol metabolism is also disturbed in peroxisomal disorders, such as rhizomelic chondroplasia punctata and Zellweger syndrome. Little is known about the mechanism of disease in Sjogren-Larsson syndrome, and treatment is not available.

## References

1. Coates PM, Stanley CA (1992) Inherited disorders of mitochondrial fatty acid oxidation. Prog Liver Dis 10:123-38
2. Hale DE, Bennett MJ (1992) Fatty acid oxidation disorders: a new class of metabolic diseases. J Pediatr 121:1-11
3. Saudubray JM, Martin D, de Lonlay et al. (1999) Recognition and management of fatty acid oxidation defects: a series of 107 patients. J Inherit Metab Dis 22:488-502
4. Stanley CA, DeLeeuw S, Coates PM et al. (1991) Chronic cardiomyopathy and weakness or acute coma in children with a defect in carnitine uptake. Ann Neurol 30:709-716
5. Demaugre F, Bonnefont J, Mitchell G et al. (1988) Hepatic and muscular presentations of carnitine palmitoyl transferase deficiency: two distinct entities. Pediatr Res 24:308-311
6. Stanley CA, Sunaryo F, Hale DE et al. (1992) Elevated plasma carnitine in the hepatic form of carnitine palmitoyltransferase-1 deficiency. J Inherit Metab Dis 15:785-789
7. Vianey-Saban C, Mousson B, Bertrand C et al. (1993) Carnitine palmitoyl transferase I deficiency presenting as a Reye-like syndrome without hypoglycemia. Eur J Pediatr 152:334-338
8. Falik-Borenstein ZC, Jordan SC, Saudubray JM et al. (1992) Brief report: renal tubular acidosis in carnitine palmitoyltransferase type 1 deficiency. N Engl J Med 327:24-27
9. Stanley CA, Hale DE, Berry GT et al. (1992) Brief report: a deficiency of carnitine-acylcarnitine translocase in the inner mitochondrial membrane. N Engl J Med 327:19-23
10. Pande SV, Brivet B, Slama A et al. (1993) Carnitine-acylcarnitine translocase deficiency with severe hypoglycemia and auriculo ventricular block. Translocase assay in permeabilized fibroblasts. J Clin Invest 91:1247-1252
11. Chalmers RA, Stanley CA, English N, Wigglesworth JS (1997) Mitochondrial carnitine-acylcarnitine translocase deficiency presenting as sudden neonatal death. J Pediatr 131:220-225
12. Morris AAM, Olpin SE, Brivet M et al. (1998) A patient with carnitine-acylcarnitine translocase deficiency with a mild phenotype. J Pediatr 132:514-516
13. DiMauro S, DiMauro PMM (1973) Muscle carnitine palmityltransferase deficiency and myoglobinuria. Science 182:929-931
14. Demaugre F, Bonnefont JP, Colonna M et al. (1991) Infantile form of carnitine palmitoyltransferase II deficiency with hepatomuscular symptoms and sudden death. Physiopathological approach to carnitine palmitoyltransferase II deficiencies. J Clin Invest 87:859-864
15. Taroni F, Verderio E, Garavaglia B et al. (1992) Biochemical and molecular studies of carnitine palmitoyltransferase II deficiency with hepatocardiomyopathic presentation. Prog Clin Biol Res 375:521-531
16. Taroni F, Verderio E, Dworzak F et al. (1993) Identification of a common mutation in the carnitine palmitoyltransferase II gene in familial recurrent myoglobinuria patients. Nat Genet 4:314-320
17. Hale DE, Batshaw ML, Coates PM et al. (1985) Long-chain acyl coenzyme A dehydrogenase deficiency: an inherited cause of nonketotic hypoglycemia. Pediatr Res 19:666-671
18. Vianey-Saban C, Divry P, Brivet M et al. (1998) Mitochondrial very-long-chain acyl-coenzyme A dehydrogenase deficiency: clinical characteristics and diagnostic considerations in 30 patients. Clin Chim Acta 269:43-62
19. Mao LF, Chu C, Luo MJ et al. (1995) Mitochondrial beta-oxidation of 2-methyl fatty acids in rat liver. Arch Biochem Biophys 321:221-228
20. Wanders RJ, Denis S, Ruiter JP, Ijlst L, Dacremont G (1998) 2,6-Dimethylheptanoyl-CoA is a specific substrate for long-chain acyl-CoA dehydrogenase (LCAD): evidence for a major role of LCAD in branched-chain fatty acid oxidation. Biochim Biophys Acta 1393:35-40
21. Stanley CA, Hale DE, Coates PM et al. (1983) Medium-chain acyl-CoA dehydrogenase deficiency in children with non-ketotic hypoglycemia and low carnitine levels. Pediatr Res 17:877-884
22. Wilson CJ, Champion MP, Collins JE, Clayton PT, Leonard JV (1999) Outcome of medium chain acyl-CoA dehydrogenase deficiency after diagnosis. Arch Dis Child 80:459-462
23. Yokota I, Coates PM, Hale DE, Rinaldo P, Tanaka K (1992) The molecular basis of medium chain acyl-CoA dehydrogenase deficiency: survey and evolution of 985A-G transition, and identification of five rare types of mutation within the medium chain acyl-CoA dehydrogenase gene. Prog Clin Biol Res 375:425-440
24. Yokota I, Coates PM, Hale DE, Rinaldo P, Tanaka K (1991) Molecular survey of a prevalent mutation, 985A-to-G transition, and identification of five infrequent mutations in the medium-chain Acyl-CoA dehydrogenase (MCAD) gene in 55 patients with MCAD deficiency. Am J Hum Genet 49:1280-1291
25. Bhala A, Willi SM, Rinaldo P et al. (1995) Clinical and biochemical characterization of short-chain acyl-coenzyme A dehydrogenase deficiency. J Pediatr 126:910-915
26. Coates PM, Hale DE, Finocchiaro G, Tanaka K, Winter SC (1988) Genetic deficiency of short-chain acyl-coenzyme A dehydrogenase in cultured fibroblasts from a patient with muscle carnitine deficiency and severe skeletal muscle weakness. J Clin Invest 81:171-175
27. Gregersen N, Winter VS, Corydon MJ et al. (1998) Identification of four new mutations in the short-chain acyl-CoA dehydrogenase (SCAD) gene in two patients: one of the variant alleles, 511C $\rightarrow$ T, is present at an unexpectedly high frequency in the general population, as was the case for 625G $\rightarrow$ A, together conferring susceptibility to ethylmalonic aciduria. Hum Mol Genet 7:619-627
28. Poll-The BT, Bonnefont JP, Ogier H et al. (1988) Familial hypoketotic hypoglycaemia associated with peripheral neuropathy, pigmentary retinopathy and C6-C14 hydroxydicarboxylic aciduria. A new defect in fatty acid oxidation? J Inherit Metab Dis 2:183-185
29. Tyni T, Pihko H (1999) Long-chain 3-hydroxyacyl-CoA dehydrogenase deficiency. Review article. Acta Paediatr 88:237-245
30. Dionisi-Vici C, Burlina AB, Bertini E et al. (1991) Long-chain 3-hydroxyacyl-CoA dehydrogenase deficiency: clinical and therapeutic considerations. J Pediatr 118:744-746
31. Rocchiccioli F, Wanders RJ, Aubourg P et al. (1990) Deficiency of long-chain 3-hydroxyacyl-CoA dehydrogenase: a cause of lethal myopathy and cardiomyopathy in early childhood. Pediatr Res 28:657-662
32. Jackson S, Kler RS, Bartlett K et al. (1992) Combined defect of long-chain 3-hydroxyacyl-CoA dehydrogenase, 2-enoyl-CoA hydratase and 3-oxoacyl-CoA thiolase. Prog Clin Biol Res 375:327-337
33. Treem WR, Rinaldo P, Hale DE et al. (1994) Acute fatty liver of pregnancy and long-chain 3-hydroxyacyl-coenzyme A dehydrogenase deficiency. Hepatology 19:339-345
34. Treem WR, Shoup ME, Hale DE et al. (1996) Acute fatty liver of pregnancy, hemolysis, elevated liver enzymes, and low platelets syndrome, and long chain 3-hydroxyacyl-coenzyme

A dehydrogenase deficiency [see comments]. Am J Gastroenterol 91:2293-2300
35. Tein I, De VDC, Hale DE et al. (1991) Short-chain L-3-hydroxyacyl-CoA dehydrogenase deficiency in muscle: a new cause for recurrent myoglobinuria and encephalopathy. Ann Neurol 30:415-419
36. Bennett MJ, Weinberger MJ, Kobori JA, Rinaldo P, Burlina AB (1996) Mitochondrial short-chain L-3-hydroxyacyl-coenzyme A dehydrogenase deficiency: a new defect of fatty acid oxidation. Pediatr Res 39:185-188
37. Kamigo T, Indo Y, Souri M et al. (1997) Medium chain 3-ketoacyl-coenzyme A thilase deficiency: a new disorder of mitochondrial fatty acid beta-oxidation. Pediatr Res 42: 569-576
38. Roe C, Millington D, Norwood D et al. (1990) 2,4-Dienoyl-coenzyme A reductase deficiency: a possible new disorder of fatty acid oxidation. J Clin Invest 85:1703-1707
39. Freeman FE, Goodman SI (1985) Deficiency of electron transfer flavoprotein or electron transfer flavoprotein:ubiquinone oxidoreductase in glutaric acidemia type II fibroblasts. Proc Natl Acad Sci USA 82:4517-4520
40. Thompson GN, Hsu BY, Pitt JJ, Treacy E, Stanley CA (1997) Fasting hypoketotic coma in a child with deficiency of mitochondrial 3-hydroxy-3-methylglutaryl-CoA synthase. N Engl J Med 337:1203-1207
41. Gibson K, Breuer J, Nyhan W (1988) 3-Hydroxy-3-methylglutaryl-coenzyme A lyase deficiency: review of 18 reported cases. Eur J Pediatr 148:180-186
42. Stanley CA (1995) Carnitine disorders. Adv Pediatr 42: 209-242
43. Millington DS, Terada N, Chace DH et al. (1992) The role of tandem mass spectrometry in the diagnosis of fatty acid oxidation disorders. In: Coates PM, Tanaka K (eds) New developments in fatty acid oxidation. Progress in clinical and biochemical research. New York, John Wiley-Liss, pp 339-354
44. Ziadeh R, Hoffman EP, Finegold DM et al. (1995) Medium chain acyl-CoA dehydrogenase deficiency in Pennsylvania: neonatal screening shows high incidence and unexpected mutation frequencies. Pediatr Res 37:675-678
45. Stanley CA, Berry GT, Bennett MJ et al. (1993) Renal handling of carnitine in secondary carnitine deficiency disorders. Pediatr Res 34:89-97
46. Rinaldo P, O'Shea JJ, Coates PM et al. (1988) Medium-chain acyl-CoA dehydrogenase deficiency. Diagnosis by stable-isotope dilution measurement of urinary n-hexanoylglycine and 3-phenylpropionylglycine [published erratum appears in N Engl J Med 1989, 320(18):1227]. N Engl J Med 319:1308-1313
47. Runsby G, Seakins J, Leonard J (1886) A simple screening test for medium chain acyl-CoA dehydrogenase deficiency. Lancet 2:467
48. Parini R, Garavaglia B, Saudubray J et al. (1991) Clinical diagnosis of long-chain acyl-coenzyme A dehydrogenase deficiency: use of stress and fat-loading tests. J Pediatr 119:77-80
49. Treem WR, Witzleben CA, Piccoli DA et al. (1986) Medium-chain and long-chain acyl CoA dehydrogenase deficiency: clinical, pathologic and ultrastructural differentiation from Reye's syndrome. Hepatology 6:1270-1278
50. Lieu YK, Hsu BY, Price WA, Corkey BE, Stanley CA (1997) Carnitine effects on coenzyme A profiles in rat liver with hypoglycin inhibition of multiple dehydrogenases. Am J Physiol 272:E359-366
51. Odaib AA, Schneider BL, Bennett MJ et al. (1998) A defect in the transport of long-chain fatty acids associated with acute liver failure. N Engl J Med 339:1752-1757
52. Rizzo WB (1998) Inherited disorders of fatty alcohol metabolism. Minireview. Mol Genet Metab 65:63-73

CHAPTER 12

## Ketogenesis and Ketolysis

During fasting, ketone bodies are an important fuel for many tissues, including cardiac and skeletal muscle. They are particularly important for the brain, which cannot oxidise fatty acids. The principal ketone bodies, acetoacetate and 3-hydroxybutyrate, are maintained in equilibrium by 3-hydroxybutyrate dehydrogenase; acetone is formed non-enzymatically from acetoacetate and is eliminated in breath. Ketone bodies are formed in liver mitochondria, predominantly from fatty acids, but also from certain amino acids, such as leucine. For use as fuel, ketone bodies are converted to acetyl-coenzyme A (CoA) in the mitochondria of extrahepatic tissues. One of the ketolytic enzymes, methylacetoacetyl-CoA thiolase, is also involved in the breakdown of isoleucine (Fig. 12.1).

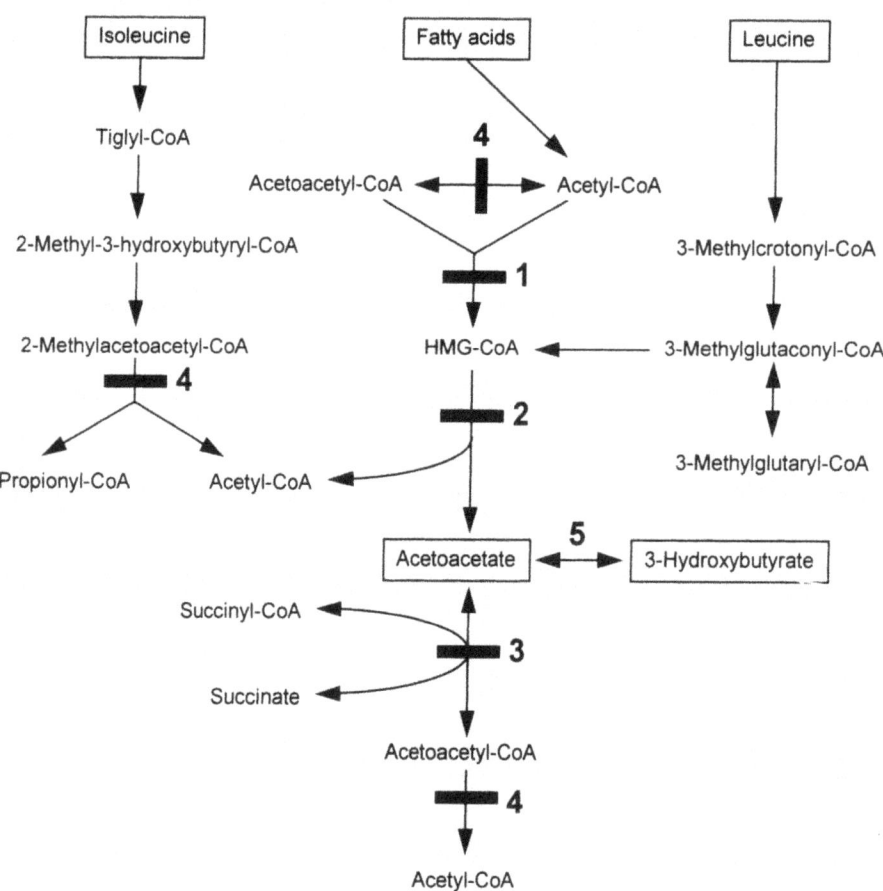

**Fig. 12.1.** Biochemical pathways involving enzymes of ketogenesis and ketolysis. *HMG-CoA*, 3-hydroxy-3-methylglutaryl coenzyme A; 1, mitochondrial HMG-CoA synthase; 2, HMG-CoA lyase; 3, succinyl-CoA oxoacid transferase; 4, methylacetoacetyl-CoA thiolase; 5, 3-hydroxybutyrate dehydrogenase. The enzyme defects are depicted by *solid bars* across the arrows

# Disorders of Ketogenesis and Ketolysis

A.A.M. Morris

CONTENTS

Clinical Presentation ........................... 153
   Mitochondrial 3-Hydroxy-3-Methylglutaryl-
   Coenzyme A Synthase Deficiency ................ 153
   HMG-CoA Lyase Deficiency ..................... 153
   Succinyl-CoA Oxoacid Transferase Deficiency ....... 153
   Methylacetoacetyl-CoA Thiolase Deficiency ......... 154
Metabolic Derangement ......................... 154
Diagnostic Tests ............................... 154
   3-Hydroxy-3-methylglutaryl (HMG)-CoA Synthase
   Deficiency .................................. 154
   3-Hydroxy-3-methylglutaryl (HMG)-CoA Lyase
   Deficiency .................................. 154
   Succinyl CoA Oxoacid Transferase (SCOT) Deficiency 154
   Methylacetoacetyl-CoA Thiolase (MAT) Deficiency ... 155
Treatment and Prognosis ........................ 155
Genetics ...................................... 155
Cytosolic Acetoacetyl-CoA Thiolase Deficiency ......... 155
References .................................... 156

---

Disorders of ketone-body metabolism present either in the first few days of life or later in childhood, during an infection or some other metabolic stress. In defects of ketogenesis, decompensation leads to encephalopathy with vomiting and a reduced level of consciousness, often accompanied by hepatomegaly. The biochemical features (hypoketotic hypoglycaemia with or without hyperammonaemia) resemble those seen in fatty-acid-oxidation disorders. In defects of ketolysis, the presentation is dominated by severe ketoacidosis. This is often accompanied by decreased consciousness and dehydration.

## Clinical Presentation

### Mitochondrial 3-Hydroxy-3-Methylglutaryl-Coenzyme A Synthase Deficiency

Two patients with mitochondrial 3-hydroxy-3-methylglutaryl-coenzyme A (mHMG-CoA)-synthase deficiency have recently been reported [1, 2]. They presented with coma following gastroenteritis, at the ages of 6 years and 16 months, respectively. The older child suffered a brief generalised seizure, and the younger child had hepatomegaly. Investigations showed hypoglycaemia with normal blood lactate and ammonia concentrations; urine was negative for ketones. Both recovered promptly with intravenous glucose and suffered no long-term complications.

### 3-Hydroxy-3-Methylglutaryl-CoA Lyase Deficiency

Approximately half of these patients present by 5 days of age, after a short, initial, symptom-free period. Most of the other patients present later in the first year, when they have fasted or suffer infections [3]. A few patients remain asymptomatic for a number of years [4].

Typical clinical features include vomiting, hypotonia and a reduced level of consciousness. All patients are acidotic and almost all have hypoglycaemia, which is frequently profound [3]. Ketone-body levels are inappropriately low, but blood lactate concentrations may be markedly elevated, particularly in neonatal-onset cases [5]. Many patients have hepatomegaly, abnormal liver-function tests and hyperammonaemia and, in the past, cases may have been misdiagnosed as Reye's syndrome [6]. Pancreatitis is a recognised complication [7], as in other branched-chain organic acidaemias. With appropriate treatment, most patients recover from their initial episode of metabolic decompensation, but a number suffer neurological sequelae, including epilepsy, hemiplegia and intellectual handicap [3, 6]. Even in asymptomatic patients, magnetic resonance imaging (MRI) shows multiple abnormal foci within the cerebral white matter. The cause of these changes is unknown, but there does not appear to be progressive demyelination [4].

### Succinyl-CoA Oxoacid Transferase Deficiency

Succinyl-CoA oxoacid transferase (SCOT) deficiency is characterised by recurrent episodes of severe ketoacidosis. Tachypnoea is often accompanied by hypotonia or coma. As in HMG-CoA lyase deficiency, half the patients become symptomatic within a few days of birth, and most of the others present later in the first year [8]. Two patients had cardiomegaly at the time of presentation [9, 10]. Blood glucose, lactate and ammonia concentrations are generally normal, though high and slightly low glucose levels have both been recorded [8, 10]. Since ketosis and acidosis are common in sick children, SCOT deficiency enters the differential diagnosis for a large number of patients.

### Methylacetoacetyl-CoA Thiolase Deficiency

Methylacetoacetyl-CoA thiolase (MAT) deficiency patients also have recurrent episodes of ketoacidosis, but a wider clinical spectrum is recognised [11]. Neonatal onset is rare [12, 13]. Most patients present during the first 2 years, but one asymptomatic adult has been diagnosed [14].

Episodes of decompensation generally start with vomiting and tachypnoea, followed by dehydration and a falling level of consciousness [11]. Some patients have seizures or haematemesis [15]. Investigations show a severe metabolic acidosis with ketonuria. Blood glucose, lactate and ammonia concentrations are normal in most cases, but hyper- and hypoglycaemia have been reported [10, 16]. The high acetoacetate levels in blood and urine may cause screening tests for salicylate to give false positive results [17]. Most patients make a full recovery from episodes of decompensation, but there are several reports of developmental delay, ataxia or other neurological problems [16, 18]. In three patients, MRI showed abnormal signal bilaterally in the posterior lateral part of the putamen [18].

### Metabolic Derangement

Ketone bodies are synthesised in hepatic mitochondria, primarily using acetyl-CoA derived from fatty-acid oxidation (Fig. 12.1). mHMG-CoA synthase catalyses the condensation of acetoacetyl-CoA and acetyl-CoA to form HMG-CoA, which is cleaved by HMG-CoA lyase to release acetyl-CoA and acetoacetate. HMG-CoA can also be derived from the ketogenic amino acid leucine. Thus, mHMG-CoA synthase and HMG-CoA lyase deficiencies both impair ketogenesis, but HMG-CoA-lyase deficiency also causes the accumulation of intermediates of the leucine catabolic pathway. The hypoglycaemia seen in these defects may result from impaired gluconeogenesis or from excessive glucose consumption due to the lack of ketone bodies.

Ketone-body utilisation occurs in extrahepatic mitochondria, starting with the transfer of CoA from succinyl-CoA to acetoacetate (catalysed by SCOT). This forms acetoacetyl-CoA, which is converted to acetyl-CoA by MAT. SCOT is not expressed in liver and has no role other than ketolysis; ketosis is, therefore, the only consistent biochemical abnormality in SCOT deficiency. In contrast, MAT is expressed in liver and has three different roles. Whereas MAT promotes ketolysis in extrahepatic tissues, in liver it participates in ketogenesis by converting acetyl-CoA to acetoacetyl-CoA. Finally, MAT cleaves methylacetoacetyl-CoA in the degradation pathway of isoleucine. Patients with MAT deficiency present with ketoacidosis, implying that the enzyme is more crucial in ketolysis than in ketogenesis; they also excrete intermediates of isoleucine catabolism.

### Diagnostic Tests

HMG-CoA lyase and MAT deficiencies are generally diagnosed by detecting abnormal urinary organic acids. Recognition of the other defects is more difficult, and it is likely that many cases remain undiagnosed.

### 3-Hydroxy-3-Methylglutaryl-CoA Synthase Deficiency

This diagnosis should be suspected when there is grossly impaired ketogenesis during fasting or after a long-chain-fat load, but there is normal fatty-acid-oxidation flux in vitro (Chaps. 2, 11). Blood acylcarnitine analysis was normal in both reported cases, and urine organic acids were normal or showed non-specific abnormalities [1, 2]. Since the enzyme is only expressed at high levels in liver, further studies must be performed on this tissue. Some residual activity is to be expected in enzyme assays, since approximately 10% of total activity can be attributed to a cytoplasmic isoenzyme involved in cholesterol synthesis. There may also be a peroxisomal isoenzyme [19].

### 3-Hydroxy-3-Methylglutaryl-CoA Lyase Deficiency

Patients with this condition excrete increased quantities of 3-hydroxy-3-methylglutaric, 3-hydroxyisovaleric, 3-methylglutaconic and 3-methylglutaric acids (Fig. 12.1); 3-methylcrotonylglycine may also be present [3]. It is important to confirm the diagnosis by enzymology, since a similar pattern of urinary organic acids has been found in patients with normal HMG-CoA lyase activity [20]. Assays can be undertaken on leukocytes, cultured fibroblasts or liver. Currently, the diagnosis is generally made following an acute presentation. There is, however, the potential to diagnose cases by neonatal screening, since 3-methylglutarylcarnitine accumulates in blood and can be detected by tandem mass spectrometry [21].

### Succinyl-CoA Oxoacid Transferase Deficiency

This deficiency should be suspected in patients with persistent ketonuria and a circulating concentration of ketone bodies (acetoacetate and 3-hydroxybutyrate) above 0.2 mmol/l, even in the fed state [10]. If a diagnostic fast is undertaken, there is an excessive rise in blood ketone-body levels (sometimes to over 10 mmol/l) without hypoglycaemia [10]. Urinary organic-acid analysis reveals high concentrations of 3-hydroxybutyrate, acetoacetate and sometimes 3-hydroxyisovalerate, but reveals no specific abnormalities. Proof of the diagnosis requires measurement of enzyme activity in lymphocytes or cultured fibroblasts.

### Methylacetoacetyl-CoA Thiolase Deficiency

Patients with MAT deficiency usually excrete 2-methyl-3-hydroxybutyric acid and tiglylglycine (Fig. 12.1); 2-methylacetoacetate may also be present, particularly during episodes of ketosis [11]. In some cases, however, the organic aciduria may be hard to detect [16]. Moreover, similar abnormalities have been reported in a few patients with normal MAT activity [22]. Sensitivity and specificity can be improved by demonstrating increased 2-methylacetoacetate excretion after an isoleucine load [23]. Nevertheless, enzymologic confirmation using leukocytes or cultured fibroblasts is essential. Assays are complicated by the presence of three other thiolases that act on acetoacetyl-CoA (cytosolic and peroxisomal acetoacetyl-CoA thiolases and mitochondrial 3-oxoacyl-CoA thiolase). Methylacetoacetyl-CoA is a specific substrate for MAT, but it is difficult to synthesise [24]. One solution is to measure acetoacetyl-CoA thiolysis in the presence and absence of potassium, which enhances the activity of MAT but not the other enzymes [17]. MAT deficiency can sometimes be detected by analysis of acylcarnitine species in blood spots, but it is not yet clear how consistently abnormalities are present [21].

### Treatment and Prognosis

All these patients can decompensate rapidly in early childhood. To prevent this, fasting must be avoided, and a high carbohydrate intake must be maintained during any metabolic stress, such as surgery or infection (Chap. 3). Drinks containing carbohydrate should be started at the first sign of illness; if these are not tolerated, hospital admission is needed for an intravenous infusion of glucose. Patients with disorders of ketolysis quickly become severely acidotic and should be started on bicarbonate prior to admission if they develop heavy ketonuria. Dehydration is also common in the ketolysis disorders and requires intravenous administration of fluids, with careful attention to electrolytes.

A moderate protein restriction is usually recommended in HMG-CoA lyase, MAT and SCOT deficiencies, since these enzymes are directly or indirectly involved in amino acid catabolism. A low-fat diet has also been recommended [25]. Protein and fat should certainly be avoided during illness. At other times, however, dietary restriction is unnecessary in some patients [26, 27]. Carnitine supplements are often given if serum levels are low, though their value is unproven.

Patients with these disorders can die or suffer irreversible neurological damage during episodes of metabolic decompensation. Outcomes have been least good for neonatal-onset cases of HMG-CoA lyase deficiency, such as those from Saudi Arabia [5]. Once the diagnosis has been made, the outlook is much improved. Patients, particularly those with ketolysis defects, become more stable with age. Late complications are rare. In one case, cardiomyopathy was reported in MAT deficiency, but this case was atypical, and the diagnosis was not established by a direct enzyme assay [28]. Cerebral MRI findings are discussed in the section "Clinical Presentation".

### Genetics

All four disorders are inherited as autosomal-recessive traits. Their prevalence is unknown, but HMG-CoA-lyase deficiency is relatively common in Saudi Arabia [5]. The cDNA sequences are known for all the enzymes, and pathogenic mutations have been identified in patients with deficiencies of each disorder [13, 26, 29, 30 and unpublished data]. There appears to be no common mutation in MAT deficiency. In contrast, two common mutations have been found in HMG-CoA-lyase deficiency, one in the Saudi population [30] and the other in Mediterranean patients [13].

Prenatal diagnosis is currently possible for all these disorders, except for mHMG-CoA synthase deficiency. Molecular techniques are used when the mutations are known or there are informative polymorphisms [29, 31]. In other cases, enzymology can be undertaken. HMG-CoA lyase activity can be measured in chorionic villi, but cultured amniocytes are recommended for SCOT assays [32].

## Cytosolic Acetoacetyl-CoA Thiolase Deficiency

Cytosolic acetoacetyl-CoA thiolase (CAT) is primarily involved in the synthesis of isoprenoid compounds, such as cholesterol, rather than in ketone-body metabolism. Two patients with CAT deficiency have been reported [33, 34]. Both presented with mental retardation after apparently normal early development. One patient developed severe ketoacidosis on a ketogenic diet, whilst the other had persistent ketonuria that resolved on a low-fat diet. No treatment had any effect on the neurological problems. Assays in liver and fibroblasts showed reduced CAT activity and, in one case, reduced cholesterol synthesis. The human CAT cDNA has been cloned, but mutations have not yet been reported [35].

## References

1. Thompson GN, Hsu BY, Pitt JJ, Treacy E, Stanley CA (1997) Fasting hypoketotic coma in a child with deficiency of mitochondrial 3-hydroxy-3-methylglutaryl-CoA synthase. N Engl J Med 337:1203–1207
2. Morris AAM, Lascelles CV, Olpin SE, Lake BD, Leonard JV, Quant PA (1998) Hepatic mitochondrial 3-hydroxy-3-methylglutaryl-CoA synthase deficiency. Pediatr Res 44:392–396
3. Gibson KM, Breuer J, Nyhan WL (1988) 3-Hydroxy-3-methylglutaryl-CoA lyase deficiency: review of 18 reported patients. Eur J Pediatr 148:180–186
4. van der Knaap MS, Bakker HD, Valk J (1998) MR imaging and proton spectroscopy in 3-Hydroxy-3-methylglutaryl CoA lyase deficiency. Am J Neuroradiol 19:378–382
5. Ozand PT, Al Aqeel A, Gascon G, Brismar J, Thomas E, Gleispach H (1991) 3-Hydroxy-3-methylglutaryl-CoA lyase deficiency in Saudi Arabia. J Inherit Metab Dis 14:174–188
6. Leonard JV, Seakins J, Griffin NK (1979) β-hydroxy-β-methylglutaric aciduria presenting as Reye's syndrome. Lancet I:680
7. Wilson WG, Cass MB, Søvik O, Gibson KM, Sweetman L (1984) A child with acute pancreatitis and recurrent hypoglycemia due to 3-hydroxy-3-methylglutaryl-CoA lyase deficiency. Eur J Pediatr 142:289–291
8. Niezen-Koning KE, Wanders RJ, Ruiter JP, IJlst L et al. (1997) Succinyl-CoA:acetoacetate transferase deficiency: identification of a new patient with a neonatal onset and review of the literature. Eur J Pediatr 156:870–873
9. Tildon JT, Cornblath M (1972) Succinyl-CoA: 3-ketoacid CoA transferase deficiency. J Clin Invest 51:493–498
10. Saudubray J-M, Specola N, Middleton B et al. (1987) Hyperketotic states due to inherited defects of ketolysis. Enzyme 38:80–90
11. Søvik O (1993) Mitochondrial 2-methylacetoacetyl-CoA thiolase deficiency: an inborn error of isoleucine and ketone body metabolism. J Inherit Metab Dis 16:46–54
12. Hillman RE, Keating JP (1974) β-Ketothiolase deficiency as a cause for 'ketotic hyperglycinaemia syndrome'. Pediatrics 53:221–225
13. Casale C, Casals N, Pié J et al. (1998) A nonsense mutation in the exon 2 of the 3-hydroxy-3-methylglutaryl CoA lyase gene producing 3 mature mRNAs is the main cause of 3-hydroxy-3-methylglutaric aciduria in European Mediterranean patients. Arch Biochem Biophys 349:129–137
14. Schutgens RB, Middleton B, van der Blij JF et al. (1982) β-Ketothiolase deficiency in a family confirmed by in vitro assay in fibroblasts. Eur J Pediatr 139:39–42
15. Daum RS, Scriver CR, Mamer OA, Devlin E, Lamm P, Goldman H (1973) An inherited disorder of isoleucine metabolism causing accumulation of 2-methylacetoacetate and 2-methyl-3-hydroxybutyrate and intermittent acidosis. Pediatr Res 7:149–160
16. Leonard JV, Middleton B, Seakins JW (1987) Acetoacetyl-CoA thiolase deficiency presenting as ketotic hypoglycaemia. Pediatr Res 21:211–213
17. Robinson BH, Sherwood WG, Taylor J, Balfe JW, Mamer OA (1979) Acetoacetyl CoA thiolase deficiency: a cause of severe ketoacidosis in infancy simulating salicylism. J Pediatr 95:228–233
18. Ozand PT, Rashed M, Gascon GG et al. (1994) 3-Ketothiolase deficiency: a review and four new patients with neurological symptoms. Brain Dev 16 [Suppl]:38–45
19. Krisans S, Rusnak N, Keller G, Edwards P (1988) Localisation of 3-hydroxy-3-methylglutaryl-CoA synthase in rat liver peroxisomes. J Cell Biol 107:122A
20. Hammond J, Wilken B (1984) 3-Hydroxy-3-methylglutaric, 3-methylglutaconic and 3-methylglutaric acids can be non-specific indicators of metabolic disease. J Inherit Metab Dis 7 [Suppl 2]:117–118
21. Rashed MS, Ozand PT, Bucknall MP, Little D (1995) Diagnosis of inborn errors of metabolism from blood spots by acylcarnitine and amino acids profiling using automated electrospray tandem mass spectrometry. Pediatr Res 38:324–331
22. Iden P, Middleton B, Robinson BH, Sherwood WG, Gibson KM, Sweetman L, Søvik O (1990) 3-Oxothiolase activities and $^{14}C$-2-methylbutanoic acid incorporation in cultured fibroblasts from 13 cases of suspected 3-oxothiolase deficiency. Pediatr Res 28:518–522
23. Aramaki S, Lehotay D, Sweetman L, Nyhan WL, Winter SC, Middleton B (1991) Urinary excretion of 2-methylacetoacetate, 2-methyl-3-hydroxybutyrate and tiglylglycine after isoleucine loading in the diagnosis of 2-methylacetoacetyl-CoA thiolase deficiency. J Inherit Metab Dis 14:63–74
24. Middleton B, Bartlett K (1983) The synthesis and characterisation of 2-methylacetoacetyl-CoA and its use in the identification of the site of the defect in 2-methylacetoacetic and 2-methyl-3-hydroxybutyric aciduria. Clin Chim Acta 128:291–305
25. Thompson GN, Chalmers RA, Halliday D (1990) The contribution of protein catabolism to metabolic decompensation in 3-hydroxy-3-methylglutaric aciduria. Eur J Pediatr 149:346–350
26. Kassovska-Bratinova S, Fukao T, Song X-Q et al. (1996) Succinyl-CoA: 3-oxoacid CoA transferase: human cDNA cloning, human chromosomal mapping to 5p13 and mutation detection in a SCOT deficient patient. Am J Hum Genet 59:519–528
27. Fukao T, Kodama A, Aoyanagi N et al. (1996) Mild form of β-ketothiolase deficiency (mitochondrial acetoacetyl-CoA thiolase deficiency) in two Japanese siblings. Clin Genet 50:263–266
28. Henry CG, Strauss AW, Keating JP, Hillman RE (1981) Congestive cardiomyopathy associated with β-ketothiolase deficiency. J Pediatr 99:754–757
29. Fukao T, Yamaguchi S, Orii T, Hashimoto T (1995) Molecular basis of β-ketothiolase deficiency: mutations and polymorphisms in the human mitochondrial acetoacetyl-CoA thiolase gene. Hum Mutat 5:113–120
30. Mitchell GA, Ozand PT, Robert M-F et al. (1998) HMG CoA lyase deficiency: identification of 5 causal point mutations in codons 41 and 42, including a frequent Saudi Arabian mutation, R41Q. Am J Hum Genet 62:295–300
31. Mitchell GA, Jacobs C, Gibson KM et al. (1995) Molecular prenatal diagnosis of 3-hydroxy-3-methylglutaryl CoA lyase deficiency. Prenat Diagn 15:725–729
32. Fukao T, Song X-Q, Watanabe H et al. (1996) Prenatal diagnosis of succinyl-CoA:3-ketoacid CoA transferase deficiency. Prenat Diagn 16:471–474
33. Bennett MJ, Hosking GP, Smith MF, Gray RG, Middleton B (1984) Biochemical investigations on a patient with a defect in cytosolic acetoacetyl-CoA thiolase, associated with mental retardation. J Inherit Metab Dis 7:125–128
34. De Groot CJ, Luit-De Haan G, Hulstaert CE, Hommes FA (1977) A patient with severe neurological symptoms and acetoacetyl-CoA thiolase deficiency. Pediatr Res 11:1112–1116
35. Song X-Q, Fukao T, Yamaguchi S, Miyazawa S, Hashimoto T, Orii T (1994) Molecular cloning and nucleotide sequence of cDNA for human hepatic cytosolic acetoacetyl-CoA thiolase. Biochem Biophys Res Commun 201:478–485

# CHAPTER 13

## The Respiratory Chain

The respiratory chain is divided into five functional units or complexes embedded in the inner mitochondrial membrane. *Complex I* [reduced nicotinamide adenine dinucleotide (NADH)–coenzyme Q (CoQ) reductase] carries reducing equivalents from NADH to CoQ and consists of 25–28 different polypeptides, seven of which are encoded by mitochondrial DNA. *Complex II* (succinate–CoQ reductase) carries reducing equivalents from reduced flavin adenine dinucleotide ($FADH_2$) to CoQ and contains five polypeptides, including the flavin adenine dinucleotide-dependent succinate dehydrogenase and a few non-heme-iron–sulfur centers. *Complex III* (reduced-CoQ-cytochrome-c reductase) carries electrons from CoQ to cytochrome c and contains 11 subunits. *Complex IV* (cytochrome-c oxidase) catalyzes the transfer of reducing equivalents from cytochrome c to molecular oxygen. It is composed of two cytochromes (a and $a_3$), two copper atoms, and 13 different protein subunits.

The respiratory chain catalyzes the oxidation of fuel molecules by oxygen and the concomitant energy transduction into ATP. During the oxidation process, electrons are transferred to oxygen via the energy-transducing complexes: complexes I, III, and IV for succinate and complexes III and IV for $FADH_2$ derived from the β-oxidation pathway via the electron-transfer flavoprotein (ETF) and the ETF-CoQ oxidoreductase system. CoQ (a lipoidal quinone) and cytochrome c (a low-molecular-weight hemoprotein) act as "shuttles" between the complexes.

The flux of electrons is coupled to the translocation of protons ($H^+$) into the intermembrane space at three coupling sites (complexes I, III, and IV). This creates a transmembrane gradient. *Complex V* (ATP synthase) allows protons to flow back into the mitochondrial matrix and uses the released energy to synthesize ATP. Three molecules of ATP are generated for each molecule of NADH oxidized.

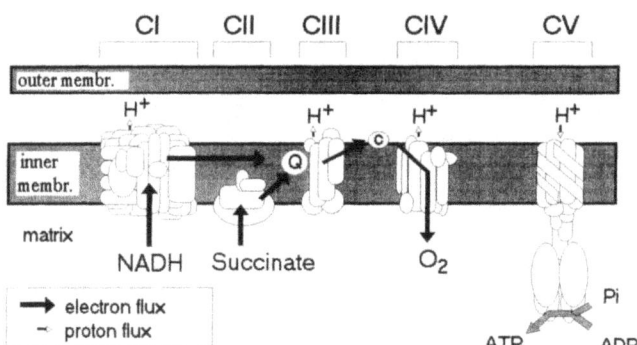

**Fig. 13.1.** The mitochondrial respiratory chain. *ADP*, adenosine diphosphate; *ATP*, adenosine triphosphate; *c*, cytochrome c; *CI*, complex I (NADH–coenzyme-Q reductase); *CII*, complex II (succinate–coenzyme-Q reductase); *CIII*, complex III (reduced-coenzyme-Q–cytochrome-c reductase); *CIV*, complex IV (cytochrome-c oxidase); *CV*, complex V (ATP synthase); *NADH*, reduced nicotinamide adenine dinucleotide; *Pi*, inorganic phosphate; *Q*, coenzyme Q

# CHAPTER 13

# Defects of the Respiratory Chain

Arnold Munnich

CONTENTS

| | |
|---|---|
| Clinical Presentation | 159 |
| Neonates | 160 |
| Infants | 161 |
| Children and Adults | 161 |
| Metabolic Derangement | 162 |
| Diagnostic Tests | 162 |
| Screening Tests | 162 |
| Enzyme Assays | 163 |
| Histopathological Studies | 165 |
| Magnetic Resonance Spectroscopy | 165 |
| Treatment and Prognosis | 165 |
| Genetics | 165 |
| Genetic Counseling and Prenatal Diagnosis | 167 |
| References | 167 |

Respiratory chain deficiencies have long been regarded as neuromuscular diseases. However, *oxidative phosphorylation* (i.e., ATP synthesis by the respiratory chain) is not restricted to the neuromuscular system but proceeds in all cells that contain mitochondria (Fig. 13.1). Most non-neuromuscular organs and tissues are, therefore, also dependent upon mitochondrial energy supply. Therefore, due to the twofold genetic origin of respiratory enzymes [nuclear DNA and mitochondrial (mtDNA)] a respiratory chain deficiency can theoretically give rise to any symptom in any organ or tissue at any age and with any mode of inheritance.

The diagnosis of a respiratory chain deficiency is difficult to consider initially when only one abnormal symptom is present. In contrast, this diagnosis is easier to consider when two or more seemingly unrelated symptoms are observed. The treatment, mainly dietetic, does not markedly influence the usually unfavorable course of the disease.

## Clinical Presentation

Due to the ubiquitous nature of oxidative phosphorylation, a defect of the mitochondrial respiratory chain should be considered in patients presenting (1) with an unexplained association of neuromuscular and/or non-neuromuscular symptoms, (2) with a rapidly progressive course, and (3) with symptoms involving seemingly unrelated organs or tissues. The disease may begin at virtually any age. Table 13.1 summarizes the most frequently observed symptoms. Whatever the age of onset and the presenting symptom, the major feature is the increasing number of tissues affected in the course of the disease. This progressive organ involvement is constant, and the central nervous system is almost consistently involved in the late stage of the disease.

While the initial symptoms usually persist and gradually worsen, they may occasionally improve or even disappear as other organs become involved. This is particularly true for bone marrow and gut. Indeed, remarkable remissions of pancytopenia or watery diarrhea have been reported in infants who later developed other organ involvements. Moreover, several patients whose disease apparently started in childhood or adulthood were retrospectively shown to have experienced symptoms (transient sideroblastic anemia, neutropenia, chronic watery diarrhea, or failure to thrive) of unexplained origin in early infancy. Similarly, a "benign" reversible infantile myopathy with hypotonia, weakness, macroglossia, respiratory distress, and spontaneous remission within 1–2 years has been described.

Certain clinical features or associations are more frequent at certain ages and have occasionally been identified as distinct entities, suggesting that these associations are not fortuitous. However, considerable overlap in clinical features leads to difficulties in the classification of many patients, and the nature, clinical course, and severity of symptoms vary among (and even within) affected individuals. It is more useful to bear in mind that the diagnosis of respiratory chain deficiency should be considered regardless of the age of onset and the nature of the presenting symptom when presented with an unexplained association of signs with a progressive course involving seemingly unrelated organs or tissues. The non-exhaustive list of clinical profiles listed below illustrates the diversity of presentations (Table 13.1).

**Table 13.1.** The most frequently observed symptoms in defects of the respiratory chain

*Neonatal period (0–1 month)*
  Central nervous system
    Iterative apnea, lethargy, drowsiness, near-miss, limb and trunk hypotonia
    Congenital lactic acidosis
    Ketoacidotic coma
  Muscle
    Myopathic presentation
    Muscular atrophy, hypotonia
    Stiffness, hypertonia
    Recurrent myoglobinuria
    Poor head control, poor spontaneous movement
  Liver
    Hepatic failure, liver enlargement
  Heart
    Hypertrophic cardiomyopathy (concentric)
  Kidney
    Proximal tubulopathy (De Toni-Debré-Fanconi syndrome)

*Infancy (1 month–2 years)*
  Central nervous system
    Recurrent apneas, near-miss
    Recurrent ketoacidotic coma
    Poor head control, limb spasticity
    Psychomotor regression, mental retardation
    Cerebellar ataxia
    "Stroke-like" episodes
    Myoclonus, generalized seizures
    Subacute necrotizing encephalomyopathy (Leigh syndrome)
    Progressive infantile poliodystrophy (Alpers syndrome)
  Muscle
    Myopathic features
    Muscular atrophy
    Limb weakness, hypotonia
    Myalgia, exercise intolerance
    Recurrent myoglobinuria
  Liver
    Progressive liver enlargement
    Hepatocellular dysfunction
    Valproate-induced hepatic failure
  Heart
    Hypertrophic cardiomyopathy (concentric)
  Kidney
    Proximal tubulopathy (De Toni-Debré-Fanconi syndrome)
    Tubulo-interstitial nephritis (mimicking nephronophtisis)
    Nephrotic syndrome
    Renal failure
    Hemolytic uremic syndrome
  Gut
    Recurrent vomiting
    Chronic diarrhea, villous atrophy
    Exocrine pancreatic dysfunction
    Failure to thrive
    Chronic interstitial pseudo-obstruction
  Endocrine
    Short stature, retarded skeletal maturation
    Recurrent hypoglycemia
    Multiple hormone deficiency

  Bone marrow
    Sideroblastic anemia
    Neutropenia, thrombopenia
    Myelodysplastic syndrome, dyserythropoiesis
  Ear
    Hearing loss
    Sensorineural deafness (brain stem or cochlear origin)
  Eye
    Optic atrophy
    Diplopia
    Progressive external ophthalmoplegia
    Limitation of eye movements (all directions, upgaze)
    "Salt-and-pepper" retinopathy, pigmentary retinal degeneration
    Lid ptosis
    Cataract
  Skin
    Mottled pigmentation of photo-exposed areas
    Trichothiodystrophy
    Dry, thick and brittle hair

*Childhood (>2 years) and adulthood*
  Central nervous system
    Myoclonus
    Seizures (generalized, focal, drop attacks, photosensitivity, tonicoclonus)
    Cerebellar ataxia
    Spasticity
    Psychomotor regression, dementia, mental retardation
    "Stroke-like" episodes
    Hemicranial headache, migraine
    Recurrent hemiparesis, cortical blindness or hemianopsia
    Leukodystrophy, cortical atrophy
    Peripheral neuropathy
  Muscle
    Progressive myopathy
    Limb weakness (proximal)
    Myalgia, exercise intolerance
    Recurrent myoglobinuria
  Heart
    Concentric hypertrophic or dilated cardiomyopathy
    Different types of heart block
  Endocrine
    Diabetes mellitus (insulin- and non-insulin dependent)
    Growth-hormone deficiency
    Hypoparathyroidism
    Hypothyroidism
    Adrenocoticotrophin deficiency
    Hyperaldosteronism
    Infertility (ovarian failure or hypothalamic dysfunction)
  Eye
    Lid ptosis
    Diplopia
    Progressive external ophthalmoplegia
    Limitation of eye movements (all directions, upgaze)
    "Salt-and-pepper" retinopathy, pigmentary retinal degeneration
    Cataract, corneal opacities
    Leber hereditary optic neuroretinopathy
  Ear
    Sensorineural deafness
    Aminoglycoside-induced ototoxicity (maternally inherited)

## Neonates

In the neonate (age less than 1 month), the following clinical profiles are seen:

- Ketoacidotic coma with recurrent apneas, seizures, severe hypotonia, liver enlargement, and proximal tubulopathy in the neonatal period, with or without a symptom-free period [1]
- Severe neonatal sideroblastic anemia (with or without hydrops fetalis), with neutropenia, thrombopenia, and exocrine pancreatic dysfunction of unexplained origin (Pearson marrow-pancreas syndrome) [2]

- Concentric hypertrophic cardiomyopathy and muscle weakness with an early onset and a rapidly progressive course (dilated cardiomyopathies are exceptional) [3]
- Concentric hypertrophic cardiomyopathy with profound central neutropenia and myopathic features in males (Barth syndrome) [4]
- Hepatic failure with lethargy, hypotonia, and proximal tubulopathy of unknown origin with neonatal onset [1]

## Infants

In infancy (1 month–2 years), the clinical profiles include the following:

- Failure to thrive, with or without chronic watery diarrhea and villous atrophy; unresponsiveness to gluten-free and cow's milk-protein-free diet [5].
- Recurrent episodes of acute myoglobinuria, hypertonia, muscle stiffness and elevated plasma levels of enzymes unexplained by an inborn error of glycolysis, glycogenolysis, fatty acid oxidation or muscular dystrophy [6].
- Proximal tubulopathy (de Toni-Debré-Fanconi syndrome) with recurrent episodes of watery diarrhea, rickets and mottled pigmentation of photo-exposed areas.
- A tubulo-interstitial nephritis mimicking nephronophtisis, with the subsequent development of renal failure and encephalomyopathy with leukodystrophy [7].
- Severe trunk and limb dwarfism unresponsive to growth-hormone administration, with subsequent hypertrophic cardiomyopathy, sensorineural deafness, and retinitis pigmentosa.
- Early-onset insulin-dependent diabetes mellitus with diabetes insipidus, optic atrophy, and deafness (Wolfram syndrome) [8].
- Rapidly progressive encephalomyopathy with hypotonia, poor sucking, weak crying, poor head control, cerebellar ataxia, pyramidal syndrome, psychomotor regression, developmental delay, muscle weakness, and respiratory insufficiency; occasionally associated with proximal tubulopathy and/or hypertrophic cardiomyopathy.
- Subacute necrotizing encephalomyopathy (Leigh's disease). This is a devastating encephalopathy characterized by recurrent attacks of psychomotor regression with pyramidal and extrapyramidal symptoms, leukodystrophy, and brainstem dysfunction (respiratory abnormalities). The pathological hallmark consists of focal, symmetrical, and necrotic lesions in the thalamus, brain stem, and the posterior columns of the spinal cord. Microscopically, these spongiform lesions show demyelination, vascular proliferation, and astrocytosis [9].

## Children and Adults

In childhood (above 2 years) and adulthood, the neuromuscular presentation is the most frequent:

- Muscle weakness with myalgia and exercise intolerance, with or without progressive external ophthalmoplegia [9].
- Progressive sclerosing poliodystrophy (Alpers disease) associated with hepatic failure [10].
- Encephalomyopathy with myoclonus, ataxia, hearing loss, muscle weakness, and generalized seizures (myoclonus epilepsy, ragged red fibers, MERRF) [9].
- Progressive external ophthalmoplegia (PEO) ranging in severity from pure ocular myopathy to Kearns-Sayre syndrome (KSS). KSS is a multisystem disorder characterized by the triad (1) onset before age 20 years, (2) PEO, and (3) pigmentary retinal degeneration, plus at least one of the following: complete heart block, CSF protein levels above 100 mg/dl, or cerebellar ataxia [9].
- Mitochondrial encephalomyopathy with lactic acidosis and stroke-like episodes (MELAS). This syndrome is characterized by onset in childhood, with intermittent hemicranial headache, vomiting, proximal limb weakness, and recurrent neurological deficit resembling strokes (hemiparesis, cortical blindness, hemianopsia), lactic acidosis, and ragged red fibers (RRFs) in the muscle biopsy. Computed tomography (CT) brain scanning shows low-density areas (usually posterior), which may occur in both white and gray matter but do not always correlate with the clinical symptoms or the vascular territories. The pathogenesis of stroke-like episodes in MELAS has been ascribed to either cerebral blood-flow disruption or acute metabolic decompensation in biochemically deficient areas of the brain [9].
- Leber's hereditary optic neuroretinopathy (LHON). This disease is associated with rapid loss of bilateral central vision due to optic nerve death. Cardiac dysrythmia is frequently associated with the disease, but no evidence of skeletal muscle pathology or gross structural mitochondrial abnormality has been documented. The median age of vision loss is 20–24 years, but it can occur at any age between adolescence and late adulthood. Expression among maternally related individuals is variable, and more males are affected [9].
- Neurogenic muscle weakness, ataxia, retinitis pigmentosa (NARP), and variable sensory neuropathy

with seizures and mental retardation or dementia [11].
- Mitochondrial myopathy and peripheral neuropathy, encephalopathy, and gastrointestinal disease manifesting as intermittent diarrhea and intestinal pseudo-obstruction (myo-neuro-gastro intestinal encephalopathy, MNGIE) [9].

## Metabolic Derangement

As the respiratory chain transfers electrons to oxygen, a disorder of oxidative phosphorylation should result in (1) an increase in the concentration of reducing equivalents in both mitochondria and cytoplasm and (2) the functional impairment of the citric acid cycle, due to the excess of reduced nicotinamide adenine dinucleotide (NADH) and the lack of nicotinamide adenine dinucleotide (NAD). Therefore, an increase in the ketone body (3-hydroxybutyrate/acetoacetate) and lactate/pyruvate molar ratios (L/P) with a secondary elevation of blood lactate might be expected in the plasma of affected individuals. This is particularly true in the post-absorptive period, when more NAD is required to adequately oxidize glycolytic substrates.

Similarly, as a consequence of the functional impairment of the citric-acid cycle, ketone body synthesis increases after meals, due to the channeling of acetyl-coenzyme A (CoA) towards ketogenesis. The elevation of the total level of ketone bodies in a fed individual is paradoxical, as it should normally decrease after meals, due to insulin release (paradoxical hyperketonemia).

Yet, the position of the block might differentially alter the metabolic profile of the patient. A block at the level of complex I impairs the oxidation of the 3 mol of NADH formed in the citric acid cycle. In theory, at least, oxidation of reduced flavin adenine dinucleotide ($FADH_2$) derived from succinate producing substrates (methionine, threonine, valine, isoleucine, and odd-numbered fatty acids) should not be altered, because it is mediated by complex II. Similarly, oxidation of $FADH_2$ derived from the first reaction of the β-oxidation pathway should occur normally, because it is mediated by the electron-transfer-flavoprotein–coenzyme-Q-reductase system. However, complex-II deficiency should not markedly alter the redox status of affected individuals fed a carbohydrate-rich diet. A block at the level of complex III should impair the oxidation of both NAD-linked and FAD-linked substrates. Finally, given the crucial role of complex IV in the respiratory chain, it is not surprising that severe defects of cytochrome c oxidase (COX) activity cause severe lactic acidosis and markedly alter redox status in plasma.

## Diagnostic Tests

### Screening Tests

Screening tests include the determination of lactate, pyruvate, ketone bodies, and their molar ratios in plasma as indices of oxidation/reduction status in cytoplasm and mitochondria, respectively (Table 13.2). Determinations should be made before and 1 h after meals throughout the day. Blood glucose and non-esterified fatty acids should be simultaneously monitored (Chap. 2). The observation of a *persistent hyperlactatemia* (>2.5 mM) with elevated L/P and ketone body molar ratios (particularly in the post-absorptive period) is highly suggestive of a respiratory chain deficiency. In addition, investigation of the redox status in plasma can help discriminate between the different causes of congenital lactic acidosis based on L/P and ketone body molar ratios in vivo [12]. Indeed, an impairment of oxidative phosphorylation usually results in L/P ratios above 20 and ketone body ratios above 2, whereas a defect of the pyruvate dehydrogenase (PDH) complex results in low L/P ratios (<10). Although little is known regarding tricarboxylic acid cycle disorders, it appears that these diseases also result in high L/P ratios, but ketone body molar ratios are lower in these conditions (<1) than in respiratory-chain defects (as also observed in pyruvate carboxylase deficiency; Chap. 10) [13, 14]. However, the above diagnostic tests may fail to detect any disturbance of the redox status in plasma. Pitfalls of metabolic screening are the following:

- Hyperlactatemia may be latent in basal conditions and may be revealed by a glucose loading test only (2 g/kg orally) or by determination of the redox status in the CSF. The measurement of lactatorachia and/or the L/P ratio in the CSF is useless when the redox status in plasma is altered.
- Proximal tubulopathy may lower blood lactate and increase urinary lactate. In this case, gas chromatography–mass spectrometry can detect urinary lactate and citric acid-cycle intermediates.
- Diabetes mellitus may hamper the entry of pyruvate into the citric acid cycle.
- Tissue-specific isoforms may be selectively impaired, barely altering the redox status in plasma (this may be particularly true for hypertrophic cardiomyopathies).
- The defect may be generalized but partial; the more those tissues with higher dependencies on oxidative metabolism (such as brain and muscle) suffer, the more the oxidation/reduction status in plasma is impaired.
- The defect may be confined to complex II, barely altering (in principle) the redox status in plasma.

**Table 13.2.** Screening of the respiratory chain

**Standard screening tests** (at least four determinations per day in fasted and 1-h-fed individuals)
  Plasma lactate
  Lactate/pyruvate molar ratio: redox status in cytoplasm
  Ketonemia ("paradoxical" elevation in fed individuals)
  β-hydroxy butyrate/acetoacetate molar ratio: redox status in the mitochondria
  Blood glucose and free fatty acids
  Urinary organic acids (GC-MS): lactate, ketone bodies, citric acid cycle intermediates

**Provocative tests** (when standard tests are inconclusive)
  Glucose test (2 g/kg orally) in fasted individuals, with determination of blood glucose, lactate, pyruvate,
    ketone bodies and their molar ratios just before glucose administration, and then every 30 min for 3–4 h (Chap. 2)
  Lactate/pyruvate molar ratios in the CSF (only when no elevation of plasma lactate is observed)
  Redox status in plasma following exercise

**Screening for multiple organ involvement**
  Liver: hepatocellular dysfunction
  Kidney: proximal tubulopathy, distal tubulopathy, proteinuria, renal failure
  Heart: hypertrophic cardiomyopathy, heart block (ultrasound, ECG)
  Muscle: myopathic features (CK, ALAT, ASAT, histological anomalies, RRF)
  Brain: leukodystrophy, poliodystrophy, hypodensity of the cerebrum, cerebellum and the brainstem, multifocal areas
    of hyperintense signal (MELAS), bilateral symmetrical lesions of the basal ganglia and brain stem (Leigh)
    (EEG, NMR, CT scan)
  Peripheral nerve: distal sensory loss, hypo- or areflexia, distal muscle wasting (usually subclinical), reduced motor
    nerve conduction velocity (NCV) and denervation features (NCV, EMG, peripheral nerve biopsy showing axonal
    degeneration and myelinated-fiber loss)
  Pancreas: exocrine pancreatic dysfunction
  Gut: villous atrophy
  Endocrine: hypoglycemia, hypocalcemia, hypoparathyroidism, growth hormone deficiency (stimulation tests)
  Bone marrow: anemia, neutropenia, thrombopenia, pancytopenia, vacuolization of marrow precursors
  Eye: PEO, ptosis, optic atrophy, retinal degeneration (fundus, ERG, visually evoked potentials)
  Ear: sensorineural deafness (auditory evoked potentials, brain-stem-evoked response)
  Skin: trichothiodystrophy, mottled pigmentation of photo exposed areas

*ALAT*, alanine aminotransferase; *ASAT*, aspartate aminotransferase; *CK*, creatine kinase; *CT*, computed tomography; *ECG*, electrocardiogram; *EEG*, electroencephalogram; *EMG*, electromyogram; *ERG*, electroretinogram; *GC-MS*, gas chromatography–mass spectrometry; *MELAS*, mitochondrial encephalopathy with lactic acidosis and stroke-like episodes; *NCV*, nerve conduction velocity; *NMR*, nuclear magnetic resonance; *PEO*, progressive external ophthalmoplegia; *RRF*, ragged red fiber

When diagnostic tests are negative, the diagnosis of a respiratory chain deficiency may be missed, especially when only the onset symptom is present. By contrast, the diagnosis is easier to consider when seemingly unrelated symptoms are observed. For this reason, the investigation of patients at risk (whatever the onset symptom) includes the systematic screening of all target organs, as multiple organ involvement is an important clue to the diagnosis of this condition (Table 13.2).

## Enzyme Assays

The observation of an abnormal redox status in plasma and/or the evidence of multiple organ involvement prompts one to carry out further enzyme investigations. These investigations include two entirely distinct diagnostic procedures that provide independent clues to respiratory-chain deficiencies: *polarographic* studies and *spectrophotometric* studies.

POLAROGRAPHIC STUDIES. *Polarographic studies* consist of the measurement of oxygen consumption by mitochondria-enriched fractions in a Clark electrode in the presence of various oxidative substrates (malate with pyruvate, malate with glutamate, succinate, palmitate, etc.). In the case of complex I deficiency, polarographic studies show impaired respiration with NADH-producing substrates, whilst respiration and phosphorylation are normal with FADH-producing substrates (succinate). The opposite is observed in the case of complex II deficiency, whereas a block at the level of complexes III or IV impairs oxidation of both NADH- and FADH-producing substrates. In complex-V deficiency, respiration is impaired with various substrates, but adding the uncoupling agent 2,4-dinitrophenol or calcium ions returns the respiratory rate to normal, suggesting that the limiting step involves phosphorylation rather than the respiratory chain [15].

It is worth remembering that polarographic studies detect not only disorders of oxidative phosphorylation but also PDH deficiencies, citric acid cycle enzyme deficiencies, and genetic defects of carriers, shuttles, and substrates (including cytochrome c, cations, and adenylate), as these conditions also impair the production of reducing equivalents in the mitochondrion. In these cases, however, respiratory enzyme activities are expected to be normal.

While previous techniques required gram-sized amounts of muscle tissue, the scaled-down procedures available now allow the rapid recovery of mitochondria-enriched fractions from small skeletal muscle biopsies (100–200 mg, obtained under local anesthetic), thus making polarography feasible in infants and children [16]. Polarographic studies on intact circulating lymphocytes (isolated from 10 ml of blood on a Percoll cushion) or detergent-permeabilized cultured cells (lymphoblastoid cell lines, skin fibroblasts) are also feasible and represent a less invasive and easily reproducible diagnostic test [17]. The only limitation of these techniques is the absolute requirement of fresh material: no polarographic studies are possible on frozen material.

SPECTROPHOTOMETRIC STUDIES. *Spectrophotometric studies* consist of the measurement of respiratory enzyme activities separately or in groups, using specific electron acceptors and donors. They do not require the isolation of mitochondrial fractions and can be carried out on tissue homogenates. For this reason, the amount of material required for enzyme assays (1–20 mg) is very small and can easily be obtained by needle biopsies of liver and kidney, and even by endomyocardial biopsies [3]. Similarly, a 25-ml flask of cultured skin fibroblasts or a lymphocyte pellet derived from a 10-ml blood sample are sufficient for extensive spectrophotometric studies. Samples should be frozen immediately and kept dry in liquid nitrogen (or at −80 °C).

Since conclusive evidence of respiratory chain deficiency is given by enzyme assays, the question of which tissue should be investigated deserves particular attention. In principle, the relevant tissue is the one that clinically expresses the disease. When the skeletal muscle expresses the disease, the appropriate working material is a microbiopsy of the deltoid. When the hematopoietic system expresses the disease (i.e., Pearson syndrome), tests should be carried out on circulating lymphocytes, polymorphonuclear cells, or bone marrow. However, when the disease is predominantly expressed in the liver or heart, gaining access to the target organ is far less simple. Yet, a needle biopsy of the liver or an endomyocardial biopsy are usually feasible. If not, or when the disease is mainly expressed in a barely accessible organ (brain, retina, endocrine system, smooth muscle), peripheral tissues (including skeletal muscle, cultured skin fibroblasts, and circulating lymphocytes) should be extensively tested. Whichever the expressing organ, it is essential to take skin biopsies from such patients (even postmortem) for subsequent investigations on cultured fibroblasts.

It should be borne in mind, however, that the in vitro investigation of oxidative phosphorylation remains difficult regardless the tissue tested. Several pitfalls should be considered:

- A normal respiratory enzyme activity does not preclude mitochondrial dysfunction even when the tissue tested clinically expresses the disease. One might be dealing with a kinetic mutant, tissue heterogeneity, or cellular mosaicism (heteroplasmy; see below). In this case, one should carry out extensive molecular genetic analyses, test other tissues, and (possibly) repeat investigations later.

- A deficient respiratory enzyme activity does not imply that oxidative phosphorylation is primarily impaired. We are now aware of deficient respiratory enzyme activities secondary either to myoadenylate deaminase deficiency or to inborn errors of intra-mitochondrial β-oxidation (long-chain and 3-hydroxy long-chain acyl-CoA dehydrogenase deficiency). It is important to carry out in vitro investigations of β-oxidation when the clinical presentation is also compatible with an inborn error of fatty acid oxidation (cardiomyopathy, hepatic failure).

- The scattering of control values occasionally hampers the recognition of enzyme deficiencies, as normal values frequently overlap those found in the patients. It is helpful to express results as ratios, especially as the normal functioning of the respiratory chain requires a constant ratio of enzyme activities [18]. Under these conditions, patients whose absolute activities are in the low normal range can be unambiguously diagnosed as enzyme deficient, although this expression of results may fail to recognize generalized defects of oxidative phosphorylation.

- No reliable method for the assessment of complex I activity in circulating or cultured cells is presently available, because oxidation of NADH-generating substrates by detergent-treated or freeze-thawed control cells is variable, and the rotenone-resistant NADH cytochrome c reductase activity is very high in this tissue.

- The phenotypic expression of respiratory enzyme deficiencies in cultured cells is unstable, and activities return to normal values when cells are grown in a standard medium [19]. The addition of uridine (200 μM) to the culture medium avoids counterselection of respiratory enzyme-deficient cells and allows them to grow normally, thereby stabilizing the mutant phenotype (uridine, which is required for nucleic-acid synthesis, is probably limited by the secondary deficiency of the respiratory chain-dependent dehydro-orotate dehydrogenase activity) [20].

- Discrepancies between control values may indicate faulty experimental conditions. Activities dependent on a single substrate should be consistent when tested under non-rate-limiting conditions. For example, normal succinate cytochrome c reductase activity should be twice as high as normal succinate-quinone dichlorophenolindophenol (DCPIP)

reductase activity (because one electron is required to reduce cytochrome c, while two are required to reduce DCPIP).
- Incorrect freezing may result in a rapid loss of quinone-dependent activities, probably due to peroxidation of membrane lipids. Tissue samples fixed for morphological studies are inadequate for subsequent respiratory enzyme assays.

## Histopathological Studies

The histological hallmark of mitochondrial myopathy is the RRF, which is demonstrated using the modified Gomori trichrome stain and contains peripheral and inter-myofibrillar accumulations of abnormal mitochondria. Although the diagnostic importance of pathological studies is undisputed, the absence of RRFs does not rule out the diagnosis of mitochondrial disorder [9]. Different histochemical stains for oxidative enzymes are used to analyze the distribution of mitochondria in the individual fibers and to evaluate the presence or absence of the enzymatic activities. Histochemical staining permits an estimation of the severity and heterogeneity of enzyme deficiency in the same muscle section. Myofibrillar integrity and the predominant fiber type and distribution can be evaluated with the myofibrillar adenosine triphosphatase stain. Studies using polyclonal and monoclonal antibodies directed against COX subunits are carried out in specialized centers. For analysis, the muscle specimen taken under local anesthetic must be frozen immediately in liquid-nitrogen-cooled isopentane.

## Magnetic Resonance Spectroscopy of Muscle and Brain

Phosphorus magnetic resonance spectroscopy (MRS) allows the study of muscle and brain-energy metabolism in vivo. Inorganic phosphate (Pi), phosphocreatine (PCr), AMP, adenosine diphosphate (ADP) or ATP, and intracellular pH may be measured. The Pi/PCr ratio is the most useful parameter and may be monitored at rest, during exercise, and during recovery. An increased ratio is found in most patients, and MRS is becoming a useful tool in the diagnosis of mitochondrial diseases and in the monitoring of therapeutic trials. However, the observed anomalies are not specific to respiratory enzyme deficiencies, and no correlation between MRS findings and the site of the respiratory enzyme defect can be made [9].

## Treatment and Prognosis

No satisfactory therapy is presently available for respiratory chain deficiency. Treatment remains largely symptomatic and does not significantly alter the course of the disease. It includes symptomatic treatments, supplementation with cofactors, prevention of oxygen-radical damage to mitochondrial membranes, dietary recommendations, and avoidance of drugs and procedures known to have a detrimental effect.

It is advisable to avoid sodium valproate and barbiturates, which inhibit the respiratory chain and have occasionally been shown to precipitate hepatic failure in respiratory enzyme-deficient children [10]. Tetracyclines and chloramphenicol should also be avoided, as they inhibit mitochondrial protein synthesis. Due to the increasing number of tissues affected in the course of the disease, it is recommended that organ transplantations (bone marrow, liver, heart) be avoided.

- Symptomatic treatments include: slow infusion of sodium bicarbonate during acute exacerbation of lactic acidosis, pancreatic extract administration in cases of exocrine pancreatic dysfunction, and repeated transfusions in cases of anemia or thrombopenia.
- Sustained improvement has been reported in patients with complex III deficiency given coenzyme Q10 (5–10 mg/kg/day). Treatment with riboflavin (100 mg/day) has been associated with improvement in a few patients with complex I deficiency myopathy. Carnitine is suggested in patients with secondary carnitine deficiency. Dichloroacetate or 2-chloropropionate administration has been proposed to stimulate PDH activity and has occasionally reduced the level of lactic acid [21], but detrimental effects of dichloroacetate have recently been reported.
- The dietary recommendation are a high-lipid, low-carbohydrate diet in patients with complex I deficiency. Indeed, a high-glucose diet is a metabolic challenge for patients with impaired oxidative phosphorylation, especially as glucose oxidation is largely aerobic in the liver. Based on our experience, we suggest avoiding a hypercaloric diet and parenteral nutrition and recommend a low-carbohydrate diet in addition to the symptomatic treatment. Succinate (6 g/day), succinate producing amino acids (isoleucine, methionine, threonine, and valine) or propionyl carnitine have occasionally been given to patients with complex I deficiency, as these substrates enter the respiratory chain via complex II.

## Genetics

Any mode of inheritance may be observed in mitochondrial diseases: autosomal recessive, dominant, X-linked, maternal, or sporadic. This variability is due to the high number of genes that encode the respiratory chain proteins, most of which are located in the cell nuclei, and 13 of which are located in the mitochondria. mtDNA encodes seven polypeptides of complex I, one of complex III (the apoprotein of cytochrome b), three of complex IV, and two of complex V. The mtDNA molecules are small (16.5 kb), double-stranded, circular, and contain no introns (Fig. 13.2). mtDNA has a number of unique genetic features [22]:

- It is maternally inherited, and its mutations are, therefore, transmitted by the mother
- It has a very high mutation rate involving both nucleotide substitutions and insertions/deletions
- During cell division, mitochondria are randomly partitioned into daughter cells

This means that, if normal and mutant mtDNA molecules are present in the mother's cells (heteroplasmy), some lineages will have only abnormal mtDNA (homoplasmy), others will have only normal mtDNA (wild type), and still others will have both normal and abnormal mtDNA. In these last cells, the phenotype will reflect the proportion of abnormal mtDNA.

## Mutations of Mitochondrial DNA

Pathological alterations of mtDNA fall into three major classes: point mutations, rearrangements, and depletions of the number of copies.

- Point mutations result in amino acid substitutions and modifications of mRNA and tRNA. Most are heteroplasmic, maternally inherited, and associated with a striking variety of clinical phenotypes (LHON, MERRF, MELAS, NARP, Leigh syndrome, diabetes, and deafness) [23].
- Rearrangements comprise deletions/duplications that markedly differ in size and position from patient to patient but usually encompass several coding and tRNA genes. They are usually sporadic, heteroplasmic, and unique and probably arise de novo during oogenesis or in early development (KSS, Pearson syndrome, PEO, diabetes, and deafness [23]). Occasionally, maternally transmitted mtDNA rearrangements are found [7]. In other cases, autosomal dominant transmission of multiple mtDNA deletions occurs, suggesting mutation of a

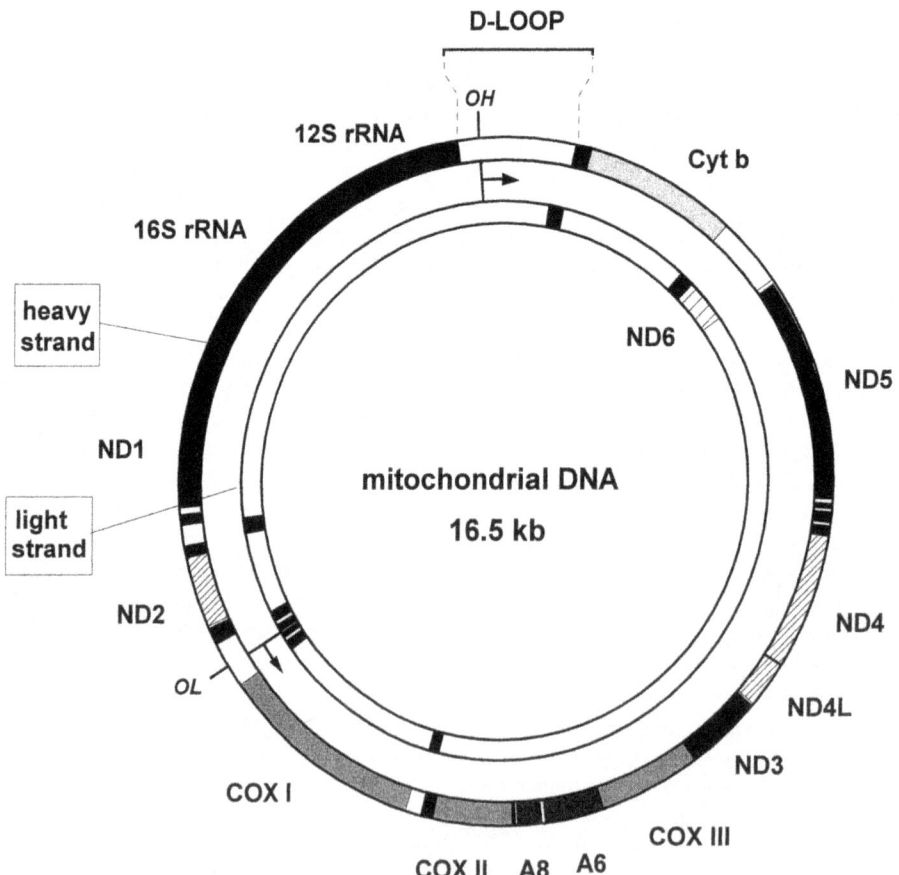

**Fig. 13.2.** Map of the mitochondrial genome. Regions encoding cytochrome b (*cyt b*), various subunits of reduced nicotinamide adenine dinucleotide–coenzyme Q reductase (*ND*), cytochrome oxidase (*COX*), and adenosine triphosphatase (*A*), and *rRNAs* are indicated. Replication of the heavy strand starts in this displacement (*D*) loop at the heavy-strand origin (*OH*), and that of the light strand at *OL*

nuclear gene essential for the function of the mitochondrial genome [24].

- Depletions of the number of copies of mtDNA, consistent with autosomal recessive inheritance, have been reported in rare cases of lethal infantile respiratory, muscle, liver, or kidney failure [25].

### Mutations in Nuclear DNA

A few of the numerous disease-causing nuclear genes have been recently identified, including the gene of Barth syndrome (tafazzin) [26], nuclear genes for complex-I (18-kDa subunit) [27], complex-II (Fp subunit of succinate dehydrogenase) [28], and complex-IV (SURF-1) [29, 30] deficiencies, and a MNGIE gene (thymidine phosphorylase) [31].

### Genetic Analysis of Respiratory Chain Deficiencies

An extensive family history, with documentation of minor signs in relatives, is of paramount importance in recognizing the mode of inheritance and in deciding on the molecular studies to be performed. Maternal inheritance indicates mtDNA mutations, autosomal dominant inheritance indicates multiple mtDNA deletions, and sporadic cases and autosomal recessive inheritance (consanguineous parents) indicate mtDNA deletions/duplications and nuclear gene mutations, respectively.

Investigations require a highly specialized, experienced laboratory and should take into account the following points:

- The distribution of mutated mtDNA molecules may differ widely among tissues, accounting for the variable clinical expression and requiring investigation of the tissue that actually expresses the disease
- mtDNA rearrangements are unstable and gradually disappear in cultured cells unless uridine is included in the culture medium, thus precluding growth under standard conditions [19]
- Negative results neither rule out an mtDNA mutation nor provide a clue that a nuclear mutation is involved
- Certain clinical presentations hint at particular nuclear genes: SURF-1 and the flavoprotein of complex II in Leigh disease are associated with succinate dehydrogenase and COX deficiencies, respectively, and the thymidine phosphorylase gene is associated with MNGIE

### Genetic Counseling and Prenatal Diagnosis

The identification of certain clinical phenotypes, listed above, allows some prediction with respect to their inheritance. Moreover, it should be borne in mind that, in cases of maternal inheritance of a mtDNA mutation, risk is absent for the progeny of an affected male but is high for that of a carrier female. In this case, determination of the proportion of mutant mtDNA on chorionic villi or amniotic cells is a rational approach. Nevertheless, its predictive value remains uncertain, owing to incomplete knowledge of the tissue distribution of abnormal mtDNA, its change during development, and its quantitative relationship to disease severity.

In the absence of detectable mtDNA mutations, the measurement of the activities of respiratory enzymes in cultured amniocytes or choriocytes provides the only possibility of prenatal diagnosis, particularly since few nuclear mutations have been identified. Unfortunately, relatively few enzyme deficiencies are expressed in cultured fibroblasts of probands, even when grown with uridine. For this reason, the ongoing identification of disease-causing nuclear genes will certainly help in delivering accurate prenatal diagnoses of respiratory chain deficiencies in the future.

## References

1. Cormier V, Rustin P, Bonnefont JP et al. (1991) Hepatic failure in neonatal - onset disorders of oxidative phosphorylation. J Pediatr 119:951–954
2. Rötig A, Cormier V, Blanche S et al. (1990) Pearson's marrow-pancreas syndrome: a multisystem mitochondrial disorder in infancy. J Clin Invest 86:1601–1608
3. Rustin P, LeBidois J, Chretien D et al. (1994) Endomyocardial biopsies for early detection of mitochondrial disorders in hypertrophic cardiomyopathies. J Pediatr 124:224–228
4. Bolhuis PA, Hensels GW, Hulsebos TJM, Baas F, Barth PG (1991) Mapping of the locus for X-linked cardioskeletal myopathy with neutropenia and abnormal mitochondria (Barth syndrome) to Xq28. Am J Hum Genet 48:481–485
5. Cormier-Daire V, Bonnefont JP, Rustin P et al. (1994) Deletion-duplication of the mitochondrial DNA presenting as chronic diarrhea with villous atrophy. J Pediatr 124:63–70
6. Saunier P, Chretien D, Wood C et al. (1995) Cytochrome c oxidase deficiency presenting as recurrent neonatal myoglobinuria. Neuromuscular Disorders 5:285–289
7. Rötig A, Bessis JL, Romero N et al. (1991) Maternally inherited duplication of the mitochondrial DNA in proximal tubulopathy with diabetes mellitus. Am J Hum Genet 50:364–370
8. Rötig A, Cormier V, Chatelain P et al. (1993) Deletion of the mitochondrial DNA in a case of early-onset diabetes mellitus, optic atrophy and deafness (DIDMOAD, Wolfram syndrome). J Clin Invest 91:1095–1098
9. Hammans SR, Morgan-Hughes JA (1994) Mitochondrial myopathies: Clinical features, investigation, treatment and genetic counselling. In: Schapira AHV, DiMauro S (eds) Mitochondrial disorders in neurology. Butterworth-Enemann, Stoneham, MA, pp 49–74
10. Chabrol B, Mancini J, Chretien D et al. (1994) Cytochrome c oxidase defect, fatal hepatic failure and valproate: a case report. Eur J Pediatr 153:133–135
11. Holt IJ, Harding AE, Petty RKH, Morgan-Hugues JA (1990) A new mitochondrial disease associated with mitochondrial DNA heteroplasmy. Am J Hum Genet 46:428–433
12. Poggi-Travert F, Martin D, Billette de Villeneur T et al. (1996) Metabolic intermediates in lactic acidosis: compounds, samples and interpretation. J Inherit Metab Dis 19:478–488

13. Bonnefont JP, Chretien D, Rustin P et al. (1992) 2-ketoglutarate dehydrogenase deficiency: a rare inherited defect of the Krebs cycle. J Pediatr 121:255-258
14. Saudubray JM, Marsac C, Cathelineau C (1989) Neonatal congenital lactic acidosis with pyruvate carboxylase deficiency in two siblings. Acta Paediatr Scand 65:717-724
15. Estabrook RW (1967) Mitochondrial respiratory control and the polarographic measurement of ADP/O ratios. Methods Enzymol 10:41-47
16. Rustin P, Chretien D, Gérard B et al. (1994) Biochemical, molecular investigations in respiratory chain deficiencies. Clin Chim Acta 220:35-51
17. Bourgeron T, Chretien D, Rötig A, Munnich A, Rustin P (1992) Isolation and characterization of mitochondria from human B lymphoblastoid cell lines. Biochem Biophys Res Commun 186:16-23
18. Rustin P, Chretien D, Bourgeron T et al. (1991) Assessment of the mitochondrial respiratory chain. Lancet 338:60
19. Gérard B, Bourgeron T, Chretien D et al. (1992) Uridine preserves the expression of respiratory enzyme deficiencies in cultured fibroblasts. Eur J Pediatr 152:270
20. Bourgeron T, Chretien D, Rötig A, Munnich A, Rustin P (1993) Fate and expression of the deleted mitochondrial DNA differ between heteroplasmic skin fibroblast and Epstein-Barr virus-transformed lymphocyte cultures. J Biol Chem 268:19369-19376
21. Stacpoole P, Harman EM, Curry SH, Baumgartner TG, Misbin RI (1983) Treatment of lactic acidosis with dichloroacetate. N Engl J Med 309:390-396
22. Clayton DA (1991) Replication and transcription of vertebrate mitochodnrial DNA. Annu Rev Cell Biol 7:453-478
23. Wallace DC (1992) Diseases of the mitochondrial DNA. Annu Rev Biochem 61:1175-1212
24. Zeviani M, Servidei S, Gellera C et al. (1989) An autosomal dominant disorder with multiple deletions of mitochondrial DNA starting at the D-loop region. Nature 339:309-311
25. Moraes CT, Shanske S, Trischler HJ et al. (1991) mtDNA depletion with variable tissue expression: a novel genetic abnormality in mitochondrial diseases. Am J Hum Genet 48:492-501
26. Bione S, D'Adamo P, Maestrini E et al. (1996) A novel X-linked gene, G4.5 is responsible for Barth syndrome. Nature Genet 12:385-389
27. van der Heuvel L, Ruitenbeek W, Smeets R et al. (1998) Demonstration of a new pathogenic mutation in human complex I deficiency: a 5-bp duplication in the nuclear gene encoding the 18-kD (AQDQ) subunit. Am J Hum Genet 62:262-268
28. Bourgeron T, Rustin P, Chretien D et al. (1995) Mutation of a nuclear succinate dehydrogenase gene results in mitochondrial respiratory chain deficiency. Nature Genet 11:144-149
29. Zhu Z, Yao J, Johns T, Fu K et al. (1998) SURF1, encoding a factor involved in the biogenesis of cytochrome c oxidase, is mutated in Leigh syndrome. Nature Genet 20:337-343
30. Tiranti V, Hoertnagel K, Carrozzo R et al. (1998) Mutation of SURF-1 in Leigh disease associated with cytochrome c oxidase deficiency. Am J Hum Genet 63:1609-1621
31. Nishino I, Spinazzola A, Hirano M (1999) Thymidine phosphorylase gene mutations in MNGIE, a human mitochondrial disorder. Science 283:689-692

# PART IV
# DISORDERS OF AMINO ACID METABOLISM AND TRANSPORT

## Phenylalanine Metabolism

Phenylalanine is an essential amino acid making up between 4% and 6% of the amino acid content of proteins. Normally, a high proportion of ingested phenylalanine is converted to tyrosine in the liver by phenylalanine hydroxylase, the rate-controlling enzyme of phenylalanine homeostasis (Fig. 14.1, reaction 4) [1]. Phenylalanine hydroxylase (and tyrosine hydroxylase, tryptophan hydroxylase and nitric oxide synthase; Fig. 14.2) requires tetrahydrobiopterin as a cofactor. This compound is formed from guanosine triphosphate, converted into tetrahydropterin carbinolamine during catalysis and recycled via quinonoid dihydrobiopterin. All defects that reduce the rate of conversion of phenylalanine into tyrosine increase the concentration of phenylalanine relative to that of tyrosine in blood and other body fluids. Due to an initial transamination to phenylpyruvic acid followed by reduction and decarboxylation, there is also a parallel increase in the production and excretion of phenylketones and phenylamines. In defects of biopterin metabolism, neopterins, primapterin or biopterins accumulate (Fig. 14.1), depending on the location of the enzyme deficiency.

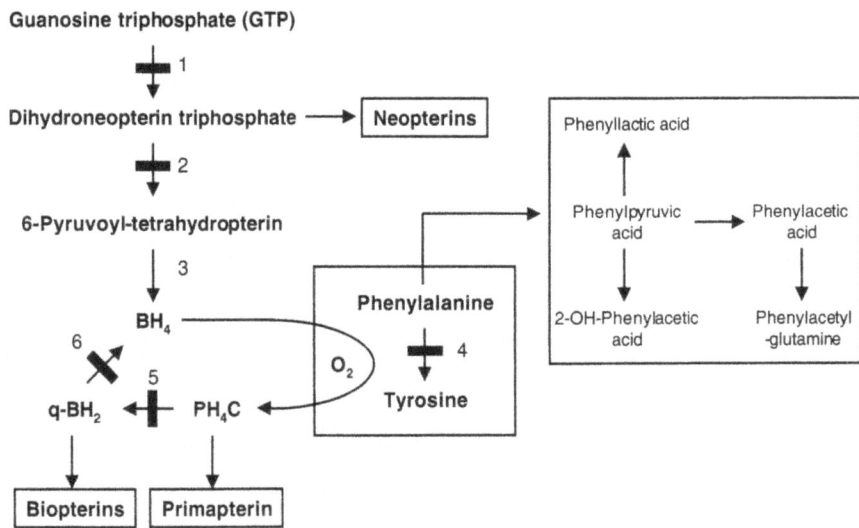

**Fig. 14.1.** Hydroxylation of phenylalanine into tyrosine, with synthesis and recycling of pterins in liver. $BH_4$, tetrahydrobiopterin; $PH_4C$, tetrahydropterin carbinolamine; $q\text{-}BH_2$, quinonoid dihydrobiopterin; 1, guanosine triphosphate cyclohydrolase; 2, 6-pyruvoyltetrahydropterin synthase; 3, sepiapterin reductase; 4, phenylalanine hydroxylase; 5, tetrahydropterin carbinolamine dehydratase; 6, dihydropteridine reductase. Enzyme defects are depicted by *solid bars*

CHAPTER 14

# The Hyperphenylalaninaemias

Isabel Smith and Philip Lee

CONTENTS

Introduction ................................. 171
Phenylalanine Hydroxylase Deficiency,
Phenylketonuria (PKU) ...................... 172
   Clinical Presentation........................ 172
   Metabolic Derangement .................... 172
   Diagnostic Tests ........................... 172
      Detection by Screening and Confirmation of Diagnosis 172
      Counselling Parents ...................... 173
   Treatment and Prognosis ................... 173
      Implementing the Diet and Phenylalanine Control in Young Children ........................ 174
      Dietary Management in Older Children, Adolescents and Adults ..................... 175
      Prescription of Phenylalanine-Containing Foods..... 175
      Prescription of Manufactured and Natural Phenylalanine-Low Foods ...................... 176
      Managing Illness and Feeding Problems ........... 176
      Risks of Nutrient Deficiency .................... 176
      Prognosis.................................... 177
   Genetics ................................... 177
Maternal Phenylketonuria........................ 178
Defects of Biopterin Metabolism................... 180
References...................................... 183

Primary inherited hyperphenylalaninaemia is probably best defined as a ratio of blood phenylalanine to tyrosine persistently greater than 3 (approximate normal ranges for blood concentrations: phenylalanine 35–120 μmol/l; tyrosine 40–130 μmol/l). Hyperphenylalaninaemia arises due to a block in the conversion of phenylalanine into tyrosine due to a defect in either the enzyme phenylalanine hydroxylase (98% of subjects) or in the metabolism of the cofactor tetrahydrobiopterin ($BH_4$). Increased production of phenylketones [hence phenylketonuria (PKU)] occurs in parallel with the rise in phenylalanine level. $BH_4$ is also the cofactor required for conversion of tyrosine into L-dopa, of tryptophan into 5-hydroxy-tryptophan and of arginine into nitric oxide (NO). In Caucasians, persistent hyperphenylalaninaemia occurs in between 1 in 4000 and 1 in 40,000 newborns. Although there is a considerable variation in genotype and phenotype, the great majority of affected subjects are at risk of intellectual and neurological impairments and handicaps, which can be prevented by introduction of strictly controlled treatment within a few weeks of birth (a phenylalanine-restricted diet with the addition of drug therapy for subjects with cofactor defects). Neonatal screening detects all but a few affected subjects. Definitive diagnosis relies on the measurement of blood phenylalanine and tyrosine and the exclusion of cofactor defects by measurement of pterins and dihydropteridine reductase (DHPR). The foetus, like the newborn infant, is at risk of damage proportionate to the degree of hyperphenylalaninaemia. The offspring of affected and untreated mothers present with low birth weight, reduced longitudinal pre- and postnatal growth, microcephaly, facial dysmorphism and a range of other congenital anomalies, notably in the heart and large vessels. It is of great importance that potential mothers with hyperphenylalaninaemia receive well-controlled treatment from before conception to the end of pregnancy. Within one generation, failure to achieve proper management for maternal PKU will result in loss of the benefits of neonatal screening.

## Introduction

How do we define "hyperphenylalaninaemia"? Normally, fasting plasma phenylalanine concentrations range from around 35 μmol/l (0.6 mg/100 ml) to 100 μmol/l (1.7 mg/100 ml). Following a protein-rich meal, concentrations rise by as much as two to three times, reaching a peak 2–3 h after a meal. Tyrosine concentrations are generally at or just above those of phenylalanine, giving a ratio of between 0.6 and 1.5. During pregnancy, or in women taking the pill, the ratio tends to be at the higher end of the range. In heterozygous carriers of PKU, the ratio is usually between 1.2 and 2.5, with blood phenylalanine levels occasionally reaching 200 µmol/l (2.3 mg/100 ml). A phenylalanine-to-tyrosine ratio of 3 or more provides us with a working definition of "hyperphenylalaninaemia". The term "phenylketonuria" is reserved, rather illogically, for the more severe forms of phenylalanine hydroxylase deficiency, when urinary phenylketones are easy to detect by simple chemical methods although, in practice, the disorders form a continuum of biochemical disturbance, with parallel increases in blood phenylalanine and urine phenylketones.

## Phenylalanine Hydroxylase Deficiency, Phenylketonuria (PKU)

### Clinical Presentation

PKU was described over 60 years ago by Ashborn Folling. In late-detected subjects with the more severe forms of hyperphenylalaninaemia [plasma phenylalanine concentrations greater than 1200 µmol/l (20 mg/100 ml)], retarded development and a mousy odour due to excretion of phenylacetic acid are the most constant clinical features. Infantile spasms with hypsarrythmia on electroencephalography (EEG), often with microcephaly, begin after the first few months in approximately one-third of infants. The majority are lightly pigmented (eyes, hair and skin) in comparison with parents and unaffected siblings, and eczema occurs in 20–40%. In older patients, mental handicap and disturbed behaviour is common, with hyperactivity, destructiveness, self-injury, autistic features and episodes of excitement. Schizophrenia-like symptoms have also been described. About 25% have seizures of grand-mal type at some time, and abnormalities in the EEG are almost invariable. Other neurological features include pyramidal signs (increased tone, hyperreflexia and exaggerated extensor plantar responses), tremor and overt parkinsonian features, abnormalities of gait, posturing and tics. Generally, the level of disability is stable beyond early childhood, although intellectual regression and neurological deterioration in adulthood are described in association with neuropathological evidence of demyelination [2].

Lesser degrees of biochemical disturbance are associated with a lower risk of mental handicap, and blood phenylalanine concentrations below 600 µmol/l are widely considered to be harmless (hence the term "benign hyperphenylalaninaemia"). However, handicap has been reported in subjects whose blood phenylalanine concentrations are only just above this range (700–1000 µmol/l), and there is evidence that the threshold for intellectual impairment may lie below 600 µmol/l [2].

### Metabolic Derangement

In addition to increasing the production of phenylketones and phenylamines, hyperphenylalaninaemia competitively inhibits transport of aromatic and other large neutral and dibasic amino acids across cell membranes (in both directions), including the blood–brain barrier and choroid plexus [3]. High brain phenylalanine concentrations inhibit the rate of protein synthesis, which may affect early dendritic proliferation and myelinisation, increase myelin turnover and inhibit synthesis of serotonin, dopamine and norepinephrine due to reduced intraneuronal amino acid concentrations and competitive inhibition of tyrosine and tryptophan hydroxylation [2]. Synthesis of dopamine in the prefrontal cortex (in contrast to the basal ganglia) is physiologically sensitive to small changes in tyrosine concentration [4], making this an area of the brain where dopamine turnover may be impaired by relatively small increases in phenylalanine. Which of the above events (or others) are critical for brain development and function remains uncertain, an extraordinary fact after 60 years of investigation.

It has been suggested that the small amino acid changes occurring in utero in the offspring of heterozygous mothers may cause foetal damage. Reports, in some (but not all) studies [2, 5], of a small reduction in birth weight, changes in the metaphyses of the long bones soon after birth and an increase in the frequency of cardiac anomalies add support to this view.

### Variation in Genotype and in Biochemical and Clinical Phenotype

Phenylalanine hydroxylase deficiency is genetically and phenotypically diverse. In subjects with little or no hydroxylase activity, blood phenylalanine concentrations rise to more than 30 times normal but, even in the biochemically severe forms of the disorder, there is wide variation in clinical phenotype. Untreated, one in six to seven subjects achieves an intellectual status adequate for education in a normal school [2].

### Diagnostic Tests

#### Detection by Screening and Confirmation of Diagnosis

Neonatal screening depends on the detection of raised blood phenylalanine concentrations, which can occur due to a variety of non-genetic and genetic causes (Table 14.1). Where testing is carried out within 48 h of birth, the chosen limit of phenylalanine for a positive result is likely to be 120–150 µmol/l, and those tested within 24 h of birth must be re-tested before 3 weeks of age. Where the recommended time of testing is later (often 3-5 days), the chosen limit will be higher (150-240 µmol/l) [6]. Screening laboratories and paediatricians responsible for the clinical care of infants with positive results need to work according to shared guidelines [7] on responsibilities and actions following a positive result.

It is important to have (1) confirmation of the rise in phenylalanine, (2) a measurement of tyrosine for the original specimen, (3) an evaluation to identify prematurity, failure to thrive or jaundice, (4) a review of milk intake to identify unusually high or low protein or energy intake and (5) a repeat measurement of phenylalanine and tyrosine. If tyrosine concentrations are normal, exclusion of defective biopterin metabolism (see "Defects of Biopterin Metabolism" and Table 14.2) is essential. If tyrosine concentrations are elevated, the infant is most unlikely to have a defect in hydroxylation and, except in premature or sick infants (Table 14.1), transient hyperphenylalaninaemia without tyrosinaemia is rare. Healthy infants with phenylalanine/tyrosine ratios of 3 or more, even if phenylalanine levels are only just above the normal range, nearly always continue to exhibit hyperphenylalaninaemia to some degree.

Subjects presenting with symptoms require developmental assessment and an EEG as a basis for judging prognosis and response to treatment.

The place of molecular genetics in diagnosis is still being explored [8]. Mutation detection does appear to be a good predictor of biochemical phenotype and, therefore, of the degree of phenylalanine restriction required for treatment. However, given the evidence of a close association between clinical outcome and blood phenylalanine profiles in early life, mutation detection in those with milder forms of enzyme deficiency seems unlikely to replace serial phenylalanine measurements as the means of deciding who to treat (see "Planning Treatment").

### Counselling Parents

Identification of hyperphenylalaninaemia in a screening programme is a crisis for the parents. The way in which the positive result is explained and the style and promptness of subsequent diagnostic investigations, counselling and advice on management are of vital importance to the family's well being, including those whose child has mild or transient hyperphenylalaninaemia. Parents want to know everything about their child's disorder, and they need a reasoned account of why particular decisions are made. This helps them come to terms with the limits of scientific knowledge and the differences in professional opinion that they may encounter later. Written materials are needed to support oral communications.

## Treatment and Prognosis

### Planning Treatment

Because of the association between age at diagnosis and long-term outcome, planning treatment should swiftly follow confirmation of the diagnosis (see "Prognosis"). Decisions on who should receive dietary treatment are not based on randomised controlled trials, and there remain differences of opinion and variations in practice regarding the level of phenylalanine at which to initiate treatment [7, 9]. The authors' practice is to start a phenylalanine-low diet in infants with diagnostic phenylalanine concentrations greater than 360 µmol/l (6 mg/100 ml) if the phenylalanine/tyrosine ratio is 4 or more. This does not preclude a withdrawal of treatment if, subsequently, it is found that therapeutic levels (120-360 µmol/l; 2-6 mg/100 ml; see below) can be consistently achieved with a normal protein intake. Infants with blood phenylalanine concentrations below 360 µmol/l and a phenylalanine/tyrosine ratio of 3 or more are monitored monthly for the first year of life; blood phenylalanine concentrations commonly vary by as much as 300-400 µmol/l (5-6.7 mg/100 ml) in these infants and may increase substantially at weaning.

The above approach would be considered unduly interventionist by those using 600 µmol/l (10 mg/100 ml) as the limit for treatment [9]. However, the choice of the lower limit is consistent with the choice of therapeutic phenylalanine levels, provides a "margin of safety" for variations in blood phenylalanine (often 300-400 µmol/l) and avoids the problems

**Table 14.1.** Causes of neonatal hyperphenylalaninaemia

*Primary, inherited*
  Phenylalanine hydroxylase deficiency
    Severe, less than 1% enzyme activity
      (classical or typical phenylketonuria)
    Moderate, 1-5% enzyme activity
      (non-classical, atypical phenylketonuria)
    Mild, more than 5% enzyme activity
  Defects of biopterin metabolism
    Guanosine-triphosphate-cyclohydrolase deficiency
    6-Pyruvoyltetrahydropterin-synthase deficiency
    Dihydropteridine reductase deficiency
    Tetrahydropterin carbinolamine dehydratase deficiency

*Secondary, sporadic or inherited*
  With hypertyrosinaemia
    Transient neonatal, often with prematurity
    High protein intake
    Liver disease including galactosaemia and tyrosinaemia
  Without hypertyrosinaemia
    Transient neonatal, often with prematurity
    Drug related (methotrexate, trimethoprim)
    Severe inflammatory response
    Renal disease

**Table 14.2.** Protocol of diagnostic biochemical studies in subjects with hyperphenylalaninaemia; normal values for children (units in parenthesis)

| Investigations | Approximate normal ranges |
|---|---|
| *Routine investigations following positive neonatal screening test* | |
| 1. Plasma phenylalanine and tyrosine concentrations (µmol/l) on a normal protein intake (3 g/kg in infancy) | Phenylalanine: 35–180<br>Tyrosine: 50–180<br>Ratio: 1.1[a] |
| 2. DHPR activity in dried blood spots or red cells or white cells | Depends on assay |
| 3. (a) Urine total biopterin and neopterin[b] concentrations (mmol/mol creatinine), | Biopterin: 0.4–2.5<br>Neopterin: 0.1–5.0<br>BNCR: 9–200<br>Biopterin percentage: 20–80 |
| (b) plasma (or dried blood spot) total biopterins (ng/ml; *Crithidia fasciculata*), or | Plasma biopterin: 1.4–3<br>Blood biopterin: <2.4–6 |
| (c) $BH_4$ load (Chap. 2) | No change (or small fall) in phenylalanine, tyrosine |
| *Additional investigations for suspected defect of biopterin metabolism* | |
| 4. CSF concentrations of HVA and 5HIAA (nmol/ml)[b] | HVA: 400–1000<br>5HIAA: 200–400 |
| 5. CSF total biopterin and neopterin (nmol/l)[b] | Biopterin: 12–40<br>Neopterin: 10–30 |
| 6. Percentage of total biopterins as $BH_4$ in urine and CSF | Urine: 60–80<br>CSF: 90–98 |
| 7. Total folates (ng/ml; *Lactobacillus casei*) in serum, red cells and CSF (DHPR deficiency only) | Serum: 3–12<br>RBC: 150–500<br>CSF: 25–50 |
| 8. Phenylalanine load. 100 mg/kg given orally after an overnight fast. Blood specimens pre-load and 1, 2, 4 and 6 h post-load | Phenylalanine: 4–5×<br>Tyrosine: 2–3×<br>Biopterin: 4–5×<br>Baseline values by 6 h |
| 9. Combined phenylalanine/biopterin load (see 8 and Chap. 2) | Compare response to phenylalanine load |

*5HIAA*, 5-hydroxy-indoleacetic acid; $BH_4$, tetrahydrobiopterin; *BNCR*, biopterin:neopterin creatinine ratio; *DHPR*, dihydropteridine reductase; *HVA*, homovanillic acid; *RBC*, red blood cells

[a] Women taking birth-control pills and carriers of hydroxylase deficiency or biopterin defects may have ratios of up to 2.5. A ratio of 3 or higher is a useful definition of hyperphenylalaninaemia

[b] Neonates have the higher values

(and anxiety) that accompany a decision to start treatment following late rises of phenylalanine to above 600 µmol/l.

### Principles of Phenylalanine-Restricted Diet

The diet replaces a measured proportion of phenylalanine-containing foods with synthetic, phenylalanine-low substitutes to provide nutrients (amino acids, vitamins, minerals, fat and carbohydrate) at a level recommended for the patient's age and size. There are now a wide range of good manufactured products low in phenylalanine [10] but, as in any severely restricted diet, nutrient deficiency and imbalance is a constant risk. It is essential to undertake regular reviews of nutritional status and intake, feeding and eating styles, and new knowledge about dietary requirements. Wide variation in the severity of hyperphenylalaninaemia, eating patterns, culture, health services and the composition of manufactured foods means that management has to be tailored to individual circumstances.

### Implementing the Diet and Phenylalanine Control in Young Children

At the start of treatment in infants with blood phenylalanine levels above 1200 µmol/l (20 mg/100 ml), a period of phenylalanine-free milk brings blood levels down at a rate of around 400 µmol/l (6.7 mg/100 ml) per day. As levels approach the therapeutic range, phenylalanine is added at a rate of 75 mg/kg/day (1.5 g protein/kg/day). Infants with lesser degrees of phenylalanine accumulation need less rigorous restriction, and smooth control is easier to achieve. The prescription of phenylalanine is adjusted until serial blood levels have stabilised.

Serial monitoring of blood phenylalanine levels (weekly during the first 2 years, declining to monthly by school age) is an essential element of treatment [7, 9]. Levels rise in response to minor events, such as intercurrent illness, decline in energy intake or in growth rate, reduction in the amount of protein substitute and rise in phenylalanine intake. The choice of therapeutic limits varies. The UK guidelines recommend 120 to 360 μmol/l (2-6 mg/100 ml) [7] up to school age, based on the need to avoid both phenylalanine deficiency and excess, whereas the recent German guidelines [9] choose 40-200 μmol/l (0.7-3.3 mg/100 ml) even whilst recommending 600 μmol/l (10 mg/100 ml) as the cut-off for decisions on treatment. Given the diurnal variation of 30-250 μmol/l (0.5-4.2 mg/100 ml) in phenylalanine levels and the level of accuracy achieved by the assay of phenylalanine in microspecimens [11], many will regard a lower limit of 40 μmol/l (0.7 mg/100 ml) as posing an undue risk of phenylalanine deficiency.

## Dietary Management in Older Children, Adolescents and Adults

Given the scale of intervention and the practical difficulties involved in sustaining a strict low-phenylalanine diet, many clinics plan to allow a relaxation of the diet at some point before adolescence. No randomised trials of relaxation have been conducted. One trial of stopping treatment at 6 years of age reported a worsened outcome (school progress and intellectual status) 2 years later in children who had stopped treatment [12]. In the UK, it is recommended that families and older children be appraised of current findings and be offered the opportunity to remain on a diet that aims to keep blood phenylalanine concentrations at or below a limit of 700 μmol/l (11.7 mg/100 ml) after mid-childhood and into adulthood [2, 7]. In severe forms of PKU, this requires restriction of phenylalanine to at least 600 mg/day (12 g natural protein).

The new German guidelines [9] recommend an upper limit of 900 μmol/l (15 mg/100 ml) for patients 10-15 years of age and 1200 μmol/l (20 mg/100 ml) for older subjects (which means continuing with phenylalanine-low supplements in subjects with severe biochemical disease). What is essential is a full explanation to the patients and their families concerning the basis for any recommendations (see "Prognosis"). This explanation should include the limits of our knowledge and the need for the active involvement of patients and their families in decision making. In terms of psychosocial development, adolescence is a vulnerable period and is an important time for careful guidance aiming at a smooth transition to full adult responsibility for decisions about treatment (Chap. 4).

## Prescription of Phenylalanine-Containing Foods

The amount of dietary phenylalanine required to provide essential requirements and control blood phenylalanine concentrations depends on the chosen "therapeutic" range, the severity of the underlying defect, the growth rate and the amount and distribution of the protein substitute. In the early months (during the phase of rapid growth), even infants with the severe disease may need 300-400 mg of phenylalanine (6-8 g protein) per day. Once into the phase of more linear growth, such children need ~200 mg phenylalanine/day to keep phenylalanine in the range of 120-360 μmol/l (2-6 mg/100 ml). The protein and, to a lesser extent, the amino acid contents of foods have been fairly well studied [13]. One gram of protein contains approximately 50 mg phenylalanine. However, there is wide variation between the protein content of different batches of foods (including vegetables, fruit and cereals), and the very approximate nature of the figures has to be borne in mind.

In bottle-fed infants, phenylalanine is prescribed and adjusted in measured amounts of milk powder (or volumes of liquid milk) and is given first at each feeding. It is divided between five feedings and spread over the 24 h. The amount should be the maximum consistent with phenylalanine control. The phenylalanine-low milk is given, to appetite, in a separate bottle and is also spread evenly over the 24 h. The amount given is calculated to cover the remaining energy needs with "a little to spare". In breast-fed infants, the pattern is reversed; the phenylalanine-low milk is prescribed in measured amounts and is given first at each of four or five feedings, and breast-feeding is offered to appetite. There is some evidence that the advantages of breast feeding for infant development in the general population are also seen in children with PKU [14].

Once addition of solids begins, the prescription and adjustment of phenylalanine intake (at the normal time, around 4 months of age) is based on food lists giving the average phenylalanine content of useful portions of food. One approach is to prescribe foods in measured portions equivalent to 15 mg of phenylalanine. Another approach, which simplifies the arithmetic, makes communication easier and takes into account the approximate nature of the figures for phenylalanine content, is to divide foods into three groups, as follows:

- High-phenylalanine foods (meat, fish, cheese, egg, pulses, flour and other refined wheat products), which are either totally excluded or are prescribed in useful portions equivalent to 300 mg phenylalanine (6 g protein).
- Medium-phenylalanine foods [milk, yoghurt, cream, rice, corn (maize), potato, spinach, broccoli, breakfast cereals (whole wheat, corn or rice)], which provide meal-appropriate portions (one half or one exchange) equivalent to 25 mg or 50 mg phenylalanine.
- Low-phenylalanine foods (refined fat and carbohydrate, fruit and many vegetables), which provide

less than 25 mg phenylalanine in meal-appropriate portions. These are ignored for the purpose of calculating the daily phenylalanine intake but need to be incorporated into the exchanges if portion sizes increase.

Fruit, cereals and vegetables are a valuable source of fibre, vitamins and minerals (and, given in sufficient amounts, can provide the requirements for nitrogen). However, the recommended requirements for certain nutrients (for example, vitamin $B_{12}$, selenium, cholesterol, long-chain polyunsaturated fat, calcium and phosphate) are impossible to meet without animal products or by replacement with synthetic substitutes when using a phenylalanine-low diet. Unsupplemented low-protein diets pose a serious risk of nutrient deficiency and have no place in management.

### Prescription of Manufactured and Natural Phenylalanine-Low Foods

Manufactured foods vary in composition from "complete" (other than phenylalanine) infant milks to single-category-nutrient mixtures (minerals, vitamins, amino acids, fats or carbohydrates). Amino acids, vitamins and minerals are usually provided in "pharmaceutical" form. The products are unpalatable and are a frequent cause of difficulty with the diet, although there are now some amino-acid-containing fruit bars. Energy is provided in the form of protein-free flour, bread, biscuits, cakes, milk substitute, chocolate, pasta and rice. Currently, these products remain deficient in amino acids, minerals and vitamins and should only be used in combination with appropriate supplements. The latest amino acid products for infants are adequately supplemented with vitamins, minerals (including selenium) and even long-chain, polyunsaturated fat, but there remain concerns about deficiencies and imbalances (for example, of cholesterol [15]). The supplements should be spread as evenly as possible over the day [11].

It is our practice to maintain the total amino acid intake at the equivalent of 3 g/kg/day until the end of the first year and at 2 g/kg until mid-childhood; thereafter, we aim to maintain the amino acid intake at 1 g/kg into adulthood. The amino acid supplements make an essential contribution in keeping blood levels of phenylalanine under control [11] and in maintaining the plasma levels of the other amino acids (including tyrosine) in a range that will help enhance the transport of amino acids across the blood–brain barrier [3]. A careful history of the pattern of consumption should be taken at regular intervals from infancy onwards to help avoid the problem of inadvertent or covert reduction in intake. Biochemical monitoring, especially of blood levels of urea and vitamin $B_{12}$, is helpful as a means of judging the adequacy of supplement intake.

### Managing Illness and Feeding Problems

During illness, most children cannot take their prescribed diet. High-energy fluids (fruit-flavoured glucose polymer with or without fat emulsion) will help reduce catabolism and are more acceptable to sick children than protein supplements, which are usually refused until recovery begins. As anabolism takes over, it is important to reintroduce the phenylalanine allowance to avoid phenylalanine deficiency as diet is re-established. Parents should be instructed *never* to force their children to take the diet during illness (or at other times); children recover rapidly from illness but take months or years to recover from the experience of force feeding.

Diets always risk causing feeding problems, especially if parents are anxious or angry about the diagnosis, are poorly supported or if the diet prescription is inappropriate (energy too high, amino acids too low). Prescriptions too high in energy, inflexible instructions ("he must eat all of it!"), parents blamed when problems arise and failure to acknowledge the immense pressure on the family are all potent causes of feeding problems. When problems do occur, careful analysis of the areas of difficulty followed by modification in small, graded steps within a "desensitisation" framework usually works very well. The aid of behavioural psychologists experienced in managing feeding problems and phobias is invaluable (Chap. 4).

### Risks of Nutrient Deficiency

It is of fundamental importance that all professionals caring for subjects with PKU, and the families themselves, fully understand the nutrient requirements, the compositions of the products being used, the limitations in nutrient content of certain products and the consequences of failing to take adequate amounts. Deficiencies can easily arise from failure to consume appropriate supplements. The risk of subacute combined degeneration of the cord and dementia due to deficiency of vitamin $B_{12}$, and the effects of nitrous oxide anaesthesia in subjects with borderline cobalamin status, should never be forgotten [16, 17]. All phenylalanine-free amino acid, mineral and vitamin substitutes are unpalatable. As children get older, there is a tendency [10] to reduce the prescription of supplements, especially if phenylalanine restrictions have been relaxed deliberately or if blood levels are regularly high due to non-compliance. However, if natural foods are chosen from a restricted range of foods, and especially if a high intake of refined fats and carbohydrates is maintained, certain nutrients (see "Prescription of Phenylalanine-Containing Foods") may become deficient even when overall protein and energy intake is adequate. Older subjects (over 10 years of age) unable to sustain the regular intake of phenylalanine-low supplements should be advised of the risks and, if the problem persists, should be encouraged to eat a wider range of normal food. After years of dietary restriction, this can be surprisingly hard to achieve, and monitoring of nutrient status should continue.

### Prognosis

In severe forms of phenylalanine-hydroxylase deficiency, early diagnosis has reduced the frequency of mental handicap from around 80-90% to 4-8% (compared with a general population risk of around 2%) [2, 8]. However, subtle impairments persist in early-treated subjects; the scale relates to the quality of phenylalanine control during the pre-school years (the age at start of treatment, the average phenylalanine concentrations and the duration of low phenylalanine levels). A reduction of 4-8 points in the mean intelligence quotient (IQ) and an increased risk of behavioural disturbance have been reported from centres across the world [2, 9]. It is likely that emotional stress and neurobiological factors both play a part in the aetiology.

In severe PKU, stopping or relaxing treatment before mid-childhood is associated with cumulative impairment but, after the age of 10 years, IQ appears to remain stable. However, there is evidence that, at all ages, raising blood phenylalanine concentrations is associated with reversible impairments in neuropsychological performance. In the UK, the choice of 700 μmol/l (11.7 mg/100 ml) as the recommended upper limit of phenylalanine for adolescents and adults was based on a range of findings. Despite a lack of change in IQ when the diet stops or is relaxed, phenylalanine levels much above 700 μmol/l are associated with impairments [2, 9] (usually reversible and in proportion to phenylalanine status) in sustained attention, calculation speeds and planning skills. There are also changes in the brain white matter on imaging (an increased water content and reduced glucose consumption, changes in amino acid transport, brain protein turnover, dopamine and serotonin turnover) and changes in EEG, visually evoked responses and pattern recognition in the retina [18]. However, it is universally recognised that the diet is demanding and is not without risk (emotional, social and physical).

Apart from the above effects, adults who stopped the diet in childhood have been reported to exhibit unusually brisk tendon jerks (including ankle clonus, finger and jaw jerks) and intention tremor [2, 19, 20].

These neurological features are not present in well-controlled children or in adolescents who have continued treatment or in those who stopped the diet but have milder forms of PKU, with phenylalanine levels below 700 μmol/l. In addition, there have been a few reports of overt neurological deterioration in both early- and late-treated adolescents and adults (limb weakness, increased tone, tremor and fits occurring with a prevalence of around 0.5-1%), some of whom had blood phenylalanine levels of around 1000 μmol/l (16.7 mg/100 ml) when on a normal diet. Improvement or stabilisation occurred when a strict diet was re-implemented.

These findings have echoes of the clinical features of untreated PKU and have to be considered alongside the evidence of the biochemical effects of phenylalanine on myelin and the older literature showing demyelination in some subjects post-mortem [2]. Given the above information, the authors encourage adolescents and adults with PKU to participate fully in making their own decisions on treatment, including the choice of blood phenylalanine concentrations. We advise stopping treatment only if the risks and problems of continuing clearly outweigh the possible benefits for the individual or if blood phenylalanine concentrations remain around or below 700 μmol/l on a normal, mixed diet.

### Genetics

In most populations of European origin, phenylalanine hydroxylase deficiency is by far the commonest form of inherited hyperphenylalaninaemia, with a prevalence of between 1 in 4000 and 1 in 40,000 births; Iceland and Ireland have the highest frequency, and Finland has the lowest frequency. Over 400 mutations in the phenylalanine hydroxylase gene have been described to date [8], and most subjects with hyperphenylalaninaemia are compound heterozygotes rather than homozygotes, although different groups of mutations and polymorphisms do predominate in certain populations (for example, Yemenite Jews). The genetic background explains the wide and continuous spectrum of biochemical phenotype long observed in clinical practice. The genotype, assessed by expression studies, correlates well with biochemical severity.

Using a combination of mutation analysis and polymorphisms in the hydroxylase gene, it is now possible to undertake reliable prenatal diagnosis in the vast majority of couples (or inbred families) with an affected child. It is still not usual to provide a molecular genetics service to detect carriers in the general population (for example, for partners of PKU subjects), although this is likely to become more common when the technology advances. Simple biochemical tests, such as a phenylalanine/tyrosine ratio, produce a broad overlap between normal subjects and carriers.

## Maternal Phenylketonuria

### Clinical Presentation

Over 90% of infants born to mothers with blood phenylalanine concentrations above 1200 μmol/l exhibit evidence of foetal damage, low birth weight, microcephaly, dysmorphic facies, slow postnatal growth and development and long-term intellectual impairment (Table 14.3) [21]. Facial features resemble those of the foetal alcohol syndrome, with small palpebral fissures, epicanthic folds, long philtrum and reduced upper lip. Malformations occur in around 10–20% of subjects, most often in the heart or great vessels but also in other organs, causing, for example, oesophageal atresia and tracheo-oesophogeal fistula, bowel malrotation, bladder extrophy and other urogenital anomalies, coloboma or cataract of eye and cleft lip and palate. As always in PKU, wide variations in phenotype occur and, even in severe PKU, the foetus may escape significant damage. The risk of abnormalities in the offspring of women with mild to moderate hyperphenylalaninaemia appears to diminish linearly in proportion to maternal phenylalanine concentrations, reaching a minimal prevalence as maternal phenylalanine concentrations approach the normal range [22, 23].

### Metabolic Derangement

Due to active net transport from mother to foetus, foetal phenylalanine concentrations are about double those in the mother [24]. Phenylalanine competes with other neutral amino acids for placental transport and, just as in postnatal life, is likely to affect foetal chemistry in a variety of ways, though which are critical to foetal development is not known.

**Table 14.3.** Percentage of mothers with phenylketonuria whose offspring have abnormalities. The sample size used to calculate the percentage is in brackets [21]

| | Maternal phenylalanine concentrations mg/100 ml[a] | | | |
| --- | --- | --- | --- | --- |
| | 20 | 16–19 | 11–15 | 3–10 |
| Mental retardation | 92 (172) | 73 (37) | 22 (23) | 21 (29) |
| Microcephaly | 73 (138) | 68 (44) | 35 (23) | 24 (21) |
| Congenital heart disease | 12 (225) | 15 (46) | 6 (33) | 0 (44) |
| Birth weight < 2500 g | 40 (89) | 52 (33) | 56 (9) | 13 (16) |

[a] Multiply by 60 to convert to μmol/l

## Treatment and Prognosis

### Counselling

Counselling begins with the parents of newborns with PKU and continues at intervals during the course of long-term management of girls with PKU. Very young children can understand a simple explanation of maternal PKU and, by late childhood, counselling should be directed to the girls themselves in order to:

- Ensure a basic understanding of conception and PKU
- Explain that a baby in utero is exposed to the mother's high phenylalanine levels and is likely to develop abnormally from conception onwards
- Stress the necessity for strict diet from before conception
- Distinguish the very high risk of foetal damage during pregnancy from the relatively small chance of a baby actually inheriting PKU
- Explain that the risk to the foetus is proportional to mother's blood phenylalanine control
- Stress the importance of continued follow-up

The key to managing pregnancy [25] lies in ensuring that girls with PKU are, from an early age, educated and counselled thoroughly and are, as adolescents and adults, followed up consistently by a committed clinic team whom they know and trust. The team must have the expertise required to provide the girls with the skills needed to re-implement a strict phenylalanine-low diet. Contraception should be discussed with teenage girls and should be reviewed at intervals. The risks of unprotected intercourse need to be emphasised, as must the importance of reporting a missed period as soon as possible when receiving a normal diet. The difficulties of implementing a synthetic diet during early pregnancy and the option of termination of pregnancy within this context should be discussed. When appropriate, the patient should bring her partner to the clinic so that they share the counselling. Girls who do not arrive for follow-up are a "high-risk" group.

### Re-Starting a Strict Diet

A patient who has not been on a diet for several years will either need to be admitted to a hospital or will need to receive very close supervision in her own home to ensure that she is able to consume (and knows how to use) the prescribed amounts of phenylalanine and the dietary substitutes. This also ensures that blood phenylalanine concentrations fall rapidly, giving a sense of achievement and encouragement. Every aspect of the diet needs to be rehearsed, including taking the protein substitute and other supplements, choosing foods and calculating exchange foods, preparing food, blood collection and dispatch to the laboratory and responding to blood-test results.

It is a medical emergency when a woman with PKU presents already pregnant. Present evidence does not accurately define the risk to the foetus (see "Prognosis") if diet is successfully introduced within a few weeks of conception, although dysmorphic facies, other congenital malformations and microcephaly may still occur [26]. The uncertainties and the fact that it is too late to completely avoid these risks must be explained. Foetal ultrasound examination of the heart, other organs and skull diameter may provide further information, but women who have conceived with phenylalanine levels greater than or equal to 700 μmol/l (11.7 mg/100 ml) have a strong case for termination. The patient who requires time to reach a decision on termination should re-start a diet without delay.

### Control of Blood Phenylalanine Concentrations

Keeping foetal phenylalanine below 500 μmol/l (8.3 mg/100 ml) requires maternal concentrations below 250 μmol/l (4.2 mg/100 ml). It is recommended [7] that treatment be introduced in all women with blood phenylalanine concentrations greater than 300 μmol/l (6 mg/100 ml) and that maternal phenylalanine values be maintained between 60 μmol/l and 250 μmol/l (1–4.2 mg/100 ml) from conception onward, with monitoring at least twice weekly. In subjects with severe disease, natural protein intake will need to be reduced to a total of 6 g/day or less, and at least 70 g of supplemental amino acids per day will be required to cover nutritional requirements. Although direct evidence for any benefit from extra tyrosine is lacking, plasma tyrosine levels often fall below 30 μmol/l (0.4 mg/100 ml) and, from around 16 weeks gestation, it is our practice to add tyrosine to provide 3–4 g/day up to a maximum of 8 g/day. Unless the infant has PKU, phenylalanine tolerance shows a consistent and progressive increase (often to around 25–30 g natural protein per day) from around 20–22 weeks gestation, which presumably relates to increases in phenylalanine-hydroxylase activity in foetal liver.

### Antenatal and Obstetric Care

The patient is seen monthly for nutritional monitoring, antenatal care and consultation with the clinician and the dietician. Routine ultrasound examination of the foetus is carried out at 12 weeks and 20 weeks. Admission may be required for poor weight gain, vomiting, poor phenylalanine control or other problems. There is nothing to suggest that the birth of the baby needs anything other than normal obstetric consideration at delivery. One would not expect additional free phenylalanine (which would be present in the mother's milk if she returned to a free diet) to be harmful unless the neonate is homozygous for PKU.

### Prognosis

In severe forms of PKU, the risk to the foetus during untreated pregnancies is high (Table 14.3). Treatment with a well-controlled phenylalanine-low diet from before conception or early in the first trimester reduces the risk of growth retardation, microcephaly and malformations [25, 27]. The information on untreated and treated pregnancies is consistent, providing good evidence of a graded effect of maternal phenylalanine [21–23, 25, 27]. Birth weight, head circumference, the risk of congenital anomalies, postnatal developmental progress and intelligence have all been shown to relate to maternal blood phenylalanine levels in the crucial first trimester. In women with milder disease [phenylalanine levels less than 700 μmol/l (11.7 mg/100 ml)], the risks remain proportionate to the phenylalanine level down to the normal range [23].

Despite the evidence for benefits of dietary treatment in maternal PKU, there remains concern about foetal outcome. Even the offspring of mothers who received a strict diet starting before conception [25] or during the first trimester [27] have shown a degree of neurological and intellectual impairment, with IQ results some 8–10 points below control values. However, many PKU mothers were treated late in childhood or had poor phenylalanine control. As a group, their intellectual status is below that of the general population, and this is likely to influence the development of their offspring. The outcome that can be achieved under optimal circumstances remains to be established.

## Defects of Biopterin Metabolism

Tetrahydrobiopterin ($BH_4$) is the cofactor required by phenylalanine, tyrosine and tryptophan hydroxylases and by NO synthase [1, 19, 28]. Defects in cofactor metabolism are generally much rarer (Japan is an exception) than phenylalanine hydroxylase deficiency, occurring in around 1 in 500,000 to 1 in 1,000,000 births [6].

## Clinical Presentation

The most severe types of disorder lead to microcephaly, developmental delay and progressive neurological deterioration leading to death in childhood, although some patients stabilise or make slow developmental progress. The neurological features are highly characteristic, although L-dopa decarboxylase deficiency (Chapter 26) can present in a similar way, and include infantile Parkinsonism (hypokinesis, drooling, swallowing difficulty, gastro-oesophageal reflux, sweating, pinpoint pupils, oculogyric crises, truncal hypotonia, increased limb tone, blank facies with relative preservation of smiling), myoclonus, choreic or dystonic limb movements, very brisk tendon jerks and sometimes infantile spasms, grandmal fits, hyperpyrexia and disturbance of sleep pattern. In milder forms of disorder, symptoms may consist simply of minor developmental delays or movement disorders; they may also be absent or intermittent or may be precipitated by a phenylalanine load or acute infection.

Some forms of familial, dopa-responsive dystonia have been shown to be due to inborn errors of pterin metabolism [heterozygous guanosine triphosphate (GTP)-cyclohydrolase deficiency or partial 6-pyruvoyl-$BH_4$-synthase deficiency]. Truncal or limb dystonia often with marked diurnal fluctuation that responds rapidly to small daily doses of L-dopa is a characteristic feature. Tyrosine hydroxylase deficiency can lead to a very similar picture. In subjects with severely defective synthesis of $BH_4$, damage may occur in utero and contributes to neurological impairment.

Symptoms of progressive demyelination occur in Dihydropteridine reductase (DHPR) deficiency due to defective folate metabolism resulting in progressive upper motor neurone damage, paraplegia, bulbar palsy, long-tract sensory loss and deterioration in cortical function. Magnetic resonance imaging, cranial tomography and histology of the brain reveal myelin loss with a characteristic microvascular calcification, appearances that closely resemble those of methotrexate toxicity [19].

## Metabolic Derangement

Defective synthesis or recycling of $BH_4$ causes hyperphenylalaninaemia (Fig. 14.1), although this may be very mild or manifest only in a reversal of the phenylalanine/tyrosine ratio. Defective synthesis of dopamine, serotonin, noradrenaline and adrenaline (Fig. 14.2) causes many of the characteristic symptoms. Dopamine deficiency is likely to be the explanation for the raised levels of prolactin reported in patients with defects of $BH_4$ synthesis. 5-Hydroxytryptophan is the precursor of melatonin in the pineal gland, and it seems probable that sleep disturbances are due to impaired turnover of this hormone. In CSF, concentrations of homovanillic acid and 5-hydroxy-indoleacetic acid, the major metabolites of dopamine and serotonin in the human nervous system, provide a useful measure of the degree of impairment of amine turnover in vivo.

$BH_4$ is also the essential cofactor for synthesis of NO from arginine and for oxidation of the ether bond in plasmalogens (cell membrane lipids that are oxidised in peroxisomes). No obvious symptoms relating to disturbed NO or plasmalogen turnover have been identified in subjects with pterin defects, probably because these enzymes operate at a lower residual $BH_4$ concentration. However, studies of NO turnover in the mouse model of GTP-cyclohydrolase deficiency have revealed detectable biochemical impairment.

### Defects of Synthesis

When $BH_4$ metabolism is intact, phenylalanine concentrations are a controlling factor in GTP-cyclohydrolase activity. Hyperphenylalaninaemia increases the production of $BH_4$ (Fig. 14.1) [1], and a failure of

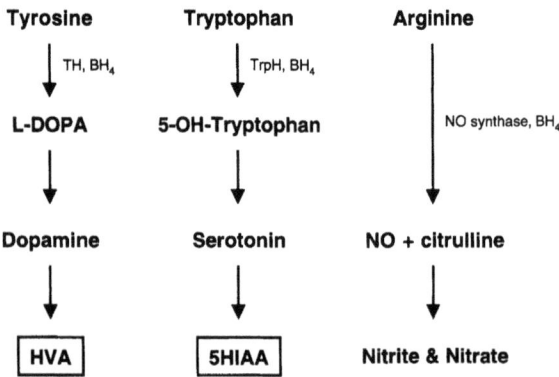

**Fig. 14.2.** Synthesis of dopamine, serotonin and nitric oxide (NO). Tyrosine hydroxylase (*TH*), tryptophan hydroxylase (*TrpH*) and NO synthase are rate-controlling enzymes requiring tetrahydropbiopterin ($BH_4$) as a cofactor. Homovanillic acid (*HVA*) and 5-hydroxyindoleacetic acid (*5HIAA*) in CSF are measures of amine turnover

biopterins to rise in blood and urine in response to hyperphenylalaninaemia is typical of defects of synthesis. This characteristic is exploited in diagnosis (Table 14.2). In GTP-cyclohydrolase deficiency (Fig. 14.1), the concentrations of biopterins and neopterins are low in urine, blood and CSF; in defects of 6-pyruvoyltetrahydropterin synthase (Fig. 14.1), biopterins are low but neopterins are raised. Administration of $BH_4$ (Table 14.2; Chap. 2) corrects the phenylalanine accumulation, increases plasma tyrosine concentrations and restores the pterin profile to normal.

Subjects with "partial" defects in synthesis have less marked changes in pterin concentration and smaller or intermittent reductions in CSF amine metabolites. However, the dopa-responsive dystonias due to heterozygous GTP-cyclohydrolase deficiency exhibit obvious CSF amine changes [29]. For a discussion of GTP-cyclohydrolase deficiency, see Chap. 26. Obligate heterozygotes for 6-pyruvoyltetrahydropterin synthase deficiency exhibit persistent neopterin accumulation but, to date, they have not been known to develop dystonias. Subjects without evidence of central amine disturbance are said to have "peripheral" defects but, as in phenylalanine hydroxylase deficiency, the disorders present a continuous spectrum of severity rather than falling into discrete groups.

### Defects of Recycling

Phenylalanine hydroxylase requires two enzymes to successfully recycle $BH_4$ in the liver: tetrahydropterin carbinolamine dehydratase and DHPR (Fig. 14.1). A defect in either enzyme may cause hyperphenylalaninaemia but, in the dehydratase deficiency, this is associated with a failure of the biopterins to rise whereas, in DHPR deficiency, biopterins and neopterins increase in proportion to the hyperphenylalaninaemia.

Subjects with dehydratase deficiency show hyperphenylalaninaemia (usually mild) and excrete an excess of primapterin, oxoprimapterin, anapterin and neopterin in the urine. There is a reduction of biopterin in blood and urine, and the hyperphenylalaninaemia responds to administration of $BH_4$. To date, all subjects reported have had a "peripheral" type of defect.

DHPR deficiency blocks the normal recycling of quinonoid dihydrobiopterin, leading to accumulation of dihydropterins in blood, urine and CSF. Whilst total biopterins are increased in CSF, concentrations of active cofactor ($BH_4$) are reduced to about half that expected for the phenylalanine level (though remaining well above the levels found in defects of synthesis). Most subjects show a fall in phenylalanine and a rise in tyrosine levels with large enough doses of $BH_4$, although the response is variable [28]. "Partial" defects do occur.

### Defective Folate Metabolism in DHPR Deficiency

Defective folate metabolism in DHPR deficiency is responsible for progressive neurological damage and probably occurs due to the effects of accumulating dihydropterins. Megaloblastic changes are unusual despite very low folate concentrations in serum, red cells, CSF and brain. Histological and neuroradiological changes in the brain are similar to those due to congenital folate malabsorption, 5,10-methylenetetrahydrofolate-reductase deficiency (Chap. 25) or methotrexate toxicity (see "Clinical Presentation") [19].

### Foetal Effects of Pterin Defects

In defects of synthesis, the pterin and amine changes occurring after birth are mirrored in the amniotic fluid of affected foetuses, providing an explanation for the prenatal neurological damage and low birth weight. Reduced receptor density and increased sensitivity due to reduced exposure to neurotransmitter amines in utero may also explain some of the problems of treatment in subjects having this form of disorder. There is no evidence of a foetal effect of DHPR deficiency.

## Diagnostic Tests

### Routine Testing of Infants with Hyperphenylalaninaemia

All subjects with hyperphenylalaninaemia should be screened for disorders of biopterin metabolism (Table 14.2). Testing for defects of synthesis can be carried out either by pterin measurements on urine (total biopterins and neopterins) or by total biopterins in dried blood spots. In order to achieve maximum discrimination, testing must be carried out before phenylalanine concentrations have been brought under control. A $BH_4$ load can also be used. Currently, Milupa (Friedreichsdorf, Germany) holds a license for the use of $BH_4$ as a test substance. DHPR deficiency should be excluded by an enzyme assay. Use of dried blood spots to screen all hyperphenylalaninaemic infants by measuring DHPR activity and total biopterins [2, 6] has the great advantage that specimens are easy to obtain, are stable under a wide range of conditions and can be posted to the specialist laboratory.

### Diagnostic Investigation

Transient increases in the neopterin/biopterin ratio can occur in neonates, but all infants with positive routine tests for biopterin defects require further investigation (Table 14.2) including measurement of

CSF pterins and amine metabolites. The pattern of pterin and amine-metabolite changes in the presence of even mild hyperphenylalaninaemia discriminates between the various defects. However, in subjects without obvious hyperphenylalaninaemia, a phenylalanine load (which may precipitate acute symptoms) has been used in combination with a $BH_4$ load to identify pterin defects (Chap. 2). Apart from DHPR assay (in dried blood spots or whole blood), enzyme studies are not necessary for diagnosis but can be useful preliminary to prenatal diagnosis [28, 30, 31]. However, molecular genetics is now supplanting enzyme studies, including the diagnosis of heterozygous GTP-cyclohydrolase deficiency as a cause of the dominant dystonias [29].

Careful assessment of development and neurological features with imaging and electrophysiological investigation are important for assessing prognosis and progress. Gastroenterological investigation to identify oesophageal dysmotility or sphincter dysfunction may also be required.

### Prenatal Diagnosis

Molecular-genetic methods are now available to an increasing proportion of families [28–31]. It is possible to confirm deficiencies of 6-pyruvoyl-$BH_4$ synthase and GTP cyclohydrolase using analysis of amine metabolites and pterins in amniotic fluid.

## Treatment and Prognosis

In the authors' experience, within the limits imposed by existing neurological damage, treatment of biopterin-metabolism defects has proved to be rather straightforward and remarkably effective compared with treatment of many metabolic disorders. Due to the side effects of amine replacement therapy (and probably persisting receptor imbalance) in subjects who have been chronically amine deficient, difficulties can occur, especially at the start of treatment and in defects of synthesis. These problems seem to improve over time, provided that slow, graded exposure to therapy is sustained. Gastroenterological problems (vomiting and feeding difficulties) can be exacerbated by treatment, and gastrostomy feeding may be needed. In most patients, treatment will be required throughout life.

### Control of Phenylalanine Accumulation

Strict control of plasma phenylalanine concentrations (60–180 µmol/l; 1–3 mg/100 ml) promotes residual endogenous amine synthesis and the uptake of administered L-dopa and 5-hydroxytryptophan (see below). In DHPR deficiency, hyperphenylalaninaemia stimulates production of the dihydropterins, which probably block folate metabolism. In defects of synthesis, phenylalanine control is best achieved by administration of a single dose of $BH_4$ daily (1 mg/kg) but has to be given on a "named-patient" basis. The cost is comparable to that of a phenylalanine-low diet. $BH_4$ is not recommended for DHPR deficiency, because the doses required are large and exacerbate the accumulation of dihydropterins. A low-phenylalanine (15–20 g/day) diet with relatively generous tolerance of natural protein is an effective means of phenylalanine control in both groups of disorder.

### Correction of Amine Deficiency

In subjects with severely defective amine synthesis, administration of L-dopa and a decarboxylase inhibitor (carbidopa) in a ratio of 1:4, combined with 5-hydroxytryptophan at doses of 10–12 mg/kg/day L-dopa and 8–10 mg/kg/day of 5-hydroxytryptophan, will restore CSF amine metabolite concentrations to the normal range and will control symptoms. Especially in defects of synthesis, it is advisable to start with very low doses (1–2 mg/kg/day) spread through the day, increasing the amounts stepwise every 2–3 days to avoid side-effects, such as vomiting, excessive fidgetiness (akathisia) or other abnormal movements. If side effects are troublesome, it is important that every effort be made to continue replacement therapy, even if at lower doses, because only in this way can the hypersensitivity of the end-organs be overcome. The amine precursors are given together, divided into at least four doses and given 30 min before meals. The lowest dose of commercially available L-dopa comes as 50-mg tablets with carbidopa (Sinemet LS, 4:1 ratio); these tablets can be halved. Ten milligram capsules of 5-hydroxytryptophan are available, but this is not a registered drug. For infants, pharmacists need to formulate both drugs individually in very small doses so that they can be spread through the day.

In subjects with a milder biochemical phenotype, lower doses of amine replacement therapy will be sufficient and, in some subjects, controlling the hyperphenylalaninaemia alone is sufficient to improve neurological symptoms and to restore amine metabolite concentrations to normal. This probably explains the beneficial neurological effects of $BH_4$ administration seen in some patients. Large doses of $BH_4$ (20 mg/kg) are required to penetrate the blood–brain barrier, and such doses have no place in routine treatment. In some subjects with mild disease, reduced amine turnover in infancy gradually resolves over the first year or two, probably because the high amine turnover in infants makes more demands on the pterin pathways. Familial dystonias due to heterozygous GTP-cyclohydrolase deficiency are especially responsive to small doses of L-dopa (1–2 mg/kg/day) [29].

Once treatment has been stabilised, it is important to measure CSF amine metabolites at intervals. Final adjustment of the timing and balance of L-dopa/carbidopa and 5-hydroxytryptophan will need to be made according to individual response. The doses need to be regularly reviewed and updated (every 3–6 months in early childhood) according to response, weight and CSF amine metabolite concentrations. The "on/off" phenomenon, familiar in adult Parkinson's disease, may occur but, in our experience, it is very uncommon if the above approach is used. The use of dopamine agonists (such as bromocriptine) and inhibitors of monoamine oxidase-B (such as selegiline) to replace or supplement L-dopa requires further exploration.

### Folate Therapy in DHPR Deficiency

Although not all subjects with DHPR deficiency develop brain folate deficiency, in those who do, the risk of irreversible damage to white matter is high. Administration of tetrahydrofolate (5-formyl-tetrahydrofolate, folic acid) in amounts sufficient to keep CSF concentrations in the high to normal range (30–40 ng/ml in infants, 20–30 ng/ml in older children) prevents demyelination and halts the demyelinating process in those who already show abnormalities. As in subjects with methotrexate toxicity, the use of folic acid may cause acute neurological deterioration and should be avoided. If control of plasma phenylalanine concentrations is strictly maintained, 15 mg of folic acid orally each day will maintain CSF folate concentrations in the normal range during infancy and early childhood; 3 mg/day is insufficient. Larger doses (up to 60 mg/day) are needed in older subjects and those with poorly controlled phenylalanine concentrations. The authors have come to the conclusion that long-term folic acid treatment should be given to all subjects with DHPR deficiency, with careful serial monitoring of CSF folate concentrations.

### Prognosis

In infants detected by routine testing, changes in feeding, expression, tone, posture and mobility may be observed within days of starting treatment, even when symptoms initially seem to be minor or absent. Some infants with defective synthesis, particularly if they are of low birth weight and exhibit neurological symptoms early in life, may show developmental delay and persistent movement disorders even when treatment is started early. In DHPR deficiency, the major determinant of outcome is whether or not neurological damage due to the folate disturbance has occurred.

Long-term progress, at least up to early adult life, is well maintained, and prognosis seems to be predictable on the basis of early events – provided treatment is maintained. In patients who present with severe neurological disease, the response to therapy varies, but in our experience, marked clinical improvement occurs in most and is usually dramatic.

## Genetics

The genes have been cloned, and mutations causing defective metabolism have been identified for all the enzymes involved in $BH_4$ metabolism. These and polymorphisms within the genes are being exploited in prenatal diagnosis [31].

## References

1. Kaufman S (1987) Tetrahydrobiopterin and hydroxylation systems in health and disease. In: Lovenberg W, Levine RA (eds) Unconjugated pteridines in neurobiology: basic and clinical aspects. Taylor and Francis, London, pp 1–28
2. Medical Research Council Working Party on Phenylketonuria (1993) Phenylketonuria due to phenylalanine hydroxylase deficiency: an unfolding story. Br Med J 306:115–119
3. Pardridge WM (1998) Blood brain barrier carrier mediated transport and brain metabolism of amino acids. Neurochem Res 23:635–644
4. Bannon MJ, Bunney EB, Roth RH (1981) Mesocortical dopamine neurons: rapid transmitter turnover compared to other brain catecholamine systems. Brain Res 218:376–382
5. Tillotson SL, Costello PM, Smith I (1995) No reduction in birth weight in phenylketonuria. Eur J Paediatr 154:847–849
6. Smith I, Cook B, Beasley M (1991) Review of neonatal screening programme for phenylketonuria. Br Med J 303:333–335
7. Report of Medical Research Council Working Party on Phenylketonuria (1993) Recommendations on the dietary management of phenylketonuria. Arch Dis Child 68:426–427
8. Guldberg P, Rey F, Zschocke J et al. (1998) A European multicentre study of phenylalanine hydroxylase deficiency: classification of 105 mutations and a system of genotype-based prediction of metabolic phenotype. Am J Hum Genet 63:71–79
9. Burgard P, Bremer HJ, Buhrdel P et al. (1999) Rationale for the German recommendations for phenylalanine level control in phenylketonuria 1997. Eur J Paediatr 158:46–54
10. Prince AP, McMurray HP, Buist NR (1997) Treatment products and approaches for phenylketonuria: improved palatability and flexibility demonstrate safety, efficacy and acceptance in US clinical trials. J Inherit Metab Dis 20:486–498
11. MacDonald A, Rylance GW, Asplin D, Hall SK, Booth IW (1998) Does a single plasma phenylalanine predict quality of control in phenylketonuria? Arch Dis Child 78:122–126
12. Azen C, Koch R, Friedman E, Wenz E, Fishler K (1996) Summary of findings from the United States Collaborative Study of Children treated for Phenylketonuria. Eur J Paediatr 155 [Suppl 1]:S29–S32
13. Paul AA, Southgate DAT (1992) McCance and Widdowson's. The composition of foods. Royal Society of Chemistry and Ministry of Agriculture, Fisheries and Foods, Cambridge
14. Riva E, Agostoni C, Biasucci G et al. (1996) Early breast feeding is linked to higher intelligent quotient scores in dietary treated phenylketonuric children. Acta Paediatr 85:56–58

15. Galli C, Agostini C, Mosconi C, Riva E, Salari P, Giovannini H (1991) Reduced plasma and C-22 polyunsaturated fatty acids in children with phenylketonuria during dietary intervention. J Pediatr 1191:562-567
16. Hanley WB, Feigenbaum AS, Clarke JT, Schoonheyt WE, Austin VJ (1996) Vitamin B12 deficiency in adolescents and young adults with phenylketonuria. Eur J Paediatr 155 [Suppl]:S145-147
17. Lee P, Smith I, Piesowicz A, Brenton DP (1999) Spastic paresis after anaesthesia. Lancet 353:554
18. Diamond A, Herzberg C (1996) Impaired sensitivity to visual contrast in children treated early and continuously for phenylketonuria. Brain 19:523-538
19. Smith I (1985) The hyperphenylalaninaemias. In: Lloyd JK, Scriver CR (eds) Genetic and metabolic disease. Butterworth, London, pp 166-210
20. McDonnell GV, Esmond TF, Hadden DR, Morrow JI (1998) A neurological evaluation of adult phenylketonuria in Northern Ireland. Eur Neurol 39:38-43
21. Lenke RL, Levy HL (1980) Maternal phenylketonuria and hyperphenylalaninaemia. N Engl J Med 303:1202-1208
22. Smith I, Glossop J, Beasley M (1990) Fetal damage due to maternal phenylketonuria: effects of dietary treatment and maternal phenylalanine concentrations around the time of conception. J Inherit Metab Dis 13:651-657
23. Levy HL, Waisbren SE, Lobbregt D et al. (1994) Maternal mild hyperphenylalaninaemia: an international survey of offspring outcome. Lancet 344:1589-1594
24. Soltesz G, Harris D, Mackenzie IZ, Aynsley-Green A (1985) The metabolic and endocrine milieu of the human fetus and mother at 18-21 weeks of gestation. 1. Plasma amino acid concentrations. Pediatr Res 19:91-92
25. Brenton DP, Lilburn M (1996) Maternal phenylketonuria. A study from the United Kingdom. Eur Paediatr 155 [Suppl 1]:S177-180
26. Rouse B, Azen C, Koch R et al. (1997) Maternal Phenylketonuria Collaborative Study (MPKUCS) offspring: facial anomalies, malformations and early neurological sequelae. Am J Med Genet 69:89-95
27. Hanley WB, Koch R, Levy HL et al. (1996) The North American Maternal Phenylketonuria Collaborative Study, developmental assessment of the offspring: preliminary report. Eur J Paediatr 155 [Suppl 1]:S169-172
28. Blau N, Thony B, Heizmann CW, Dhondt J-L (1993) Tetrahydrobiopterin deficiency: from phenotype to genotype. Pteridines 4:1-10
29. Bandemann O, Valente EM, Holmans P et al. (1998) Dopa-responsive dystonia: a clinical and molecular genetic study. Ann Neurol 44:649-656
30. Thony B, Neuheiser F, Kierat L et al. (1998) Mutations in the pterin 4 alphacarbinolamine dehydratase (PCBD) gene cause a benign form of hyperphenylalaninaemia. Hum Genet 103:162-167
31. Blau N, Duran M, Blaskovics ME (eds) (1996) Physician's guide to the laboratory diagnosis of metabolic diseases. Chapman and Hall Medical, London

CHAPTER 15

## Tyrosine Metabolism

Tyrosine is one of the least soluble amino acids and forms characteristic crystals upon precipitation. It derives from two sources, diet and hydroxylation of phenylalanine (Fig. 15.1). Tyrosine is both glucogenic and ketogenic, since its catabolism, which proceeds predominantly in the liver cytosol, results in the formation of fumarate and acetoacetate. The first step of tyrosine catabolism is conversion into 4-hydroxyphenylpyruvate by cytosolic tyrosine aminotransferase. Transamination of tyrosine can also be accomplished in the liver and in other tissues by mitochondrial aspartate aminotransferase, but this enzyme plays only a minor role under normal conditions. The penultimate intermediates of tyrosine catabolism, maleylacetoacetate and fumarylacetoacetate, can be reduced to succinylacetoacetate, followed by decarboxylation to succinylacetone. The latter is the most potent known inhibitor of the heme biosynthetic enzyme, δ-aminolevulinic acid dehydratase (prophobilinogen synthase, see Fig. 32.1).

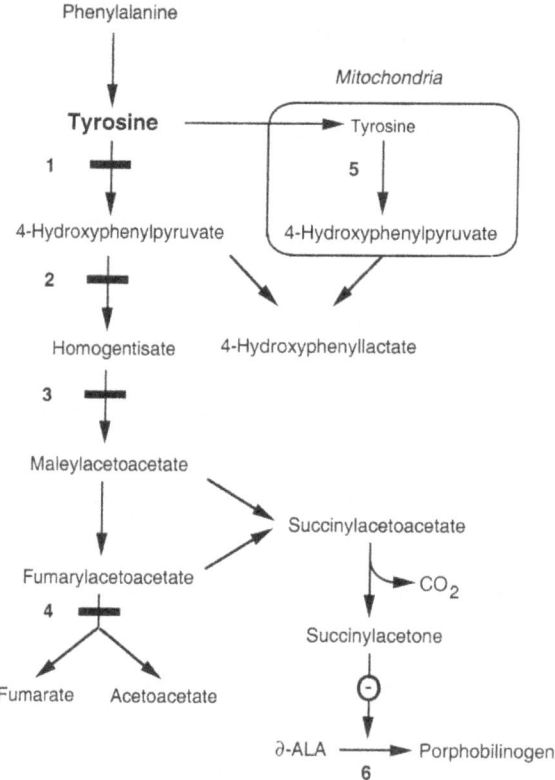

**Fig. 15.1.** The tyrosine catabolic pathway. 1, Tyrosine aminotransferase (deficient in tyrosinaemia type II); 2, 4-hydroxyphenylpyruvate dioxygenase (deficient in tyrosinaemia type III and in hawkinsinuria, also site of inhibition by NTBC); 3, homogentisate dioxygenase (deficient in alkaptonuria); 4, fumarylacetoacetase (deficient in tyrosinemia type I); 5, aspartate aminotransferase; 6, δ-aminolevulinic acid (δ-ALA) dehydratase (porphobilinogen synthase). Enzyme defects are depicted by *solid bars* across the arrows

# CHAPTER 15

# Disorders of Tyrosine Metabolism

Eli Anne Kvittingen and Elisabeth Holme

CONTENTS

Hereditary Tyrosinaemia Type I.................... 187
   Clinical Presentation........................... 187
   Metabolic Derangement ......................... 188
   Diagnostic Tests .............................. 188
   Treatment and Prognosis ....................... 188
      NTBC Treatment ........................... 188
      Liver Transplantation....................... 189
      Dietary Treatment .......................... 189
      Supportive Treatment ...................... 190
   Genetics ..................................... 190
Hereditary Tyrosinaemia Type II ................... 190
Hereditary Tyrosinaemia Type III .................. 191
Hawkinsinuria..................................... 192
Alkaptonuria...................................... 192
References........................................ 193

Five inherited disorders of tyrosine catabolism are known and depicted in Fig. 15.1. Hereditary tyrosinaemia type I is characterized by progressive liver disease and renal tubular dysfunction with rickets. Hereditary tyrosinaemia type II (Richner-Hanhart syndrome) presents with keratitis and blisterous lesions of the palms and soles. Tyrosinaemia type III may be asymptomatic or associated with mental retardation. Hawkinsinuria may be asymptomatic or presents with failure to thrive and metabolic acidosis in infancy. In alkaptonuria symptoms of osteoarthritis usually appear at an advanced age. Other inborn errors of tyrosine metabolism include oculocutaneous albinism caused by a deficiency of melanocyte-specific tyrosinase, converting tyrosine into DOPA-quinone; the deficiency of tyrosine hydroxylase, the first enzyme in the synthesis of dopamine from tyrosine; and the deficiency of aromatic L-amino acid decarboxylase, which also affects tryptophan metabolism. The latter two disorders are covered in Chap. 26.

## Hereditary Tyrosinaemia Type I

### Clinical Presentation

The clinical heterogeneity of tyrosinaemia is wide, and patients may present at almost any age from infancy to adulthood. In the most acute forms of the disease, patients present within weeks of birth with severe liver failure and often present with vomiting, diarrhoea, jaundice, hypoglycaemia, oedema, ascites and particularly a bleeding diathesis. Sepsis is common and patients may also have early hypophosphataemic bone disease secondary to renal tubular dysfunction. Later in infancy and childhood, patients may present with liver failure, rickets or more non-specific problems, such as bleeding tendency, failure to thrive or hepatosplenomegaly. Although the acute forms of tyrosinaemia present with a picture of liver failure, the jaundice is moderate even in severely affected patients; furthermore, the transaminases are rarely exceedingly high. The coagulopathy of tyrosinaemia type I is striking; blood coagulation may be highly abnormal even with few other signs of liver disease. Very occasionally, children may present with neurological crisis, hypotonia secondary to rickets or a relapsing polyneuropathy [1]. The neurological problems often complicate the course of the disease, and the attacks resemble those of acute intermittent porphyria, with abdominal pain and peripheral neuropathy with muscle weakness and hypertension [2] (Chap. 32).

In the chronic forms of tyrosinaemia, the symptoms and signs are more subtle. There may be only slight enlargement of the liver, mild growth retardation and sub-clinical rickets. A history of bruising may be apparent on careful inquiry. In some patients, symptoms of liver disease are minimal, and the liver involvement is only evident on careful biochemical investigation or imaging. Computed tomography and ultrasound of the liver usually demonstrate abnormal structure [3], but the techniques do not detect all abnormalities, even with overt regeneration nodules. Most (but not all) patients with chronic tyrosinaemia have renal tubular dysfunction and rickets. Nephromegaly and nephrocalcinosis may also be observed. Some develop renal failure that requires renal transplant [4]. In the absence of rickets, and with few (if any) symptoms of liver disease, some patients with chronic disease easily elude diagnosis until liver cirrhosis is evident or hepatocellular carcinoma has developed. Hepatocellular carcinoma is a major complication of all types of tyrosinaemia type I [5], even those with few symptoms and signs of liver disease. Carcinoma may develop during infancy but is more common in late childhood or adolescence. The incidental finding of any abnormality of the kidneys and/or liver function or of a bone disorder in a child or adolescent should prompt investigation for tyrosinaemia type I.

## Metabolic Derangement

Tyrosinaemia type I is due to a deficiency of fumarylacetoacetase (Fig. 15.1, enzyme 4). The enzyme block results in accumulation of fumarylacetoacetate and possibly maleylacetoacetate, which are alkylating agents thought to cause the hepatorenal damage. These metabolites are the precursors of succinylacetoacetate and succinylacetone, which accumulates in the plasma and urine of the patients. Succinylacetone is a potent inhibitor of δ-aminolevulinate dehydratase. This results in elevation of δ-aminolevulinate (δ-ALA) and may explain the porphyria-like symptoms that occur in some patients. Other intermediates located between tyrosine and the enzyme defect also accumulate, namely 4-hydroxyphenylpyruvate and its derivative, 4-hydroxyphenyllactate.

## Diagnostic Tests

The hallmark of the diagnosis is demonstration of elevated levels of succinylacetone (and/or its precursors). The method employed should be sufficiently sensitive to detect urine succinylacetone at a level of 0.4 µmol/l [6] and plasma succinylacetone at a level of 0.1 µmol/l. Some patients with tyrosinaemia type I accumulate only small amounts of succinylacetone, and whenever succinylacetone is only marginally elevated, the diagnosis should be confirmed by assay of fumarylacetoacetase in lymphocytes or fibroblasts or by mutation analysis. The diagnosis of tyrosinaemia should not be based on fumarylacetoacetase assay alone because of a genetic variant in healthy individuals that is responsible for low activity of fumarylacetoacetase in fibroblasts and lymphocytes that is close to the range of activity in affected patients [7]. Enzyme diagnosis in the liver tissue of tyrosinaemia patients may also be misleading because of frequent reversion of the genetic defect, resulting in a mosaic pattern of fumarylacetoacetase in liver tissue [8]. The mechanism behind this frequently observed reversion of the genetic defect in tyrosinaemia is not known, but it is believed to be due to a high mutation rate in tyrosinaemia in combination with a positive selection pressure resulting in clonal expansion of the normally functioning cells.

Apart from the diagnostic tests discussed above, a number of other biochemical abnormalities are present. Serum tyrosine is elevated and, in acutely ill patients, methionine is also elevated. Erythrocyte δ-aminolevulinate-dehydratase activity is most often low, and there is a high excretion of δ-ALA in urine. Urinary excretion of 4-hydroxyphenylpyruvate and 4-hydroxyphenyllactate is often high. If tubulopathy is present, a full Fanconi syndrome (hyperaminoaciduria, glucosuria and hyperphosphaturia) may be present. Biochemical findings of liver cell damage and failure of protein synthesis are present to varying degrees. Vitamin-K-dependent coagulation factors are highly abnormal in acute cases and are also usually abnormal in patients with the chronic form of the disease. γ-Glutamyltransferase may be elevated in patients in whom other liver enzymes are within the reference range. The α-fetoprotein level is often highly elevated in acute patients but may be normal in chronically ill patients.

## Treatment and Prognosis

Tyrosinaemia type I has traditionally been treated with a tyrosine- and phenylalanine-restricted diet and, later, by liver transplantation. In 1992, a new drug was introduced, 2-(2-nitro-4-trifluoro-methylbenzoyl)-1,3-cyclohexanedione (NTBC), which is a potent inhibitor of 4-hydroxyphenylpyruvate dioxygenase (Fig. 15.1) [9]. It has proved to be very effective and is the treatment of choice for tyrosinaemia type I. It will be discussed first, as other forms of therapy were influenced by this new approach.

### NTBC Treatment

The rationale for NTBC treatment is to stop tyrosine degradation at an early stage to prevent production of the downstream metabolites that accumulate in tyrosinaemia type I (Fig. 15.1). The effect was demonstrated in a pilot study of five patients, in whom disappearance of succinylacetone and normalisation of porphyrin metabolism (with a concomitant increase in the levels of tyrosine and phenolic tyrosine metabolites) was found [9]. During the study period of 6–8 months, there was a marked improvement of the clinical condition of the patients, of whom one was an infant with acute liver failure. There was normalisation of prothrombin time, and α-fetoprotein concentrations decreased, except in one 7-year-old patient, in whom liver transplantation for hepatocellular carcinoma was required. The tyrosine level was kept below 500 µmol/l by a phenylalanine- and tyrosine-restricted diet in order to prevent the known adverse effects of a high tyrosine level (see "Hereditary Tyrosinaemia Type II"). Since the initial study, another 260 patients have been included in a multicentre study conducted from Gothenburg, and the drug has become commercially available. Adverse events have been few. Transient thrombocytopenia and neutropenia have been reported in four cases, and transient eye symptoms (burning/photophobia/corneal erosion/corneal clouding) have been reported in 13 cases [10].

NTBC is now recommended in a dose of 1 mg/kg body weight per day. Individual dose adjustment is then based on the biochemical response and NTBC concentration. Results from the NTBC study confirm the initial finding of a marked effect on the disease course, especially in patients with acute tyrosinaemia [10]. The acute illness has resolved in 90% of the infants, and the need for urgent liver transplantation has been reduced accordingly. In non-responders, there is increasing jaundice and a persistent coagulopathy, with a continuous need for plasma transfusion. NTBC has also been shown to be effective in the treatment of porphyria-like symptoms [1] and in the treatment of long-standing renal tubular dysfunction [11]. However, the real long-term value of NTBC treatment has yet to be determined, particularly with respect to the risk of developing liver cancer. The international NTBC study has now been going for 6 years, and data indicating a decreased risk for early development of hepatocellular carcinoma in patients in whom NTBC treatment is started at an early age have emerged. There are now more than 100 patients aged 2–8 years who have started NTBC treatment before 2 years of age and have been on treatment for more than a year, and no case of hepatocellular carcinoma has been recognised after 1 year of treatment [10]. However, liver cancer was recognised at the time of diagnosis in a 1-year-old girl and after a few months of treatment in an infant who was given NTBC at 5 months of age [12]. These cases emphasise that there is no safe age with respect to occurrence of hepatocellular carcinoma in tyrosinaemia, although the risk increases considerably with age. It should be noted that the risk for hepatocellular carcinoma at the start of NTBC treatment (and for an unknown period of time thereafter) must be considered to be of the same magnitude as in patients receiving diet alone, especially for older patients. This is in accordance with the results of the NTBC study, where hepatocellular carcinoma was recognised 0.5–4 years after start of NTBC treatment in 11 out of 64 patients who started NTBC treatment after 2 years of age (range: 2–21 years of age). In all patients with hepatocallular carcinoma, there was either (1) an increasing concentration of α-fetoprotein or (2) a lack of reduction in α-fetoprotein concentration in response to NTBC. It is obvious that patients with a late diagnosis and advanced disease have a considerable risk of developing liver cancer, so the decision of whether or not to transplant (and, if transplantation is chosen, the timing of the procedure) is crucial in these cases, as discussed below.

### Liver Transplantation

Liver transplantation for tyrosinaemia has been performed for a decade [13, 14]. Since the recipient retains the genotype of the donor, the patient no longer has the enzyme defect in liver tissue. The blood biochemical abnormalities are corrected, as are the renal tubular defects. Transplanted patients do not need dietary treatment. However, increased urinary excretion of succinylacetone, presumably originating from the kidneys, persists. The long-term prognosis of kidney function after liver transplantation remains unclear.

The immediate prognosis of tyrosinaemia patients after liver transplantation is related to the immunosuppressive treatment and the procedure itself. Current results for liver transplantation in children indicate that the 1-year survival may be as high as 90%, with the 5- to 8-year survival rate ranging from 60% to 80% [15]. NTBC treatment has greatly reduced the need for urgent liver transplantation due to acute liver failure or end-stage liver disease. The current indications for liver transplantation include only acute patients in whom there is no clinical response to NTBC and patients in whom there is suspicion of hepatocellular carcinoma. In our experience, frequent monitoring of α-fetoprotein and imaging studies are useful for the detection of early lesions curable by liver transplantation. However, it is known that hepatocellular carcinoma is not always accompanied by an increase in α-fetoprotein concentration, and liver imaging might not be informative in all cases. There will also be patients with rapid development of malignant growth in whom the malignancy will be detected too late to be cured by liver transplantation. These cases seem to be few, but we need to continuously follow the outcome of a strategy (as outlined above) in order to get reliable risk estimates to compare with the risk encountered in liver transplantation.

### Dietary Treatment

Dietary treatment with restriction of tyrosine and phenylalanine intake has been employed in tyrosinaemia type I for 30 years [16]. Although dietary treatment may relieve the acute symptoms of tyrosinaemia and may completely resolve the renal tubular dysfunction, the prognosis of patients receiving dietary treatment remains poor [17], and dietary treatment is now mainly used together with NTBC treatment. When diet is given alone, the aim is to reduce phenylalanine and tyrosine intake as much as possible whereas, in NTBC-treated patients, the aim of the diet is to keep the plasma tyrosine level at least below 500 μmol/l. In both cases, the principle of the diet is the same: the protein requirement for normal growth is provided by a protein hydrolysate or amino acid mixture free of phenylalanine and tyrosine, together with a limited intake of natural protein.

### Supportive Treatment

In the acutely ill patient, supportive treatment is essential. The patient is often depleted of potassium and phosphate. Clotting factors, albumin, calcium, phosphate, electrolytes and acid/base balance should be closely monitored and corrected as necessary. Tyrosine/phenylalanine intake should be kept to a minimum during acute decompensation. Addition of vitamin D, preferably 1,25-hydroxy vitamin $D_3$ or an analogue, may be required to heal the rickets. Infections should be treated intensively.

## Genetics

Hereditary tyrosinaemia type I is inherited as an autosomal-recessive trait. The gene for fumarylacetoacetase is located at 15q 23–25. More than 20 mutations causing tyrosinaemia have been reported [18–21], and the most common mutation, IVS12 + 5(g-a), is only found in about 25% of the alleles, except in the French-Canadian population, where this mutation is found in 86% of the alleles [22]. There is no clear-cut genotype–phenotype correlation, although some of the mutations seem to predispose for acute forms and others for more chronic forms of tyrosinaemia [20]. When the mutation(s) is known, gene analysis is useful for prenatal diagnosis. Prenatal diagnosis is also available by determination of succinylacetone in amniotic-fluid supernatant [6] or by determination of fumarylacetoacetase in cultured amniotic-fluid cells or in chorionic villus material [23]. In families in which either parent has a compound genotype for tyrosinaemia and a genetic variant ("pseudo-deficiency"), prenatal diagnosis by enzyme determination may not be feasible. Before embarking on enzyme-activity-based prenatal diagnosis, the enzyme activity of both parents should be determined to decide if either carries the genetic variant.

## Hereditary Tyrosinaemia Type II

### Clinical Presentation

Hereditary tyrosinaemia type II is characterised by eye lesions (in about 75% of the cases), skin disease (80%) or neurological complications (60%), or any combination of these [24]. The disorder usually presents in infancy but may become manifest at any age.

Eye symptoms are often the presenting problem and may start in the first months of life [25]. The patient usually has photophobia and lacrimation, and intense burning pain is common. The conjunctivae are inflamed and, on split-lamp examination, herpetic-like corneal ulcerations are found. However, in contrast to herpetic ulcers, which are usually unilateral, the lesions in tyrosinaemia type II are bilateral. The lesions stain poorly with fluorescein. Neovascularisation may be prominent. If untreated, serious damage may occur, including corneal scarring, visual impairment, nystagmus and glaucoma.

The skin lesions are limited to the palms and soles and especially affect the pressure areas [26]. They begin as blisters or erosions with crusts and become hyperkeratotic with an erythematous rim. The lesions are painful and may range in diameter from 2 mm to 3 cm.

The neurological complications are highly variable. Some patients are normal, whilst others have defects of fine co-ordination and language. The problems may also be more severe, including microcephaly, self-mutilation and gross retardation. As the eye and/or skin problems may be the only symptoms, any patient with bilateral keratitis (particularly if it does not respond to routine treatment) or hyperkeratotic skin lesions of palms and soles should be investigated for tyrosinaemia type II. It should be noted that the diagnosis of tyrosinaemia type II has only been confirmed by enzymatic and/or molecular genetic analysis in a minority of cases, and it is possible that some of these patients actually are suffering from tyrosinaemia type III.

### Metabolic Derangement

Tyrosinaemia type II is due to a defect of hepatic cytosol tyrosine aminotransferase (Fig. 15.1, enzyme 1). As a result of the metabolic block, the tyrosine concentration in serum and cerebrospinal fluid is markedly elevated. The accompanying increased production of the phenolic acids 4-hydroxyphenyl-pyruvate, -lactate and -acetate may be a consequence of the direct deamination of tyrosine in the kidneys or of the metabolism of tyrosine by mitochondrial aminotransferase (Fig. 15.1). The eye and skin damage probably results from the intracellular formation of tyrosine crystals as concentrations exceed saturation. The crystals may resolve when plasma tyrosine concentrations fall. The aetiology of the neurological problems is not known but is probably related to the high tyrosine concentration.

### Diagnostic Tests

A plasma tyrosine concentration above 1200 µmol/l can be regarded as diagnostic in this disorder. When the tyrosinaemia is less pronounced, a diagnosis of tyrosinaemia type III might be considered (see "Hereditary Tyrosinaemia Type III"). Urinary excretion of the phenolic acids 4-hydroxyphenyl-pyruvate, -lactate and -acetate is highly elevated, and N-acetyltyrosine and 4-tyramine are also increased. Enzyme determination requires a liver biopsy and is generally not considered justified. However, until diagnostic mutation analysis becomes available, this is the only way to differentiate between tyrosinaemia types II and III in cases with moderately increased plasma tyrosine concentrations. Irrespective of the primary cause of the tyrosinaemia, there should be a rapid resolution of the eye and skin symptoms on treatment (see below).

### Treatment and Prognosis

A diet with restriction of phenylalanine and tyrosine intake alleviates the eye disorder within a week and alleviates the skin lesions after a few months. If the patient has general symptoms, such as failure to thrive, improvement is evident within days of treatment. Traditionally, the diet is based on a tyrosine/phenylalanine-free amino acid mixture. The intake of tyrosine and phenylalanine is adjusted to allow for appropriate growth. A diet based on low protein intake (1.5 g/kg body weight) has also proved successful and is easier to manage [26]. Generally, there are no eye and skin lesions at tyrosine levels less than 800 µmol/l. Neurological development has also been satisfactory with this treatment [27]. However, since the basis for the neurological damage is not understood, a safe tyrosine level cannot be fixed unambiguously and may have to be decided on an individual basis. As the impact of hypertyrosinaemia on the developing brain is not known, strict dietary control may be indicated in pregnancy of women with tyrosinaemia type II [28].

### Genetics

Tyrosinaemia type II is of autosomal-recessive inheritance. The gene is located at 16q22.1-q22.3. At present, twelve different mutations have been identified in the tyrosine-aminotransferase gene [29, 30].

## Hereditary Tyrosinaemia Type III

### Clinical Presentation

There are only a few cases of hereditary tyrosinaemia type III known, and the clinical spectrum has not been defined. The patients described to date have presented with various neurological symptoms or have been detected by the finding of a high tyrosine concentration on neonatal screening [31–35]. There are no signs of liver disease. Eye and skin lesions have not been reported but, since eye symptoms may occur at the tyrosine levels found in this disorder, it is reasonable to be aware of this possibility.

### Metabolic Derangement

Tyrosinaemia type III is due to deficiency of 4-hydroxyphenylpyruvate dioxygenase (Fig. 15.1, enzyme 2), which occurs in liver and kidney. As a result of the enzyme block, there is an increased plasma tyrosine concentration and increased excretion of 4-hydroxyphenylpyruvate and its derivatives 4-hydroxyphenyllactate and 4-hydroxyphenylacetate in urine. The aetiology of the neurological symptoms is not known but may be related to the high tyrosine concentration (see "Hereditary Tyrosinaemia Type II").

### Diagnostic Tests

A plasma tyrosine concentration of 500–1200 µmol/l is compatible with a diagnosis of tyrosinaemia type III. An elevated excretion of 4-hydroxyphenyl-pyruvate, -lactate and -acetate in urine accompanies the increased plasma tyrosine concentration. Diagnosis is confirmed by enzyme determination in liver or kidney biopsy specimens.

### Treatment and Prognosis

It is not known if dietary treatment with a phenylalanine- and tyrosine- restricted diet is actually needed in this condition. However, since it is possible that the mental retardation and neurological symptoms are caused by the increased tyrosine concentration, it seems reasonable to reduce the tyrosine level. In tyrosinaemia type II, neurological development has been reported to be satisfactory at tyrosine levels less than 800 µmol/l [27]. However, a safe tyrosine level cannot be fixed and, for NTBC-treated patients (in whom the enzyme deficient in tyrosinaemia type III is inhibited; Fig. 15.1), we recommend a tyrosine level less than 500 µmol/l.

### Genetics

Tyrosinaemia type III is of autosomal-recessive inheritance. The gene has been localised to 12q24-qter [36], and the gene has been characterised [37]. Mutations have not been published.

## Hawkinsinuria

### Clinical Presentation

This rare condition, which has only been described in four families [38–40] is characterised by failure to thrive and metabolic acidosis confined to infancy. After the first year of life the condition is asymptomatic. Early weaning from breastfeeding seems to precipitate the disease and the condition may be asymptomatic in breastfed children.

### Metabolic Derangement

The abnormal metabolites produced in hawkinsinuria (hawkinsin: 2-cysteinyl-1,4-dihydroxycyclohexenylacetate and 4-hydroxycyclohexylacetate) are thought to derive from an incomplete conversion of 4-hydroxyphenylpyruvate to homogentisate caused by a defect of 4-hydroxyphenylpyruvate dioxygenase (Fig. 15.1, enzyme 2). Hawkinsin is thought to be the product of a reaction of an epoxide intermediate with glutathione, which may be depleted. The acidosis is thought to be due to 5-oxoprolinuria, which is secondary to glutathione depletion.

### Diagnostic Tests

Identification of urine hawkinsin or 4-hydroxycyclohexylacetate by gas chromatography–mass spectrometry is diagnostic [40]. Hawkinsin is a ninhydrin-positive compound, which appears between urea and threonine in ion-exchange chromatography of urine amino acids [41]. Increased excretion of 4-hydroxycyclohexylacetate is revealed in analysis of organic acids in urine. In addition to hawkinsinuria, moderate tyrosinaemia, increased urine 4-hydroxyphenylpyruvate and 4-hydroxyphenyllacetate, metabolic acidosis and 5-oxoprolinuria are seen during infancy; 4-hydroxycyclohexylacetate appears with time and is not detected during infancy.

### Treatment and Prognosis

Return to breastfeeding or a diet restricted in tyrosine and phenylalanine may be required in infancy. The condition is asymptomatic after the first year of life and symptomatic infants have developed normally.

### Genetics

Unlike most other inborn errors of metabolism, hawkinsinuria is of autosomal dominant inheritance. It is believed that one deficient allele of 4-hydroxyphenylpyruvate dioxygenase (Fig. 15.1, enzyme 2) is responsible for the production of the abnormal metabolite of 4-hydroxyphenylpyruvate. Neither the enzymatic defect, nor molecular genetics of hawkinsinuria have been studied.

## Alkaptonuria

### Clinical Presentation

Alkaptonuria may be detected in infancy by the darkening of urine when exposed to air. However, clinical symptoms first appear late in adulthood. A greyish coloration (ochre on microscopic examination, thus the name ochronosis), first seen in the sclera and cartilage of the ears, usually appears at approximately 30 years of age. Later, dark coloration of the skin (particularly over the nose, cheeks and in the axillar and pubic areas) may become evident. In elderly patients, the pigmentation seen at operation or post-mortem is striking in cartilage and fibrous tissues and, to a lesser degree, in endothelium and various organs [37, 38]. The arthritis following the ochronosis may occur at an earlier age and may be more severe in males than in females [44]. Symptoms usually start in the large weight-bearing joints – the hip and knee joints – and, occasionally, in the shoulders. Acute periods of inflammation may resemble rheumatoid arthritis. The arthritis can be severe and can lead to marked limitation of movement or, in later years, to a completely bedridden existence [45, 46]. Ankylosis in the lumbosacral region is common. The roentgenographic appearance of the lumbar spine may be pathognomonic, with degeneration and calcification of the intervertebral discs and fusion of the vertebral bodies [47]. Heart disease, particularly mitral and aortic valvulitis is more common in alkaptonurics than in the general population [48]. Alkaptonuria was the first disease to be interpreted as an inborn error of metabolism in 1902 by Garrod [49].

### Metabolic Derangement

The disorder is due to a defect of the enzyme homogentisate dioxygenase (Fig. 15.1, enzyme 3), which exists primarily in the liver and kidneys. Due to the enzyme block, there is accumulation of homogentisate and its oxidised derivative, benzoquinone acetic acid, the putative toxic metabolite and immediate precursor to the dark pigment, which is deposited in various tissues. The relationship between the pigment deposits and the arthritis is not known. It has been proposed that the pigment deposit may act as a chemical irritant [50] and that inhibition of the enzymes involved in cartilage metabolism by the ochronotic pigment, homogentisate or benzoquinone acetic acid might be part of the pathogenic mechanism [51, 52].

## Diagnostic Tests

Alkalisation of urine from alkaptonuric patients results in immediate dark brown coloration of the urine. Excessive homogentisate also results in a positive test for urinary reducing substances. A positive identification is obtained by gas-chromatography–mass-spectrometry-based organic-acid screening methods, which may also be useful for quantification. Homogentisate may also be quantified by HPLC [53] and by specific enzymatic methods [54].

## Treatment and Prognosis

No definite treatment has been reported to be successful in reducing the accumulation of homogentisic acid or preventing the late effects of the disorder. Dietary restriction of phenylalanine and tyrosine intake reduces homogentisate formation, but compliance is a major problem, since symptoms appear late in adult life [50]. Ascorbic acid prevents the binding of $^{14}$C-homogentisic acid to connective tissue in rats [51] and reduces the excretion of benzoquinone acetic acid in urine [52], but no long-term experience with ascorbic acid supplementation in alkaptonuric patients exits. Administration of the drug NTBC (see "Hereditary Tyrosinaemia Type I") most likely would prevent homogentisate formation. Complete inhibition of 4-hydroxyphenylpyruvate dioxygenase would require dietary adjustment and give problems with compliance. However, if NTBC could be dosed to give a substantial reduction in homogentisate production and an acceptable increase in tyrosine level, NTBC may become an option for alkaptonuric patients.

## Genetics

Alkaptonuria is an autosomal-recessive disease. The gene for homogentisate oxidase has been mapped to chromosome 3q2, and several mutations have been identified [58, 59].

*Acknowledgement.* We acknowledge the contributions of James Leonard and Peter Clayton to the previous edition of this chapter, on which the present chapter is based.

## References

1. Gibbs TC, Payan J, Brett EM, Lindstedt S, Holme E, Clayton PT (1993) Peripheral neuropathy as the presenting feature of tyrosinaemia type I and effectively treated with an inhibitor of 4-hydroxyphenylpyruvate dioxygenase. J Neurol Neurosurg Psychiatr 56:1129-1132
2. Mitchell G, Larochelle J, Lambert M, Michaud J, Grenier A, Ogier H, Gauthier M, Lacroix J, Vanasse M, Larbrisseau A, Paradis K, Weber A, Lefevre Y, Melancon S, Daillaire L (1990) Neurologic crisis in hereditary tyrosinemia. N Engl J Med 322:432-437
3. Macvicar D, Dicks-Mireaux C, Leonard JV, Wight DG (1990) Hepatic imaging with computed tomography of chronic tyrosin-aemia type I. Br J Radiol 63:605-608
4. Kvittingen EA, Talseth T, Halvorsen S, Jakobs C, Hovig T, Flatmark A (1991) Renal failure in adult patients with hereditary tyrosinaemia type I. J Inherit Metab Dis 14:53-62
5. Weinberg AG, Mize CE, Worthen HG (1976) The occurrence of hepatoma in the chronic form of hereditary tyrosinemia. J Pediatr 88:434-438
6. Jakobs C, Dorland L, Wikkerink B, Kok RM, de Jong APJM, Wadman SK (1988) Stable isotope dilution analysis of succinylacetone using electron capture negative ion mass fragmentography: an accurate approach to the pre- and neonatal diagnosis of hereditary tyrosinemia type I. Clin Chim Acta 223:223-232
7. Kvittingen EA (1991) Tyrosinaemia type I - an update. J Inherit Metab Dis 14:554-562
8. Kvittingen EA, Rootwelt H, Brandtzaeg P, Bergan A, Berger R (1993) Hereditary tyrosinemia type I. Self-induced correction of the fumarylacetoacetase defect. J Clin Invest 91:1816-1821
9. Lindstedt S, Holme E, Lock EA, Hjalmarson O, Strandvik B (1992) Treatment of hereditary tyrosinaemia type I by inhibition of 4-hydroxyphenylpyruvate dioxygenase. Lancet 340:813-817
10. Holme E, Lindstedt S (1998) Tyrosinaemia type I and NTBC (2-(2-nitro-4-trifluoromethylbenzoyl)-1,3-cyclohexanedione). J Inherit Metab Dis 21:507-517
11. Pronicka E, Rowinska E, Bentkowski Z, Zawadski J, Holme E, Lindstedt S (1996) Treatment of two children with hereditary tyrosinaemia type I and long-standing renal disease with a 4-hydroxyphenylpyruvate dioxygenase (NTBC). J Inherit Metab Dis 19:234-238
12. Dionisi-Vici C, Boglino C, Marcellini M, De Sio L, Inserra A, Cotugno G, Sabetta G, Donfrancesco A (1997) Tyrosinemia type I with early metastatic hepatocellular carcinoma: combined treatment with NTBC, chemotherapy and surgical mass removal. J Inherit Metab Dis 20 [Suppl 1]:15
13. Mieles LA, Esquivel CO, van Thiel DH, Koneru B, Makowka L, Tzakis AG, Starzl TE (1990) Liver transplantation for tyrosinemia. A review of 10 cases from the University of Pittsburgh. Dig Dis Sci 35:153-157
14. Burdelski M, Rodeck B, Latta A, Brodehl J, Ringe B, Pichlmayr R (1991) Treatment of inherited metabolic disorders by liver transplantation. J Inherit Metab Dis 14:604-618
15. Kelly DA (1998) Current results and evolving indications for liver transplantation in children. J Pediatr Gastroenterol Nutr 27:214-221
16. Halvorsen S, Gjessing LR (1964) Studies on tyrosinosis: 1. Effect of low-tyrosine and low-phenylalanine diet. Br Med J 2:1171-1173
17. van Spronsen FJ, Thomasse Y, Smit GP, Leonard JV, Clayton PT, Fidler V, Berger R, Heymans HS (1994) Hereditary tyrosinemia type I: a new clinical classification with difference in prognosis on dietary treatment. Hepatology 20:1187-1191
18. Grompe M, Al-Dhalimy M (1993) Mutations of the fumarylacetoacetate hydrolase gene in four patients with tyrosinemia type I. Hum Mutat 2:85-93
19. Labelle Y, Phaneuf D, Leclerc B, Tanguay RM (1993) Characterization of the human fumarylacetoacetate hydrolase gene and identification of a missense mutation abolishing enzymatic activity. Hum Mol Genet 2:941-946
20. Rootwelt H, Hoie K, Berger R, Kvittingen EA (1996) Fumarylacetase mutations in tyrosinaemia type I. Hum Mutat 7:239-243
21. Bergman AJ, van den Berg IE, Brink W, Poll-The BT, Ploos van Amstel JK, Berger R (1998) Spectrum of mutations in the fumarylacetoacetate hydrolase gene of tyrosinemia type 1 patients in northwestern Europe and Mediterranean countries. Hum Mutat 12:19-26

22. Grompe M, St-Louis M, Demers SI, al-Dhalimy M, Leclerc B, Tanguay RM (1994) A single mutation of the fumarylacetoacetate hydrolase gene in French Canadians with hereditary tyrosinemia type I. N Engl J Med 331:353-357
23. Kvittingen EA, Brodtkorb E (1986) The pre- and post-natal diagnosis of tyrosinemia type I and the detection of the carrier state by assay of fumarylacetoacetase. Scand J Clin Lab Invest 46 [Suppl 184]:35-40
24. Buist NRM, Kennaway NG, Fellman JH (1985) Tyrosinaemia type II. In: Bickel H, Wachtel U (eds) Inherited diseases of amino-acid metabolism. Thieme, Stuttgart, pp 203-235
25. Heidemann DG, Dunn SP, Bawle EV, Shepherd DM (1989) Early diagnosis of tyrosinemia type II. Am J Ophthalmol 107:559-560
26. Paige DG, Clayton P, Bowron A, Harper JI (1992) Richner-Hanhart syndrome (oculocutaneous tyrosinaemia type II). J R Soc Med 85:759-760
27. Barr DGD, Kirk JM, Laing SC (1991) Outcome of tyrosinaemia type II. Arch Dis Child 66:1249-1250
28. Francis DEM, Kirby DM, Thompson GN (1992) Maternal tyrosinaemia type II: management and successful outcome. Eur J Pediatr 151:196-199
29. Natt E, Kida K, Odievre M, Di Rocco M, Scherer G (1992) Point mutations in the tyrosine aminotransferase gene in tyrosinemia type II. Proc Natl Acad Sci USA 89:9297-9301
30. Huhn R, Stoermer H, Klingele B, Bausch E, Fois A, Farnetani M, Di Rocco M, Boue J, Kirk JM, Coleman R, Scherer G (1998) Novel and recurrent tyrosine aminotransferase gene mutations in tyrosinemia type II. Hum Genet 102:305-313
31. Endo F, Kitano A, Uehara I, Nagata N, Matsuda I, Shinka T, Kuhara T, Matsumoto I (1983) Four-hydroxyphenylpyruvic acid oxidase deficiency with normal fumarylacetoacetase: a new variant form of hereditary hypertyrosinemia. Pediatr Res 17:92-96
32. Giardini O, Cantani A, Kennaway NG, D'Eufemia P (1983) Chronic tyrosinemia associated with 4-hydroxyphenylpyruvate dioxygenase deficiency with acute intermittent ataxia and without visceral and bone involvement. Pediatr Res 17:25-29
33. Preece MA, Rylance GW, MacDonald A, Green A, Gray RGF (1996) A new case of tyrosinaemia type III detected by neonatal screening. J Inherit Metab Dis 19 [Suppl 1]:32
34. Cerone R, Holme E, Schiaffino MC, Caruso U, Maritano L, Romano C (1997) Tyrosinemia type III: diagnosis and ten-year follow-up. Acta Paediatr 86:1013-1015
35. Standing SJ, Dunger D, Ruetschi U, Holme E (1998) Tyrosinaemia type III detected by neonatal screening. J Inherit Metab Dis 21 [Suppl 2]:25
36. Stenman G, Roijer E, Ruetschi U, Dellsen A, Rymo L, Lindstedt S (1995) Regional assignment of the human 4-hydroxyphenylpyruvate dioxygenase gene (HPD) to 12q24 → qter by fluorescence in situ hybridization. Cytogenet Cell Genet 71:374-376
37. Ruetschi U, Rymo L, Lindstedt S (1997) Human 4-hydroxyphenylpyruvate dioxygenase gene (HPD). Genomics 44:292-299
38. Niederwieser A, Matasovic A, Tippett P, Danks DM (1977) A new sulfur amino acid, named Hawkinsin, identified in a baby with transient tyrosinemia and her mother. Clin Chim Acta 76:345-356
39. Wilcken B, Hammond JW, Howard N et al. (1981) Hawkinsinuria: a dominantly inherited defect of tyrosine metabolism with severe effects in infancy. New Eng J Med 305:865-869.
40. Borden M, Holm J, Leslie J et al. (1992) Hawkinsinuria in two families. Am J Med Genet 44:52-56
41. Nyhan WL (1984) Hawkinsinuria. Nyhan WL (ed) Abnormalities in amino acid Metabolism in Clinical Medicine, Norwalk, CT, Appleton-Century-Crofts, p 187-188
42. Osler W (1904) Ochronosis: the pigmentation of cartilage, sclerotics, and skin in alkaptonuria. Lancet 1:10
43. Lichtenstein L, Kaplan L (1954) Hereditary ochronosis: Pathological changes observed in two necropsied cases. Am J Pathol 30:99
44. Harrold AJ (1956) Alkaptonuric arthritis. J Bone Surg (Br) 38:532
45. O'Brien WWM, La Due BN, Bunim JJJ (1963) Biochemical, patholgical and clinical aspects of alkaptonuria: ochronosis and ochronotic arthropathy. Am J Med 34:813
46. O.Brien WM, Bansfield WG, Sokoloff L (1961) Studies on the pathogenesis of ochronotic arthropathy. Artheritis Rheum 4:137
47. Pomeranz MM, Friedman LJ, Tunick IS (1941) Roentgen findings in alkaptonuric ochronosis. Radiology 37:295
48. Vlay SC, Hartmann AR, Culliford AT (1986) Alkaptonuria and aortic stenosis. Ann Intern Med 104:446
49. Garrod AE (1902) The insidence of alkaptonuria: a study in chemical individuality. Lancet 2:1616
50. Crissy RE, Day AJ (1950) Ochronosis: a case report. J Bone Joint Surg (Am) 32:688
51. Dihlmann W, Greiling H, Kisters R, Stuhlsatz IW (1970) Biochemische und radiologische Untersuchungen zur Pathogenese der Alkaptonurie. Dtsch Med Wochenschr 95:839
52. Murray JC, Lindberg KA, Pinnell SR (1977) In vitro inhibition of chick embryo lysyl hydroxylase by homogentisic acid. A proposed connective tissue defect in alkaptonuria. J Clin Invest 59:1071-1079
53. Bory C, Boulieu R, Chantin C, Mathieu M (1990) Diagnosis of alcaptonuria: Rapid analysis of homogentisic acid by HPLC. Clin Chim Acta 189:7
54. Fernandez-Canon JM, Penalva MA (1997) Spectrophotometric determination of homogentisate using Aspergillus nidulans homogentisate dioxygenase. Anal Biochem 245:218-221
55. de Haas V, Carbasius Weber EC, de Klerk JB, Bakker HD, Smit GP, Huijbers WA, Duran M, Poll-The BT (1998) The success of dietary protein restriction in alkaptonuria patients is age-dependent. J Inherit Metab Dis 21:791-798
56. Lustberg TJ, Schulman JD, Seegmiller JE (1970) Decreased binding of $^{14}C$-homogentisic acid induced by ascorbic acid in connective tissue of rats with experimental alkaptonuria. Nature 228:770
57. Wolff JA, Barshop B, Nyhan WL, Leslie J, Seegmiller JE, Gruber H, Garst M, Winter S, Michals K, Matalon R (1989) Effects of ascorbic acid in alkaptonuria: alterations in benzoquinone acetic acid and an ontogenic effect in infancy. Pediatr Res 26:140-144
58. Beltran-Valero de Bernabe D, Granadino B, Chiarelli I, Porfirio B, Mayatepek E, Aquaron R, Moore MM, Festen JJ, Sanmarti R, Penalva MA, de Cordoba SR (1998) Mutation and polymorphism analysis of the human homogentisate 1, 2-dioxygenase gene in alkaptonuria patients. Am J Hum Genet 62:776-784
59. Gehrig A, Schmidt SR, Muller CR, Srsen S, Srsnova K, Kress W (1997) Molecular defects in alkaptonuria. Cytogenet Cell Genet 76:14-16

CHAPTER 16

## Catabolism of Branched-Chain Amino Acids

The three essential branched-chain amino acids (BCAAs) – leucine, isolecuine, and valine – are initially catabolized by a common pathway. This first involves reversible transamination to 2-oxo (or keto) acids, which occurs mainly in muscle, followed by oxidative decarboxylation into coenzyme A (CoA) derivatives by branched-chain oxo- (or keto-) acid dehydrogenase. The latter enzyme is similar in structure to pyruvate dehydrogenase (Fig. 10.2). Subsequently, the degradative pathways of BCAA diverge. Leucine is catabolized to acetoacetate and acetyl-CoA, which enters the Krebs' cycle. The final step in the catabolism of isoleucine involves cleavage into acetyl-CoA and propionyl-CoA, which also enters the Krebs cycle via conversion into succinyl-CoA. Valine is also ultimately metabolized to propionyl-CoA. Methionine, threonine, fatty acids with odd numbers of carbons, the side chain of cholesterol, and bacterial gut activity also contribute to the formation of propionyl-CoA (Fig. 16.1).

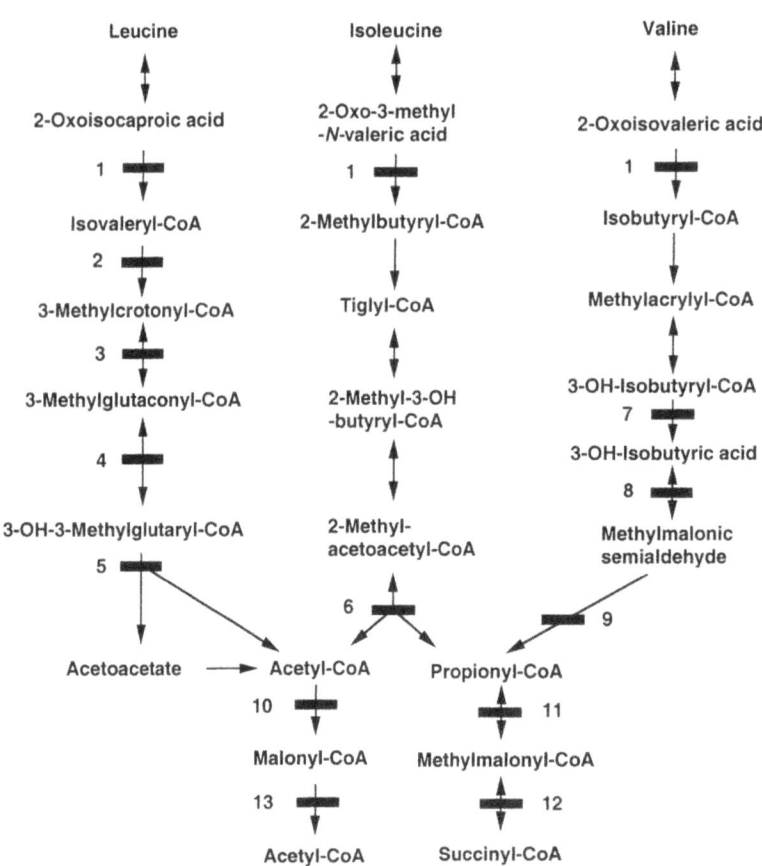

**Fig. 16.1.** Pathways of branched-chain amino acid catabolism. 1, Branched-chain oxo- (or keto-) acid dehydrogenase; 2, isovaleryl-coenzyme A (CoA) dehydrogenase; 3, 3-methylcrotonyl-CoA carboxylase; 4, 3-methylglutaconyl-CoA hydratase; 5, 3-hydroxy-3-methylglutaryl-CoA lyase; 6, 2-methylacetoacetyl-CoA thiolase; 7, 3-hydroxybutyryl-CoA deacylase; 8, 3-hydroxyisobutyric-acid dehydrogenase; 9, methylmalonylsemialdehyde dehydrogenase; 10, acetyl-CoA carboxylase (in cytosol), propionyl-CoA carboxylase (in mitochondria); 11, propionyl-CoA carboxylase; 12, methylmalonyl-CoA mutase; 13, malonyl-CoA decarboxylase. Enzyme defects are indicated by *solid bars*

# CHAPTER 16

# Branched-Chain Organic Acidurias

H. Ogier de Baulny and Jean-Marie Saudubray

CONTENTS

Maple Syrup Urine Disease,
Isovaleric Aciduria, Propionic Aciduria,
Methylmalonic Aciduria ........................ 197
    Clinical Presentation ........................... 197
        Severe Neonatal-Onset Form ................... 197
        Acute Intermittent Late-Onset Form .............. 198
        Chronic, Progressive Forms .................... 199
        Complications ............................ 199
    Metabolic Derangement ....................... 200
    Diagnostic Tests ............................. 201
    Treatment and Prognosis ...................... 202
        Principles of Long-Term Dietary Treatment ........ 202
        Specific Adjustments ........................ 203
        Management of Intercurrent Decompensations ...... 205
    Genetics .................................. 206
3-Methylcrotonyl Glycinuria ...................... 207
Malonic Aciduria................................ 208
3-Methylglutaconic Aciduria Type I ................ 209
3-Hydroxyisobutyric Aciduria..................... 210
References ..................................... 210

Branched-chain organic acidurias are a group of disorders that result from an abnormality of specific enzymes involving the catabolism of branched-chain amino acids (BCAAs). Collectively, maple syrup urine disease (MSUD), isovaleric aciduria (IVA), propionic aciduria (PA) and methylmalonic aciduria (MMA) represent the most commonly encountered abnormal organic acidurias. They can present clinically as a severe neonatal onset form of metabolic distress, an acute, intermittent, late-onset form, or a chronic progressive form presenting as hypotonia, failure to thrive, and developmental delay. 3-Methylcrotonyl glycinuria, 3-methylglutaconic (3-MGC) aciduria, 3-hydroxyisobutyric aciduria and malonic aciduria are other rare diseases involving leucine, isoleucine, and valine catabolism. All these disorders can be diagnosed by identifying acylcarnitines and other organic acid compounds in plasma and urine by gas chromatography–mass spectrometry (MS) or tandem MS-MS.

## Maple Syrup Urine Disease, Isovaleric Aciduria, Propionic Aciduria, Methylmalonic Aciduria

### Clinical Presentation

Children with MSUD, IVA, PA, or MMA have many clinical and biochemical symptoms in common. They can be divided into three schematic presentations:

1. A severe neonatal-onset form with metabolic distress
2. An acute, intermittent, late-onset form
3. A chronic, progressive form presenting as hypotonia, failure to thrive, and developmental delay

In addition, prospective data gathered by some newborn-screening programs and the systematic screening of siblings have demonstrated the relative frequency of asymptomatic forms.

#### Severe Neonatal-Onset Form

**General Presentation**

The general presentation of this form can be summarized as a neurologic distress of the intoxication type with either ketosis or ketoacidosis; it belongs to type I or II in the classification of the neonatal inborn errors of metabolism (Chap. 1). An extremely evocative clinical setting is that of a full-term baby born after a normal pregnancy and delivery who, after an initial symptom-free period, undergoes relentless deterioration that has no apparent cause and that is not responsive to symptomatic therapy. The interval between birth and clinical symptoms may range from hours to weeks, depending on the nature of the defect, and may be linked to the time schedule of the sequential catabolism of carbohydrates, proteins, and fats. Typically, the first signs are poor feeding and drowsiness, after which the newborn sinks into an unexplained progressive coma. It may display cerebral edema with bulging fontanel, arousing suspicion of CNS infection. At a more advanced stage, neurovegetative dysregulation with respiratory distress, hiccups, apneas, bradycardia, and hypothermia may appear. In the comatose state, most patients have characteristic changes in muscle tone and exhibit involuntary movements. Generalized hypertonic episodes with opisthotonus, boxing, or pedaling movements and slow limb elevations, spontaneously or upon stimulation, are frequently observed. Another pattern is that of axial hypotonia and limb hypertonia with large-amplitude tremors and myoclonic jerks, which are often mistaken for convulsions. In contrast, true convulsions occur late and inconsistently. The electroencephalogram may show a burst-suppression pattern. In addition to neurologic signs, patients may present with dehydration and mild hepatomegaly. Once clinical suspicion of an organic aciduria has been aroused, general laboratory investigations (Chap. 1; Table 1.1) and the storage of adequate amounts of plasma, urine, and CSF must be undertaken.

## Specic Signs

### Maple Syrup Urine Disease

Concomitantly with the onset of the symptoms, the patient emits an intensive (sweet, malty, caramel-like) maple-syrup-like odor. In general, neonatal MSUD does not display pronounced abnormalities on routine laboratory tests. Patients are not severely dehydrated, have no metabolic acidosis, no hyperammonemia or only a slight elevation (<130 µmol/l), no blood lactate accumulation, and the blood cell count is normal. The main laboratory abnormality is the presence of 2-oxo acids detected in urine with the 2,4-dinitrophenylhydrazine (DNPH) test.

### Isovaleric Aciduria, Propionic Aciduria and Methylmalonic Aciduria

In contrast, dehydration is a frequent finding in patients with IVA, PA, or MMA, and moderate hepatomegaly may be observed. They have metabolic acidosis (pH < 7.30) with increased anion gap and ketonuria (Acetest 2–3 positive). However, ketoacidosis can be moderate and is often responsive to symptomatic therapy. Hyperammonemia is a constant finding. When the ammonia level is very high (>800 µmol/l), it can induce respiratory alkalosis and can lead to the erroneous diagnosis of an urea-cycle disorder. Moderate hypocalcemia (<1.7 mmol/l) and hyperlactacidemia (3–6 mmol/l) are frequent symptoms. The physician should be wary of attributing marked neurologic dysfunction merely to these findings. Blood glucose can be normal, reduced, or elevated. When hyperglycemia is very high (≥20 mmol/l) and is associated with glucosuria, ketoacidosis, and dehydration, it may mimic neonatal diabetes. Neutropenia, thrombocytopenia, non-regenerative anemia, and pancytopenia are findings frequently confused with sepsis. Among these disorders, IVA is easily recognized by its unpleasant "sweaty feet" odor.

In some cases, the combination of vomiting, abdominal distension, and constipation may suggest gastrointestinal obstruction. Cerebellar hemorrhages have been described in a few neonates, a complication that may be linked to inappropriate correction of acidosis and may explain some poor neurological outcomes [1, 2].

## Acute Intermittent Late-Onset Form

In approximately one third of the patients, the disease presents with late onset after a symptom-free period commonly longer than 1 year, or may even arise in adolescence or adulthood. Recurrent attacks are frequent and, between them, the child may seem entirely normal. Onset of an acute attack may arise during catabolic stress, such as during infections or following increased intake of protein-rich foods, but sometimes occurs without overt cause.

## Neurologic Presentation

Recurrent attacks of either coma or lethargy with ataxia are the main presentations of these acute late-onset forms. The most frequent variety of coma is that presenting with ketoacidosis. Exceptionally, ketosis may be absent.

Hypoglycemia is frequent in patients with MSUD while, in other disorders, blood glucose levels are low, normal, or high. Mild hyperammonemia can be present in IVA, PA, and MMA patients. Although most recurrent comas are not accompanied by focal neurologic signs, some patients may present with acute hemiplegia, hemianopsia, or symptoms and signs of cerebral edema mimicking encephalitis, a cerebrovascular accident, or a cerebral tumor.

These acute neurologic manifestations have frequently been preceded by other premonitory symptoms that had been missed or misdiagnosed. They include acute ataxia, unexplained episodes of dehydration, persistent and selective anorexia, chronic vomiting with failure to thrive, hypotonia, and progressive developmental delay.

## Hepatic Forms

Some patients may present with a Reye-syndrome-like illness characterized by onset of coma, cerebral edema, hepatomegaly, liver dysfunction, hyperammonemia, and even macro- or microvesicular fatty infiltration of the liver. These observations stress the importance of complete metabolic investigations in such situations.

## Hematological and Immunological Forms

Severe hematologic manifestations are frequent, mostly concomitant with ketoacidosis and coma, sometimes as the presenting problem. Neutropenia is regularly observed in both neonatal and late-onset forms of IVA, PA, and MMA. Thrombocytopenia occurs only in infancy, and anemia occurs only in the neonatal period. Various cellular and humoral immunological abnormalities have been described in patients presenting with recurrent infections, leading to erroneous diagnosis and management [3-5].

### Chronic, Progressive Forms

### Digestive Presentation

Persistent anorexia, chronic vomiting, failure to thrive, and osteoporosis (sign of a long-term digestive disturbance) are frequent revealing signs. In infants, this presentation is easily misdiagnosed as gastroesophageal reflux, cow's-milk-protein intolerance, celiac disease, late-onset chronic pyloric stenosis, or hereditary fructose intolerance, particularly if these symptoms start after weaning and diversifying food intake. Later in life, recurrent vomiting with ketosis is easily diagnosed. Most of these patients remain undetected until an acute neurologic crisis with coma leads to the diagnosis [4].

### Chronic Neurologic Presentation

Some patients present with severe hypotonia, muscular weakness, and poor muscle mass and can simulate congenital neurologic disorders or myopathies [4]. Nonspecific developmental delay, progressive psychomotor retardation, dementia, seizures, and movement disorders can also be observed during the course of the disease. However, these rather unspecific symptoms are rarely the only presenting symptoms [6].

### Complications

### Neurologic Complications

*Maple Syrup Urine Disease*
Cerebral edema and its sequelae are well-recognized complications of untreated MSUD in the newborn period. Later, during acute metabolic decompensation, cerebral edema and brain-stem compression may cause unexpected death in the hours following intensive rehydration [7, 8]. Increased intracranial pressure may also develop slowly due to long-standing elevations of BCAA [9]. Additionally, chronic dysmyelination proportional to the average plasma leucine values has been found by brain computed tomography (CT) and magnetic resonance imaging (MRI) in patients on a relaxed treatment. These changes mainly involve the periventricular areas, the deep cerebellar white matter, the dorsal part of the brain stem, the cerebral peduncles, the dorsal limb of the internal capsule, and the basal ganglia. These changes are reversible with appropriate treatment [10, 11].

*Propionic Aciduria and Methylmalonic Aciduria*
An increasing number of patients with MMA and PA present with an acute or progressive extrapyramidal syndrome due to bilateral necrosis of the basal ganglia. These lesions mainly affect the globus pallidus in MMA patients and both the lentiform nuclei and the caudate heads in PA [12]. In addition, MRI studies indicate cerebral atrophy and delayed myelination [13, 14]. For both MSUD and organic acidurias, these dramatic complications are arguments for adequate life-long dietary control, even if the patient is free of symptoms.

### Renal Complications

Tubular acidosis associated with hyperuricemia may be an early and presenting sign in some late-onset MMA patients. This condition partially improves with metabolic control, but chronic renal failure is increasingly recognized in patients older than 10 years [15]. The renal lesion is a tubulo-interstitial nephritis with type-4 tubular acidosis and adaptative changes secondary to the reduced glomerular filtration rate [16]. The course of the disease is usually indolent, but end-stage renal failure may develop, and renal transplantation by the end of the second decade is likely to be necessary in many patients [17, 18]. If this nephritis is the complication of a chronic glomerular hyperfiltration secondary to excessive MMA excretion, prevention of renal injury requires strict metabolic control.

### Skin Disorders

Frequently, large, superficial desquamation, alopecia, and corneal ulcerations may develop in the course of late and severe MSUD, PA, or MMA decompensations. It has been described as a staphylococcal scalded-skin syndrome with epidermolysis or as acrodermatitis-enteropathica-like syndrome. In many cases, these complications occur together with diarrhea and can be ascribed to acute protein malnutrition, especially to L-isoleucine deficiency [19-22].

### Pancreatitis

Pancreatitis has been described in patients affected with MSUD and other organic acidurias. It has been the revealing symptom in two patients with late-onset forms of IVA. In other cases, pancreatic involvement in the course of acute episodes has been associated with hyperglycemia and hypocalcemia. The increase of serum lipase and amylase is variable. Enlargement and even calcification or hemorrhage of the gland can be visualized by abdominal ultrasonography [23].

### Cardiomyopathy

Acute cardiac failure due to cardiomyopathy, may be responsible for rapid deterioration or death in cases of MMA and PA. This complication may develop as part of the presenting illness or may occur during an intercurrent metabolic decompensation [24]. The pathogenesis is unclear. Diverse factors, such as carnitine or micronutrient deficiencies or intercurrent viral infections, may play additional roles. Acute energy deprivation due to the propionyl-coenzyme-A (CoA)-oxidation defect is another possibility.

## Metabolic Derangement

### Maple Syrup Urine Disease

This disorder is caused by a deficiency of branched-chain oxo- (or keto-) acid dehydrogenase (BCKD), the second common step in the catabolism of the three BCAAs (Fig. 16.1, enzyme 1). Like the other 2-ketoacid dehydrogenases, BCKD is composed of three components (Fig. 10.2): a decarboxylase (E1), composed of E1$\alpha$ and E1$\beta$ subunits and requiring thiamine pyrophosphate as a coenzyme, a dihydrolipoyl acyltransferase (E2), and a dihydrolipoamide dehydrogenase (E3). A deficiency of any of these components can cause MSUD.

The enzyme defect results in marked increases of the branched-chain 2-oxo (or keto) acids in plasma, urine, and CSF. Owing to the reversibility of the initial transamination step, the BCAAs also accumulate. Reduction of the 2-oxo acids results in the appearance of smaller amounts of the respective 2-hydroxy acids, especially 2-hydroxyisovaleric acid (2-HIVA). Tautomerization of isoleucine results in the formation of allo-isoleucine, which is invariably found in blood in MSUD, owing to its very slow renal excretion. A deficiency of the E3 component produces a syndrome with congenital lactic acidosis, branched-chain 2-oxoaciduria and 2-oxoglutaric aciduria, which sometimes becomes obvious only after protein loading [25, 26].

Among the BCAA metabolites, the leucine/2-oxoisocaproic-acid pair appears to be the most toxic. Always present in approximately equimolar concentrations, they may cause acute brain dysfunction when their plasma concentrations rise above 1 mmol/l. Isoleucine and valine are of lesser clinical significance. Their 2-oxo-acid-to-amino-acid ratios favor the less toxic amino acids, and cerebral symptoms do not occur even when the blood levels of both amino acids are extremely high.

### Isovaleric Aciduria

Isovaleric aciduria is caused by a deficiency of isovaleryl-CoA dehydrogenase (Fig. 16.1, enzyme 2), a mitochondrial flavoprotein which, similar to the acyl-CoA dehydrogenases (Fig. 11.1), transfers electrons to the respiratory chain via the electron-transfer flavoprotein (ETF). IVA is caused by a defect of the isovaleryl-CoA dehydrogenase apoenzyme. Deficiencies of ETF result in multiple acyl-CoA-dehydrogenase deficiencies or glutaric aciduria type II (Chap. 11).

The enzyme defect results in the accumulation of derivatives of isovaleryl-CoA, including: free isovaleric acid, which is usually increased both in plasma and urine (although normal levels have been reported); 3-hydroxyvaleric acid; and N-isovalerylglycine. This conjugate is the major derivative of isovaleryl-CoA, owing to the high affinity of the latter for glycine N-acylase. Conjugation with carnitine (catalyzed by carnitine N-acylase) results in the formation of isovaleryl-carnitine. These two alternative pathways are of major therapeutic interest, because they allow the transformation of highly toxic isovaleric acid into nontoxic byproducts that are rapidly excreted in urine due to their high renal clearance.

### Propionic Aciduria

Propionic aciduria is caused by a deficiency of propionyl-CoA carboxylase (PCC; Fig. 16.1, enzyme 11), a mitochondrial biotin-dependent enzyme composed of $\alpha$ subunits (which bind biotin) and $\beta$ subunits. Therefore, a biotin-responsive form of PA (in addition to the common, biotin-unresponsive form) may theoretically exist, independent of multiple carboxylase deficiency.

PA is characterized by greatly increased concentrations of free propionate in blood and urine. However, this sign may be absent and, in that case, the diagnosis is based upon the presence of multiple organic acid byproducts, among which propionylcarnitine, 3-hydroxypropionate, and methylcitrate are the major diagnostic metabolites. The first is formed by acylation to carnitine, mostly in muscles and liver. The second is formed by either $\beta$- or $\omega$-oxidation of propionyl-CoA. Methylcitrate arises by condensation of propionyl-CoA with oxaloacetate, which is catalyzed by citrate synthase. During ketotic episodes, 3-HIVA is formed by condensation of propionyl-CoA with acetyl-CoA, followed by reduction. Low levels of organic acids derived from a variety of intermediates of the isoleucine catabolic pathway, such as tiglic acid, tiglylglycine,

2-methyl-3-hydroxybutyrate, 3-hydroxybutyrate, propionylglycine, and methylmalonate, can also be found. Due to abnormal biotin metabolism, propionyl-CoA accumulation also occurs in multiple carboxylase deficiency, resulting in defective activity of all biotin-dependent carboxylases (Chap. 24).

### Methylmalonic Aciduria

Methylmalonic aciduria (MMA) is caused by a deficiency of methylmalonyl-CoA mutase (MCM; Fig. 16.1, enzyme 12), a vitamin $B_{12}$-dependent enzyme. Deficient activity of the apoenzyme leads to MMA. Because the apomutase requires adenosylcobalamin (AdoCbl), defects of AdoCbl metabolism lead to variant forms of MMA (Chap. 25).

The deficiency of MCM leads to the accumulation of methylmalonyl-CoA, resulting in greatly increased amounts of methylmalonic acid in plasma and urine. Owing to secondary inhibition of PCC, propionic acid also accumulates, and other propionyl-CoA metabolites, such as propionylcarnitine, 3-hydroxypropionic acid, methylcitrate, and 3-HIVA, are usually also found in urine. Through neonatal screening, some "healthy" patients who have MMA levels 10- to 50-fold higher than normal have been identified, but no metabolites derived from propionyl-CoA have been found in their urine. Other screened subjects have elevated excretion of MMA associated with propionate derivatives. These patients are probably at risk in cases of catabolic stress, even if they have been asymptomatic for several years [27]. Vitamin-$B_{12}$ deficiency must be excluded when excessive urinary MMA is found, particularly in infants who are breast fed by a mother who is either a strict vegetarian or suffers from subclinical pernicious anemia.

### Secondary Metabolic Disturbances

The accumulation of propionyl-CoA and related compounds also results in inhibitory effects on various pathways of intermediary metabolism, in carnitine deficiency, and in the synthesis of abnormal fatty acids. Inhibition of the pyruvate dehydrogenase complex, N-acetyl-glutamate synthetase, and the glycine-cleavage system by propionyl-CoA and inhibition of pyruvate carboxylase by methylmalonyl-CoA may explain some clinical features reported in both PA and MMA, such as hypoglycemia, mild hyperlactacidemia, hyperammonemia, and hyperglycinemia [28]. Moreover, inhibition of Krebs' cycle activity may result in reduced synthesis of ATP.

Owing to the activity of carnitine N-acylase, which can remove accumulating propionyl-CoA from the mitochondria to restore the essential pool of free CoA, patients with PA or MMA have increased levels of acylcarnitines (particularly propionyl-carnitine) in their blood and urine. Thus, a relative carnitine deficiency, which can be corrected by L-carnitine supplements, may appear. L-Carnitine supplements will further increase propionyl-CoA removal and will thereby relieve the intramitochondrial inhibitions caused by propionyl-CoA accumulation.

When propionyl-CoA accumulates, it can compete with acetyl-CoA for conversion into malonyl-CoA by acetyl-CoA carboxylase (Fig. 16.1, enzyme 10), the first, rate-limiting step of de novo fatty acid synthesis. This will result in the synthesis of non-physiological, odd-numbered (instead of even-numbered) long-chain fatty acids. Similarly, accumulating methylmalonyl-CoA will compete with malonyl-CoA for conversion into long-chain fatty acids, resulting in the formation of methyl-branched long-chain fatty acids. The abnormal fatty acids can be incorporated into lipids throughout pre- and postnatal life. In turn, lipolysis during catabolic conditions can release sizable amounts of toxic propionyl-CoA. Thus, measurement of odd-numbered and methylated fatty acids in erythrocytes, for instance, is a potentially useful means of long-term assessment of PA and MMA [29].

It has been estimated by stable isotope studies in PA and MMA that amino acid catabolism accounts for approximately 50% of propionate production, oxidation of odd-numbered fatty acids accounts for 25%, and gut bacterial activity accounts for 25% [30, 31]. Only the amino acid catabolism source gives rise to the simultaneous excretion of urea [32]. Furthermore, propionyl-carnitine does not derive from gut bacterial activity [33].

### Diagnostic Tests

In this group of disorders, the final diagnosis is made by identifying specific abnormal metabolites. The classical means use amino acid chromatography, gas–liquid chromatography and MS (GLC-MS). New technologies to screen BCAA disorders are becoming available, such as proton nuclear magnetic resonance spectroscopy applied to urine samples to detect various organic acids [34] and MS-MS to detect abnormal acylcarnitine or amino acid profiles in blood or plasma [35, 36]. Only MSUD can be diagnosed by using amino acid chromatography alone. IVA, PA and MMA are diagnosed by their specific organic acid profiles, while amino acid chromatography displays nonspecific abnormalities, such as hyperglycinemia and hyperalaninemia. Some patients with PA and MMA may present with pseudo-cystinuria–lysinuria [37]. Whatever the acute or chronic clinical presentation, the diagnosis can be made by sending fresh or frozen urine samples, 5 ml of fresh heparinized whole blood or 1–2 ml of fresh or frozen plasma, and blood sampled on a "Guthrie" card to an experienced laboratory. Specific loading tests are not necessary, even in cases of chronic intermittent forms, which most often display a biochemical pattern sufficiently evocative to allow diagnosis.

Enzymatic studies are useful for diagnostic confirmation, for better delineation of the enzymatic group and, combined with molecular analysis, for determination of phenotype–genotype relationships. In each disease, these studies can be performed in cultured fibroblasts. For rapid diagnosis, fresh peripheral leukocytes can be used, with the possibility of evaluating the response to vitamins in vitamin-responsive disorders.

At the 12th to 14th week of gestation, reliable and fast prenatal diagnosis of IVA, PA, and MMA can be performed through the direct measurement of metabolites in amniotic fluid collected using GLC-MS, stable-isotope-dilution techniques, or MS-MS [38–40]. Direct enzymatic assay can also be performed in fresh or cultured chorionic villi or in cultured amniotic cells. Prenatal diagnosis of MSUD relies on enzymatic assays in cultured amniocytes or in chorionic villi.

## Treatment and Prognosis

Over the past decades, several hundred patients have been treated. Evidence is accumulating that the CNS dysfunction can be prevented by early diagnosis and emergency treatment followed by compliance with the restricted diet. This aspect is important in order to avoid more radical therapeutic approaches, such as liver transplantation and gene therapy, which represent a real hope for patients with the most severe forms of PA and MMA, who have to deal with recurrent life-threatening episodes of metabolic acidosis.

The heterogeneity of clinical manifestations is reflected in the different management strategies. Neonatal-onset forms require early toxin removal. Thereafter, the restricted food pattern essential to limit formation of organic acid byproducts is applied to survivors of the difficult newborn period and to patients affected with the late-onset form. In both, prevention and early treatment of recurrent episodes of metabolic imbalance is crucial. At any age, each metabolic derangement is potentially life threatening, and parents must be taught to recognize early warning signs and have an immediate plan for intervention.

### Principles of Long-Term Dietary Treatment

Long-term dietary treatment is aimed at reducing accumulated toxic metabolites while at the same time maintaining normal development and nutritional status and preventing catabolism. Variability of treatment has been reported. Some patients tolerate normal foods; others need only minimal restriction or can even regulate the diet themselves. However, many need very specific food allowances, implying stringent dietary restrictions that will likely be a life-long necessity.

The treatment involves limiting one or more essential amino acids which, if present in excess, are toxic or precursors of organic acids. This means that protein is highly restricted, which might interfere with normal growth if the diet is not otherwise enriched. Therefore, the amount of protein in relation to total nutrient requirements, energy intake, and the distribution of protein must be carefully planned.

Precise prescriptions are established for the daily intake of amino acids, protein, and energy. The diet is checked for the recommended daily allowance (RDA) and for the estimated safe and adequate daily dietary intakes of minerals and vitamins. In order to prevent dehydration in infants, the osmolality of synthetic or semisynthetic formulas must be estimated.

### Amino Acid Prescriptions

Requirements for BBCAs vary widely from patient to patient and in the same patient, depending on the nature and severity of the disorder, other therapies prescribed (stimulation of an alternate pathway), growth rate, state of health, and feeding difficulties. Individual requirements must be estimated for each child by frequent monitoring of clinical and metabolic status. Balancing between protein malnutrition and metabolic disequilibrium is difficult and needs regular control, especially after an acute intercurrent imbalance or after a change in the diet.

The prescribed amounts of amino acids are provided by natural foods. Infant formula is used in young infants. For toddlers or children, solids are introduced, using specific serving lists. Apart from milk, high-protein foods (eggs, meat, dairy products) are generally avoided, because the lower percentage of amino acids in vegetable protein (as compared with that in animal protein) makes it easier to satisfy the appetites of children.

### Protein Prescriptions

Limitation of essential amino acids to or even below the minimum requirements necessitates the use of synthetic amino acid mixtures that do not contain any of the potentially toxic amino acid(s). Although still controversial, the goal is to supply some additional nitrogen and other essential and nonessential amino acids in order to promote a protein-sparing anabolic effect. Some studies show that the addition of a special amino acid mixture to a severely restricted diet has no effect on growth or metabolic status and that these amino acids are mostly broken down and excreted as urea [32].

However, in MSUD patients, normal growth requires the use of an amino acid mixture. Food companies have developed specific formulas for each disorder.

In theory, an amino acid mixture is added to the natural protein in an amount sufficient to meet the protein RDA for the patient's age. From a practical point of view, this is rarely possible, first because of the bad taste of these synthetic mixtures, and second because a satisfactory nutritional status can be reached with much less protein than the RDA suggests. In patients with high tolerance, it is even possible to restrict protein in the diet without adding amino acid mixtures.

## Energy Prescriptions

Energy requirements vary widely and may be greater than normal to ensure that essential amino acids are not degraded to provide energy or nitrogen for the biosynthesis of nitrogenous metabolites. Reduction of energy intake below the individual's requirements results in a decreased growth rate and a metabolic imbalance. The energy requirement is met through natural foods, special amino acid formulas, and additional fat and carbohydrates from other sources. Distribution of energy intakes from protein, carbohydrates, and lipids should approach the recommended percentages.

## Micronutrient Prescriptions

The diet must be checked for minerals, vitamins, and trace elements. If incomplete, the diet must be supplemented with an appropriate commercial preparations.

## Water Prescriptions

Enough water must be added to prevent dehydration of these patients, who may have a low renal concentrating capacity and may not tolerate hyperosmolar formulas. The appropriate concentration of formula mixtures is approximately 0.7–0.9 kcal/ml, and the measured or calculated osmolarity should be less than 450 mosmol/kg.

## Design of a Low-Protein Diet

The design of a low-protein diet is illustrated by the following guidelines:

1. Calculate the amount of infant formula and/or servings of solid food required to meet the desired intake of the amino acid concerned
2. Calculate the amount of protein provided by each natural food used, subtract it from the protein prescription, and provide the remainder via a special amino acid mixture
3. Calculate the energy provided by natural foods and the amino acid mixture and fulfill the additional energy requirement with carbohydrates, fat, and/or specialized low-protein products
4. Check the other nutrients of the diet, particularly minerals and vitamins
5. Add sufficient water to meet liquid requirements

For young infants, total alimentation is provided by the prescribed formula. Introduction of solid foods must be planned carefully, depending on the infant's appetite and metabolic stability. During this introduction, parents are taught how to use serving lists in order to introduce variety in the diet and promote appetite. Milk intake is progressively reduced, whereas fruits and vegetables are increased. Because of its high biologic value, maintenance of a certain percentage (25%) of natural protein intake as dairy protein is recommended.

## Evaluation of Nutritional Status

This comprises monthly evaluation of length, weight, and head circumference, which should follow growth percentiles appropriate for the patient's age. Nutritional status is also judged by blood cell count, hemoglobin and hematocrit, plasma protein and albumin, iron and ferritin, calcemia, phosphatemia, and alkaline phosphatase. The metabolic and nutritional statuses are both evaluated weekly during the first month of therapy, then every 3–6 months. In patients treated with a low-protein diet without an added amino acid mixture, evaluation of urea excretion is an easy means to evaluate anabolism [32]. Regular assessment of developmental progress provides the opportunity for psychological support, as social and emotional needs are major elements of the overall therapy of the affected child and of the well being of the family (Chap. 4).

## *Specific Adjustments*

## Maple Syrup Urine Disease

*Toxin Removal Procedures*
In order to protect the neonatal brain from permanent damage, the acutely ill newborn needs exogenous toxin removal, because a high-energy enteral or parenteral nutrition alone, is insufficiently effective to rapidly lower plasma leucine levels [41, 42]. Continuous blood exchange transfusion, hemodialysis, and hemofiltration are efficient methods that allow high-energy dietary treatment within hours as soon as the plasma leucine level is reduced to 1 mmol/l or less

[43]. During the recovery interval, the BCAA intake has to be adjusted according to the plasma levels, which are monitored every day until the optimal equilibrium is attained. During this stage, plasma concentrations of isoleucine and valine may decrease too much and become rate limiting for protein synthesis, a situation which requires supplements in doses of 100–200 mg/day.

*Dietary Therapy*
The objective of life-long maintenance therapy is to maintain 2–3 h postprandial plasma BCAA at near-normal concentrations (leucine: 80–200 μmol/l; isoleucine 40–90 μmol/l; valine 200–425 μmol/l). Because leucine is the most toxic precursor, the diet can be based on leucine requirement; isoleucine and valine are provided in proportion. In the classical severe form, the leucine requirement is 300–400 mg/day, which is approximately 50–60% of the leucine intake in the healthy newborn. Minimum isoleucine and valine requirements are approximately 200–250 mg/day. Intakes must frequently be titrated against plasma concentrations. Occasionally, small amounts of free valine and isoleucine must be added to the amounts provided by natural protein, because the tolerance for leucine is lower than for the other two. Under conditions of high leucine and low valine and isoleucine levels, a rapid fall of plasma leucine can be achieved only by combining a reduced leucine intake with a temporary supplement of valine and isoleucine. Apart from considerable interindividual variation, children and adolescents with the classical form of MSUD tolerate about 500–700 mg of leucine per day. Individuals with variant forms tolerate higher amounts, and some do well on a low-protein diet. Nevertheless, constant care is indicated, especially during intercurrent episodes.

*Vitamin Therapy*
Pharmacologic doses of thiamine (5 mg/kg/day) for a minimum of 3 weeks may improve BCAA tolerance in some patients. However, normal leucine tolerance has never been restored, and the degree of response is often difficult to assess [44]. The original thiamine-responsive patient has a normal long-term outcome without obvious cerebral impairment on MRI [8].

*Prognosis*
Patients with MSUD are now expected to survive; they are generally healthy between episodes of metabolic imbalance, and some attend regular schools and have a normal intelligence-quotient score. However, on the whole, the average intellectual performance is far below normal. This intellectual outcome is inversely related to the amount of time after birth that plasma leucine levels remained above 1 mmol/l, and is dependent on the quality of long-term metabolic control [45]. In addition, timely evaluation and intensive treatment of minor illnesses at any age is essential, as late death attributed to recurrence of metabolic crises with infections has occurred [8].

## Isovaleric Aciduria

*Toxin Removal Procedures*
Exogenous toxin removal, such as blood exchange transfusion, may be needed in newborns, who are often in a poor clinical condition, precluding the effective use of alternate pathways. Oral L-glycine (250–600 mg/kg/day) and intravenous L-carnitine (100–400 mg/kg/day) therapies are effective means of treatment. Glycine can be provided as a 100 mg/ml water solution delivered in four to eight separate doses.

*Dietary Therapy*
Goals of nutritional support are to keep the urine free of IVA and 3-hydroxy-IVA. A special amino acid mixture free of leucine is useful if a stringent protein-restricted diet is maintained. In comparison with MSUD, IVA patients supplemented with glycine and carnitine have a high leucine tolerance. During the first year of life, leucine intake can be gradually increased to 800 mg/day, which represents an important provision of natural proteins (>8 g/day). Subsequently, higher amounts can be tested, and the strict low-leucine diet may be replaced by a low-protein diet, which requires less monitoring. Most children can tolerate approximately 20–30 g of protein per day, which is sufficient to assure normal growth and development without amino-acid-mixture supplements.

*Glycine and Carnitine Therapy*
Patients can be cared for with either oral L-carnitine (50–100 mg/kg/day) or oral L-glycine (150–300 mg/kg/day). During the steady state, the need for both supplementations is still controversial, but it can be useful during metabolic crises when toxic acyl-CoA accumulation increases the need for detoxifying agents [46].

*Prognosis*
Once they have passed the neonatal period, patients need careful nutritional support. However, prognosis is better than for any other organic aciduria. Intellectual prognosis depends on early diagnosis and treatment and, subsequently, on long-term compliance.

## Propionic Aciduria and Methylmalonic Aciduria

*Toxin Removal Procedures*
The urinary excretion of propionic acid is negligible, and no alternate urinary pathway is sufficient to

effectively detoxify newborns with PA; therefore, they need exogenous toxin-removal procedures. In contrast, the efficient removal of toxin in MMA takes place via urinary excretion. The high clearance of methylmalonic acid ($22 \pm 9$ ml/min per $1.73$ m$^2$) allows excretion as high as 4–6 mmol/day. Thus, emergency treatment of the MMA newborn mainly comprises rehydration and promotion of anabolism. Simultaneously, most neonatal cases of MMA benefit from rapid toxin removal, such as a blood exchange transfusion, which is successful in ensuring a partial removal of methylmalonic acid accumulated in blood.

*Dietary Therapy*
Special amino acid mixtures for PA and MMA patients are available. They are free of isoleucine, valine, methionine, and threonine. Since valine is one of the more direct precursors of propionyl-CoA, the diet can be based on valine intake; other amino acids are provided in proportion. In the neonatal period during the refeeding phase, valine intake is progressively increased to 220–250 mg/day over a period of 5–7 days, depending on clinical status, weight gain, and biochemical results. Thereafter, the individual child's tolerance should be tested. Subsequently, the valine intake for the following years is quite homogenous – between 350–700 mg/day, which represents approximately 5–10 g natural protein per day. The stringent protein restriction may require additional intake of special amino acid mixtures to prevent protein deficiency. In general, the entire artificial diet supplement must be delivered during a nocturnal gastric feeding. Apart from the natural foods that provide the required amounts of valine, low-protein products may be offered during the day (more for social, psychological, and developmental reasons than for nutritional reasons). This practice prevents chronic malnutrition, catabolism, and prolonged fasting periods and allows a more rapid and effective adaptation in case of intercurrent crisis.

Most patients with a late-onset form are easier to manage. Individual tolerance is quite high, and the diet may be based on the protein intake rather than on the daily valine intake. By the age of 2 years, they can tolerate more than 12 g of natural protein per day, and supplementary amino acid mixtures are no longer necessary. Even though their individual tolerance allows a less rigid protein restriction and leads to lower risk of malnutrition, these patients must be taught to immediately reduce their protein intake during intercurrent illness in order to prevent metabolic imbalance.

*Vitamin Therapy*
Some late-onset forms (and, more rarely, neonatal-onset forms) of MMA are vitamin B$_{12}$-responsive; thus, parenteral vitamin therapy, starting with hydroxy-cobalamin 1000–2000 µg/day for a few days, must be carefully tried. During this period, 24-h urine samples are collected for an organic acid analysis. Vitamin-B$_{12}$ responsiveness leads to a prompt and sustained decrease of propionyl-CoA byproducts. However, as biochemical results may be difficult to assess, they must later be confirmed by in vitro studies. Most of the B$_{12}$-responsive patients need only mild protein restriction or none at all; vitamin B$_{12}$ is either given orally once per day or is administered once a week (1000–2000 µg i.m.). In some cases, i.m. vitamin therapy can be kept in reserve for intercurrent infections.

*Carnitine Therapy*
Chronic oral administration of L-carnitine (100 mg/kg/day) appears to be effective not only in preventing carnitine depletion but also in allowing urinary propionyl-carnitine excretion and then reducing propionate toxicity [33].

*Metronidazole Therapy*
Microbial propionate production can be suppressed by antibiotics. Metronidazole, an antibiotic that inhibits anaerobic colonic flora has been found to be specifically effective in reducing urinary excretion of propionate metabolites by 40% in MMA and PA patients. Long-term metronidazole therapy (at a dose of 10–20 mg/kg once daily for ten consecutive days each month) may be of significant clinical benefit [30, 33]. This alternate administration may prevent the known side effects of the drug, such as leukopenia, peripheral neuropathy, and pseudomembranous colitis.

*Prognosis*
Vitamin B$_{12}$-responsive MMA patients have a mild disease and good outcome. Conversely, both the vitamin-B$_{12}$-unresponsive MMA and the PA patients have severe disease and many encephalopathic episodes, mainly due to intercurrent infections. The early-onset patients have the poorest survival rate. Survivors of both the early- and late-onset forms have many nutritional problems, with poor growth and neurologic sequelae with various degree of developmental delay and neurological impairment [17, 47–50], and many patients present with chronic renal failure.

This hazardous long-term prognosis associated with the high risk of complications raises the question of other therapeutic means, such as liver transplantation, for those patients difficult to manage. However, this procedure carries its own, often severe complications despite its proven metabolic efficacy [18, 51].

## Management of Intercurrent Decompensations

Acute intercurrent episodes are prevented by being aware of those situations that may induce protein catabolism. These include intercurrent infections, immunization, trauma, anesthesia and surgery, and dietary indiscretion. In all cases, the main adaptation comprises a more reduced protein intake.

Some of these situations are predictable and require a planned scheme of dietetic adaptation depending on the basal clinical and metabolic status. The most frequent situations, such as those occurring during intercurrent infections, are unpredictable. The related management is, in fact, the most difficult. Multiple protein reduction, if not necessary, may lead to malnutrition and chronic metabolic imbalance; however, a patient's stress tolerance is so unpredictable that the same child could tolerate an acute febrile infection well and develop a life-threatening coma secondary to an apparently trivial illness. Therefore, parents need guidelines in order to avoid any protein loads and to recognize early situations of impending protein catabolism and metabolic imbalance.

From a practical point of view, parents must have at their disposal a semi-emergency diet in which natural protein intakes are reduced by half and an emergency diet in which natural proteins are suppressed. In both, energy supply is reinforced using carbohydrates and lipids. The composition and modes of administration are regularly re-evaluated to ensure that they are in accordance with dietary requirements based on the patient's age and with the usual diet. Alternatively, solutions based on 80056 powder or on a mixture of glucose polymer and lipids diluted in an oral rehydration solution can be used. For children treated with specific amino acid mixtures, the usual supplements can be added, though one should be aware that they increase osmolarity and that their taste renders nasogastric tube feeding quite unavoidable. Their use is contraindicated in MMA and PA in cases of severe hyperammonemia. At home, the solution is given in small, frequent drinks at intervals of 2–4 h each day and night. In these situations, children receiving a nocturnal feeding can benefit from total nasogastric tube feeding [52, 53]. After 24–48 h, if the child is doing well, the usual diet is resumed within 2 days or 3 days. The glucose/fat diet should not be continued for more than 48 h, because it does not provide adequate nutrition and carries the risk of inducing sustained metabolic imbalance.

In cases of clinical deterioration with anorexia and/or gastric intolerance or if the child is obviously unwell, the patient must be hospitalized to evaluate the clinical status and metabolic imbalance, to search for and treat intercurrent disease, and to halt protein catabolism. Emergency therapy depends on the presence of dehydration, acidosis, ketosis, and hyperammonemia. However, attention must be paid to the fact that large discrepancies between the clinical and biological status may occur. Parents who feel that their child is not well are most often right, even if there is "no cause for worry" after the first investigation, and therapeutic measures should be anticipated until the clinical situation stabilizes.

Most often, intravenous rehydration for 12–24 h would allow some clinical improvement and progressive renutrition with continuous enteral feeding. During this step, it is better to introduce some natural protein in the enteral solution to at least cover the minimal dietary requirements. The energy intakes are supplied with carbohydrates and lipids, applying the same rules as for the treatment of late-onset forms. During this management, regular metabolic evaluation is closely monitored, as the acute decompensation may rapidly deteriorate, requiring adjustment of the therapy. Conversely, if the patient's condition improves quickly, basal feeding should be restored without delay.

In cases of severe decompensation or worsening of the clinical and metabolic statuses, the use of total parenteral nutrition appears to be an effective means to relieve metabolic imbalance and should prevent the need to use extra-renal removal procedures. The decision to use such therapeutic means is an emergency judgement for which no objective criteria are available. Selection of the form of procedure depends ultimately on local resources and experience, keeping in mind that intervention started too late is likely to fail.

## Genetics

### Maple Syrup Urine Disease

MSUD is inherited in an autosomal-recessive mode, with an incidence of 1 in 120,000 to 1 in 500,000. About 75% of those affected suffer from a severe classical form, and the remainder suffers from intermediate or intermittent variants. Over 50 different causal mutations scattered among the three E1α, E1β, and E2 genes give rise to either classical or intermediate clinical phenotypes. Despite the implication of the E1α subunit in thiamine-dependent decarboxylation, the E2 subunit is mutated in the thiamine-responsive patient. This vitamin responsiveness is still poorly understood. Gene therapy is available in experimental systems [25, 26].

### Isovaleric Aciduria

IVA is an autosomal-recessive inherited disorder. With respect to the two clinical phenotypes (neonatal-onset and the intermittent late-onset forms) no interallelic complementation has been defined, and various point mutations and deletions of the gene (chromosome 15) have been described in both clinical phenotypes [54].

## Propionic Aciduria

PA is an autosomal recessive disorder with an incidence of less than 1 in 100,000. Irrespective the clinical phenotype, severe reduction (but not complete absence) of PCC activity (1–5%) has been found in cultured fibroblasts. Two distinct genotypic forms are distinguished by cell complementation: *pcc*A, resulting from defects in the α (PCCA) gene, and *pcc*B, resulting from defects in the β (PCCB) gene. PCCA and PCCB cDNA clones have been obtained, and mutations in both PCCA and PCCB have been identified [55]. Interallelic complementation has been described among cell lines from patients with mutations in the PCCB gene. This result may partially explain the clinical heterogeneity observed due to β-subunit defects [56]. In addition, DNA-mediated gene-transfer developments might allow somatic gene therapy in the future [57].

## Methylmalonic Aciduria

MMA is an autosomal-recessive disorder. The incidences of benign and severe forms are each about 1 in 50,000. Genetic defects are categorized by somatic cell complementation as either *mut* defects, due to mutations in the gene encoding MCM, or *Cbl* defects, due to mutations in genes required for provision of the cobalamin cofactors. Approximately one half of patients have a mutase apoenzyme defect, which is further divided into *mut*° and *mut*⁻ groups. *Mut*° lines show no detectable enzyme activity even when hydroxycobalamin is provided in excess; *mut*⁻ lines have an enzyme with detectable activity when stimulated by a high concentration of hydroxycobalamin. Thirty mutations at the *mut* locus on the short arm of chromosome 6 have been described. They involve either the C-terminal cobalamin-binding domain in *mut*⁻ or the active site in *mut*° [58]. Gene therapy is now available in experimental systems [59].

The remaining patients are cobalamin variants. Among them, CblA is due to a defect in mitochondrial cobalamin reductase, and CblB is due to defective AdoCbl transferase. Their corresponding genes have not been identified.

## 3-Methylcrotonyl Glycinuria

Less than 30 patients with isolated and biotin-resistant 3-methylcrotonylglycinuria (3-MCG) have been reported. In addition, about ten asymptomatic individuals, most of whom are sibs of symptomatic patients, have been found, and a few have been recognized by neonatal screening.

### Clinical Presentation

Clinical signs are highly variable but, on the whole, symptomatic patients present with either an early onset in the neonatal period or a late onset in childhood. Some neonates present with intractable seizures since the first days of life [60]. Other newborns present, within the first weeks after weaning, with feeding difficulties, poor growth, and hypotonia [61], and some have recurrent seizures resulting in microcephaly and developmental delay [62]. All the routine laboratory tests are normal, and this clinical presentation belongs to type IV in the classification of the neonatal inborn errors of metabolism (Chap. 1).

Most patients, however, present with a Reye-like syndrome following intercurrent illness or a protein-enriched diet within the first 2 years of life. They have metabolic acidosis associated with ketonuria, hyperammonemia, and often hypoglycemia [63].

In a few patients, chronic muscular signs were prominent, with hypotonia and poor growth [64]. In addition, two adult women diagnosed via neonatal screening of their newborns complained of muscle weakness [65]. Finally, in a few patients affected with unspecific developmental delay and recurrent seizures, the diagnosis was the result of a systematic metabolic work-up. In these two groups, routine laboratory investigations were normal [66].

### Metabolic Derangement

3-MCG is an inborn error of leucine catabolism due to 3-methylcrotonyl-CoA carboxylase deficiency (Fig. 16.1, enzyme 3). This enzyme is one of the biotin-dependent carboxylases, and variant forms of 3-MCG are secondary to defective biotin metabolism, resulting in multiple carboxylase deficiency, which is much more common than isolated biotin-resistant 3-MCG (Chap. 24).

Due to the block, 3-methylcrotonyl-CoA and 3-methylcrotonic acid accumulate. However, most of this accumulated acyl-CoA is conjugated with glycine into 3-methylcrotonylglycine. In contrast, acylation of 3-methylcrotonyl-CoA by carnitine appears to be a minor pathway.

3-Hydroxyisovalerate (3-HIV), another major metabolite, is derived through the action of a crotonase on 3-methylcrotonyl-CoA and the subsequent hydrolysis of the CoA-ester. 3-HIV-glycine has not been found in this condition, while acylation with carnitine leads to 3-HIV-carnitine formation. This latter compound is the major abnormal acylcarnitine found in the plasma by MS-MS techniques [67].

### Diagnostic Tests

The diagnosis relies on a characteristic urinary profile of organic acids, with huge excretion of 3-HIV and 3-methylcrotonylglycine and without the lactate, methylcitrate, and tiglylglycine found in multiple carboxylase deficiency (MCD). Supplementation with pharmacological doses of biotin does not alter this pattern. Total and free carnitine concentrations in plasma are very low. The presence of 3-HIV-carnitine in plasma is diagnostic for 3-MCG deficiency, since it is not found in IVA. In other disorders, such as MCD, propionylcarnitine is also seen and, in 3-hydroxy-3-methylglutaryl-CoA-lyase deficiency, glutarylcarnitine is the major signal [67].

Enzymatic activity is present in lymphocytes and cultured fibroblasts. In patients, residual activity is higher in lymphocytes than in fibroblasts, without any correlation between the residual activity and the various-onset forms of the disease.

### Treatment and Prognosis

Long-term treatment based on a mildly restricted protein diet (0.75–2 g/kg/day) results in a general improvement and a reduction in the number of exacerbations. It is effective in lowering the abnormal organic-acid excretion which, however, never disappears.

Glycine and carnitine therapies directed at increasing the excretion of glycine and carnitine conjugates are complementary rather than competitive means of detoxification. Glycine supplementation (175 mg/kg/day) increases the excretion of 3-MCG. Carnitine supplementation (100 mg/kg/day) corrects carnitine deficiency and increases the excretion of 3-HIV [63, 67].

The poor prognosis described in early-onset forms presenting with neonatal seizures could be due to late diagnosis and treatment. In acute late-onset forms presenting with Reye-like syndrome all patients but one fully recovered.

### Genetics

This rare disorder is recessively inherited, but the human genes of the two α and β subunits that comprise the enzyme have not yet been mapped or cloned.

## Malonic Aciduria

Only a few patients with malonic aciduria have been described. This condition can be divided into two groups. The first group has clear defective mitochondrial malonyl CoA decarboxylase deficiency expressed in fibroblasts and/or leukocytes; the other has normal enzyme activity despite similar clinical and biological presentation.

### Clinical Presentation

A neonatal form has been described in two patients, who displayed progressive lethargy, hypotonia, and hepatomegaly associated with metabolic acidosis and mild hyperammonemia. Hypoglycemia was present in one newborn, and hyperlactacidemia was present in the other [68, 69].

In the late-onset forms, most patients present with acute episodes of gastroenteritis, febrile seizures, unexplained lethargy associated with metabolic acidosis, and hypoglycemia. Some of these patients were previously known to be affected with a mild and unspecific psychomotor retardation [70–73]. Other children have been diagnosed following a systematic screening indicated mental retardation and hypotonia [74, personal observations]. Cardiomyopathy was present in three patients [72, 74, personal observations].

### Metabolic Derangement

Malonic aciduria is due to deficiency of malonyl CoA decarboxylase (Fig. 16.1, enzyme 13; Fig. 16.2, enzyme 1). The role of this mitochondrial enzyme is unclear. It is thought to remove mitochondrial malonyl-CoA, which is produced by mitochondrial PCC (Fig. 16.1, enzyme 11; Fig. 16.2, enzyme 3), not (like cytosolic malonyl-CoA) by the action of acetyl-CoA carboxylase on acetyl-CoA (Fig. 16.1, enzyme 10; Fig. 16.2, enzyme 2). Accumulating mitochondrial malonyl-CoA is both hydrolyzed to free malonic acid and esterified to malonylcarnitine. The latter can leave the mitochondria by the carnitine shuttle and can be transesterified to malonyl-CoA in the cytosol. Since cytosolic malonyl-CoA is a potent inhibitor of carnitine palmitoyl transferase I (a component of the carnitine cycle that allows entry of long-chain fatty acids into the mitochondria; Fig. 11.1), harmful, secondary inhibition of fatty-acid β-oxidation can occur [72]. This may explain the dicarboxylic aciduria found in patients during catabolic episodes. However, this inhibition seems only partial, since patients display ketonuria during acute decompensations and exhibit normal ketogenesis on acute fat-loading tests [personal observations]. Moreover, the accumulation of malonyl-CoA inhibits methylmalonyl-CoA mutase, succinyl-CoA dehydrogenase, and glutaryl-CoA dehydrogenase, leading to mild urinary excretion of, respectively, methylmalonic, succinic, and glutaric acids [70].

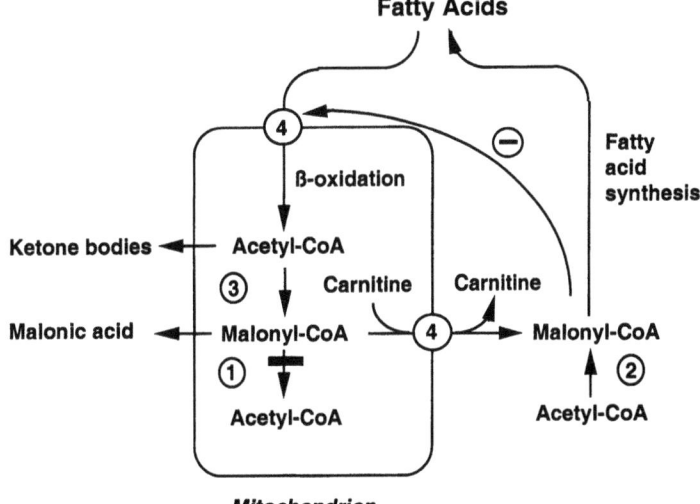

**Fig. 16.2.** Metabolism of malonyl-coenzyme A (CoA). 1, Malonyl-CoA decarboxylase; 2, acetyl-CoA carboxylase; 3, propionyl-CoA carboxylase; 4, carnitine shuttle. The enzyme defect is depicted by a *solid bar*, and (−) denotes inhibition

## Diagnostic Tests

Diagnosis relies on a characteristic profile of urinary organic acids, in which malonic and methylmalonic acids are constant findings. In all patients but one, malonic acid was much higher than methylmalonic acid [73]. Abnormal succinic aciduria has been found in about half of the cases, as have various dicarboxylic and glutaric acidurias.

Total and free carnitine concentrations in plasma are low, and accumulation of malonylcarnitine has been documented in a few patients [72, personal observations]. Despite little experience, acute fat-loading tests can be helpful investigations. They clearly increase malonic acid accumulation [70, personal observations].

Mitochondrial malonyl CoA decarboxylase has been measured in cultured fibroblasts and in the leukocytes of seven out of 11 defective cell lines. Patients with normal enzyme activity may suffer from another enzyme defect or may have a tissue-specific isoenzyme alteration not expressed in fibroblasts.

Prenatal diagnosis should be possible, as malonic acid can be measured in normal amniotic fluid, and malonyl CoA decarboxylase activity is present in cultured amniotic cells. To date, no affected fetuses have been diagnosed [69].

## Treatment and Prognosis

No rules for treatment and prognosis have been established. Due to the results of function tests performed in a few patients [70, 73], a low-fat, high-carbohydrate diet is usually used. However, combined malonic and methylmalonic excretion varies spontaneously and is probably linked to the patients' nutritional status. In addition, during these function tests, it has never been demonstrated that a low-fat, high-carbohydrate diet is clinically more effective than a normal well-balanced diet. Carnitine supplementation (100 mg/kg/day) corrects the carnitine deficiency and has improved the cardiomyopathy and muscle weakness in two patients [74, personal observations].

Long-term prognosis is unknown. Except for the two patients who have developed extrapyramidal signs following an acute crisis, most patients have residual mild developmental delay [71, 75].

## Genetics

Malonic aciduria is inherited as an autosomal-recessive disorder, but there is no available information on the human gene.

## 3-Methylglutaconic Aciduria Type I

3-Methylglutaconic aciduria type I has only been identified in eight patients, who presented with a wide spectrum of clinical signs ranging from mild neurologic impairment to severe encephalopathy with basal-ganglia involvement [76]. 3-MGC-CoA is metabolized to 3-hydroxy-3-methylglutaryl-CoA by 3-MGC-CoA hydratase (Fig. 16.1, enzyme 4). Defective activity leads to 3-MGC aciduria type I, characterized by urinary excretion of 3-MGC and 3-methylglutaric acids. Both metabolites derive from accumulated 3-methylglutaconyl-CoA through hydrolysis and dehydrogenation, respectively. The combined urinary excretion of 3-MGC and 3-methylglutaric acids ranges from 500 to 1000 mmol/mol creatinine, of which 3-methylglutaric acid represents about 1%. The metabolic pattern also includes 3-HIVA and a normal amount of 3-

hydroxy-3-methylglutaric acid. In fibroblasts, the 3-MGC-CoA-hydratase activity is evaluated using the production of acyl-CoA and 3-hydroxybutyric acid from radiolabeled 3-methylcrotonyl-CoA [77].

This disease must be distinguished from many other conditions associated with 3-MGC aciduria of unknown origin. They include Barth's syndrome (3-MGC aciduria type II, an X-linked disorder described in male patients with cardiomyopathy), neutropenia, and growth retardation [78]. Type III is reported in Iraqi Jewish patients who showed early optic atrophy, motor dysfunction, choreiform movement disorder, dysarthria and later development of spasticity, ataxia, and cognitive deficit [79]. Finally, among a large heterogeneous group of patients suffering from variable, multisystemic diseases (3-MGC aciduria type IV, unspecified diseases), some respiratory-chain disorders have been described [80]. In these latter conditions, the urinary excretion of 3-MGC and 3-methylglutaric acid is low (10–200 mmol/mol creatinine), without elevation of 3-HIV and 3-hydroxy-3-methylglutaric acids.

No clear therapeutic regimen has been described. Carnitine supplementation may have beneficial effects.

## 3-Hydroxyisobutyric Aciduria

A few patients with increased excretion of 3-hydroxyisobutyric acid (3-HIBA), an intermediate of the catabolic pathways of valine and thymidine, have been identified. This condition is linked to various enzymatic defects. Unfortunately, in most described cases, the enzymatic diagnosis has been speculative, with only three identified defects in eight families [81]. In addition, two recently reported children with strong evidence of defective valine catabolism in fibroblasts do not excrete 3-HIBA [82, 83].

Clinical presentation is heterogeneous [81]. Some patients present in infancy, with acute episodes of vomiting, lethargy, ketoacidosis, and (sometimes) associated hypoglycemia or hyperlactacidemia [84, 85]. Muscle involvement can be a prominent sign, with marked hypotonia and even myopathic features [85]. In addition, hypertrophic cardiomyopathy has been reported in two cases [82, 86]. CNS involvement is highly variable, ranging from patients with normal development [84] to neonates with brain dysgenesis and other congenital malformations and children with microcephaly, hypotonia, and seizures [81].

Several enzyme defects have been found to underlie 3-hydroxyisobutyric aciduria. A combined deficiency of semialdehyde dehydrogenase (Fig. 16.1, enzyme 9) has been confirmed in two patients and is suspected in three additional cases [81, 84]. This defect results in the accumulation not only of 3-hydroxyisobutyric acid but also of β-aminoisobutyrate, which is formed from methylmalonic semialdehyde. In addition, β-alanine, its derivative 3-hydroxypropionate, and 2-ethylhydracrylic acid (an intermediate of the catabolism of alloisoleucine) are also found in urine, because semialdehyde dehydrogenases intervene in the catabolism of β-alanine and alloisoleucine. The residual activities of the various semialdehyde dehydrogenases are estimated by measuring the conversion of radiolabeled valine and β-alanine into $CO_2$ in fibroblasts [81]. Two recently reported patients show deficiency of 3-hydroxyisobutyrate dehydrogenase (Fig. 16.1, enzyme 8) and isolated semialdehyde, respectively, in their fibroblasts. The first patient did not exhibit any abnormal organic aciduria. The second patient was diagnosed following investigations for mild and unexplained MMA [82, 83]. An unique patient with 3-hydroxyisobutyryl-CoA deacylase deficiency (Fig. 16.1, enzyme 7) in liver and fibroblasts was identified [87]. He did not exhibit any organic aciduria, but cysteine and cysteamine conjugates of methylacrylic acid, S-(2-carboxypropyl)-cysteine and S-(2-carboxypropyl)-cysteamine were found.

## References

1. Dave P, Curless R, Steinman L (1984) Cerebellar hemorrhage complicating methylmalonic and propionic acidemia. Arch Neurol 41:1293–1296
2. Orban T, Mpofu C, Blackensee D (1994) Severe CNS bleeding followed by a good clinical outcome in the acute neonatal form of isovaleric acidaemia. J Inherit Metab Dis 17:755–756
3. Raby RB, Ward JC, Herrod HG (1994) Propionic acidaemia and immunodeficiency. J Inherit Metab Dis 17: 250–251
4. Ozand PT, Rashed M, Gascon GG et al. (1994) Unusual presentation of propionic acidemia. Brain Dev 16 [Suppl]: 46–57
5. Wajner M, Schlottfeld JL, Ckless K, Wannmacher MD (1995) Immunosuppressive effects of organic acids accumulating in patients with maple syrup urine disease. J Inher Metab Dis 18:165–168
6. Gascon GC, Ozand PT, Brismar J (1994) Movement disorders in childhood organic acidurias clinical, neuroimaging, and biochemical correlations. Brain Dev 16:94–103
7. Riviello JJ, Rezvani I, DiGeorge AM, Foley CM (1991) Cerebral edema causing death in children with maple syrup urine disease. J Pediatr 119:42–45
8. Treacy E, Clow CL, Reade TR et al. (1992) Maple syrup urine disease: interrelations between branched-chain amino-, oxo- and hydroxyacids; implications for treatment; associations with CNS dysmyelination. J Inherit Metab Dis 15:121–135
9. Levin ML, Scheimann A, Lewis RA, Beaudet AL (1993) Cerebral edema in maple syrup urine disease. J Pediatr 122:167–168
10. Brismar J, Aqeel A, Brismar G et al. (1990) Maple syrup urine disease: findings on CT and MR scans of the brain in 10 infants. Am J Neuroradiol 11:1219–1228
11. Taccone A, Schiaffino MC, Cerone R, Fondelli MP, Romano C (1992) Computed tomography in maple syrup urine disease. Eur J Radiol 14:207–212

12. Pérez-Cerda C, Merinero B, Marti M et al. (1998) An unusual late onset case of propionic acidaemia: biochemical investigations, neuroradiological findings and mutation analysis. Eur J Pediatr 157:50–52
13. Brismar J, Ozand PT (1994) CT and MR of the brain in disorders of the propionate and methylmalonate metabolism. Am J Neuroradiol 15:1459–1473
14. Hamilton RL, Haas RH, Nyhan WL et al. (1995) Neuropathology of propionic acidemia: a report of two patients with basal ganglia lesions. J Child Neurol 10:25–30
15. Baumgartner ER, Viardot et al. (1995) Long-term follow-up of 77 patients with isolated methylmalonic acidaemia. J Inherit Metab Dis 18:138–142
16. Rutledge SL, Geraghty M, Mroczek E, Rosenblatt D, Kohout E (1993) Tubulointersticial nephritis in methylmalonic acidemia. Pediatr Nephrol 7:81–82
17. Leonard JV (1995) The management and outcome of propionic and methylmalonic acidaemia. J Inherit Metab Dis 18:430–434
18. van't Hoff WG, Dixon M, Taylor J et al. (1998) Combined liver-kidney transplantation in methylmalonic acidemia. J Pediatr 132:1043–1044
19. Giacoia GP, Berry GT (1993) Acrodermatitis enteropathica-like syndrome secondary to isoleucine deficiency during treatment of maple syrup urine disease. Am J Dis Child 147:954–956
20. Bodemer C, de Prost Y, Bachollet B et al. (1994) Cutaneous manifestations of methylmalonic and propionic acidaemia: a description based on 38 cases. Br J Dermatol 131:93–98
21. De Raeve L, De Meirleir L, Ramet J, Vandenplas Y, Gerlo E (1994) Acrodematitis enteropathica-like cutaneous lesions in organic aciduria. J Pediatr 124:416–420
22. Tornqvist K, Tornqvist H (1996) Corneal deepithelialization caused by acute deficiency of isoleucine during treatment of a patient with maple syrup urine disease. Acta Ophthalmol Scand 74 [Suppl 219]:48–49
23. Burlina AB, Dionisi-Vici C, Piovan S et al. (1995) Acute pancreatitis in propionic acidaemia. J Inherit Metab Dis 18:169–172
24. Massoud AF, Leonard JV (1993) Cardiomyopathy in propionic acidaemia. Eur J Pediatr 152:441–445
25. Danner DJ, Doering CB (1998) Human mutations affecting branched chain α-ketoacid dehydrogenase. Front Biosci 3:d517–524
26. Chuang DT (1998) Maple syrup urine disease: it has come a long way. J Pediatr 132:S17–S23
27. Treacy E, Clow C, Mamer A, Scriver R (1993) Methylmalonic acidemia with a severe chemical but benign clinical phenotype. J Pediatr 122:428–429
28. Chalmers RA, Lawson AM (1982) Disorders of propionate and methylmalonate metabolism. In: Organic acids in man. Chapman and Hall, London, pp 296–331
29. Wendel U, Eissler A, Perl W, Schaedewaldt P (1995) On the differences between urinary metabolite excretion and odd-numbered fatty acid production in propionic and methylmalonic acidaemias. J Inherit Metab Dis 18:584–591
30. Leonard JV (1996) Stable isotope studies in propionic and methylmalonic acidaemia. Eur J Pediatr 156 [Suppl 1]:S67–S69
31. Sbai D, Narcy C, Thompson GN et al. (1994) Contribution of odd-chain fatty acid oxidation to propionate production in disorders of propionate metabolism. Am J Clin Nutr 59:1332–1337
32. Saudubray JM, Poggi-Travert F, Martin D et al. (1996) Management and long-term follow-up of organic acidemias: criteria for therapeutic decisions. Jpn J Inherit Metab Dis 12:9–18
33. Burns Sp, Iles RA, Saudubray JM, Chalmers RA (1996) Propionylcarnitine excretion is not affected by metronidazole administration to patients with disorders of propionate metabolism. Eur J Pediatr 155:31–35
34. Holmes E, Foxall PJ, Spraul M et al. (1997) 750 Mhz 1H NMR spectroscopy characterisation of the complex metabolic pattern of urine from patients with inborn errors of metabolism: 2-hydroxyglutaric aciduria and maple syrup urine disease. J Pharm Biomed Anal 15:1647–1659
35. Chace DH, Hillman SL, Millington DS, Kahler SG, Roe CR, Naylor EW (1995) Rapid diagnosis of maple syrup urine disease in blood spots from newborns by tandem mass spectrometry. Clin Chem 41:62–68
36. Johnson AW, Mills K, Clayton PT (1996) The use of automated electrospray ionization tandem MS for the diagnosis of inborn errors of metabolism from dried blood spots. Biochem Soc Trans 24:932–938
37. Parvy P, Bardet J, Rabier D, Kamoun P (1988) Pseudo cystinuria lysinuria in neonatal propionic acidemia. Clin Chem 34:1258
38. Lehnert W, Sperl W, Suormala T, Baumgartner ER (1994) Propionic acidaemia: clinical biochemical and therapeutic aspects (experience in 30 patients). Eur J Pediatr 153 [Suppl 1]:S68–S80
39. Jakobs C, Ten Brink HG, Stellaard F (1991) Prenatal diagnosis of inherited metabolic disorders by quantification of characteristic metabolites in amniotic fluid: facts and future. Prenat Diagn 10:265–271
40. Shigematsu Y, Hata I, Nakai A et al. (1996) Prenatal diagnosis of organic acidemias based on amniotic fluid levels of acylcarnitines. Pediatr Res 39:680–684
41. Berry GT, Heindenreich R, Kaplan P et al. (1991) Branched chain amino acid-free parenteral nutrition in the treatment of acute metabolic decompensation in patients with maple syrup urine disease. N Engl J Med 324:175–179
42. Parini R, Sereni LP, Bagozzi DC, Corbetta C, Rabier R, Narcy C, Hubert P, Saudubray JM (1993) Nasogastric drip feeding as the only treatment of neonatal maple syrup urine disease. Pediatrics 92:280–283
43. Jouvet P, Poggi F, Rabier D et al. (1997) Continuous venovenous haemofiltration in the acute phase of neonatal maple syrup urine disease. J Inherit Metab Dis 20:463–472
44. Elsas LJ, Ellerine NP, Klein PD (1993) Practical methods to estimate whole body leucine oxidation in maple syrup urine disease. Pediatr Res 33:445–451
45. Hilliges C, Awiszus D, Wendel U (1993) Intellectual performance of children with maple urine disease. Eur J Pediatr 152:144–147
46. Fries MH, Rinaldo P, Schmidt-Sommerfeld E, Jurecki E, Packman S (1996) Isovaleric acidemia: response to a leucine load after three weeks of supplementation with glycine, L-carnitine, and combined glycine-carnitine therapy. J Pediatr 129:449–452
47. Surtees RAH, Matthews EE, Leonard JV (1992) Neurologic outcome of propionic acidemia. Pediatr Neurol 5:334–337
48. van der Meer SB, Poggi F, Spada M et al. (1994) Clinical outcome of long-term management of patients with vitamin-$B_{12}$ unresponsive methylmalonic acidemia. J Pediatr 125:903–908
49. van der Meer SB, Poggi F, Spada M et al. (1996) Clinical outcome and long term management of 17 patients with propionic acidaemia. Eur J Pediatr 155:205–210
50. Nicolaides P, Leonard J, Surtees R (1998) Neurological outcome of methylmalonic acidaemia. Arch Dis Child 78:508–512
51. Schlenzig JS, Poggi-Travert F et al. (1995) Liver transplantation in two cases of propionic acidaemia. J Inherit Metab Dis 18:448–461
52. Dixon MA, Leonard JV (1992) Intercurrent illness in inborn errors of intermediary metabolism. Arch Dis Child 67:1387–1391
53. Thompson GN, Francis DEM, Halliday D (1991) Acute illness in maple syrup urine disease: dynamics of protein metabolism and implications for management J Pediatr 119:35–41
54. Moshen AA, Anderson BD, Volchenboum SL et al. (1998) Characterization of molecular defects in isovaleryl-CoA dehydrogenase in patients with isovaleric acidemia. Biochemistry 37:10325–10335
55. Rodriguez-Pombo P, Hoenicka J, Muro S et al. (1998) Human propionyl-CoA carboxylase beta subunit gene: exon-intron definition and mutation spectrum in spanish and latin

american propionic acidemia patients. Am J Genet 63: 360-369
56. Loyer M, Leclerc D, Gravel RA (1995) Interallelic complementation of β-subunit defects in fibroblasts of patients with propionyl-CoA carboxylase deficiency microinjected with mutant cDNA constructs. Hum Mol Genet 4:1035-1039
57. Lamhonwah AM, Leclerc D, Loyer M, Clarizio R, Gravel RA (1994) Correction of the metabolic defect in propionic acidemia fibroblasts by microinjection of a full-length cDNA or RNA transcript encoding the propionyl-CoA carboxylase β-subunit. Genomics 19:500-505
58. Adjalla CE, Hosack AR, Gilfix BM et al. (1998) Seven novel mutations in mut methylmalonic aciduria. Hum Mutat 11:270-274
59. Sawada T, Ledley FD (1992) Correction of methylmalonyl-CoA mutase deficiency in mut° fibroblasts and constitution of gene expression in primary human hepatocytes by retroviral-mediated genetransfer. Somat Cell Mol Genet 18:507-516
60. Bannwart C, Wermuth B, Baumgartner R, Suormala T, Wiesmann UN (1992) Isolated biotin-resistant deficiency of 3-methylcrotonyl-CoA carboxylase presenting as a clinically severe form in a newborn with fatal outcome. J Inherit Metab Dis 15:863-868
61. Tuchman M, Berry SA, Thuy LP, Nyhan WL (1993) Partial methylcrotonyl-Coenzyme A carboxylase deficiency in an infant with failure to thrive, gastrointestinal dysfunction, and hypertonia. Pediatrics 91:664-666
62. Wiesmann UN, Suormala T, Pfenninger J, Baumgartner ER (1998) Partial 3-methylcrotonyl-CoA carboxylase deficiency in an infant with fatal outcome due to progressive respiratory failure. Eur J Pediatr 157:225-229
63. Rutledge SL, Berry GT, Stanley CA, van Hove JLK, Millington D (1995) Glycine and L-carnitine therapy in 3-methylcrotonyl-CoA carboxylase deficiency. J Inherit Metab Dis 18: 299-305
64. Elpeleg ON, Havkin S, Barash V et al. (1992) Familial hypotonia of childhood caused by isolated 3-methylcrotonyl-coenzyme A carboxylase deficiency. J Pediatr 121:407-410
65. Gibson KM, Bennettt MJ, Naylor EW, Morton DH (1998) 3-methylcrotonyl-coenzyme A carboxylase deficiency in Amish/Mennonite adults identified by detection of increased acylcarnitines in blood spots of their children. J Pediatr 132:519-523
66. Yap S, Monavari AA, Thornton P, Naughteen E (1998) Late-infantile 3-methylcrotonyl-CoA carboxylase deficiency presenting as global developmental delay. J Inherit Metab Dis 21:175-176
67. van Hove JLK, Rutledge SL, Nada MA, Khaler SG, Millington DS (1995) 3-hydroxyisovalerylcarnitine in 3-methylcrotonyl-CoA carboxylase deficiency. J Inherit Metab Dis 18:592-601
68. Krawinkel MB, Oldigs HD, Santer R et al. (1994) Association of malonyl-CoA decarboxylase deficiency and heterozygote state for haemoglobin C disease. J Inherit Metab Dis 17: 636-637
69. Buyukgebiz B, Jakobs C, Scholte HR, Huijmans JGM, Kleijer WJ (1998) Fatal neonatal malonic aciduria. J Inherit Metab Dis 21:76-77
70. Haan EA, Scholem RD, Croll HB, Brown GK (1986) Malonyl coenzyme A decarboxylase deficiency. Clinical and biochemical findings in a second child with a more severe enzyme defect. Eur J Pediatr 144:567-570
71. MacPhee GB, Logan RW, Mitchell JS et al. (1993) Malonyl coenzyme A decarboxylase deficiency. Arch Dis Chil 69: 433-436
72. Yano S, Sweetman L, Thoburn DR, Modifi S, Williams JC (1997) A new case of malonyl coenzyme A decarboxylase deficiency presenting with cardiomyopathy. Eur J Pediatr 156:382-383
73. Gregg AR, Warman AW, Thoburn DR, O'Brien WE (1998) Combined malonic and methylmalonic aciduria with normal malonyl-coenzyme A decarboxylase activity: a case supporting multiple aetiologies. J Inherit Metab Dis 21:382-390
74. Matalon R, Michaels K, Kaul R et al. (1993) Malonic aciduria and cardiomyopathy. J Inherit Metab Dis 16:571-573
75. Ozand PT, Nyhan WL, Al Aqeel A, Christodoulou J (1994) Malonic aciduria. Brain Dev 16:7-11
76. Gibson KM, Wappner RS, Jooste S et al. (1998) Variable clinical presentation in three patients with 3-methylglutaconyl-coenzyme A hydratase deficiency. J Inherit Metab Dis 21:631-638
77. Gibson KM, Lee CF, Wappner RS (1992) 3-Methylglutaconyl-Coenzyme-A hydratase deficiency: a new case. J Inherit Metab Dis 15:363-366
78. Johnston J, Kelley RI, Feigenbaum A et al. (1997) Mutation characterization and genotype-phenotype correlation in Barth syndrome. Am J Hum Genet 61:1053-1058
79. Costeff H, Apter N, Elpeleg ON, Prialnic M, Böhles HJ (1998) Ineffectiveness of oral coenzyme Q10 supplementation in 3-methylglutaconic aciduria type 3. Brain Dev 20:33-35
80. Broid E, Elpeleg O, Lahat E (1997) Type IV 3-methylglutaconic aciduria: a new case presenting with hepatic dysfunction. Pediatr Neurol 17:353-355
81. Gibson KM, Lee CF, Bennett MJ, Holmes B, Nyhan WL (1993) Combined malonic, methylmalonic and ethylmalonic acid semialdehyde dehydrogenase deficiencies: an inborn error of β-alanine, L-valine and L-alloisoleucine metabolism. J Inherit Metab Dis 16:563-567
82. Roe CR, Cederbaum SC, Roe DS, Mardach R, Gaiindo A (1998) Isolated isobutyryl-CoA dehydrogenase deficiency: a new defect of human valine metabolism. J Inherit Metab Dis 21 [Suppl 2]:54
83. Roe CR, Struys E, Kok R et al. (1998) Methylmalonic semialdehyde dehydrogenase deficiency: psychomotor delay and methylmalonic aciduria without metabolic decompensation. Mol Genet Metab 65:35-43
84. Boulat O, Benador N, Girardin E, Bachman C (1995) 3-Hydroxyisobutyric aciduria with mild clinical course. J Inherit Metab Dis 18:204-206
85. Sasaki M, Kimura M, Sugai K, Hashimoto T, Yamaguchi S (1998) 3-Hydroxyisobutyric aciduria in two brothers. Pediatr Neurol 18:253-255
86. Ko FJ, Nyhan WL, Wolff J, Barshop B, Sweetman L (1991) 3-Hydroxyisobutyric aciduria: an inborn error of valine metabolism. Pediatr Res 30:322-326
87. Brown GK, Huint SM, Scholem R et al. (1982) β-Hydroxyisobutyryl-Coenzyme A deacylase deficiency: a defect in valine metabolism associated with physical malformations. Pediatrics 70:532-538

# CHAPTER 17

# The Urea Cycle

The urea cycle which, in its complete form, is only present in the liver, is the main pathway for the disposal of excess of ammonium nitrogen. This cycle sequence of reactions, localised in part in the mitochondria and in part in the cytosol, converts the toxic ammonia molecule into the non-toxic product, urea, which is excreted in the urine. There are genetic defects of each of the enzymes of the urea cycle which lead to hyperammonaemia. Some genetic defects of other important metabolic pathways may lead to secondary inhibition of the urea cycle. Alternative pathways for nitrogen excretion, namely conjugation of glycine with benzoate and of glutamine with phenylacetate can be exploited in the treatment of patients with defective ureagenesis.

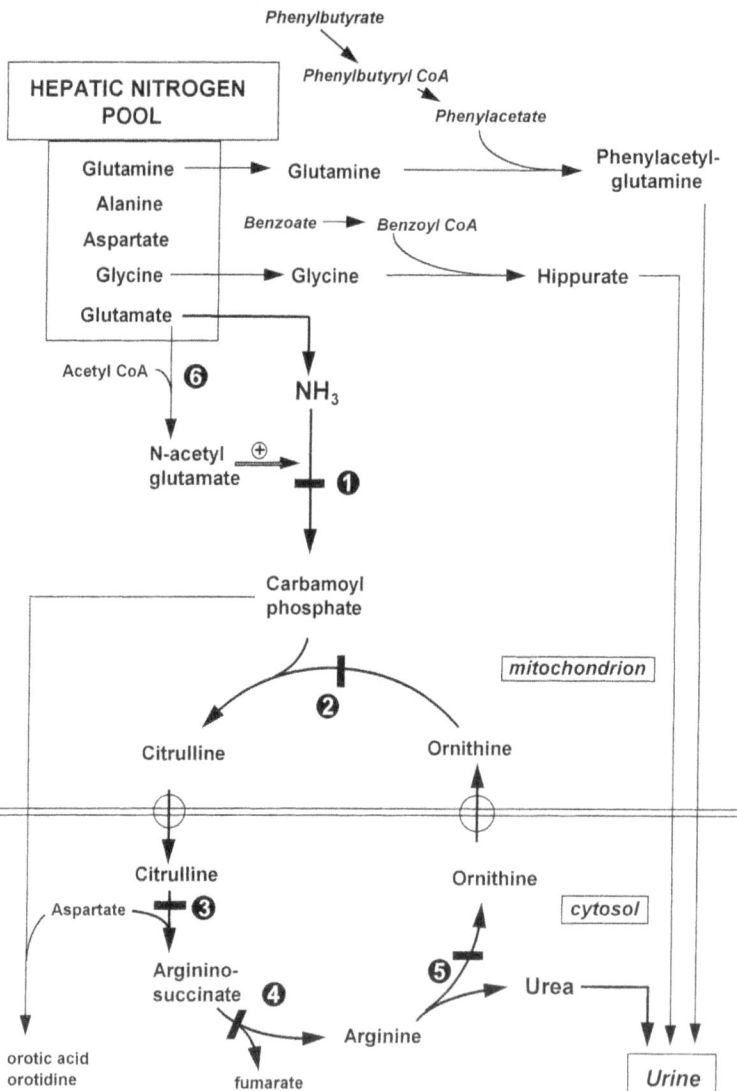

**Fig. 17.1.** The urea cycle and alternative pathways of nitrogen excretion. Enzymes: 1, carbamoyl phosphate synthetase; 2, ornithine transcarbamoylase; 3, argininosuccinate synthetase; 4, argininosuccinate lyase; 5, arginase; 6, $N$-acetylglutamate synthetase. Enzyme defects are depicted by *solid bars* across the arrows

# CHAPTER 17

# Disorders of the Urea Cycle

J.V. Leonard

## CONTENTS

| | |
|---|---|
| Clinical Presentation | 215 |
| Neonatal Presentation | 215 |
| Infantile Presentation | 216 |
| Children and Adults | 216 |
| Metabolic Derangement | 216 |
| Toxicity | 217 |
| Diagnostic Tests | 217 |
| Biochemical Tests | 217 |
| Imaging | 218 |
| Differential Diagnosis | 218 |
| Treatment | 218 |
| Low-Protein Diet | 218 |
| Essential Amino Acids | 219 |
| Alternative Pathways for Nitrogen Excretion | 219 |
| General Aspects of Therapy | 219 |
| Assessment for Treatment | 220 |
| Emergency Treatment | 220 |
| Prognosis | 221 |
| Genetics and Prenatal Diagnosis | 221 |
| References | 221 |

Five inherited disorders of the urea cycle are now well described. These are characterised by hyperammonaemia and disordered amino-acid metabolism. The presentation is highly variable: those presenting in the newborn period usually have an overwhelming illness that rapidly progresses from poor feeding, vomiting, lethargy or irritability and tachypnoea to fits, coma and respiratory arrest. In infancy, the symptoms are less severe and more variable. Poor developmental progress, behavioural problems, hepatomegaly and gastrointestinal symptoms are usually observed. In children and adults, chronic neurological illness is characterised by behavioural problems, confusion, irritability and cyclic vomiting, which deteriorates to acute encephalopathy during metabolic stress. Arginase deficiency shows more specific symptoms, such as spastic diplegia, dystonia, ataxia and fits. All urea-cycle disorders have autosomal-recessive inheritance except ornithine carbamoyl transferase deficiency, which is X-linked.

## Clinical Presentation

Patients with urea-cycle disorders may present at almost any age. However, there are certain times at which they are more likely to develop symptoms because of metabolic stress, such as infection precipitating protein catabolism. These are:

- The neonatal period.
- During late infancy. Children are vulnerable during this period because of the slowing of growth, the change to cow's milk and weaning foods and the declining maternal antibody and consequent development of intercurrent infections.
- Puberty. The changing growth rate and psychosocial factors may precipitate decompensation.

However, it must be emphasised that many patients may present outside these periods. The patterns of the clinical presentation of hyperammonaemia are rather characteristic and are broadly similar for all the disorders except arginase deficiency, which is discussed separately. The early symptoms are often non-specific and initially, therefore, the diagnosis is easily overlooked. The most important points in diagnosing hyperammonaemia are to think of it during diagnosis and to measure the plasma ammonia concentration.

### Neonatal Presentation

Most babies with urea cycle disorders are of normal birthweight and are initially healthy but, after a short interval that can be less than 24 h, they become unwell. Common early symptoms are poor feeding, vomiting, lethargy and/or irritability and tachypnoea. The initial working diagnosis is almost invariably sepsis. Rather characteristically, these babies may have a transient mild respiratory alkalosis, which can be a useful diagnostic clue at this stage. Usually, they deteriorate rapidly, with more obvious neurological and autonomic problems, including changes of tone with loss of normal reflexes, vasomotor instability and hypothermia, apnoea and fits. The baby may soon become totally unresponsive and may require full intensive care. Untreated, most babies will die, often with complications, such as cerebral or pulmonary haemorrhage, the underlying metabolic cause for which may not be recognised. Some survive neonatal hyperammonaemia but are invariably handicapped to some degree.

### Infantile Presentation

In infancy, the symptoms are generally rather less acute and more variable than in the neonatal period and include anorexia, lethargy, vomiting and failure to thrive, with poor developmental progress. Irritability and behavioural problems are also common. The liver is often enlarged but, as the symptoms are rarely specific, the illness is initially attributed to many different causes that include gastrointestinal disorders (gastro-oesophageal reflux, cow's milk protein intolerance), food allergies, behaviour problems or hepatitis. The correct diagnosis is often only established when the patient develops a more obvious encephalopathy with changes in consciousness level and neurological signs (see below).

### Children and Adults

At these ages, the patients commonly present with a more obviously neurological illness.

ACUTE ENCEPHALOPATHY. Whilst older patients often present with episodes of acute metabolic encephalopathy, they may also have chronic symptoms. Usually, symptoms develop following metabolic stress precipitated by infection, anaesthesia or protein catabolism, such as that produced by the rapid involution of the uterus in the puerperium [1]. However an obvious trigger is not always apparent. The patients first become anorexic, lethargic and unwell. Sometimes they are agitated and irritable, with behaviour problems or confusion. Vomiting and headaches may be prominent, suggesting migraine or cyclical vomiting. Others may be ataxic as though intoxicated. On examination, hepatomegaly may be present, particularly in those with argininosuccinic aciduria. The patients may then recover completely but, if not, they may then develop neurological problems, including a fluctuating level of consciousness, fits and (sometimes) focal neurological signs, such as hemiplegia [2] or cortical blindness. Untreated, they continue to deteriorate, becoming comatose, and they may die. Alternatively, they may recover with a significant neurological deficit. The cause of death is usually cerebral oedema.

Between episodes, the patients are usually relatively well, although some, particularly younger ones, may continue to have problems, such as vomiting or poor developmental progress. Some patients may voluntarily restrict their protein intake. In addition to those disorders already mentioned, the illness may be attributed to a wide variety of other disorders, including Reye's syndrome, encephalitis, poisoning and psychosocial problems.

CHRONIC NEUROLOGICAL ILLNESS. Learning difficulties or more obvious mental retardation are common, and some patients, particularly those with argininosuccinic aciduria, may present with relatively few symptoms apart from mental retardation and fits. About half the patients with argininosuccinic acid have brittle hair (trichorrhexis nodosa). Patients may present with chronic ataxia, which is worse during intercurrent infections [3].

ARGINASE DEFICIENCY. Arginase deficiency commonly presents with spastic diplegia and, initially, a diagnosis of cerebral palsy is almost always suspected. However, the neurological abnormalities appear to be slowly progressive, although it may be difficult to distinguish this from an evolving cerebral palsy. During the course of the disease, fits, ataxia and dystonia may develop. Occasionally, patients may present with an acute encephalopathy or anticonvulsant-resistant fits [4].

### Metabolic Derangement

The urea cycle is the final common pathway for the excretion of waste nitrogen in mammals. The steps in the urea cycle are shown in Fig. 17.1. Ammonia is probably derived principally from glutamine and glutamate and is converted to carbamoyl phosphate by *carbamoyl phosphate synthetase* (CPS). This enzyme requires an allosteric activator, N-acetylglutamate, for full activity. This compound is formed by the condensation of acetyl coenzyme A (acetyl CoA) and glutamate in a reaction catalysed by *N-acetyl glutamate synthetase*. Carbamoyl phosphate condenses with ornithine to form citrulline in a reaction catalysed by *ornithine transcarbamoylase*. The product, citrulline, condenses with aspartate to produce argininosuccinate in a reaction catalysed by *argininosuccinate synthetase*, and the argininosuccinate is then hydrolysed to arginine and fumarate by *argininosuccinate lyase*. The arginine is itself cleaved by *arginase*, releasing urea and re-forming ornithine. Within the urea cycle itself, ornithine acts as a carrier; it is neither formed nor lost.

Each molecule of urea contains two atoms of waste nitrogen, one derived from ammonia and the other from aspartate. Regulation of the urea cycle is not fully understood, and it is likely that there are several mechanisms controlling flux through this pathway [5].

These include enzyme induction, the concentrations of substrates, intermediates and N-acetyl glutamate, and hormonal effects. Defects of each step have now been described and are listed in Table 17.1.

The plasma *ammonia* concentration is raised as a result of metabolic blocks in the urea-cycle. The degree to which it is elevated depends on several factors, including the enzyme involved and its residual activity, the protein intake and the rate of endogenous protein catabolism, particularly if this is increased because of infection, fever or other metabolic stresses. The values may also be falsely elevated if the specimen is not collected and handled correctly.

The concentrations of the amino acids in the metabolic pathway immediately proximal to the enzyme defect will increase, and those beyond the block will decrease (Table 17.1). In addition, plasma alanine and particularly *glutamine* accumulate in all the disorders. The concentration of *citrulline* is often helpful, but it may not always be reliable during the newborn period [6].

*Orotic acid and orotidine* are excreted in excess in the urine if there is a metabolic block distal to the formation of carbamoyl phosphate, as is the case in ornithine transcarbamoylase (OTC) deficiency, citrullinaemia, argininosuccinic aciduria and arginase deficiency (Fig. 17.1). In these disorders, carbamoyl phosphate accumulates, leaves the mitochondrion and, once in the cytosol, enters the pathway for the de novo synthesis of pyrimidines. The urea cycle is also closely linked to many other pathways of intermediary metabolism, particularly the citric-acid cycle.

## Toxicity

Ammonia increases the transport of tryptophan across the blood-brain barrier, which then leads to an increased production and release of serotonin [7].

Some of the symptoms of hyperammonaemia can be explained on this basis, and the dietary tryptophan restriction has reversed anorexia in some patients with urea cycle disorders [8]. Ammonia induces many other electrophysiological, vascular and biochemical changes in experimental systems, but it is not known to what extent all of these are relevant to the problems of clinical hyperammonaemia in man [9].

Using proton nuclear magnetic resonance spectroscopy, glutamine can also be shown to accumulate at high concentrations, both in experimental models and in man in vivo [10]. The concentrations are such that the increase in osmolality could be responsible for cellular swelling and cerebral oedema.

## Diagnostic Tests

### Biochemical Tests

Routine tests are not helpful for establishing the diagnosis of hyperammonaemia. Plasma transaminases may be elevated; combined with hepatomegaly, this may lead to the erroneous diagnosis of hepatitis.

The most important diagnostic test in urea cycle disorders is measurement of the plasma ammonia concentration. Normally, this is less than 50 µmol/l but may be slightly raised as a result of a high protein intake, exercise, struggling or a haemolysed blood sample. Generally, patients who are acutely unwell with urea cycle disorders have plasma ammonia concentrations greater than 150 µmol/l, and often significantly higher. However, the concentrations may be near normal when patients are well, are early in an episode of decompensation or if they have been on a low-protein, high-carbohydrate intake for some time.

Healthy neonates have slightly higher values [11]. If they are ill (sepsis, perinatal asphyxia, etc.), plasma

**Table 17.1.** Urea-cycle disorders: biochemical and genetic details

| Disorder | Alternative names | Plasma amino acid concentrations | Urine orotic acid | Tissue for enzyme diagnosis | Genetics (chromosome localisation) |
|---|---|---|---|---|---|
| CPS deficiency | CPS deficiency | ↑ Glutamine; ↑ alanine; ↓ citrulline; ↓ arginine | N | Liver | AR (chromosome 2p) |
| OTC deficiency | OTC deficiency | ↑ Glutamine; ↑ alanine; ↓ citrulline; ↓ arginine | ↑↑ | Liver | X-linked (Xp21.1) |
| Argininosuccinic synthetase deficiency | Citrullinaemia | ↑↑ Citrulline; ↓ arginine | ↑ | Liver/fibroblasts | AR (chromosome 9q) |
| Argininosuccinic lyase deficiency | Argininosuccinic aciduria | ↑ Citrulline; ↑ argininosuccinic acid; ↓ arginine | ↑ | RBC/liver/fibroblasts | AR (chromosome 7q) |
| Arginase deficiency | Hyperargininaemia | ↑ Arginine | ↑ | RBC/liver | AR (chromosome 6q) |
| NAGS deficiency | NAGS deficiency | ↑ Glutamine; ↑ alanine | N | Liver | AR (not confirmed) |

↑, increased; ↓, decreased; AR, autosomal recessive; CPS, carbamyl phosphate synthetase; N, normal; NAGS, N-acetylglutamate synthetase; OTC, ornithine transcarbamoylase; RBC, red blood cell

ammonia concentrations may increase to 180 μmol/l. Patients with inborn errors presenting in the newborn period usually have concentrations greater than 200 μmol/l, often very much greater. In that case, further investigations (particularly of the plasma amino acid and urine organic acid levels) are urgent. The following investigations should be performed:

- Blood pH and gases
- Plasma chemistry: sodium, urea and electrolytes, glucose and creatinine
- Liver-function tests and clotting studies
- Plasma amino acids
- Urine organic acids, orotic acid and amino acids
- Plasma free and acyl carnitines

In all urea-cycle disorders, there is accumulation of glutamine and alanine and, in citrullinaemia, argininosuccinic aciduria and arginase deficiency, the changes in the amino acids are usually diagnostic (Table 17.1). Orotic aciduria with raised plasma glutamine and alanine concentrations suggests OTC deficiency. The diagnosis of this and the other disorders can be confirmed by measuring enzyme activity in appropriate tissue (Table 17.1). The enzyme diagnosis of *N*-acetyl glutamate synthetase deficiency is not straightforward, and the response to a load of *N*-carbamyl glutamate, an orally active analogue of *N*-acetyl glutamate, may be helpful both diagnostically and for treatment.

### *Imaging*

Patients who present with an acute encephalopathy commonly receive brain imaging at an early stage. This may show no abnormality, a localised area of altered signal or, if the patient is very seriously ill, widespread cerebral oedema [12].

Focal areas of altered signal may be identified and need to be distinguished from herpes simplex encephalitis. A careful history revealing previous episodes of encephalopathy, albeit mild, may provide vital clues. Imaging in patients who have recovered from a severe episode of hyperammonaemia usually show cerebral atrophy that may be focal, particularly in those areas in which there were altered signals during the acute illness.

### *Differential Diagnosis*

The differential diagnosis of hyperammonaemia is wide, and the most common conditions are summarised in Table 17.2. In the neonatal period, the most common differential diagnoses are organic acidaemias, particularly propionic and methylmalonic acidaemia. Patients with these disorders may have had marked hyperammonaemia with minimal metabolic acidosis

**Table 17.2. Differential diagnosis of hyperammonaemia**

**Inherited disorders**
*Urea cycle enzyme defects*
  Carbamoyl phosphate synthetase deficiency
  Ornithine transcarbamoylase deficiency
  Argininosuccinate synthetase deficiency (citrullinaemia)
  Argininosuccinate lyase deficiency (arginosuccinic aciduria)
  Arginase deficiency
  *N*-acetylglutamate synthetase deficiency
*Transport defects of urea cycle intermediates*
  Lysinuric protein intolerance
  Hyperammonemia–hyperornithinemia–homocitrullinuria syndrome
*Organic acidurias*
  Propionic acidaemia
  Methylmalonic acidaemia and other organic acidaemias
  Fatty acid oxidation disorders
  Medium-chain acyl-CoA dehydrogenase deficiency
  Systemic carnitine deficiency
  Long-chain fatty acid oxidation defects and other related disorders
*Other inborn errors*
  Pyruvate carboxylase deficiency (neonatal form)
**Acquired**
  Transient hyperammonemia of the newborn
  Reye's syndrome
  Liver failure, any cause (both acute and chronic)
  Valproate therapy
  Infection with urease-positive bacteria (particularly with stasis in the urinary tract)
  Leukaemia therapy, including treatment with asparaginase
  Severe systemic illness, particularly in neonates

*CoA*, coenzyme A

or ketosis. Although babies with transient hyperammonaemia of the newborn are often born prematurely, with early onset of symptoms [13], it may be difficult to distinguish between urea-cycle disorders and transient hyperammonaemia of the newborn. All patients in whom a tentative diagnosis of Reye's syndrome is made should be investigated in detail for inherited metabolic disorders, including urea-cycle disorders.

### Treatment

The aim of treatment is to correct the biochemical disorder and to ensure that all the nutritional needs are met. The major strategies used are to reduce protein intake, to utilise alternative pathways of nitrogen excretion and to replace nutrients that are deficient.

### *Low-Protein Diet*

Most patients require a low-protein diet. The exact quantity will depend mainly on the age of the patient and the severity of the disorder. Many published regimens suggest severe protein restriction but, in early infancy, patients may need 1.8–2 g/kg/day or more during phases of very rapid growth. The protein intake usually decreases to approximately 1.2–1.5 g/kg/day during pre-school years and 1 g/kg/day in late childhood. After puberty, the quantity of natural protein may be less than 0.5 g/kg/day. However, it must be emphasised that there is considerable variation in the needs of individual patients.

### Essential Amino Acids

In the most severe variants, it may not be possible to achieve good metabolic control and satisfactory nutrition with restriction of natural protein alone. Other patients will not take their full protein allowance. In both these groups of patients, some of the natural protein may be replaced with an essential amino acid mixture, giving up to 0.7 g/kg/day. Using this, the requirements for essential amino acids can be met; in addition, waste nitrogen is re-utilised to synthesise non-essential amino acids, hence reducing the load of waste nitrogen.

### Alternative Pathways for Nitrogen Excretion

In many patients, additional therapy is necessary. A major advance in this field has been the development of compounds that are conjugated to amino acids and rapidly excreted [14, 15]. The effect of the administration of these substances is that nitrogen is excreted in compounds other than urea; hence, the load on the urea cycle is reduced (Fig. 17.1). The first compound introduced was sodium benzoate. *Benzoate* is conjugated with glycine to form hippurate, which is rapidly excreted. For each mole of benzoate given, 1 mol of nitrogen is removed. Sodium benzoate is usually given in doses up to 250 mg/kg/day but, in acute emergencies, this can be increased to 500 mg/kg/day. The major side effects are nausea, vomiting and irritability. In neonates, conjugation may be incomplete, with increased risk of toxicity [C. Bachmann, personal communication].

The next drug used was phenylacetate, but this has now been superseded by *phenylbutyrate*, because the former has a peculiarly unpleasant, clinging, mousy odour. In the liver phenylbutyrate is oxidised to phenylacetate, which is then conjugated with glutamine. The resulting phenylacetylglutamine is rapidly excreted in urine; hence, 2 mol of nitrogen are lost for each mole of phenylbutyrate given. Phenylbutyrate is usually given as the sodium salt in doses of 250 mg/kg/day, but has been given in doses of up to 650 mg/kg/day [16]. In a recent study of the side effects [17], there was a high incidence of menstrual disturbance in females. Other problems included anorexia, but it was not easy to distinguish between the effects of the disorder and those of the medicine. Patients are often reluctant to take the medicine, and great ingenuity is sometimes needed to ensure that the patient takes it.

ARGININE AND CITRULLINE. Arginine is normally a nonessential amino acid, because it is synthesised within the urea cycle. For this reason, all patients with urea-cycle disorders (except those with arginase deficiency) are likely to need a supplement of arginine to replace that which is not synthesised [18]. The aim should be to maintain plasma arginine concentrations between 50 µmol/l and 200 µmol/l. For OTC and CPS deficiencies, a dose of 100–150 mg/kg/day appears to be sufficient for most patients. However, in severe variants of OTC and CPS, citrulline may be substituted for arginine in doses up to 170 mg/kg/day, as this will utilise an additional nitrogen molecule. Patients with citrullinaemia and argininosuccinic aciduria have a higher requirement, because ornithine is lost as a result of the metabolic block; this is replaced by administering arginine. Doses of up to 700 mg/kg/day may be needed, but this does have the disadvantage of increasing the concentrations of citrulline and argininosuccinate, respectively. The consequences of this are thought to be less important than those caused by the accumulation of ammonia and glutamine.

OTHER MEDICATION. *Citrate* has long been used to provide a supply of Krebs-cycle intermediates [19]. It is known to reduce postprandial elevation of ammonia and may be helpful in the management of argininosuccinic aciduria [20].

*N-carbamyl glutamate* can be used in N-acetylglutamate synthetase deficiency to replace the missing compound, as it is active orally. The dose is 100–300 mg/kg/day [21]. Patients who respond may require treatment with this compound only. *Anticonvulsants* may be needed for patients with urea-cycle disorders, but sodium valproate should *not* be used, as this drug may precipitate fatal decompensation, particularly in OTC patients [22].

### General Aspects of Therapy

All treatment must be monitored with regular quantitative estimation of plasma ammonia and amino acids, paying particular attention to the concentrations of glutamine and essential amino acids. The aim is to keep plasma ammonia levels below 80 µmol/l and plasma glutamine levels below 800 µmol/l [23]. In practice, a glutamine concentration of 1000 µmol/l together with concentrations of essential amino acids within the normal range (see the algorithm, Fig. 17.2) is probably more realistic. All diets must, of course, be nutritionally complete and must meet requirements for growth and normal development.

The concept of balance of diet and medicine is important. The protein intake of the patients varies considerably, and the figures that have been given should be regarded only as a guide. The variation

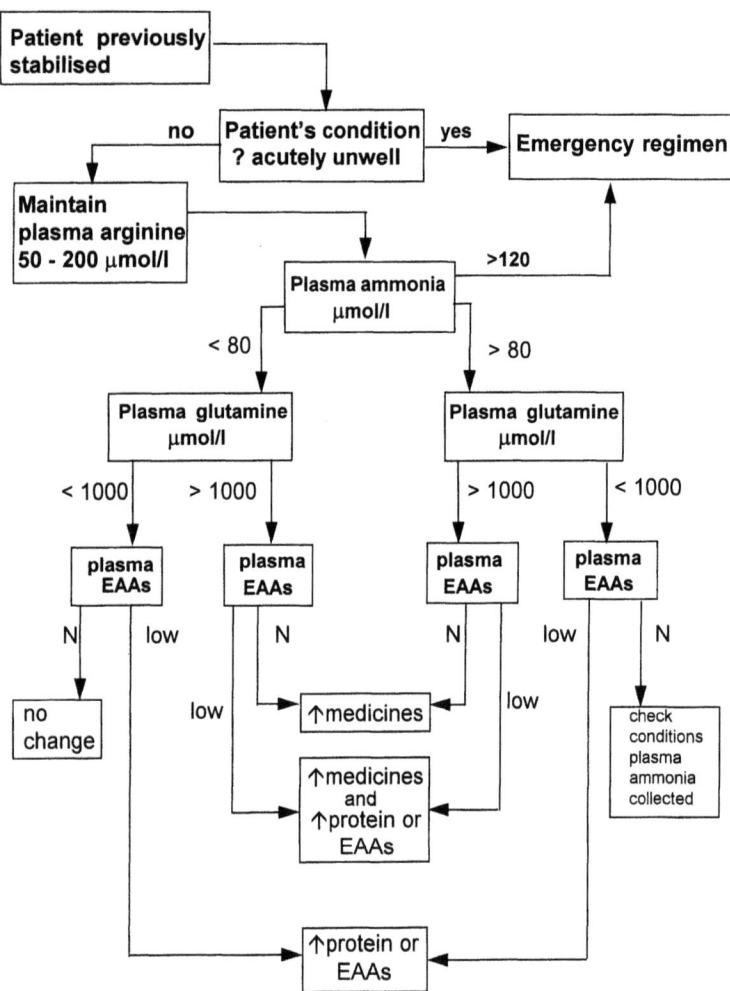

Fig. 17.2. Guidelines for the management of patients with urea-cycle disorders (except arginase deficiency). This is intended for use in patients who have been stabilised previously and should only be regarded as a guide, as some patients may have individual requirements. For more detail and information about doses, please refer to the text. *EAAs*, essential amino acids

reflects not only the residual enzyme activity but also many other factors, including appetite and growth rate. Some patients have an aversion to protein, so it can be difficult to get them to take even their recommended intake. Consequently, they are likely to need smaller doses of sodium benzoate and phenylbutyrate. Others prefer to take more protein, and this has to be balanced by an increase in the dosages of benzoate and phenylbutyrate. Some will not take adequate quantities of sodium benzoate or sodium phenylbutyrate and, therefore, their protein intakes necessarily have to be stricter than would be needed if they took the medicines. Hence, for each patient, a balance must be found between their protein intake and the dose of their medicines to achieve good metabolic control.

### Assessment for Treatment

All patients with urea cycle disorders are at risk of acute decompensation with acute hyperammonaemia. This can be precipitated by different kinds of metabolic stress, such as fasting, a large protein load, infection, anaesthesia or surgery. For this reason, all patients should have detailed instructions of what to do when they are at risk. We routinely use a three-stage procedure [24]. If the patient is off-colour, the protein is reduced, and more carbohydrate is given. If symptoms continue, protein should be stopped and a high-energy intake given with their medication by day and night. However, if they cannot tolerate oral drinks and medicines, are vomiting or are becoming progressively encephalopathic, they should go to a hospital for assessment and intravenous therapy without delay. For further practical details, see Dixon and Leonard [24]. Patients should also have a high carbohydrate intake before any anaesthesia or surgery.

For patients who are seriously ill with hyperammonaemia, treatment is urgent. The steps are listed below, and early treatment is essential (Chap. 3).

### Emergency Treatment

The volumes which are given are related to age and the condition of the patient. Fluid volumes should be restricted if there is any concern about cerebral oedema.

- Stop protein intake.
- Give a high energy intake.
  1. Orally: (a) 10-20% soluble glucose polymer or (b) protein-free formula or
  2. Intravenously: (a) 10% glucose by peripheral infusion or (b) 10-25% glucose by central venous line
- Give sodium benzoate up to 500 mg/kg/day orally or intravenously.
- Give sodium phenylbutyrate up to 600 mg/kg/day.
- Give L-arginine:
  - Up to 700 mg/kg/day in citrullinaemia and arginosuccinic aciduria.
  - Up to 150 mg/kg/day in OTC and CPS deficiencies.
  - For the emergency treatment of hyperammonaemia before diagnosis is known, this may be replaced by L-arginine 300 mg/kg/24 h and L-carnitine 200 mg/kg/24 h. Both can be given orally or intravenously.
- Dialysis. If hyperammonaemia is not controlled or the medicines are not immediately available, haemofiltration (or haemodialysis/haemodiafiltration) should be started without delay. Alternatively, peritoneal dialysis can be used, but this is a less effective method for reducing hyperammonaemia.
- Treat other conditions (sepsis, fits, etc.).
- Monitor intracranial pressure with the usual measures to reduce raised pressure and maintain perfusion pressure.

## Prognosis

The prognosis in these disorders is closely related to the age of the patient and their condition at the time of diagnosis. For those patients who present with symptomatic hyperammonaemia in the newborn period, the outlook is very poor. Even with the most aggressive treatment, the majority of the survivors will be handicapped. Those who are treated prospectively do better, but there may still be significant complications [25]. For these patients, there remains a serious risk of decompensation, and careful consideration should be given to early liver transplantation, which may offer the hope of a better long-term outlook [26]. Of those who present later, their neurological problems at the time of diagnosis are critical, as most will have already suffered neurological damage. At best, this may apparently resolve, but almost all are left with some degree of learning and neurological problems. Patients who have widespread cerebral oedema almost all die or survive with severe handicaps. By contrast, those who are treated prospectively have a better outcome.

## Genetics and Prenatal Diagnosis

The genes for urea-cycle enzymes (except N-acetylglutamate synthetase) have been mapped, isolated and fully characterised [27]. Many mutations have been described. The most common urea cycle disorder is OTC deficiency, which is an X-linked disorder in which molecular genetic studies are particularly helpful. When the diagnosis of OTC deficiency is established, it is necessary to take a careful family history and for the mother's carrier status to be assessed. Currently, if the mutation is not known, the most convenient investigation is the allopurinol test, which is used to detect increased de novo synthesis of pyrimidines (see "Metabolic Derangement"). It appears to have good sensitivity and specificity [28, 29]. This is easier than the protein- or alanine-loading tests and carries no risk of hyperammonaemia. Prenatal diagnosis using a gene probe for mutation detection or to identify informative polymorphisms can help most families. However, whilst the phenotype of the males can be predicted, that of the females cannot because of the random inactivation of the X chromosome. This presents a problem when counselling families, but the prognosis for females who are treated prospectively from birth is good.

All the other conditions have autosomal-recessive inheritance, and prenatal diagnosis is possible for all disorders except N-acetyl glutamate synthetase deficiency. For CPS deficiency, prenatal diagnosis using closely linked gene markers is now possible for a substantial proportion of families. If the molecular-genetic studies are uninformative, prenatal liver biopsy is a possible alternative. Citrullinaemia and argininosuccinic aciduria can both be diagnosed on chorionic villus biopsy. Arginase deficiency can be diagnosed either with molecular-genetic studies or, if they are not informative, with a foetal blood sample.

## References

1. Arn PH, Hauser ER, Thomas GH et al. (1990) Hyperammonemia in women with a mutation at the ornithine carbamoyltransferase locus. A cause of postpartum coma. N Engl J Med 322:1652-1655
2. Christodoulou J, Qureshi IA, McInnes RR, Clarke JT (1993) Ornithine transcarbamylase deficiency presenting with strokelike episodes. J Pediatr 122:423-425
3. Fowler GW (1979) Intermittent ataxia in heterozygote ornithine transcarbamylase deficiency. Ann Neurol 6:185-186
4. Patel JS, Van't Hoff W, Leonard JV (1994) Arginase deficiency presenting with convulsions. J Inherit Metab Dis 17:254
5. Newsholme EA, Leech AR (1983) Biochemistry for the medical sciences. Wiley, Chichester, pp 491-494
6. Batshaw ML, Brusilow SW (1978) Asymptomatic hyperammonaemia in low birthweight infants. Pediatr Res 12:221-224
7. Bachmann C, Colombo JP (1983) Increased tryptophan uptake into the brain in hyperammonaemia. Life Sci 33: 2417-2424

8. Hyman SL, Porter CA, Page TJ et al. (1987) Behavior management of feeding disturbances in urea cycle and organic acid disorders. J Pediatr 111:558-562
9. Surtees RJ, Leonard JV (1989) Acute metabolic encephalopathy. J Inherit Metab Dis 12 [Suppl 1]:42-54
10. Connelly A, Cross JH, Gadian DG et al. (1993) Magnetic resonance spectroscopy shows increased brain glutamine in ornithine carbamoyl transferase deficiency. Pediatr Res 33:77-81
11. Batshaw ML, Berry GT (1991) Use of citrulline as a diagnostic marker in the prospective treatment of urea cycle disorders. J Pediatr 118:914-917
12. Kendall B, Kingsley DPE, Leonard JV, Lingam S, Oberholzer VG (1983) Neurological features and computed tomography of the brain in children with ornithine carbamyl transferase deficiency. J Neurol Neurosurg Psychiatr 46:28-34
13. Hudak ML, Jones MD, Brusilow SW (1985) Differentiation of transient hyperammonaemia of the newborn and urea cycle enzyme defects by clinical presentation. J Pediatr 107:712-719
14. Brusilow SW, Valle DL, Batshaw ML (1979) New pathways of nitrogen excretion in inborn errors of urea synthesis. Lancet II:452-454
15. Feillet F, Leonard JV (1998) Alternative pathway therapy for urea cycle disorders. J Inherit Metab Dis 21 [Suppl 1]:101-111
16. Brusilow SW (1991) Phenylacetylglutamine may replace urea as a vehicle for waste nitrogen excretion. Pediatr Res 29:147-150
17. Wiech NL, Clissold DM, MacArthur RB (1997) Safety and efficacy of buphenyl (sodium phenylbutyrate) tablets and powder (abstract). Advances in inherited urea cycle disorders, satellite to the 7th international congress for inborn errors of metabolism, Vienna, p 25
18. Brusilow SW (1984) Arginine, an indispensible aminoacid for patients with inborn errors of urea synthesis. J Clin Invest 74:2144-2148
19. Levin B, Russell A (1967) Treatment of hyperammonaemia. Am J Dis Child 113:142-144
20. Iafolla AK, Gale DS, Roe CR (1990) Citrate therapy in arginosuccinate lyase deficiency. J Pediatr 117:102-105
21. Bachmann C, Colombo JP, Jaggi K (1982) N-acetylglutamate synthetase (NAGS) deficiency: diagnosis, clinical observations and treatment. Adv Exp Med Biol 153:39-45
22. Tripp JH, Hargreaves T, Anthony PP et al. (1981) Sodium Valproate and ornithine carbamyl transferase deficiency (letter). Lancet 1:1165-1166
23. Maestri NE, McGowan KD, Brusilow SW (1992) Plasma glutamine concentration: a guide to the management of urea cycle disorders. J Pediatr 121:259-261
24. Dixon MA, Leonard JV (1992) Intercurrent illness in inborn errors of intermediary metabolism. Arch Dis Child 67:1387-1391
25. Maestri NE, Hauser ER, Bartholomew D, Brusilow SW (1991) Prospective treatment of urea cycle disorders. J Pediatr 119:923-928
26. Todo S, Starzl TE, Tzakis A et al. (1992) Orthotopic liver transplantation for urea cycle enzyme deficiency. Hepatology 15:419-422
27. Saudubray J-M, Tonati G, DeLonlay P et al. (1999) Liver transplantation in urea cycle disorders. Eur J Pediatr 158 [Suppl 2]:S55-59
28. Hauser ER, Finkelstein JE, Valle D, Brusilow SW (1990) Allopurinol-induced orotidinuria. A test for mutations at the ornithine carbamoyltransferase locus in women. N Engl J Med 322:1641-1645
29. Sebesta I, Fairbanks LD, Davies PM, Simmonds HA, Leonard JV (1994) The allopurinol loading test for the identification of carriers for ornithine carbamoyl transferase deficiency: studies in a healthy control population and females at risk. Clin Chim Acta 224:45-54

# CHAPTER 18

## Metabolism of the Sulfur-Containing Amino Acids

Methionine, homocysteine and cysteine are linked by the methylation cycle (Fig. 18.1, left) and the trans-sulfuration pathway (Fig. 18.1, right). Conversion of methionine into homocysteine proceeds via methionine S-adenosyltransferase (enzyme 4). This yields S-adenosylmethionine, the methyl-group donor in a wide range of transmethylation reactions. These reactions also produce S-adenosylhomocysteine, which is cleaved to adenosine and homocysteine. Depending on a number of factors, about 50% of available homocysteine is recycled into methionine. This involves methyl transfer from either 5-methyl-tetrahydrofolate (THF), catalyzed by cobalamin-requiring 5-methyl THF-homocysteine methyltransferase (enzyme 2; see also Fig. 25.2), or betaine, catalyzed by betaine-homocysteine methyltransferase (enzyme 3). Homocysteine can also be condensed with serine to form cystathionine via a reaction catalyzed by pyridoxal-phosphate-requiring cystathionine-β-synthase (enzyme 1). Cystathionine is cleaved to cysteine and α-ketobutyrate by another pyridoxal-phosphate-dependent enzyme, γ-cystathionase (enzyme 5). The last step of the trans-sulfuration pathway converts sulfite to sulfate and is catalyzed by sulfite oxidase (enzyme 6), which requires a molybdenum cofactor.

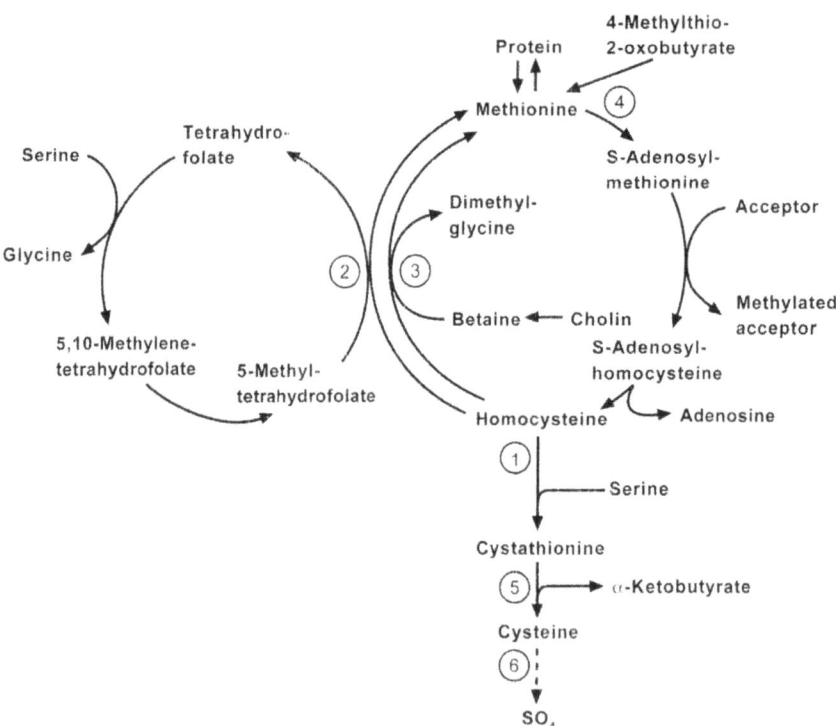

**Fig. 18.1.** Metabolism of the sulfur-containing amino acids. 1, Cystathionine β-synthase; 2, 5-methyltetrahydrofolate-homocysteine methyltransferase; 3, betaine-homocysteine methyltransferase; 4, methionine adenosyltransferase; 5, γ-cystathionase; 6, sulfite oxidase

# Disorders of Sulfur Amino Acid Metabolism

Generoso Andria, Brian Fowler, and Gianfranco Sebastio

CONTENTS

Homocystinuria due to Cystathione-β-Synthase
Deficiency ................................... 225
   Clinical Presentation ......................... 225
      Eye ....................................... 225
      Skeleton ................................. 225
      Central Nervous System ................... 226
      Vascular System .......................... 226
   Metabolic Derangement ..................... 226
   Diagnostic Tests ............................ 227
      Prenatal Diagnosis ....................... 227
      Heterozygotes ............................ 227
   Treatment and Prognosis .................... 228
   Genetics ..................................... 228
Methionine-S-Adenosyltransferase Deficiency ......... 229
γ-Cystathionase Deficiency ....................... 229
Isolated Sulfite Oxidase Deficiency ................. 230
References ..................................... 230

Several defects can exist in the conversion of the sulfur-containing amino acid methionine to cysteine and the ultimate oxidation of cysteine to inorganic sulfate (Fig. 18.1). Cystathionine-β-synthase (CBS) deficiency is the most important. It is associated with severe abnormalities of four organs or organ systems: the eye (dislocation of the lens), the skeleton (dolichostenomelia and arachnodactyly), the vascular system (thromboembolism), and the central nervous system (mental retardation, cerebrovascular accidents). A low-methionine, high-cystine diet, pyridoxine, folate, and betaine in various combinations, and antithrombotic treatment may halt the otherwise unfavorable course of the disease. Methionine adenosyltransferase deficiency and γ-cystathionase deficiency usually do not require treatment. Isolated sulfite oxidase deficiency leads (in its severe form) to refractory convulsions, lens dislocation, and early death. No effective treatment exists.

## Homocystinuria due to Cystathione-β-Synthase Deficiency

### Clinical Presentation

The eye, skeleton, central nervous system, and vascular system are all involved in the typical presentation. The patient is normal at birth and, if not treated, progressively develops the full clinical picture.

### Eye

*Ectopia lentis* (the dislocation of the ocular lens), myopia, and glaucoma (sometimes with pupil entrapment of the dislocated lens) are frequent, severe and characteristic complications. Retinal detachment and degeneration, optical atrophy, and cataracts eventually appear. Myopia may precede lens dislocation, and worsens afterwards. Ectopia lentis is detected in most untreated patients after a few years during the first decade of life and in nearly all patients by the end of the fourth decade; it is often the clue to diagnosis. The dislocation is generally downwards, whereas it is usually upwards in Marfan syndrome. Once ectopia lentis has occurred, a peculiar trembling of the iris (iridodonesis) following eye or head movement may be evident.

### Skeleton

Osteoporosis is almost invariably detected, at least after childhood. Frequent consequences are scoliosis and a tendency towards pathological fractures and vertebral collapse. As in Marfan syndrome, homocystinuric patients tend to be tall, with thinning and elongation (dolichostenomelia) of long bones near puberty and enlarged metaphyses and epiphyses, especially at the knees. Arachnodactyly, defined as a metacarpal index (the average ratio of length to breadth for metacarpals II–V) above 8.5, is present in about half the patients. Other bone deformities include *genu valgum* with knobby knees, *pes cavus*, and *pectus carinatum* or *excavatum*. Restricted joint mobility, particularly at the extremities, contrasts with the joint laxity observed in Marfan syndrome. Abnormal X-ray findings include biconcavity and flattening of the intervertebral discs, growth arrest lines in the distal tibia, metaphyseal spicules in the hands and feet, enlarged carpal bones, retarded lunate development, and shortening of the fourth metacarpal.

### Central Nervous System

Developmental delay and mental retardation affect about 60% of patients to a variable degree of severity. Seizures, electroencephalogram abnormalities, and psychiatric disturbances have also been reported in approximately half of cases. Focal neurologic signs may be a consequence of cerebrovascular accidents (see below).

### Vascular System

Thromboembolic complications, occurring in arteries and veins of all parts of the body, constitute the major cause of morbidity and mortality. The prognosis is influenced by the site and extent of the vascular occlusion. Thrombophlebitis and pulmonary embolism are the commonest vascular accidents. Thrombosis of large- and medium-sized arteries, particularly carotid and renal arteries, is a frequent cause of death. Ischemic heart disease has also been reported, though it does not represent a prominent feature of homocystinuria. Noninvasive methods, such as echo-Doppler techniques, can detect abnormalities of the vessels at a presymptomatic stage [1]. Even a relatively small increase of plasma homocyst(e)ine, measured fasting or after methionine loading, has been demonstrated to be associated with a higher risk of premature vascular disease [2, 3].

### Clinical Variability and Natural History

The spectrum of clinical abnormalities is wide, and mild cases may not be recognized until later severe complications, such as thromboembolic accidents, occur. Using a questionnaire survey, detailed information on 629 patients has been assembled by Mudd et al. [4]. Time-to-event curves were calculated for the main clinical manifestations and mortality. Each abnormality occurred significantly earlier and at a higher rate in untreated pyridoxine-nonresponsive individuals than in untreated pyridoxine-responsive ones. Interestingly, the risk of thromboembolic accidents in patients undergoing surgery was relatively small, complications (six of which were lethal) being recorded in only 25 patients following 586 operations.

In an Italian multicenter survey [5], a strong correspondence of ectopia lentis, mental retardation, seizures, dolichostenomelia, and thrombotic accidents was found among 11 couples of affected siblings, supporting a prominent role of genetic factors in determining the phenotype. Nevertheless, rare cases of intrafamilial variability have been reported [6].

Pyridoxine-responsive women can carry out pregnancies without a significant risk of malformations in the offspring. However, this issue is not definitely settled. The outcome of pregnancies in non-responsive women seems to be worse [4].

### Metabolic Derangement

CBS activity can be found in many tissues, including liver (from which it has been purified), brain, pancreas, and cultured fibroblasts [7]. In addition to the coenzyme pyridoxal phosphate, CBS also binds two other ligands: the activator S-adenosylmethionine and a heme moiety whose function is not yet clear [6]. Fibroblasts from patients with homocystinuria can be classified into three groups: those exhibiting no residual activity; those exhibiting reduced activity with normal affinity for pyridoxal phosphate; and those exhibiting residual activity with markedly reduced affinity for pyridoxal phosphate [8]. In vivo responsiveness to pharmacological doses of pyridoxine, present in approximately 50% of homocystinuric patients, is generally associated with the presence of a small amount of residual enzymatic activity, although exceptions to this rule are known.

Deficiency of CBS leads to tissue accumulation of methionine, homocysteine, and their S-adenosyl derivatives, with lack of cystathionine and low levels of cysteine. The –SH group of homocysteine can easily react with the –SH group of a second homocysteine molecule or of other molecules, leading to the formation of a number of disulfide compounds, such as homocystine, homocysteine–cysteine mixed disulfide or protein-bound homocysteine.

The pathophysiology of CBS deficiency has not yet been completely elucidated, but accumulation of homocysteine probably plays a major role in determining some of the most relevant clinical manifestations, including generalized vascular damage and thromboembolic complications. This view is supported by the observation that patients with homocystinuria (due to defects in the remethylation pathway) but without accumulation of methionine show similar lesions in blood vessels [9]. Thromboembolism has been suggested to be the end-point of homocysteine-induced abnormalities of platelets, endothelial cells, and coagulation factors.

Homocysteine may also cause abnormal cross-linking of collagen, leading to abnormalities of the skin, joints, and skeleton in patients. This mechanism seems unlikely to cause damage of the non-collagenous zonular fibers of the lens; instead, disturbed fibrillin structure has been proposed.

## Diagnostic Tests

Screening of urine with the cyanide–nitroprusside test often yields positive results, but it can give false negative results and lacks specificity. A modified nitroprusside reagent is more specific for homocystine [10].

Initial diagnosis is best achieved by ion-exchange chromatography of plasma, which must be immediately deproteinized to prevent loss of disulfide amino acids by binding to protein. Increased levels of methionine, homocystine and cysteine–homocysteine disulfide, with low cystine and no increase of cystathionine, is typical of CBS deficiency [10]. As some responsive patients are extremely sensitive to very low doses of pyridoxine (as contained in multivitamin tablets), false-negative results may be obtained.

A useful adjunct to this approach and for monitoring treatment is the determination of total homocysteine after treatment of the plasma sample with reducing agents (70–80% of homocysteinyl moieties are bound to protein). Normal plasma total homocyst(e)ine values are less than 15 μmol/l [11], whereas most untreated CBS patients exhibit levels above 200 μmol/l.

Hyperhomocyst(e)inemia also occurs in remethylation defects, due to 5,10-methylene-tetrahydrofolate (THF)-reductase deficiency and 5-methyl-THF-homocysteine-methyltransferase deficiency either isolated or due to defects in cytosolic cobalamin metabolism (Chap. 25). These disorders can mostly be distinguished from CBS deficiency by the very low to normal plasma methionine level. It must be noted that, in CBS deficiency, methionine concentrations tend to decrease with age and may even be normal in some older patients; this decrease may be contributed to by folate depletion and a consequent reduced capacity for remethylation.

A wide range of non-genetic causes of hyperhomocyst(e)inemia are known: for example, renal insufficiency and administration of 6-azauridine triacetate, methotrexate, isonicotinic acid hydroxide, and colestipol combined with niacin. Other biochemical sequelae include a slight increase in plasma concentrations of ornithine, copper, and ceruloplasmin and decreased serine levels.

Definitive diagnosis requires demonstration of greatly reduced CBS activity, usually assayed in cultured skin fibroblasts [7, 8] but also possible in phytohemagglutinin-stimulated lymphocytes [12] and liver biopsies [13]. Exceptional patients may have significant residual activity of CBS in fibroblast extracts but have the typical abnormalities of the disease. The molecular diagnosis of CBS deficiency now provides a powerful additional approach to the diagnosis (see "Genetics"). In many countries, newborn mass-screening programs based on detection of hypermethioninemia have been implemented, but they yield an unacceptably high rate of false-negative results, since many pyridoxine-responsive patients are missed [14].

### Prenatal Diagnosis

Prenatal diagnosis of homocystinuria has been performed in some at-risk pregnancies by assaying CBS in extracts of cultured amniocytes [15]. CBS activity is very low in uncultured chorionic villi from control subjects and can only be measured after culturing [16]. In families where the mutation(s) is known, direct analysis of the CBS gene will allow rapid prenatal diagnosis and, in other cases, DNA linkage to the CBS locus may have diagnostic value.

### Heterozygotes

On a *group basis*, differences between obligate heterozygotes and control subjects have been clearly demonstrated using either measurements of homocyst(e)ine in plasma after methionine loading [17] or assay of CBS in liver biopsies, cultured skin fibroblasts, or phytohemagglutinin-stimulated lymphocytes [18]. However, with the methods described so far, a considerable number of obligate heterozygotes exhibit values that overlap those at the lower end of the control range. Therefore, the diagnostic power of these two approaches for heterozygote testing in *individual subjects* is limited.

Molecular analysis of established mutations will allow heterozygote detection in individual families, and the most common CBS mutations (see "Genetics") might be considered in population screening for CBS heterozygotes. A CBS activity compatible with a heterozygous condition was found in a significant number of vascular patients with hyperhomocyst(e)inemia [19, 20]. In another study, obligate heterozygotes for CBS deficiency showed early vascular lesions (detected by non-invasive ultrasound methods) in a pre-symptomatic stage at a significant higher rate than was exhibited by controls [1].

Since the molecular basis of CBS deficiency has been elucidated, evidence against the role of heterozygosity for CBS deficiency for hyperhomocyst(e)inemia has been suggested. In fact, the search for CBS-mutation disease failed to demonstrate a causative role of CBS heterozygosity in patients affected by premature vascular disease [21]. However, a major role for the C677T mutation of 5,10-methylene-THF reductase has been established for moderate hyperhomocyst(e)inemia and premature vascular disease [22].

## Treatment and Prognosis

The aim of treatment is to reduce plasma total homocyst(e)ine levels to as close to normal as possible while maintaining the normal growth rate. Plasma cystine should be kept within the normal range (67 ± 20 μmol/l) and should be supplemented accordingly (up to 200 mg/kg/day), since it becomes an essential amino acid in methionine-restricted diets. Homocysteine levels can be lowered in a number of ways, and the best approach or combination for the individual patient will depend on the nature of the defect and social factors.

About half of patients with CBS deficiency respond, often only partially, to large oral doses of pyridoxine. In fully responsive patients, fasting plasma homocystine disappears, hypermethioninemia decreases, and cystinemia increases to values within the control range, following a variable period of up to a few weeks of daily administration of between a few milligrams and 1000 mg of pyridoxine.

Response to the vitamin is also influenced by folate depletion, which may be due to pyridoxine administration itself. Therefore, folic acid (5–10 mg/day) should be added to the treatment [23].

Doses higher than 1000 mg/day should be avoided, since megadoses taken for other disorders have been associated with sensory neuropathy. In patients who do not respond to pyridoxine, a low-methionine/high-cystine diet must be introduced [24]. A less strict low-methionine diet may also be necessary to achieve adequate control in pyridoxine-responsive patients. Synthetic methionine-free amino acid mixtures are commercially available and are especially useful for infants (Hominex-1, Hominex-2: Ross Products Division, USA; Analog RVHB, Maxamaid RVHB, Maxamum RVHB, Albumaid Methionine low, Methionine-free amino acid mix: Scientific Hospital Supplies, UK; HOM 1, HOM 2: Milupa AG, Germany). The requirement for methionine is met by small amounts of infant formula; supplements of essential fatty acids and carbohydrates are also required. After infancy, foods low in methionine (relative to their protein contents) can be introduced, including gelatin and pulses (such as lentils and soybeans). However, it should be noted that soya-modified formulas are usually enriched with methionine. In homocystinuria, a more strict regimen is not necessary during infections and catabolism, unlike the case for other amino acid disorders. In addition to pyridoxine, folate and, possibly vitamin $B_{12}$, the usual vitamin and mineral supplements are recommended.

Betaine given orally at a maximum dose of 150 mg/kg/day is another important homocysteine-lowering agent that can be useful in addition to the other treatments, especially when it is difficult to obtain good compliance to diet. Betaine remethylates homocysteine [25], often leading to very high methionine concentrations, but these apparently do not influence the pathophysiology of the disease.

Dipyridamole (100 mg four times per day) either alone or combined with aspirin (100 mg/day with 1 g aspirin/day) has been proposed for the primary prevention of thromboembolic complications, but no clear-cut evidence of its effectiveness has been obtained. Once either venous or arterial thrombotic events have occurred, anticoagulant and antiplatelet therapy (as used for such complications due to other causes) should be started. As the suggested aspirin dose is potentially dangerous, a rationale for testing the efficacy and safety of low-dose aspirin has been reported [26].

Whatever the combination of regimens employed, it is very difficult to achieve the aim of virtually normal total homocysteine levels in most patients. Nevertheless, lifelong treatment is clearly needed to prevent the severe clinical abnormalities associated with this disorder, and considerable impact on outcome has been achieved in patients for whom the criterion of treatment was removal of free-disulfide homocystine from plasma [27]. The results of the international survey provide a firmly established baseline for the evaluation of past and future therapeutic regimens [4]. When the low-methionine diet was started in the newborn period, mental retardation was prevented, the start and progression of lens dislocation were delayed, and the incidence of seizures decreased. When late-diagnosed, responsive subjects received pyridoxine treatment, the first thromboembolic episode occurred later.

Since the treatment is more successful when the diagnosis is made early, mass newborn-screening programs should probably be implemented as soon as the present technical pitfalls are solved; screening could be performed by the detection of homocysteine by tandem mass spectrometry. It is still unknown whether preventive treatment with vitamins is indicated in heterozygotes for homocystinuria or in other subjects with homocyst(e)inemia from various causes.

## Genetics

Homocystinuria due to CBS deficiency is inherited as an autosomal-recessive trait. Clinical and biochemical variations, such as pyridoxine responsiveness, are also genetically determined and related to specific mutations.

CBS is a tetramer of identical 63-kDa subunits but can also exist as a more active dimer of 48-kDa subunits. The CBS gene is located on chromosome 21 (21q22.3), and both the cDNA and the gene have been characterized [28, 29]. Molecular studies on CBS patients have led to the characterization of more than

90 mutations [30], most of which are private. Only three mutations appear to be of epidemiologic relevance. In some patients, a double mutational event has been observed on a single allele. This makes systematic screening for mutations of the entire coding region of the CBS gene a prerequisite for reliable establishment of genotype/phenotype correlation.

In at least 5% of Caucasian alleles, exon 8 displays a 68-bp duplication of the 5′ intron–exon junction. Whether this polymorphic mutation plays a role in mild hyperhomocysteinemia as a risk factor and in multifactorial diseases related to this condition remains to be established [31, 32].

## Methionine-S-Adenosyltransferase Deficiency

### Clinical Presentation

Of a total of 30 patients described, 27 have been symptom-free, indicating a benign disorder [33]. However, two subjects developed neurological abnormalities and demyelination of the brain attributed to deficient formation of S-adenosylmethionine, the methyl donor in the synthesis of major myelin phospholipids.

### Metabolic Derangement

This disorder is characterized by a deficiency of the hepatic form (but not the extrahepatic form) of the enzyme leading to elevated methionine concentrations in tissues and physiological fluids; this enzyme is coded by different genes. The product of this enzyme reaction, S-adenosylmethionine, appears not to be deficient in most cases. Alternative metabolism of methionine seems to occur above a threshold plasma methionine concentration of about 300 µM, resulting in the formation of the transamination product 4-methylthio-3-oxobutyrate and dimethyl sulfide, the latter resulting in a distinct odor of the breath.

### Diagnostic Tests

High methionine in plasma and urine (detected by usual chromatographic methods) without increased homocyst(e)ine is suggestive of this defect, but several other causes of hypermethioninemia are possible and must be excluded.

### Treatment and Prognosis

Treatment is generally not indicated but, in patients with evidence of demyelination, administration of S-adenosylmethionine corrects deficiency of this compound. If the postulated association between specific mutations leading to a severe enzyme deficiency holds true [34, 35], treatment with S-adenosylmethionine may be advisable in such cases.

### Genetics

Three forms of methionine S-adenosyltransferase (MAT) are known: MAT-I, -II, and -III. MAT-I and -III are encoded by the same gene, MAT1, and correspond to tetrameric and dimeric forms of a single α1 subunit, respectively. MAT-II is encoded by a separate gene, mainly expressed in fetal liver and in kidney, brain, testis, and lymphocytes. Mutations of the MAT1 gene account for both autosomal-recessive and autosomal-dominant hypermethioninemia [36, 37]. The rarer, latter form is caused by mutation on a single allele with a dominant-negative effect.

## γ-Cystathionase Deficiency

### Clinical Presentation

This is considered to be a benign disorder. Subjects detected without ascertainment bias are mainly asymptomatic subjects with mental retardation having healthy siblings with the same defect or without the defect but showing the same symptoms.

### Metabolic Derangement

Deficiency of the pyridoxal-phosphate-requiring γ-cystathionase leads to tissue accumulation of cystathionine. Increased plasma concentrations and markedly increased excretion of cystathionine occur, and N-acetylcystathionine is also excreted.

### Diagnostic Tests

High urinary excretion of cystathionine without homocystine and with normal plasma methionine points to this defect. Transient cystathioninuria in newborns is due to known secondary causes, such as vitamin-$B_6$ deficiency, generalized liver disease, thyrotoxicosis, and neural tumors. Milder increases of plasma and urine levels of cystathionine can also occur in the remethylation defects due to overproduction of this metabolite. While γ-cystathionase activity is certainly expressed in cultured skin fibroblasts, the level of activity is probably too small to allow reliable measurement by specific enzyme assay [38].

### Treatment and Prognosis

Most subjects respond to administration of about 100 mg of pyridoxine daily though, as a benign disorder, it is debatable whether treatment is needed.

## Genetics

Inheritance is autosomal recessive. The cystathionase gene was cloned [39], but no molecular studies on individuals with this condition have been reported.

## Isolated Sulfite Oxidase Deficiency

### Clinical Presentation

Characteristic findings in the severe form of this enzyme deficiency, whether isolated (nine patients reported) or due to molybdenum cofactor deficiency (more than 50 described), are early refractory convulsions, severe psychomotor retardation, failure to thrive, microcephaly, hypotonia passing into hypertonia, lens dislocation, and early death [40]. Milder presentation has been reported.

### Metabolic Derangement

Sulfite oxidase catalyses the last step in the oxidation of the sulfur atom of cysteine into inorganic sulfate (Fig. 18.1). Its deficiency results in accumulation of the suspected toxic compound sulfite together with its detoxification products, S-sulfocysteine and thiosulfate, with reduced formation of sulfate.

### Diagnostic Tests

Increased sulfite can be detected with urine-test strips, but samples must be fresh and, in one case, no increased sulfite with normal sulfate excretion was reported. S-sulfocysteine is a stable, ninhydrin-positive diagnostic parameter that can be detected by electrophoresis or chromatography and can be quantified in urine and plasma by classical ion-exchange techniques. Cystine levels are always very low. Using a thin-layer-chromatography technique, specific excretion of thiosulfate can also be searched for. The absence of xanthinuria distinguishes the isolated deficiency from the molybdenum-cofactor defect, in which xanthinuria is observed (Chap. 31). Sulfite oxidase activity must be determined in the liver.

### Treatment and Prognosis

So far, no successful treatment has been reported. Attempts to remove sulfite by binding to penicillamine were unsuccessful [41].

### Genetics

This autosomal-recessive disease has been explained at the molecular level by cloning of the gene and characterization of mutations in a few patients affected by the isolated sulfite oxidase deficiency [42]. The gene for molybdenum cofactor deficiency has been localized to chromosome 6.

## References

1. Rubba P, Faccenda F, Pauciullo P, Carbone L, Mancini M, Strisciuglio P, Carrozzo R, Sartorio R, Del Giudice E, Andria G (1990) Early signs of vascular disease in homocystinuria: a noninvasive study by ultrasound methods in eight families with cystathionine β-synthase deficiency. Metabolism 39:1191–1195
2. Kang S-S, Wong PWK, Malinow MR (1992) Hyperhomocyst(e)inemia as a risk factor for occlusive vascular disease. Annu Rev Nutr 12:279–288
3. Boushey CJ, Beresford SA, Omenn GS, Motulsky AG (1995) A quantitative assessment of plasma homocysteine as a risk factor for vascular disease. Probable benefits of increasing folic acid intakes. JAMA 274:1049–1057
4. Mudd SH, Skovby F, Levy HL, Pettigrew KD, Wilcken B, Pyeritz RE, Andria G, Boers GHJ, Bromberg IL, Cerone R, Fowler B, Grobe H, Schmidt H, Schweitzer L (1985) The natural history of homocystinuria due to cystathionine β-synthase deficiency. Am J Hum Genet 37:1–31
5. de Franchis R, Sperandeo MP, Sebastio G, Andria G. The Italian Collaborative Study Group on Homocystinuria (1998) Clinical aspects of cystathionine β-synthase deficiency: how wide is the spectrum? Eur J Pediatr 157:S67–70
6. Kraus JP (1994) Molecular basis of phenotype expression in homocystinuria. J Inherited Metab Dis 17:383–390
7. Uhlendorf BW, Mudd SH (1968) Cystathionine β-synthase in tissue culture derived from human skin: enzyme defect in homocystinuria. Science 160:1007–1009
8. Fowler B, Kraus J, Packman S, Rosenberg LE (1978) Homocystinuria: evidence for three distinct classes of cystathionine β-synthase mutants in cultured fibroblasts. J Clin Invest 61:645–653
9. Baumgartner R, Wick H, Ohnacker H, Probst A, Maurer R (1980) Vascular lesions in two patients with congenital homocystinuria due to different defects of remethylation. J Inherited Metab Dis 3:101–103
10. Fowler B, Jakobs C (1998) Post- and prenatal diagnostic methods for the homocystinurias. Eur J Pediatr 157:S88–93
11. ASHG/ACMG Statement (1998) Measurement and use of total plasma homocysteine. Am J Hum Genet 63:1541–1543
12. Goldstein JL, Campbell BK, Gartler SM (1972) Cystathionine β-synthase activity in human lymphocytes: induction by phytohemagglutinin. J Clin Invest 51:1034–1037
13. Finkelstein JD, Mudd SH, Irreverre F, Laster L (1964) Homocystinuria due to cystathionine β-synthase deficiency: the mode of inheritance. Science 146:785–787
14. Naughten ER, Yap S, Mayne PD (1998) Newborn screening for homocystinuria: Irish and world experience. Eur J Pediatr 157:S84–87
15. Fowler B, Borresen AL, Boman N (1982) Prenatal diagnosis of homocystinuria. Lancet 2:875
16. Poenaru L (1987) First trimester prenatal diagnosis of metabolic diseases: a survey in countries from the European Community. Prenat Diagn 7:333–342
17. Sardharwalla IB, Fowler B, Robins AJ, Komrower GM (1974) Detection of heterozygotes for homocystinuria: study of sulfur-containing amino acids in plasma and urine after L-methionine loading. Arch Dis Child 49:553–559
18. McGill JJ, Mettler G, Rosenblatt DS, Scriver CR (1990) Detection of heterozygotes for recessive alleles. Homocyst(e)inemia: paradigm of pitfalls in phenotypes. Am J Med Genet 36:45–52
19. Boers GHJ, Smals AGH, Trijbels FJM, Fowler B, Bakkeren JAJM, Schoonderwaldt HC, Kleijer WJ, Kloppenborg PWC

(1985) Heterozygosity for homocystinuria in premature peripheral and cerebral occlusive arterial disease? N Engl J Med 313:709-715
20. Clarke R, Daly L, Obinson K, Naughten E, Cahalane S, Fowler B, Graham I (1991) Homocysteinemia: a risk factor for vascular disease. N Engl J Med 324:1149-1155
21. Kozich V, Kraus E, de Franchis R, Fowler B, Boers GH, Graham I, Kraus JP (1995) Hyperhomocysteinemia in premature arterial disease: examination of cystathionine beta-synthase alleles at the molecular level. Hum Mol Genet 4:623-629
22. Kluijtmans LAJ, van der Heuvel LPWJ, Boers GHJ, Frosst P, Stevens EMB, van Oost BA, den Heijer M, Trijbels FJ, Rozen R, Blom HJ (1996) Molecular genetic analysis in mild hyperhomocysteinemia: a common mutation in the methylenetetrahydrofolate reductase gene is a genetic risk factor for cardiovascular disease. Am J Hum Genet 58:35-41
23. Wilcken B, Turner B (1973) Homocystinuria: reduced folate levels during pyridoxine treatment. Arch Dis Child 48:58-62
24. The Ross Metabolic Formula System (1997) Nutrition Support Protocols. Ross Products Division, Columbus
25. Wilcken DE, Wilcken B, Dudman NPB, Tyrrell PA (1983) Homocystinuria - the effect of betaine in the treatment of patients not responsive to pyridoxine. N Engl J Med 309:448-453
26. Di Minno G, Davi G, Margaglione M, Cirillo F, Grandone E, Ciabattoni G, Catalano I, Strisciuglio P, Andria G, Patrono C, Mancini M (1993) Abnormally high thromboxane biosynthesis in homozygous homocystinuria. Evidence for platelet involvement and probucol-sensitive mechanism. J Clin Invest 92:1400-1406
27. Wilcken DE, Wilcken B (1997) The natural history of vascular disease in homocystinuria and the effects of treatment. J Inherit Metab Dis 20:295-300
28. Kraus JP, Le K, Swaroop M, Ohura T, Tahara T, Rosenberg LE, Roper MD, Kozich V (1993) Human cystathionine β-synthase cDNA: sequence, alternative splicing and expression in cultured cells. Hum Mol Genet 2:1633-1638
29. Kraus JP, Oliveriusova J, Sokolova J, Kraus E, Vlcek C, de Franchis R, Maclean KN, Bao L, Bukovska G, Patterson D, Paces V, Ansorge W, Kozich V (1998) The human cystathionine β-synthase (CBS) gene: complete sequence, alternative splicing and polymorphism. Genomics 52:312-324
30. Kraus JP, Janosik M, Kozich V, Mandell R, Shih V, Sperandeo MP, Sebastio G, de Franchis R, Andria G, Kluijtmans AJ, Blom H, Boers GHJ, Gordon RB, Kamoun P, Tsai MY, Kruger WD, Koch HG, Ohura T, Gaustadnes (1999) M Cystathionine β-synthase mutations in homocystinuria. Hum Mutat 13:368-375
31. Sebastio G, Sperandeo MP, Panico M, de Franchis R, Kraus JP, Andria G (1995) The molecular basis of homocystinuria due to cystathionine β-synthase deficiency in Italian families and report of four novel mutations. Am J Hum Genet 56:1324-1333
32. Sperandeo MP, de Franchis R, Andria G, Sebastio G (1996) A 68 bp insertion found in a homocystinuric patient is a common variant and is skipped by alternative splicing of the cystathionine β-synthase mRNA. Am J Hum Genet 59:1391-1393
33. Mudd SH, Levy HL, Tangerman A, Boujet C, Buist N, Davidson-Mundt A, Hudgins L, Oyanagi K, Nagao M, Wilson WG (1995b) Isolated persistent hypermethioninemia. Am J Hum Genet 57:882-892
34. Chamberlin ME, Ubagai T, Mudd SH, Wilson WG, Leonard JV, Chou JY (1996) Demyelination of the brain is associated with methionine adenosyltransferase I/III deficiency. J Clin Invest 98:1021-1027
35. Hazelwood S, Barnardini I, Shotelersuk V, Tangerman A, Guo J, Mudd H, Gahl WA (1998) Normal brain myelination in a patient homozygous for a mutation that encodes a severely truncated methionine adenosyltransferase I/III. Am J Med Genet 75:395-400
36. Ubagai T, Lei K-J, Huang S, Mudd SH, Levy HL, Chou JY (1995) Molecular mechanisms of an inborn error of methionine pathway: methionine adenosyltransferase deficiency. J Clin Invest 96:1943-1947
37. Chamberlin ME, Ubagai T, Mudd SH, Levy HL, Chou JY (1997) Dominant inheritance of isolated hypermethioninemia is associated with a mutation in the human methionine adenosyltransferase 1 A gene. Am J Hum Genet 60:540-546
38. Fowler B (1982) Transsulphuration and methylation of homocysteine in control and mutant human fibroblasts. Biochim Biophys Acta 721:201-207
39. Lu Y, Odowd BF, Orrego H, Israel Y (1992) Cloning and nucleotide sequence of human liver cDNA encoding for cystathionine gamma-lyase. Biochem Biophys Res Commun 189:749-758
40. Rupar CA, Gillett J, Gordon BA et al. (1996) Isolated sulfite oxidase deficiency. Neurpediatrics 27:299-304
41. Tardy P, Parvy P, Charpentier C, Bonnefont JP, Saudubray JM, Kamoun P (1989) Attempt at therapy in sulphite oxidase deficiency. J Inherit Metab Dis 12:94-95
42. Garrett RM, Johnson JL, Graf TN, Feigenbaum A, Rajagopalan KV (1998) Human sulfite oxidase R160Q: identification of the mutation in a sulfite oxidase-deficient patient and expression of the mutant enzyme. Proc Natl Acad Sci USA 95:6394-6398

## Ornithine and Creatine Metabolism

Ornithine is an important intermediate in several metabolic pathways. The pyridoxal-phosphate-requiring enzyme ornithine-δ-aminotransferase (OAT) plays a pivotal role in its metabolism. During the neonatal period, the flux of the OAT reaction is in the direction of ornithine synthesis, and arginine, an essential nutrient for young infants, is synthesized from ornithine via citrulline. At a later age, OAT catabolizes excess ornithine generated from arginine. Ornithine also plays an important role in urea synthesis (Fig. 17.1), in which it is utilized by ornithine transcarbamoylase (OTC) to form citrulline and is regenerated by hydrolytic cleavage of arginine. Since both OTC and OAT are mitochondrial matrix enzymes, ornithine produced in the cytoplasm must be transported to the mitochondrial matrix by a specific, energy-requiring transport system. A small fraction of ornithine is decarboxylated in the cytosol to form putrescine for polyamine biosynthesis.

Creatine is synthesized in the liver and pancreas. Guanidinoacetate is first formed from arginine and glycine by arginine:glycine amidinotransferase. Subsequently, guanidinoacetate is converted into creatine by guanidinoacetate methyltransferase. In muscle and brain, creatine is phosphorylated by creatine kinase into creatine phosphate, which serves as a high-energy phosphate store with which ATP is regenerated. Creatine and creatine phosphate are non-enzymatically converted (at a constant daily turnover of 1.5% of body creatine) into a cyclic compound, creatinine, which is mainly excreted in urine. Thus, daily excretion of creatinine is directly proportional to total body creatine.

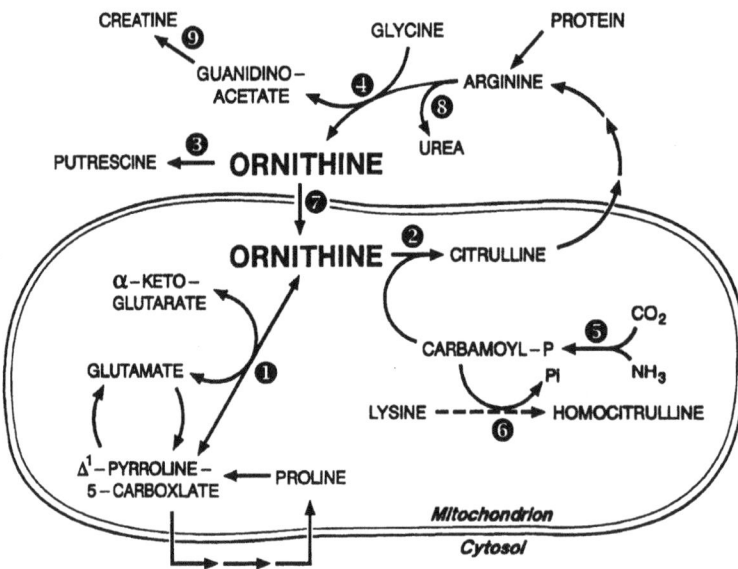

**Fig. 19.1.** Ornithine and creatine metabolic pathways. *Pi* inorganic phosphate. 1, ornithine-δ-aminotransferase; 2, ornithine transcarbamoylase; 3, ornithine decarboxylase; 4, arginine:glycine transamidinase; 5, carbamoylphosphate synthetase; 6, lysine transcarbamoylase; 7, mitochondrial ornithine transporter; 8, arginase; 9, guanidinoacetate methyltransferase. The step indicated by the *broken line* is not well defined

# CHAPTER 19

# Disorders of Ornithine and Creatine Metabolism

Vivian E. Shih and Sylvia Stöckler-Ipsiroglu

CONTENTS

Hyperornithinemia Due to Ornithine Aminotransferase
Deficiency (Gyrate Atrophy of the Choroid and Retina) .. 233
The Hyperornithinemia, Hyperammonemia,
Homocitrullinemia Syndrome . . . . . . . . . . . . . . . . . . . . . . . 236
Disorders of Creatine Metabolism . . . . . . . . . . . . . . . . . . . 237
References . . . . . . . . . . . . . . . . . . . . . . . . . . . . . . . . . . . . . . . 239

*Hyperornithinemia* due to ornithine aminotransferase (OAT) deficiency is associated with gyrate atrophy (GA) of the choroid and retina. Patients usually become virtually blind by age 55 years. Treatment includes pharmacological doses of pyridoxine (vitamin $B_6$) and/or a low-arginine diet. Preliminary results are encouraging.

In the *hyperornithinemia, hyperammonemia, homocitrullinuria (HHH) syndrome*, there is a wide spectrum of clinical manifestations, most of which are related to hyperammonemia. Progressive spastic paraparesis is often a late complication. Patients have a marked elevation of plasma ornithine associated with hyperammonemia and increased urinary excretion of homocitrulline, a derivative of lysine. HHH results from a defect in the importation of ornithine into the mitochondrion. Treatment is aimed at preventing ammonia toxicity.

*Guanidinoacetate-methyltransferase (GAMT) deficiency* is a newly recognized inborn error of creatine biosynthesis manifesting in infancy with severe neurologic symptoms, high urinary excretion of guanidinoacetate, low urinary excretion of creatinine, and depletion of creatine in brain and muscle. Symptoms are partly reversible with oral supplementation of creatine monohydrate.

## Hyperornithinemia Due to Ornithine Aminotransferase Deficiency (Gyrate Atrophy of the Choroid and Retina)

### Clinical Presentation

Night blindness and myopia in early childhood are usually the first symptoms of hyperornithinemia due to OAT deficiency. Ocular findings include myopia, constricted visual fields, elevated dark-adaptation thresholds, and very small or non-detectable electroretinographic responses. Retinopathy can be detected before visual disturbances. Fundoscopic appearances of chorioretinal atrophy are illustrated in Fig. 19.2. Patients develop posterior subcapsular *cataracts* by late in the second decade of life and usually become virtually blind between the ages of 40 years and 55 years due to extensive chorioretinal atrophy. Pyridoxine-responsive patients often have a milder course and maintain adequate visual acuity at older ages. Even within the same family, considerable heterogeneity exists in the appearance of the fundus, and siblings at the same age can show substantial differences in the severity of the ocular disease. Vitreous hemorrhage causing sudden loss of vision is a rare complication [1].

Histopathologic study of the eye obtained postmortem from a pyridoxine-responsive patient showed focal areas of photoreceptor atrophy with adjacent retinal-pigment epithelial hyperplasia [2]. Electronmicroscopic studies revealed abnormal mitochondria in the corneal endothelium and the non-pigmented ciliary epithelium and similar (but less severe) abnormalities in the photoreceptors.

In addition to the ocular findings, systemic abnormalities have been reported in some patients. Most patients have normal intelligence. The common finding on electroencephalography (EEG) has been diffuse slowing. Muscle pathology includes tubular aggregates and type-2 fiber atrophy [3], but only a small number of patients have clinical evidence of muscle weakness. Abnormal ultrastructure of hepatic mitochondria has been described [4]. Peculiar fine, sparse, straight hair with microscopic abnormalities has been found in some patients [5].

### Metabolic Derangement

Patients with GA of the choroid and retina have marked hyperornithinemia due to a deficiency of OAT (also known as ornithine ketoacid transaminase, OKT) activity [6]. The enzyme deficiency has been demonstrated in liver, muscle, hair roots, cultured skin

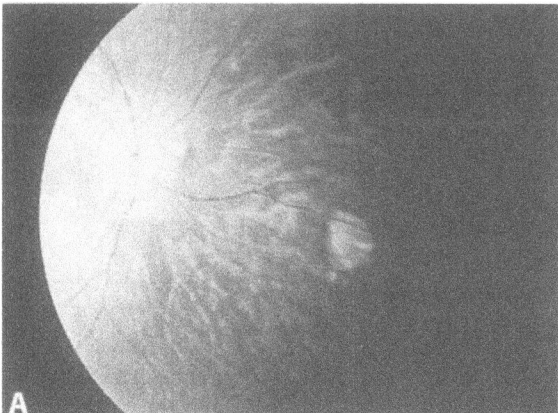
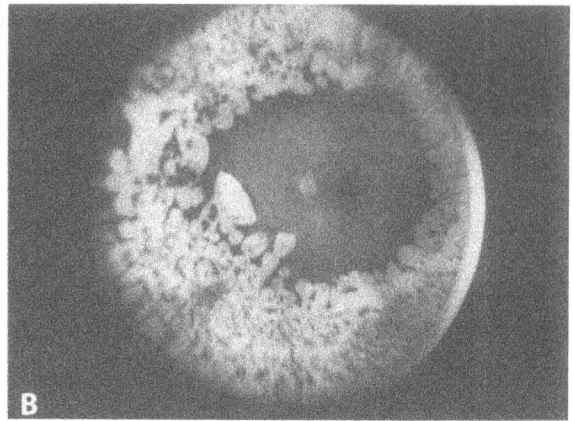

**Fig. 19.2A,B.** Fundoscopic appearances of the chorioretinal atrophy showing early changes (A), or advanced changes (B)

fibroblasts, and lymphoblasts. The pathophysiological mechanism of the retinal degeneration is unclear. OAT requires pyridoxal phosphate (PLP) as a cofactor. In a small number of patients, the OAT activity increased substantially when measured in the presence of high concentrations of PLP. Most of these patients showed a partial reduction of plasma ornithine when given pharmacological doses of pyridoxine (vitamin $B_6$). On the basis of in vitro and in vivo responses, at least two variants – pyridoxine responsive and pyridoxine nonresponsive – have been described. In rare cases, the in vivo response is inconsistent with the in vitro response to pyridoxine. GA patients have low levels of creatine and its precursor, guanidinoacetate, in the blood and urine, in addition to low levels of creatine and creatine phosphate in skeletal muscle [7], most likely as a result of ornithine inhibition of glycine transamidase and the subsequent reduction of creatine biosynthesis (Fig. 19.1).

## Diagnostic Tests

The main biochemical finding is a plasma ornithine concentration 5- to 20-fold above normal. Patients with the pyridoxine-responsive variant tend to have lower plasma ornithine levels than those with the pyridoxine-nonresponsive variant. Urinary excretion of ornithine, lysine, arginine, and cystine is often increased when the concentration of plasma ornithine is 400 μmol/l or greater. These changes are secondary to competitive inhibition by ornithine for the common renal transport shared by these amino acids. Small amounts of ornithine methyl ester and γ-glutamylornithine are sometimes detected in the urine. The absence of hyperammonemia and homocitrullinuria differentiate this disorder from HHH syndrome (Table 19.1). Hyperornithinemia may not be present in affected neonates; thus, newborn screening may give false negative results.

For confirmation of OAT deficiency, skin fibroblasts and lymphoblasts are suitable. Direct assay of OAT activity is performed in cell extracts, and the production of pyrroline-5-carboxylate from ornithine is monitored by either a colorimetric or a radioisotopic technique. An indirect assay measures the isotopic incorporation from $^{14}$C-labelled L-ornithine into macromolecules by intact cells [8]. Since ornithine transcarbamoylase (OTC) is not expressed in cultured cells, cells lacking OAT activity cannot convert $^{14}$C-L-ornithine into protein amino acids (i.e. $^{14}$C-L-proline and $^{14}$C-L-glutamate) for incorporation. Cells from patients with the pyridoxine-responsive variant may incorporate as much as 45% of the control value. This incorporation assay uses fewer cells than the direct assay and is applicable to genetic complementation analysis and prenatal diagnosis but is not useful for heterozygote identification.

## Treatment and Prognosis

The goal of treatment has been to correct the amino acid abnormalities. Megavitamin and/or diet therapy has been used to reduce hyperornithinemia. In addition, the administration of proline or creatine has been tried.

Pharmacological dosage of pyridoxine hydrochloride has resulted in plasma ornithine reduction in a small number of patients. Doses between 15 mg/day and 600 mg/day lowered plasma ornithine levels from 25% to 60% [9–11]. A 2-week trial of pyridoxine treatment (300–600 mg/day) is recommended for all newly diagnosed patients to determine their responsiveness. Reduction of hyperornithinemia can also be achieved by restriction of protein intake and arginine (a precursor of ornithine in foods) [7, 12, 13]. On average, food proteins contain 4–6% arginine (nuts and seeds have higher arginine contents). The low-

Table 19.1. Differential diagnosis of disorders involving ornithine metabolism

|  | OAT deficiency | HHH syndrome | OTC deficiency | GAMT deficiency |
| --- | --- | --- | --- | --- |
| Major clinical findings | Gyrate atrophy of the choroid and retina | Mental retardation, episodic lethargy and ataxia, seizures, coagulopathy | Severe form: neonatal onset of coma, early death | Autistic/self-injurious behavior, hypotonia, extrapyramidal movements, basal-ganglia involvement, seizures |
|  |  |  | Milder form: mental retardation, aversion to protein-rich foods, episodic lethargy and ataxia |  |
| Inheritance | Autosomal recessive | Autosomal recessive | X-linked | Autosomal recessive |
| Major biochemical changes | Increased blood ornithine | Increased blood ornithine, increased urine homocitrulline | Normal blood ornithine, decreased blood citrulline | Increased blood ornithine in some cases, increased urine guanidinoacetate, creatine deficiency, low blood and urine creatinine |
| Orotic acid levels | Normal | Increased | Increased |  |

*GAMT*, guanidinoacetate methyltransferase; *HHH*, hyperornithinemia, hyperammonemia, and homocitrullinuria; *OAT*, ornithine-δ-aminotransferase; *OTC*, ornithine transcarbamoylase

protein diet may be supplemented with a synthetic mixture of essential amino acids; this mixture can be used to provide up to one half of the nitrogen intake. Commercial products containing no arginine have been developed for this purpose. Products designated for patients with urea-cycle disorders can also be used. It is cautioned that severe arginine depletion can result in hyperammonemic complication [13].

Compliance with diet restriction can be a major problem, and long-term commitment and motivation are important factors. At least 30 patients have been given a low arginine diet, some in combination with pharmacological doses of pyridoxine, for periods up to 10 years. The results have been mixed. In the early reports, most patients were teenagers or adults with already advanced retinal disease, and the results were disappointing. Despite the fact that plasma ornithine was kept at near-normal levels, progression in the retinal atrophy and/or in the loss of visual function occurred in most patients, including three patients whose treatment was initiated in the first decade of life [7, 12–15]. Stabilization of visual function has been reported in two adult patients during 10 years of dietary treatment with good biochemical control [15]. The results of an arginine-restricted diet during 5–7 years in two sets of siblings are encouraging and showed that the younger and earlier-treated sibling in each pair had less retinal atrophy [16]. The vision of one 14-year-old patient was preserved, and his retinopathy was unchanged during 3 years of treatment [Shih, unpublished observation].

It has been hypothesized that, rather than ornithine toxicity, insufficient de novo formation of proline from ornithine in retinal pigment epithelium and ciliary bodies is responsible for the development of retinal degeneration [10]. Five patients with GA were given proline supplements of 2–10 g/day for periods of 2–4 years. The youngest patient had been detected by routine amino acid screening and had only minimal fundoscopic changes and had no visual complaints. Although this patient was started on 3 g of proline daily at age 4 years and remained asymptomatic at age 8 years, the effectiveness of proline supplement has not been established. In five other patients, creatine administration (0.75–1.5 g/day) corrected muscle histopathology but did not halt the progression of the retinal degeneration [7].

The long-term effects of the above therapeutic measures have yet to be assessed. Recently, a transgenic mouse model for GA was created, and a trial of dietary arginine restriction prevented the appearance of retinopathy at the age when untreated mice developed GA [17]. This observation supports the importance of early diagnosis and early treatment. However, it should be noted that, during the neonatal period, ornithine is the precursor of arginine. This metabolic flux is the opposite of that in older individuals, in whom arginine is the precursor of ornithine. Thus, arginine intake in patients less than 3 months of age should not be restricted until plasma ornithine begins to increase.

### Genetics

GA is an autosomal-recessive disorder and has been described in patients from various ethnic backgrounds, but its incidence is highest in the Finnish population [18]. Intermediate levels of OAT activity have been observed in skin fibroblasts from obligate heterozygotes for both pyridoxine-nonresponsive and pyridoxine-responsive variants. Heterozygotes for the pyridoxine-responsive variant can be distinguished by a doubling of OAT activity when assayed with and without PLP.

The human gene for OAT has been mapped to chromosome 10, with pseudogenes on the X-chromosome. Many different mutations in the OAT structural gene have been defined in GA patients of varied ethnic origins [19].

## The Hyperornithinemia, Hypercitrullinemia, Hyperammonemia Syndrome

### Clinical Presentation

HHH syndrome [20] is rare, and only approximately 50 cases are known. There is wide spectrum of clinical manifestations, most of which are related to hyperammonemia (Table 19.1). Ocular abnormalities are notably absent. Intolerance to protein feeding, vomiting, seizures, and developmental delay beginning during infancy are common complaints. Neonatal onset of lethargy, hypotonia, seizures with progression to coma, and death have been observed in the most severe form [21]. Progressive spastic paraparesis is often a late complication. Abnormal cranial computed tomography showing diffuse white matter density and cerebellar vermis atrophy was described in one patient. Coagulopathy, especially factor-VII and -X deficiencies, has been reported in several patients [22].

Mildly affected adult patients may present with hepatitis-like hyperammonemia and apparently normal intelligence. One adult man with HHH syndrome came to the attention of the physician because of episodic ataxia and personality change, and his sister was diagnosed by family survey [23]. She had a history of protein intolerance, refusal of high-protein foods in childhood, and poor school performance. One other patient was misdiagnosed with multiple sclerosis [Shih, unpublished observation].

Electron-microscopic examination of biopsied hepatic tissue showed an increased number of mitochondria, and many were large or bizarrely shaped, with unusually long tubular internal structure [24]. The structural changes seen in fibroblast mitochondria are, in some respects, similar to those found in hepatic mitochondria.

### Metabolic Derangement

Patients with the HHH syndrome have a marked elevation of plasma ornithine associated with hyperammonemia and increased urinary excretion of homocitrulline, a derivative of lysine. Activities of the two major ornithine-metabolizing enzymes, OTC and OAT, measured in liver homogenate and fibroblast extracts, respectively, are normal [25]. In contrast, the utilization of ornithine by intact fibroblasts is impaired [8]. These findings suggest that HHH syndrome is a disorder of compartmentation, and its defect is in the import of ornithine into the mitochondrion (Fig. 19.1), resulting in a functional deficiency of both OTC and OAT activities (Table 19.1). Results of in vitro studies using isolated mitochondria support this concept [26, 27]. Presumably, the intramitochondrial deficiency of ornithine results in the utilization of carbamoylphosphate via alternate pathways: formation of homocitrulline from lysine (Fig. 19.1) and formation of orotic acid (Chap. 31; Fig. 31.3).

### Diagnostic Tests

The HHH syndrome can be differentiated from other hyperammonemic syndromes, including the urea-cycle enzymopathies, by laboratory findings (Table 19.1). The triad of hyperornithinemia, hyperammonemia, and homocitrullinuria is pathognomonic. Plasma ornithine concentration is three to ten times higher than normal and tends to be somewhat lower than that seen in GA patients. Despite a functional deficiency of OTC activity, plasma citrulline is often normal in HHH syndrome.

Urine amino acid screening shows increased ornithine and homocitrulline when the plasma ornithine concentration is above 400 µmol/l. As seen in GA, ornithine methyl ester and γ-glutamylornithine are sometimes increased. At lower plasma ornithine concentrations, homocitrullinuria may be the only urine amino acid abnormality. Due to transformation of lysine to homocitrulline during manufacture, excessive homocitrulline excretion is known to occur in infants being fed canned formula. Persistent homocitrullinuria without dietary source is abnormal. Increased urinary homocitrulline has also been detected in hyperlysinemia. Homocitrulline may escape detection when present in small quantities, since it co-elutes with methionine in some amino acid analyzing buffer systems. However, its presence can be revealed by a pink color reaction with Ehrlich reagent (2% acidic p-dimethylaminobenzaldehyde in acetone) when used to overstain isatin- or ninhydrin-treated thin layer or paper chromatograms. Orotic aciduria is common, and increased excretion of orotic acid and orotidine can be induced by allopurinol challenge [23], as can be done in patients with primary OTC deficiency (Chap. 17).

The metabolic defect can best be confirmed by $^{14}$C-L-ornithine incorporation assay using fibroblast monolayers [8]. Frozen tissue is not suitable for this study. The compartmentation of ornithine in HHH fibroblasts prevents the conversion of ornithine to proline and glutamate (Fig. 19.1) and results in very low incorporation of radioactivity into trichloroacetic-acid-precipitable macromolecules. Direct measurement of OAT activity in cell extracts is normal. In GA fibroblasts, both tests are abnormal. The $^{14}$C-ornithine-incorporation assay is a sensitive test that has been used for prenatal diagnosis of HHH syndrome [21] but it is not useful for heterozygote identification.

### Treatment and Prognosis

Treatment is aimed at preventing ammonia toxicity and follows the principles outlined for the urea-cycle disorders (Chap. 17). Based upon the observation that ornithine given with protein loading prevented hyperammonemic response [28], several HHH patients were put on oral administration of ornithine in the range of 0.5–1.0 mmol/kg/day and were found to show lower blood ammonia levels and increased nitrogen tolerance. Presumably, the higher blood ornithine levels increased the transport of ornithine into the mitochondria, leading to improved urea-cycle functioning. However, not all patients benefited from ornithine supplementation. A low dose of arginine decreased the blood ammonia concentration in one patient [29] and citrulline supplementation in combination with a low-protein diet reduced plasma glutamine and homocitrulline excretion in two patients [22]. These supplements also corrected low creatine excretion in the two patients. One patient with neonatal onset of moderate hyperammonemia responded well to treatment and had normal growth and development at 18 months of age [30]. In general, a low-protein diet has been effective in achieving biochemical control for most patients. This treatment results in improved growth and development in children but has not prevented the late development of spastic gait.

The report of women with urea-cycle disorders who developed hyperammonemic coma after childbirth [31] suggests that women with HHH syndrome may also be at risk for such complications. Indeed, one HHH patient developed hyperammonemia 1 day post-partum in two instances [Shih, unpublished observation]. Thus, it is advisable to exercise caution in the postpartum dietary management of HHH patients, whose protein tolerance is expected to be lower than during pregnancy. Offspring from both women and men with HHH syndrome have been apparently normal.

### Genetics

The HHH syndrome is a rare inborn error of metabolism but is more frequently seen in French-Canadians than in other ethnic groups. Inheritance is autosomal recessive. A recent study has identified the gene encoding both the transporter protein and a common mutation in HHH patients of French-Canadian origin [32]. The mode of clinical presentation and responses to treatment suggest heterogeneity among HHH patients. Obligate heterozygotes are clinically normal and cannot be identified via biochemical studies.

## Disorders of Creatine Metabolism

### Clinical Presentation

The first patient with guanidinoacetate-methyltransferase (GAMT) deficiency was described in 1994 [33]. This boy was considered to be normal until 4 months of age, when he was noted to have developmental arrest. He gradually developed severe extrapyramidal movements, hypotonia, frequent vomiting, and difficulties in handling secretions. His EEG showed very slow background activity and multifocal-spike slow waves. Magnetic resonance imaging revealed bilateral abnormalities of the globus pallidus as hypointensities in T1-weighted images and as hyperintensities in T2-weighted images. In two other patients, severe epilepsy and early global developmental delay were predominant symptoms [34, 35]. The most recently diagnosed patient presented with moderate developmental delay and autistic features [36]. The clinical phenotype varies widely, from predominance of extrapyramidal encephalopathy and intractable epilepsy to moderate mental retardation only (Table 19.1).

While it has been postulated that the retinopathy in hyperornithinemia due to OAT deficiency may be attributed to impairment of creatine synthesis, retinal changes have not been observed in patients with GAMT deficiency. Patients with GAMT deficiency do not have signs of skeletal or cardiac myopathy, although muscle tissue is a main site of creatine depletion.

### Metabolic Derangement

Patients with GAMT deficiency have systemic depletion of creatine and creatine phosphate due to impairment of de novo creatine biosynthesis. The main affected organs are muscle and brain, which contain the bulk of the body creatine pool. In vivo proton magnetic resonance spectroscopy has shown that patients with GAMT deficiency have extremely low brain creatine concentrations, while guanidinoacetate, the immediate precursor of creatine and substrate of the deficient enzyme activity, accumulates in unusually high concentrations. In vivo phosphorus magnetic resonance spectroscopy of the brain has further shown that reduced availability of creatine to the action of creatine kinase leads to a deficiency of creatine phosphate. Instead, abundant guanidinoacetate in the brain is phosphorylated by creatine kinase and represents the major proportion of high-energy phosphate. In one patient, the creatine concentration measured in a muscle biopsy was found to be low [33]. As a consequence of systemic depletion of creatine and creatine phosphate, daily urinary creatinine excretion is low, as are creatinine concentrations in plasma and CSF [34, 35, 37].

Hyperornithinemia in the presence of low plasma arginine levels was a pronounced finding in the first reported patient with GAMT deficiency [33]. It was explained by abnormal de-repression of arginine:glycine amidinotransferase in the creatine-deficient state and by overproduction of guanidinoacetate and ornithine and simultaneous equimolar consumption of arginine and glycine [33]. These findings were not reported in the other patients with GAMT deficiency and, therefore, seem to be nonspecific. Moderate hyperammonemia and hyperuricemia have been reported occasionally [33–35].

## Diagnostic Tests

Guanidinoacetate is the most specific marker of GAMT deficiency, and its accumulation in tissues and body fluids is pathognomonic for the disease [37]. Guanidinoacetate can be detected by several chromatographic and colorimetric methods [34, 37]. The most efficient way to identify patients at risk is the determination of guanidinoacetate in urine by gas chromatography/mass spectrometry [36, 38]. Separate derivatization of urine samples prior to gas chromatography/mass spectrometry is needed [38], as current procedures for organic acid screening are not suitable for the determination of guanidino compounds. Laboratories using thin-layer chromatography or high-voltage electrophoresis for semiquantitative amino acid determination might also screen their samples for GAMT deficiency by application of the Sakaguchi stain [39], a color reaction of monounsaturated guanidines, including guanidinoacetate [40]. Due to the nonspecific nature of this reaction, a positive result must be further evaluated by determination of guanidinoacetate by gas chromatography/mass spectrometry [39] or ion-exchange chromatography [41].

Creatine deficiency is another diagnostic marker of GAMT deficiency. Creatine has a prominent proton magnetic spectrum in the brain in vivo, and its absence cannot be overlooked. Complete lack of creatine in the presence of a normal spectral pattern of the remaining metabolites is a striking and unique pattern, allowing diagnosis of creatine deficiency by in vivo proton magnetic resonance spectroscopy of the brain. As this technique is not yet commonly available and deep sedation is required in pediatric patients, in vivo magnetic resonance spectroscopy is not a primary screening tool in patients at risk.

Creatinine excretion is directly related to body creatine, and assessment of 24-h urine creatinine may be helpful in the diagnosis of creatine-deficient states. However, this test may not be reliable in newborns and very young infants, since the maternal intrauterine creatine supply provides the newborn with a full creatine pool, and depletion will follow as a slow process over several weeks. However, low creatinine levels are also observed in patients with vanishing muscle mass; therefore, low creatinine levels can be a nonspecific finding in several myopathies and muscular dystrophies. Urinary organic acids may be nonspecifically elevated in random urine samples when calculated in relation to urinary creatinine concentrations.

GAMT deficiency is confirmed enzymatically by determination of GAMT activity. Absence of GAMT activity was first demonstrated in liver biopsy [42], but it can also be reliably measured in fibroblasts and virus (Epstein-Barr virus)-transformed lymphoblasts [Stöckler-Ipsiroglu, unpublished observation]. The GAMT activity in cultured amniotic cells is in the same range as that in lymphoblasts, suggesting that prenatal diagnosis is feasible. Characterization of the molecular defect may complete the diagnosis.

## Treatment and Prognosis

Systemic creatine deficiency can be corrected by oral supplementation of creatine monohydrate. Dosages from 350 mg/kg/day to 2 g/kg/day have been used [34, 35, 43]. The dose level of 350 mg/kg/day is about 20 times the daily creatine requirement and has been reported not to induce side effects in healthy volunteers [44]. A marked increase of plasma creatine concentrations (with peak values after 1 h of creatine-monohydrate ingestion) and normalization of urinary creatinine excretion reflect intestinal creatine absorption and replenishment of intracellular, mainly muscular creatine pools [37, 43]. Restoration of brain creatine and creatine phosphate can be monitored by in vivo proton and phosphorus magnetic resonance spectroscopy. During a 25-month period of treatment, nearly complete recovery of brain creatine was demonstrated in one patient [43]. The time course of creatine increase was characterized by a fast process during the first 3 months, resulting in 40% restoration of brain creatine; the increase was then characterized by a slower process during the following 22 months, leading to 90% restoration. The rate of increase was not accelerated when the initial creatine dose (350 mg/kg/day) was doubled. Clinical responses to oral creatine supplementation include resolution of extrapyramidal signs and symptoms, substantial developmental progress, and improvement of epilepsy and general condition [34, 35, 43]. Although creatine supplementation leads to substantial clinical benefit, none of the patients has achieved normal development.

In contrast to creatine deficiency, accumulation of guanidinoacetate cannot be sufficiently corrected by therapeutic means. Guanidinoacetate is synthesized from arginine and glycine by the activity of arginine:glycine amidinotransferase, the regulatory enzyme of creatine biosynthesis. The gene expression of this enzyme is mainly controlled by a creatine-dependent negative feedback mechanism. Investigations in one patient with GAMT deficiency have shown that repression of (highly expressed) arginine:glycine amidinotransferase activity by exogenous creatine leads to a decrease (but not to normalization) of guanidinoacetate in body fluids [37]. Further reduction of guanidinoacetate concentrations via competitive inhibition of arginine:glycine amidinotransferase activity by additional substitution with high-dose ornithine failed [37]. Restriction of dietary arginine, which is the immediate precursor of guanidinoacetate and is the substrate for arginine:glycine amidinotransferase activity, has also failed to lower guanidinoacetate levels [45]. Substantial improvement of clinical outcome might be achieved by early recognition and presymptomatic introduction of creatine supplementation.

## Genetics

Inheritance of GAMT is autosomal recessive. The four known cases are from families of German [33], Kurdish [34], Welsh [35], and Dutch [36] origins. Direct sequence analysis performed in the first two patients (German and Kurdish origin) revealed compound heterozygosity for 327G-A/309ins13 and homozygosity for 327G-A mutations, respectively [42]. Both mutant alleles are located in exon 2 of the GAMT gene and encode for truncated, non-functional enzyme proteins. DGGE analysis and subsequent sequence analysis of two new patients of Welsh and Dutch origin revealed homozygosity for 327G-A mutation [Stöckler-Ipsiroglu, unpublished observation]. GAMT activity was below detection limits in all four patients, while their parents and siblings had values within the range expected for heterozygous values. Prenatal diagnosis should be possible by determination of GAMT activity and of mutant alleles in cultivated amniotic cells.

## References

1. Takahashi O, Hayasaka S, Kiyosawa M, Mizuno K, Saito T et al. (1985) Gyrate atrophy of choroid and retina complicated by vitreous hemorrhage. Jpn J Ophthalmol 29:170-176
2. Wilson DJ, Weleber RG, Green WR (1991) Ocular clinicopathologic study of gyrate atrophy. Am J Ophthalmol 111:24-33
3. Sipila I, Simell R, Rapola J, Sainio K, Tuuteri I (1979) Gyrate atrophy of the choroid and retina with hyperornithinemia: tubular aggregates and type 2 fiber atrophy in muscle. Neurology 29:996-1005
4. Arshinoff SA, McCullock JC, Matuk Y, Phillips MJ, Gordon BA, et al. (1979) Amino-acid metabolism and liver ultrastructure in hyperornithinemia with gyrate atrophy of the choroid and retina. Metabolism 28:979-928
5. Kaiser-Kupfer MI, Kuwabara T, Askanas V (1981) Systemic manifestations of gyrate atrophy of the choroid and retina. Ophthalmology 88:918-928
6. Trijbels JMF, Sengers RCA, Bakkaren JAJM et al. (1977) L-Ornithine-ketoacidtransaminase deficiency in cultured fibroblasts of a patient with hyperornithineaemia and gyrate atrophy of the choroid and retina. Clin Chim Acta 79:371
7. Vannas-Sulonen K, Simell O, Sipila I (1987) Gyrate atrophy of the choroid and retina. The ocular disease progresses in juvenile patients despite normal or near normal plasma ornithine concentration. Ophthalmology 94:1428-1433
8. Shih VE, Mandell R, Herzfeld A (1982) Defective ornithine metabolism in cultured skin fibroblasts from patients with the syndrome of hyperornithinemia, hyperammonemia and homocitrullinuria. Clin Chim Acta 118:149
9. Kennaway NG, Weleber RG, Buist NRM (1980) Gyrate atrophy of the choroid and retina with hyperornithinemia: biochemical and histologic studies and repsonse to vitamin B6. Am J Hum Genet 32:529-541
10. Hayasaka S, Saito T, Nakajima H, Takahashi O, Mizuno K et al. (1985) Clinical trials of vitamin B6 and proline supplementation for gyrate atrophy of the choroid and retina. Br J Ophthalmol 69:283-290
11. Shih VE, Berson EL, Gargiulo M (1981) Reduction of hyperornithinemia with a low protein, low arginine diet and pyridoxine in patients with a deficiency of ornithine- ketoacid transaminase (OKT) activity and gyrate atrophy of the choroid and retina. Clin Chim Acta 113:243-251
12. Valle D, Walser M, Brusilow SW, Kaiser-Kupfer M (1980) Gyrate atrophy of the choroid and retina: amino acid metabolism and correction of hyperornithinemia with an arginine-deficient diet. J Clin Invest 65:371-378
13. McInnes R, Arshinoff FS, Bell L, Marliss E, McCulloch J (1981) Hyperornithinaemia and gyrate atrophy of the retina. Improvement of vision during treatment with a low-arginine diet. Lancet 1:513
14. Berson EL, Hanson AH, Rosner B et al. (1982) A two year trial of low protein, low arginine diets or vitamin B6 for patients with gyrate atrophy. Birth Defects xviii:209
15. Kaiser-Kupfer MI, Valle DL (1987) Clinical, biochemical and therapeutic aspects of gyrate atrophy. In: Osborne N, Chader J (eds) Progress in retinal research, vol 6. Pergamon, Elmsford, NY, pp 179-206
16. Kaiser-Kupfer MI, Caruso RC, Valle D (1991) Gyrate atrophy of the choroid and retina. Long-term reduction of ornithine slows retinal degeneration. Arch Ophthalmol 109:1539-1548
17. Wang T, Lawler AM, Steel G, Sipila I, Milam AH et al. (1995) Mice lacking ornithine-δ-aminotransferase have paradoxical neonatal hypoornithinaemia and retinal degeneration. Nat Genet 11:185-190
18. Takki K (1974) Gyrate atrophy of the choroid and retina associated with hyperornithinaemia. Br J Ophthalmol 58:3
19. Ramesh V, Gusella JF, Shih VE (1991) Molecular pathology of gyrate atrophy of the choroid and retina due to ornithine aminotransferase deficiency. Mol Biol Med 8:81-93
20. Shih VE, Efron ML, Moser HW (1969) Hyperornithinemia, hyperammonemia, and homocitrullinuria: a new disorder of amino acid metabolism associated with myoclonic seizures and mental retardation. Am J Dis Child 117:83
21. Shih VE, Laframboise R, Mandell R, Pichette J (1992) Neonatal form of the hyperornithinemia, hyperammonemia and homocitrullinuria (HHH) syndrome and prenatal diagnosis. Prenat Diagn 12:717-723
22. Dionisi Vici C, Bachmann C, Gambarara M, Colombo JP, Sabetta G (1987) Hyperornithinemia-hyperammonemia-homocitrullinuria syndrome: low creatine excretion and effect of citrulline, arginine, or ornithine supplement. Pediatr Res 22:364-367

23. Tuchman M, Knopman DS, Shih VE (1990) Episodic hyperammonemia in adult siblings with hyperornithinemia, hyperammonemia, and homocitrullinuria syndrome. Arch Neurol 47:1134-1137
24. Haust MD, Gordon BA (1987) Possible pathogenetic mechanism in hyperornithinemia, hyperammonemia, and homocitrullinuria syndrome. Birth Defects 23:17-45
25. Oyanagi K, Tsuchiyama A, Itakkura Y, Sogawa H, Wagatsuma K et al. (1983) The mechanism of hyperammonaemia and hyperornithinaemia in the syndrome of hyperornithinaemia, hyperammonaemia with homocitrullinuria. J Inherited Metab Dis 6:133-134
26. Hommes FA, Ho CK, Roesel RA et al. (1982) Decreased transport of ornithine across the inner mitochondrial membrane as a cause of hyperornithinaemia. J Inherit Metab Dis 5:41
27. Inoue I, Saheki T, Kayanuma K, Uono M, Nakajima M et al. (1988) Biochemical analysis of decreased ornithine transport activity in the liver mitochondria from patients with hyperornithinemia, hyperammonemia and homocitrullinuria. Biochim Biophys Acta 964:90-95
28. Gordon BA, Gatfield DP, Haust MD (1987) The hyperornithinemia, hyperammonemia, homocitrullinuria syndrome: an ornithine transport defect remediable with ornithine supplements. Clin Invest Med 10:329-336
29. Hommes FA, Roesel RA (1986) Studies on a case of HHH-syndrome (hyperornithinemia, hyperammonemia and homocitrullinuria). Neuropediatrics 17:48-52
30. Zammarchi E, Ciani R, Pasquini E, Bonocore G, Shih VE et al. (1997) Neonatal Onset of Hyperornithinemia-Hyperammonemia-Homocitrullinuria Syndrome with Favourable Outcome. J Pediatr 131:440-443
31. Arn PH, Hauser ER, Thomas GH, Herman G, Hess D et al. (1990) Hyperammonemia in women with a mutation at the ornithine carbamoyltransferase locus: a cause of postpartum coma. N Engl J Med 322:1652-1655
32. Camacho J, Biery B, Mitchell G, Almashanu S, Hu C-A et al. (1998) Identification and molecular analysis of the gene responsible for the hyperornithinemia-hyperammonemia-homocitrullinuria (HHH) syndrome. Am J Hum Genet 63:A14
33. Stockler S, Holzbach U, Hanefeld F, Marquardt I, Helms G et al. (1994) Creatine deficiency in the brain: a new, treatable inborn error of metabolism. Pediatr Res 36:409-413
34. Schulze A, Hess T, Wevers R, Mayatepek E, Bachert P et al. (1997) Creatine deficiency syndrome caused by guanidinoacetate methyltransferase deficiency: diagnostic tools for a new inborn error of metabolism. J Pediatr 131:626-631
35. Ganesan V, Johnson A, Connelly A, Eckhardt S, Surtees RA (1997) Guanidinoacetate methyltransferase deficiency: new clinical features. Pediatr Neurol 17:155-157
36. van der Knaap MS, Verhoeven NM, Stuys E, Powels PJW, Jacobs C (1998) Mental retardation and autism as presenting signs in creatine synthesis defect. J Inherit Metab Dis 21:136
37. Stockler S, Marescau B, De Deyn PP, Trijbels JM, Hanefeld F (1997) Guanidino compounds in guanidinoacetate methyltransferase deficiency, a new inborn error of creatine synthesis. Metabolism 46:1189-1193
38. Hunneman DH, Hanefeld F (1997) GC-MS determination of guanidinoacetate in urine and plasma. J Inherit Metab Dis 20:450-452
39. Bremer HJ, Duran M, Kamerling JP, Przyrembel H, Wadman SK (1981) Sakaguchi reaction. Disturbances of amino acid metabolism: clinical chemistry and diagnosis. Urban and Schwarzenberg, Baltimore, p 439
40. Jepson JB, Smith I (1953) "Multiple dipping" procedures in paper chromatography: a specific test for hydroxyproline. Nature 172:1100-1101
41. Marescau B, Deshmukh DR, Kockx M, Possemiers I, Quereshi EA et al. (1992) Guanidinocompounds in serum, urine, liver, kidney, and brain of man and some uretelic animals. Metabolism 41:526-532
42. Stockler S, Isbrandt D, Hanefeld F, Schmidt B, von Figura K (1996) Guanidinoacetate methyltransferase deficiency: the first inborn error of creatine metabolism in man. Am J Hum Genet 58:914-922
43. Stockler S, Hanefeld F, Frahm J (1996) Creatine replacement therapy in guanidinoacetate methyltransferase deficiency, a novel inborn error of metabolism. Lancet 348:789-790
44. Greenhaff PL, Casey A, Short AH, Harris R, Soderlund K et al. (1993) Influence of oral creatine supplementation of muscle torque during repeated bouts of maximal voluntary exercise in man. Clin Sci (Colch) 84:565-571
45. Schulze A, Mayatepek E, Bachert P, Marescau B, De Deyn PP et al. (1998) Therapeutic trial of arginine restriction in creatine deficiency syndrome (letter). Eur J Pediatr 157:606-671

# CHAPTER 20

## Catabolism of Lysine, Hydroxylysine and Tryptophan

Lysine, hydroxylysine and tryptophan are degraded intramitochondrially via initially separate pathways that converge in a common pathway starting with 2-aminoadipic and 2-oxoadipic acids (Fig. 20.1). The initial catabolism of lysine proceeds mainly via the bifunctional enzyme 2-aminoadipic-4-semialdehyde synthase (1). A small amount of lysine is catabolized via pipecolic acid and the peroxisomal enzyme pipecolic acid oxidase (2). 2-Aminoadipic acid is converted into glutaryl-coenzyme A (CoA) by a second bifunctional enzyme, 2-aminoadipate aminotransferase/2-oxoadipate dehydrogenase (3). Glutaryl-CoA is converted into crotonyl-CoA by a third bifunctional enzyme, glutaryl-CoA dehydrogenase/glutaconyl-CoA decarboxylase (4). This enzyme transfers electrons to flavin adenine dinucleotide and, hence, to the respiratory chain (Fig. 11.1) via electron-transfer flavoprotein (ETF)/ETF-dehydrogenase (ETF-DH). Of the four distinct enzyme deficiencies identified in the degradation of lysine, only the last has proven relevance as a neurometabolic disorder. Glutaric aciduria type I is caused by the isolated deficiency of glutaryl-CoA dehydrogenase/glutaconyl-CoA decarboxylase. Glutaric aciduria type II, caused by ETF/ETF-DH deficiencies, is discussed in Chap. 11. Pipecolic-acid-oxidase deficiency is discussed in Chap. 37.

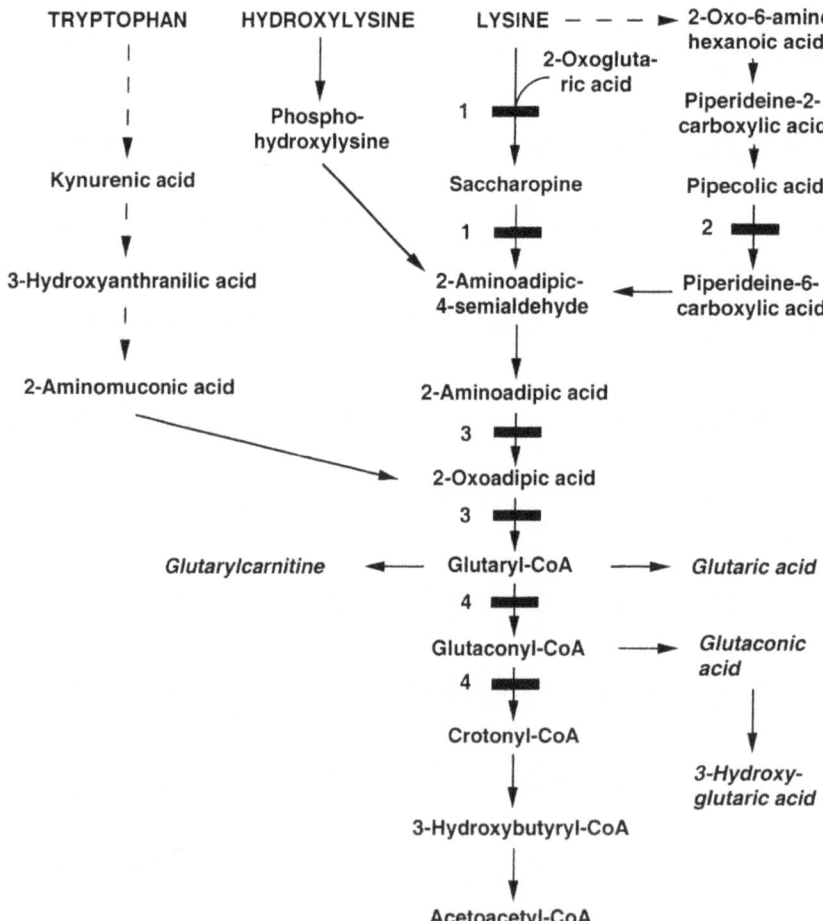

**Fig. 20.1.** Tryptophan, hydroxylysine and lysine catabolic pathways. *CoA*, coenzyme A. Enzyme deficiencies are indicated by *solid bars*

CHAPTER 20

# Disorders of Lysine Catabolism and Related Cerebral Organic-Acid Disorders

G.F. Hoffmann

CONTENTS

Introduction................................... 243
Hyperlysinemia/Saccharopinuria................. 244
2-Amino-/2-Oxoadipic Aciduria.................. 244
Glutaric Aciduria Type I....................... 245
    Clinical Presentation........................ 245
    Metabolic Derangement ...................... 248
    Diagnostic Tests ............................ 248
Treatment and Prognosis....................... 249
    Emergency Treatment ........................ 249
    Oral Supplementation with Carnitine and Riboflavin . 249
    Diet ....................................... 249
    Neuropharmaceutical Agents.................. 249
    Nonspecific Multiprofessional Support .......... 250
    Genetics ................................... 250
L-2-Hydroxyglutaric Aciduria ................... 250
D-2-Hydroxyglutaric Aciduria ................... 251
References..................................... 252

---

Five inborn errors are known in the pathway of lysine catabolism, which is shared by tryptophan and hydroxylysine degradation.

- Hyperlysinemia-saccharopinuria is a non-disease.
- 2-Amino-/2-oxoadipic aciduria may also have no clinical significance, but some patients are retarded and show variable neurological abnormalities.
- Glutaric aciduria type 1 is a severe neurological disease. A very characteristic macrocephalus, progressive cerebral atrophy, subdural hemorrhages, encephalopathic crises during stress, a dystonic-dyskinetic syndrome and, ultimately, often a fatal outcome occur in untreated patients. However, restriction of protein and lysine, administration of L-carnitine, timely vigorous treatment during intercurrent illness and neuropharmaceutical agents may prevent or at least halt the unfavorable course of the disease.
- L-2-hydroxyglutaric aciduria is associated with gradual onset of ataxia, extrapyramidal signs and progressive mental retardation, and with macrocephaly in some patients. The enzyme defect is not known and there is no rational treatment
- D-2-Hydroxyglutaric aciduria mostly causes severe early-onset epileptic encephalopathy with neonatal seizures, lack of psychomotor development and early death. Some patients exhibit milder neurological symptoms like mild developmental delay, delayed speech and febrile convulsions.

## Introduction

Neurological manifestations are very common and are sometimes the leading and/or presenting feature in organic-acid disorders [1]. A group of organic acid disorders presents exclusively with characteristic (progressive) neurological symptoms of ataxia, epilepsy, myoclonus, extrapyramidal symptoms, metabolic stroke, and macrocephaly. These "cerebral" organic acid disorders include glutaric aciduria type I (GAI), L-2-hydroxyglutaric aciduria, D-2-hydroxyglutaric aciduria and 2-oxoglutaric aciduria. Strikingly, in all these disorders, the pathological compound is a five-carbon organic acid. There is evidence of a common pathophysiology through specific interference of the accumulating metabolites with the main excitatory neurotransmission system via the glutaminergic N-methyl-D-aspartate (NMDA) receptor [2].

Cerebral organic acid disorders may often remain undiagnosed. Abnormalities, such as hypoglycemia, metabolic acidosis or lactic acidemia, the usual concomitants of disorders of organic acid metabolism, are generally absent. Furthermore, elevations of diagnostic metabolites may be small and, therefore, may be missed on "routine" organic acid analysis, e.g. in GAI. Therefore, the correct diagnosis requires an increased awareness of these disorders by the referring physician and the biochemist in the metabolic laboratory. Diagnostic clues can be derived from neuroimaging findings (Fig. 20.2). Progressive disturbances of myelination, cerebellar atrophy, frontotemporal atrophy, hypodensities and/or infarcts of the basal ganglia and any symmetrical (fluctuating) pathology apparently independent of defined regions of vascular supply are suggestive of cerebral organic acid disorders [1].

## Hyperlysinemia/Saccharopinuria

### Clinical Presentation

Hyperlysinemia/saccharopinuria appears to be a rare "non-disease". About half of the patients described were detected incidentally and are healthy [3, 4]. Symptoms described as associated with the disorder include psychomotor retardation, epilepsy, spasticity, ataxia and short stature. Single patients with joint laxity and spherophakia were described. These observations suggest that the condition is not deleterious and that the combination of symptoms is accounted for by sampling bias.

### Metabolic Derangement

Hyperlysinemia/saccharopinuria is caused by deficiency of the bifunctional protein 2-aminoadipic semialdehyde synthase, the first enzyme of the main pathway of lysine degradation. The two functions of the enzyme, lysine:2-oxoglutarate reductase and saccharopine dehydrogenase, may be differently affected by mutations. In most cases, both activities are severely reduced, resulting predominantly in hyperlysinemia and hyperlysinuria, accompanied by relatively mild saccharopinuria. Failure to remove the ε-amino group results in an overflow of the minor pathway, with removal of the α-amino group by oxidative deamination. The oxoacid cyclizes and is reduced to pipecolic acid. As a consequence, hyperpipecolatemia is regularly observed in hyperlysinemia. Hyperlysinuria without hyperlysinemia can be assumed to result from renal tubular defects, often as part of a transport defect of dibasic amino acids.

In one adult retarded woman and a second incompletely characterized patient, saccharopinuria with significantly less hyperlysinuria was the predominant biochemical abnormality [5, 6]. Fibroblasts of the first patient revealed a high residual activity of lysine:2-oxoglutarate reductase, while saccharopine dehydrogenase was virtually absent.

### Diagnostic Tests

The initial observation in patients with hyperlysinemia/saccharopinuria is an impressive lysinuria up to 15,000 mmol/mol creatinine (controls < 70 mmol/mol creatinine). Detailed amino acid analysis reveals additional elevations of saccharopine (up to 80 mmol/mol creatinine, controls not detectable), homoarginine (up to 90 mmol/mol creatinine, controls not detectable), 2-aminoadipic acid (up to 160 mmol/mol creatinine, controls < 20 mmol/mol creatinine) and pipecolic acid (3–6 mmol/mol creatinine, controls not detectable) [7]. The patient with predominant saccharopinuria excretes less than 1000 mmol lysine/mol creatinine and more than 2000 mmol saccharopine/mol creatinine. The same metabolites can be documented in other body fluids (like plasma and CSF), with elevated lysine as the predominant abnormality (up to 1700 µmol/l in plasma, controls < 200 µmol/l, and up to 270 µmol/l in CSF, controls < 28 µmol/l).

2-Aminoadipic semialdehyde synthase is widely distributed in human tissues, and the enzymatic defect can be ascertained in fibroblasts and tissue biopsies by determining the overall degradation of [1-$^{14}$C]-lysine to $^{14}CO_2$. Specific assays for lysine:2-oxoglutarate reductase and saccharopine dehydrogenase have also been described, the latter requiring a substrate commercially not available [3]. Molecular diagnosis has not yet been achieved.

### Treatment and Prognosis

Specific interventions during intercurrent illnesses do not appear to be necessary. Long-term dietary restriction of lysine has no proven benefit.

### Genetics

Both sexes are affected, and consanguinity has been reported, indicating autosomal-recessive inheritance.

## 2-Amino-/2-Oxoadipic Aciduria

### Clinical Presentation

Like hyperlysinemia/saccharopinemia, 2-amino-/2-oxoadipic aciduria may have no clinical significance. Over 20 patients are known, of whom more than half are asymptomatic [7]. Symptoms include psychomotor retardation, muscular hypotonia, epilepsy, ataxia and failure to thrive. Single patients with dysphagia, dysmorphy and heart defects were described. In one girl, intermittent metabolic acidosis was documented [8].

### Metabolic Derangement

The metabolic profile is heterogeneous, with most patients showing elevations of all three metabolites, whereas some excrete only 2-aminoadipic acid [9]. Normally, 2-aminoadipic acid is deaminated to

2-oxoadipic acid by a mitochondrial 2-aminoadipate aminotransferase. 2-Oxoadipic acid is also formed from the degradation of tryptophan. 2-Oxoadipic acid is further metabolized to glutaryl-coenzyme A (CoA) by oxidative decarboxylation. It is still debated whether 2-oxoadipate dehydrogenase is partially or totally identical to the 2-oxoglutarate dehydrogenase complex. In my opinion, the benign phenotype and the lack of significant elevations of 2-oxoglutaric acid in patients with 2-oxoadipic aciduria suggest that the two enzymes are different.

It could be assumed that isolated 2-aminoadipic aciduria without significant 2-oxoadipic aciduria is caused by a deficiency of 2-aminoadipate aminotransferase, whereas combined 2-amino/2-oxoadipic aciduria would be caused by a deficiency of the 2-oxoadipate-dehydrogenase complex. However, the biochemical profile of the reported patients overlap, loading studies were inconclusive [9], and a deficiency of either enzyme has not been shown directly. Enzymatic studies in fibroblasts have been performed in patients with combined 2-amino/2-oxoadipic aciduria demonstrating an inability to oxidize both 2-amino-[1-$^{14}$C] adipic and 2-oxo-[1-$^{14}$C] adipic acids to $^{14}CO_2$ [10, 11].

2-Aminoadipic acid shows a complex excitatory-amino-acid synaptic pharmacology, especially at the metabotropic excitatory-amino-acid receptors [12]; this pharmacology could be related to neurological symptoms. Unfortunately, CSF studies in affected patients have not been reported.

### Diagnostic Tests

Patients are diagnosed by demonstrating elevated 2-aminoadipic acid levels on amino acid chromatography or of 2-oxoadipic and 2-hydroxyadipic acid levels on urinary organic acid analysis. Elevations of metabolites in urine and plasma are variable. 2-aminoadipic acid levels are 50–2000 mmol/mol of creatinine in urine (controls < 20 mmol/mol of creatinine) and 2–120 µmol/l in plasma (controls < 5 µmol/l). 2-oxoadipic acid levels are up to 360 mmol/mol of creatinine (controls < 10 mmol/mol of creatinine), and 2-hydroxyadipic acid levels are up to 180 mmol/mol of creatinine (controls < 10 mmol/mol of creatinine) [7]. Plasma lysine may be two times the normal level, and urinary glutaric acid may be up to 50 mmol/mol of creatinine (controls < 9 mmol/mol of creatinine) [13]. Glutaric acid results from spontaneous decarboxylation of 2-oxoadipic acid in patients with significant elevations of 2-oxoadipic acid.

Oral loading with lysine or tryptophan (100 mg/kg body weight) increases the pathological metabolites two- to sixfold [9]. A decreased ability to oxidize either 2-amino-[1-$^{14}$C] adipic or 2-oxo-[1-$^{14}$C] adipic acids to $^{14}CO_2$ can be demonstrated in fibroblasts [10, 11]. Molecular analyses have not yet been achieved.

### Treatment and Prognosis

As patients with hyperlysinemia/saccharopinuria do not suffer from metabolic decompensations, specific interventions during intercurrent illnesses do not appear to be necessary. Administration of pharmaceutical doses of vitamins $B_1$ and $B_6$ had no effect on the levels of pathological metabolites [9]. Dietary restriction of lysine also failed to correct the biochemical abnormalities in some patients [9] and has no proven long-term benefit.

### Genetics

Autosomal-recessive inheritance is implied by the pedigrees and because parents can not be biochemically differentiated from controls.

## Glutaric Aciduria Type I (Glutaryl-CoA Dehydrogenase Deficiency)

### Clinical Presentation

If there is megalencephaly in an infant together with progressive atrophic changes on computed tomography or nuclear magnetic resonance (Fig. 20.2a–d) and/or acute profound dyskinesia or subacute motor delay accompanied by increasingly severe choreoathetosis and dystonia, GAI should have a high priority in the differential diagnosis [14–19]. In many patients, macrocephalus is present at or shortly after birth and precedes the severe neurological disease. An important clue to early diagnosis is not so much the finding of macrocephalus at birth but the observation of pathologically increased head growth peaking at the age of 3–6 months. Furthermore, affected babies often present additional "soft" neurological symptoms of hypotonia, with prominent head lag, irritability, jitteriness and feeding difficulties. Neonatal posture and tone may persist until 4–8 months of age. During febrile illnesses or after immunizations, hypotonia is often inadequately aggravated, and unusual movements and postures of hands appear. All these symptoms are still reversible and are of little prognostic significance. Neuroimaging investigations have been performed retrospectively in a number of patients during this "presymptomatic" period, revealing the characteristic findings of frontotemporal atrophy and delayed myelination (Fig. 20.2a). The clinical significance of enlarged subdural-fluid spaces in infants with GAI is the crossing of these spaces by bridging veins. Infants with GAI are prone to suffer acute subdural

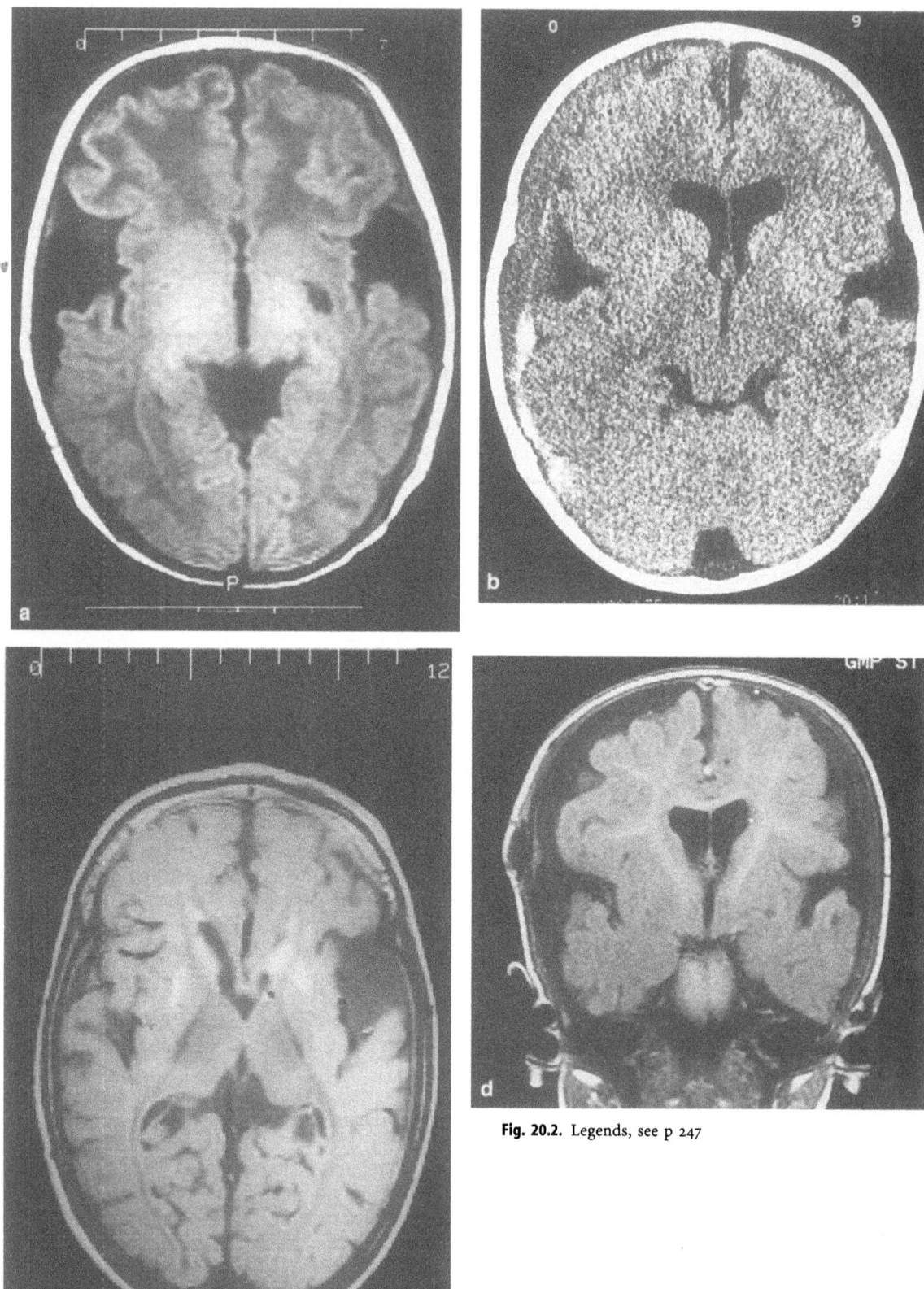

**Fig. 20.2.** Legends, see p 247

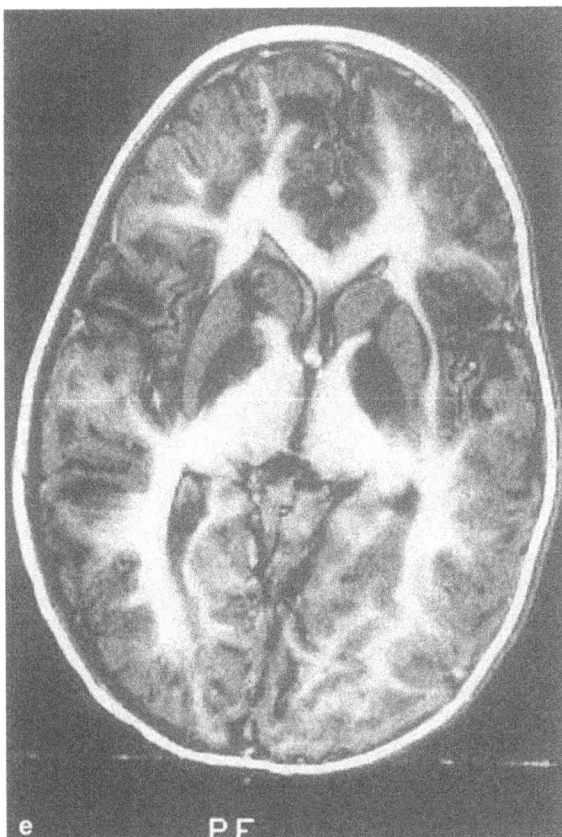 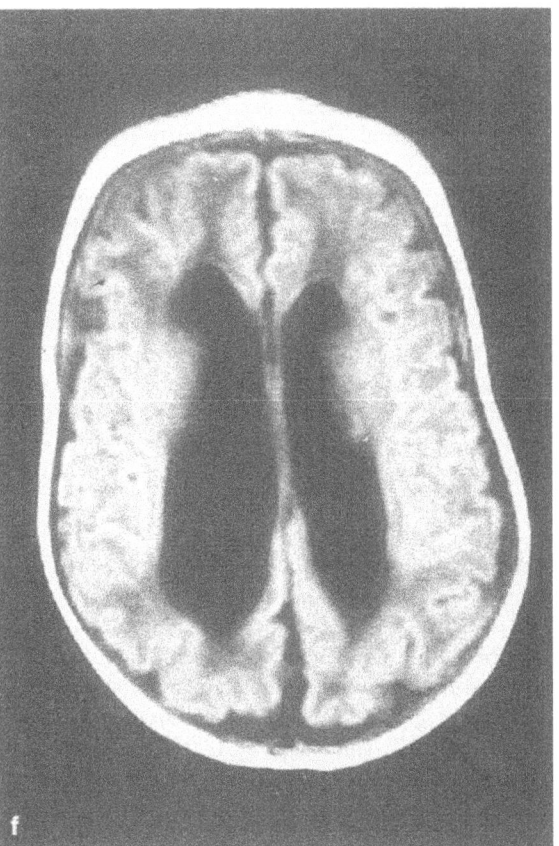

Fig. 20.2. Neuroimaging findings, which are suggestive of cerebral organic acid disorders. a Transverse magnetic resonance image (MRI) of a 2-month-old presymptomatic boy with glutaric aciduria type-I deficiency, showing enlargement of CSF spaces anterior to the temporal lobes, with marked extension of sylvian fissures (frontotemporal atrophy). Spin-echo technique (1.0 T); time of repetition: 660 ms; time of echo: 20 ms; slice thickness: 5.5 mm. b Computed tomography (CT) scan of a 9-month-old presymptomatic boy with glutaric aciduria type I. In addition to frontotemporal atrophy, subdural effusions and hematomas causing midline shift are visible. There is no pathology of the basal ganglia, and the child continued to develop normally. c Transverse MRI of a 2-year-old boy with glutaric aciduria type I, who suffered an encephalopathic crisis at the age of 11 months, illustrating regression of the temporal lobes, more pronounced on the left side, with dilatation of the sylvian fissures anterior to them and dilated insular cisterns (frontotemporal atrophy), delayed myelination and bilateral hypodensities of the basal ganglia. Spin-echo technique (1.0 T); time of repetition: 2660 ms; time of echo: 100 ms; slice thickness: 5 mm. d Coronal section of the central nervous system of an 8-month-old girl with glutaric aciduria type I, with dilatation of CSF spaces prominent at the sylvian fissures on MRI. At the age of 6 months, a ventricularperitoneal shunt was inserted in an attempt to drain "bilateral subdural hygromas" without knowledge of the underlying metabolic disorder. The catabolic stress following the operation resulted in an encephalopathic crisis, consecutive destruction of the basal ganglia and a severe dystonic/dyskinetic syndrome. Spin echo technique (1.0 T); time of repetition: 380 ms; time of echo: 25 ms; slice thickness: 6 mm. e Transverse MRI image of a 3-year-old boy with L-2-hydroxyglutaric aciduria, illustrating characteristic involvement of the subcortical white matter (U fibers), the nucleus caudati and the putamina. In this boy, the periventricular white matter was also affected early on. Spin-echo technique (1.0 T); time of repetition: 2660 ms; time of echo: 100 ms; slice thickness: 5 mm. f Transverse MRI image of a 2-month-old girl with D-2-hydroxyglutaric aciduria. Please note delayed myelination and considerable occipitally pronounced enlargement of lateral ventricles. Spin-echo technique (1.0 T); time of repetition: 618 ms; time of echo: 15 ms; slice thickness: 6 mm

hemorrhages (including retinal hemorrhages) after minor head trauma, i.e. commonly around the first birthday, when starting to walk (Fig. 20.2b). Repeatedly, parents were wrongly charged with child abuse because of chronic or acute subdurals and/or hemorrhages [20]. On average, at the age of 14 months, 75% of patients suffer an acute brain injury, mostly associated with an upper respiratory and/or gastrointestinal infection, but the encephalopathic crisis may also develop in association with fasts required for surgery, after routine immunizations, or following minor head traumas. Infants then present acute loss of neurological functions, such as ability to sit, ability to pull to a standing position, head control and suck and swallow reflexes. They appear to be alert, with profound hypotonia of the neck and trunk, stiff arms and legs, and twisting (athetoid) movements of hands and feet. There may also be generalized seizures. Mostly, there are no or only mild metabolic symptoms. After recovery, the children have lost motor skills. A severe dys- or hypotonic movement disorder develops. At this point, the distinctive clinical picture of a severe dystonic/dyskinetic syndrome in an alert-looking child with relatively well-preserved intellectual functions and a prominent forehead may be recognized. If the underlying metabolic disorder remains undiagnosed, additional cerebral systems are slowly but progressively affected. A generalized cerebral atrophy emerges (Fig. 20.2c,d), giving rise to pyramidal tract signs and, finally, mental retardation.

Although the majority of patients present with characteristic symptoms and disease course, the natural history of GAI can be variable even within families [14–18, 21]. A minority of patients (25%) presents with developmental delay from birth and (progressive) dystonic cerebral palsy. Finally, there have been single cases who were older than 6 years of age or were even adults but who never developed neurological disease [14, 16, 22].

## Metabolic Derangement

GAI is caused by a deficiency of glutaryl-CoA dehydrogenase, a mitochondrial flavin-adenine-dinucleotide-requiring enzyme that catalyzes the dehydrogenation of glutaryl-CoA and the subsequent decarboxylation of glutaconyl-CoA to crotonyl-CoA (Fig. 20.1). Part of the accumulating glutaryl-CoA is esterified with carnitine by carnitine acyltransferase, leading to increased ratios of acylcarnitines to free carnitine in plasma and urine. Glutaryl-carnitine is excreted, which contributes to secondary carnitine deficiency [15, 17]. Patients with GAI often show increased urinary excretion of dicarboxylic acids and increased excretion of 2-oxoglutarate and succinate, indicative of disturbed mitochondrial function [15, 17]. Secondary carnitine deficiency is probably a major causative factor of metabolic crises. These can present with severe hypoglycemia and variable metabolic acidosis and quickly progress to a Reye-like syndrome (14% in our series). Metabolic crises seem to be different from the above-described encephalopathic crises. They can develop at any age and respond well to intravenous therapy with glucose, carnitine and bicarbonate (Table 20.1) [15, 19, 21].

The mechanism of age-specific destruction of specific cerebral structures in GA-1 has been subject of intense debates and different hypotheses [23, 24]. Substantial evidence points to an excitotoxic sequence. Both glutaric and 3-hydroxyglutaric acids exhibit structural similarities to the excitotoxic amino acid glutamate. *Post-mortem* examination of the basal ganglia of patients has revealed postsynaptic vacuolation similar to that in glutamate-mediated damage [25].

There is now strong evidence that glutaric and 3-hydroxyglutaric acids mediate neuronal damage specifically via the NMDA 2B-subtype receptor, whose dynamic spatio-temporal expression pattern in the brain matches the neuropathological events and disease course [2, 26]. Massive activation of glutaminergic neurons from the cortex to the putamen via the nucleus caudatus can be anticipated during encephalopathic crises. The inhibitory output of the thalamus would explain the massive muscular hypotonia and the temporary improvement of dystonia during further catabolic episodes as a consequence of overstimulation of the remaining glutaminergic neurons.

## Diagnostic Tests

Patients with GAI are generally diagnosed by urinary organic acid analysis [15, 24]. Diagnostic elevations of glutaryl-carnitine in the body fluids of patients can be detected through acylcarnitine analysis in blood spots (Guthrie cards) using fast atom bombardment tandem mass spectrometry (MS) or electrospray tandem MS. However, individuals with deficiency of glutaryl-CoA dehydrogenase and severe characteristic neurological disease but without (or with only slight or inconsistent) elevations of glutaric acid or glutaryl-carnitine have been diagnosed in increasing numbers [15, 18, 21, 27]. Furthermore, elevated urinary excretion of glutaric acid can also be found in a number of other disease states, mostly related to mitochondrial dysfunction [13]. Repeated and quantitative urinary organic acid analyses may be necessary for diagnosis [13, 21]. Additional diagnostic hints can be obtained by finding carnitine deficiency in serum and/or a pathologically increased ratio of acylcarnitines to free carnitine in serum and urine. Differentiation between patients and healthy children can be further improved by alkaline hydrolysis of glutaryl conjugates (such as glutaryl-carnitine and glutaryl-glycine) prior to quantification. However, an overlap of urinary levels of glutaric acid among patients with deficiency of glutaryl-CoA dehydrogenase and nonspecific disease states remains. Recently, a clear discrimination could be achieved by determining 3-hydroxyglutaric acid with a stable-isotope dilution assay in urine. 3-Hydroxyglutaric acid proved to be permanently and reliably elevated in patients with deficiency of glutaryl-CoA dehydrogenase [13].

Ultimately, analysis of the enzyme glutaryl-CoA dehydrogenase and mutation analysis are the only methods in use that can establish the diagnosis of GAI with certainty. Glutaryl-CoA dehydrogenase activity can be determined in tissues, cultured fibroblasts, peripheral leukocytes, amniocytes and chorionic villi cells [15, 24, 28]. This procedure is justified in family studies or whenever there is very strong clinical suspicion [15, 16, 27].

*Carrier detection* is possible by enzyme assay (though the results are not always unequivocal [28]) and by molecular means in families in which the mutations are already known.

**Table 20.1.** Therapeutic recommendations for patients with glutaric aciduria type I

| Measures | Infants | Children under 6 years of age | Children over 6 years of age | Adults |
|---|---|---|---|---|
| **Emergency measures (all age groups)** | | | | |
| Stop protein supply[a] | | | | |
| Glucose infusion (8–15 mg/kg/min; with insulin, if necessary)[a] | | | | |
| Carnitine (200 mg/kg/day intravenously)[a] | | | | |
| **Diet** | | | | |
| Natural protein (g/kg/day) | 1.0–1.2 | 0.8–1.0 | 1.0–1.5 | 0.8–1.0 |
| Amino acid mixture (g/kg/day) | 1.0 | 1.0 | n.a. | n.a. |
| Lysine (mg/kg/day) | 90–100 | 60–70 | n.a. | n.a. |
| Tryptophan (mg/kg/day) | 20 | 12 | n.a. | n.a. |
| Energy (kcal/kg/day) | 120 | 80–95 | 60–70 | 40–50 |
| **Supplements** | | | | |
| L-Carnitine[a] (mg/day) | 100 | 50–100 | 50–100 | 50 |
| Riboflavine (mg/day) | 100–200 | 100–200 | 100–200 | 100–200 |
| **Neuropharmaceuticals[b] (for patients with neurological disease; all age groups)** | | | | |
| Baclofen[a] (1–2 mg/kg/day), Vigabatrin, Clonazepam[a], Diazepam[a], Trihexyphenidyl, Memantine, Haloperidol, L-Dopa/Levodopa, Glutamine | | | | |
| **Multiprofessional support of patient and family[a] (all age groups)** | | | | |

*n.a.*, not applicable
[a] Therapeutic measures whose benefit has been convincingly demonstrated
[b] Do not use valproic acid

*Prenatal diagnosis* can be offered by enzyme assay [28], determination of glutaric acid by stable-isotope-dilution gas chromatography (GC)-MS assay in amniotic fluid [13] and by molecular analysis.

## Treatment and Prognosis

Early diagnosis and treatment of the asymptomatic child is essential, as current therapy has little effect upon the brain-injured child. Therapy prevents brain degeneration in more than 90% of affected infants who are treated prospectively [21]. Without treatment, more than 90% of affected children will develop severe neurological disabilities. Five different therapeutic measures are generally employed (Table 20.1).

### Emergency Treatment

Emergency treatment during intercurrent illnesses (especially gastrointestinal infections), consisting of frequent feedings, high carbohydrates and zero protein intake, followed by high-dose intravenous glucose and carnitine, if necessary. Clomethiazole was found useful in severe cases of hyperpyrexia. All patients with GAI should be supplied with an emergency card [15].

### Oral Supplementation with Carnitine and Riboflavin

Carnitine should be supplemented in any patient who is carnitine deficient (free carnitine < 15 µmol/l). The rationale for carnitine supplementation is not to enhance the elimination of glutaric acid but to prevent secondary metabolic crises. The amount of glutaric acid, which is excreted as glutaryl-carnitine, can usually not be raised higher than 5% [17]. Riboflavin responsiveness in patients with GAI is an extreme rarity and should be investigated by (1) giving riboflavin in doses increasing from 50 mg to 300 mg and (2) monitoring total glutaric acid in 24-h urine samples. In evaluating the response, unrelated high daily variations of the urinary excretion of glutarate must be taken into account [17, 18].

### Diet

Many patients with GAI are treated with either a diet low in protein or a strict reduction of the intake of lysine supplemented with a lysine-free amino acid mixture. The intake of tryptophan should not be equally reduced. Tryptophan contributes only 20% or less to total body glutarate production, and plasma tryptophan levels cannot be reliably monitored by regular amino acid analysis. Furthermore, concentrations of tryptophan directly modulate the production of serotonin in the CNS. Using diets low in tryptophan, we observed side effects like sleeplessness, ill temper, irritability and loss of appetite, which were improved by tryptophan supplementation [21].

In neurologically symptomatic patients, dietary treatment results in no major clinical improvement. All patients remain severely handicapped [19, 21]. Nevertheless, the combination of modest protein restriction with carnitine supplementation at least halts the course of the disease.

### Neuropharmaceutical Agents

Several neuropharmaceutical agents have been used to ameliorate neurological symptoms in patients with GAI. In our experience, baclofen (Lioresal, 1–2 mg/kg/

day) or benzodiazepines (Diazepam, 0.1–1 mg/kg/day) reduce involuntary movements and improve motor function. In some patients, use and dosage is limited by worsening of truncal hypotonia. Valproic acid should not be given, as it effectively competes with glutaric acid for esterification with L-carnitine and may promote disturbances in the mitochondrial acetyl-CoA/CoA ratio [17]. Considering the severe neurological disease, surprisingly little information on the effects of other neuropharmaceutical agents is available, and all medications listed in Table 20.1 could be empirically employed.

The finding of the specific NMDA R2B-subunit-mediated toxicity of glutaric acid and 3-hydroxyglutaric acid opens new prospects for therapy [2]. High-dose application of NMDA-receptor antagonists, calcium-channel blockage during intercurrent illnesses and (possibly) continuous therapy between the ages of 6 months and 18 months may be beneficial for children with GAI and should be tested in clinical studies.

### Nonspecific Multiprofessional Support

Nonspecific multiprofessional support is of the utmost importance. It must be kept in mind that, considering the severe motor handicap, intellectual functions are well preserved until late in the disease course. Using Bliss boards and (especially) language computers, the social integration of patients can be greatly improved. As involuntary movements of orofacial muscles may be severe, feeding difficulties can become a major problem. Increased muscular tension and sweating, common findings in GAI, require a high intake of energy and water. Percutaneous gastrostomy often leads to a dramatic improvement of nutritional status, a marked relief of psychological tension and care load in the families and even reduction of dystonic/dyskinetic symptoms [19]. As a final remark, neurosurgical interventions on subdural hygromas and hematomas in infants and toddlers with GAI should be avoided if possible (Fig. 20.2d).

Our current policy of dietary and drug therapy for patients with GAI is specified in Table 20.1 [15]. As the risk of encephalopathic crises subsides after 4–5 years of age, patients should not be treated with severe protein restriction beyond 6 years of age. Carnitine supplementation and emergency measures during intercurrent illnesses (especially gastrointestinal infections) have to be followed throughout life, as in other disorders of amino acid degradation.

### Genetics

GAI is an autosomal-recessive disorder. An incidence around 1:30,000 has been suggested [15, 24]. The disease is particularly frequent in certain communities, such as the Amish people in Pennsylvania (incidence of 1:400) and the Saulteaux/Ojibway Indians in Canada [16, 18]. More than 70 different disease-causing mutations in the glutaryl-CoA dehydrogenase gene have been identified so far [29, 30, and unpublished data]. There is a correlation between genotype and biochemical phenotype in that specific mutations with significant residual enzyme activity may be associated with low excretions of metabolites in heterozygous patients who carry a severe mutation on the other allele. However, no correlation between genotype and clinical phenotype has yet been found. Single common mutations are found in genetically homogenous communities, such as the Amish of Pennsylvania [31], but GAI in general is quite heterogeneous: the most frequent mutation in Caucasians, R402W, has been identified on 10–20% of alleles [29, 30, and unpublished data]. Apart from three short or single-nucleotide deletions, the great majority of mutations are single base changes that are frequently found at hypermutable CpG sites in the gene.

## L-2-Hydroxyglutaric Aciduria

### Clinical Presentation

The initial description of L-2-hydroxyglutaric aciduria [32] was followed by a number of reports from all over the world, illustrating previous mis- and underdiagnoses. Most patients with L-2-hydroxyglutaric aciduria follow a characteristic disease course [32–34]. In infancy and early childhood, mental and psychomotor development of the patients appears to be normal or only slightly retarded. Thereafter, progressive ataxia, seizures, slight extrapyramidal signs and progressive mental retardation become the most obvious clinical findings. Macrocephaly is present in about half of the patients. The intelligence quotient during the second decade of life is 40–50. The oldest known patients are over 30 years of age. They are bedridden and severely retarded. Sometimes, mental deterioration is rapidly progressive, and a single patient with fatal neonatal outcome has been described [35].

In L-2-hydroxyglutaric aciduria, the neuroimaging findings are very specific [33]. The subcortical white matter appears to be mildly swollen, with some effacement of gyri. The progressive loss of arcuate fibers is combined with a severe cerebellar atrophy and increased signal densities of dentate nucleus, nucleus caudatus and putamen (Fig. 20.2e).

### Metabolic Derangement

Despite a number of loading and fasting studies, the origin of L-2-hydroxyglutaric acid remains unknown.

Recently, the enzyme L-2-hydroxyglutaric acid dehydrogenase was tentatively identified in human liver [33], but defective function could not be demonstrated in L-2-hydroxyglutaric aciduria [36]. Quantitative analysis of organic acids in CSF revealed higher elevations of L-2-hydroxyglutaric acid in CSF than in plasma [37]. In addition, a number of hydroxydicarboxylic acids (glycolate, glycerate, 2,4-dihydroxybutyrate, citrate and isocitrate) were only found elevated in CSF, pointing to a specific disturbance of brain metabolism.

### Diagnostic Tests

On organic acid analysis, L-2-hydroxyglutarate is found to be elevated in all body fluids [32–34, 38]. In addition, lysine is slightly elevated in CSF and protein in the absence of pleocytosis. Differentiation between the two isomers of 2-hydroxyglutarate is indispensable for diagnosis.

*Prenatal diagnosis* may be possible utilizing accurate determination of L-2-hydroxyglutarate by stable-isotope-dilution GC-MS assay in amniotic fluid [38].

### Treatment

To date, no rational therapy exists for L-2-hydroxyglutaric aciduria. Epilepsy can generally be controlled by standard medications.

### Genetics

The observation of up to three affected children in families and a similar clinical picture in males and females strongly supports an autosomal-recessive mode of inheritance [32–34]. Heterozygotes display no detectable clinical or biochemical abnormalities related to L-2-hydroxyglutaric aciduria.

## D-2-Hydroxyglutaric Aciduria

### Clinical Presentation

Patients with D-2-hydroxyglutaric aciduria exhibit a more variable phenotype than patients with L-2-hydroxyglutaric aciduria. The clinical spectrum varies from neonatal onset, severe seizures, lack of psychomotor development and early death [39] to mild developmental delay and no symptoms at all [40]. A recent international survey of 17 patients revealed a continuous spectrum within these extremes, with most patients suffering from a severe early-onset epileptic encephalopathy, while a substantial group shows mild symptoms or is even asymptomatic [41]. Clinical and neuroradiological symptoms of the severely affected patients were quite uniform. Severe, often intractable seizures started in early infancy. The babies were severely hypotonic. Consciousness levels varied from irritability to stupor. Cerebral visual failure was uniformly present. Psychomotor development appeared to be almost absent. A third of the severely affected patients suffered from cardiomyopathy. Less severely affected patients exhibited mostly mild neurological symptoms like mild developmental delay, delayed speech and febrile convulsions.

In severely affected patients, neuroimaging uniformly revealed disturbed and delayed gyration, myelination and opercularization, ventriculomegaly, cysts over the head of the caudate nucleus and more pronounced occipital horns (Fig. 20.2f). Enlarged prefrontal spaces and subdural effusions in some patients were further similarities with the neuroimaging findings in GAI.

### Metabolic Derangement

Hitherto no specific biochemical function or pathway involving D-2-hydroxyglutaric acid has been proven in humans. D-2-Hydroxyglutaric acid can be assumed to be an intermediate in the conversion of 5-aminolevulinic acid to 2-oxoglutaric acid [39, 40]. A possible defect would be the conversion of D-2-hydroxyglutaric acid into 2-oxoglutaric acid by D-2-hydroxyglutaric acid dehydrogenase. A further enzyme, D-2-hydroxyglutaric-acid transhydrogenase, described in rat tissues, converts D-2-hydroxyglutaric acid into 2-oxoglutaric acid, combined with the formation of 4-hydroxybutyric acid from succinic semialdehyde, an intermediate in the degradation of γ-aminobutyric acid (GABA). In agreement with this hypothesis is the elevation of GABA in CSF and Krebs-cycle intermediates in urine in some patients with D-2-hydroxyglutaric aciduria [41].

### Diagnostic Tests

On organic acid analysis, D-2-hydroxyglutaric acid levels are 120–26,000 mmol/mol of creatinine (controls < 17 mmol/mol of creatinine) in urine, 3–660 μmol/l (control < 0.9 μmol/l) in plasma and 3–320 μmol/l in CSF (controls < 0.34 μmol/l) [38, 41]. In addition, GABA was found elevated in cerebrospinal fluid and intermediates of energy metabolism in urine (lactic, succinic, malic and 2-oxoglutaric acids) in some patients. Differentiation between the two isomers of 2-hydroxyglutarate is indispensable for diagnosis. *Prenatal diagnosis* may be possible utilizing accurate determination of D-2-hydroxyglutarate by stable-isotope-dilution GC-MS assay in amniotic fluid [38]. D-2-

Hydroxyglutaric acid can also be elevated in multiple acyl-CoA-dehydrogenase deficiency (glutaric aciduria type II), which can be distinguished by the classical urine organic acid profile of the latter disorder.

## Treatment and Prognosis

To date, there is no rational therapy for D-2-hydroxyglutaric aciduria. Attempts of riboflavin and L-carnitine supplementation resulted in no benefits. Seizures can be very difficult to control, and patients have died with profound developmental delay. In general, the clinical course does not appear to be progressive per se and, if affected children do not develop an early-onset epileptic encephalopathy, D-2-hydroxyglutaric aciduria can to be a relatively benign condition.

## Genetics

The observation of affected siblings and a similar clinical picture and equal numerical distribution for males and females strongly supports an autosomal-recessive mode of inheritance [41]. Heterozygotes display no detectable clinical or biochemical abnormalities related to D-2-hydroxyglutaric aciduria.

## References

1. Hoffmann GF, Gibson KM (1996) Disorders of organic acid metabolism. In: Moser HW (ed) Handbook of clinical neurology: neurodystrophies and neurolipidoses, vol 66. Elsevier Science, Amsterdam, pp 639–660
2. Kölker S, Ahlemeyer B, Krieglstein J, Hoffmann GF (1999) 3-Hydroxyglutaric and glutaric acids are neurotoxic through NMDA receptors in vitro. J Inherit Metab Dis 22:259–262
3. Dancis J, Hutzler J, Cox RP (1979) Familial hyperlysinemia: enzyme studies, diagnostic methods, comments on terminology. Am J Hum Genet 31:290–299
4. Dancis J, Hutzler J, Ampola MG et al. (1983) The prognosis of hyperlysinemia: an interim report. Am J Hum Genet 35:438–442
5. Carson NAJ (1969) Sacccharopinuria: a new inborn error of lysine metabolism. In: Allen JD, Holt KS, Ireland JT, Pollitt RJ (eds) Enzymopenic anemias, lysosomes, and other papers, Proc 6th Symposium of SSIEM, Livingstone, Edinburgh, pp 163–173
6. Simell O, Johansson T, Aula P (1973) Enzyme defect in sacccharopinuria. J Pediatr 82:54
7. Przyrembel H (1996) Disorders of ornithine, lysine and tryptophan. In: Blau N, Duran M, Blaskovics M (eds) Physician's guide to the laboratory diagnosis of inherited metabolic disease. Chapman and Hall, London, pp 223–245
8. Przyrembel H, Bachmann D, Lombeck I et al. (1975) Alpha-ketoadipic aciduria, a new inborn error of lysine metabolism; biochemical studies. Clin Chim Acta 58:257–269
9. Casey RE, Zaleski WA, Philp M, Mendelson IS (1978) Biochemical and clinical studies of a new case of α-aminoadipic aciduria. J Inherit Metab Dis 1:129–135
10. Duran M, Beemer FA, Wadman SK, Wendel U, Janssen B (1974) A patient with α-ketoadipic and α-aminoadipic aciduria. J Inherit Metab Dis 7:61
11. Wendel U, Rüdiger HW, Przyrembel H, Bremer HJ (1975) Alpha-ketoadipic aciduria: degradation studies with fibroblasts. Clin Chim Acta 58:271–276
12. Brauner-Osborne H, Slok FA, Skjaerbaek N et al. (1996) A new highly selective metabotropic excitatory amino acid agonist: 2-amino-4-(3-hydroxy-5-methylisoxazol-4-yl)butyric acid. J Med Chem 39:3188–3194
13. Baric I, Wagner L, Feyh P et al. (1999) Sensitivity and specificity of free and total glutaric and 3-hydroxyglutaric acids measurements by stable isotope dilution assays for the diagnosis of glutaric aciduria type I. J Inherit Metab Dis 22:867–882
14. Amir N, Elpeleg ON, Shalev RS, Christensen E (1989) Glutaric aciduria type I: enzymatic and neuroradiologic investigations of two kindreds. J Pediatr 114:983–989
15. Baric I, Zschoke J, Christensen E et al. (1998) Diagnosis and management of glutaric aciduria type I. J Inherit Metab Dis 21:326–340
16. Haworth JC, Booth FA, Chudley AE et al. (1991) Phenotypic variability in glutaric aciduria type I: report of fourteen cases in five Canadian Indian kindreds. J Pediatr 118:52–58
17. Hoffmann GF, Trefz FK, Barth P et al. (1991) Glutaryl-CoA dehydrogease deficiency: a distinct encephalopathy. Pediatrics 88:1194–1203
18. Morton DH, Bennett MJ, Seargeant LE, Nichter CA, Kelley RI (1991) Glutaric aciduria type I: a common cause of episodic encephalopathy and spastic paralysis in the Amish of Lancaster county, Pennsylvania. Am J Med Genet 41:89–95
19. Kyllerman M, Skjeldal OH, Lundberg M et al. (1994) Dystonia and dyskinesia in glutaric aciduria type I: clinical heterogeneity and therapeutic considerations. Mov Disord 9:22–30
20. Muntau AC, Röschinger W, Pfluger T, Enders A, Hoffmann GF (1997) Subdurale Hygrome und Hämatome im Säuglingsalter als Initialmanifestation der Glutarazidurie Typ I: Folgenschwere Fehldiagnose als Kindesmißhandlung. Monatsschr Kinderh 145:646–651
21. Hoffmann GF, Athanassopoulos S, Burlina A et al. (1996) Clinical course, early diagnosis, treatment and prevention of disease in glutaryl-CoA dehydrogenase deficiency. Neuropediatrics 27:115–123
22. Christensen E, Brandt NJ, Rosenberg T, Bömers K, Jakobs C (1994) The segregation of glutaryl-CoA dehydrogenase deficiency and Refsum syndrome in a family. J Inherit Metab Dis 17:287–290
23. Heyes MP (1987) Hypothesis: a role for quinolinic acid in the neuropathology of glutaric aciduria type I. Can J Neurol Sci 14:441–443
24. Superti-Furga A, Hoffmann GF (1997) Glutaric aciduria type 1 (glutaryl-CoA-dehydrogenase deficiency): advances and unanswered questions. Eur J Pediatr 156:821–828
25. Goodman SI, Norenberg MD, Shikes RH, Breslich DJ, Moe PG (1977) Glutaric aciduria type 1: biochemical and morphological considerations. J Pediatr 90:746–750
26. Monyer H, Burnashev N, Laurie DJ, Sakmann B, Seeburg PH (1994) Developmental and regional expression in the rat brain and functional properties of four NMDA receptors. Neuron 12:529–540
27. Bergman I, Finegold D, Gartner JC et al. (1989) Acute profound dystonia in infants with glutaric acidemia. Pediatrics 83:228–234
28. Christensen E (1993) A fibroblast glutaryl-CoA dehydrogenase assay using detritiation of ³H-labelled glutaryl-CoA: application in the genotyping of the glutaryl-CoA dehydrogenase locus. Clin Chim Acta 220:71–80
29. Goodman SI, Stein DE, Schlesinger S et al. (1998) Glutaryl-CoA dehydrogenase mutations in glutaric acidemia (type I): review and report of thirty novel mutations. Hum Mutat 12:141–144
30. Schwarz M, Christensen E, Superti-Furga A, Brandt NJ (1998) The human glutaryl-CoA dehydrogenase gene: report of intronic sequences and of thirteen novel mutations causing glutaric aciduria type 1. Hum Genet 102:452–458
31. Biery BJ, Stein DE, Morton DH, Goodman SI (1996) Gene structure and mutations of glutaryl-CoA dehydrogenase:

impaired association of enzyme subunits that is due to an A421 V muatation causes glutaric acidemia type I in the Amish. Am J Hum Genet 59:1006–1011

32. Barth PG, Hoffmann GF, Jaeken J et al. (1992) L-2-Hydroxyglutaric acidemia: a novel inherited neurometabolic disease. Ann Neurol 32:66–71
33. Barth PG, Hoffmann GF, Jaeken J et al. (1993) L-2-Hydroxyglutaric acidemia: clinical and biochemical findings in 12 patients and preliminary report on L-2-hydroxyacid dehydrogenase. J Inherit Metab Dis 16:753–761
34. de Klerk JBC, Huijmans JGM, Stroink H et al. (1997) L-2-Hydroxyglutaric aciduria: clinical heterogeneity versus biochemical homogeneity in a sibship. Neuropediatrics 28:314–317
35. Chen E, Nyhan WL, Jakobs C et al. (1996) L-2-Hydroxyglutaric aciduria: neuropathological correlations and first report of severe neurodegenerative disease and neonatal death. J Inherit Metab Dis 19:335–343
36. Wanders RJA, Vilharinho L, Hartung HP et al. (1997) L-2-Hydroxyglutaric aciduria: normal L-2-hydroxyglutarate dehydrogenase activity in liver from two new patients. J Inherit Metab Dis 20:725–726
37. Hoffmann GF, Jakobs C, Holmes B et al. (1995) Organic acids in cerebrospinal fluid and plasma of patients with L-2-hydroxyglutaric aciduria. J Inherit Metab Dis 18:189–193
38. Gibson KM, Schor DSM, Kok RM et al. (1993) Stable-isotope dilution analysis of D- and L-2-hydroxyglutaric acid: application to the detection and prenatal diagnosis of D- and L-2-hydroxyglutaric acidemias. Pediatr Res 34:277–280
39. Gibson KM, Craigen W, Herman GE, Jakobs C (1993) D-2-Hydroxyglutaric aciduria in a newborn with neurological abnormalities: a new neurometabolic disorder? J Inherit Metab Dis 16:497–500
40. Nyhan WL, Shelton GD, Jakobs C et al. (1995) D-2-Hydroxyglutaric aciduria. J Child Neurol 10:137–142
41. van der Knaap MS, Jakobs C, Hoffmann GF et al. (1999) D-2-Hydroxyglutaric aciduria. Biochemical marker or clinical disease entity? Ann Neurol 45:111–119

## Glycine Metabolism

Glycine, the simplest of the amino acids, is abundant in nearly all animal proteins and enters into more biosynthetic routes than any other amino acid. Formation of glycine conjugates plays an important role in the detoxification of various compounds, including those that accumulate in certain inborn errors of metabolism. The catabolism of glycine involves several pathways, among which the glycine-cleavage system is of major importance. This multienzyme complex degrades glycine into $NH_3$ and $CO_2$ and, thereby, also converts tetrahydrofolate into 5,10-methylene tetrahydrofolate. The latter compounds also intervene in the interconversion of serine and glycine, catalyzed by serine hydroxymethyl transferase.

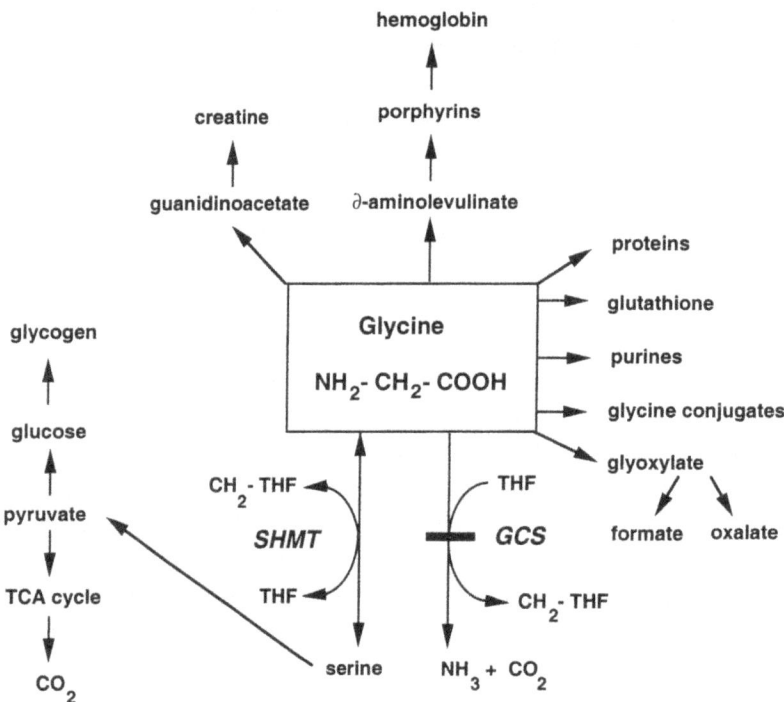

**Fig. 21.1.** Pathways of glycine metabolism. *CH₂-THF*, 5,10-methylene tetrahydrofolate; *GCS*, glycine cleavage system; *SHMT*, serine hydroxymethyl transferase; *TCA*, tricarboxylic acid; *THF*, tetrahydrofolate. Enzyme defect is depicted by *solid bar*

# CHAPTER 21

# Nonketotic Hyperglycinemia

K. Tada

## CONTENTS

Introduction.................................. 255
  Clinical Presentation....................... 255
  Metabolic Derangement...................... 256
  Diagnostic Tests........................... 257
    Prenatal Diagnosis....................... 257
  Treatment and Prognosis.................... 257
  Genetics................................... 258
References.................................... 258

---

Nonketotic hyperglycinemia (NKH) is an autosomal recessive disorder that is characterized by rapidly progressing neurological symptoms, such as muscular hypotonia, seizures, apneic attacks, and lethargy or coma in early infancy, mostly in the neonatal period. Most patients die within a few weeks, whereas the survivors show severe psychomotor retardation. Increased glycine concentrations in plasma, urine, and cerebrospinal fluid are biochemical features of the disorder. The primary lesion was found to be a defect in the glycine cleavage system (GCS) (Fig. 21.1). No specific treatment is available. Prenatal diagnosis is feasible by determining the activity of GCS in chorionic villi.

---

## Introduction

Hyperglycinemia represents a group of disorders characterized by elevated concentrations of glycine in body fluids. There are two types of hyperglycinemia: the nonketotic type and the ketotic type [1]. Nonketotic hyperglycinemia (NKH) is a disorder of glycine degradation due to a primary defect in the glycine-cleavage system (GCS). NKH is a relatively frequent metabolic cause of overwhelming illness in infancy. In the ketotic type, the most striking feature is ketoacidosis, which begins early in life and in which hyperglycinemia is secondarily associated with organic acidemias, such as methylmalonic acidemia, propionic acidemia, and isovaleric acidemia.

## Clinical Presentation

NKH is usually classified into two types from a clinical point of view: neonatal and late-onset types. The *neonatal type* is the common type. Most affected infants appear normal at birth except for patients who present prenatally with in utero brain-damage-like dysgenesis of the corpus callosum and gyral malformations [2]. After a short interval (seldom longer than 48 h), the patient develops rapidly progressing neurological symptoms, such as muscular hypotonia, depressed Moro response, seizures, apneic attacks, and lethargy or coma. Most patients die within a few weeks; the survivors show severe psychomotor retardation. Convulsive seizures range from myoclonic to grand mal convulsions. Hiccuping is often seen. During the first few weeks of life, a characteristic electroencephalogram (EEG) pattern with bursts of high complex waves of 1–3 s arising periodically from a hypoactive background is seen (Fig. 21.2). This so-called burst-suppression pattern is present immediately after birth [3], preceding clinical symptoms. It disappears at the end of the first month and changes to hypsarrythmia. Muscular hypotonia is prominent during the neonatal period but, thereafter, spasticity develops gradually. In our experience of 32 cases of NKH, 28 (87.5%) were of the neonatal type. Among them, 24 died at between 6 days and 5 years of age. The remaining four patients survived but were severely retarded [4].

*Transient neonatal NKH* has been described in six newborns with symptoms indistinguishable from those of the neonatal phenotype. Plasma and CSF glycine concentrations initially are elevated to those seen in neonatal NKH but return to normal by 8 weeks of age. Five of the six patients had no neurological sequelae after 6 months to 13 years of follow-up. However, one patient had severe developmental delay at 9 months of age [5, 6].

In the *late-onset type*, the patient has no abnormal symptoms or signs in the neonatal period and, thereafter, nonspecific neurological symptoms develop to various degrees. Onset ranges from infancy to adolescence.

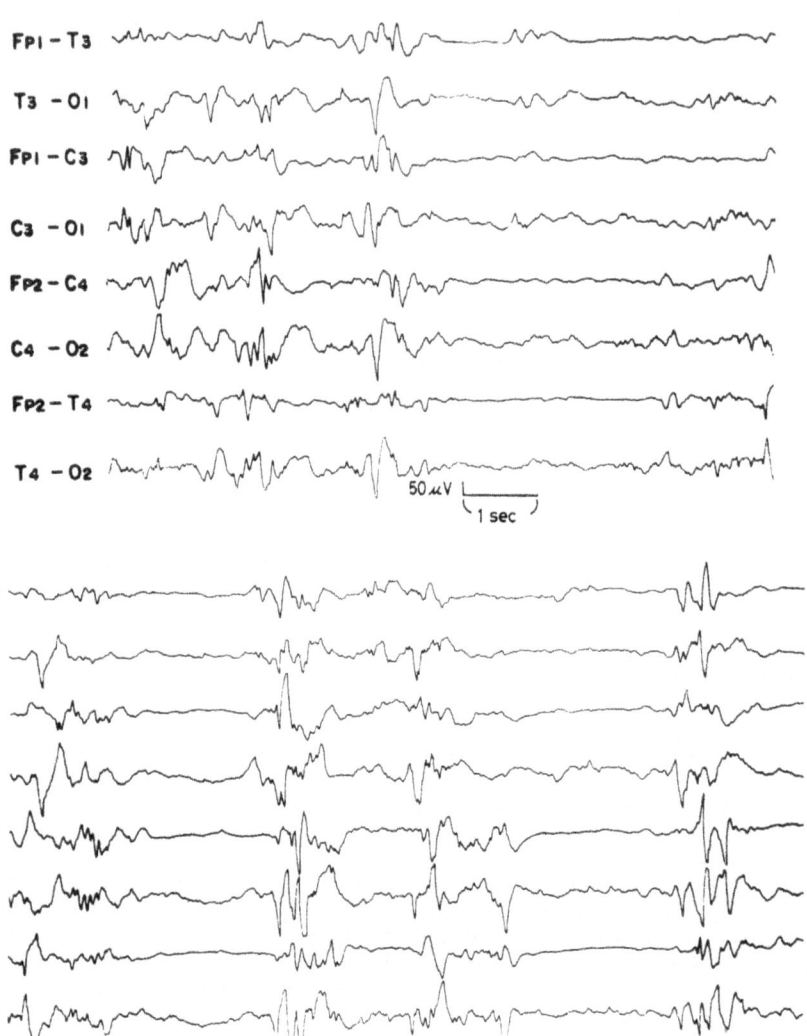

Fig. 21.2. Electroencephalogram of a 6-week-old patient with nonketotic hyperglycinemia, showing "burst-suppression" pattern

## Metabolic Derangement

The defect of NKH is in the GCS, which catalyzes the transformation of glycine and tetrahydrofolate into $CO_2$, $NH_3$, and methylene tetrahydrofolate (Fig. 21.1) [7, 8]. The accumulation of glycine is due not only to the metabolic block itself but also to the overproduction of glycine from serine via the serine hydroxymethyl transferase reaction (Fig. 21.1). The GCS is a multienzyme complex composed of four protein components: P-protein (a pyridoxal phosphate dependent glycine decarboxylase), H-protein (a lipoic acid containing protein), T-protein (a tetrahydrofolate requiring enzyme), and L-protein (lipoamide dehydrogenase) [9]. The overall activity of the GCS in the livers of 30 patients was not detectable or was extremely low in the neonatal type whereas, in the late-onset type, some residual activity of the GCS was seen [4]. Thus, the clinical phenotypes do seem to relate to the degree of the defect in the GCS. The mechanism for transient NKH is unclear, because of the lack of enzymatic data. Immaturity of one or more of the four components of the GCS, or deficiency of any of the GCS cofactors, is postulated. The accumulation of glycine is due not only to the metabolic block itself (which suppresses glycine degradation), but also to the overproduction of glycine from serine via the serine hydroxymethyl transferase reaction. The glycine content of the brain is elevated in NKH in contrast to the content in ketotic hyperglycinemia [1, 10]. Consistent with this, the activity of GCS in the brain is undetectable in NKH and normal in ketotic hyperglycinemia [1]. The majority of the NKH patients (26 out of 30; 87%) appeared to have a specific defect in the P-protein, and the remaining had a specific defect in the T-protein. The component analysis was also made in the brains of seven autopsied patients. The sites of the defects in these cases were identical in both the brain and liver. This suggests that the GCS proteins in the liver and brain are controlled by the same gene.

With respect to secondary ketotic hyperglycinemia, GCS activity was normal in liver obtained at biopsy from a patient with propionic acidemia who was under adequate dietary treatment and had a normal plasma glycine level [10]. In contrast, GCS activity was significantly reduced in the livers of patients with propionic acidemia or methylmalonic acidemia, who died in a hyperglycinemic state. This suggests that, in ketotic hyperglycinemia, the elevation of glycine occurs as a result of a secondary suppression of GCS or as a result of the excess of some organic acids or their coenzyme-A derivatives [11].

Elevation of glycine in the brain is thought to be responsible for clinical symptoms in NKH. Recently, several observations have provided a new insight into the relationship of glycine and excitotoxicity of the brain. Although glycine has been considered to act as an inhibitory neurotransmitter at a strychnine-sensitive receptor, it has also been found to have an excitatory property that potentiates the glutaminergic N-methyl-D-aspartate (NMDA) receptors [12]. It is suggested that glycine enhances NMDA-mediated responses at a site closely associated with the NMDA receptor. Glycine administration enhanced NMDA-induced seizures in mice whose classic glycine receptor had been blocked with strychnine. Furthermore, it was shown that the developing brain has enhanced susceptibility to NMDA-mediated brain injury, and high levels of glycine may be particularly devastating to the central nervous system of the neonate [13]. Also, in immunohistochemical studies, the sites of the GCS coincided with a region rich in NMDA receptors in rat brain [14]. Thus, the elevated concentrations of glycine in the brain may contribute to the pathophysiology of NKH by overstimulating NMDA receptors via an action at the associated glycine-modulatory site.

## Diagnostic Tests

When infants develop seizures, muscular hypotonia, and somnolence or lethargy, and these symptoms can not readily be explained by infection, trauma, hypoxia, or other commonly encountered pediatric problems, NKH should be considered and plasma amino acids analyzed. Differentiation from ketotic hyperglycinemia is sometimes not easy. Absence of ketoacidosis, as reflected by normal plasma bicarbonate levels, normal arterial or capillary blood pH, and exclusion of organic acidemia by gas-chromatographic analyses of urine or plasma are crucial. In NKH, the glycine level in CSF is elevated, and the ratio of CSF glycine concentration to plasma glycine concentration is above 0.09 whereas, under normal circumstances and in ketotic hyperglycinemia, it is below 0.04 [1].

As the GCS is expressed in liver, kidney, and brain, liver biopsy is performed for the enzymatic diagnosis of NKH. However, a new method using peripheral blood is now available. It is based on the fact that the GCS is induced in B lymphocytes by infection and transformation using Epstein-Barr virus [15]. This method is useful for differential diagnosis between NKH and ketotic hyperglycinemia and for carrier detection of NKH.

## Prenatal Diagnosis

There is a strong demand for prenatal diagnosis, since no effective treatment is available for NKH. Cultured amniotic cells do not have GCS activity and are, therefore, not useful. However, we found GCS in chorionic villi of placenta and suggested prenatal diagnosis of NKH by chorionic villi sampling [16]. Accordingly, 20 pregnant women who had children with NKH were investigated at the 8th–16th week of gestation [17]. In 15 out of 20 cases, GCS activity was found to be normal. Their pregnancies were continued, and healthy babies were born after full-term pregnancies. In the remaining five cases, GCS activity was undetectable, suggesting that the fetus was affected with NKH. The pregnancies were terminated in accordance with the parents' desire. The GCS activities in the livers and brains of the aborted fetuses were nearly undetectable. DNA diagnosis, too, is possible when the mutation of the family is known [18].

## Treatment and Prognosis

To lower the glycine concentration in NKH patients, many therapeutic approaches have been attempted, including: protein restriction; a synthetic diet devoid of glycine and its precursor, serine; promotion of renal clearance by benzoate; administration of ursodeoxycholic acid, which conjugates with glycine and is excreted in bile; and exchange transfusion. These treatments were effective for lowering plasma levels (but not CSF levels) of glycine and did not alter appreciably the clinical course of the disease. Tentative treatment with strychnine [19], diazepam (a competitor of glycine receptors), or one-carbon donors such as methionine, leucovorin, choline, or formate, did not improve or only ambiguously improved the clinical symptoms.

Recently, a tentative treatment using NMDA antagonists, in keeping with the excitotoxicity hypothesis suggested by glycine, has been tried. Oral administration (8 mg/kg/day, in four divided doses) of ketamine, an anesthetic NMDA-channel blocker, to a 7-month-old patient with NKH brought a partial improvement of neurological symptoms and EEG findings [20]. The regimen of ketamine and strychnine [21] or sodium benzoate [22] was reported to have a beneficial effect on clinical symptoms of NKH patients, but their developmental milestones were delayed. The general use of ketamine is limited by its short duration of action and its route of administration.

Dextromethorphan (DM) is converted into its active metabolite, dextrorphan, which is a noncompetitive inhibitor of the NMDA-receptor-channel complex. Hamosh et al. [23] reported the results of long-term use of high-dose benzoate and DM for the treatment of NKH. They administered benzoate (at doses of 500–750 mg/kg/day) and DM (at doses of 3.5–22.5 mg/kg/day) to four infants with NKH, with follow-ups of 3 months to 6 years. Benzoate reduced the glycine concentration in plasma to normal and substantially reduced (but did not normalize) the glycine concentration in CSF. DM was a potent anticonvulsant in some but not all patients. Three patients are living (ages ranging from 4 years to 6 years) and are moderately to severely developmentally delayed; two are free of seizures. One patient died of intractable seizures at 3 months. These outcomes suggest that benzoate and DM are not uniformly effective in NKH but, for some patients, they decrease or eliminate seizures, and the patients show developmental progress. Further trials with additional patients and other receptor-channel blockers are warranted [23, 24].

## Genetics

NKH is transmitted as an autosomal-recessive trait. The prevalence is not firmly known. In northern Finland, it is estimated to be 1:12,000.

cDNAs encoding the human P-protein and T-protein of the GGS were cloned by our group [4, 25, 26], and many missense, nonsense, or frame-shift mutations have been identified [27–29]. The majority of NKH patients in Finland were found to carry a common G → T substitution, resulting in a S564I amino acid change in the P-protein of the GCS [18].

## References

1. Tada K (1987) Nonketotic hyperglycinemia: clinical and metabolic aspects. Enzyme 38:27–35
2. Dobyns WB (1989) Agenesis of the corpus callosum and gyral malformations are frequent manifestations of nonketotic hyperglycinemia. Neurology 39:817–820
3. Markand ON, Garg BP, Brondt IK (1982) Nonketotic hyperglycinemia: electroencephalographic and evoked potential abnormalities. Neurology 32:151–156
4. Tada K, Kure S (l983) Nonketotic hyperglycinemia: molecular lesion, diagnosis and pathophysiology. J Inherit Metab Dis 16:691–703
5. Luder AS, Davidson A, Goodman SI, Green CL (1989) Transient nonketotic hyperglycinemia in neonates. J Pediatr 114:1013–1015
6. Zammarchi E, Donati MA, Ciani F (1995) Transient neonatal nonketotic hyperglycinemia: a 13-year follow-up. Neuropediatrics 26:328–330
7. Tada K, Narisawa K, Yoshida T, Konno T, Yokoyama Y et al. (1969) Hyperglycinemia: defect in glycine cleavage reaction. Tohoku J Exp Med 98:289–296
8. Kikuchi G (1983) The glycine cleavage system: composition, reaction mechanism, and physiological significance. Mol Cell Biochem 1:169–187
9. Hayasaka K, Tada K, Kikuchi G, Winter S, Nyhan WL (1983) Nonketotic hyperglycinemia: two patients with primary defects of P-protein and T-protein, respectively, in the glycine cleavage system. Pediatr Res 17:967–970
10. Hayasaka K, Narisawa K, Satoh T, Tateda H, Metoki K et al. (l982) Glycine cleavage system in ketotic hyperglycinemia. Pediatr Res 16:5–7
11. Hayasaka K, Tada K (1983) Effects of the metabolites of the branched-chain aminoacids and cysteamine on the glycine cleavage system. Biochem Int 6:225–230
12. Johnson JW, Ascher P (1987) Glycine potentiates the NMDA response in cultured mouse brain neurons. Nature 325:529–531
13. McDonald JW, Johnstone MV (1990) Physiological and pathophysiological roles of excitatory amino acids during central nervous system development. Brain Res Rev 15:41–70
14. Sato K, Yoshiada S, Fujiwara K, Tada K, Tohyama M (1991) Glycine cleavage system in astrocytes. Brain Res 567:64–70
15. Kure S, Narisawa K, Tada K (1992) Enzymatic diagnosis of nonketotic hyperglycinemia with lymphoblasts. J Pediatr 120:95–98
16. Hayasaka K, Tada K, Fueki N et al. (1987) Feasibility of prenatal diagnosis of nonketotic hyperglycinemia: existence of the glycine cleavage system in placenta. J Pediatr 110:124–126
17. Hayasaka K, Tada K, Fueki N, Aikawa D (1990) Prenatal diagnosis of nonketotic hyperglycinemia: enzymatic analysis of the glycine cleavage system in chorionic villi. J Pediatr 16:444–445
18. Kure S, Takayanagi M, Narisawa K, Tada K, Leisti J (1992) Identification of a common mutation in Finnish patients with nonketotic hyperglycinemia. J Clin Invest 90:160–164
19. Gitzelmann R, Cuenod M, Otten A, Steinmann B, Dumermuth G (1977) Nonketotic hyperglycinemia treated with strychnin. Pediatr Res 11:1016
20. ohya Y, Ochi N, Mizutani N, Hayakawa C, Watanabe K (1991) Nonketotic hyperglycinemia: treatment with NMDA antagonist and consideration of neuropathogenesis. Pediatr Neurol 7:65–68
21. Tegtmeyer-Metzdorf H, Roth B, Gunther M et al. (l995) Ketamine and strychnine treatment of an infant with nonketotic hyperglycinemia. Eur J Pediatr 154:649–653
22. Boneh A, Degni Y, Harari M (1996) Prognostic clues and outcome of early treatment of nonketotic hyperglycinemia. Pediatr Neurol 15:137–147
23. Hamosh A, Maber JF, Bellus GA, Rasmussen SA, Johnston MV (1998) Long-term use of high-dose benzoate and dextromethorphan for the treatment of nonketotic hyperglycinemia. J Pediatr 132:709–713
24. Deutsch SI, Rosse RB, Mastropaolo J (1998) Current status of NMDA antagonist interventions in the treatment of nonketotic hyperglycinemia. Clin Neuropharmacol 21:71–79
25. Kure S, Narisawa K, Tada K (1991) Structural and expression analyses of normal and mutant mRNA encoding glycine decarboxylase: three base deletion in mRNA causes nonketotic hyperglycinemia. Biochem Biophys Res Commun 174:1176–1182
26. Hyasaka K, Nanao K, Takada G, Ikeda KO, Motokawa Y (1993) Isolation and sequence determination of cDNA encoding human T-protein of the glycine cleavage system. Biochem Biophys Res Commun 192:766–771
27. Nanao K, Ikeda KO, Motokawa Y, Danks DM, Hayasaka K et al. (1994) Identification of the mutations in the T-protein gene causing typical and atypical nonketotic hyperglycinemia. Hum Genet 93:655–658
28. Kure S, Shinka T, Sakata Y, Narasaki O, Takayanagi M et al. (1998) A one-base deletion (l83delC) and a missense mutation (D276H) in the T-protein gene from a Japanese family with nonketotic hyperglycinemia. J Hum Genet 43:135–137
29. Kure S, Mandel H, Rolland MO, Sakata Y, Shinka T et al. (1998) A missense mutation (His42Arg) in the T-protein gene from a large Israeli-Arab kindred with nonketotic hyperglycinemia. Hum Genet 102:430–434

# CHAPTER 22

## Proline and Serine Metabolism

Proline and serine are non-essential amino acids. Unlike all other amino acids, *proline* has no primary amino group ("imino acid") and uses, therefore, a specific system of enzymes for its metabolism (Fig. 22.1). $\Delta^1$-Pyrroline 5-carboxylate (P 5-C) is both the immediate precursor and the degradation product of proline. The P 5-C/proline cycle transfers reducing/oxidizing potential between cellular organelles. Due to its pyrrolidine ring, proline (together with hydroxyproline) contributes to the structural stability of proteins, particularly collagen, with its high proline and hydroxyproline content.

*Serine* also has important functions besides its role in protein synthesis. It is a precursor of the synthesis of a number of compounds, including glycine, cysteine, serine phospholipids, sphingomyelins, and cerebrosides. Moreover, it is a major source of methylenetetrahydrofolate and of other one-carbon donors that are required for the synthesis of purines and thymidine. Serine is synthesized de novo from a glycolytic intermediate, 3-phosphoglycerate (Fig. 22.2), and can also be synthesized from glycine by reversal of the reaction catalyzed by serine hydroxymethyltransferase.

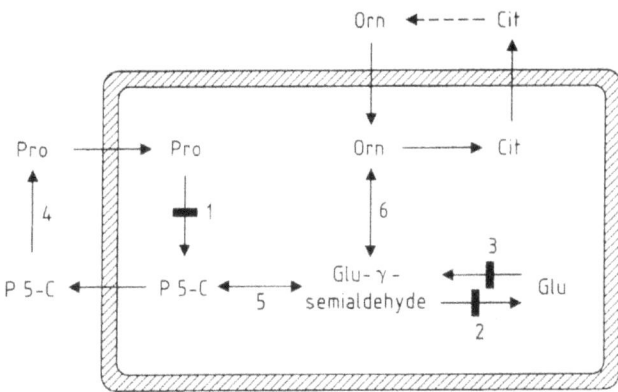

**Fig. 22.1.** Proline metabolism. *Shaded area* represents mitochondrial membrane. *1*, Proline oxidase (deficient in hyperprolinemia type 1); *2*, pyrroline 5-carboxylate (P 5-C) dehydrogenase (deficient in hyperprolinemia type 2); *3*, P 5-C synthase; *4*, P 5-C reductase; *5*, non-enzymatic reaction; *6*, ornithine aminotransferase (deficient in gyrate atrophy). *Bars* across *arrows* indicate defects of proline metabolism

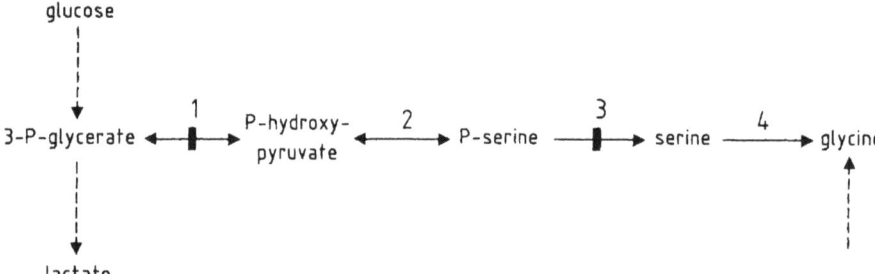

**Fig. 22.2.** Pathway of de novo serine synthesis. *1*, 3-Phosphoglycerate dehydrogenase; *2*, 3-phosphohydroxypyruvate transaminase; *3*, 3-phosphoserine phosphatase. Serine can also be synthesized from glycine by reversal of the reaction catalysed by serine hydroxymethyltransferase (enzyme 4). The *vertical bars* indicate the known defects in serine synthesis. Glycine is synthesized from serine, but also from other sources (*broken arrow*)

# CHAPTER 22

# Disorders of Proline and Serine Metabolism

J. Jaeken

CONTENTS

Inborn Errors of Proline Metabolism.............. 261
   Proline Oxidase Deficiency (Hyperprolinemia Type I).. 261
   $\Delta^1$-Pyrroline 5-Carboxylate Dehydrogenase Deficiency
   (Hyperprolinemia Type II)...................... 262
   $\Delta^1$-Pyrroline 5-Carboxylate Synthase Deficiency....... 262

Inborn Errors of Serine Metabolism ................ 262
   3-Phosphoglycerate Dehydrogenase Deficiency........ 262
   Phosphoserine Phosphatase Deficiency.............. 263
   Serine Deficiency with Ichthyosis
   and Polyneuropathy ........................... 263

References...................................... 263

Three disorders of *proline* metabolism are known: two in its catabolism (hyperprolinemia type I due to proline oxidase deficiency and hyperprolinemia type II due to $\Delta^1$-pyrroline 5-carboxylate dehydrogenase deficiency) and one in its synthesis ($\Delta^1$-pyrroline 5-carboxylate synthetase deficiency). Hyperprolinemia type I is a non-disease, while hyperprolinemia type II seems to be associated with a disposition to recurrent seizures. A defect in a proline-synthesizing enzyme, $\Delta^1$-pyrroline 5-carboxylate synthetase, was demonstrated in siblings with joint hyperlaxity, skin hyperelasticity, cataract, mental retardation with hyperammonemia, and low citrulline, ornithine and proline.

Three disorders of *serine* metabolism are known. One is in its biosynthesis: namely, 3-phosphoglycerate dehydrogenase deficiency. Patients are affected with congenital microcephaly, psychomotor retardation and intractable seizures and partially respond to L-serine or L-serine and glycine. One patient with an association of Williams syndrome and phosphoserine phosphatase deficiency has been reported. Another, unexplained serine disorder has been reported in a patient with decreased serine in body fluids, ichthyosis and polyneuropathy but no central nervous system manifestations. There was a spectacular response to L-serine.

## Inborn Errors of Proline Metabolism

### Proline Oxidase Deficiency (Hyperprolinemia Type I)

### Clinical Presentation

Hyperprolinemia type I is a very rare "non-disease" that is detected by neonatal metabolic screening [1–3].

### Metabolic Derangement

Hyperprolinemia type I is caused by a deficiency of proline oxidase (a mitochondrial inner-membrane enzyme), which catalyses the conversion of proline into P 5-C (Fig. 22.1, enzyme 1). Hence, in hyperprolinemia type I, there are increased levels of proline in plasma (usually not above 2000 µM; normal range ~100–450 µM), urine and CSF. Hyperprolinemia (as high as 1000 µM) is also observed as a secondary phenomenon in hyperlactacidemia, possibly due to inhibition of proline oxidase by lactic acid. Remarkably, and contrary to the case for hyperprolinemia type II, heterozygotes have hyperprolinemia.

### Diagnostic Tests

The diagnosis is made by amino acid analysis. Direct enzyme assay is not possible, since the enzyme is not present in leukocytes or skin fibroblasts. Mutation analysis of patients is in progress [4].

### Treatment and Prognosis

Since the prognosis is excellent, dietary treatment is not indicated.

### Genetics

The mode of inheritance is autosomal recessive. Two different forms of the human gene (HsPOX) have been identified: HsPOX1 was mapped to 19q13.1 and HsPOX2 to 22q11.2 [4, 5].

## $\Delta'$-Pyrroline 5-Carboxylate Dehydrogenase Deficiency (Hyperprolinemia Type II)

### Clinical Presentation

This is a relatively benign disorder, though a disposition to recurrent seizures is highly likely [2].

### Metabolic Derangement

Hyperprolinemia type II is caused by a deficiency of P 5-C dehydrogenase (a mitochondrial inner-membrane enzyme), which catalyses the conversion of proline into glutamate (Fig. 22.1, enzyme 2). Hence, in hyperprolinemia type II, there are increased levels of proline in plasma (usually exceeding 2000 μM; normal range ~100-450 μM), urine and CSF as well as of P 5-C. Heterozygotes do not have hyperprolinemia.

### Diagnostic Tests

The accumulation of P 5-C in physiological fluids differentiates type-II and type-I hyperprolinemias. This compound can be qualitatively identified by its reactivity with *ortho*-aminobenzaldehyde and can be quantitatively measured by several specific assays [2]. P 5-C-dehydrogenase activity can be measured in skin fibroblasts and leukocytes.

### Treatment and Prognosis

The benign character of the disorder does not justify dietary treatment (which, in any case, would be very difficult).

### Genetics

This is an autosomal-recessive disease. Mutations have recently been reported in four patients (two frameshift mutations and two missense mutations) [6].

## $\Delta'$-Pyrroline 5-Carboxylate Synthetase Deficiency

### Clinical Presentation

Only two patients (siblings, aged 4 years and 12 years) have been reported; they showed bilateral cataract, psychomotor retardation, joint hyperlaxity and skin hyperelasticity [7, 8].

### Metabolic Derangement

P 5-C-synthetase deficiency is one of the few examples of a defect in the biosynthesis of amino acids, namely proline and ornithine. In addition to the low plasma levels of proline and ornithine, there is also hypocitrullinemia, which is secondary to the hypo-ornithinemia (Fig. 22.1). However, this hypo-ornithinemia causes a moderate, apparently paradoxical hyperammonemia, since it is elicited by fasting and decreases after feeding. This observation is easily explained by the normal provision of dietary arginine, the source of ornithine.

### Diagnostic Tests

Plasma amino acid analysis shows the specific amino acid pattern described above, which is associated with fasting hyperammonemia. Indirect confirmation of the enzyme defect can be obtained using incorporation of $C^{14}$ from $C^{14}$-glutamate into ornithine and proline in fibroblasts.

### Treatment and Prognosis

No reported data are available. The disease is expected to be treatable with excess dietary proline and ornithine (or arginine).

### Genetics

Inheritance is autosomal recessive. The gene has been mapped to 10q24.3. Recently a L396S mutation has been identified in patients [8].

# Inborn Errors of Serine Metabolism

## 3-Phosphoglycerate Dehydrogenase Deficiency

### Clinical Presentation

At least nine patients belonging to four families with this disease (first reported in 1996) are known [9]. They presented at birth with microcephaly and developed pronounced psychomotor retardation, severe spastic tetraplegia, nystagmus and intractable seizures (including hypsarrythmia).

In addition, one patient showed congenital bilateral cataract, two siblings showed growth retardation and hypogonadism, and two other siblings showed megaloblastic anemia. Magnetic resonance imaging of the brain revealed cortical and subcortical hypotrophy and evidence of disturbed myelination [9-11].

### Metabolic Derangement

3-Phosphoglycerate dehydrogenase deficiency is due to a defect in the first step of the serine biosynthesis catalyzed by 3-phosphoglycerate dehydrogenase (Fig. 22.2, enzyme 1). This causes decreased concentrations of serine and, to a lesser extent, of glycine in CSF and in fasting plasma. Serine thus becomes an essential amino acid in these patients. It is unlikely that there is a significant accumulation of the substrate 3-phosphoglycerate, since it is an intermediate of the glycolytic pathway. Therefore, the deficiency of brain serine seems to be the main determinant of the disease. Serine plays a major role in the synthesis of important brain and myelin constituents, such as proteins, glycine, cysteine, serine phospholipids, sphingomyelins and cerebrosides.

In the two patients with megaloblastic anemia, a decreased methyltetrahydrofolate was found in CSF. This can be explained by the fact that serine is converted into glycine in the same reaction that forms methylenetetrahydrofolate, which is further reduced to methyltetrahydrofolate (Chap. 25) [10].

### Diagnostic Tests

The diagnosis should be suspected in patients with encephalopathy comprising congenital microcephaly. Plasma amino acids have to be measured in the fasting state, because serine and glycine levels can be normal after feeding. In CSF, serine levels are always decreased, as are glycine levels, to a lesser extent. The diagnosis is confirmed by finding a deficient activity of 3-phosphoglycerate dehydrogenase in fibroblasts (reported residual activities from 6% to 22%).

### Treatment and Prognosis

Treatment with L-serine has a beneficial effect on the convulsions, spasticity, feeding and behavior of these patients. Oral L-serine treatment (up to 500 mg/kg/day in six divided doses) corrected the biochemical abnormalities in all reported patients and abolished the convulsions in most patients. In two patients, convulsions stopped only after adding glycine (200 mg/kg/day) [10]. In order to prevent microcephaly and the other neurological abnormalities, prenatal treatment of the mother with serine would be a possible approach, but no experience is available.

### Genetics

This is an autosomal-recessive disease. The gene for 3-phosphoglycerate dehydrogenase has not yet been localized. Mutation analysis is under way. Prenatal diagnosis by enzyme analysis of amniocytes or chorionic villi should be possible but has not yet been performed.

### Phosphoserine Phosphatase Deficiency

Decreased serine levels were found in the plasma (53–80 µM; normal range 70–187 µM) and CSF (18 µM; control range 27–57 µM) of one patient with Williams syndrome [12]. Phosphoserine phosphatase activity in lymphoblasts and fibroblasts amounted to about 25% of normal values (Fig. 22.2; enzyme 3). Oral serine normalized the plasma and CSF levels of this amino acid and seemed to have some clinical effect. This observation suggests that the phosphoserine phosphatase gene is closely linked to the elastin gene region (7q11.23).

### Serine Deficiency with Ichthyosis and Polyneuropathy

A remarkable new serine-deficiency syndrome has recently been discovered by De Klerk et al. in a 15-year-old girl [13]. She had ichthyosis since the first year of life and growth retardation from the age of 6 years. Since the age of 14 years, she presented symptoms of axonal polyneuropathy with walking difficulties and areflexia. Psychomotor development and magnetic resonance imaging of the brain were normal. Fasting plasma and CSF serine levels were decreased, but the CSF glycine level was slightly increased. Oral ingestion of serine (400 mg/kg/day) cured the skin lesions and the polyneuropathy. It is hypothesized that this patient exhibits an increased conversion of serine into glycine, possibly due to hyperactivity of serine hydroxymethyltransferase (Fig. 22.2, enzyme 4).

### References

1. Scriver CR, Schafer IA, Efron ML (1961) New renal tubular amino acid transport system and a new hereditary disorder of amino acid metabolism. Nature 192:672
2. Phang JM, Yeh GC, Scriver CR (1995) Disorders of proline and hydroxyproline metabolism. In: Scriver CR, Beaudet AL, Sly WS, Valle D (eds) The metabolic and molecular bases of inherited disease, 7th edn. McGraw-Hill, New York, pp 1125–1146
3. Aral B, Kamoun P (1997) The proline biosynthesis in living organisms. Amino Acids 13:189–217
4. Lin WW, Hu C-A, Valle D (1997) Cloning and characterization of cDNAs encoding human proline oxidase, the enzyme deficient in type I hyperprolinemia. Pediatr Res 41:105A
5. Jaeken J, Goemans N, Fryns J-P, François I, de Zegher F (1996) Association of hyperprolinaemia type I and heparin cofactor II deficiency with CATCH 22 syndrome: evidence for a contiguous gene syndrome locating the proline oxidase gene. J Inherit Metab Dis 19:275–277
6. Geraghty MT, Vaughn D, Nicholson AJ, Lin WW, Jimenez-Sanchez G, Obie C, Flynn MP, Valle D (1998) Mutations in the delta 1-pyrroline 5-carboxylate dehydrogenase gene cause type II hyperprolinemia. Hum Mol Genet 7:1411–1415
7. Rabier D, Nuttin C, Poggi F, Padovani JP, Abdo K, Bardet J, Parvy P, Kamoun P, Saudubray JM (1992) Familial joint hyperlaxity, skin hyperelasticity, cataract and mental retarda-

tion with hyperammonemia and low citrulline, ornithine and proline. A new disorder of collagen metabolism? Abstracts of the 30th annual symposium of the Society for the Study of Inborn Errors of Metabolism. Leuven, 8–11 Sept 1992
8. Kamoun P, Aral B, Saudubray J-M (1998) A new inherited metabolic disease: $\Delta^1$-pyrroline 5-carboxylate synthetase deficiency. Bull Acad Natl Med 182:131–139
9. Jaeken J, Detheux M, Van Maldergem L, Foulon M, Carchon H, Van Schaftingen E (1996) 3-Phosphoglycerate dehydrogenase deficiency: an inborn error of serine biosynthesis. Arch Dis Child 74:542–545
10. De Koning TJ, Duran M, Dorland L, Gooskens R, Van Schaftingen E, Jaeken J, Blau N, Berger R, Poll-The BT (1998) Beneficial effects of L-serine and glycine in the management of seizures in 3-phosphoglycerate dehydrogenase deficiency. Ann Neurol 44:261–265
11. Pineda M, Vilaseca MA, Artuch R, Santos S, Garcia Gonzalez MM, San I, Jaeken J (1998) Serine deficiency in West syndrome. J Inherit Metab Dis 22 [Suppl 2]:55
12. Jaeken J, Detheux M, Fryns J-P, Collet J-F, Alliet P, Van Schaftingen E (1997) Phosphoserine phosphatase deficiency in a patient with Williams syndrome. J Med Genet 34:594–596
13. De Klerk JBC, Huijmans JGM, Catsman-Berrevoets CE, Lindhout D, den Hollander JC, Duran M (1996) Disturbed biosynthesis of serine; a second phenotype with axonal polyneuropathy and ichthyosis. Abstracts of the annual meeting of the Erfelijke Stofwisselingsziekten Nederland. Maastricht, 13–15 Oct 1996

# CHAPTER 23

## Amino Acid Transport at the Cell Membrane

Several amino acid transport systems exist at the cell membrane of epithelia, including renal tubules and intestinal brush border. Cystine and the structurally related dibasic amino acids lysine, arginine and ornithine, are transported into cells by an apical (luminal) system (Fig. 23.1). In the intestine, amino acids are absorbed not only as free amino acids but also as small peptides. These peptides are cleaved to free amino acids inside the epithelial cells. The dibasic amino acids are then transported through the basolateral (antiluminal) dibasic amino acid transporter and into the body. A neutral amino acid transporter that is expressed only at the luminal border of the epithelial cells in the renal tubuli and the intestinal mucosa transports alanine, asparagine, citrulline, glutamine, histidine, isoleucine, leucine, phenylalanine, serine, threonine, tryptophan, tyrosine and valine. Cystinuria, lysinuric protein intolerance and Hartnup disease are caused by defects of the luminal cystine/dibasic amino acid transporter, the antiluminal dibasic amino acid transporter and the neutral amino acid transporter, respectively.

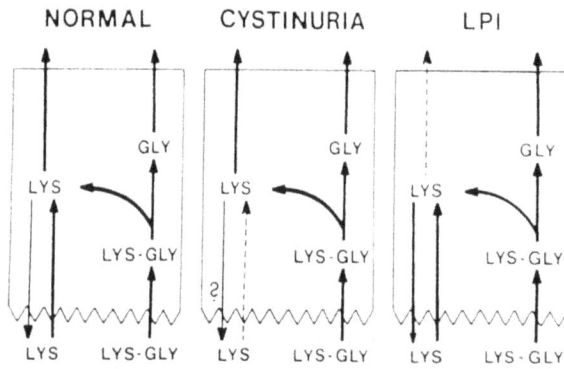

**Fig. 23.1.** Absorption of diamino acids (here lysine in free and dipeptide forms) by brush-border cells of jejunal mucosa. Sites of the defect in cystinuria (*bottom*, at the apical, luminal membrane) and in lysinuric protein intolerance (*top*, at the basolateral, antiluminal membrane) are indicated by *dashed arrows* [52]. *GLY*, glycine; *LPI*, lysinuric protein intolerance; *LYS*, lysine; *LYS-GLY*, lysyl-glycine dipeptide

# CHAPTER 23

# Transport Defects of Amino Acids at the Cell Membrane: Cystinuria, Hartnup Disease, and Lysinuric Protein Intolerance

O. Simell, K. Parto, and K. Näntö-Salonen

CONTENTS

Cystinuria .................................... 267
Hartnup Disease ............................. 268
Lysinuric Protein Intolerance .................. 270
References ................................... 272

Inherited defects in amino acid transport at the cell membrane are usually expressed as selective renal aminoaciduria, i.e., the concentration of the affected amino acids is high in the urine but is normal or low in plasma. Intestinal absorption of the affected amino acids is also almost always impaired. The clinical symptoms may be due to substrate excess in the urine or deficiencies of the substrates in the body.

Consequently, the symptoms in *cystinuria* are caused by renal stones due to the poor solubility of cystine. In *Hartnup disease*, the pellagra-like dermatitis and ataxia are due to a deficiency of tryptophan, the precursor of niacin synthesis.

In *lysinuric protein intolerance* (LPI), the transporter defect for the cationic (dibasic) amino acids leads to poor absorption and urinary loss of arginine, ornithine and, particularly, lysine. Deficiencies of arginine and ornithine, intermediates of the urea cycle, cause hyperammonemia and protein intolerance, and insufficient supply of lysine probably plays a major role in the growth retardation and skeletal and immunological manifestations of LPI.

## Cystinuria

### Clinical Presentation

Subjects with cystinuria have a high risk for *urolithiasis* [1]. Occasionally, patients never develop symptoms, but others already have recurrent symptomatic stones in early childhood. Severe acute episodes of abdominal or lower-back pain, hematuria, pyuria or spontaneous passing of stones may be the presenting sign. Cystine stones are usually radio-opaque and are visible on ultrasonography.

### Metabolic Derangement

In cystinuria, the luminal transport of *cystine and dibasic amino acids* in the kidney and small intestine is defective. Several cell-membrane transport systems have been described for cystine and the structurally related dibasic amino acids lysine, arginine and ornithine [2]. The defect in cystinuria is expressed on the luminal membrane of the epithelial cells and leads to poor absorption of cystine in the intestine and the renal tubules (Fig. 23.1). The intratubular cystine concentration may then exceed the threshold for cystine solubility, and crystals and stones may be formed.

Amino acids are absorbed from the intestine not only as free molecules but also as small peptides, which are then cleaved into free amino acids inside the epithelial cell and are exported to the body across the antiluminal membrane. In cystinuria, sufficient amounts of cystine are apparently absorbed as cystine-containing oligopeptides, since no definite signs of cystine deficiency have been described.

### Diagnostic Tests

A positive urinary nitroprusside test and analysis of urinary amino acids lead to the diagnosis. Homozygotes with cystinuria excrete more than 0.1 mmol cystine/mmol creatinine (250 mg cystine/g creatinine) in urine; this corresponds to more than 800 µmol cystine/24 h per 1.73 m² body surface area (>200 mg cystine/24 h per 1.73 m² body surface area), but the excreted amount varies markedly. Chemical analysis of the stones alone may be misleading, because mixed stones are not uncommon in patients with homozygous cystinuria, and some stones may contain no cystine at all. Plasma concentrations of cystine and the dibasic amino acids are normal or slightly decreased.

Cystinuria is divided into three subtypes on the basis of urinary amino acid excretion in obligate heterozygotes: parents of *type I* patients excrete normal amounts of cystine, those with *type III* excrete intermediate amounts, and those with *type II* excrete the greatest amount of cystine (0–11, 11–68, 112–197 µmol/g creatinine or 0–48, 48–144, 238–418 mg/g creatinine, respectively). After an oral cystine load, the plasma cystine concentration remains unchanged in cases of cystinuria types I and II but increases slowly in type III. Practically all patients with cystinuria type I carry two mutated alleles of the SCL3A1 gene [3].

## Treatment and Prognosis

Crystallization of cystine can be counteracted by decreasing its intratubular concentration and by increasing its solubility. Excessive hydration (in order to dilute the urine) and moderate sodium restriction are recommended, as reduced sodium intake decreases cystine excretion [4, 5]. Adults should consume more than 3000 ml fluid/24 h, 500 ml of this before bedtime and, if possible, 500 ml during the night. Because cystine is much more soluble in alkaline urine, permanent alkalinization of urine by sodium bicarbonate ($\sim$100 mmol/24 h or 8.4 g/24 h for adults) or potassium and sodium citrate should be encouraged. Restriction of methionine intake to limit endogenous cystine synthesis is difficult but may be helpful [6].

Thiol compounds form water-soluble disulfides with cystine and are thus suitable for stone dissolving [7, 8] and prophylaxis of stone formation. Daily D-penicillamine in doses of 2 g/1.73 m$^2$ body surface area is well tolerated by most patients but may cause hypersensitivity reactions, glomerulopathy or nephrotic syndrome and production of antinuclear antibodies. α-Mercaptopropionylglycine (tiopronin) is better tolerated than D-penicillamine but has occasionally caused glomerulopathy or hyperlipidemia [9, 10]. The daily dose (500–3000 mg in adults) is adjusted individually, aiming at maintaining urinary cysteine concentration below 1200 µmol/l [11]. The drug has also been successfully used (one dose every two days) as stone prophylaxis [12]. Captopril is well tolerated but may not be as effective as thiol compounds [13–15].

Percutaneous nephrolithotomy and extracorporeal shock-wave lithotripsy are seldom effective in stone removal, because cystine stones are extremely hard. New, minimally invasive urological techniques, e.g., ureteroscopic lithotripsy, laser probes [16] and in situ litholysis via percutaneous nephrostomies [17], minimize the need for open surgery. Surgical procedures should always be combined with conservative preventive therapy. Regular follow-up of the patient is mandatory in order to detect the stones as early as possible and to monitor renal function, which is frequently impaired as a result of stone-forming cystinuria. Early detection of the disease by screening the family members of the patient is also essential. Homozygotes with cystinuria type I frequently develop stones during the first decade of life and should perhaps be treated prophylactically beginning at an early age. Other subtypes may have a milder course [18].

## Genetics

The combined incidence of the three major genetic subtypes of cystinuria varies from 1 in 2000 to 1 in 15,000. They are genetically heterogeneous. Autosomal-recessive type I cystinuria is caused by over 20 different mutations in the SLC3A1 (originally called RBAT) gene on the short arm of chromosome 2 [19]. The gene product is a 90-kDa, type-II glycoprotein capable of transporting neutral amino acids, cystine and dibasic amino acids. SCL3A1 expression is highly restricted to the straight segment of the proximal tubule [20]. The gene(s) of cystinuria types II and III are located in the long arm of chromosome 19, according to linkage analysis [21]. The gene(s) have not been cloned but presumably code for a cystine transporter at a nephron site different than that of the SCL3A1 gene. Genetic compounds, such as type I/III also occur.

## Hartnup Disease

### Clinical Presentation

Since the first description of the syndrome in several members of the Hartnup family in 1956 [22], an extensive number of patients who fulfil the biochemical diagnostic criteria have been reported. The classical clinical symptoms – pellagra-like dermatitis and neurological involvement – closely resemble those of nutritional niacin deficiency. They probably reflect deficient production of the essential tryptophan metabolites, particularly of nicotinamide. However, most patients are asymptomatic, possibly because the necessary amount of tryptophan is absorbed in oligopeptide form or because their niacin intake is sufficient.

The skin lesions and neurological symptoms usually appear in early childhood [23] and usually become milder with increasing age. Exposure to sunlight, fever, diarrhoea, inadequate diet or psychological stress may precipitate the symptoms. *Pellagra-like skin changes* are found on light-exposed areas of the face, neck, forearms and dorsal aspects of the hands and legs. The skin hardens and becomes scaly, rough and hyperpigmented. *Cerebellar ataxia, attacks of headache, muscle pain* and *weakness*, resemble the symptoms of *porphyria* or the porphyria-like crises of type 1 tyrosinemia. Occasionally, patients present with *mental retardation* or *seizures*. *Psychiatric symptoms* ranging from emotional instability to delirium may occur. Growth and developmental outcome of the patients is generally normal, although low academic scores have been reported.

## Metabolic Derangement

The *hyperaminoaciduria* in Hartnup disease is characteristic (Fig. 23.2). The affected *neutral amino acids* share a common transporter, which is expressed only at the luminal border of the epithelial cells in the renal tubuli and intestinal epithelium. Alanine, serine, threonine, asparagine, glutamine, valine, leucine, isoleucine, phenylalanine, tyrosine, tryptophan, histidine and citrulline are excreted in excess in the urine, and their plasma concentrations are decreased or are low in the normal range. The renal clearance values of other amino acids are within the normal range [24].

The transport defect is also expressed in the intestine [25]. Stools of the patients contain increased amounts of free amino acids, reflecting closely the urinary excretion pattern. After oral tryptophan loads, the patients show smaller plasma tryptophan peaks than controls and excrete less kynurenine and N-methylnicotinamide (tryptophan metabolites produced in the body) in the urine [26]. Oral loads with histidine-containing dipeptides and incubation of intestinal biopsy specimens with histidine and with glycylhistidine show that the affected amino acids are readily absorbed as short oligopeptides but not as free amino acids [27, 28].

The unabsorbed amino acids in the colon are exposed to bacterial degradation. Degradation of tryptophan produces large amounts of indican, indole acetic acid, indole acetylglutamine, indolylacylglycine and several other compounds, which are then excreted in the urine. Urinary indole excretion may be normal during fasting but is always increased after a tryptophan load. Oral administration of antibiotics may decrease tryptophan-induced indole excretion.

## Diagnostic Tests

The characteristic *excess of neutral monoamino-monocarboxylic acids* in the urine and the normal or low normal concentrations in plasma confirm the diagnosis. Urinary excretion of indole compounds may be within the normal range if the patient consumes normal or low amounts of dietary protein, but an oral load of L-tryptophan (100 mg/kg) always leads to a supranormal increase in indole excretion.

## Treatment and Prognosis

Dermatitis and neurological symptoms usually disappear with *oral nicotinamide* (40–200 mg/day) in 1–2 weeks. Oral neomycin reduces intestinal degradation of tryptophan and decreases indole production; however, the role of the indole compounds in the disease has been poorly characterized. Sunlight should be avoided, and exposed areas should be protected by sun-blocking agents. Psychosis, ataxia, or other neurological symptoms disappear rapidly during nicotinamide supplementation. A patient who presented with severe intermittent dystonia failed to respond to nicotinic acid, but his condition improved with trihexyphenidyl (1–2 mg/kg/day) [29]. An adequate supply of high-quality protein is probably important for prevention of the symptoms in patients detected in newborn screening. Recently, tryptophan

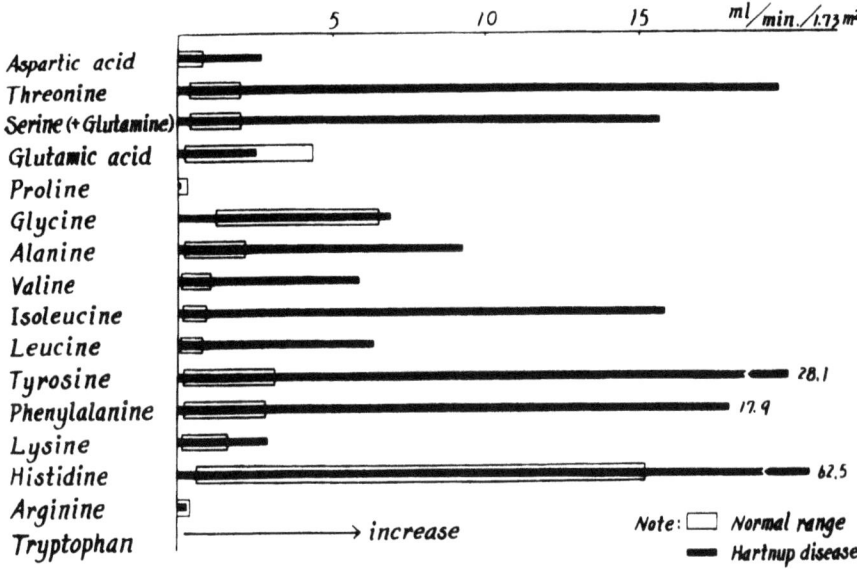

**Fig. 23.2.** Renal clearance of amino acids in Hartnup Disease

ethyl ester has been successfully used to circumvent the transport defect [30].

## Genetics

Hartnup disease follows an autosomal-recessive pattern of inheritance. Heterozygotes excrete normal amounts of amino acids in the urine, but tests of intestinal amino acid absorption have been inconclusive. The incidence in newborns screened for aminoaciduria has varied from 1 in 18 000 in Massachusetts to 1 in 200 000 in Australia, with a mean value of 1 in 24 000 [31, 32].

Neither the gene responsible for the defect nor the transporter have been characterized in man, but a candidate mouse model was recently described. Homozygotes of the mutant strain HPH2 show selective hyperaminoaciduria, normal plasma amino acid concentrations and symptoms of niacin deficiency, all features resembling those of human Hartnup disease [33].

## Lysinuric Protein Intolerance

### Clinical Presentation

About 120 patients with LPI, 46 of them from Finland, have been reported or are known to us. The full natural history of LPI has not yet been characterized, as most of the oldest patients are still in their forties. Newborns and older infants are usually asymptomatic if they are only fed breast milk (protein content ~9 g/l). Postprandial episodes of *hyperammonemia* usually emerge when the infants begin receiving formula (protein content of most brands = 13–20 g/l) or supplementary high-protein foods [34, 35]. Hyperammonemia may present as refusing to eat, vomiting, stupor, coma and, finally, death (caution: avoid forced tube feeding). Strong *aversion to high-protein foods* with *failure to thrive* usually develops around the age of 1 year. The liver and spleen are moderately enlarged.

In toddlers and school-aged children, the presenting signs are most often *growth failure* and *hepato-* and *splenomegaly*. The children are usually hypotonic, and muscular endurance is decreased. The neurological development is normal if severe or prolonged hyperammonemia has been avoided. Because of marked *osteoporosis*, the children may have fractures after minor traumas. Their bone maturation is retarded [36, 37]. In childhood, some patients have *anemia* and *leukopenia*, and their *platelet count may be decreased*. The erythrocytes show poikilocytosis, anisocytosis, and mild macrocytosis [38]. Serum *ferritin* concentration is constantly two to 20 times higher than the reference values and may exceed the normal range 50- to 100-fold during severe infections [38].

The clinical heterogeneity is obvious at adult age. Most patients are of moderately short stature, but some are of normal height. They have abundant subcutaneous fat on a square-shaped trunk, but their extremities are thin. They may have marked hepatomegaly with or without splenomegaly. Two thirds of the adults have *skeletal abnormalities*, e.g., osteoporosis, abnormal cortices of the long bones, scoliosis, vertebral end-plate deformities, early cartilage destruction or metacarpal cortical thickening [37]. Pathologic fractures seldom occur in adults. Radiological signs of pulmonary fibrosis are common, but few patients suffer from clinical interstitial lung disease [39]. Mental capacity varies from normal to moderate retardation depending on previous periods of hyperammonemia. Pregnancies of LPI patients have been complicated by toxemia, anemia or bleeding problems during the delivery.

Many patients show signs of *disturbed immune function*, reflected as leukopenia, high serum immunoglobulin-G concentration and abnormalities of the distribution of lymphocyte subpopulations [40]. *Varicella* infections are usually severe and can be fatal [41]. Systemic lupus erythematosus [42] has also been reported. Bone-marrow involvement with erythrophagocytosis, interstitial pulmonary disease with alveolar proteinosis and slowly progressing renal insufficiency are other rare complications [43–48]. A few patients have died in childhood after a very uniform course: they develop an acute or less aggressively progressing multiorgan failure. This almost always comprises interstitial lung involvement with alveolar proteinosis, progressive glomerulonephritis that leads to renal insufficiency, and severe bleeding diathesis [45].

### Metabolic Derangement

In LPI, transport of the cationic (dibasic) amino acids lysine, arginine and ornithine is impaired at the cell membrane (Fig. 23.1). In normal intestine, free amino acids and short oligopeptides first cross the luminal membrane of the epithelial cells to reach the cytoplasm. Oligopeptides are hydrolyzed to free amino acids, which are able to cross the antiluminal membrane of the cell. In LPI, the cationic amino acids pass through the antiluminal membrane of the epithelial cells poorly [49–51], and their reabsorption in kidney tubuli is also decreased, resulting in massive loss of lysine and moderate losses of arginine and ornithine in the urine [52]. In cultured skin fibroblasts and other parenchymal cells, the transport defect is expressed as deficient amino acid efflux [53], leading to increased intracellular concentration of the cationic amino acids. Poor intestinal absorption and increased renal loss of cationic amino acids lead to a negative net ionic balance in the body.

The symptoms of LPI result from the decreased availability of lysine, arginine and ornithine. Deficiency of arginine and ornithine, which are essential intermediates in the urea cycle, delays ammonia elimination after protein ingestion and causes hyperammonemia. Thus, the clinical symptoms resemble those of urea cycle-enzyme deficiencies. As arginine is also the rate-limiting precursor of nitric-oxide synthesis, arginine deficiency may also result in persistently low nitric-oxide concentrations and may influence, e.g., the immunological function of the patient. Reduced availability of lysine, an essential amino acid, probably has a prominent role in the poor growth and skeletal and immunological manifestations of the patients.

## Diagnostic Tests

The diagnosis of LPI is based on the following:

- Urinary excretion of the cationic amino acids, especially of lysine, is increased. The excretion rates of the cationic amino acids have, in a few instances, been within the reference range when protein intake in the diet has been extremely low and plasma concentrations of the cationic amino acids have been exceptionally low. Cyst(e)ine excretion may be slightly increased, resembling the situation in heterozygotes with cystinuria type II or type III. Glutamine, alanine, serine, proline, glycine and citrulline are excreted in slight excess, because their plasma concentrations in the patients are slightly increased.
- Plasma lysine concentration is usually less than 80 µmol/l, the arginine level is less than 40 µmol/l, and the ornithine level is less than 30 µmol/l. Plasma glutamine, alanine, serine, proline, citrulline and glycine concentrations are 1.5–10 times higher than the upper reference values.
- Blood ammonia concentration increases after protein-rich meals or an intravenous L-alanine load (6.6 mmol/kg during a 90-min load of a 5% aqueous solution; samples are taken at 0, 120, 270 and 360 min). The serum concentration of urea rises slowly. Orotic aciduria occurs, at least in the first 2-h urine collection after the load.
- Serum lactate-dehydrogenase activity and the concentrations of ferritin and thyroxin-binding globulin are increased.

## Treatment and Prognosis

The treatment of LPI aims to prevent hyperammonemia and to provide a sufficient supply of protein and essential amino acids for normal metabolism and growth. The protein tolerance can be improved if citrulline, a neutral amino acid and intermediate in the urea cycle, is given as a daily supplement (0.1–0.5 g/kg before or during 3–5 meals) [54]. Citrulline is readily absorbed and is partly converted to arginine and ornithine. All three amino acids improve the function of the urea cycle. If supplemented with citrulline, children usually tolerate 1.0–1.5 g/kg and adults 0.5–0.8 g/kg of protein daily without hyperammonemia or increases in orotic acid excretion. There is marked day-to-day variation in the protein tolerance of the patients, and interindividual differences are even larger. Infections, pregnancy and lactation may alter protein tolerance extensively. Frequent monitoring of blood ammonia at home and follow-up of urinary orotic acid excretion (fasting morning urine and postprandial evening urine samples are sent to a laboratory by regular mail at 2-week to 12-week intervals) is necessary for optimization of the diet.

Correction of lysine deficiency with oral lysine supplement or lysine-containing oligopeptides is complicated by the poor absorption of lysine and the resulting osmotic diarrhea. Experimentally, a neutral lysine analogue, ε-$N$-acetyl-lysine, increased the plasma lysine concentration, while homocitrulline, another potential lysine supplier, had no influence [55]. Because of high price and poor availability, ε-$N$-acetyl-lysine has not been used in long-term supplementation. In one patient, plasma lysine concentration increased during carnitine substitution; further clinical trials of the effects of carnitine are warranted [56].

In *acute hyperammonemic crisis*, all protein- and nitrogen-containing substances should be removed from the nutrition and sufficient energy should be supplied as intravenous glucose. Intravenous infusion of ornithine, arginine or citrulline, beginning with a priming dose of 0.5–1.0 mmol/kg in 5–10 min and continuing with 0.5–1.0 mmol/kg/h, will clear hyperammonemia rapidly. Sodium benzoate and sodium phenylbutyrate may also be used as alternate pathways of ammonia elimination (Chap. 17) [57].

Hyperammonemia and mental retardation are avoidable with the current therapy. However, several features and complications of LPI are not caused by hyperammonemia; they are more likely to be due to deficiency of the essential amino acid lysine or may be caused by abnormalities in, e.g., arginine or nitric-oxide metabolism. The patients should be immunized against *Varicella*, and non-immunized patients should be treated with acyclovir if they contact the infection [41].

## Genetics

LPI is an autosomal-recessive disease. An amino-acid-transporter gene, y$^+$ LAT-1, located within the LPI locus detected by linkage analysis in the long arm of chromosome 14, shows different mutations in different LPI populations [58-61].

## References

1. Lindell A, Denneberg T, Granerus G (1997) Studies on renal function in patients with cystinuria. Nephron 77:76-85
2. Seriver CR (1986) Cystinuria. N Engl J Med 1:1155-1157
3. Chesney RW (1998) Mutational analysis of patients with cystinuria detected by a genetic screening network: powerful tools in understanding the several forms of the disorder. Kidney Int 54:279-280
4. Jeager P, Portmann L, Saunders A, Rosenberg LE, Thier SO (1986) Anticystinuric effects of glutamine and of dietary sodium restriction. N Engl J Med 1120-1123
5. Pewes R, Sanchez L, Gorostidi M, Alvarez J (1991) Effects of variation in sodium intake on cystinuria. Nephron 57:421-423
6. Kolb FO, Earil JM, Harper HA (1967) "Disappearance" of cystinuria in a patient treated with prolonged low methionine diet. Metabolism 16:378-381
7. Pak CYC, Fuller C, Sakhaee K et al. (1986) Management of cystine nephrolithiasis with alphamercaptopropionyl glycine. J Urol [Suppl] 136:1003-1008
8. Stephens AD (1989) Cystinuria and its treatment: 25 years experience at St Bartholomew's Hospital. J Inherited Metab Dis 12:197-209
9. Lindell A, Denneberg T, Enestrom S, Fich C, Skogh T (1990) Membranous glomerulonephritis induced by 2-mercaptopropionylglycine (2-MPG). Clin Nephrol 34:108-115
10. Siskind MS, Popovtzer MM (1992) Hyperlipidemia associated with alphamercapto-propionyl-glycine therapy for cystinuria. Am J Kidney Dis 19:179-180
11. Lindell A, Denneberg T, Hellgren E, Jeppson JO, Tiselius HG (1995) Clinical course and cystine stone formation during tiopronin treatment. Urol Res 23:111-117
12. Berio A, Piazzi A (1998) Prophylaxis of cystine calculi by low dose of alpha mercaptopropionylglycine administered every other day. Panminerva Med 40:244-246
13. Perazella MA, Buller GK (1993) Successful treatment of cystinuria with captopril. Am J Kidney Dis 21:504-507
14. Streem SB, Hall P (1989) Effect of captopril on urinary cystine excretion in homozygous. J Uro1 142:1522-1524
15. Chow GK, Streem SB (1996) Medical treatment of cystinuria: results of contemporary clinical practice. J Urol 156:1576-1578
16. Maes K, Goethuys H, Baert L (1988) A cystine lower pole renal calculus treated with holmium: YAG laser using a flexible 9.5F transurethral ureteroscope. Acta Urol Belg 66:29-32
17. Benitez-Navio J, Tudela-Pavon P, Laguna-Urraca G et al. (1995) The dissolution of cystine lithiasis with n-acetylcysteine. Arch Esp Urol 48:944-948
18. Goodyer P, Saadi I, Ong P, Elkas G, Rozen R (1998) Cystinuria subtype and the risk of nephrolithiasis. Kidney Int 54:56-61
19. Calonge MJ, Gasparini P, Chillaron J et al. (1994) Cystinuria caused by mutations in RBAT, a gene involved in the transport of cystine. Nat Genet 6:420-425
20. Saadi I, Chen X-Z, Hediger M et al. (1998) Molecular genetics of cystinuria: mutation analysis of SLC3A1 and evidence for another gene in the Type I (silent) phenotype. Kidney Int 54:48-55
21. Bisceglia L, Calonge MJ, Totaro A et al. (1997) Localisation, by linkage analysis, of the cystinuria type III gene to chromosome 19q13.1. Am J Hum Genet 60:611-616
22. Baron DN, Dent CE, Harris H, Hart EW, Jepson JB (1956) Hereditary pellagra-like skin rash with temporary cerebellar ataxia, constant renal amino aciduria and other bizarre biochemical features. Lancet 2:421-428
23. Seriver CR, Mahon B, Levy HL et al. (1987) The Hartnup phenotype: mendelian transport disorder, multifactorial disease. Am J Hum Genet 40:401-402
24. Tada K, Hirono H, Arakawa T (1967) Endogenous renal clearance rates of free amino acids in probinuric and Hartnup patients. Tohoku J Exp Med 278:57-61
25. Scriver CR (1965) Hartnup disease. N Engl J Med 273:530-532
26. Tada K, Morikawa T, Arakawa T (1966) Tryptophan load and uptake of tryptophan by leukocytes in Hartnup disease. Tohoku J Exp Med 90:337-346
27. Asatoor AM, Cheng B, Edwards KDG et al. (1970) Intestinal absorption of two dipeptides in Hartnup disease. Gut 11: 380-387
28. Tarlow MJ, Seakins JW, Lloyd JK et al. (1972) Absorption of amino acids and peptides in a child with a variant of Hartnup disease and coexistent coeliac disease. Arch Dis Child 47: 798-803
29. Darras BT, Ampola MG, Dietz WH, Gilmore HE (1989) Intermittent dystonia in Hartnup disease. Pediatr Neurol 5:118-120
30. Jonas AJ, Butler IJ (1989) Circumvention of defective neutral amino acid transport in Hartnup disease using tryptophan ethyl ester. J Clin Invest 84:200-204
31. Levy HL (1973) Genetic screening. Adv Hum Genet 4:1-104
32. Wileken B, Smith A, Gaha TJ, McLeay AC, Brown DA (1973) Screening for metabolic diseases in New South Wales. Med J Aust 1:1129-1133
33. Symula DJ, Shedlovsky A, Guillery EN, Dove WF (1997) A candidate mouse model for Hartnup disorder deficient in neutral amino acid transport. Mamm Genome 8:102-107
34. Perheentupa J, Visakorpi JK (1965) Protein intolerance with deficient transport of basic amino acids: another inborn error of metabolism. Lancet 2:813-815
35. Simell o, Perheentupa J, Rapola J, Visakorpi JK, Eskelin L-E (1975) Lysinuric protein intolerance. Am J Med 59:229-240
36. Carpenter TO, Levy HL, Holtrop ME, Shih VE, Anast CS (1985) Lysinuric protein intolerance presenting as childhood osteoporosis. Clinical and skeletal response to citrulline therapy. N Engl J Med 312:290-294
37. Svedström E, Parto K, Marttinen M, Virtama P, Simell O (1993) Skeletal manifestations of lysinuric protein intolerance. A follow-up study of 29 patients. Skeletal Radiol 22:11-16
38. Rajantie J, Simell O, Perheentupa J, Siimes M (1980) Changes in peripheral blood cells and serum ferritin in lysinuric protein intolerance. Acta Paediatr Scand 69:741-745
39. Parto K, Svedström E, Majurin M-L, Härkönen R, Simell O (1993) Pulmonary manifestations in lysinuric protein intolerance. Chest 104:1176-1182
40. Yoshida Y, Machigashira K, Suehara M et al. (1995) Immunological abnormality in patients with lysinuric protein intolerance. J Neurol Sci 134:178-182
41. Lukkarinen M, Näntö-Salonen K, Ruuskanen O et al. (1998) Varicella and varicella immunity in patients with lysinuric protein intolerance. J Inherit Metab Dis 21:103-111
42. Dionisi-Vici C, DeFelice L, el Hachem M et al. (1998) Intravenous immunoglobulin in lysinuric protein intolerance. J Inherit Metab Dis 21:95-102
43. Kerem E, Elpelg ON, Shalev RS et al. (1993) Lysinuric protein intolerance with chronic interstitial lung disease and pulmonary cholesterol granulomas at onset. J Pediatr 123: 275-278
44. DiRocco M, Garibotto G, Rossi GA et al. (1993) Role of haematological, pulmonary and renal complications in the long-term prognosis of patients with lysinuric protein intolerance. Eur J Pediatr 152:437-440
45. Parto K, Kallajoki M, Aho H, Simell O (1994) Pulmonary alveolar proteinosis and glomerulonephritis in lysinuric protein intolerance: case reports and autopsy findings of four pediatric patients. Hum Pathol 25:400-407
46. Santamaria F, Parenti G, Guidi G et al. (1996) Early detection of lung involvement in lysinuric protein intolerance: role of high-resolution computed tomography and radioisotopic methods. Am J Respir Crit Care Med 153:731-735
47. Parenti G, Sebastio G, Strisciuglio P et al. (1995) Lysinuric protein intolerance characterised by bone marrow abnormalities and severe clinical course. J Pediatr 126:246-251

48. Doireau V, Fenneteau O, Duval M et al. (1996) Lysinuric protein intolerance: characteristic aspects of bone marrow involvement. Arch Pediatr 3:877–880
49. Rajantie J, Simell O, Perheentupa J (1980) Intestinal absorption in lysinuric protein intolerance: impaired for diamino acids, normal for citrulline. Gut 21:519–524
50. Desjeux J-F, Rajantie J, Simell O, Dumontier A-M, Perheentupa J (1980) Lysine fluxes across the jejunal epithelium in lysinuric protein intolerance. J Clin Invest 65:1382–1387
51. Rajantie J, Simell O, Perheentupa J (1980) Basolateral membrane transport defect for lysine in lysinuric protein intolerance. Lancet 1:1219–1221
52. Rajantie J, Simell O, Perheentupa J (1981) Lysinuric protein intolerance. Basolateral transport defect in renal tubuli. J Clin Invest 67:1078–1082
53. Smith DW, Seriver CR, Tenenhouse HS, Simell O (1987) Lysinuric protein intolerance mutation is expressed in the plasma membrane of cultured skin fibroblasts. Proc Natl Acad Sci USA 84:7711–7715
54. Rajantie J, Simell O, Rapola J, Perheentupa J (1980) Lysinuric protein intolerance: a two-year trial of dietary supplementation therapy with citrulline and lysine. J Pediatr 97:927–932
55. Rajantie J, Simell O, Perheentupa J (1983) Oral administration of ε-N-acetyllysine and homocitrulline for lysinuric protein intolerance. J Pediatr 102:388–390
56. Takada G, Goto A, Komatsu K, Goto R (1987) Carnitine deficiency in lysinuric protein intolerance: lysine-sparing effect of carnitine. Tohoku Exp Med 153:331–334
57. Brusilow SW, Danney M, Waber LJ et al. (1984) Treatment of episodic hyperammonemia in children with inborn errors of urea synthesis. N Engl J Med 310:1630–1634
58. Lauteala T, Sistonen P, Savontaus ML et al. (1997) Lysinuric protein intolerance (LPI) gene maps to the long arm of chromosome 14. Am J Hum Genet 60:1479–1486
59. Torrents D, Estevez R, Pineda M et al. (1998) Identification and characterisation of a membrane protein (y + L amino acid transporter-1) that associates with 4F2hc to encode the amino acid transport activity y + L. A candidate gene for lysinuric protein intolerance. J Biol Chem 273:32437–32445
60. Torrents D, Mykkänen J, Pineda M et al. (1999) Identification of SLC7A7, encoding y + LAT-1, as the lysinuric protein intolerance gene. Nature Genet 21:293–296
61. Borsani G, Bassi MT, Sperandeo MP et al. (1999) SLC7A7, encoding a putative permease-related protein, is mutated in patients with lysinuric protein intolerance. Nature Genet 21:297–301

# PART V
# VITAMIN-RESPONSIVE DISORDERS

## The Biotin Cycle and Biotin-Dependent Enzymes

Biotin is a water-soluble vitamin that is widely present in small amounts in natural foodstuffs, in which it is mostly protein bound. It is the coenzyme of four important carboxylases, which intervene in gluconeogenesis, fatty-acid synthesis, and the catabolism of several amino acids (Fig. 24.1) [1]. Binding of biotin to the four inactive apocarboxylases, catalyzed by *holocarboxylase synthetase*, is required to generate the active holocarboxylases (Fig. 24.2). Regeneration of biotin first involves proteolytic degradation of the holocarboxylases, yielding biotin bound to lysine (biocytin) or short biotinyl peptides. *Biotinidase* releases biotin from the latter compounds, which are derived from endogenous and from dietary sources.

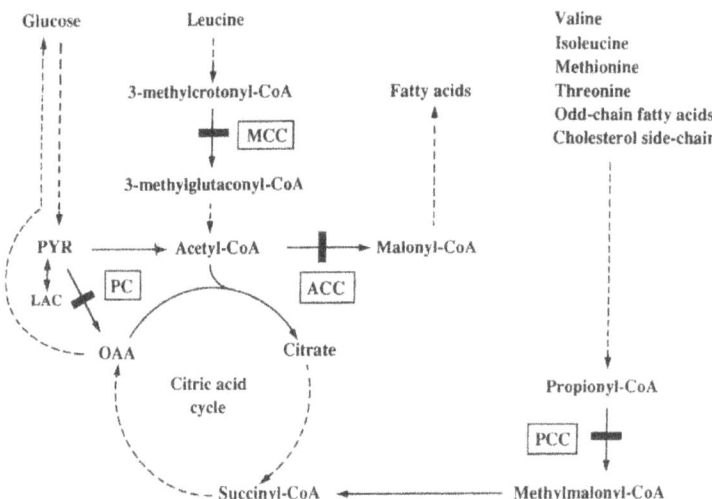

**Fig. 24.1.** Location of biotin-dependent carboxylases in intermediary metabolism. *ACC*, acetyl-CoA carboxylase; *CoA*, coenzyme A; *LAC*, lactate; *MCC*, 3-methylcrotonyl-CoA carboxylase; *OAA*, oxaloacetate; *PC*, pyruvate carboxylase; *PCC*, propionyl-CoA carboxylase; *PYR*, pyruvate. *Full lines* indicate one enzyme, and *dotted lines* indicate that several enzymes are involved. Sites of the enzyme defects are indicated by *solid bars*

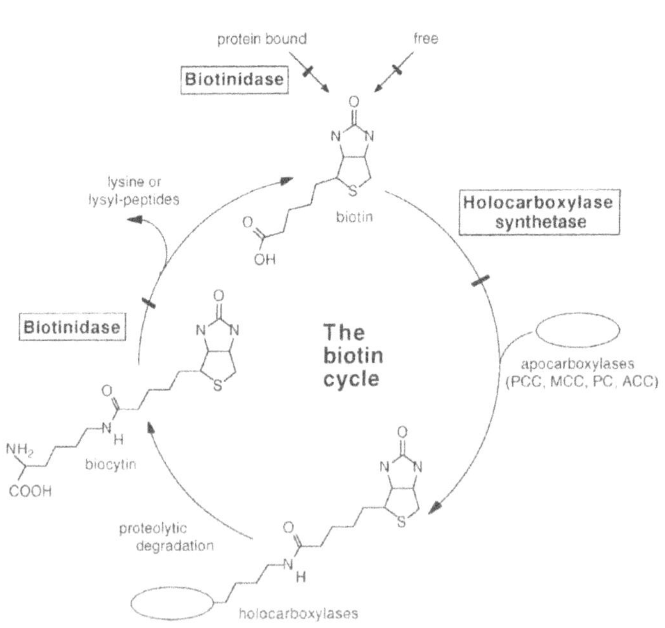

**Fig. 24.2.** The biotin cycle. For definitions of abbreviations, see Fig. 24.1 [2]

# CHAPTER 24

# Biotin-Responsive Multiple Carboxylase Deficiency

E. Regula Baumgartner and Terttu Suormala

CONTENTS

Clinical Presentation ............................ 277
   Holocarboxylase Synthetase Deficiency ............ 277
   Biotinidase Deficiency ......................... 278
Metabolic Derangement ......................... 278
Diagnostic Tests ............................... 278
   Holocarboxylase Synthetase Deficiency ............ 279
   Biotinidase Deficiency ......................... 279
   Acquired Biotin Deficiency ..................... 279
   Prenatal Diagnosis ........................... 279
Treatment and Prognosis ....................... 279
   Holocarboxylase Synthetase Deficiency ............ 279
   Biotinidase Deficiency ......................... 280
Genetics ..................................... 281
References ................................... 281

Two inherited defects in biotin metabolism are known: holocarboxylase synthetase (HCS) deficiency and biotinidase deficiency. Both lead to multiple carboxylase deficiency (MCD). In HCS deficiency, the formation of holocarboxylases is impaired. In biotinidase deficiency, biotin depletion ensues from the inability to recycle endogenous biotin and to utilize protein-bound biotin from the diet. As the carboxylases play an important role in the catabolism of several amino acids, in gluconeogenesis and in fatty-acid synthesis, their deficiency provokes multiple, life-threatening metabolic derangements, eliciting characteristic organicaciduria and neurologic symptoms. The clinical presentation is extremely variable in both disorders. Characteristic manifestations of MCD are metabolic acidosis, hypotonia, seizures, ataxia, impaired consciousness and cutaneous symptoms, such as skin rash and alopecia. Both disorders respond dramatically to oral therapy with pharmacological doses of biotin. Acquired biotin deficiency, which also causes MCD, is extremely rare.

## Clinical Presentation

The characteristic manifestation of MCD is metabolic acidosis associated with neurologic abnormalities and skin disease. The expression of the clinical and biochemical features is variable in both inherited disorders [1]. While patients with HCS deficiency commonly present with the characteristic manifestations of MCD, patients with biotinidase deficiency show a less consistent clinical picture, particularly during the early stage of the disease. The onset in biotinidase deficiency may be insidious, and the manifestation is usually very variable, neurologic symptoms often being prominent without markedly abnormal organic-acid excretion or metabolic acidosis. Later-onset forms of HCS deficiency cannot be clinically distinguished from biotinidase deficiency, necessitating confirmation of the diagnosis by enzyme assay.

### Holocarboxylase Synthetase Deficiency

Although HCS deficiency was initially termed early-onset MCD, recent experience shows that the age of onset of symptoms varies widely, from a few hours after birth to 6 years of age [1, 3]. Nevertheless, about half of the patients have presented acutely in the first days of life, with symptoms very similar to those observed in other severe organic acidurias, i.e., lethargy, hypotonia, vomiting, seizures and hypothermia. The most common initial clinical features consist of respiratory difficulties, such as tachypnea or Kussmaul breathing. Increasingly severe metabolic acidosis, ketosis and hyperammoniaemia may lead to coma and early death. Patients with a less severe defect and later onset may also present with recurrent life-threatening attacks of metabolic acidosis and typical organic aciduria [4, 5]. Untreated early-onset patients and patients with a less severe defect may additionally develop psychomotor retardation, hair loss and skin lesions. The latter include an erythematous, scaly skin rash that spreads over the whole body but is particularly prominent in the diaper and intertriginous areas; alternatively, the rash may resemble seborrheic dermatitis or ichthyosis. Superinfection with *Candida* may occur. Disorders of immune function have been observed with decreased T cell count and impaired in vitro and in vivo response to *Candida* antigen. Episodes of acute illness are often precipitated by catabolism during intercurrent infections or by a higher protein intake.

### Biotinidase Deficiency

Important features are the gradual development of symptoms and episodes of remission, which may be related to increased free biotin in the diet. The full clinical picture has been reported as early as 7 weeks, but discrete neurologic symptoms may occur much earlier, even in the neonatal period [6]. Neurologic manifestations (lethargy, muscular hypotonia, grand mal and myoclonic seizures, ataxia) are the most frequent initial symptoms. In addition, respiratory abnormalities, such as stridor, episodes of hyperventilation and apnoea, frequently occur and may also be of neurologic origin [7]. Skin rash and/or alopecia are hallmarks of the disease; however, they may develop late or not at all [8, 9]. Skin lesions are usually patchy, erythematous/exudative and typically localized periorificially. Eczematoid dermatitis or an erythematous rash covering large parts of the body has also been observed, as has keratoconjunctivitis. Hair loss is usually discrete but may, in severe cases, become complete, including the eyelashes and eyebrows. Immunologic dysfunction may occur in acutely ill patients. Because of the variability and unspecificity of clinical manifestations, there is a great risk of a delay in diagnosis [7, 10]. Late-diagnosed patients often have psychomotor retardation and permanent neurologic deficits, such as hearing loss and optic atrophy, which may be irreversible [8–10]. The outcome may even be fatal. One patient died at the age of 22 months, with features of Leigh's syndrome proven by histopathology [7].

Metabolic acidosis and the characteristic organic aciduria of MCD are frequently lacking in the early stages of the disease. Plasma lactate and 3-hydroxyisovalerate may be only slightly elevated, whereas in CSF their levels may be significantly higher [11]. This fact and the finding of severely decreased carboxylase activities in brain but moderately deficient activity in liver and kidney in a patient with lethal outcome [7] are in accordance with the predominance of neurologic symptoms and show that, in biotinidase deficiency, the brain is affected earlier and more severely than other organs. The threat of irreversible brain damage demands that this disorder be considered in all children with neurologic problems, even without obvious organic aciduria and/or cutaneous findings.

### Metabolic Derangement

In *HCS deficiency*, a decreased affinity of the enzyme for biotin and/or a decreased maximal velocity lead to reduced formation of the four holocarboxylases from their corresponding inactive apocarboxylases at physiological biotin concentrations [12, 13, 40]. In *biotinidase deficiency*, biotin cannot be released from biocytin and biotinyl peptides. Thus, patients with biotinidase deficiency are unable to either recycle endogenous biotin or to use protein-bound dietary biotin [1]. Consequently, biotin is lost in the urine, mainly in the form of biocytin [6, 14], and progressive biotin depletion occurs. Depending on the amount of free biotin in the diet and the severity of the enzyme defect, the disease becomes clinically manifest during the first months of life or later in infancy or childhood.

Biotinidase was recently shown to have biotinyltransferase activity in addition to its hydrolase activity [15]. In the presence of biocytin, but not of biotin, biotin is bound to biotinidase and can be transferred to nucleophilic acceptors, such as histones. Although the physiological significance of this function is still unknown, it may play a role in the transport of biotin. Patients with deficient hydrolase activity usually also have deficiency of the transferase activity [16–18].

Deficient activity of carboxylases in both HCS and biotinidase deficiencies (Fig. 24.2) results in accumulation of lactic acid and derivatives of 3-methylcrotonyl-coenzyme A (CoA) and propionyl-CoA (see "Diagnostic Tests").

*Acquired biotin deficiency* is rare. It occurs under the following conditions: excessive consumption of raw egg white (binding to avidin makes biotin unavailable), malabsorption due to short-bowel syndrome, long-term total parenteral nutrition or hemodialysis without biotin supplementation, long-term anticonvulsant therapy [1].

*Isolated inherited deficiencies* of each of the three mitochondrial carboxylases [propionyl-CoA carboxylase (PCC, the most common), 3-methylcrotonyl-CoA carboxylase (MCC; for both, see Chap. 16), and pyruvate carboxylase (PC; Chap. 10)] are also known. A single patient with an isolated defect of hepatic and fibroblast acetyl-CoA carboxylase (ACC, cytosolic) has been reported [1]. These isolated deficiencies are due to absence or abnormal structure of the apoenzyme and do not respond to biotin therapy.

### Diagnostic Tests

Characteristic organic aciduria due to systemic deficiency of the carboxylases is the key feature of MCD. In severe cases, an unpleasant odor of the urine (cat's urine) may even be suggestive of the defect. MCD is reflected in elevated urinary and plasma concentrations of organic acids as follows:

- Deficiency of MCC: 3-hydroxyisovaleric acid in high concentrations, 3-methylcrotonylglycine in smaller amounts

- Deficiency of PCC: methylcitrate, 3-hydroxypropionate, propionylglycine, tiglylglycine, propionic acid in small to moderate amounts
- Deficiency of PC: lactate in high concentrations, pyruvate in smaller amounts

The majority of HCS-deficient patients excrete all of the typical organic acids in elevated concentrations, provided that the urine sample has been taken during an episode of acute illness whereas, in biotinidase deficiency, elevated excretion of only 3-hydroxyisovalerate may be found, especially in early stages of the disease.

The measurement of carboxylase activities in lymphocytes provides direct evidence of MCD. These activities are low in HCS deficiency but may be normal in biotinidase deficiency, depending on the degree of biotin deficiency [3, 19].

The two inherited disorders can easily be distinguished by assay of biotinidase in serum. Today, this assay is included in the neonatal screening programs in many countries worldwide.

### Holocarboxylase Synthetase Deficiency

- Biotin concentrations in plasma and urine are normal.
- Carboxylase activities in lymphocytes are deficient and cannot be activated by in vitro preincubation with biotin [20].
- Direct measurement of HCS activity requires an apocarboxylase or an apocarboxyl carrier protein of ACC as one of the substrates [12, 21]; therefore, it is not routinely performed.
- HCS deficiency can be diagnosed indirectly by demonstrating severely decreased carboxylase activities in fibroblasts cultured in a medium with low biotin concentration ($10^{-10}$ mol/l) and by normalization (or, at least, increase) of the activities in culture media supplemented with high biotin concentrations ($10^{-6}$–$10^{-5}$ mol/l) [3, 12]. It must be noted that fibroblasts of some late-onset patients may exhibit normal levels of carboxylase activities when cultured in standard media supplemented with 10% fetal calf serum, which results in a final biotin concentration of about $10^{-8}$ mol/l [3, 5].

### Biotinidase Deficiency

- Biotinidase activity in plasma is absent or decreased [1, 19, 22]. Many patients have measurable residual activity and should be evaluated for the presence of Km mutations (see below).
- Symptomatic patients usually have decreased biotin concentrations in plasma and urine [6, 19], provided that a method that does not detect biocytin is used [23]. In addition, carboxylase activities in lymphocytes are usually decreased but are normalized within hours after either a single dose of oral biotin [6] or in vitro preincubation with biotin [19, 20].
- Patients excrete biocytin in urine [14], the concentration being dependent on the level of residual biotinidase activity [19].
- Carboxylase activities in fibroblasts cultured in low-biotin medium are similar to those in control fibroblasts, and are always normal in fibroblasts cultured in standard medium.

### Acquired Biotin Deficiency

- Biotinidase activity is normal in plasma
- Biotin concentrations are low in plasma and urine
- Carboxylase activities in lymphocytes are decreased and are promptly normalized after a single dose of oral biotin or after preincubation with biotin in vitro [20]

### Prenatal Diagnosis

Prenatal diagnosis of HCS deficiency is possible by enzymatic studies in cultured chorionic villi or amniotic fluid cells or by organic-acid analysis of amniotic fluid. However, organic acid analysis in milder forms of HCS deficiency may fail to show an affected fetus, necessitating enzymatic investigation in these cases [5]. Prenatal diagnosis allows rational prenatal therapy, preventing severe metabolic derangement in the early neonatal period [1, 5, 24]. Biotinidase can be measured in chorionic villi or cultured amniotic fluid cells but, in our opinion, this is not warranted, because prenatal treatment is not necessary.

## Treatment and Prognosis

Both inherited disorders can be treated effectively with pharmacologic doses of biotin. Restriction of protein intake is not necessary except in very severe cases of HCS deficiency. Acutely ill patients with metabolic decompensation require general emergency treatment in addition to biotin therapy (Chap. 3).

### Holocarboxylase Synthetase Deficiency

The required dose of biotin is dependent on the severity of the enzyme defect and has to be assessed individually [20]. Most patients have shown a good clinical response to 10–20 mg/day, although some may require higher doses, i.e. 40–100 mg/day [3, 20, 24, 25]. In spite of apparently complete clinical recovery, some patients continue to excrete abnormal metabolites (particularly 3-hydroxyisovalerate), a finding that correlates inversely with the actual level of carboxylase activity in lymphocytes. Exceptionally, persistent clinical and biochemical abnormalities have been observed despite treatment with very high doses of biotin [20, 24, 25]. One severely retarded patient showing some biochemical but no clinical response to 20 mg of biotin per day died at the age of 6 years [26]. Mutations causing totally unresponsive HCS deficiency are probably lethal in utero [13].

To date, the prognosis for most surviving, well-treated patients with HCS deficiency seems to be good, with the exception of those who show only a partial response [20, 24, 25]. Careful follow-up studies are needed to judge the long-term outcome. In one patient, followed for 9 years and treated prenatally and from the age of 3.5 months with 6 mg biotin/day, some difficulties in fine motor tasks were obvious at the age of 9 years [27]. In five Japanese patients (four families), the intelligence quotient (IQ) at the age of 5-10 years varied between 64 and 80 [24]. Four of these patients had a severe neonatal onset form, and one of them (IQ = 64) was treated prenatally. Three of these patients showed recurrent respiratory infections, metabolic acidosis and organic aciduria despite high-dose (20-60 mg/day) biotin therapy. However, irreversible neurologic auditory-visual deficits, as described for biotinidase deficiency, have not been reported. Successful prenatal treatment (10 mg/day), preventing acute neonatal symptoms, has been reported in four pregnancies [1, 5, 24].

### Biotinidase Deficiency

Introduction of neonatal screening programs has resulted in the detection of asymptomatic patients with residual biotinidase activity [22]. Based on measurement of plasma biotinidase activity, the patients are classified into three main groups.

1. Patients with *profound biotinidase deficiency*, with less than 10% of mean normal serum biotinidase activity. Using a sensitive method with the natural substrate biocytin, we classify these patients further into those with complete deficiency (undetectable activity, limit of detection ~0.05% of the mean normal) and those with residual biotinidase activity up to 10% [19].
2. Patients with *partial biotinidase deficiency*, with 10-30% residual activity.
3. Patients with decreased affinity of biotinidase for biocytin, i.e. *Km variants* [28].

### Group 1

In early-diagnosed children with *complete biotinidase deficiency*, 5-10 mg of oral biotin per day promptly reverse or prevent all clinical and biochemical abnormalities. For chronic treatment, the same dose is recommended. No adverse effects have been observed from such therapy and, importantly, there is no accumulation of biocytin in body fluids [14], which was previously suspected to be a possible risk.

Under careful clinical and biochemical control, it may be possible to reduce the daily dose of biotin to 2.5 mg. However, biotin has to be given throughout life and regularly each day, since biotin depletion develops rapidly [6]. In one patient, 1 mg/day was insufficient during infections [9].

Neonatal screening for biotinidase deficiency [22] allows early diagnosis and effective treatment. In such patients, the diagnosis must be confirmed by quantitative measurement of biotinidase activity. Treatment should be instituted without delay, since patients may become biotin deficient within a few days after birth [6].

In patients who are diagnosed late, irreversible brain damage may have occurred before the commencement of treatment. In particular, auditory and visual deficits often persist in spite of biotin therapy [8-10], and intellectual impairment and ataxia have been observed as long-term complications [9].

Patients with *residual activity up to 10%*, usually detected by neonatal screening, may remain asymptomatic for several years or even until adulthood [19, 29]. According to our experience with 61 such patients (52 families), however, they show a great risk of becoming biotin deficient and should be treated with, e.g., 2.5 mg of biotin per day [1, 19, 22].

### Group 2

Patients with *partial biotinidase deficiency* (10-30% residual activity) are mostly detected by neonatal screening and in family studies and usually remain asymptomatic. One infant with about 30% enzyme activity developed hypotonia, skin rash and hair loss during an episode of gastroenteritis at 6 months of age. This was reversed by biotin therapy [30]. Recently, we showed that, among 24 patients with 14-25% serum biotinidase activity studied at the age of 8 months-8 years, 16 patients had a subnormal biotin concentration in at least one plasma sample, with a tendency toward lower values with increasing age [31]. Therefore, it seems necessary to regularly control patients with 10-30% of residual activity and to supplement patients with borderline abnormalities with small doses of biotin, e.g., 2.5-5 mg/week.

### Group 3

Among 201 patients (176 families), we found ten patients (eight families) with a Km mutation. In the routine colorimetric biotinidase assay with 0.15 mmol/l biotinyl-*p*-aminobenzoate as substrate, six of these patients (five families) showed profound deficiency (0.9–4.3% residual activity), whereas four patients (three families) showed partial deficiency (18–20% residual activity). The index patient in all five families with profound deficiency presented with a severe clinical illness [10, 28], and one of the patients with partial deficiency, although apparently asymptomatic, had marginal biotin deficiency at the age of 2 years [28]. These results show the importance of testing all patients with residual biotinidase activity for a Km mutation. They all seem to have a high risk of becoming biotin deficient and, therefore, must be treated with biotin.

## Genetics

Both disorders are inherited as autosomal-recessive traits. HCS deficiency seems to be rarer than biotinidase deficiency. The incidences of profound (<10% residual activity) and partial (10–30% residual activity) biotinidase deficiencies are, on average, 1:112,000 and 1:129,000, respectively [22]. The cDNAs for both human HCS [32, 33] and biotinidase [34] have been cloned, and the corresponding genes were mapped to human chromosomes 21q22.1 [32] and 3p25 [35], respectively. In both genes, several mutations have been identified, in accordance with the variability of the clinical picture.

### Holocarboxylase Synthetase Deficiency

So far, fifteen different mutations have been reported in 21 families [13, 24, 32, 36, 39, 40]. Six mutations are within the putative biotin-binding region of the enzyme, probably accounting for the observed decrease of the affinity of HCS for biotin [13, 36, 40] and the in vivo responsiveness to biotin therapy of these patients. The degree of abnormality of the Km of HCS for biotin correlated well with the time of onset and severity of illness [12]. Other mutations, located outside the biotin-binding region, most likely decrease catalytic activity without changing the affinity for biotin, although in vivo responsiveness to biotin remains present, as demonstrated for the L327P mutation [13].

### Biotinidase Deficiency

About 30 different mutations have been found in patients with profound or partial biotinidase deficiency [16–18, 37, 38]. In 30 symptomatic patients with profound deficiency, the most common mutations, accounting for half of the alleles, were a deletion/insertion (G98:d713) and a R538C mutation [17]. All mutations found in symptomatic patients totally abolished biotinyl-transferase activity. In contrast, in patients with profound biotinidase deficiency detected by newborn screening, the most common mutations were Q456H and a double-mutation allele A171T + D444H [37]. Strikingly, these mutations were not found in any of the symptomatic patients [17], and an asymptomatic adult, homozygous for the double mutation, has been identified [29]. The lack of symptoms was explained by the finding of trace levels of biotinyl-transferase activity in these mutations. In many patients with partial biotinidase deficiency, mutation D444H is found in compound-heterozygote form with a mutation causing profound biotinidase deficiency [38].

## References

1. Wolf B (1995) Disorders of biotin metabolism. In: Scriver CR, Beaudet AL, Sly WS, Valle D (eds) The metabolic and molecular bases of inherited disease, 7th edn. McGraw-Hill, New York, pp 3151–3177
2. Wolf B, Heard GS, Jefferson LG et al. (1986) Newborn screening for biotinidase deficiency. In Carter TP, Willey AM (eds) Genetic disease: screening and management. Liss, New York, pp 175–182
3. Suormala T, Fowler B, Duran M et al. (1997) Five patients with a biotin-responsive defect in holocarboxylase formation: evaluation of responsiveness to biotin therapy in vivo and comparative studies in vitro. Pediatr Res 41:666–673
4. Sherwood WG, Saunders M, Robinson BH, Brewster T, Gravel RA (1982) Lactic acidosis in biotin-responsive multiple carboxylase deficiency caused by holocarboxylase synthetase deficiency of early and late onset. J Pediatr 101:546–550
5. Suormala T, Fowler B, Jakobs C et al. (1998) Late-onset holocarboxylase synthetase-deficiency: pre- and post-natal diagnosis and evaluation of effectiveness of antenatal biotin therapy. Eur J Pediatr 157:570–575
6. Baumgartner ER, Suormala TM, Wick H, Bausch J, Bonjour JP (1985) Biotinidase deficiency associated with renal loss of biocytin and biotin. Ann NY Acad Sci 447:272–286
7. Baumgartner ER, Suormala TM, Wick H et al. (1989) Biotinidase deficiency: a cause of subacute necrotizing encephalomyelopathy (Leigh syndrome). Report of a case with lethal outcome. Pediatr Res 26:260–266
8. Wolf B, Heard GS, Weissbecker KA et al. (1985) Biotinidase deficiency: initial clinical features and rapid diagnosis. Ann Neurol 18:614–617
9. Wastell HJ, Bartlett K, Dale G, Shein A (1988) Biotinidase deficiency: a survey of 10 cases. Arch Dis Child 63:1244–1249
10. Ramaekers VTH, Suormala TM, Brab M et al. (1992) A biotinidase Km variant causing late onset bilateral optic neuropathy. Arch Dis Child 67:115–119
11. Duran M, Baumgartner ER, Suormala TM et al. (1993) Cerebrospinal fluid organic acids in biotinidase deficiency. J Inherit Metab Dis 16:513–516
12. Burri BJ, Sweetman L, Nyhan WL (1985) Heterogeneity in holocarboxylase synthetase in patients with biotin-responsive multiple carboxylase deficiency. Am J Hum Genet 37:326–337
13. Aoki Y, Suzuki Y, Li X et al. (1997) Characterization of mutant holocarboxylase synthetase (HCS): a Km for biotin was not elevated in a patient with HCS deficiency. Pediatr Res 42:849–854
14. Suormala TM, Baumgartner ER, Bausch J, Holiock W, Wick H (1988) Quantitative determination of biocytin in urine of patients with biotinidase deficiency using high-

performance liquid chromatography (HPLC). Clin Chim Acta 177:253–270
15. Hymes J, Fleischhauer K, Wolf B (1995) Biotinylation of histones by human serum biotinidase: assessment of biotinyltransferase activity in sera from normal individuals and children with biotinidase deficiency. Biochem Mol Med 56:76–83
16. Norrgard KJ, Pomponio RJ, Swango KL et al. (1997) Mutation (Q456H) is the most common cause of profound biotinidase deficiency in children ascertained by newborn screening in the United States. Biochem Mol Med 61:22–27
17. Pomponio RJ, Hymes J, Reynolds TR et al. (1997) Mutation in the human biotinidase gene that causes profound biotinidase deficiency in symptomatic children: molecular, biochemical, and clinical analysis. Pediatr Res 42:840–848
18. Wolf B, Pomponio RJ, Norrgard KJ et al. (1998) Delayed-onset profound biotinidase deficiency. J Pediatr 132:362–365
19. Suormala TM, Baumgartner ER, Wick H, Scheibenreiter S, Schweitzer S (1990) Comparison of patients with complete and partial biotinidase deficiency: biochemical studies. J Inherit Metab Dis 13:76–92
20. Baumgartner ER, Suormala T (1997) Multiple carboxylase deficiency: inherited and acquired disorders of biotin metabolism. Int J Vit Nutr Res 67:377–384
21. Suzuki Y, Aoki Y, Sakamoto O et al. (1996) Enzymatic diagnosis of holocarboxylase synthetase deficiency using apocarboxyl carrier protein as a substrate. Clin Chim Acta 251:41–52
22. Wolf B (1991) Worldwide survey of neonatal screening for biotinidase deficiency. J Inherit Metab Dis 14:923–927
23. Baur B, Suormala T, Bernoulli C, Baumgartner ER (1998) Biotin determination by three different methods: specificity and application to urine and plasma ultrafiltrates of patients with and without disorders in biotin metabolism. Int J Vit Nutr Res 68:300–308
24. Aoki Y, Suzuki Y, Sakamoto O et al. (1995) Molecular analysis of holocarboxylase synthetase deficiency: a missense mutation and a single base deletion are predominant in Japanese patients. Biochim Biophys Acta 1272:168–174
25. Wolf B, Hsia YE, Sweetman L et al. (1981) Multiple carboxylase deficiency: clinical and biochemical improvement following neonatal biotin treatment. Pediatrics 68:113–118
26. Velazquez A, von Raesfeld D, Gonzalez-Noriega A et al. (1986) Pyruvate carboxylase responsive to ketosis in a multiple carboxylase deficient patient. J Inherit Metab Dis 9 [suppl 2]:300–302
27. Michalski AJ, Berry GT, Segal S (1989) Holocarboxylase synthetase deficiency: 9-year follow-up of a patient on chronic biotin therapy and a review of the literature. J Inherit Metab Dis 12:312–316
28. Suormala T, Ramaekers VTH, Schweitzer S et al. (1995) Biotinidase Km-variants: detection and detailed biochemical investigations. J Inherit Metab Dis 18:689–700
29. Wolf B, Norrgard KJ, Pomponio RJ et al. (1997) Profound biotinidase deficiency in two asymptomatic adults. Am J Med Genet 73:5–9
30. Secor McVoy JR, Levy HL, Lawler M et al. (1990) Partial biotinidase deficiency: clinical and biochemical features. J Pediatr 116:78–83
31. Bernoulli C, Suormala T, Baur B, Baumgartner ER (1998) A sensitive method for the determination of biotin in plasma and CSF, and application to partial biotinidase deficiency. J Inherit Metab Dis 21 [suppl 2] 46:92
32. Suzuki Y, Aoki Y, Ishida Y et al. (1994) Isolation and characterization of mutations in the human holocarboxylase synthetase cDNA. Nature Genet 8:122–128
33. Leon-Del-Rio A, Leclerc D, Akerman B, Wakamatsu N, Gravel RA (1995) Isolation of cDNA encoding human holocarboxylase synthetase by functional complementation of a biotin auxotroph of Escherichia coli. Proc Natl Acad Sci USA 92:4626–4630
34. Cole H, Reynolds TR, Lockyer JM et al. (1994) Human serum biotinidase. cDNA cloning, sequence, and characterization. J Biol Chem 269:6566–6570
35. Cole H, Weremovicz S, Morton CC, Wolf B (1994) Localization of serum biotinidase (BTD) to human chromosome 3 in Band p25. Genomics 22:662–663
36. Dupuis L, Leon-Del-Rio A, Leclerc D et al. (1996) Clustering of mutations in the biotin-binding region of holocarboxylase synthetase in biotin-responsive multiple carboxylase deficiency. Hum Mol Genet 5:1011–1016
37. Norrgard KJ, Pomponio RJ, Swango KL et al. (1997) Double mutation (A171 T and D444H) is a common cause of profound biotinidase deficiency in children ascertained by newborn screening in United States. Hum Mutat, mutations in brief no. 128, on-line
38. Swango KL, Demirkol M, Huner G et al. (1998) Partial biotinidase deficiency is usually due to the D444H mutation in the biotinidase gene. Hum Genet 102:571–575
39. Aoki Y, Li X, Sakamoto O et al. (1999) Identification and characterization of seven mutations in patients with holocarboxylase synthetase deficiency. Hum Genet 104:143–148
40. Sakamoto O, Suzuki Y, Li X et al. (1999) Relationship between kinetic properties of mutant enzyme and biochemical and clinical responsiveness to biotin in holocarboxylase synthetase deficiency. Pediatr Res 46:671–676

# CHAPTER 25

## Cobalamin Transport and Metabolism

Cobalamin (Cbl or vitamin $B_{12}$) is a cobalt-containing, water-soluble vitamin that is synthesized by lower organisms but not by higher plants and animals. In the human diet, its only source is animal products in which it has accumulated by microbial synthesis. Cbl is needed for only two reactions in man, but its metabolism involves complex absorption and transport systems and multiple intracellular conversions (Fig. 25.1). As methylcobalamin, it is a cofactor of the cytoplasmic enzyme methionine synthase. As adenosylcobalamin, it is a cofactor of the mitochondrial enzyme methylmalonyl-coenzyme A mutase, which is involved in the catabolism of valine, threonine and odd-chain fatty acids into succinyl-CoA, an intermediate of the Krebs' cycle.

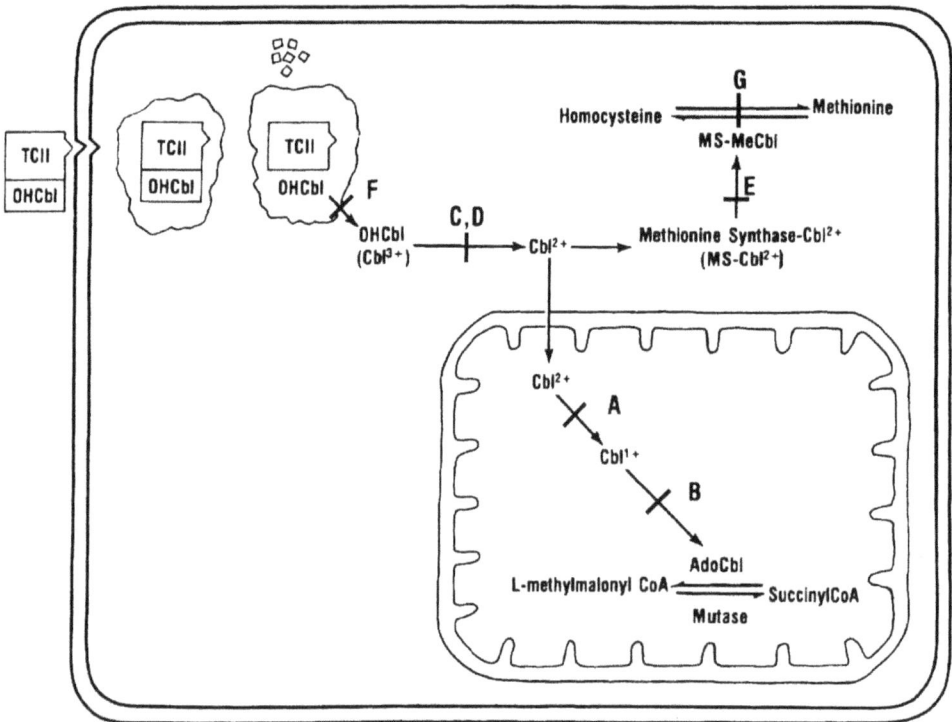

**Fig. 25.1.** Cobalamin (Cbl) endocytosis and intracellular metabolism. The cytoplasmic, lysosomal, and mitochondrial compartments are indicated. *AdoCbl*, adenosyl-Cbl; *CoA*, coenzyme A; *MeCbl*, methyl-Cbl; *MS*, methionine synthase; *OHCbl*, hydroxy-Cbl; *TCII*, transcobalamin II. $1^+$, $2^+$ and $3^+$ refer to the oxidation state of the central cobalt of Cbl. *Letters* refer to the sites of blocks [5]. Enzyme defects are indicated by *solid bars*

# CHAPTER 25

# Disorders of Cobalamin and Folate Transport and Metabolism

David S. Rosenblatt

CONTENTS

Cobalamin................................... 285
   Disorders of Absorption and Transport of Cobalamin.. 285
      Hereditary Intrinsic Factor Deficiency............ 285
      Defective Transport of Cbl by Enterocytes
      (Imerslund-Gräsbeck Syndrome)................ 286
      Transcobalamin I (R Binder) Deficiency.......... 287
      Transcobalamin II Deficiency................... 287
   Disorders of Intracellular Utilization of Cbl.......... 288
      Combined Deficiencies of AdoCbl and MeCbl....... 288
      AdoCbl Deficiency............................ 290
      MeCbl Deficiency............................ 290

Folate...................................... 293
   Disorders of Absorption and Metabolism of Folate.... 293
      Hereditary Folate Malabsorption................. 293
      Glutamate-Formiminotransferase Deficiency........ 294
      Methylenetetrahydrofolate Reductase Deficiency..... 294

References................................... 296

---

Patients with inherited disorders affecting cobalamin (Cbl) absorption or metabolism show elevations of homocysteine or methylmalonic acid, either alone or in combination. For those disorders that affect methylcobalamin (MeCbl) formation, the major manifestations include megaloblastic anemia secondary to folate deficiency and neurological abnormalities presumably secondary to methionine deficiency or homocysteine elevation. For those disorders that affect adenosylcobalamin (AdoCbl) formation, the main findings are secondary to methylmalonic-acid elevations and resultant acidosis.

---

## COBALAMIN

Inherited disorders of Cbl metabolism are classified as those involving absorption and transport and those involving intracellular utilization [1-5].

### Disorders of Absorption and Transport of Cobalamin

Absorption of dietary Cbl involves first binding to a glycoprotein (R binder, haptocorrin, TCI) in the saliva. In the intestine, TCI is digested by proteases, allowing the Cbl to bind to intrinsic factor (IF), which is produced in the stomach by parietal cells. Using a specific receptor, the IF-Cbl complex enters the enterocyte. Cbl bound to transcobalamin II (TCII), the physiologically important circulating Cbl-binding protein, slowly enters the portal vein after release. Inherited defects of several of these steps are known.

### Hereditary Intrinsic Factor Deficiency

#### Clinical Presentation

Megaloblastic anemia is the main finding [6-8]. It usually appears after the first year of life but before the age of 5 years. In cases of partial deficiency, clinical presentation has been delayed until adolescence or adulthood. The patients present with failure to thrive, often with vomiting and alternating diarrhea and constipation, anorexia and irritability. They are anemic and may have hepatosplenomegaly, stomatitis or atrophic glossitis, developmental delay, and myelopathy or peripheral neuropathy.

#### Metabolic Derangement

IF is either absent or nonfunctional. Some patients produce no IF whereas, in others, it may be detectable immunologically. There have been reports of IF with reduced affinity for Cbl, reduced affinity for the Cbl receptor or increased susceptibility to proteolysis [7-9].

#### Diagnostic Tests

The hematological abnormalities in the defects of Cbl absorption and transport should be detected by a measurement of red blood cell indices, a complete blood count and a bone-marrow examination. Megaloblastic anemia and a low serum Cbl are present. Homocystinuria and methylmalonic aciduria may be present. A deoxyuridine-suppression test on marrow cells is useful but is not easily available in most clinical laboratories. This test measures the incorporation of label from thymidylate (dTMP) in a trichloroacetic-acid precipitate before and after incubation of washed bone-marrow cells in an excess of deoxyuridine. In the presence of folate or Cbl deficiency, this preincubation reduces incorporation to only 30–40% of that observed in the absence of deoxyuridine, as compared with about 10% when there is no folate or Cbl deficiency.

In hereditary IF deficiency, in contrast to acquired forms of pernicious anemia, there is normal gastric acidity and normal gastric cytology. Cbl absorption, as measured by the Schilling test, is abnormal but is normalized when the labeled Cbl is mixed with a source of normal IF, such as gastric juice from an unaffected individual. Some (but not all) patients may have a lack of immunologically reactive IF.

### Treatment and Prognosis

IF deficiency can be treated initially with hydroxycobalamin (OHCbl, 1 mg/day intramuscularly) to replete body stores until biochemical and hematological values normalize. The subsequent dose of OHCbl required to maintain normal values may be as low as 0.25 mg (250 µg) every 3 months. Whereas the hematological and biochemical findings may be completely reversed, if treatment is delayed, some neurological abnormalities may persist.

### Genetics

At least 45 patients of both sexes have been reported, and inheritance is autosomal recessive. A cDNA has been characterized, and the gene is localized on chromosome 11q13 [10, 11]; no mutations are known.

## Defective Transport of Cobalamin by Enterocytes (Imerslund-Gräsbeck Syndrome)

### Clinical Presentation

Megaloblastic anemia also presents once fetal hepatic Cbl stores have been depleted. The disease usually appears between the ages of 1 year and 5 years, but onset may be even later [12–18]. Most patients have proteinuria and, in a few cases, this was of the tubular type, with all species of proteins represented rather than albumin alone. The literature on the renal pathology has been reviewed [19]. Although patients who excreted protein during childhood continued to excrete protein in adulthood, the renal lesions were not progressive [13]. Neurological abnormalities, such as spasticity, truncal ataxia and cerebral atrophy, may be present as a consequence of the Cbl deficiency.

### Metabolic Derangement

Although the defect in this disorder is thought to be in the IF receptor, in some patients, homogenates of ileal biopsies bound IF–Cbl normally, suggesting that the defect does not lie in the absence of receptors. In other patients, there appears to be an absent receptor. In the canine model of this disease, Cbl malabsorption may be due to a mutation in the receptor, which results in its retention in an early biosynthetic compartment; this results in inefficient expression in the intestine and kidney [20]. Cubilin has been purified as the IF–Cbl receptor from the proximal renal tubule [21–23]. It copurifies with a larger receptor, megalin, and the two proteins are found together in both intestinal and renal epithelia [23].

### Diagnostic Tests

In contrast to patients with IF deficiency, the Schilling test is not corrected by providing a source of human IF with the labeled Cbl [1]. The diagnosis is aided by finding low serum Cbl levels, megaloblastic anemia and proteinuria. As with hereditary IF deficiency, gastric morphology and pancreatic function are normal, and there are no IF autoantibodies. IF levels are normal.

### Treatment and Prognosis

Treatment with systemic OHCbl corrects the anemia and the neurologic findings, but not the proteinuria. As with hereditary IF deficiency, once Cbl stores are replete, low doses of systemic OHCbl may be sufficient to maintain normal hematological and biochemical values.

### Genetics

About 250 cases have been reported, and inheritance is autosomal recessive [18]. Most patients are found in Finland, Norway, Saudi Arabia, Turkey and among Sephardic Jews. The gene was mapped to a 6-cm region of chromosome 10p using microsatellite markers in Finnish and Norwegian families [24]. Cubilin has been mapped to the same region and is thus the prime candidate gene for this disease [22]. The sequence of cubilin is characterized by eight EGF repeats followed by 27 contiguous, 110-amino-acid CUB domains [21, 22]. In recent years, there has been a decrease in the number of new cases in Norway and Finland, suggesting that there may be environmental factors affecting expression [24].

## Transcobalamin I (R Binder) Deficiency

### Clinical Presentation

It is not clear that this entity has a distinct phenotype. As adults, the patients have been diagnosed with predominantly neurological rather than hematological findings [25]. One man in the fifth decade of life had findings consistent with subacute combined degeneration of the spinal cord. Another had optic atrophy, ataxia, long-tract signs and dementia.

### Metabolic Derangement

The role of TCI is uncertain but may be in the scavenging of toxic Cbl analogs.

### Diagnostic Tests

Serum Cbl levels are low, because most circulating Cbl is bound to TCI. TCII–Cbl levels are normal, and there are no hematologic findings of Cbl deficiency. A deficiency or absence of TCI is found in plasma, saliva and leukocytes.

### Treatment and Prognosis

It has not been possible to reliably assign a phenotype to the biochemical finding of low or absent TCI. It is uncertain whether treatment is warranted.

### Genetics

The TCI gene has been cloned and mapped to chromosome 11q11-q12 [26, 27]. No mutations have been described in any patient with TCI deficiency.

## Transcobalamin II Deficiency

### Clinical Presentation

TCII-deficient patients usually develop symptoms much earlier than patients with other causes of Cbl malabsorption, usually within the first few months of life. Even though the only TCII in cord blood is of fetal origin, patients are not sick at birth. Presenting findings include pallor, failure to thrive, weakness and diarrhea. Although the anemia is usually megaloblastic, patients with pancytopenia or isolated erythroid hypoplasia have been described. Leukemia may be mistakenly diagnosed because of the presence of immature white cell precursors in an otherwise hypocellular marrow. Neurologic disease is not found at diagnosis but may develop with prolonged duration of untreated illness, treatment with folate in the absence of Cbl, or inadequate Cbl treatment [28]. This may consist of developmental delay, neuropathy, myelopathy and encephalopathy. Defective granulocyte function with both defective humoral and cellular immunity has been seen.

### Metabolic Derangement

The majority of patients have no immunologically detectable TCII, although others have some detectable protein or a TCII that is able to bind Cbl but does not function normally [1, 29, 30].

### Diagnostic Tests

Serum Cbl levels are usually normal, because the majority of serum Cbl is bound to TCI and not TCII. Cbl bound to TCII, as reflected by the unsaturated vitamin-$B_{12}$-binding capacity, is low. This test must be performed before Cbl treatment is started. Most patients have no immunologically detectable circulating TCII. Since TCII is involved in the transcytosis of Cbl through the enterocyte, the Schilling test may be abnormal in TCII-deficient patients [1]. In those patients in whom the Schilling test is normal, immunoreactive TCII is found. Methylmalonic aciduria and homocystinuria have been described, but the levels are not usually as high as in defects affecting the intracellular synthesis of Cbl cofactors. In patients who do not synthesize TCII, both diagnosis and prenatal diagnosis are possible by studying TCII synthesis in cultured fibroblasts or amniocytes [31]. In families in which the molecular defect has been identified, DNA testing is possible for both diagnosis and heterozygote detection.

### Treatment and Prognosis

To treat successfully, it is necessary to keep serum Cbl levels in the range of 1000–10,000 pg/ml. These levels have been achieved with doses of oral or systemic OHCbl or cyanocobalamin (CNCbl) of 500–1000 µg twice weekly. Initially, it is reasonable to begin with a daily dose of systemic OHCbl and then to decrease the dose to once or twice weekly once the hematological profile has responded. Intravenous Cbl is not suggested, because of the rapid loss of vitamin in the urine. Folic acid or folinic acid can reverse the megaloblastic anemia and has been used in doses up to 15 mg orally four times daily. Folates must never be given as the only therapy in TCII deficiency, because of the danger of hematological relapse and neurological deterioration.

## Genetics

Inheritance is autosomal recessive; there have been at least 36 cases, including both twins and siblings [1, 28]. The TCII gene is on chromosome 22, and defects have included deletions and nonsense mutations [32, 33]. In addition to the pathogenic mutations, a number of polymorphic variants are known.

## Disorders of Intracellular Utilization of Cbl

### Combined Deficiencies of Adenosylcobalamin and Methylcobalamin

Three disorders are associated with functional defects in both methylmalonyl-coenzyme A (CoA) mutase and methionine synthase. As such, they are characterized by both methylmalonic aciduria and homocystinuria.

### Cobalamin F

*Clinical Presentation*
Of the five known patients with *cblF* disease, four presented in the first year of life. The original infant girl had glossitis and stomatitis in the first week of life [34, 35]. She had severe feeding difficulties requiring tube feeding. Tooth abnormalities and dextrocardia were present. Other clinical findings have included anemia, failure to thrive, recurrent infections, developmental delay, lethargy, hypotonia, aspiration pneumonia, hepatomegaly and encephalopathy. One infant died suddenly at home in the first year of life. One boy developed juvenile rheumatoid arthritis at the age of 4 years and a pigmented skin abnormality at 10 years.

*Diagnostic Tests*
Precise diagnosis of the inborn errors of Cbl metabolism requires tests in cultured fibroblasts. The incorporation of [$^{14}$C]propionate in macromolecules is a good screen for the integrity of the methylmalonyl-CoA mutase reaction, and the incorporation of [$^{14}$C]methyltetrahydrofolate is a good screen for the function of methionine synthase. The total incorporation of [$^{57}$Co]CNCbl by fibroblasts can differentiate a number of the disorders. The conversion of labeled CNCbl to both MeCbl and AdoCbl can be measured following hot ethanol extraction using high-performance liquid chromatography. Complementation analysis is used to define the specific mutant class. Cells from the undiagnosed patient are co-cultivated with cells from patients with known defects, and replicate cultures are either treated or not treated with polyethylene glycol, which acts as a fusing agent. If cells belong to the same class, incorporation of labeled substrate (methyltetrahydrofolate or propionate) will not increase in fused cells but, if the cell lines belong to different classes, fusion results in a partial correction of the defect in incorporation.

In fibroblasts from *cblF* patients, total incorporation of labeled CNCbl is elevated, but there is no conversion of CNCbl to either AdoCbl or MeCbl. All the label is found as free CNCbl in lysosomes. There is decreased incorporation of both labeled propionate and labeled methyltetrahydrofolate. In *cblF*, a complete blood count and bone-marrow examination may reveal megaloblastic anemia, neutropenia and thrombocytopenia. The serum Cbl level may be low, and the Schilling test has been abnormal in all patients tested. Usually, both homocystinuria and methylmalonic aciduria are found, although homocystinuria was not detected in the original patient.

*Metabolic Derangement*
The defect in *cblF* appears to be due to a failure of Cbl to be transported across the lysosomal membrane following degradation of TCII in the lysosome. As a result, Cbl cannot be converted to either AdoCbl or MeCbl. The inability of *cblF* patients to absorb oral Cbl suggested that IF–Cbl also has to pass through a lysosomal stage in the enterocyte before Cbl is released into the portal circulation.

*Treatment and Prognosis*
Treatment with parenteral OHCbl (first daily and then biweekly) at a dose of 1 mg/day seems to be effective in correcting the metabolic and clinical findings. Despite the fact that two Schilling tests showed an inability to absorb Cbl with or without IF, the original patient responded to oral Cbl before being switched to parenteral Cbl.

*Genetics*
As both male and female patients of unaffected parents have been reported, inheritance is presumed to be autosomal recessive. The gene responsible for *cblF* has not been identified.

### Cobalamin C

*Clinical Presentation*
This is the most frequent inborn error of Cbl metabolism, and more than 100 patients are known [2, 36–39]. Many were acutely ill in the first month of life, and most were diagnosed within the first year. The *early-onset group* shows feeding difficulties, hypotonia and lethargy, followed by progressive neurological deterioration. Patients may go into coma, show hypotonia,

hypertonia or both and develop abnormal movements or seizures. There is a severe pancytopenia or a non-regenerative anemia, which is not always associated with macrocytosis and hypersegmented neutrophils but which is megaloblastic on bone-marrow examination. Many patients develop multisystem pathology, such as renal failure, hepatic dysfunction, cardiomyopathy, interstitial pneumonia or the hemolytic uremic syndrome caused by widespread microangiopathy. An unusual retinopathy consisting of perimacular hypopigmentation surrounded by a hyperpigmented ring and a more peripheral salt-and-pepper retinopathy has been described; this is sometimes accompanied by nystagmus. Both microcephaly and hydrocephalus occur. A small number of *cblC* patients were not diagnosed until after the first year of life and as late as the end of the second decade of life. The earlier-diagnosed patients in this group had overlapping findings with those found in the first year of life. Major clinical findings in this *late-onset cblC* group included confusion, disorientation and gait abnormalities. Macrocytic anemia was seen in only about a third of the oldest patients; therefore, it is important both to consider the diagnosis of *cblC* in the presence of neurological findings alone and to order metabolite levels.

*Metabolic Derangement*
In both the *cblC* and *cblD* disorders, patients are thought to have a defect in the reduction of the oxidation state of the central cobalt of Cbl from $3^+$ to $2^+$ after efflux of the Cbl from the lysosome. Decreased activities of microsomal Cbl-$3^+$ reductase, CNCbl β-ligand transferase and a mitochondrial, reduced-nicotinamide-adenine-dinucleotide-linked aquacobalamin reductase have been described in fibroblast extracts [40-42]. If the reduction of Cbl does not occur, Cbl does not bind to the two intracellular enzymes and leaves the cell. Neither AdoCbl nor MeCbl is formed.

*Diagnostic Tests*
Homocystinuria and methylmalonic acidemia are the biochemical hallmarks of this disease. In general, the levels seen are lower than those found in methylmalonyl-CoA mutase deficiency but higher than those seen in the Cbl-transport defects. Plasma methionine levels are either normal or decreased but are not elevated. A complete blood count and bone-marrow examination will detect the hematologic abnormalities.

Fibroblast studies show decreased incorporation of label from propionate, methyltetrahydrofolate and CNCbl, and there is decreased synthesis of both AdoCbl and MeCbl. Cells fail to complement those of other *cblC* patients.

*Treatment and Prognosis*
Treatment with 1 mg/day OHCbl (parenteral) decreases the elevated metabolite levels, but these are not usually completely normalized. In one comprehensive study, oral OHCbl was found to be insufficient, and both folinic acid and carnitine were ineffective. Daily oral betaine (250 mg/kg/day) with twice weekly systemic OHCbl (1 mg/day) resulted in normalization of methionine and homocysteine levels and decreased methylmalonic aciduria [43].

Of a group of 44 patients with onset in the first year of life, 13 died, and only one patient was neurologically intact, with other survivors described as having severe or moderate impairment. Survival with mild to moderate disability was found in the patients who had a later onset [39].

*Genetics*
The gene responsible for *cblC* has not been isolated. Inheritance is autosomal recessive. Prenatal diagnosis can be performed by measuring the incorporation of labeled propionate or labeled methyltetrahydrofolate, by the synthesis of MeCbl and AdoCbl in cultured chorionic villus cells and amniocytes and by looking at methylmalonic acid levels in amniotic fluid. Heterozygote detection is not possible.

## Cobalamin D

*Clinical Presentation*
There is only one sibship of two males with *cblD* [44-46]. The elder sibling was diagnosed with behavioral problems and mild mental retardation at the age of 14 years. He had ataxia and nystagmus.

*Metabolic Derangement*
Although *cblD* is a distinct complementation group, many of the differences between it and *cblC* appear to be quantitative. All the biochemical findings are similar to those in *cblC*, and the defect is thought to be in the reduction of Cbl-$3^+$.

*Diagnostic Tests*
Homocystinuria, methylmalonic aciduria and hyperglycinemia may be found. Although the original patient did not have megaloblastic anemia, the deoxyuridine-suppression test was abnormal. There are only minor biochemical differences between *cblD* and *cblC*, and *cblD* cell lines show decreased incorporation of labeled methyltetrahydrofolate, propionate and CNCbl and show decreased synthesis of both AdoCbl and MeCbl. As *cblC* and *cblD* cells belong to distinct complementation classes, complementation analysis is required to make a specific diagnosis.

*Genetics*

As both siblings with *cblD* are male, the possibility of sex linkage has not been disproved.

### Adenosylcobalamin Deficiency; cblA, cblB

**Clinical Presentation**

These two disorders are characterized by Cbl-responsive methylmalonic aciduria (Chap. 16) [2]. The phenotype resembles methylmalonyl-CoA mutase deficiency. Most patients have an acidotic crisis in the first year of life, many in the neonatal period. Symptoms are related to methylmalonic-acid accumulation and include vomiting, dehydration, tachypnea, lethargy, failure to thrive, developmental retardation, hypotonia and encephalopathy. The toxic levels of methylmalonic acid may result in bone-marrow abnormalities and produce anemia, leukopenia and thrombocytopenia. Hyperammonemia, hyperglycinemia and ketonuria may be found.

**Metabolic Derangement**

The defect in *cblA* presumably lies in the reduction of the central cobalt of Cbl from the $2^+$ to the $1^+$ oxidation state in mitochondria. A patient with all the clinical and biochemical features of *cblA* has been described, but cells from this patient complement those from other *cblA* patients. This implies that more than one step may be involved in the intramitochondrial reduction of Cbl or that intragenic complementation may occur among *cblA* lines [47]. The defect in *cblB* is in an adenosyltransferase, the final intramitochondrial catalyst in the synthesis of AdoCbl [48].

**Diagnostic Tests**

Total serum Cbl is usually normal, and there is massive methylmalonic aciduria (0.8–1.7 mmol/day; normal <0.04 mmol/day) but no homocystinuria. A decrease in the level of methylmalonic-acid excretion in response to Cbl therapy is useful in distinguishing these disorders from methylmalonyl-CoA-mutase deficiency. The differentiation of *cblA* and *cblB* from mutase deficiency can be made by finding normal levels of methylmalonyl-CoA mutase in fibroblast extracts or by the failure of intact *cblA* or *cblB* fibroblasts to increase labeled propionate incorporation following transfection by a vector containing cloned mutase cDNA. Neither of the above tests are available clinically. In cultured fibroblasts, the incorporation of labeled propionate is decreased but is responsive to the addition of OHCbl to the culture medium. There is decreased synthesis of AdoCbl following incubation in labeled CNCbl. Both *cblA* and *cblB* cell lines complement both one another and lines from methylmalonyl-CoA-mutase patients. Cells lines do not complement those from patients in the same complementation group.

**Treatment and Prognosis**

These patients respond to protein restriction and to OHCbl treatment, either 1 mg orally daily or intramuscularly one or twice weekly. For details of the planning of a protein-restricted diet, see Chap. 16. Some patients appear to become resistant to Cbl treatment. Therapy with AdoCbl has been attempted in *cblB* with and without success, and it may be that AdoCbl does not reach the target enzyme intact. There have been reports of prenatal therapy with Cbl in AdoCbl deficiency. Most (90%) *cblA* patients improve on Cbl therapy, with 70% doing well long term. Only 40% of *cblB* patients respond to Cbl, and their long-term survival is poorer [49].

**Genetics**

Male and female patients with either *cblA* or *cblB* have been described, and parents of *cblB* patients have decreased adenosyltransferase activity. Autosomal-recessive inheritance is presumed, but neither gene product has been isolated.

### MethylCobalamin Deficiency; cblE, cblG

**Clinical Presentation**

The most common clinical findings are megaloblastic anemia and neurological disease [50–54]. The latter include poor feeding, vomiting, failure to thrive, cerebral atrophy, developmental delay, nystagmus, hypotonia or hypertonia, ataxia, seizures and blindness. Cerebral atrophy may be seen on imaging studies of the central nervous system, and at least one *cblE* patient showed a spinal-cord cystic lesion on autopsy. Most patients are symptomatic in the first year of life, but one *cblG* patient was not diagnosed until age 21 years and carried a misdiagnosis of multiple sclerosis [55]. Another *cblG* patient, who was diagnosed during his fourth decade of life, had mainly psychiatric symptoms.

**Metabolic Derangement**

The defect in *cblE* is in a reducing enzyme, methionine-synthase reductase, required to keep Cbl in a functional state. In *cblG*, the defect is in methionine synthase itself.

## Diagnostic Tests

Homocystinuria and hyperhomocysteinemia are almost always found in the absence of methylmalonic acidemia. Only one *cblE* patient had transient unexplained methylmalonic aciduria. Hypomethioninemia and cystathioninemia may be present, and there may be increased serine in the urine. A complete blood count and bone-marrow examination will detect the hematological manifestations. Fibroblast extracts from *cblE* patients have normal activity of methionine synthase in the standard assay, but deficient activity can be found when the assay is performed under specific reducing conditions [50, 56]. Cell extracts from *cblG* patients have decreased methionine-synthase activity in the standard assay. Cultured fibroblasts from both *cblE* and *cblG* patients have decreased incorporation of labeled methyltetrahydrofolate and decreased synthesis of MeCbl following incubation in Cbl. In some *cblG* patients (*cblG* variants) no Cbl forms are bound to methionine synthase following incubation in labeled CNCbl. Complementation analysis is performed using methyltetrahydrofolate as the substrate and will distinguish *cblE* from *cblG* patients.

## Treatment and Prognosis

Both of these disorders are treated with OHCbl or MeCbl, 1 mg intramuscularly, first daily and then once or twice weekly. Although the metabolic abnormalities are nearly always corrected, it is difficult to reverse the neurologic findings once they have developed. Treatment with betaine (250 mg/kg/day) has been used, and one *cblG* patient was treated with L-methionine (40 mg/kg/day) and had neurological improvement. Despite therapy, many patients with *cblG* do not do well. In one family with *cblE*, there was successful prenatal diagnosis using cultured amniocytes, and the mother was treated with OHCbl twice per week beginning during the second trimester, and the baby was treated with OHCbl from birth. This boy has developed normally to age 14 years, in contrast to his older brother, who was not treated until after his metabolic decompensation in infancy and who is now 18 years old and has significant developmental delay.

## Genetics

There are at least 11 *cblE* and 19 *cblG* patients known. A cDNA for methionine-synthase reductase has been cloned, and mutations have been detected in *cblE* patients [57]. The methionine-synthase-reductase gene has been localized to chromosome 5p15.2-15.3. Mutations in the methionine-synthase gene have been found in *cblG* patients following cloning of the cDNA for the gene on chromosome 1q43 [58, 59]. Patients with the *cblG* variant of methionine-synthase deficiency have null mutations [60]. Where both mutations are known in a patient, molecular analysis can be used for carrier detection in the family and for prenatal diagnosis.

## Folate Metabolism

Folic acid (pteroylglutamic acid) is plentiful in foods, such as liver, leafy vegetables, legumes and some fruits. Its metabolism (Fig. 25.2) involves reduction to dihydro- (DHF) and tetrahydrofolate (THF), followed by addition of a single-carbon unit, provided by histidine or serine; this carbon unit can be in various redox states (methyl, methylene, methenyl or formyl). Transfer of this single-carbon unit is essential for the endogenous formation of methionine, thymedylate (dTMP) and formylglycineamide ribotide (FGAR) and formylaminoimidazolecarboxamide ribotide (FAICAR), two intermediates of purine synthesis. These reactions also allow regeneration of DHF and THF.

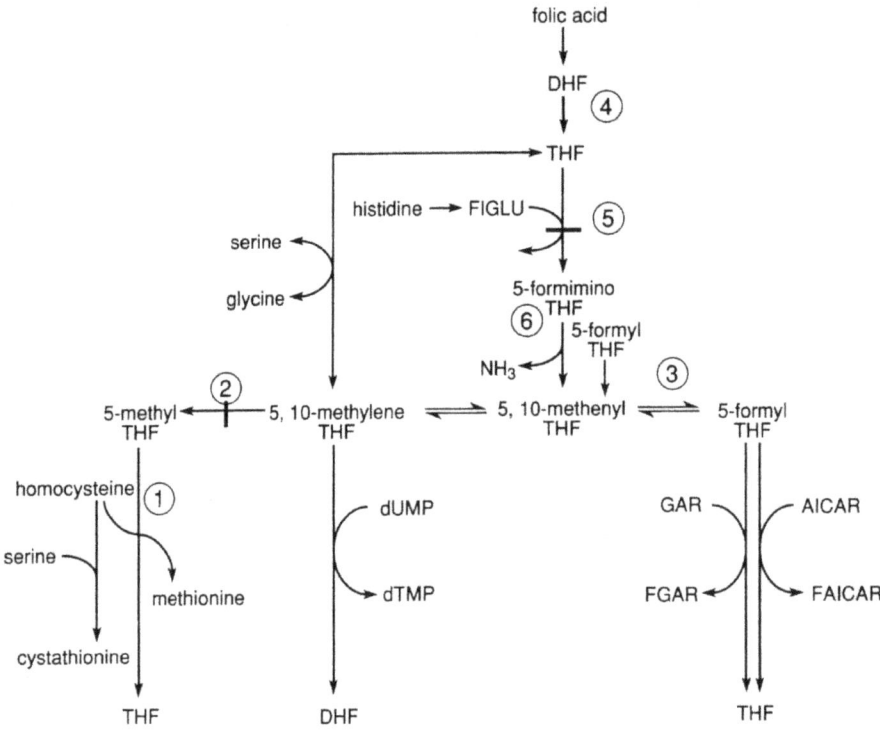

**Fig. 25.2.** Folic-acid metabolism. 1, methionine synthase; 2, methylenetetrahydrofolate reductase; 3, methenyltetrahydrofolate cyclohydrolase; 4, dihydrofolate reductase; 5, glutamate formiminotransferase; 6, formiminotetrahydrofolate cyclodeaminase; *AICAR*, aminoimidazole carboxamide ribotide; *DHF*, dihydrofolate; *dTMP*, deoxythymidine monophosphate; *dUMP*, deoxyuridine monophosphate; *FAICAR*, formylaminoimidazole carboxamide ribotide; *FGAR*, formylglycinamide ribotide; *FIGLU*, formiminoglutamate; *GAR*, glycinamide ribotide; *THF*, tetrahydrofolate. Enzyme defects are indicated by *solid bars*

Three confirmed inborn errors of folate absorption and metabolism have been described.

*Hereditary folate malabsorption* presents with severe megaloblastic anemia, due to the importance of dTMP and purine synthesis in hematopoiesis, and is usually associated with progressive neurological deterioration.

*Glutamate-formiminotransferase deficiency* has been reported in association with various degrees of psychomotor retardation and megaloblastic anemia.

*Severe methylenetetrahydrofolate-reductase deficiency* presents mainly with developmental delay, often accompanied by seizures, microcephaly and findings related to cerebrovascular events. Individuals with a *thermolabile variant* of methylenetetrahydrofolate reductase may be at increased risk of cardiovascular disease owing to elevations of plasma homocysteine. Reports of patients with deficiencies of DHF reductase deficiency, methylenetetrahydrofolate cyclohydrolase deficiency or primary methionine synthase deficiency without homocystinuria (i.e., not *cblG*) have not been confirmed and must be considered unproven.

# FOLATE

## Disorders of Absorption and Metabolism of Folate

### Hereditary Folate Malabsorption

#### Clinical Presentation

Patients present in the first months of life with severe megaloblastic anemia, diarrhea, stomatitis, failure to thrive and (usually) progressive neurological deterioration with seizures and sometimes with intracranial calcifications. Peripheral neuropathy has been seen, as have partial defects in humoral and cellular immunity.

#### Metabolic Derangement

All patients have severely decreased absorption of oral folic acid or reduced folates, such as formyltetrahydrofolic acid (folinic acid) or methyltetrahydrofolic acid. These patients provide the best evidence for the existence of a single transport system for folate at both the intestine and the choroid plexus. Transport of folates across other cell membranes is not affected in this disorder, as the hematological and gastrointestinal manifestations are corrected by relatively low levels of folate, and folate metabolism in cultured fibroblasts is not abnormal.

#### Diagnostic Tests

Measurement of serum, red blood cell and cerebrospinal fluid folate levels should be performed, as should a complete blood count and bone marrow analysis. The most important diagnostic features are the severe megaloblastic anemia in the first few months of life, together with low serum folate levels. Excretion of formiminoglutamate and orotic acid may be seen. Cerebrospinal-fluid folate levels remain low even when blood levels are high enough to correct the megaloblastic anemia [64]. Folate absorption may be directly looked for by measuring serum folate levels following an oral dose of between 5 mg and 100 mg of folic acid.

#### Treatment and Prognosis

High-dose oral folic acid (2–40 mg/day) or lower parenteral doses in the physiological range correct the hematologic findings but are less effective in correcting the neurological findings and in raising the level of folate in the cerebrospinal fluid. Both methyltetrahydrofolate or folinic acid (5 mg intramuscularly twice weekly) may be more effective in raising cerebrospinal-fluid levels and have been given in combination with high-dose (40 mg/day) oral folic acid. The clinical response to folates has varied among patients and, in some cases, seizures were worse after folate therapy was started. It is important to maintain blood and cerebrospinal-fluid folate in the normal range. Oral doses of folate may be increased to 100 mg/day if necessary and, if oral therapy does not raise cerebrospinal folate levels, parenteral therapy should be used. Consideration should be given to the use of intrathecal folate therapy if cerebrospinal-fluid levels of folate cannot be raised by other treatments, although there is no experience with the dose of folate that may be required.

#### Genetics

Of the 18 known patients, five have been female. Consanguinity has been noted in four families, and the father of one of the patients had intermediate levels of folate absorption, making autosomal-recessive inheritance likely. A cDNA for a putative intestinal folate transporter has been cloned, and it is identical to that for the reduced folate carrier [65]. To date, no mutations have been searched for in these patients. The defect in hereditary folate malabsorption is not expressed in amniocytes or chorionic villus cells.

### Glutamate-Formiminotransferase Deficiency

#### Clinical Presentation

Over a dozen patients have been described, but the clinical significance of this disorder is still unclear [4, 61–63]. In the severe form of glutamate formiminotransferase deficiency, there is both mental and physical retardation, abnormal electroencephalograms and dilatation of cerebral ventricles with cortical atrophy. Several of the patients had a folate-responsive megaloblastic anemia with macrocytosis and hypersegmentation of neutrophils. Patients ranged in age from 3 months to 42 years. Two had mental retardation, two had seizures and three had delayed speech as their presenting findings, and two were studied because they were the siblings of known patients. In the mild form of glutamate-formiminotransferase deficiency, there is no mental retardation, but there is a greater excretion of formiminoglutamate. Although mental retardation was described in most of the original patients from Japan, of the remaining eight patients, only three showed mental retardation.

#### Metabolic Derangement

Histidine catabolism is associated with a formimino-group transfer to THF, with the subsequent release of ammonia and the formation of 5,10-methenyl-tetrahydrofolate. A single octameric enzyme catalyzes two different activities: glutamate formiminotransferase and formiminotetrahydrofolate cyclodeaminase. These activities are found only in the liver and kidney, and defects in either of these activities will result in formiminoglutamate excretion. It has been suggested (without any direct enzyme measurements) that the severe form of this disease is due to a block in the cyclodeaminase activity, whereas the mild form is due to a block in the formiminotransferase activity.

#### Diagnostic Tests

Elevated formiminoglutamate excretion and elevated levels of formiminoglutamate in the blood following a histidine load help to establish the diagnosis. A complete blood count and bone-marrow examination may detect megaloblastic anemia. Normal to high serum folate levels are found. Although plasma amino acid levels are usually normal, on occasion, hyperhistidinemia and histidinuria have been reported. Two other metabolites that may be found in the urine are hydantoin-5-propionate, a stable oxidation product of the formiminoglutamate precursor, 4-imidazolone-5-propionate and 4-amino-5-imidazolecarboxamide, an intermediate of purine synthesis. Glutamate-formiminotransferase activity is expressed only in liver. It is not expressed in cultured fibroblasts, and there is doubt as to whether it is expressed in red blood cells. The residual activity that has been measured in the livers of five patients has varied from 14% to 54% of control values, and these levels are higher than would be expected for an enzymatic block causing disease.

#### Treatment and Prognosis

It is not clear whether reducing formiminoglutamate excretion is of any clinical value. Although two patients in one family responded to folate therapy by reducing excretion of formiminoglutamate, six others did not. One of two patients responded to methionine supplementation. Pyridoxine and folic acid have been used to correct the megaloblastic anemia in one infant.

#### Genetics

Glutamate-formiminotransferase deficiency has been found in both male and female children of unaffected parents. Consanguinity has not been reported, and there have been no measurements of liver enzyme levels in parents of confirmed patients. It has been presumed that the disease is inherited in an autosomal-recessive manner. Because of the lack of expression of the enzyme in cultured cells, prenatal diagnosis has not been possible, but it may be possible to look directly at formiminoglutamate levels in amniotic fluid. This has not been reported. A cDNA has been cloned for the glutamate formiminotransferase/formiminotetrahydrofolate cyclodeaminase, and molecular analysis of the patients should be possible once the human gene structure has been determined.

### Methylenetetrahydrofolate Reductase Deficiency

#### Clinical Presentation

Approximately 50 patients with severe methylenetetrahydrofolate-reductase deficiency have been described [37, 61, 63, 66–68]. Most commonly, they were diagnosed in infancy, and more than half presented in the first year of life. The most common early manifestation was progressive encephalopathy with apnea, seizures and microcephaly. However, patients became symptomatic at any time from infancy to adulthood and, in the older patients, ataxic gait, psychiatric disorders ("schizophrenia") and symptoms related to cerebrovascular events have been reported. An infant had extreme progressive brain atrophy, and the magnetic resonance image showed demyelination [69]. A 10-year-old boy had findings compatible with those of Angelman syndrome [70]. At least one adult with severe enzyme deficiency was completely asymptomatic. Autopsy findings have included dilated cerebral vessels, microgyria, hydrocephalus, perivascular changes, demyelination, gliosis, astrocytosis and macrophage infiltration. In some patients, thrombosis of both cerebral arteries and veins was the major cause of death. There have been reports of patients with findings similar to those seen in subacute degeneration of the spinal cord due to Cbl deficiency. Of note is the fact that methylenetetrahydrofolate-reductase deficiency is not associated with megaloblastic anemia. In summary: in early infancy, the clinical presentation is characterized by an acute neurological disturbance; in early childhood, a progressive encephalopathy develops, with late stage steps similar to those seen in the adult form, both forms having symptoms related to subacute degeneration of the spinal cord. Individuals with intermediate activity of methylenetetrahydrofolate reductase due to common polymorphisms may be at increased risk for hyperhomocysteinemia and premature vascular disease [71–73].

**Metabolic Derangement**

Methyltetrahydrofolate is the methyl donor for the conversion of homocysteine to methionine and, in methylenetetrahydrofolate-reductase deficiency, the result is an elevation of total plasma homocysteine levels and decreased levels of methionine. The block in the conversion of methylenetetrahydrofolate to methyltetrahydrofolate does not result in the trapping of folates as methyltetrahydrofolate and does not interfere with the availability of reduced folates for purine and pyrimidine synthesis. This explains why patients do not have megaloblastic anemia. It is not clear whether the neuropathology in this disease results from the elevated homocysteine levels, from decreased methionine and resulting interference with methylation reactions or from some other metabolic effect.

**Diagnostic Tests**

Because methyltetrahydrofolate is the major circulating form of folate, serum folate levels may sometimes be low. Homocystinuria is seen in all patients, with a mean of 130 µmol/24 h and a range of 15–667 µmol/24 h. These values are much lower than are seen in cystathionine-synthase deficiency. Plasma methionine levels range from zero to 18 µM (mean = 12 µM, range of control means from different laboratories = 23–35 µM). Although neurotransmitter levels have been measured in only a few patients, they are usually low. Direct measurement of methylenetetrahydrofolate-reductase-specific activity can be performed in liver, leukocytes, lymphocytes and cultured fibroblasts. In cultured fibroblasts, the specific activity is highly dependent on the stage of the culture cycle, with activity highest in confluent cells. There is a rough inverse correlation between the specific activity of the reductase in cultured fibroblasts and the clinical severity. There is a better inverse correlation between clinical severity and either the proportion of total cellular folate that is in the form of methyltetrahydrofolate or the extent of labeled formate incorporation into methionine. The clinical heterogeneity in methylenetetrahydrofolate-reductase deficiency can be seen at the biochemical level. Some of the patients have residual enzyme that is more thermolabile than the control enzyme [74].

Observations of thermolabile reductase have been carried out in the adult population, and it has been postulated that a thermolabile allele of methylenetetrahydrofolate reductase without severe enzyme deficiency and without any of the clinical findings in methylenetetrahydrofolate-reductase deficiency may be an independent risk factor for vascular disease and coronary heart disease. Some of these patients may be ascertained because they have elevated levels of total plasma homocysteine [75, 76]. It has subsequently been shown that, in both the general population and in some patients with severe methylenetetrahydrofolate-reductase deficiency, the thermolabile phenotype is conferred by the presence of a common polymorphism, 677C → T [71, 77]. The role of this polymorphism as a risk factor for vascular disease and for neural tube defect remains a subject of interest and debate.

**Treatment and Prognosis**

It is important to diagnose methylenetetrahydrofolate-reductase deficiency early because, in the infantile forms, the only patients that have done well have been those who have been treated from birth. Early treatment with betaine in oral doses of 3–6 g/day following prenatal diagnosis has resulted in the best outcome [78–80]. Betaine is a substrate for betaine methyltransferase, an enzyme that converts homocysteine to methionine but is mainly active in the liver. Therefore, betaine may be expected to have the doubly beneficial effect of lowering homocysteine levels and raising methionine levels. Because betaine methyltransferase is not present in the brain, the central nervous system effects must be mediated through the effects of the circulating levels of metabolites. The dose of betaine should be modified according to plasma levels of homocysteine and methionine. Other therapeutic agents that have been used in methylenetetrahydrofolate-reductase deficiency include folic acid or reduced folates, methionine, pyridoxine, Cbl and carnitine.

Most of the treatment protocols omitting betaine have not been effective.

## Genetics

Methylenetetrahydrofolate-reductase deficiency is inherited as an autosomal-recessive disorder. There have been multiple affected children of both sexes with either unaffected parents or affected families with consanguinity. Prenatal diagnosis has been reported using amniocytes, and the enzyme is present in chorionic villi. A cDNA has been isolated, and the gene coding for methylenetetrahydrofolate reductase has been localized to chromosome 1p36.3. Over 20 mutations causing severe deficiency have been described, in addition to polymorphisms that result in intermediate enzyme activity and that may contribute to disease in the general population [77, 81–83].

## References

1. Cooper BA, Rosenblatt DS (1987) Inherited defects of vitamin $B_{12}$ metabolism. Annu Rev Nutr 7:291–320
2. Fenton WA, Rosenberg LE (1995) Inherited disorders of cobalamin transport and metabolism. In: Scriver CR, Beaudet AL, Sly WS, Valle D (eds) The metabolic and molecular bases of inherited disease. McGraw-Hill, New York, pp 3129–3150
3. Rosenblatt DS, Cooper BA (1987) Inherited disorders of vitamin $B_{12}$ metabolism. Blood Rev 1:177–182
4. Whitehead VM, Rosenblatt DS, Cooper BA (1998) Megaloblastic anemia. In: Nathan DG, Orkin SH (eds) Hematology of infancy and childhood. Sanders, Philadelphia, pp 385–422
5. Rosenblatt DS, Cooper BA (1990) Inherited disorders of Vitamin $B_{12}$ utilization. Bioessays 12:331–334
6. Yang Y-M, Ducos R, Rosenberg AJ, Catrou PG, Levine JS, Podell ER, Allen RH (1985) Cobalamin malabsorption in three siblings due to an abnormal intrinsic factor that is markedly susceptible to acid and proteolysis. J Clin Invest 76:2057–2065
7. Katz M, Mehlman CS, Allen RH (1974) Isolation and characterization of an abnormal intrinsic factor. J Clin Invest 53:1274–1283
8. Rothenberg SP, Quadros EV, Straus EW, Kapelner S (1984) An abnormal intrinsic factor (IF) molecule: a new cause of "pernicious anemia" (PA). Blood 64:41a
9. Levine JS, Podell ER, Allen RH (1985) Cobalamin malabsorption in three siblings due to an abnormal intrinsic factor that is markedly susceptible to acid and proteolysis. J Clin Invest 76:2057
10. Hewitt JE, Gordon MM, Taggart RT, Mohandas TK, Alpers DH (1991) Human gastric intrinsic factor: characterization of cDNA and genomic clones and localization to human chromosome 11. Genomics 10:432–440
11. Fernandes M, Poirier C, Lespinasse F, Carle GF (1998) The mouse homologues of human GIF, DDBI, and CFL1 genes are located on chromosome 19. Mammalian Genome 9:339–342
12. Gräsbeck R (1972) Familial selective vitamin $B_{12}$ malabsorption. N Engl J Med 287:358 (letter)
13. Broch H, Imerslund O, Monn E, Hovig T, Seip M (1984) Imerslund-Gräsbeck anemia. A long-term follow-up study. Acta Paediatr Scand 73:248–253
14. el Mauhoub M, Sudarshan G, Aggarwal V, Banerjee G (1989) Imerslund-Gräsbeck syndrome in a Libyan boy. Ann Trop Paediatr 9:180–181
15. el Bez M, Souid M, Mebazaa R, Ben Dridi MF (1992) L'anemie d'Imerslund-Gräsbeck. A propos d'un cas. An Pediatr (Paris) 39:305–308
16. Salameh MM, Banda RW, Mohdi AA (1991) Reversal of severe neurological abnormalities after vitamin $B_{12}$ replacement in the Imerslund-Gräsbeck syndrome. J Neur 238 (6):349–350
17. Kulkey O, Reusz G, Sallay P, Miltenyi M (1992) Selective vitamin $B_{12}$ absorption disorder (Imerslund-Gräsbeck syndrome). Orv Hetil 133:3311–3313
18. Gräsbeck R (1997) Selective cobalamin malabsorption and the cobalamin-intrinsic factor receptor. Acta Biochim Polon 44:725–734
19. Liang DC, Hsu HC, Huang FY, Wei KN (1991) Imerslund-Gräsbeck syndrome in two brothers: renal biopsy and ultrastructure findings. Pediatr Hematol Oncol 8:361–365
20. Fyfe JC, Ramanujam KS, Ramaswamy K, Patterson DF, Seetharam B (1991) Defective brush-border expression of intrinsic factor-cobalamin receptor in canine inherited intestinal cobalamin malabsorption. J Biol Chem 266:4489–4494
21. Moestrup SK, Kozyraki R, Kristiansen M, Kaysen JH, Rasmussen HH, Brault D, Pontillon F, Goda FO, Christensen EI, Hammond TG, Verroust PJ (1998) The intinsic factor-vitamin $B_{12}$ receptor and target of teratogenic antibodies is a megalin-binding peripheral membrane protein with homology to developmental proteins. J Biol Chem 273:5235–5242
22. Kozyraki R, Kristiansen M, Silahtaroglu A, Hansen C, Jacobsen C, Tommerup N, Verroust PJ, Moestrup SK (1998) The human intrinsic factor-vitamin $B_{12}$ receptor, *Cubilin*: Molecular characterization and chromosomal mapping of the gene to 10p within the autosomal recessive megaloblastic anemia (MGA1) region. Blood 91:3593–3600
23. Birn H, Verroust PJ, Nexo E, Hager H, Jacobsen C, Christensen EI, Moestrup SK (1997) Characterization of an epithelial ~460-kDa protein that facilitates endocytosis of intrinsic factor-vitamin $B_{12}$ and binds receptor-associated protein. J Biol Chem 272:26497–26504
24. Arminoff M, Tahvanainen E, Gräsbeck R, Weissenbach J, Broch H, de la Chapelle A (1995) Selective intestinal malabsorption of vitamin $B_{12}$ displays recessive mendelian inheritance: assignment of a locus to chromosome 10 by linkage. Am J Hum Genet 57:824–831
25. Carmel R (1983) R-binder deficiency. A clinically benign cause of cobalamin pseudodeficiency. J Am Med Assoc 250:1886–1890
26. Johnston J, Bollekens J, Allen RH, Berliner N (1989) Structure of the cDNA encoding transcobalamin I, a neutrophil granule protein. J Biol Chem 264:5754–5757
27. Johnston J, Yang-Feng T, Berliner N (1992) Genomic structure and mapping of the chromosomal gene for transcobalamin I (TCN1): comparison to human intrinsic factor [published erratum appears in Genomics 1992, 14(1):208]. Genomics 12:459–464
28. Hall CA (1992) The neurologic aspects of transcobalamin II deficiency. Br J Haematol 80:117–120
29. Haurani FI, Hall CA, Rubin R (1979) Megaloblastic anemia as a result of an abnormal transcobalamin II. J Clin Invest 64:1253–1259
30. Seligman PA, Steiner LL, Allen RH (1980) Studies of a patient with megaloblastic anemia and an abnormal transcobalamin II. N Engl J Med 303:1209–1212
31. Rosenblatt DS, Hosack A, Matiaszuk N (1987) Expression of transcobalamin II by amniocytes. Prenat Diagn 7:35
32. Li N, Rosenblatt DS, Kamen BA, Seetharam S, Seetharam B (1994) Identification of two mutant alleles of transcobalamin II in an affected family. Hum Mol Genet 3:1835–1840
33. Li N, Rosenblatt DS, Seetharam B (1994) Nonsense mutations in human transcobalamin II deficiency. Biochem Biophys Res Commun 204:1111–1118
34. Rosenblatt DS, Laframboise R, Pichette J, Langevin P, Cooper BA, Costa T (1986) New disorder of vitamin $B_{12}$ metabolism (cobalamin F) presenting as methylmalonic aciduria. Pediatrics 78:51–54

35. Rosenblatt DS, Hosack A, Matiaszuk NV, Cooper BA, Laframboise R (1985) Defect in vitamin $B_{12}$ release from lysosomes: newly described inborn error of vitamin $B_{12}$ metabolism. Science 228(4705):1319–1321
36. Mitchell GA, Watkins D, Melancon SB, Rosenblatt DS, Geoffroy G, Orquin J, Homsy MB, Dallaire L (1986) Clinical heterogeneity in cobalamin C variant of combined homocystinuria and methylmalonic aciduria. J Pediatr 108: 410–415
37. Ogier de Baulny H, Gerard M, Saudubray JM, Zittoun J (1998) Remethylation defects: guidelines for clinical diagnosis and treatment. Eur J Pediatr 157 [Suppl 2]:S77–S83
38. Traboulsi EI, Silva JC, Geraghty MT, Maumenee IH, Valle D, Green WR (1992) Ocular histopathology characteristics of cobalamin-C type vitamin $B_{12}$ defect with methylmalonic aciduria and homocystinuria. Am J Opthlalmol 113: 269–280
39. Rosenblatt DS, Aspler AL, Shevell MI, Pletcher BA, Fenton WA, Seashore MR (1997) Clinical heterogeneity and progosis in combined methylmalonic aciduria and homocytinuria (cblC). J Inherit Metab Dis 20:528–538
40. Watanabe F, Saido H, Yamaji R, Miyatake K, Isegawa Y, Ito A, Yubisui T, Rosenblatt DS, Nakano Y (1996) Mitochondrial NADH- or NADP-linked aquacobalamin reductase activity is low in human skin fibroblasts with defects in synthesis of cobalamin coenzymes. J Nutr 126:2947–2951
41. Pezacka EH, Rosenblatt DS (1994) Intracellular metabolism of cobalamin. Altered activities of β-axial-ligand transferase and microsomal cob(III)alamin reducatase in cblC and cblD fibroblasts. In: Bhatt HR, James VHT, Besser GM, Bottazzo GF, Keen H (eds) Advances in Thomas Addison's Diseases (Journal of Endocrinology). Bristol, London, pp 315–323
42. Pezacka EH (1993) Identification and characterization of two enzymes involved in the intracellular metabolism of cobalamin. Cyanocobalamin beta-ligand transferase and microsomal cob(III)alamin reductase. Biochim Biophys Acta 1157 (2):167–177
43. Bartholomew DW, Batshaw ML, Allen RH, Roe CR, Rosenblatt D, Valle DL, Francomano CA (1988) Therapeutic approaches to cobalamin-C methylmalonic acidemia and homocystinuria. J Pediatr 112:32–39
44. Carmel R, Bedros AA, Mace JW, Goodman SI (1980) Congenital methylmalonic aciduria-homocystinuria with megaloblastic anemia: observations on response to hydroxocobalamin and on the effect of homocysteine and methionine on the deoxyuridine suppression test. Blood 55:570–579
45. Willard HF, Mellman IS, Rosenberg LE (1978) Genetic complementation among inherited deficiencies of methylmalonyl-CoA mutase activity: evidence for a new class of human cobalamin mutant. Am J Hum Genet 30:1–13
46. Mellman IH, Willard P, Youngdahl-Turner P, Rosenberg LE (1979) Cobalamin coenzyme synthesis in normal and mutant human fibroblasts: evidence for a processing deficiency in cblC Cells. J Biol Chem 254:11847–11853
47. Cooper BA, Rosenblatt DS, Watkins D (1990) Methylmalonic aciduria due to a new defect in adenosylcobalamin accumulation by cells. Am J Hematol 34:115–120
48. Fenton WA, Rosenberg LE (1981) The defect in the cbl B class of human methylmalonic acidemia: deficiency of cob(I)alamin adenosyltransferase activity in extracts of cultured fibroblasts. Biochem Biophys Res Commun 98:283–289
49. Matsui SM, Mahoney MJ, Rosenberg LE (1983) The natural history of the inherited methylmalonic acidemias. N Engl J Med 308:857–861
50. Rosenblatt DS, Cooper BA, Pottier A, Lue-Shing H, Matiaszuk N, Grauer K (1984) Altered vitamin B12 metabolism in fibroblasts from a patient with megaloblastic anemia and homocystinuria due to a new defect in methionine biosynthesis. J Clin Invest 74:2149–2156
51. Schuh S, Rosenblatt DS, Cooper BA, Schroeder ML, Bishop AJ, Seargeant LE, Haworth JC (1984) Homocystinuria and megaloblastic anemia responsive to vitamin $B_{12}$ therapy. An inborn error of metabolism due to a defect in cobalamin metabolism. N Engl J Med 310:686–690
52. Watkins D, Rosenblatt DS (1989) Functional methionine synthase deficiency (cblE and cblG): Clinical and biochemical heterogeneity. Am J Med Genet 34:427–434
53. Watkins D, Rosenblatt DS (1988) Genetic heterogeneity among patients with methylcobalamin deficiency: definition of two complementation groups, cblE and cblG. J Clin Invest 81:1690–1694
54. Harding CO, Arnold G, Barness LA, Wolff JA, Rosenblatt DS (1997) Functional methionine synthase deficiency due to cblG disorder: a report of two patients and a review. Amer J Med Genet 71:384–390
55. Carmel R, Watkins D, Goodman SI, Rosenblatt DS (1988) Hereditary defect of cobalamin metabolism (cblG mutation) presenting as a neurologic disorder in adulthood. N Engl J Med 318:1738–1741
56. Gulati S, Chen Z, Brody LC, Rosenblatt DS, Banerjee R (1997) Defects in auxillary redox proteins lead to functional methionine synthase deficiency. J Biol Chem 272:19171–19175
57. Leclerc D, Wilson A, Dumas R, Gafuik C, Song D, Watkins D, Heng HHQ, Rommens JM, Scherer SW, Rosenblatt DS, Gravel RA (1998) Cloning and mapping of a cDNA for methionine synthase reductase, a flavoprotein defective in patients with homocystinuria. Proc Natl Acad Sci USA 95:3059–3064
58. Gulati S, Baker P, Li YN, Fowler B, Kruger WD, Brody LC, Banerjee R (1996) Defects in human methionine synthase in cblG patients. Hum Mol Genet 5 (12):1859–1865
59. Leclerc D, Campeau E, Goyette P, Adjalla CE, Christensen B, Ross M, Eydoux P, Rosenblatt DS, Rozen R, Gravel RA (1996) Human methionine synthase: cDNA cloning and identification of mutations in patients of the cblG complementation group of folate/cobalamin disorders. Hum Mol Genet 5 (12):1867–1874
60. Wilson A, Leclerc D, Saberi F, Phillips III JA, Pfotenhauer JP, Roback E, Rosenblatt DS, Gravel RA (1997) Causal mutations in siblings with the cblG variant form of methionine synthase deficiency. Am J Hum Genet 61:A263 (abstract)
61. Erbe RW (1986) Inborn errors of folate metabolism. In: Blakley RL, Whitehead VM (eds) Nutritional, pharmacological and physiological aspects, vol 3. Wiley, New York, p 413
62. Erbe RW (1979) Genetic aspects of folate metabolism. Adv Hum Genet 9:293–354
63. Rosenblatt DS (1995) Inherited disorders of folate transport and metabolism. In: Scriver CR, Beaudet AL, Sly WS, Valle D (eds) The metabolic and molecular basis of inherited disease. McGraw-Hill, New York, pp 3111–3128
64. Urbach J, Abrahamov A, Grossowicz N (1987) Congenital isolated folic acid malabsorption. Arch Dis Child 62:78–80
65. Nguyen TT, Dyer DL, Dunning DD, Rubin SA, Grant KE, Said HM (1997) Human intestinal folate transport: cloning, expression, and distribution of complementary RNA. Gastroenterology 112:783–791
66. Visy JM, Le Coz P, Chadefaux B, Fressinaud C, Woimant F, Marquet J, Zittoun J, Visy J, Vallat JM, Haguenau M (1991) Homocystinuria due to 5,10-methylenetetrahydrofolate reductase deficiency revealed by stroke in adult siblings (see comments). Neurology 41:1313–1315
67. Haworth JC, Dilling LA, Surtees RAH, Seargeant LE, Lue-Shing H, Cooper BA, Rosenblatt DS (1993) Symptomatic and asymptomatic methylenetetrahydrofolate reductase deficiency in two adult brothers. Am J Med Genet 45:572–576
68. Fowler B (1998) Genetic defects of folate and cobalamin metabolism. Eur J Pediatr 157:S60–S66
69. Sewell AC, Neirich U, Fowler B (1998) Early infantile methylenetetrahydrofolate reductase deficiency: a rare cause of progressive brain atrophy. J Inherit Metab Dis 21:22 (abstract)
70. Arn PH, Williams CA, Zori RT, Driscoll DJ, Rosenblatt DS (1998) Methylenetetrahydrofolate reductase deficiency in a patient with phenotypic findings of Angelman syndrome. Am J Med Genet 77:198–200
71. Frosst P, Blom HJ, Milos R, Goyette P, Sheppard CA, Matthews RG, Boers GJH, den Heijer M, Kluijtmans LAJ,

van den Heuvel LP, Rozen R (1995) A candiate genetic risk factor for vascular disease: a common methylenetetrahydrofolate reductase mutation causes thermoinstability. Nat Genet 10:111–113
72. van der Put NMJ, Gabreels F, Stevens EMB, Smeitink JAM, Trijbels FJM, Eskes TKAB, van der Heuvel LP, Blom HJ (1998) A second common mutation in the methylenetetrahydrofolate reductase gene: an additional risk factor for neural-tube defects? Am J Hum Genet 62:1044–1051
73. Weisberg I, Tran P, Christensen B, Sibani S, Rozen R (1998) A second genetic polymorphism in methylenetetrahydrofolate reductase (MTHFR) associated with decreased enzyme activity. Mol Genet Metab 64:169–172
74. Rosenblatt DS, Lue-Shing H, Arzoumanian A, Low-Nang L, Matiaszuk N (1992) Methylenetetrahydrofolate reductase (MR) deficiency: thermolability of residual MR activity, methionine synthase activity, and methylcobalamin levels in cultured fibroblasts. Biochem Med Met Biol 47(3):221–225
75. Kang S, Zhou J, Wong P, Kowalisyn J, Strokosch G (1988) Intermediate homocysteinemia: a thermolabile variant of methylenetetrahydrofolate reductase. Am J Hum Genet 43:414–421
76. Kang S-S, Wong PWK, Susmano A, Sora J, Norusis M, Ruggie N (1991) Thermolabile methylenetetrahydrofolate reductase: an inherited risk factor for coronary artery disease. Am J Hum Genet 48:536–545
77. Goyette P, Christensen B, Rosenblatt DS, Rozen R (1996) Severe and mild mutations in cis for the methylenetetrahydrofolate reductase (MTHFR) gene, and description of 5 novel mutations in MTHFR. Am J Hum Genet 59:1268–1275
78. Wendel U, Bremer HJ (1983) Betaine in the treatment of homocystinuria due to 5,10-methylene THF reductase deficiency. J Pediatr 103:1007
79. Holme E, Kjellman B, Ronge E (1989) Betaine for treatment of homocystinuria caused by methylenetetrahydrofolate reductase deficiency. Arch Dis Child 64:1061–1064
80. Ronge E, Kjellman B (1996) Long term treatment with betaine in methylenetetrahydrofolate reductase deficiency. Arch Dis Child 74:239–241
81. Goyette P, Sumner JS, Milos R, Duncan AM, Rosenblatt DS, Matthews RG, Rozen R (1994) Human methylenetetrahydrofolate reductase: isolation of cDNA, mapping and mutation identification. Nat Genet 7(2):195–200
82. Goyette P, Frosst P, Rosenblatt DS, Rozen R (1995) Seven novel mutations in the methylenetetrahydrofolate reductase gene and genotype/phenotype correlations in severe methylenetetrahydrofolate reductase deficiency. Am J Hum Genet 56:1052–1059
83. Kluijtmans LAJ, Wendel U, Stevens EMB, van den Heuvel LPWJ, Trijbels FJM, Blom HJ (1998) Indentification of four novel mutations in severe methylenetetrahydrofolate reductase deficiency. Eur J Hum Genet 6:257–265

# PART VI
# NEUROTRANSMITTER AND SMALL PEPTIDE DISORDERS

## Neurotransmitters

The neurotransmitter systems can be divided into mainly inhibitory aminoacidergic [γ-aminobutyric acid (GABA) and glycine], excitatory aminoacidergic (aspartate and glutamate), cholinergic (acetylcholine), monoaminergic (mainly adrenaline, noradrenaline, dopamine, and serotonin), and purinergic (adenosine and adenosine mono-, di-, and triphosphate). A rapidly growing list of peptides are also considered putative neurotransmitters.

GABA is formed from glutamic acid by glutamic acid decarboxylase. It is catabolized into succinic acid through the sequential action of two mitochondrial enzymes, GABA transaminase and succinic semialdehyde dehydrogenase. All these enzymes require pyridoxal phosphate as a coenzyme.

A major inhibitory neurotransmitter, GABA is present in high concentrations in the central nervous system, predominantly in the gray matter. GABA mainly modulates brain activity locally, by release from interneurons. On depolarization, neuronal GABA undergoes calcium-dependent release into the synaptic cleft. At the postsynaptic membrane, GABA binds to sodium-independent, high-affinity, mostly $GABA_A$ receptors.

GLYCINE, a non-essential amino acid, is not only an essential intermediate in many metabolic processes but is also one of the major inhibitory neurotransmitters in the central nervous system. In addition, glycine has an excitation function at the N-methyl-D-aspartate receptor. The inhibitory *glycine receptors* are mostly found in the brain stem and spinal cord. This receptor is a member of the ligand-gated ion-channel superfamily, which also includes the $GABA_A$ receptor, serotonin receptors, and the excitatory nicotinic-acetylcholine-receptor superfamily. The receptor complex consists of five subunits (three α subunits and two β subunits) that form a ring with a central chloride-conducting channel. Each subunit comprises a N-terminal extracellular domain with the ligand binding sites, and four membrane-spanning domains.

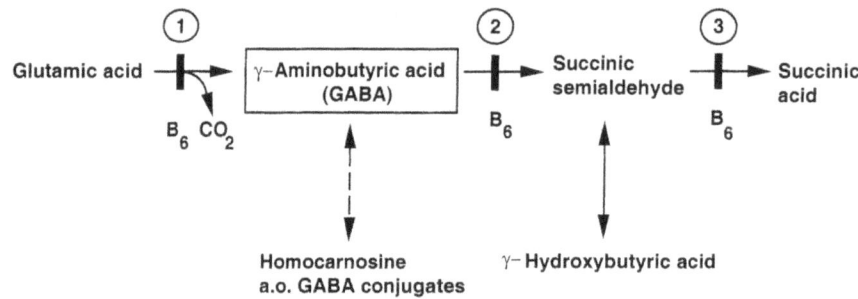

**Fig. 26.1.** Brain metabolism of γ-aminobutyric acid (GABA). $B_6$ pyridoxal phosphate. 1, Glutamic acid decarboxylase; 2, GABA transaminase; 3, succinic semialdehyde dehydrogenase. *Dotted arrow* indicates reactions postulated. Enzyme defects are depicted by *solid bars*

CHAPTER 26, PART 1

# Disorders of Neurotransmission

J. Jaeken, C. Jakobs, and R. Wevers

CONTENTS

Inborn Errors of Gamma Amino Butyric
Acid Metabolism.................................. 301
   Pyridoxine-Responsive and -Unresponsive Putative
   Glutamic Acid Decarboxylase Deficiency............ 301
   Gamma Amino Butyric Acid Transaminase
   Deficiency...................................... 302
   Succinic Semialdehyde Dehydrogenase Deficiency..... 303

Inborn Errors of Glycine Receptors................. 304
   Hyperekplexia.................................. 304
Inborn Errors of Monoamine Metabolism............ 307
   Tyrosine Hydroxylase Deficiency.................. 307
   Aromatic L-Aminoacid Decarboxylase Deficiency...... 307
   Monoamine Oxidase-A Deficiency................. 308
   Guanosine Triphosphate Cyclohydrolase I Deficiency .. 308
Folinic Acid Responsive Seizures................... 309
References....................................... 309

> This review deals with (1) hereditary diseases in the metabolism of γ-aminobutyric acid (GABA) and of a glycine receptor and (2) some inborn errors of monoamine metabolism. Disorders of the metabolism of glycine are treated in Chap. 21. A putative deficiency of glutamic-acid decarboxylase is a rare cause of early or late infantile *pyridoxine-responsive convulsions*. Two other defects of GABA catabolism are: the very rare, severe, and untreatable *GABA transaminase deficiency* and the much more frequent *succinic semialdehyde dehydrogenase (SSADH) deficiency* which, to some extent, responds to GABA transaminase inhibition. *Hyperekplexia* is a dominantly inherited defect of the $\alpha_1$ subunit of the glycine receptor; it is a cause of excessive startle responses treatable with clonazepam.
>
> Four disorders of monoamine metabolism are discussed: *monoamine-oxidase-A (MAO-A) deficiency* mainly causes behavioral disturbances; no efficient treatment is known. *Tyrosine-hydroxylase (TH) deficiency* is an extrapyramidal disorder responding very well to dihydroxyphenylalanine (L-Dopa). The two other disorders of monoamine metabolism involve both catecholamine and serotonin metabolism. *Guanosine triphosphate (GTP)-cyclohydrolase-I (GTPCH-I) deficiency* is a defect upstream of L-Dopa and 5-hydroxytryptophan (5-HTP) and, therefore, is efficiently treatable with these compounds. *Aromatic-amino-acid-decarboxylase (AADC) deficiency* is located downstream of these intermediates. Its treatment is more difficult and less efficient.

## Inborn Errors of γ-Aminobutyric Acid Metabolism

Three genetic diseases due to a defect in brain GABA metabolism have been reported: a putative glutamic-acid-decarboxylase deficiency and two defects in GABA catabolism, GABA-transaminase deficiency and SSADH deficiency (Fig. 26.1).

## Pyridoxine-Responsive and -Unresponsive Putative Glutamic Acid Decarboxylase Deficiency

Pyridoxine-responsive putative glutamic acid decarboxylase deficiency (*pyridoxine-responsive convulsions*) was first reported in 1954 [1]. It is a rare cause of convulsions in early childhood, with less than 100 probands having been reported [2, 3].

Recently, indirect evidence was presented for pyridoxine-unresponsive putative glutamic acid decarboxylase deficiency in infants with a "stiff baby-like" syndrome and convulsions. This is not further discussed in this chapter.

### Clinical Presentation

The clinical picture of typical pyridoxine-responsive convulsions has to be differentiated from the more recently identified atypical presentation. The typical form satisfies the following criteria:

- Onset of convulsions before or shortly after birth
- Rapid response to pyridoxine

- Refractoriness to other anticonvulsants
- Dependence on a maintenance dose
- Absence of pyridoxine deficiency

The disease may start as intrauterine convulsions as early as in the fifth month of pregnancy. Some patients suffered from peripartal asphyxia, probably as a consequence of this disorder. The seizures are intermittent at onset but may proceed to status epilepticus. All types of seizures can be observed; most are long-lasting seizures and repeated status epilepticus, but brief convulsions (generalized or partial), atonic attacks, and infantile spasms can also occur. There is pronounced hyperirratibility that can alternate with flaccidity. Abnormal eye movements are often reported (nystagmus, "rolling" eyes, miosis, and/or poor reaction of the pupils to light). The atypical presentation [4] differs from the typical one as follows:

- Later onset of the attacks (up to the age of about 18 months)
- Prolonged seizure-free intervals without pyridoxine (for as long as 5 months)
- The need for larger pyridoxine doses in some patients

### Metabolic Derangement

Pyridoxine-responsive convulsions are considered to be due, at least in some patients, to brain GABA deficiency resulting from a genetic defect at the pyridoxal phosphate coenzyme-binding site of glutamic acid decarboxylase, the rate-limiting enzyme in GABA synthesis. Brain and CSF GABA were found to be low in the rare cases where it was measured [5, 6]. The putative substrate, glutamate, has been reported to be increased [7] or normal [8] in CSF. In those patients in whom CSF glutamate levels are normal, the basic defect is probably not at the level of glutamate decarboxylase.

### Diagnostic Tests

The diagnosis rests on the clinical response to *pyridoxine*. A trial of pyridoxine should be performed in all unclear seizure disorders with onset before the age of about 18 months. Results of CSF free-GABA determinations should not be waited for, as these are mostly not readily available.

### Treatment and Prognosis

The disease promptly responds to pyridoxine (an intravenous test dose of 100 mg is recommended [3]) but is refractory to other antiepileptic medications.. The recommended daily maintenance dose is 5–10 mg/kg. It should be noted that oral or intravenous test doses of pyridoxine may result in prolonged neurological and respiratory depression [3]. Therefore, it is recommended that these patients be monitored in an intensive-care unit. Treatment with isoniazid increases the minimum effective dose. The convulsions cease within a few minutes when pyridoxine is administered parenterally and within a few hours when it is given orally. The effect of a single dose (mostly 2–5 days) remains constant in the same patient. When treatment is interrupted, the seizures return, although there might be exceptions to this rule (due to delayed maturation of enzyme activity?) [6]. In the case of (suspected) *intrauterine convulsions*, treatment of the mother with pyridoxine is effective (~100 mg/day). In the absence of early appropriate treatment, severe psychomotor retardation is the rule and, if untreated, the disease runs a fatal course, at least in the neonatal form.

### Genetics

In its typical form, the disease has an autosomal-recessive inheritance, and there is evidence that this also holds true for the later-onset presentation.

## γ Amino Butyric Acid Transaminase Deficiency

GABA transaminase deficiency was first reported in 1984 in a brother and sister from a Flemish family [9]. Only one other patient has been identified [W. Nyhan, personal communication].

### Clinical Presentation

The two siblings showed feeding difficulties from birth, often necessitating gavage feeding. They had a pronounced axial hypotonia and generalized convulsions. A high-pitched cry and hyperreflexia were present during the first 6–8 months. Further evolution was characterized by lethargy and psychomotor retardation (the developmental level of 4 weeks was never attained). Corneal reflexes and the reaction of the pupils to light remained normal. A remarkable, continued acceleration of length growth was noted from birth until death. This was explained by increased fasting plasma growth-hormone levels (8–39 ng/ml; normal < 5 ng/ml); these could be suppressed by oral glucose. In one of the patients, head circumference showed a rapid increase during the last 6 weeks of life (from the 50th to the 97th percentiles). Postmortem examination of the brain showed a spongiform leukodystrophy.

The third patient presented with seizures from birth, severe hypotonia, hyperreflexia, and brain-stem dysfunction. Growth was normal. Electroencephalography showed a burst-suppression pattern, and magnetic resonance imaging (MRI) showed cerebellar hypoplasia and agenesis of the corpus callosum.

### Metabolic Derangement

The CSF and plasma concentrations of GABA, GABA conjugates, and β-alanine in one of the siblings are shown in Table 26.1. Liver GABA and β-alanine concentrations were normal. This metabolite pattern could be explained by a decrease in GABA transaminase activity in the liver (and lymphocytes). Intermediate levels were found in the healthy sibling, the father, and the mother. It can be assumed that the same enzymatic defect exists in the brain, since GABA transaminases of human brain and of peripheral tissues have the same kinetic and molecular properties. β-Alanine seems to be an alternative substrate for GABA transaminase, hence its increase in this disease.

In this context, it can be mentioned that the antiepileptic drug γ-vinyl-GABA (*vigabatrin*) causes an irreversible inhibition of GABA transaminase, leading to two- to threefold increases in CSF free GABA. Interestingly, this drug also significantly decreases serum glutamic pyruvic transaminase but not glutamic oxaloacetic transaminase activity.

### Diagnostic Tests

The diagnosis requires amino acid analysis of the CSF. Due to enzymatic homocarnosine degradation, free-GABA levels in the CSF show artifactual increases unless samples are deep-frozen (at −20°C) within a few minutes when analysis is performed within a few weeks and at −70°C if the time until analysis is longer. Control CSF free-GABA levels range from about 40 nmol/l to 150 nmol/l after the age of 1 year and are lower in younger children. Because of these low levels, sensitive techniques, such as ion-exchange chromatography and fluorescence detection [10] or a stable-isotope-dilution technique [11], have to be used. Enzymatic confirmation can be obtained in lymphocytes, lymphoblasts, and liver. As for prenatal diagnosis, GABA transaminase activity is not expressed in fibroblasts, but activity is present in chorionic villus tissue [12].

### Treatment and Prognosis

We obtained no clinical or biochemical response after administration of either pharmacological doses of pyridoxine (the precursor of the coenzyme of GABA transaminase) or with picrotoxin, a potent, noncompetitive GABA antagonist. The siblings died at the ages of 1 year and 2 years and 7 months, respectively.

### Genetics

Inheritance is autosomal recessive. A missense mutation has been identified in one of the siblings, resulting in substitution of arginine 220 by lysine [13]. The father of the patient (but not the mother) carried this allele, indicating that the patient was a compound heterozygote. The gene for GABA transaminase has not yet been assigned.

## Succinic Semialdehyde Dehydrogenase Deficiency

SSADH deficiency was first reported as γ-hydroxybutyric aciduria in 1981 [14]. It has been documented in at least 150 patients, with approximately 60 of them reported [15].

### Clinical Presentation

The clinical presentation is nonspecific and varies from mild to severe. It comprises psychomotor retardation, delayed speech development, hypotonia, ataxia and, less frequently, hyporeflexia, convulsions, aggressive behavior, hyperkinesis, oculomotor apraxia, choreoathetosis, and nystagmus. Ataxia, when present, may resolve with age. MRI showed basal-ganglia abnormalities in some patients.

### Metabolic Derangement

The key feature is an accumulation of γ-hydroxybutyrate in urine, plasma, and CSF (Fig. 26.1). γ-Hydroxybutyrate and GABA are neuropharmacologically active compounds. The accumulation of γ-hydroxybutyrate tends to decrease with age. Metabolites indicative of the β- and α-oxidation of γ-hydroxybutyric acid may be variably detected in the urine of SSADH-deficient patients. The identification of other metabolites in the urine of SSADH-deficient patients related to pathways of fatty acid, pyruvate, and glycine metabolism suggests that the deficiency has metabolic consequences beyond the pathway of GABA metabolism.

### Diagnostic Tests

Diagnosis is made by organic-acid analysis of urine, plasma, and/or CSF. Pitfalls in this diagnosis are the instability of γ-hydroxybutyrate in urine and the variable excretion pattern of this compound which, in some patients, is only marginally increased. The enzyme deficiency can be demonstrated in lymphocytes and lymphoblasts. Residual SSADH activity measured in extracts of cultured cells has been less than 5% of control values in all patients, and parents have intermediate levels of enzyme activity [16]. SSADH activity is expressed in normal human fibroblasts (low activity), liver, kidney, and brain, and SSADH deficiency in these tissues has been demonstrated. Prenatal diagnosis can be accurately performed using both (1) isotope-dilution mass spectrometry to measure γ-hydroxybutyric acid levels in amniotic fluid and (2) determination of SSADH activity in amniocytes or chorionic villus tissue.

### Treatment and Prognosis

In an attempt to reduce the accumulation of γ-hydroxybutyrate, we introduced a novel treatment principle: inhibition of the preceding enzymatic step. This was realized by giving vigabatrin, an irreversible inhibitor of GABA transaminase, in doses of 50–100 mg/kg/day (divided into two daily doses) [17]. This treatment was shown to reduce CSF γ-hydroxybutyrate levels and, in the majority of patients, it was associated with variable improvement particularly of ataxia, behavior and manageability [18, 19]. However, long-term administration of vigabatrin should be monitored closely, because this drug increases CSF (and probably also brain) GABA levels.

As to prognosis, this disease can manifest a mild to severe neurological course. Some patients have died, although there was no evidence for metabolic acidosis or decompensation.

### Genetics

The mode of inheritance is autosomal recessive. The human gene maps to chromosome 6p22. In four patients from two families, two exon-skipping mutations have been identified [20].

## Inborn Errors of Glycine Receptors

### Hyperekplexia

### Clinical Presentation

Hyperekplexia, or "startle disease" seems to have been reported first in 1958 [21]. Three main symptoms are required for the diagnosis [22]. The first is a generalized stiffness immediately after birth, which normalizes during the first years of life; the stiffness increases with handling and disappears during sleep. The second feature is an excessive startle reflex to unexpected stimuli (particularly auditory stimuli) from birth on. This causes frequent falling. The third is a short period of generalized stiffness (during which voluntary movements are impossible) following the startle response. Associated features may occur, particularly periodic limb movements during sleep and hypnagogic myoclonus. Other symptoms are inguinal, umbilical, or epigastric herniations, congenital hip dislocation, and epilepsy.

Psychomotor development is mostly normal. MRI has provided evidence for frontal neuronal dysfunction [23].

### Metabolic Derangement

Most patients show defects in the glycine receptor, especially in the $\alpha_1$ subunit [24]. In patients with unexplained hyperekplexia, defects in the other subunits are being looked for but have not yet been found [25]. The finding of mutations in the β subunit of the glycine receptor in a mouse model of hyperekplexia suggests the existence of further defects in man [26]. In a few patients, decreased CSF GABA levels have been reported. The relationship to the glycine receptor defect is not clear [27].

### Diagnostic Tests

The diagnosis is based on the response to medication: the benzodiazepine clonazepam reduces the frequency and magnitude of startle responses and diminishes the frequency of falls. This drug binds to the benzodiazepine site of the $GABA_A$ receptor [28].

### Treatment and Prognosis

The stiffness decreases during the first years of life, but the excessive startle responses remain. Clonazepam significantly reduces the frequency and magnitude of the startle responses but has less effect on the stiffness. The mechanism of the beneficial effect of clonazepam is not known.

### Genetics

Hyperekplexia has, in the great majority of the patients, an autosomal-dominant inheritance with nearly complete penetrance and variable expression in most pedigrees. The gene for the glycine $\alpha_1$-subunit receptor is located on chromosome 5q33. At least six mutations have been identified, most of them point mutations [22, 27, 29].

# CHAPTER 26, PART 2

## Monamines

The monoamines, adrenaline, noradrenaline, dopamine, and serotonin, are metabolites of the amino acids tyrosine and tryptophan. The first step in their formation is catalyzed by amino-acid-specific hydroxylases, which require tetrahydrobiopterin ($BH_4$) as a cofactor. $BH_4$ is also a cofactor of phenylalanine hydroxylase (Fig. 14.1). Its synthesis from GTP is initiated by the rate-limiting GTPCH-I, which forms dihydroneopterin triphosphate. L-Dopa and 5-HTP are metabolized by a common $B_6$-dependent AADC into, respectively, dopamine (the precursor of the catecholamines, adrenaline and noradrenaline) and serotonin (5-hydroxytryptamine). Adrenaline and noradrenaline are catabolized into vanillylmandelic acid (VMA) and 3-methoxy-4-hydroxyphenylethyleneglycol (MHPG) via MAO-A. This enzyme also intervenes in the catabolism of both dopamine into homovanillic acid (HVA) via 3-methoxytyramine and of serotonin into 5-hydroxyindoleacetic acid (5-HIAA). Dopaminergic modulation of ion fluxes regulates emotion, activity, behavior, nerve conduction, and the release of a number of hormones via G-protein-coupled cell-surface dopamine receptors. Serotinergic neurotransmission modulates body temperature, blood pressure, endocrine secretion, appetite, sexual behavior, movement, emesis, and pain.

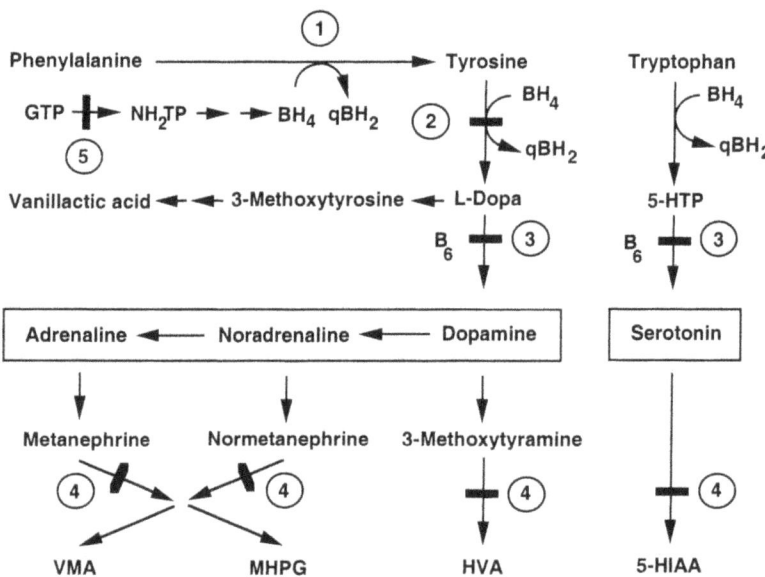

**Fig. 26.2.** Metabolism of adrenaline, noradrenaline, dopamine, and serotonin. *5-HIAA*, 5-hydroxyindoleacetic acid; *5-HTP*, 5-hydroxytryptophan; $B_6$, pyridoxal phosphate; $BH_4$, tetrahydrobiopterin; *L-Dopa*, L-dihydrophenylalanine; *GTP*, guanosine triphosphate; *HVA*, homovanillic acid; *MHPG*, 3-methoxy-4-hydroxyphenylethyleneglycol; $NH_2TP$, dihydroneopterin triphosphate; $qBH_2$, quininoid dihydrobiopterin; *VMA*, vanillylmandelic acid. *1*, Phenylalanine hydroxylase; *2*, tyrosine hydroxylase; *3*, aromatic L-amino acid decarboxylase; *4*, monoamine oxidase A; *5*, GTP cyclohydrolase 1. Enzyme defects covered in this chapter are depicted by *solid bars*

## Inborn Errors of Monoamine Metabolism

### Tyrosine Hydroxylase Deficiency

#### Clinical Presentation

To date, only a few patients with genetically confirmed TH deficiency have been reported [30–35]. All presented in the first year of life with extrapyramidal signs and symptoms. This can either be a dystonia in the lower extremities or a severe hypokinesia with generalized rigidity (a hypokinetic-rigid Parkinsonian syndrome). A few patients were labeled as having cerebral palsy before diagnosis. A diurnal fluctuation of the clinical symptoms can be observed in some patients.

#### Metabolic Derangement

TH converts tyrosine into L-Dopa, the direct precursor of catacholamine biosynthesis (Fig. 26.2). This enzymatic step is rate-limiting in the biosynthesis of the catecholamines. The enzyme is expressed in the brain and in the adrenals. $BH_4$ is a cofactor. The biochemical hallmarks of the disease are low CSF levels of homovanillic acid (HVA) and 3-methoxy-4-hydroxyphenylethyleneglycol (MHPG) (the catabolites of dopamine and norepinephrine, respectively) with normal 5-hydroxyindoleacetic acid (5-HIAA) levels in the CSF. The serotonin metabolism is unaffected.

#### Diagnostic Tests

The most important diagnostic test is the measurement of HVA, MHPG, and 5-HIAA in the CSF [34]. As there is a lumbosacral gradient in the concentration of HVA and 5-HIAA, care should be taken that the measurements are carried out in a standardized CSF volume fraction in a laboratory that is experienced in this field. Urinary measurements of HVA and 5-HIAA are not reliable in diagnosing the defect. Determination of amino acids in body fluids (including CSF) does not contribute diagnostically, as phenylalanine and tyrosine generally are normal in all body fluids of patients [34]. Direct enzyme measurement is not a diagnostic option, as there is no enzyme activity detectable in body fluids, blood cells and fibroblasts. The finding of elevated prolactin in blood as a measure of dopamine deficiency may be helpful but it is unclear whether normal prolactin levels actually exclude the diagnosis.

#### Treatment and Prognosis

Patients can be treated successfully with low-dose L-Dopa (initial dose 3 mg/kg/day in three divided doses) together with an L-Dopa-decarboxylase inhibitor. Several patients on this therapy have been followed for up to 3 years [unpublished observation]. The motor performance improved very significantly in most cases. There are no data available yet on the long-term effects of life-long low-dose L-Dopa treatment in these patients.

#### Genetics

TH deficiency is inherited as an autosomal-recessive trait. The TH gene is located on chromosome 11p15.5. Four different mutations have been described in the TH gene [30–33, 35]. One of these is a "common" mutation in the Dutch population [35].

### Aromatic L-Aminoacid Decarboxylase Deficiency

#### Clinical Presentation

Symptoms of AADC deficiency often present in the first months of life. Patients show an extrapyramidal movement disorder with generalized hypotonia and often with oculogyric crises and developmental delay [36–38].

#### Metabolic Derangement

AADC is implicated in two metabolic pathways: the biosynthesis of catecholamines and of serotonin (Fig. 26.2). The activity of the homodimeric enzyme requires pyridoxal phosphate as a cofactor. A deficiency of the enzyme results in a deficiency of the catecholamines and of serotonin. The concentrations of the catabolites (HVA from dopamine, 5-HIAA from serotonin, and MHPG in the central nervous system from norepinephrine) are severely reduced in the CSF. Another biochemical hallmark of the disease is the increased concentration of metabolites upstream of the metabolic block: L-Dopa, 3-methoxytyrosine, vanillactic acid, and 5-HTP.

#### Diagnostic Tests

L-Dopa, 3-methoxytyrosine, vanillactic acid, and 5-HTP can be found elevated in urine, CSF and plasma. Also, low concentrations of HVA, 5-HIAA, and MHPG in CSF may lead to the diagnosis [37]. Just as in TH deficiency, these CSF measurements require the cooperation of a laboratory with experience in this field. AADC deficiency can easily be confirmed at the enzyme level, as all patients have also shown a deficiency of the enzyme in plasma.

## Treatment and Prognosis

Treatment with pyridoxine, the AADC cofactor, seems to have no clinical effect, but it may lower the abnormal metabolite levels. Treatment with bromocriptine (a dopamine-D2 agonist) or tranylcypromine (a MAO inhibitor) stopped the abnormal eye movements [36]. A combined treatment of pyridoxine, bromocriptine, and tranylcypromine produced a sustained improvement in muscle tone and voluntary movements. Only moderate clinical improvement was observed in a patient that started the therapy (pyridoxine, selegiline as a MAO inhibitor, and bromocriptine) during the fifth year of life [38]. Patients who were treated since the first year of life seem to react better to the therapy.

## Genetics

AADC deficiency is inherited as an autosomal-recessive trait. Mutations in the gene have been reported in an abstract [39].

# Monoamine Oxidase-A Deficiency

## Clinical Presentation

MAO-A deficiency has been identified in five generations in one Dutch family [40, 41]. Only males were affected. They showed borderline mental retardation with prominent behavioral disturbances, including aggressive and sometimes violent behavior, arson, attempted rape, and exhibitionism. The patients were non-dysmorphic and showed a tendency towards stereotyped hand movements, such as hand wringing, plucking, or fiddling. Growth was normal. All females in the family functioned normally.

MAO exists as two isoenzymes (A and B). The genes encoding for both isoenzymes are located on the X-chromosome. Patients with contiguous gene syndrome affecting both the MAO-A and -B genes and also the gene responsible for Norrie disease have been described [42]. They are severely mentally retarded and blind. Patients with only the MAO-B and the Norrie genes affected were also found. These patients are not mentally retarded and do not have abnormalities in catecholamine metabolites in the urine. Elevated excretion of phenylethylamine as a specific MAO-B substrate is a consistent finding in patients where the MAO-B gene is involved in the contiguous-gene syndrome. Here, we address only the isolated deficiency of MAO-A.

## Metabolic Derangement

In patients with MAO-A deficiency, marked elevations were noted in the MAO substrates serotonin, normetanephrine, 3-methoxytyramine, and tyramine in urine (Fig. 26.2). The concentrations of the metabolites downstream of the metabolic block, VMA, HVA, 5-HIAA, and MHPG, were reduced [43]. As platelet MAO-B activity was found to be normal, these results are consistent with a deficiency of MAO-A, the isoenzyme found in neural tissue.

## Diagnostic Tests

A characteristically abnormal excretion pattern of biogenic amine metabolites is present in random urine samples of the patients [43]. The deficiency can be diagnosed at the metabolite level by finding elevated urinary serotonin, normetanephrine, metanephrine, and 3-methoxytyramine [43]. The ratios in urine of normetanephrine to VMA or normetanephrine to MHPG are abnormally high in patients with the defect [43].

The ratio HVA/VMA in urine (patients > 4) may also form a first indication for this diagnosis. In such cases, subsequent measurement of normetanephrine remains essential. The discovery of this disorder suggests that it might be worthwhile to perform systematic urinary monoamine analysis when diagnosing unexplained significant *behavior disturbances*, particularly when these occur in several male family members.

## Treatment and Prognosis

No effective treatment is known at present. Both the borderline mental retardation and the behavioral abnormalities seem to be stable with time. No patient has been institutionalized because of the mental retardation.

## Genetics

The locus for this X-linked inherited disease has been assigned to Xp11.21. A point mutation in the eighth exon of the MAO-A gene, causing a premature truncation of the protein, has been found in this family [40, 41].

# Guanosine Triphosphate Cyclohydrolase-I Deficiency

## Clinical presentation

In 1994, GTPCH-I was identified as the first causative gene for Dopa-responsive dystonia [44]. Patients with this deficiency start symptomatology during the first decade of life. Dystonia in the lower limbs is generally considered to be the initial and most prominent symptom. Unless treated with L-Dopa, the dystonia becomes generalized. Diurnal fluctuation of the symptoms with improvement after sleep is a feature in most patients. The clinical phenotype is still expanding [45]. Recently, onset of symptomatology in the first week of life was reported. The disease may also present with focal dystonia, dystonia with a relapsing and remitting course, or hemidystonia.

### Metabolic Derangement

GTPCH-I is the initial and rate-limiting step in the biosynthesis of $BH_4$, the essential cofactor of various aromatic amino acid hydroxylases, such as phenylalanine hydroxylase, TH, and tryptophan hydroxylase (Fig. 26.2). Therefore, the deficiency state of the enzyme is characterized by defective biosynthesis of serotonin and the catecholamines. For the other defects of biopterin metabolism, see Chap. 14.

### Diagnostic Tests

Some patients with the recessively inherited form of the disease may be diagnosed through the high phenylalaninemia found on neonatal screening. Patients with dominantly inherited GTPCH-1 deficiency, however, escape detection during the newborn screening, as they have normal phenylalanine levels in body fluids. Several tests may play a role in reaching the correct diagnosis in such patients. These are:

1. Measurement of urinary pterins (biopterin and neopterin).
2. Measurement of CSF HVA and 5-HIAA. The determination should be carried out by a laboratory with experience in the field, having age-related and CSF volume-fraction related reference values. A normal or slightly low CSF HVA in combination with a low or slightly low 5-HIAA may be meaningful.
3. An oral phenylalanine-loading test.
4. Measurement of the enzyme in a liver biopsy.

### Treatment and Prognosis

Optimal treatment of patients would be life-long supplementation of the defective cofactor $BH_4$. However, in most cases, this is not feasible, because of the high price of this treatment. Instead, patients have been treated successfully with a combination of low-dose L-Dopa, a Dopa-decarboxylase inhibitor, and 5-hydroxy-tryptophan. There is normally a complete or near-complete response of motor problems soon after the start of the therapy. Even when the therapy is started after a diagnostic delay of several years, the results are satisfactory.

### Genetics

The gene for GTPCH-I is located on chromosome 14q22.1-q22.2. GTPCH-I deficiency can be inherited as an autosomal-dominant trait with 30% penetrance. The female:male ratio is approximately 3:1. Several mutations have been found as a cause for the dominant form of the disease. These often occur as heterozygous mutations in the patient. However, there are also patients with mutations on both alleles [46, 47]. In such patients, the defect has been inherited as an autosomal-recessive trait. The residual activity of the enzyme may be so low in such cases that they actually present with neonatal hyperphenylalaninemia. Some cases, however, escape detection during neonatal screening for phenylketonuria [47].

## Folinic Acid Responsive Seizures

Although this is not strictly a neurotransmission disorder, it resembles pyridoxine-responsive convulsions in that the convulsions start very early, and their response to the medication (here folinic acid) is specific, rapid, and complete. The reported patient showed generalized seizures from the age of 8 h. There was no response to vitamin B6, and seizure control under high-dose phenobarbital was unsatisfactory. Biochemical studies revealed a slight but significant increase of glutamine, histidine, leucine, methionine, tyrosine, tryptophan, and valine; these increases were seen only in the CSF. CSF analyses for biogenic amines revealed high concentrations of an unknown compound. After initiation of folinic acid at 7 weeks (2.5 mg twice daily), seizures ceased, and the patient's psychomotor development improved dramatically. An older brother had suffered from severe neonatal epileptic encephalopathy unresponsive to conventional antiepileptic therapy. There was no psychomotor development. He died in status epilepticus at 8 months of age. He probably had the same disease his sister has [48, 49].

## References

1. Hunt AD, Stokes J, McCrory WW, Stroud HH (1954) Pyridoxine dependency: report of a case of intractable convulsions in an infant controlled by pyridoxine. Pediatrics 13:140-145
2. Minns R (1980) Vitamin B6 deficiency and dependency. Dev Med Child Neurol 22:795-799

3. Gospe SM (1998) Current perspectives on pyridoxine-dependent seizures. J Pediatr 132:919–923
4. Goutières F, Aicardi J (1985) Atypical presentations of pyridoxine-dependent seizures: a treatable cause of intractable epilepsy in infants. Ann Neurol 17:117–120
5. Lott IT, Coulombe T, Di Paolo RV, Richardson EP Jr, Levy HL (1978) Vitamin B6-dependent seizures: pathology and chemical findings in brain. Neurology 28:47–54
6. Kurlemann G, Löscher W, Dominick HC, Palm GD (1987) Disappearance of neonatal seizures and low CSF GABA levels after treatment with vitamin B6. Epilepsy Res 1: 152–154
7. Baumeister FAM, Gsell W, Shin YS, Egger J (1994) Glutamate in pyridoxine-dependent epilepsy: neurotoxic glutamate concentration in the cerebrospinal fluid and its normalization by pyridoxine. Pediatrics 94:318–321
8. Kure S, Maeda T, Fukushima N et al. (1998) A subtype of pyridoxine-dependent epilepsy with normal CSF glutamate concentration. J Inherit Metab Dis 21:431–432
9. Jaeken J, Casaer P, De Cock P, Corbeel L, Eeckels R et al. (1984) Gamma-aminobutyric acid-transaminase deficiency: a newly recognized inborn error of neurotransmitter metabolism. Neuropediatrics 15:165–169
10. Carchon HA, Jaeken J, Jansen E, Eggermont E (1991) Reference values for free gamma-aminobutyric acid determined by ion-exchange chromatography and fluorescence detection in the cerebrospinal fluid of children. Clin Chim Acta 201:83–88
11. Kok RM, Howells DW, van den Heuvel CCM et al. (1993) Stable isotope dilution analysis of GABA in CSF using simple solvent extraction and electron-capture negative-ion mass fragmentography. J Inherited Metab Dis 16:508–512
12. Sweetman FR, Gibson KM, Sweetman L, Nyhan WL, Chin H et al. (1986) Activity of biotin-dependent and GABA metabolizing enzymes in chorionic villus samples: potential for 1st trimester prenatal diagnosis. Prenat Diagn 6:187–194
13. Medina-Kauwe LK, Nyhan WL, Gibson KM, Tobin AJ (1998) Identification of a familial mutation associated with GABA-transaminase deficiency disease. Neurobiol Dis 5: 89–96
14. Jakobs C, Bojasch M, Monch E et al. (1981) Urinary excretion of gamma-hydroxybutyric acid in a patient with neurological abnormalities. The probability of a new inborn error of metabolism. Clin Chim Acta 111:169–117
15. Jakobs C, Jaeken J, Gibson KM (1993) Inherited disorders of GABA metabolism. J Inherited Metab Dis 16:704–715
16. Chambliss KL, Lee CF, Ogier H et al. (1993) Enzymatic and immunological demonstration of normal and defective succinic semialdehyde dehydrogenase activity in fetal brain, liver and kidney. J Inherit Metab Dis 16:523–526
17. Jaeken J, Casaer P, De Cock P, François B (1989) Vigabatrin in GABA metabolism disorders. Lancet 1:1074
18. Jakobs C, Michael T, Jaeger E, Jaeken J, Gibson KM (1992) Further evaluation of vigabatrin therapy in 4-hydroxybutyric aciduria. Eur J Pediatr 151:466–468
19. Gibson KM, Jakobs C, Ogier H et al. (1995) Vigabatrin therapy in six patients with succinic semialdehyde dehydrogenase deficiency. J Inherit Metab Dis 18:143–146
20. Chambliss KL, Hinson DD, Trettel F et al. (1998) Two exon-skipping mutations as the molecular basis of succinic semialdehyde dehydrogenase deficiency (4-hydroxybutyric aciduria). Am J Hum Genet 63:399–408
21. Kirstein I, Silfverskiold BP (1958) A family with emotionally precipitated "drop seizures". Acta Psychiatr Neurol Scand 33:471–476
22. Tijssen MAJ, van Dijk JG, Roos RAC, Padberg GW (1995) "Startle disease": van schrik verstijven. Ned Tijdschr Geneeskd 139:1940–1944
23. Bernasconi A, Cendes F, Shoubridge EA et al. (1998) Spectroscopic imaging of frontal neuronal dysfunction in hyperekplexia. Brain 121:1507–1512
24. Shiang R, Ryan SG, Zhu Y-Z et al. (1993) Mutations in the α1 subunit of the inhibitory glycine receptor cause the dominant neurologic disorder, hyperekplexia. Nature Genet 5:351–358
25. Vergouwen MN, Tijssen MA, Shiang R et al. (1997) Hyperekplexia-like syndromes without mutations in the GLRA$_1$ gene. Clin Neurol Neurosurg 99:172–178
26. Simon ES (1997) Phenotypic heterogeneity and disease course in three murine strains with mutations in genes encoding for alpha 1 and beta glycine receptor subunits. Mov Disord 12:221–228
27. Berthier M, Bonneau D, Desbordes JM et al. (1994) Possible involvement of a gamma-hydroxybutyric acid receptor in startle disease. Acta Paediatr 83:678–680
28. Tijssen MA, Schoemaker HC, Edelbroek PJ et al. (1997) The effects of clonazepam and vigabatrin in hyperekplexia. J Neurol Sci 149:63–67
29. Rees MI, Andrew M, Jawad S, Owen MJ (1994) Evidence for recessive as well as dominant forms of startle disease (hyperekplexia) caused by mutations in the α$_1$ subunit of the inhibitory glycine receptor. Hum Mol Genet 12:2175–2179
30. Lüdecke B, Knappskog PM, Clayton PT et al. (1996) Recessively inherited L-Dopa-responsive parkinsonism in infancy caused by a point mutation (L205P) in the tyrosine hydroxylase gene. Hum Mol Genet 5:1023–1028
31. Knapskogg PM, Flatmark T, Mallet J, Lüdecke B, Bartholomé K (1995) Recessively inherited L-Dopa-responsive dystonia caused by a point mutation (Q381K) in the tyrosine hydroxylase gene. Hum Mol Genet 4:1209–1212
32. Lüdecke B, Dworniczak B, Bartholomé K (1995) A point mutation in the tyrosine hydroxylase gene associated with Segawa's syndrome. Hum Genet 95:123–125
33. Wevers RA, de Rijk-van Andel JF, Bräutigam C et al. (1999) A review on biochemical and molecular genetic aspects of tyrosine hydroxylase deficiency including a novel mutation (291delC). J Inherit Metab Dis 22:364–373
34. Bräutigam C, Wevers RA, Jansen RJT et al. (1998) Biochemical hallmarks of tyrosine hydroxylase deficiency. Clin Chem 44:1897–1904
35. Van den Heuvel LPWJ, Luiten B, Smeitink JAM et al. (1998) A common point mutation in the tyrosine hydroxylase gene in autosomal recessive L-Dopa-responsive dystonia (DRD) in the Dutch population. Hum Genet 102:644–646
36. Hyland K, Surtees RAH, Rodeck C, Clayton PT (1988) Aromatic L-amino acid decarboxylase deficiency: clinical features, diagnosis, and treatment of a new inborn error of neurotransmitter amine synthesis. Neurology 42:1980–1988
37. Hyland K, Clayton PT (1992) Aromatic L-amino acid decarboxylase deficiency: diagnostic methodology. Clin Chem 38:2405–2410
38. Korenke GC, Christen H-J, Hyland K, Hunneman DH, Hanefeld F (1997) Aromatic L-amino acid decarboxylase deficiency: an extrapyramidal movement disorder with oculogyric crises. Eur J Pediatr Neurol 2/3:67–71
39. Chang YT, Mues G, McPherson JD et al. (1998) Mutations in the human aromatic L-amino acid decarboxylase gene. J Inherit Metab Dis 21 [Suppl 2]:4
40. Brunner HG, Nelen MR, Breakefield XO, Ropers HH, van Ost BA (1993) Abnormal behavior associated with a point mutation in the structural gene for monoamine oxidase A. Science 262:578–580
41. Brunner HG, Nelen MR, van Zandvoort P et al. (1993) X-linked borderline mental retardation with prominent behavioral disturbance: phenotype, genetic localization, and evidence for disturbed monoamine metabolism. Am J Hum Genet 52:1032–1039
42. Lenders JWM, Eisenhofer G, Abeling NGGM et al. (1966) Specific genetic deficiencies of the A and B isoenzymes of monoamine oxidase are characterized by distinct neurochemical and clinical phenotypes. J Clin Invest 97:1010–1019
43. Abeling NGGM, van Gennip AH, van Cruchten AG, Overmars H, Brunner HG (1998) Monoamine oxidase A deficiency: biogenic amine metabolites in random urine samples. J Neural Transm [Suppl] 52:9–15

44. Ichinose H, Ohye T, Takahashi E et al. (1994) Hereditary progressive dystonia with marked diurnal fluctuation caused by mutations in the GTP cyclohydrolase I gene. Nature Genet 8:236-242
45. Bandmann O, Valante EM, Holmans P et al. (1998) Dopa-responsive dystonia: a clinical and molecular genetic study. Ann Neurol 44:649-656
46. Furukawa Y, Shimadzu M, Rajput AH et al. (1996) GTP-cyclohydrolase I gene mutations in hereditary progressive and Dopa-responsive dystonia. Ann Neurol 39:609-617
47. Blau N, Ichinose H, Nagatsu T et al. (1995) A missense mutation in a patient with guanosine triphosphate cyclohydrolase I deficiency missed in the newborn screening program. J Pediatr 126:401-405
48. Hyland K, Buist NRM, Powell BR et al. (1995) Folinic acid responsive seizures: a new syndrome? J Inherit Metab Dis 18:177-181
49. Hoffmann GF, Surtees RAH, Wevers RA (1998) Cerebrospinal fluid investigations for neurometabolic disorders. Neuropediatrics 29:59-71

# Glutathione and Imidazole-Dipeptide Metabolism

*Glutathione* is a tripeptide composed of glutamate, cysteine and glycine. It is present in almost all cells at substantial concentrations and plays an important role in several biological functions, such as free-radical scavenging, reducing reactions, amino acid transport, synthesis of proteins and DNA, and drug metabolism. Biosynthesis and catabolism of glutathione form the γ-glutamyl cycle. Glutathione synthesis involves the sequential actions of γ-glutamylcysteine synthetase and glutathione synthetase. Catabolism proceeds initially by way of γ-glutamyl transpeptidase. The γ-glutamyl residue is then released as 5-oxoproline, which is reconverted into glutamate. The biosynthesis of glutathione is feedback regulated, i.e., glutathione acts as an inhibitor of γ-glutamylcysteine synthetase.

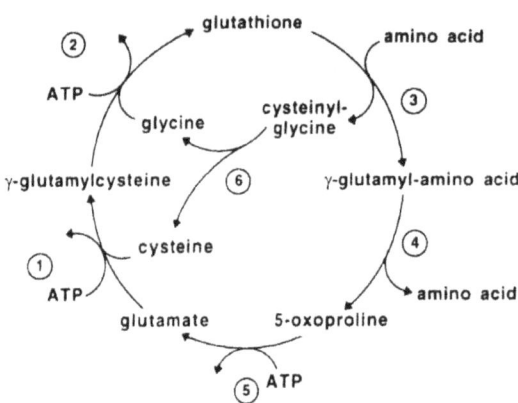

**Fig. 27.1.** The γ-glutamyl cycle. 1, γ-glutamyl-cysteine synthetase; 2, glutathione synthetase; 3, γ-glutamyl transpeptidase; 4, γ-glutamyl cyclotransferase; 5, 5-oxoprolinase; 6, cysteinyl glycinase (dipeptidase)

*Imidazole dipeptides* derive their name from the presence of the imidazole ring of histidine. *Carnosine* (β-alanyl-histidine) is found in skeletal (but not cardiac) muscle and brain, where it may be a neurotransmitter. It is hydrolyzed by two isozymes. Cytosolic carnosinase is present in most human tissues and displays a very broad dipeptidase specificity but does not hydrolyze anserine or homocarnosine. Serum carnosinase (also found in cerebrospinal fluid) hydrolyzes carnosine and anserine but hydrolyzes homocarnosine very poorly. Its activity increases gradually with age.

*Anserine* (β-alanyl-1-methylhistidine) is normally absent from human tissues and body fluids but may be derived from the diet, (particularly in infants, owing to their low serum carnosinase activity) or may be found in patients with serum carnosinase deficiency. Anserine is found is skeletal muscles of birds and certain mammals, such as rabbits. Its physiologic function is unclear.

*Homocarnosine* (γ-aminobutyryl-histidine) is a brain-specific dipeptide. It is hydrolyzed by serum but not by cytosolic carnosinase. Cerebrospinal-fluid homocarnosine concentrations are higher in children (~8 μM) than in adults (~1 μM). The physiologic function of homocarnosine is unknown. It may act as a "reservoir" for γ-aminobutyric acid in some parts of the brain (Fig. 27.1).

CHAPTER 27

# Disorders in the Metabolism of Glutathione and Imidazole Dipeptides

A. Larsson and J. Jaeken

CONTENTS

Disorders in the Metabolism of Glutathione .......... 313
 γ-Glutamyl-Cysteine-Synthetase Deficiency........... 313
 Glutathione-Synthetase Deficiency ................ 313
  Generalized Glutathione-Synthetase Deficiency ...... 314
  Erythrocyte Glutathione-Synthetase Deficiency ...... 315
 γ-Glutamyl-Transpeptidase Deficiency .............. 315
 5-Oxoprolinase Deficiency....................... 315
 Cysteinyl-Glycinase Deficiency ................... 316
 Secondary 5-Oxoprolinuria ...................... 316
Disorders of Imidazole Dipeptides.................. 316
 Serum Carnosinase Deficiency .................... 316
 Homocarnosinosis............................. 317
 Prolidase Deficiency ........................... 317
References..................................... 318

Genetic defects have been described in five of the six steps of the γ-glutamyl cycle. *Glutathione-synthetase deficiency* is the most frequently recognized disorder and, in its severe, generalized form, it is associated with hemolytic anemia, metabolic acidosis, 5-oxoprolinuria (pyroglutamic aciduria) and CNS damage. *γ-Glutamyl-cysteine-synthetase deficiency* is also associated with hemolytic anemia, and some patients with this disorder show defects of neuromuscular function and generalized aminoaciduria. *γ-Glutamyl-transpeptidase deficiency* has been found in patients with CNS involvement and glutathionuria. *5-Oxoprolinase deficiency* is associated with 5-oxoprolinuria but without a clear association with other symptoms. Finally, *cysteinyl-glycinase (or-dipeptidase) deficiency* has recently been identified in a patient with mental retardation, deafness and skeletal malformation. Serum *carnosinase deficiency* and *homocarnosinosis* are probably the same disorder. It is uncertain whether there is a relationship between the biochemical abnormalities and clinical symptoms. *Prolidase deficiency* causes skin lesions and recalcitrant ulceration (particularly on the lower legs) in addition to other features, such as impaired psychomotor development and recurrent infections. The severity of clinical expression is highly variable.

## Disorders in the Metabolism of Glutathione

### γ-Glutamyl-Cysteine-Synthetase Deficiency

*Clinical Presentation*

Six patients with this enzyme deficiency have been identified to date [1]. A common finding is hemolytic anemia. The first two patients also developed cerebellar involvement, peripheral neuropathy and myopathy [2]. Both patients had generalized aminoaciduria, but no other renal-function defect. Treatment with sulfonamide precipitated psychosis and pronounced hemolytic anemia. It remains to be established if these additional symptoms are related to the metabolic defect.

*Metabolic Derangement and Diagnostic Tests*

γ-Glutamyl-cysteine-synthetase activity was low in erythrocytes, as were the levels of glutathione in erythrocytes, leukocytes and skeletal muscle. In one adult patient, hemolytic anemia and modest decrease in the amount of glutathione in cultured lymphoblasts and fibroblasts were the only abnormalities [3]. The diagnosis is established by analysis of the relevant enzyme in erythrocytes or other tissues.

*Treatment and Prognosis*

The prognosis and treatment remain to be established.

*Genetics*

γ-Glutamyl-cysteine synthetase consists of two nonidentical subunits coded by two separate genes, located on chromosomes 1 and 6, respectively, in the human genome [4, 5]. The mode of inheritance is autosomal recessive.

### Glutathione-Synthetase Deficiency

Two forms of glutathione-synthetase deficiency can be distinguished clinically [1]. One is generalized, and the other is expressed only in erythrocytes. In the human genome, there is only one glutathione-synthetase gene. The generalized form is due to mutations affecting the catalytic properties of the enzyme, and the erythrocyte form is due to a mutation that affects the stability of the enzyme.

### Generalized Glutathione-Synthetase Deficiency

**Clinical Presentation**

Generalized deficiency of glutathione synthetase has been recognized in more than 40 patients in 35 families [1]. The clinical condition is variable and is presumably correlated to the extent of the enzyme defect. Most patients show symptoms (metabolic acidosis, jaundice and hemolytic anemia) within the first few days of life. After the neonatal period, the condition is usually stabilized. During episodes of gastroenteritis and other infections, however, the patients may become critically ill due to pronounced acidosis and electrolyte imbalance. Several patients have died during such episodes. The majority of the patients have progressive CNS damage, including mental retardation, ataxia, spasticity, and seizures. One patient died at the age of 28 years, and autopsy revealed atrophy of the granule cell layer of the cerebellum and focal lesions of the cortex. Two patients had increased susceptibility to bacterial infections due to defective granulocyte function.

**Metabolic Derangement**

Glutathione concentrations in erythrocytes and other tissues are very low, whereas γ-glutamyl cysteine, the metabolite before the enzyme defect, is produced in excess due to a lack of feedback inhibition of γ-glutamyl-cysteine synthetase. γ-Glutamyl cysteine is converted into 5-oxoproline (pyroglutamic acid) and cysteine by γ-glutamyl cyclotransferase.

5-Oxoproline is transferred to glutamate by 5-oxoprolinase, which is the rate-limiting enzyme of the γ-glutamyl cycle in many tissues. The excessive formation of 5-oxoproline exceeds the capacity of 5-oxoprolinase. Therefore, 5-oxoproline accumulates in the body fluids, causing metabolic acidosis and 5-oxoprolinuria.

**Diagnostic Tests**

The diagnosis is usually established in a newborn infant with severe metabolic acidosis but without ketosis or hypoglycemia. In the urine, massive excretion of L-5-oxoproline (up to 1 g/kg body weight/day) can be demonstrated by gas–liquid chromatography. Note that 5-oxoproline is ninhydrin negative. The distinction between the D and L forms can be made after acid hydrolysis and analysis of L-glutamate by L-glutamic-acid dehydrogenase. Decreased activity of glutathione synthetase can be demonstrated in, for instance, erythrocytes, leukocytes, or cultured skin fibroblasts. Intracellular levels of glutathione are markedly decreased.

**Treatment and Prognosis**

Treatment involves acidosis correction using parenteral administration of sodium bicarbonate initially followed by oral maintenance doses of sodium bicarbonate or citrate (up to 10 mmol/kg body weight/day). During episodes of acute infections, higher doses may be required. Vitamin E (α-tocopherol) has been shown to correct the defective granulocyte function. Therefore, vitamin E should be given in doses of about 10 mg/kg body weight/day. Recently, treatment with ascorbate (100 mg/kg body weight/day) has been postulated to be of benefit. Both α-tocopherol and ascorbate are supposed to replenish the lack of glutathione as a scavenger of free radicals. Drugs that precipitate hemolytic crises in patients with glucose-6-phosphate dehydrogenase (G6PD) deficiency should be avoided to prevent such crises. Therapeutic trials have been made in order to substitute for the lack of glutathione. Oral administration of glutathione, mercaptopropionylglycine, N-acetylcysteine and monoethyl esters of glutathione have been tested. The effect has mainly been monitored by cellular glutathione levels and excretion of 5-oxoproline. N-acetylcysteine and glutathione ethyl ester both yielded increased levels of glutathione in, for instance, leukocytes. More studies are needed before treatment with these drugs can be recommended. Dietary manipulation – including adjustment of the protein intake – has not affected the excretion of 5-oxoproline.

The prognosis of the patients is, at least in part, dependant on the measures taken during acute episodes. Especially during the neonatal period, it is essential to correct the metabolic acidosis and electrolyte imbalance, treat anemia and prevent excessive hyperbilirubinemia. CNS damage is progressive and cannot be prevented at this time. It is, however, essential to remember that generalized glutathione-synthetase deficiency is a heterogeneous condition, and it is difficult to predict the outcome for individual patients.

**Genetics**

The defective gene is transmitted by autosomal-recessive inheritance. The enzyme is a homodimer with a subunit molecular weight of 52 kDa. There is only one copy of the gene in the human genome and its location is 20.q11.2 [6]. Twelve non-related patients with generalized glutathione-synthetase deficiency have been analyzed at the DNA level, and they were found to have splice-site mutations, deletions or missense mutations [7].

### Erythrocyte Glutathione-Synthetase Deficiency

**Clinical Presentation**

At least four families with hereditary erythrocyte glutathione-synthetase deficiency have been reported [1]. The characteristic symptom is mild hemolytic anemia. Some patients had splenomegaly. No other clinical symptoms have been reported.

**Metabolic Derangement and Diagnostic Tests**

Since the mutation affects the stability of the enzyme, the concentration of glutathione [control levels in erythrocytes (mean ± SD) = 2.79 ± 0.50 mmol/l] and the activity of glutathione synthetase are decreased in erythrocytes but are considerably higher or even normal in nucleated cells. Erythrocyte glutathione S-transferase is decreased as a consequence of the lack of glutathione. Urinary levels of 5-oxoproline are normal.

**Treatment and Prognosis**

No treatment is proposed. It seems reasonable, however, that patients should avoid drugs that cause hemolytic crises in patients with G6PD deficiency. The hemolytic anemia is usually mild, and the prognosis is good.

**Genetics**

The mode of inheritance is autosomal recessive. Since the enzyme is a homodimer and there is only one copy of the subunit gene in the human genome, it is likely that the erythrocyte form is caused by mutations affecting the stability of the enzyme, whereas generalized deficiency is due to mutations affecting the catalytic properties of the enzyme [7].

### γ-Glutamyl-Transpeptidase Deficiency

**Clinical Presentation**

Five patients with γ-glutamyl-transpeptidase deficiency have been reported [1]. Three of them had CNS involvement, but two siblings have apparently no signs of CNS damage. This may reflect the fact that the first three patients were identified by screening for amino acid defects in populations of mentally retarded patients.

**Metabolic Derangement**

The patients have increased glutathione concentrations in plasma and urine, but the cellular levels are normal. In addition to glutathionuria, urinary levels of γ-glutamyl cysteine and cysteine are increased. The patients lose up to 1 g of glutathione per day via urine, which becomes a substantial source of nitrogen excretion.

**Diagnostic Tests**

Using thin-layer or paper chromatography and ninhydrin detection, the patients are identified by urinary screening for amino acid disorders. This reveals glutathionuria. Decreased activity of γ-glutamyl transpeptidase can be demonstrated in leukocytes or cultured skin fibroblasts but not in erythrocytes, which lack this enzyme under normal conditions.

**Treatment and Prognosis**

No specific treatment has been postulated. The prognosis must be considered serious if the patient presents with psychiatric or neurologic symptoms [8]. However, two affected siblings aged 11 years and 13 years have no signs of CNS involvement.

**Genetics**

γ-Glutamyl-transpeptidase deficiency is transmitted by autosomal-recessive inheritance. The human genome has at least seven different loci for γ-glutamyl transpeptidase. Five of them are located on chromosome 22 [9, 10].

### 5-Oxoprolinase Deficiency

**Clinical Presentation**

Eight patients with hereditary defects in 5-oxoprolinase have been described [1]. The clinical symptoms that led to the discovery of the patients were not necessarily related to the metabolic defect. Two brothers were investigated because of renal stone formation. They also had chronic enterocolitis but no signs of hemolytic anemia (except after salazosulfapyridine in one patient) or CNS damage. The third patient was a woman with mild mental retardation who had given birth to a child with congenital malformations. Two brothers with 5-oxoprolinase deficiency were recently reported. One had not shown any clinical symptoms, whereas the other had hypoglycemia. Two additional patients had mental retardation. In summary, two of eight patients had shown some degree of mental retardation. Whether this is a coincidence remains to be established.

## Metabolic Derangements

The patients were identified because of 5-oxoprolinuria, excreting 4–10 g of 5-oxoproline per day. They exhibited no signs of metabolic acidosis.

## Diagnostic Tests

Urinary analysis of 5-oxoproline (pyroglutamic acid) is the initial diagnostic test, followed by determination of the corresponding enzyme activity. 5-Oxoprolinase is not present in erythrocytes and, therefore, leukocytes or other nucleated cells must be used for the final diagnosis.

## Treatment and Prognosis

No specific treatment has been proposed, and prognosis remains to be established.

## Genetics

The mode of inheritance is autosomal recessive. The mammalian enzyme is a homodimer [11]. The location of the corresponding gene in the human genome remains to be established.

### Cysteinyl-Glycinase Deficiency

## Clinical Presentation

A boy with mental retardation, deafness, foot deformity and suspected peripheral neuropathy was recently diagnosed with tentative cysteinyl-glycinase (dipeptidase) deficiency [12]. It remains to be established if there is a relationship between the biochemical defect and the clinical symptoms.

## Metabolic Derangement and Diagnostic Tests

Urinary excretion of cysteinyl glycine was 49–72 mmol/mol of creatinine. The level of cysteinyl glycine in plasma was normal.

The diagnosis is made by quantitative amino acid analysis of urine. To date, the corresponding enzyme deficiency has not been established.

## Treatment and Prognosis

No treatment has been developed, and it remains to be established if therapy is necessary. The prognosis is unclear.

## Genetics

The genetics of the condition remains to be established.

### Secondary 5-Oxoprolinuria

5-Oxoprolinuria has been described in conditions other than generalized glutathione-synthetase deficiency and 5-oxoprolinase deficiency. Patients with severe burns or Stevens-Johnson syndrome and infants fed formula based on acid-hydrolyzed protein have been found to excrete increased amounts of 5-oxoproline, usually in the range of a few milligrams per day. Oxoprolinuria is also found in conditions associated with severe depression of glutathione levels. For instance, patients with homocystinuria may have increased excretion of 5-oxoproline (up to 1 g/day). Likewise, patients suffering from metabolic crises due to urea-cycle defects, e.g., ornithine transcarbamoylase deficiency, have been found to excrete a few grams of 5-oxoproline. This seems to occur as a consequence of lack of ATP in critical organs, such as liver and kidney. A few patients treated with acetaminophen (paracetamol) have been found to excrete several grams of 5-oxoproline per day.

# Disorders of Imidazole Dipeptides

## Serum Carnosinase Deficiency

### Clinical Presentation

Some 30 individuals have been reported with this disorder, first described in 1967 [13, 14]. The majority of them showed mental retardation to a variable degree. Some patients had seizures and one had congenital myopathy. A few had no symptoms at all, making the relationship between the biochemical abnormalities and the clinical picture uncertain [15].

### Metabolic Derangement

The deficiency of serum *carnosinase* activity causes a persistent carnosinuria during a meat-free diet. Several variants with abnormal kinetic properties of the enzyme have been described. In the cerebrospinal fluid of affected persons, *homocarnosine* can be normal or increased.

### Diagnostic Tests

The diagnosis is made by quantitative amino acid analysis of serum and/or urine after exclusion of meat from the diet. Anserine appears in the urine of these persons only after eating food containing the dipeptide. Normal persons excrete 1-methylhistidine after ingesting anserine; in serum carnosinase deficiency, there is little or no 1-methylhistidine excretion. The diagnosis is confirmed by measuring carnosinase activity in serum. It has to be noted that serum carnosinase activity may be low in other disorders, such as urea-cycle disorders and multiple sclerosis [16].

### Treatment and Prognosis

No efficient treatment is available. In view of the above remarks, it is uncertain whether treatment would be necessary. There is no reason to withhold meat from the diet, because the accumulating carnosine is primarily endogenous. Prognosis is variable and does not seem to correlate with the degree of enzyme deficiency.

### Genetics

Inheritance is autosomal recessive. Serum carnosinase deficiency was recently reported in a child with 18q-syndrome, suggesting a chromosomal location of this enzyme [17].

## Homocarnosinosis

### Clinical Presentation

This condition was described in 1976 in a Norwegian family (three of four siblings and their mother) [18]. The three offspring showed progressive spastic diplegia, mental retardation and retinitis pigmentosa, with onset between 6 years and 29 years of age. The mother, however, was symptom free. This makes it uncertain whether there is a relationship between the biochemical defect and the clinical symptoms.

### Metabolic Derangement and Diagnostic Tests

In the CSF of the three siblings and in that of their clinically normal mother, the homocarnosine level was 30–50 times the mean of control levels. The carnosine levels were normal. Deficiency of homocarnosinase activity was found [19]. Therefore, serum carnosinase deficiency and homocarnosinosis are probably the same disorder. The diagnosis is made by quantitative amino acid analysis of the CSF.

### Treatment and Prognosis

The remarks regarding treatment and prognosis of serum carnosinase deficiency also apply here.

### Genetics

Inheritance in the Norwegian family seems to be autosomal dominant.

## Prolidase Deficiency

### Clinical Presentation

Less than 40 individuals with prolidase deficiency have been reported since 1968 [20]. About a quarter of them were asymptomatic at the time of the report. The others had their first symptoms between birth and 22 years of age. All patients showed skin lesions, either mild (face, palms, soles) or severe, and had recalcitrant ulceration, particularly on the lower legs. Other features included a characteristic face, impaired motor or cognitive development and recurrent infections. Prolidase deficiency seems to be a risk factor for the development of systemic lupus erythematosus. Additionally, patients with systemic lupus erythematosus should, where there is a family history or presentation in childhood, be specifically investigated for prolidase deficiency [21].

### Metabolic Derangement

The hallmark biochemical finding is massive hyperexcretion of a large number of imidodipeptides (dipeptides with a N-terminal proline or hydroxyproline, particularly glycylproline). This is due to a deficiency of the exopeptidase prolidase (or peptidase D).

### Diagnostic Tests

The hyperimidodipeptiduria can be detected and quantified by partition and elution chromatography and by direct chemical-ionization mass spectrometry. The finding of low or absent prolidase activity in hemolysates or in homogenates of leukocytes or fibroblasts confirms the diagnosis.

### Treatment and Prognosis

Due to the rarity of the disease, experience with treatment is scarce. The skin ulcers improved with oral ascorbate, manganese (cofactor of prolidase), and an inhibitor of collagenase in one patient and with local applications of L-proline- and glycine-containing ointments in other patients. Skin grafts have been unsuccessful [22]. As to prognosis, the age of onset and the severity of clinical expression are highly variable.

## Genetics

Inheritance is autosomal recessive. The gene (PEPD) maps to chromosome 19p13.2. Several mutations have been identified [23].

## References

1. Larsson A, Anderson M (2000) Glutathione synthetase deficiency and other disorders of the γ-glutamyl cycle. In: Scriver CF, Beaudet AL, Sly WS, Vallee D (eds) The metabolic and molecular basis of inherited disease, 8th edn. McGraw-Hill, New York (in press)
2. Konrad PN, Richards F II, Valentine WN, Paglia D (1972) γ-Glutamylcysteine synthetase deficiency. N Engl J Med 286:557–561
3. Beutler E, Moroose R, Kramer L, Gelbart T, Forman L (1990) γ-Glutamylcysteine synthetase deficiency and hemolytic anemia. Blood 75:271–273
4. Sierra-Rivera E, Summar ML, Dasouki M et al. (1995) Assignment of the human gene (GLCLC) that encodes the heavy subunit of γ-glutamylcysteine synthetase to human chromosome 6. Cytogenet Cell Genet 70:278–279
5. Sierra-Rivera E, Dasouki M, Summar ML et al. (1996) Assignment of the human gene (GLCLR) that encodes the regulatory subunit of γ-glutamyl-cysteine synthetase to chromosome 1p21. Cytogenet Cell Genet 72:252–254
6. Webb GC, Vaska VL, Gali RR, Ford JH, Board PG (1995) The gene encoding human glutathione synthetase (GSS) maps to the long arm of chomosome 20 at band 11.2. Genomics 30:617–619
7. Dahl N, Pigg M, Ristoff E et al. (1997) Missense mutations in huam glutathione synthetase gene result in severe metabolic acidosis, 5-oxoprolinuria, hemolytic anemia and neurological dysfunction. Hum Mol Genet 6:1147–1152
8. Wright EC, Stern J, Erseer R, Patrick AD (1979) γ-glutamyl transpeptidase deficiency. J Inherit Metab Dis 2:3–7
9. Bulle F, Mattei MG, Siegrist S et al. (1987) Assignment of the human γ-glutamyl transferase gene to the long arm of chromosome 22. Hum Genet 76:283–286
10. Courtay C, Heisterkamp N, Siest G, Groffen J (1994) Expression of multiple γ-glutamyl transferase genes in man. Biochem J 297:503–508
11. Ye GJ, Breslow EB, Meister A (1996) The amino acid sequency of rat kidney 5-oxo-L-prolinase determined by cDNA cloning. J Biol Chem 271:32293–32300
12. Bellet H, Rejon F, Vallat C, Mion H, Dimeglio A (1998) Cystinyl glycinuria; a new disorder of the γ-glytamyl cycle. J Inherit Metabol Dis 21 [Suppl 2]:34
13. Perry TL, Hansen S, Tischler B, Bunting R, Berry K (1967) Carnosinemia: a new metabolic disorder associated with neurological disease and mental defect. N Engl J Med 277:1219–1227
14. Gjessing LR, Lunde HA, Morkrid L, Lenner JF, Sjasstad O (1990) Inborn errors or carnosine and homocarnosine metabolism. J Neural Transm 29 [Suppl]:91–106
15. Cohen M, Hartlage PL, Krawiecki N et al. (1985) Serum carnosinase deficiency: a non-disabling phenotype? J Ment Defic Res 29:383–389
16. Wassif WS, Sherwood RA, Amir A et al. (1994) Serum carnosinase activity in central nervous system disorders. Clin Chim Acta 225:57–64
17. Willi SM, Zhang Y, Hill JB et al. (1997) A deletion in the long arm of chromosome 18 in a child with serum carnosinase deficiency. Pediatr Res 41:210–213
18. Sjaastadt O, Berstadt J, Gjesdahl P, Gjessing L (1976) Homocarnosinosis. 2. A familial metabolic disorder associated with spastic paraplegia, progressive mental deficiency, and retinal pigmentation. Acta Neurol Scand 53:275–290
19. Lenney JF, Peppers SC, Kucera CM, Sjaastadt O (1983) Homocarnosinosis: lack of serum carnosinase is the defect probably responsible for elevated brain and CSF homocarnosine. Clin Chim Acta 132:157–165
20. Goodman SI, Solomons CC, Muschenheim F et al. (1968) A syndrome resembling lathyrism associated with iminodipeptiduria. Am J Med 45:152–159
21. Shrinath M, Walter JH, Haeney M et al. (1997) Prolidase deficiency and systemic lupus erythematosus. Arch Dis Child 76:441–444
22. Jemec GB, Moe AT (1996) Topical treatment of skin ulcers in prolidase deficiency. Pediatr Dermatol 13:58–60
23. Ledoux P, Scriver CR, Hechtman P (1996) Expression and molecular analysis of mutations in prolidase deficiency. Am J Hum Genet 59:1035–1039

# PART VII
# DISORDERS OF LIPID AND BILE ACID METABOLISM

## Lipoprotein Metabolism

Lipids are transported in plasma on lipoproteins (spherical particles that consist of a hydrophobic core of triglycerides and cholesteryl esters) surrounded by an amphiphilic coating of apolipoproteins, phospholipids and unesterified cholesterol. The human plasma lipoproteins are classified according to their density and electrophoretic mobility (Table 28.1), and several species of apolipoproteins are known (Table 28.2). Lipoprotein metabolism involves two major pathways, which are briefly summarized here and reviewed in more detail in the first section of the chapter.

The *exogenous pathway* transports dietary lipids (mainly triglycerides but also cholesterol) as chylomicrons (Fig. 28.1). Lipoprotein lipase, an enzyme on the surface of capillary endothelial cells that requires apolipoprotein C-II as a cofactor, hydrolyzes chylomicron triglycerides into free fatty acids for uptake by muscle and fat. The resultant chylomicron remnants are taken up by the liver, where they deliver dietary cholesterol.

The *endogenous pathway* transports hepatic triglycerides and cholesterol as very-low-density lipoproteins (VLDL; Fig. 28.1). In the capillaries of muscle and fat, VLDL are also hydrolyzed by lipoprotein lipase, yielding free fatty acids for uptake. Their remnants, intermediate-density lipoproteins (IDL), are in part cleared from the circulation by the liver low-density lipoprotein (LDL) receptor and in part converted into LDL. LDL are taken up via LDL receptors in a variety of extrahepatic tissues, where they supply cholesterol (mainly for membrane synthesis). Liver also takes up LDL via LDL receptors and uses their cholesterol for the synthesis of bile acids.

*Reverse cholesterol transport* involves release of unesterified cholesterol from cells into plasma, followed by binding to high-density lipoprotein (HDL), conversion (by lecithin:cholesterol acyltransferase) of unesterified cholesterol into esterified cholesterol and transfer of the latter (via cholesteryl-ester transfer protein) to VLDL and, ultimately, IDL and LDL. HDL can also deliver cholesteryl esters directly to the liver (Fig. 28.2).

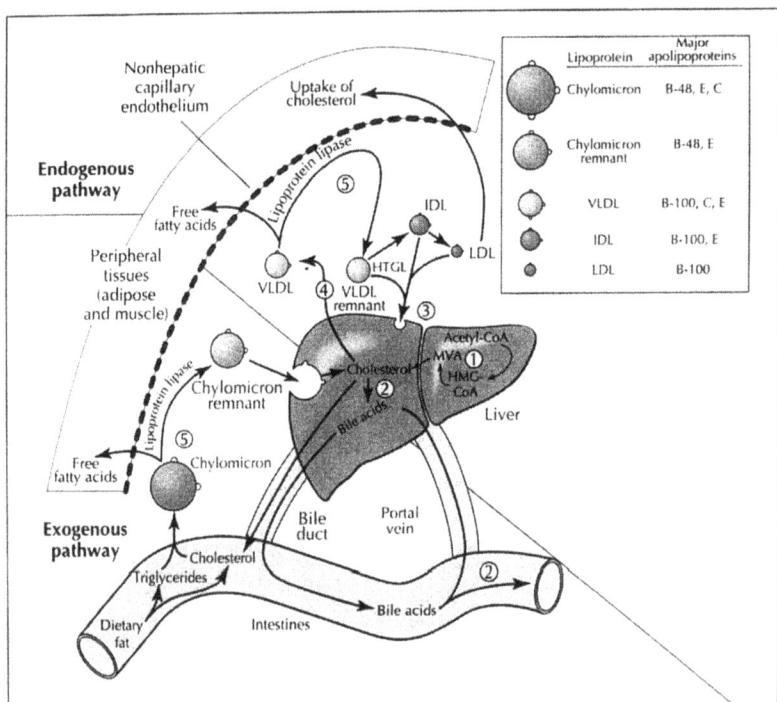

**Fig. 28.1.** Pathways of exogenous and endogenous lipoprotein metabolism. The metabolism of the apolipoprotein-B-containing lipoproteins from the intestine and the liver are depicted. The sites of action of the lipid-lowering drugs are also shown. The statins (*1*), and the bile-acid sequestrants (*2*) both induce low-density lipoprotein (LDL) receptors (*3*). Niacin (*4*), inhibits very-low-density-lipoprotein, intermediate-density-lipoprotein and LDL production. The fibric-acid derivatives (*5*) enhance lipoprotein-lipase activity. See text for abbreviations. Reproduced with permission from [40]

# CHAPTER 28

# Dyslipidemias

Annabelle Rodriguez-Oquendo and Peter O. Kwiterovich Jr

CONTENTS

Overview of Plasma Lipid and Lipoprotein Metabolism . 321
    Exogenous Lipoprotein Metabolism ............... 321
    Endogenous Lipoprotein Metabolism .............. 322
    Reverse Cholesterol Transport and HDL ............ 323
    Lipid-Lowering Drugs ......................... 324
Disorders of Exogenous Lipoprotein Metabolism....... 324
    Lipoprotein Lipase Deficiency ................... 324
    ApoC-II Deficiency .......................... 325
Disorders of Endogenous Lipoprotein Metabolism ..... 325
    Disorders of VLDL Overproduction ............... 325
        Familial Hypertriglyceridemia ................. 325
        Familial Combined Hyperlipidemia and the Small
        Dense LDL Syndromes ....................... 326
        Lysosomal Acid Lipase Deficiency ............... 327
    Disorders of LDL Removal ..................... 327
        Familial Hypercholesterolemia
        (LDL Receptor Defect) ....................... 327
        Familial Ligand-Defective ApoB ................. 328
        Sitosterolemia .............................. 328
Disorders of Endogenous and Exogenous
Lipoprotein Transport .......................... 329
    Dysbetalipoproteinemia (Type-III Hyperlipoproteinemia) 329
    Hepatic lipase Deficiency ...................... 329
Disorders of Reduced LDL-Cholesterol Levels ......... 329
    Abetalipoproteinemia ......................... 329
    Hypobetalipoproteinemia ...................... 330
    Homozygous Hypobetalipoproteinemia ............ 330
Disorders of Reverse Cholesterol Transport........... 330
    Familial Hypoalphalipoproteinemia ............... 330
    ApoA-I Mutations............................ 331
    Tangier Disease.............................. 331
    LCAT Deficiency and Fish-Eye Disease............. 331
    CETP Deficiency ............................. 331
Elevated Lipoprotein(a).......................... 332
Guidelines for the Treatment of Hyperlipidemia ....... 332
Abbreviations ................................. 335
References .................................... 336

## Overview of Plasma Lipid and Lipoprotein Metabolism

Lipoproteins play an essential role in the delivery of free fatty acids (FFA) to muscle and adipose tissue where they serve as a fuel and are stored as triglycerides (TG), respectively. Lipoproteins also intervene in the transfer of cholesterol from intestine to liver, from liver to other tissues and from the latter back to the liver. The lipoprotein structure resembles a plasma-membrane bilayer with hydrophilic phospholipids, apolipoproteins and some cholesterol on the outer surface, and hydrophobic TG and cholesteryl esters in the core. The physical-chemical properties and composition of the major human plasma proteins are given in Table 28.1.

The plasma apolipoproteins are amphipathic proteins that interact with both the polar aqueous environment of blood and the nonpolar core lipids. They serve various functions, such as ligands for receptors, cofactors for enzymes and structural proteins for packaging. The main characteristics of human plasma apolipoproteins are given in Table 28.2.

### Exogenous Lipoprotein Metabolism

The exogenous pathway of lipoprotein metabolism transports dietary fats from intestine to muscle, adipose tissue and liver. After a meal is consumed, dietary lipids, mainly TG and cholesteryl esters, are emulsified by bile acids and hydrolyzed by pancreatic lipases into their component parts: monoglyceride and FFA, and unesterified cholesterol and FFA, respectively. After absorption into the intestinal cells, the monoglycerides are reconverted into TG and incorporated (together with cholesterol) into chylomicrons that contain apolipoproteins ApoA-I, A-II, A-IV and B-48. The assembled chylomicrons are secreted into the thoracic duct, a process that requires ApoB-48. Thereafter, they enter the peripheral circulation, where they acquire ApoE and ApoC-II [derived from high-density lipoproteins (HDL)]. When they enter the capillaries of skeletal muscle and adipose tissue, the chylomicrons are exposed to the enzyme lipoprotein lipase (LPL), located on the surface of the endothelial cells (Table 28.3). ApoC-II is necessary for activation of LPL, provoking hydrolysis of the TG into FFA, which enters muscle and adipose tissue. The resulting chylomicron remnants, still containing cholesterol, ApoB-48 and ApoE (the last of these acts as a ligand for the hepatic chylomicron remnant receptor), are taken up by the liver, where they deliver dietary cholesterol (Fig. 28.1).

**Table 28.1.** Physical-chemical properties of human plasma lipoprotein. Compositions are given in percent by weight

| Class | Density (g/ml) | Electrophoretic mobility | Surface components | | | Core lipids | |
|---|---|---|---|---|---|---|---|
| | | | Cholesterol | Phospholipids | Apolipoprotein | Triglycerides | Cholesterylester |
| Chylomicrons | <0.95 | Remains at origin | 2 | 7 | 2 | 86 | 3 |
| VLDL | 0.95–1.006 | Pre-β lipoproteins | 7 | 18 | 8 | 55 | 12 |
| IDL | 1.006–1.019 | Slow pre-β lipoproteins | 9 | 19 | 19 | 23 | 29 |
| LDL | 1.019–1.063 | β-Lipoproteins | 8 | 22 | 22 | 6 | 42 |
| HDL-2[a] | 1.063–1.125 | α-Lipoproteins | 5 | 33 | 40 | 5 | 17 |
| HDL-3[a] | 1.125–1.210 | α-Lipoproteins | 4 | 35 | 55 | 3 | 13 |
| Lp(a)[b] | 1.040–1.090 | Slow pre-β-lipoproteins | | | | | |

*HDL*, high-density lipoprotein; *IDL*, intermediate-density lipoprotein; *LDL*, low-density lipoprotein; *LP(a)*, lipoprotein a; *VLDL*, very-low-density lipoprotein
[a] HDL$_2$ and HDL$_3$ are the two major subclasses of HDL
[b] Lp(a) consists of a molecule of LDL covalently attached to a molecule of apolipoprotein (a), a protein homologous to plasminogen. Its lipid composition is similar to that of LDL

**Table 28.2.** Characteristics of human plasma apolipoproteins

| Apolipoproteins | Major tissue sources | Functions | Molecular weight (Da) |
|---|---|---|---|
| Apo A-I | Liver and intestine | Co-factor of LCAT | 29,016 |
| Apo A-II | Liver and intestine | Not known | 17,414 |
| Apo A-IV | Liver and intestine | Activates LCAT | 44,465 |
| Apo B-48 | Intestine | Secretion TG from intestine | 240,800 |
| Apo B-100 | Liver | Secretion TG from liver; binding ligand to LDL receptor | 512,723 |
| Apo C-I | Liver | Activates LCAT | 6630 |
| Apo C-II | Liver | Cofactor LPL | 8900 |
| Apo C-III | Liver | Inhibits LPL | 8800 |
| Apo D | Many sources | Reverse cholesterol transport | 19,000 |
| Apo E | Liver | Ligand for uptake of chylomicron remnants and IDL | 34,145 |

*IDL*, intermediate-density lipoprotein; *LCAT*, lecithin:cholesterol acyl transferase; *LDL*, low-density lipoprotein; *LPL*, lipoprotein lipase; *TG*, triglyceride

## Endogenous Lipoprotein Metabolism

The endogenous pathway of lipoprotein metabolism transports TG and cholesteryl esters, synthesized in the liver, to the peripheral tissues. These compounds are transported in the form of very-low-density lipoproteins (VLDL) with their major apolipoproteins, ApoB-100, E and C (I, II, III). The VLDL particles are transported to tissue capillaries, where they release FFA by interacting with the same LPL that hydrolyzes chylomicrons. The resulting VLDL remnants are further hydrolyzed, generating intermediate-density lipoproteins (IDL). A portion of the IDL is cleared from the circulation via direct uptake by the liver by the binding of IDL ApoE to the low-density lipoprotein (LDL) receptor (Fig. 28.1). The remaining IDL can undergo further hydrolysis by hepatic lipase (HL; Table 28.3) to yield LDL. Most LDL are removed from the peripheral circulation by the binding of ApoB-100 to liver LDL receptors.

Liver LDL receptors are clustered in regions termed "coated pits" on the surface of the hepatocytes. In these regions, LDL particles are removed by absorptive endocytosis, a process by which both lipoprotein and receptor are internalized into endosomes. The receptor is separated from the complex and is translocated back to the plasma membrane. The cholesteryl esters are hydrolyzed to unesterified cholesterol. Overaccumulation of intrahepatic cholesterol is prevented by cholesterol-induced downregulation of the transcription of the genes for the LDL receptor and the rate-limiting enzyme of cholesterol synthesis, hydroxymethylglutaryl coenzyme A (HMG-CoA) reductase. Both downregulations are mediated by the sterol-regulatory-element-binding protein [1].

LDL also supplies cholesterol to a variety of extrahepatic parenchymal tissues (where it is used mainly for membrane synthesis) and to adrenal cortical cells, where it serves as a precursor for steroid synthesis. Like the liver, extrahepatic tissues also have

**Table 28.3.** Key enzymes and transfer proteins of plasma lipid transport

| Enzyme | Major tissue source | Functions | Molecular weight (Da) |
| --- | --- | --- | --- |
| LPL | Adipose tissue, striated muscle | Hydrolyzes triglycerides and phospholipids of chylomicrons and large VLDL | 50,394 |
| HL | Liver | Hydrolyzes triglycerides and phospholipids of small VLDL, IDL, and $HDL_2$ | 53,222 |
| LCAT | Liver | Converts free cholesterol from cell membranes to esterified cholesterol using a free fatty acid from phosphatidylcholine on nascent (pre-β) HDL | 47,090 |
| CETP | Liver, spleen, adipose tissue | Transfers cholesteryl esters from HDL to apoB-containing triglyceride-rich lipoproteins; converts α-HDL to pre-β-HDL | 74,000 |
| PTP | Placenta, pancreas, adipose tissue, lung | Transfers the majority of phospholipids in plasma; converts α-HDL to pre-β-HDL | 81,000 |

*apoB*, apolipoprotein B; *CETP*, cholesterol-ester transport protein; *HDL*, high-density lipoprotein; *HL*, hepatic lipase; *IDL*, intermediate-density lipoprotein; *LCAT*, lecithin:cholesterol acyl transferase; *LPL*, lipoprotein lipase; *PTP*, phospholipid transfer protein; *VLDL*, very-low-density lipoprotein

abundant LDL receptors. LDL cholesterol can also be removed via non-LDL-receptor mechanisms. One class of cell-surface receptors, termed scavenger receptors, takes up chemically modified LDL, such as oxidized LDL, which is generated by the release of oxygen radicals from endothelial cells. Scavenger receptors are not regulated by intracellular cholesterol levels and will continue to bind and internalize oxidized LDL and other chemically modified LDL (such as malondialdehyde LDL) residing in tissue spaces. In peripheral tissues, such as macrophages and smooth-muscle cells of the arterial wall, excess cholesterol accumulates within the plasma membrane and then is transported to the endoplasmic reticulum, where it is esterified to cholesteryl esters by the enzyme acyl-CoA cholesterol acyltransferase. It is at this stage that cytoplasmic droplets are formed and that the cells are converted into foam cells (an early stage of atherogenesis). Later on, cholesteryl esters accumulate as insoluble residues in atherosclerotic plaques.

## Reverse Cholesterol Transport and High Density Lipoproteins

Reverse cholesterol transport refers to the process by which unesterified or free cholesterol is removed from extrahepatic tissues, probably by extraction from cell membranes, and is transported on HDL [2]. HDL particles are heterogeneous and differ in their percentage of apolipoproteins (A-I, A-II, A-IV). HDL can be formed by remodeling of apolipoproteins cleaved during the hydrolysis of TG-rich lipoproteins (chylomicrons, VLDL and IDL). They can also be synthesized by intestine, liver and macrophages as nascent or pre-β-HDL particles that are relatively lipid-poor and disc-like in appearance (Fig. 28.2). Pre-β1 HDL is a molecular species of plasma HDL; it weighs approximately 67 kDa. Pre-β1 HDL contains ApoA-I, phospholipids and unesterified cholesterol and plays a major role in the retrieval of cholesterol from peripheral tissues. HDL particles possess a number of enzymes on their surface [3]. One enzyme, lecithin:cholesterol acyltransferase (LCAT), plays a significant role by catalyzing the conversion of unesterified cholesterol into esterified cholesterol (Fig. 28.2; Table 28.3). Esterified cholesterol is nonpolar and will localize in the core of the HDL particle, allowing it to remove more unesterified cholesterol from cells. Esterified cholesterol can be transferred [via the action of cholesteryl-ester transfer protein (CETP)] to VLDL and IDL particles. These TG-rich lipoproteins can be hydrolyzed to LDL, which can then be cleared by hepatic LDL receptors. Another enzyme that plays a critical role in the metabolic fate of HDL is HL, which hydrolyzes the TG and phospholipids on $HDL_2$, producing $HDL_3$. Nascent HDL particles are regenerated by the action of HL and phospholipid transfer protein (Fig. 28.2; Table 28.3). HDL may also deliver cholesteryl esters to the liver directly via the scavenger receptor SRB1 [4].

A number of epidemiological studies has shown an inverse relationship between the prevalence of cardiovascular disease and levels of HDL cholesterol. HDL are thought to be cardioprotective due to their participation in reverse cholesterol transport and

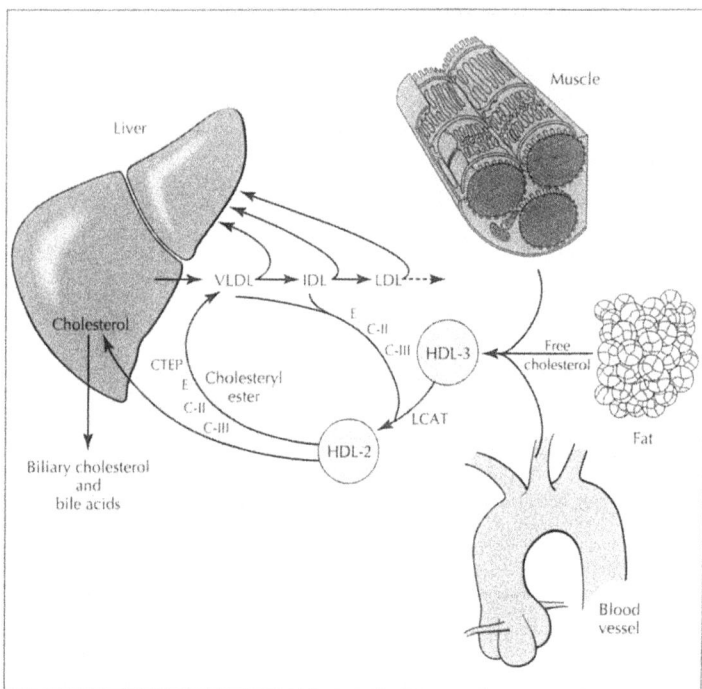

**Fig. 28.2.** The pathway for high-density-lipoprotein metabolism and reverse cholesterol transport. See text for abbreviations. Reproduced with permission from [40]

perhaps because of their roles as antioxidants. HDL impedes LDL oxidation by metal ions, an effect that may be due to the influences of several molecules on HDL, including ApoA-I, platelet-activating-factor acetylhydrolase and paraoxonase [3]. Accumulation of $HDL_2$, thought to be the most cardioprotective member of the HDL subclasses, is favored by estrogens, which negatively regulate HL. In contrast, progesterone and androgens, which positively regulate this enzyme, lead to increased production of $HDL_3$.

## Lipid-Lowering Drugs

In recent years, pharmacologic manipulation of the metabolic and cellular processes of lipid and lipoprotein metabolism has greatly improved the treatment of dyslipidemias. Inhibitors of the rate-limiting enzyme of cholesterol synthesis, HMG-CoA reductase, are called statins (Fig. 28.1); they effectively decrease intracellular cholesterol levels. This effect, in turn, leads to an upregulation of the LDL-receptor gene, with a corresponding increase in expression of the receptor protein, resulting in an increased uptake of LDL by the liver. Resins (which sequester bile acids and prevent enterohepatic recycling and re-uptake of bile acids and cholesterol) also induce LDL receptors (Fig. 28.1). Fibrates induce the LPL gene and repress the ApoC-III gene; both of these effects enhance the lipolysis of TG in VLDL (Fig. 28.1). Niacin, or vitamin $B_3$, when given at high doses, decreases the mobilization of free fatty acids and the hepatic production of ApoB-100, leading to decreased production of VLDL and, subsequently, IDL and LDL (Fig. 28.1).

## Disorders of Exogenous Lipoprotein Metabolism

### Disorders of Chylomicron Removal

#### Lipoprotein Lipase Deficiency

In the first several months of life, the patient with classic LPL deficiency presents with profound hypertriglyceridemia, often ranging between 5000 mg/dl and 10,000 mg/dl (Table 28.4). The plasma cholesterol level is usually one tenth of the TG level. This disorder is often suspected because of colic, creamy plasma on the top of a hematocrit tube, hepatosplenomegaly or eruptive xanthomas. Usually, only the chylomicrons are elevated (type-I phenotype; Table 28.5) but, occasionally, the VLDL are also elevated (type-V phenotype). The disorder can present with abdominal pain and pancreatitis later in childhood, a life-threatening complication of the massive elevation in chylomicrons. Lipemia retinalis is usually present; premature atherosclerosis is uncommon. When chylomicrons are markedly increased, they can replace water volume in plasma, producing artifactual decreases in concentrations of plasma constituents; for example, for each 1000-mg/dl increase of plasma TG, serum sodium levels decrease between 2 mEq/l and 4 mEq/l.

**Table 28.4.** Guidelines for plasma triglyceride levels in adults

| Triglyceride levels | | Category |
|---|---|---|
| mg/dl | mmol/l | |
| <200 | <2.28 | Normal |
| 200–399 | 2.28–4.55 | Borderline |
| 400–999 | 4.56–11.39 | High |
| >1000 | >11.40 | Very high |

**Table 28.5.** Lipoprotein phenotypes of hyperlipidemia

| Lipoprotein phenotype | Elevated lipoprotein |
|---|---|
| Type I | Chylomicrons |
| Type IIa | LDL |
| Type IIb | LDL, VLDL |
| Type III | Cholesterol-enriched VLDL |
| Type IV | VLDL |
| Type V | Chylomicrons, VLDL |

*LDL*, low-density lipoprotein; *VLDL*, very-low-density lipoprotein

The diagnosis is first made by a test for post-heparin lipolytic activity (PHLA). LPL is attached to the surface of endothelial cells through a heparin-binding site. After the intravenous injection of heparin (60 units/kg), LPL is released and the activity of the enzyme is assessed in plasma drawn 45 min after the injection of heparin. The mass of LPL released can also be assessed, using an enzyme-linked immunosorbent assay (ELISA). Parents of LPL-deficient patients often have LPL activity halfway between those of normal controls and those of the LPL-deficient child. The parents may or may not be hypertriglyceridemic.

Treatment is a diet very low in fat (10–15% of total calorie intake) [5]. Lipid lowering medication is ineffective and is not used. Affected infants can be given Portagen, a soybean-based formula containing medium-chain TG (MCT). MCT do not require the formation of chylomicrons for absorption, since they are directly transported from the intestine to the liver by the portal vein. A subset of LPL-deficient patients with unique, possibly post-transcriptional genetic defects respond to therapy with MCT oil or ω-3 fatty acids by normalizing fasting plasma TG; a therapeutic trial with MCT oil should, therefore, be considered in all patients presenting with the familial chylomicronemia syndrome [5]. Older children may also utilize MCT oil to improve the palatability and caloric content of their diet. Care must be taken that affected infants and children get at least 1% of their calories from the essential fatty acid linoleic acid.

Familial LPL deficiency is a rare, autosomal-recessive condition that affects about one in one million children. Parents are often consanguineous. The large amounts of chylomicrons result from mutations in the LPL gene. Over 30 mutations in the structural LPL gene have been reported to result in deficiency of LPL. These include insertion-deletion mutations, splice-site defects and nonsense mutations, particularly in the highly conserved exons (4, 5 and 6) of the LPL gene.

### ApoC-II Deficiency

Marked hypertriglyceridemia (TG > 1000 mg/dl) can also present in patients with a rare autosomal recessive disorder affecting ApoC-II, the co-factor for LPL. Affected homozygotes have been reported to have TG levels ranging from 500 mg/dl to 10,000 mg/dl (Table 28.4). ApoC-II deficiency can be expressed in childhood but is often delayed until adulthood. The disorder is suspected due to milky serum or plasma or by unexplained recurrent bouts of pancreatitis. A type-V lipoprotein phenotype (Table 28.5) is often found, but a type-I pattern may also be present. Eruptive xanthomas and lipemia retinalis may also be found. As with the LPL defect, those with ApoC-II deficiency do not get premature atherosclerosis.

The diagnosis can be confirmed by a PHLA test and by measuring ApoC-II levels in plasma using an ELISA assay. ApoC-II levels range from very low to undetectable.

The deficiency can be corrected by the addition of normal plasma to the in vitro assay for PHLA. The treatment of patients with ApoC-II deficiency is the same as that discussed above for LPL deficiency. Infusion of normal plasma in vivo into an affected patient will decrease plasma TG levels.

At least 14 families with ApoC-II deficiency have been described. Thus, it is even rarer than LPL deficiency. Frame-shift, donor splice-site and single base substitutions leading to stop codons have been found in ApoC-II deficiency. Obligate heterozygous carriers of ApoC-II mutants usually have normal plasma lipid levels despite a 50% reduction in ApoC-II levels.

## Disorders of Endogenous Lipoprotein Metabolism

### Disorders of Very Low Density Lipoprotein Overproduction

#### Familial Hypertriglyceridemia

Patients with familial hypertriglyceridemia (FHT) usually present with elevated TG levels with normal LDL-cholesterol levels (type-IV lipoprotein phenotype; Table 28.5). The diagnosis is confirmed by finding at least one (and preferably two or more) first-degree relatives with a similar type-IV lipoprotein phenotype. The VLDL levels may increase to a considerable degree, leading to hypercholesterolemia and marked hypertriglyceridemia (>1000 mg/dl) and occasionally to hyperchylomicronemia (type-V lipoprotein phenotype; Table 28.5). This extreme presentation of FHT is usually due to the presence of obesity and type-II diabetes. Throughout this spectrum of hypertriglyceridemia and hypercholesterolemia, the LDL-cholesterol levels remain normal or below normal. The LDL particles may be small and dense, secondary to the hypertriglyceridemia, but the number of these particles is not increased (see also below).

Patients with FHT often manifest hyperuricemia in addition to hyperglycemia. There is a greater propensity for peripheral vascular disease than coronary-artery disease (CAD) in FHT. A family history of premature CAD is not usually present. The unusual patient with FHT who has a type-V lipoprotein phenotype may develop pancreatitis.

The metabolic defect in FHT appears to be due to the increased hepatic production of TG, but the production of ApoB-100 is not increased. This results in the enhanced secretion of very large VLDL particles that are not hydrolyzed at a normal rate by LPL and ApoC-II. Thus, in FHT, there is not an enhanced conversion of VLDL into IDL and (subsequently) LDL (Fig. 28.1).

Diet, particularly reduction to ideal body weight, is the cornerstone of therapy in FHT. For patients with persistent hypertriglyceridemia above 400 mg/dl, treatment with fibric-acid derivatives, niacin or the statins may reduce the elevated TG by up to 50%. Management of type-II diabetes, if present, is also an important part of the management of patients with FHT (see "Guidelines for the Treatment of Hyperlipidemia").

## Familial Combined Hyperlipidemia and the Small, Dense-Low Density Lipoprotein Syndromes

### Clinical Presentation

Patients with familial combined hyperlipidemia (FCHL) may present with elevated cholesterol alone (type-IIa lipoprotein phenotype), elevated TG alone (type-IV lipoprotein phenotype) or elevations of both the cholesterol and TG (type-IIb lipoprotein phenotype; Table 28.5). The diagnosis of FCHL is confirmed by the finding of a first-degree family member who has a lipoprotein phenotype different than the proband. Other characteristics of FCHL include the presence of an increased number of *small, dense LDL particles*; this links FCHL to other disorders, including hyperapobetalipoproteinemia (hyperapoB), LDL subclass pattern B, familial dyslipidemic hypertension and syndrome X [6]. In addition to hypertension, patients with small-dense-LDL syndromes can also manifest hyperinsulinism, glucose intolerance, low HDL-C levels and increased visceral obesity (syndrome X). From a clinical perspective, FCHL and other small-dense-LDL syndromes are clearly more prevalent in families with premature CAD; as a group, these disorders are the most commonly recognized dyslipidemias associated with premature CAD and may account for one third (or more) of families with early CAD.

### Metabolic Derangement

There are three metabolic defects that have been described both in FCHL patients and in those with hyperapoB: (1) overproduction of VLDL and ApoB-100 in liver; (2) slower removal of chylomicrons and chylomicron remnants; and (3) abnormally increased FFA levels [6, 7]. The abnormal FFA metabolism in FCHL and hyperapoB subjects may reflect the primary defect in these patients. The elevated FFA levels indicate an impaired metabolism of intestinally derived TG-rich lipoproteins in the postprandial state and impaired insulin-mediated suppression of serum FFA levels. Fatty acids and glucose compete as oxidative fuel sources in muscle such that increased concentrations of FFA inhibit glucose uptake in muscle and result in insulin resistance. Finally, elevated FFA may drive hepatic overproduction of TG and ApoB.

It has been hypothesized that a cellular defect in the adipocytes of hyperapoB patients prevents the normal stimulation of FFA incorporation into TG by a small molecular-weight, basic protein (BP) called the acylation-stimulatory protein (ASP). The active component responsible for enhancement of ASP in chylomicrons of human adipocytes does not appear to be an apolipoprotein but may be transthyretin, a protein that binds retinol-binding protein and complexes thyroxine and retinol. ASP also appears to be generated in vivo by human adipocytes, a process that is accentuated postprandially, supporting the hypothesis that ASP plays an important role in the clearance of TG from plasma and in the fatty-acid storage in adipose tissue.

A defect in the adipocytes of hyperapoB patients may explain both metabolic abnormalities of TG-rich particles in hyperapoB. Following ingestion of dietary fat, chylomicron TG is hydrolyzed by LPL, producing FFA. The defect in the normal stimulation of the conversion of FFA into TG by ASP in the adipocytes of hyperapoB patients leads to increased levels of FFA that: (1) flux back into the liver, increasing VLDL ApoB production; and (2) inhibit (due to feedback) further hydrolysis of chylomicron TG by LPL [6]. Alternatively, there could be a defect in the stimulation of release of ASP by adipocytes, perhaps due to an abnormal transthyretin/retinol-binding system. In that regard, plasma retinol levels have been found to be significantly lower in FCHL patients. This may also affect the peroxisome proliferator/activator receptors, which are retinoic-acid dependent.

The Johns Hopkins group isolated and partially characterized three distinct BPs from normal human serum [6]. BP-I increases the mass of cellular triacylglycerols in cultured fibroblasts from normal subjects about twofold, while there is a 50% deficiency in such activity in cultured fibroblasts from hyperapoB patients. BP-I appears to be a distinct protein from $C_{3a}$ desArg but has similar physiologic effects [8]. In contrast, BP-II abnormally stimulates the formation of unesterified and esterified cholesterol in hyperapoB cells, an effect that might further accentuate the overproduction of ApoB and VLDL in hyperapoB patients [6]. Pilot data in hyperapoB fibroblasts indicate a deficiency in the high-affinity binding of BP-I but an enhanced high-affinity binding of BP-II [8]. These observations together suggest the existence of a receptor-mediated process for BP-I and BP-II that involves signal transduction [8]. We have postulated that a defect in a BP receptor might exist in a significant number of patients with hyperapoB and premature CAD.

### Treatment and Prognosis

The treatment of FCHL and hyperapoB starts with a diet reduced in total fat, saturated fat and cholesterol. This will reduce the burden of postprandial chylomicrons and chylomicron remnants (which may also be atherogenic). Reduction to ideal body weight may improve insulin sensitivity and decrease VLDL overproduction. Regular aerobic exercise also appears to be important. Two classes of drugs, fibric acids and nicotinic acid, lower TG and increase HDL and may also convert small, dense LDL to normal-sized LDL. The HMG-CoA-reductase inhibitors do not appear to be as effective as the fibrates or nicotinic acid in converting small, dense LDL into large, buoyant LDL. However, the "statins" are very effective in lowering LDL cholesterol and the total number of atherogenic small, dense LDL particles. In many patients with FCHL, combination therapy of a statin with either a fibrate or nicotinic acid will be required to obtain the optimal lipoprotein profile [9] (see "Guidelines for the Treatment of Hyperlipidemia"). Patients with small-dense-LDL syndromes appear to have a greater improvement in coronary stenosis severity on combined treatment. This appears to be associated with drug-induced improvement in LDL buoyancy.

### Genetics

The basic genetic defect(s) in FCHL and the other small, dense LDL syndromes are not known. FCHL and these other syndromes are clearly genetically heterogeneous, and a number of genes may influence the expression of FCHL and the small-dense-LDL syndromes (an oligogenic effect) [6, 9, 10]. FCHL in Finnish families has been linked to an area on chromosome 1 [10].

## Lysosomal Acid Lipase Deficiency: Wolman Disease and Cholesteryl-Ester Storage Disease

Wolman disease is a fatal disease that occurs in infancy [11]. Clinical manifestations include hepatosplenomegaly, steatorrhea and failure to thrive.

Patients with Wolman disease have a life span that is generally less than 1 year, while those with cholesteryl-ester storage disease (CESD) can survive for longer periods of time [12]. In some cases, patients with CESD have developed premature atherosclerosis.

Lysosomal acid lipase (LAL) is an important lysosomal enzyme that hydrolyzes LDL-derived cholesteryl esters into unesterified cholesterol. Intracellular levels of unesterified cholesterol are important in regulating cholesterol synthesis and LDL-receptor activity. In LAL deficiency, cholesteryl esters are not hydrolyzed in lysosomes and do not generate unesterified cholesterol. In response to low levels of intracellular unesterified cholesterol, cells continue to synthesize cholesterol and ApoB-containing lipoproteins.

In CESD, the inability to release free cholesterol from lysosomal cholesteryl esters results in elevated synthesis of endogenous cholesterol and increased production of ApoB-containing lipoproteins. Lovastatin reduced both the rate of cholesterol synthesis and the secretion of ApoB-containing lipoproteins, leading to significant reductions in total (–197 mg/dl) and LDL (–102 mg/dl) cholesterol and TG (–101 mg/dl) [13]. Wolman disease and CESD are autosomal-recessive disorders that are due to mutations in the LAL gene on chromosome 10.

## Disorders of Low Density Lipoprotein Removal

### Familial Hypercholesterolemia (LDL-Receptor Defect)

#### Clinical Presentation

Familial hypercholesterolemia (FH) is an autosomal-dominant disorder that presents in the heterozygous state with a twofold elevation in the plasma levels of total and LDL cholesterols [14]. Since FH is completely expressed at birth and early in childhood, it is often associated with premature CAD; by age 50 years, about half the heterozygous FH males and 25% of affected females will develop CAD. Heterozygotes develop tendon xanthomas in adulthood, often in the Achilles tendons and the extensor tendons of the hands. Homozygotes usually develop CAD in the second decade; atherosclerosis often affects the aortic valve, leading to life-threatening aortic stenosis. Virtually all FH homozygotes have planar xanthomas by the age of 5 years, notably in the webbing of fingers and toes and over the buttocks.

**Metabolic Derangement and Genetics**

FH is one of the most common inborn errors of metabolism and affects 1 in 500 children. It is due to one of more than 150 different mutant alleles at the LDL-receptor locus. If an individual inherits two mutant alleles for the LDL receptor, he/she usually presents with a four-to sixfold increase in LDL cholesterol levels (FH-homozygous phenotype). The chance that two FH heterozygotes will marry is 1/500 × 1/500, or 1/250,000; there is a one in four chance that the maternal and paternal mutant alleles will be transmitted to a child (prevalence of one in one million). Most patients inherit two different mutant alleles (genetic compounds), but some have two identical LDL-receptor mutations (true homozygotes). Mutant alleles may fail to produce LDL-receptor proteins (null alleles), may encode receptors blocked in intracellular transport between endoplasmic reticulum and Golgi (transport-defective alleles), may produce proteins that cannot bind LDL normally (binding defective) or may bind LDL normally but do not internalize LDL (internalization defects). Prenatal diagnosis of FH homozygotes can be performed by assays of LDL-receptor activity in cultured amniotic fluid cells, by direct DNA analysis of the molecular defect(s) or by linkage analysis using tetranucleotide DNA polymorphisms.

**Treatment**

Treatment of FH includes a diet low in cholesterol and saturated fat. FH heterozygotes usually respond to higher doses of HMG-CoA-reductase inhibitors. The addition of bile-acid-binding sequestrants (cholestyramine, colestipol) can produce a further complementary fall in LDL. In those FH heterozygotes who may be producing increased amounts of VLDL (leading to borderline hypertriglyceridemia and low HDL-C levels), nicotinic acid may be useful in combination therapy. Nicotinic acid can also be used to lower an elevated lipoprotein a [Lp(a)]. FH homozygotes may respond to high doses of HMG-CoA-reductase inhibitors or nicotinic acid, both of which appear to decrease the production of hepatic VLDL, leading to decreased production of LDL. Usually, FH homozygotes will require LDL apheresis every 2 weeks to effect a further lowering of LDL into a range that is less atherogenic. In the future, ex vivo gene therapy will become the treatment of choice for FH homozygotes [15].

**Familial Ligand-Defective ApoB**

Heterozygotes with familial ligand-defective ApoB (FLDB) may present with normal, moderately elevated, or markedly increased LDL-C levels [16]. Hypercholesterolemia is usually not as markedly elevated in FLDB as in patients with heterozygous FH, a difference attributed to effective removal of VLDL and IDL particles through the interaction of ApoE with the LDL receptors in FLDB. About 1/20 affected patients present with tendon xanthomas and more extreme hypercholesterolemia. This disorder represents a very small fraction of patients with premature CAD, i.e., no more than 1%.

In FLDP patients, one mutant allele produces a defective ligand-binding region of ApoB-100; the most commonly recognized mutation is due to a R3500E substitution. The frequency of FLDB heterozygotes has been estimated as 1 in 500. LDL-receptor activity is normal, but defective binding of mutant ApoB-100 reduces the clearance of LDL, leading to elevation of plasma LDL-C. Since the clearance of VLDL remnants and IDL occurs through the binding of ApoE (not ApoB) to the LDL-(B, E) receptor, the clearance of these TG-enriched particles in this disorder is not affected.

Dietary and drug treatment of FLDB is similar to that used for FH heterozygotes. Induction of LDL receptors will enhance the removal of the LDL particles that contain the normal ApoB-100 molecules and will increase the removal of VLDL remnants and IDL that utilize ApoE (but not ApoB-100) as a ligand for the LDL receptor.

**Sitosterolemia**

This is a rare, autosomal-recessive trait in which patients present with normal to moderately elevated total and LDL cholesterol levels, tendon and tuberous xanthomas and premature CAD [17]. Homozygotes manifest abnormal intestinal hyperabsorption of plant or shellfish sterols (sitosterol, campesterol and stigmasterol). In normal individuals, plant sterols are not absorbed, and plasma sitosterol levels are low (0.3–1.7 mg/dl) and are less than 1% of the total plasma sterol; in homozygotes with sitosterolemia, levels of total plant sterols are elevated (13–37 mg/dl) and represent 7–16% of the total plasma sterols.

Treatment primarily consists of avoidance of shellfish or plant foods with high fat content, such as oils and margarines. Bile-acid-binding resins, such as cholestyramine, are the drugs of choice and are effective in lowering plant sterol and LDL sterol concentrations. The statins are ineffective and should not be used to treat sitosterolemic patients. The basic defect is unknown, but the disorder has been linked to a locus on chromosome 2 [17].

## Disorders of Endogenous and Exogenous Lipoprotein Transport

### Dysbetalipoproteinemia (Type-III Hyperlipoproteinemia)

This disorder is often associated with premature atherosclerosis of the coronary, cerebral and peripheral arteries. Xanthomas are often present and usually are tuberoeruptive or planar, especially in the creases of the palms. Occasionally, tuberous and tendon xanthomas are found.

Patients with dysbetalipoproteinemia present with elevations in both plasma cholesterol and TG, usually (but not always) above 300 mg/dl. The hallmark of the disorder is the presence of VLDL that migrate as β-lipoproteins (β-VLDL), rather than as pre-β-lipoproteins (type III lipoprotein phenotype; Table 28.5). β-VLDL reflect the accumulation of cholesterol-enriched remnants of both hepatic VLDL and intestinal chylomicrons (Fig. 28.1) [18]. These remnants accumulate because of the presence of a dysfunctional ApoE, the ligand for the receptor-mediated removal of both chylomicron and VLDL remnants by the liver.

The diagnosis of dysbetalipoproteinemia includes: (1) demonstration of the presence of an $E_2E_2$ genotype; (2) performing preparative ultracentrifugation and finding the presence of β-VLDL on agarose-gel electrophoresis (floating β-lipoproteins); and (3) a cholesterol-enriched VLDL (VLDL cholesterol/TG ratio > 0.30, normal ratio < 0.30). LDL and HDL cholesterol levels are low or normal. Patients with this disorder are very responsive to therapy. A low-fat diet is important to reduce the accumulation of chylomicron remnants, and reduction to ideal body weight may decrease the hepatic overproduction of VLDL particles. The drug of choice is a fibric-acid derivative, but nicotinic acid and HMG-CoA-reductase inhibitors may also be effective. Treatment of the combined hyperlipidemia in dysbetalipoproteinemia with a fibrate will correct both the hypercholesterolemia and hypertriglyceridemia; this effect is in contrast to treatment of FCHL with fibrates alone, which usually reduces the TG level but increases the LDL cholesterol level.

There are two genetic forms of dysbetalipoproteinemia. The most common form is inherited as a recessive trait. Such patients have an $E_2E_2$ genotype. The $E_2E_2$ genotype is necessary but not sufficient for dysbetalipoproteinemia. Other genetic and metabolic factors, such as overproduction of VLDL in the liver (seen in FCHL), or hormonal and environmental conditions (such as hypothyroidism, low estrogen state, obesity or diabetes) are necessary for the full-blown expression of dysbetalipoproteinemia. The recessive form has a penetrance delayed until adulthood and a prevalence of about 1:2000. In the rarer form of the disorder, dominantly inherited and expressed as hyperlipidemia even in childhood, there is a single copy of another defective ApoE allele [18].

### Hepatic Lipase Deficiency

Patients with HL deficiency can present with features similar to dyslipoproteinemia (type-III hyperlipoproteinemia; see above), including hypercholesterolemia, hypertriglyceridemia, accumulation of TG-rich remnants, planar xanthomas and premature cardiovascular disease [19]. Recurrent bouts of pancreatitis have been described. The LDL cholesterol is usually low or normal in both disorders.

HL hydrolyzes both TG and phospholipids in plasma lipoproteins. As a result, HL converts IDL into LDL and $HDL_2$ into $HDL_3$, thus playing an important role in the metabolism of both remnant lipoproteins and HDL. HL shares a high degree of homology with LPL and pancreatic lipase. HL deficiency can be distinguished from dysbetalipoproteinemia in two ways: first, the elevated TG-rich lipoproteins have a normal VLDL cholesterol/TG ratio (less than 0.3), because the TG is not being hydrolyzed by HL; second, the HDL cholesterol often exceeds the 95th percentile in HL deficiency but is low in dysbetalipoproteinemia.

The diagnosis is made by a PHLA test (see "LPL Deficiency"). Absent HL activity is documented by measuring total PHLA activity and by measuring HL and LPL activities separately. Treatment includes a low-total-fat diet. In one report, the dyslipidemia in HL deficiency improved on treatment with lovastatin but not with gemfibrozil.

HL deficiency is a rare genetic disorder that is inherited as an autosomal-recessive trait. The frequency of this disorder is not known, but it has been identified in at least four families. Obligate heterozygotes are normal. The molecular defects described in HL deficiency include a single A → G substitution in intron I of the HL gene [20].

## Disorders of Reduced Low Density Lipoprotein-Cholesterol Levels

### Abetalipoproteinemia

Abetalipoproteinemia is a rare, autosomal-recessive disorder in patients with undetectable plasma ApoB levels [21]. Total cholesterol levels are exceedingly low

(20–50 mg/dl), and no detectable levels of chylomicrons, VLDL or LDL are present. HDL levels are measurable but low. Parents have normal lipid levels.

Patients present with symptoms of fat malabsorption and neurological problems. Fat malabsorption occurs in infancy with symptoms of failure to thrive (poor weight gain and steatorrhea). Fat malabsorption is secondary to the inability to assemble and secrete chylomicrons from enterocytes. Neurological problems begin during adolescence and include dysmetria, cerebellar ataxia and spastic gait. Other manifestations include atypical retinitis pigmentosa, anemia (acanthocytosis) and arrhythmias.

Treatment of patients with abetalipoproteinemia is difficult. Steatorrhea can be controlled by reducing the intake of fat to 5–20 g/day. This measure alone can result in marked clinical improvement and growth acceleration. In addition, the diet should be supplemented with linoleic acid (5 g corn oil or safflower oil/day). As a caloric substitute for long-chain fatty acids, MCT may produce hepatic fibrosis; thus, MCT should be used with caution, if at all. Fat-soluble vitamins should be added to the diet. Rickets can be prevented by normal quantities of vitamin D, but 200–400 IU/kg/day of vitamin A may be required to raise the level of vitamin A in plasma to normal. Enough vitamin K (5–10 mg/day) should be given to maintain normal prothrombin time. Neurologic and retinal complications may be prevented or ameliorated through oral supplementation with vitamin E (150–200 mg/kg/day). Adipose tissue (rather than plasma) may be used to assess the delivery of vitamin E.

It was initially thought that the lack of plasma ApoB levels were due to defects in the ApoB gene. Subsequent studies have demonstrated no defects in the ApoB gene. Immunoreactive ApoB-100 is present in liver and intestinal cells. Wetterau and colleagues [22] found that the defect in synthesis and secretion of ApoB is secondary to the absence of microsomal TG-transfer protein (MTP), a molecule that permits the transfer of lipid to ApoB. MTP is a heterodimer composed of both the ubiquitous multifunctional protein called protein disulfide isomerase and a unique 97-kDa subunit. Mutations that lead to the absence of the functional 97-kDa subunit cause abetalipoproteinemia. At least 13 mutant 97-kDa-subunit alleles have been described.

## Hypobetalipoproteinemia

Patients with hypobetalipoproteinemia often have both a reduced risk for premature atherosclerosis and an increased life span. These patients do not have any of the physical stigmata of dyslipidemia. The concentrations of fat-soluble vitamins in plasma are low to normal.

Most patients have low levels of LDL cholesterol (below the 5th percentile; ~40–60 mg/dl), owing to the inheritance of one normal allele and one autosomal-dominant mutant allele for a truncated ApoB. Occasionally, hypobetalipoproteinemia is secondary to anemia, dysproteinemias, hyperthyroidism, intestinal lymphangiectasia with malabsorption, myocardial infarction, severe infections and trauma.

Plasma levels of truncated ApoB are generally low and are thought to be secondary to low synthesis and secretory rates of the truncated forms of ApoB from hepatocytes and enterocytes. The catabolism of LDL in hypobetalipoproteinemia also appears to be increased. The diagnosis is confirmed by demonstrating the presence of a truncated ApoB in plasma.

No treatment is required. Neurologic signs and symptoms of a spinocerebellar degeneration similar to those of Friedreich ataxia and peripheral neuropathy have been found in several affected members.

Over 25 gene mutations (nonsense and frame-shift mutations) have been shown to affect the full transcription of ApoB and cause familial hypobetalipoproteinemia. The various gene mutations lead to the production of truncated ApoB.

## Homozygous Hypobetalipoproteinemia

The clinical presentation of children with this disorder depends on (1) whether they are homozygous for null alleles in the ApoB gene (i.e., make no detectable ApoB) or are homozygous (or are compound heterozygotes) for other alleles and (2) whether their lipoproteins contain small amounts of ApoB or a truncated ApoB [23]. Null-allele homozygotes are phenotypically similar to those with abetalipoproteinemia (see above) and may have fat malabsorption, neurologic disease and hematologic abnormalities as their prominent clinical presentation; these patients will require treatment similar to that needed by patients with abetalipoproteinemia (see above). However, the parents of these children are heterozygous for hypobetalipoproteinemia. Patients with homozygous hypobetalipoproteinemia may develop less marked ocular and neuromuscular manifestations than those with abetalipoproteinemia and may develop these symptoms at a later age. The concentrations of fat-soluble vitamins are low.

## Disorders of Reverse Cholesterol Transport

### Familial Hypoalphalipoproteinemia

Hypoalphalipoproteinemia is defined as a low level of HDL cholesterol (<5th percentile; age and sex specific) in the presence of normal lipid levels [24]. Patients with this syndrome have a significantly increased prevalence of CAD but do not manifest the clinical findings typical of other forms of HDL deficiency (see below). Low HDL cholesterol levels of this degree are most often secondary to disorders of TG metabolism (see above). Consequently, primary hypoalphalipoproteinemia, although more prevalent than rare recessive disorders including deficiencies in HDL, is relatively uncommon. In some families, hypoalphalipoproteinemia behaves as an autosomal-dominant trait, but the basic defect is unknown.

## ApoA-I Mutations

The HDL cholesterol levels are very low (0–4 mg/dl), and the ApoA-I levels are usually less than 5 mg/dl. Corneal clouding is usually present in these patients. Planar xanthomas are not infrequently described; the majority (but not all) of these patients develop premature CAD [24–26].

The *APOA-1* gene exists on chromosome 11 as part of a gene cluster with the *APOC-3* and *APOA-4* genes. A variety of molecular defects have been described in *APOA-1*, including gene inversions, gene deletions and nonsense and missense mutations.

In contrast, *APOA-1* structural variants, usually due to a single amino acid substitution, do not have (in most instances) any clinical consequences [26]. Despite lower HDL cholesterol levels (decreased by about one half), premature CAD is not ordinarily present. In fact, in one Italian variant, ApoA-I$_{Milano}$, the opposite has been observed (i.e., increased longevity in affected subjects).

## Tangier Disease

The name Tangier disease is derived from the island of Tangier in the Chesapeake Bay in Virginia, USA. HDL cholesterol levels are extremely low and have an abnormal composition (HDL Tangier or HDL$_T$). HDL$_T$ are chylomicron-like particles seen in patients on a high-fat diet; the particles disappear when the patient consumes a low-fat diet [24–26].

The characteristic clinical findings in Tangier patients include the presence of enlarged, orange–yellow tonsils, splenomegaly and a relapsing peripheral neuropathy. The finding of orange tonsils is due to the deposition of β-carotene-rich cholesteryl esters (foam cells) in the lymphatic tissue. Other sites of foam cell deposition include the skin, peripheral nerves, bone marrow and the rectum. Mild hepatomegaly, lymphadenopathy and corneal infiltration (in adulthood) may also occur.

In general, patients with Tangier disease have an increased incidence of atherosclerosis in adulthood [26]. Treatment with a low-fat diet diminishes the abnormal lipoprotein species, which are believed to be remnants of abnormal chylomicron metabolism.

The *APOA-1* gene in Tangier patients is normal, and recent studies indicate that there is hypercatabolism of HDL. In addition, other studies suggest a defect in cell signaling and decreased ApoA-I mediated cholesterol and phospholipid efflux in fibroblasts isolated from Tangier patients [27]. The disorder has been linked to chromosome 9q31 [28].

## Lecithin: Cholesterol Acyl Transferase Deficiency and Fish-Eye Disease

LCAT is an enzyme located on the surface of HDL particles and is important in transferring fatty acids from the sn-2 position of phosphatidylcholine (lecithin) to the 3-β-hydroxyl group on cholesterol (Table 28.3). In this process, lysolecithin and esterified cholesterol are generated (α-LCAT). Esterification can also occur on VLDL/LDL particles (β-LCAT).

In patients with *classic LCAT deficiency*, both α- and β-LCAT activities are missing [29]. The diagnosis should be suspected in patients presenting with low HDL-C levels, corneal opacifications and renal disease (proteinuria, hematuria). Laboratory tests include the measurement of the plasma free-cholesterol-to-total-cholesterol ratio. Levels above 0.7 are diagnostic for LCAT deficiency.

In *fish-eye disease*, only α-LCAT activity is absent. Patients present with corneal opacifications but do not have renal disease [29]. It has been hypothesized that the variability of clinical manifestations in patients with fish-eye disease compared with symptoms in patients with LCAT deficiency may reside in the amount of total plasma LCAT activity.

To date, no therapies treat the underlying genetic mutation. Patients succumb primarily to renal disease, and atherosclerosis may be accelerated by the underlying nephrosis. Thus, patients with LCAT deficiency and other lipid metabolic disorders associated with renal disease should be aggressively treated (including a low-fat diet). This includes patients with secondary dyslipidemia associated with the nephrotic syndrome, which responds to statin therapy.

LCAT deficiency is a rare, autosomal-recessive disorder. More than 20 mutations in this gene, located on chromosome 16, have been described.

## Cholesteryl Ester Transfer Protein Deficiency

The role of the CETP in atherosclerosis has not been well defined. The CETP gene is upregulated in peripheral tissues and liver in response to dietary or endogenous hypercholesterolemia. HDL particles isolated from patients with CETP deficiency have been shown to be less effective in promoting cholesterol efflux from cultured cells. This may be due to the increased concentration of cholesterol within the HDL particles and its inability to adsorb additional cholesterol from peripheral tissues. Some investigators have termed this type of HDL as being "dysfunctional".

Elevated HDL-C levels due to deficiency of CETP were first described in Japanese families, and several mutations have been found. Increased coronary heart disease (CHD) in Japanese families with CETP deficiency was primarily observed for HDL cholesterol levels in the range 41–60 mg/dl; for HDL cholesterol levels greater than 60 mg/dl, men with and without mutations had low CHD prevalence [30]. Thus, genetic CETP deficiency appears to be an independent risk factor for CHD. These effects occur despite lower levels of ApoB in CETP deficiency [31].

## Elevated Lipoprotein(a)

Lp(a) consists of one molecule of LDL whose ApoB-100 is covalently linked to one molecule of Apo(a) by a disulfide bond [32]. Apo(a) is highly homologous to plasminogen and, when the Lp(a) level is elevated (>30 mg/dl), Apo(a) interferes with the thrombolytic action of plasmin, promoting thrombosis. Apo(a) exists in a number of size isoforms, with the smaller isoforms correlating with higher plasma levels of Lp(a).

The physiological function(s) of Lp(a) are unknown. Plasma levels in whites tend to be lower than in blacks (median values: 1 mg/ml and 10 mg/ml, respectively). However, elevated plasma levels of Lp(a) do not correlate directly with the extent of cardiovascular disease in African-Americans.

Niacin and estrogen can effectively lower Lp(a) levels, while the statins and fibrates do not. To date, clinical-trial evidence is lacking regarding the effect of lowering Lp(a) on the prevalence of cardiovascular disease.

## Guidelines for the Treatment of Hyperlipidemia

### Clinical Evaluation and Treatment

#### Clinical Evaluation

The patient who is being evaluated for dyslipidemia requires a thorough family history and an evaluation of his current intake of dietary fat and cholesterol. We have found it useful to provide the patient with standard forms, which they complete beforehand. The family history is then reviewed for premature (before 60 years of age) cardiovascular disease (heart attacks, strokes, angina, peripheral vascular disease), dyslipidemia, diabetes mellitus and hypertension in grandparents, parents, siblings, children and aunts and uncles. A dietary assessment can be performed by using a questionnaire and computing a total score.

The medical history is focused on the two major complications of dyslipidemias, atherosclerotic cardiovascular disease and pancreatitis. The patient is asked about chest pain, arrhythmias, palpitations, myocardial infarction, stroke (including transient ischemic attacks), coronary-artery-bypass graft surgery and balloon angioplasty. The results of past resting and stress electrocardiograms and coronary arteriography are assessed. Any history of recurrent abdominal pain and pancreatitis is reviewed. The past and current use of lipid-lowering drugs is determined, as is a history of untoward reactions or side effects. The review of systems includes diseases of the liver, thyroid and kidney, the presence of diabetes mellitus and past operations including transplantation. For women, a menstrual history, including current use of oral contraceptives and post-menopausal estrogen-replacement therapy, is obtained.

The presence of other risk factors for premature CAD [32] are systematically assessed. All of the following are at greater risk: patients above a certain age (men over 45 years of age or females over 55 years of age or postmenopausal without estrogen-replacement therapy); current smokers; patients suffering from diabetes mellitus or hypertension (poorly controlled or treated); patients with a family history (onset of cardiovascular disease in a first-degree male relative before the age of 55 years or in a first-degree female relative before the age of 65 years); and patients with HDL-cholesterol levels below 35 mg/dl or electrocardiographic changes showing left-ventricular hypertrophy with strain.

Height and weight are determined to assess obesity using the Quetelet index [weight (kg)/height$^2$ (m$^2$)]. The physical examination includes an assessment of tendon, tuberous and planar xanthomas. The eyes are examined for the presence of xanthelasmas, corneal arcus, corneal clouding, lipemia retinalis and atherosclerotic changes in the retinal blood vessels. The cardiovascular exam includes an examination for bruits in the carotid, abdominal and femoral arteries, auscultation of the heart, assessment of peripheral pulses and measurement of blood pressure. The rest of the exam includes palpation of the thyroid, assessment of hepatosplenomegaly and deep-tendon reflexes (which are decreased in hypothyroidism).

The clinical chemistry examination includes (at the minimum) measurements of total cholesterol, total TG, LDL cholesterol and HDL cholesterol, a chemistry panel to assess fasting blood sugar and uric acid and tests of liver and kidney function and thyroid-stimulating hormone (TSH). Other tests may be ordered when clinically indicated, such as "non-traditional" risk factors for cardiovascular disease, i.e., Lp(a), homocysteine and small, dense LDL. Glycosylated hemoglobin is measured when a patient has known diabetes mellitus.

### Dietary Treatment, Weight Reduction and Exercise

The cornerstone of treatment of dyslipidemia is a diet reduced in total fat, saturated fat and cholesterol (Table 28.6) [32, 33]. This is important to reduce the burden of postprandial lipemia and to induce LDL receptors. A step-I and step-II dietary approach is often used (Table 28.6) [32], but most dyslipidemic patients will require a step-II diet. The use of a registered dietician or nutritionist is usually essential in achieving dietary goals. The addition of 400 IU of vitamin E as an antioxidant is often useful [34].

If a patient is obese (Quetelet index > 30) or moderately overweight (Quetelet index = 25–30), weight reduction will be an important part of the dietary management. This is particularly true if hypertriglyceridemia or diabetes mellitus are present.

**Table 28.6.** National cholesterol education-program diets: steps I and II

Step I
  Less than 30% of calorie intake as fat: less than 10% saturated, 10–15% monounsaturated, and up to 10% polyunsaturated
  55% of calorie intake as carbohydrates
  15–20% of calorie intake as protein
  Less than 300 mg cholesterol/day
Step II
  Less than 30% of calorie intake as fat: <7% saturated, 10–15% monounsaturated, and 10% polyunsaturated
  Less than 200 mg cholesterol/day

Regular aerobic exercise is beneficial in most patients to help control weight and dyslipidemia. The duration, intensity and frequency of exercise are critical. For an adult, a minimum of a 1000 calories/week of aerobic exercise is required. This usually translates into three or four sessions of 30-min duration or more per week, during which time the patient is in constant motion and is slightly out of breath.

### Goals for Dietary and Hygienic Therapy

Three lipid parameters are used to define abnormal levels and determine therapeutic goals: LDL cholesterol (Table 28.7), TG (Table 28.4) and HDL cholesterol (low < 35 mg/dl) [32]. If the goals for LDL cholesterol are achieved with dietary management alone, drug therapy is not recommended. The minimum goal for TG is a level below 200 mg/dl in adults; the ideal goal is below 150 mg/dl. Values above 200 mg/dl are usually associated with the presence of small, dense LDL particles. A low-HDL-cholesterol level is a value less than 35 mg/dl. The minimum treatment goal for HDL cholesterol is 40 mg/dl.

The most recent recommendations from the National Cholesterol Education Program [32] offer guidelines for assessing risk and initiating treatment in patients with hypercholesterolemia. As shown in Tables 28.6 and 28.7, dietary intervention is used initially in the treatment of patients with dyslipidemia. A more aggressive reduction in the total daily allowance of saturated fat and cholesterol is used in patients with CAD or those failing to respond to the step-I diet (Table 28.6). Patients with CAD should be placed simultaneously on the step-II diet and lipid-lowering drug therapy. Ideally, all patients should be formally counseled by a registered dietitian. Physicians should reinforce the importance of the dietary plan for their patients.

**Table 28.7.** Recommendations for initiating diet and drug therapy. Plasma low-density lipoprotein (LDL) cholesterol levels for initiating diet or drug therapy (left column) and goals for LDL cholesterol levels (right column) [32]

| Patient category | Initiation therapy[a] | LDL goal[a] |
|---|---|---|
| **Dietary therapy** | | |
| Without CAD and with fewer than two risk factors | ≥160 mg/dl | <160 mg/dl |
| Without CAD and with two or more risk factors | ≥130 mg/dl | <130 mg/dl |
| With CAD | ≥100 mg/dl | <100 mg/dl |
| **Drug therapy** | | |
| Without CAD and with fewer than two risk factors | ≥190 mg/dl | <160 mg/dl |
| Without CAD and with two or more risk factors | ≥160 mg/dl | <130 mg/dl |
| With CAD | ≥130 mg/dl | <100 mg/dl |

CAD, coronary-artery disease
[a] The LDL levels are equivalent to the following: 100 mg/dl = 2.59 mmol/l; 130 mg/dl = 3.37 mmol/l; 160 mg/dl = 4.14 mmol/l; 190 mg/dl = 4.92 mmol/l

## Low Density Lipoprotein-Lowering Drugs

Agents which will lower LDL cholesterol include: the inhibitors of HMG-CoA reductase (the statins), the bile-acid sequestrants and niacin (nicotinic acid). The fibrates can also modestly reduce LDL cholesterol levels but, in hypertriglyceridemic patients with FCHL, LDL levels may stay the same or actually increase [36].

The *statins* available in Europe and the USA include atorvastatin (Lipitor), cerivastatin (Baycol), fluvastatin (Lescol), lovastatin (Mevacor), pravastatin (Pravachol) and simvastatin (Zocor; Table 28.8) [36]. The dose ranges, average LDL reductions and costs of these statins are shown in Table 28.8. The equivalent doses are approximately: 10 mg atorvastatin = 20 mg simvastatin = 40 mg lovastatin = 40 mg pravastatin = 0.4 mg cerivastatin = 80 mg fluvastatin. Lovastatin, simvastatin and pravastatin are derived from a biological product, while atorvastatin, fluvastatin and cerivastatin are entirely synthetic products. The effects of the statins on lowering LDL cholesterol and on improving endothelial cell function and stabilizing plaques appear to be class effects [37, 38].

In general, this drug class has shown an excellent safety profile, with minimal side effects. Liver-function tests (aspartate aminotransferase, alanine aminotransferase) should be monitored at baseline, 6–8 weeks after initiating treatment and every 4 months for the first year. After that, patients on a stable dose of a statin can have their liver-function tests monitored every 6 months. Consideration should be given to reducing the dosage of drug (or discontinuing it) should the liver-function tests exceed three times the upper limit of the normal range. In clinical trials, the discontinuation rate due to elevation of transaminases was less than 2%. Between 1 in 500 and 1 in 1000 patients may develop myositis while on statin therapy; this can lead to life-threatening rhabdomyolysis. The creatinine phosphokinase (CPK) should be measured at baseline and repeated if the patient develops muscle aches and cramps. The statin is discontinued if the CPK is more than five times above the upper limit of the normal range. The CPK is not routinely measured in the asymptomatic patient, because it is not predictive of who will develop myositis.

The *bile-acid resins* (cholestryamine, colestipol) do not enter the bloodstream but bind bile acids in the intestine, preventing their reabsorption (Table 28.9). More cholesterol is converted into bile acids in the liver, decreasing the cholesterol pool, increasing the proteolytic release of sterol-regulating-element-binding protein and leading to upregulation of LDL receptors (Fig. 28.1). Side effects of the resins include constipation, heartburn, bloating and decreased serum folate levels.

*Niacin* decreases LDL by inhibiting the production of ApoB-100 in the liver, leading to decreased VLDL secretion and, subsequently, decreased IDL and LDL formation (Fig. 28.1). Nicotinic acid is commonly prescribed in those patients with mixed dyslipidemia or isolated low-HDL cholesterol. Niacin should not be used in patients with active peptic ulcer disease or liver disease and should be used with considerable caution in patients with diabetes mellitus or gout. There are a number of niacin preparations available over the counter or by prescription (Table 28.9). Immediate crystalline niacin can be purchased in most pharmacies and health-food stores. The slow release of niacin products and the new extended-release niacin (Niaspan) are available by prescription. The slow-release niacin is not associated with flushing but has been reported to increase liver-function tests. Niaspan also decreases flushing, but the prevalence of abnormal liver-function tests with Niaspan is comparable to that seen with regular niacin.

**Table 28.8.** Hydroxymethylglutaryl-coenzyme-A-reductase inhibitors ("statins") [36]

| Drugs | FDA-pproved daily dosages | Usual decrease in LDL cholesterol (%) | Costs[a] ($) |
|---|---|---|---|
| Atorvastatin – *Lipitor* (Parke-Davis) | Initial dosage: 10 mg once | 35–40 | 56.36 |
| | Maximum dosage: 80 mg once | 50–60 | 209.88 |
| Cerivastatin – *Baycol* (Bayer) | Initial dosage: 0.3 mg once | 30–40 | 39.60 |
| | Maximum dosage: 0.4 mg once | 35–40 | 39.60 |
| Fluvastatin – *Lescol* (Novartis) | Initial dosage: 20 mg once | 20–25 | 37.70 |
| | Maximum dosage: 40 mg b.i.d. | 30–35 | 75.27 |
| Lovastatin – *Mevacor* (Merck) | Initial dosage: 20 mg once | 25–30 | 69.85 |
| | Maximum dosage: 80 mg once[b] | 40–45 | 251.48 |
| Pravastatin – *Pravachol* (Bristol-Myers Squibb) | Initial dosage: 20 mg once | 25–30 | 64.95 |
| | Maximum dosage: 40 mg once | 30–35 | 106.77 |
| Simvastatin – *Zocor* (Merck) | Initial dosage: 20 mg once | 35–40 | 109.88 |
| | Maximum dosage: 80 mg once | 45–50 | 109.88 |

FDA, Food and Drug Administration; LDL, low-density lipoprotein
[a] Cost to the pharmacist for 30 days' treatment, based on average wholesale prices [41]
[b] Or divided b.i.d.

**Table 28.9.** Other lipid-lowering drugs: resins, fibrates and niacin [36]

| Drug | Daily dosage | Costs[a] ($) |
|---|---|---|
| Resins | | |
|   Cholestyramine | | |
|     Low generic price (HCFA) | 8 g resin, divided | 54.20 |
|     *Questran, Questran Light* (Bristol-Myers Squibb) | 8 g resin, divided | 61.04 |
|     *Prevalite* (Upsher-Smith) | 8 g resin, divided | 50.96 |
|   Colestipol | | |
|     *Colestid* granules (Pharmacia and Upjohn) | 10 g, divided | 59.43 |
|     *Colestid* tablets | 10 g, divided | 109.70 |
| Fibrates | | |
|   Clofibrate | | |
|     Average generic price | 1 g b.i.d. | 65.30 |
|     *Atromid S* (Wyeth-Ayerst) | 1 g b.i.d. | 123.49 |
|   Gemfibrozil | | |
|     Low generic price (HCFA) | 600 mg b.i.d. | 10.80 |
|     *Lopid* (Parke-Davis) | 600 mg b.i.d. | 81.75 |
|   Fenofibrate – *Tricor* (Abbott) | 201 mg once | 61.88 |
| Niacin | | |
|   Immediate release | | |
|     *Niacor* (Upsher-Smith) | 1 g t.i.d. | 50.71 |
|   Extended release | | |
|     *Niaspan* (Kos) | 2 g once | 44.40 |

*HCFA*, Health Care Financing Administration
[a] Cost to the pharmacist for 30 days' treatment based on average wholesale prices or HCFA listings [41, 42]

### Triglyceride-Lowering Drugs

Drugs that can effectively lower TG include nicotinic acid, fibrates and statins (particularly when used at their highest doses). A 30–50% reduction in TG is often achieved [39].

One theoretical advantage of niacin and fibrate therapy for hypertriglyceridemia is the improvement or shift of dense subfractions (pattern B) to lighter subfractions (pattern A) [35]. The measurement of dense LDL or HDL subfractions can be made by density-gradient electrophoresis or nuclear magnetic-resonance spectroscopy. These different methodologies have shown the existence of a number of lipoprotein subfractions. In vitro studies have also suggested that dense LDL is more atherogenic and that a shift to lighter subfractions may reduce the risk of CAD. To date, there have not been prospective clinical-outcome studies showing a benefit of shifting dense subfractions to lighter ones.

Fibrates can also effectively lower TG levels and raise HDL cholesterol levels [35]. The dosages of commonly prescribed agents are shown in Table 28.9.

## Abbreviations

| | |
|---|---|
| ABL: | abetalipoproteinemia |
| ACAT: | acyl-Coenzyme A: cholesterol aclytransferase |
| apo: | apolipoprotein |
| CE: | cholesteryl esters |
| CESD: | cholesteryl ester storage disease |
| CETP: | cholesteryl ester transfer protein |
| CHD: | coronary heart disease |
| CM: | chylomicron |
| CMR: | chylomicron remnant |
| FCH: | familial combined hyperlipidemia |
| FDB: | familial defective apo B |
| FED: | fish-eye disease |
| FFA: | free fatty acids |
| FHALP: | farnilial hypoalphallpoproteinemia |
| FHC: | familial hypercholesterolemia |
| FHTG: | familial hypertriglyceridemia |
| HBL: | hypobetalipoproteinemia |
| HC: | hypercholesterolemia |
| HDL: | high density lipoproteins |
| HDL-C: | HDL-cholesterol |
| HL: | hepatic lipase |
| HLP: | hyperlipoproteinemia |
| HMG-CoA: | 3-hydroxymethylglutaryl Coenzyme A |
| HTG: | hypertriglyceridemia |
| IDL: | intermediate density lipoproteins |
| IEF: | isoelectric focusing |
| LAL: | lysosomal acid lipase |
| LCAT: | Lecithin:cholesterol acyltransferase |
| LDL: | low density lipoproteins |
| LDL-C: | LDL-cholesterol |
| Lp(a): | lipoprotein(a) |
| LPL: | lipoprotein lipase |
| MI: | myocardial infarction |
| MTTP: | microsomal triglyceride transfer protein |
| PCR: | polymerase chain reaction |
| TG: | triglycerides |
| TGRL: | triglyceride-rich lipoproteins |
| UC: | unesterified cholesterol |
| VLDL: | very low density lipoproteins |

# References

1. Brown MS, Goldstein JL (1997) The SREBP pathway: regulation of cholesterol metabolism by proteolysis of a membrane-bound transcription factor. Cell 89(3):331–340
2. Tall AR (1998) An overview of reverse cholesterol transport. Eur Heart J 19:A31–A35
3. Heinecke JW, Lusis AJ (1998) Paraoxonase-gene polymorphisms associated with coronary heart disease: support for the oxidative damage hypothesis? Am J Hum Genet 62:36–44
4. Acton S, Rigotti A, Landschulz KT et al. (1996) Identification of scavenger receptor SR-BI as a high-density lipoprotein receptor. Science 271:518–520
5. Rouis M, Dugi KA, Previato L et al. (1997) Therapeutic response to medium-chain triglycerides and omega-3 fatty acids in a patient with the familial chylomicronemia syndrome. Arterioscler Thromb Biol 17(7):1400–1406
6. Kwiterovich PO Jr (1993) Genetics and molecular biology of familial combined hyperlipidemia. Curr Opin Lipidol 4(2):133–143
7. Millar JS, Packard CJ (1998) Heterogeneity of apolipoprotein B-100-containing lipoproteins: what we have learnt from kinetic studies. Curr Opin Lipidol 9(3):197–202
8. Motevalli M, Goldschmidt-Clermont, PJ, Virgil, D, Kwiterovich PO Jr (1997) Abnormal protein tyrosine phosphorylation in fibroblasts from hyperapoB subjects. J Biol Chem 272:24703–24709
9. Brown BG, Zambon A, Poulin D et al. (1998) Use of niacin, statins, and resins in patients with combined hyperlipidemia. Am J Cardiol 81(4A):52B–59B
10. Aouizerat BE, Allayee H, Bodnar J et al. (1999) Novel genes for familial combined hyperlipidemia. Curr Opin Lipidol 10:113–122
11. Wolman M (1995) Wolman disease and its treatment. Clin Pediatr 34:207–212
12. Beaudet AL, Ferry GD, Nichols BL, Rosenberg HS (1977) Cholesterol ester storage disease: clinical, biochemical, and pathological studies. J Pediatr 90:910–914
13. Ginsberg HN, Le NA, Short MP, Ramakrishnan R, Desnick RJ (1987) Suppression of apolipoprotein B production during treatment of cholesteryl ester storage disease with lovastatin. Implications for regulation of apolipoprotein B synthesis. J Clin Invest 80(6):1692–1697
14. Nicholls P, Young IS, Graham CA (1998) Genotype/phenotype correlations in familial hypercholesterolemia. Curr Opin Lipidol 9(4):313–317
15. Grossman M, Rader DJ, Muller DW et al. (1995) A pilot study of ex vivo gene therapy for homozygous familial hypercholesterolemia. Nat Med 1(11):1148–1154
16. Hansen PS (1998) Familial defective apolipoprotein B-100. Dan Med Bull 45(4):370–382
17. Patel SB, Salen G, Hidaka H et al. (1998) Mapping a gene involved in regulating dietary cholesterol absorption. The sitosterolemia locus is found at chromosome 2p21. J Clin Invest 102:1041–1044
18. Mahley RW (1996) Heparan sulfate proteoglycan/low density lipoprotein receptor-related protein pathway involved in type III hyperlipoproteinemia and Alzheimer's disease. Isr J Med Sci 32(6):414–429
19. Hegele RA, Little JA, Vezina C (1993) Hepatic lipase deficiency: clinical biochemical and molecular genetic characteristics. Arterioscler Thromb 13:720–728
20. Brand K, Dugi KA, Brunzell JD (1996) A novel A→G mutation in intron I of the hepatic lipase gene leads to alternative splicing resulting in enzyme deficiency. J Lipid Res 37(6):1213–1223
21. Rader DJ, Brewer HB (1993) Abetalipoproteinemia. New Insights into lipoprotein assembly and vitamin E metabolism from a rare genetic disease. JAMA 270:865–869
22. Wetterau JR, Aggerbeck LP, Bouma ME et al. (1992) Absence of microsomal triglyceride transfer protein in individuals with abetalipoproteinemia. Science 258:999–1001
23. Gabelli C, Bilato C, Martini S et al. (1996) Homozygous familial hypobetalipoproteinemia. Increased LDL catabolism in hypobetalipoproteinemia due to a truncated apolipoprotein B species, apoB-87Padova. Arterioscler Thromb Biol 16(9):1189–1196
24. Breslow JL (1995) Familial disorders of high-density lipoprotein metabolism. In: Scriver CR, Beaudet AL, Sly WS, Valle D (eds) The metabolic and molecular basis of inherited diseases, 7th edn, vol II. McGraw-Hill, New York, pp 2031–2052
25. Bruce C, Chouinard RA Jr, Tall AR (1998) Plasma lipid transfer proteins, high-density lipoproteins, and reverse cholesterol transport. Annu Rev Nutr 18:297–330
26. von Eckardstein A, Assmann G (1998) High density lipoproteins and reverse cholesterol transport: lessons from mutations. Atherosclerosis 137:S7–11
27. Remaley AT, Schumacher UK, Stonik JA et al. (1997) Decreased reverse cholesterol transport from Tangier disease fibroblasts. Acceptor specificity and effect of brefeldin on lipid efflux. Arterioscler Thromb Biol 17(8):1813–1821
28. Young SG, Tielding CJ (1999) The ABCs of cholesterol efflux. Nat Genet 22:316–319
29. Glomset JA, Assmann G, Gjone E, Norum KR (1995) Lecithin:cholesterol acyltransferase deficiency and fish eye disease. In: Scriver CR, Beaudet AL, Sly WS, Valle D (eds) The metabolic and molecular basis of inherited diseases, 7th edn, vol II. McGraw-Hill, New York, pp 1933–1951
30. Zhong S, Sharp DS, Grove JS et al. (1996) Increased coronary heart disease in Japanese-American men with mutations in the cholesteryl ester transfer protein gene despite increased HDL levels. J Clin Invest 97:2917–2923
31. Ikewaki K, Nishiwaki M, Sakamoto T et al. (1995) Increased catabolic rate of low density lipoproteins in humans with cholesteryl ester transfer protein deficiency. J Clin Invest 96(3):1573–1581
32. Summary of the Second Report of the National Education Program (NCEP) (1993) Expert panel on detection evaluation and treatment of high blood cholesterol in adults. JAMA 269:3015–3023
33. Dietschy JM (1997) Theoretical considerations of what regulates low-density-lipoprotein and high-density-lipoprotein cholesterol. Am J Clin Nutr 65:1581S–1589S
34. Kwiterovich PO (1997) The effect of dietary fat, antioxidants and pro-oxidants on blood lipids and lipoproteins and atherosclerosis. J Am Diet Assoc 97 [Suppl]:S31–S41
35. Fruchart J-C, Brewer HB, Leitersdorf E (1998) Consensus for the use of fibrates in the treatment of dyslipoproteinemia and coronary heart disease. Am J Cardiol 81:912–917
36. Anonymous (1998) Choice of lipid-lowering drugs. Med Lett 40:117–122
37. Ganz P, Creager MA, Fang JC et al. (1996) Pathogenetic mechanisms of atherosclerosis: effect of lipid lowering on the biology of atherosclerosis. Am J Med 101:10S–16S
38. Kwiterovich PO (1998) State-of-the-art update and review: clinical trials of lipid-lowering agents. Am J Cardiol 82: 3U–17U
39. Kwiterovich PO (1998) The antiatherogenic role of high-density lipoprotein cholesterol. Am J Cardiol 82:13Q–21Q
40. Braunwald E (ed) (1997) Essential atlas of heart diseases. Appleton and Lange, Philadelphia, p 1.28
41. Medical Economics Company (1998) Drug topics red book update, December 1998. Medical Economics, Montvale
42. Medical Economics Company (1995) Drug topics red book. Medical Economics, Montvale

# CHAPTER 29

## Cholesterol Synthesis

The initial steps in the synthesis of cholesterol involve the formation of mevalonic acid from acetyl-CoA, catalyzed by a sequence of three enzymes (Fig. 29.1). The last of these, 3-hydroxy-3-methylglutaryl-CoA reductase, is the rate-limiting step of cholesterol synthesis. The phosphorylation of mevalonic acid, catalyzed by mevalonate kinase, is followed by the synthesis of the nonsterol isoprenoids, isopentyl-, geranyl-, and farnesyl-pyrophosphate. The latter is converted into squalene, followed by cyclization into lanosterol. The final modifications of the sterol nucleus follow two distinct pathways. One ends with the reduction of 7-dehydrocholesterol by 3β-hydroxysterol-Δ7-reductase. An alternative pathway involves formation of cholesta-7,24-dien-3β-ol, reduction of the latter in desmosterol by 3β-hydroxysterol-Δ7-reductase, and conversion into cholesterol by hydroxysterol-Δ24-reductase.

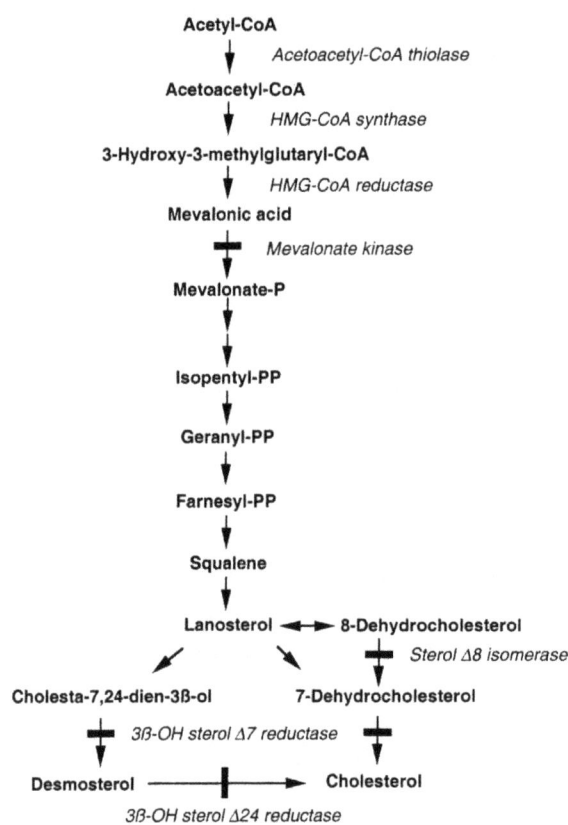

**Fig. 29.1.** Pathway of cholesterol synthesis. *CoA*, coenzyme A; *HMG*, 3-hydroxy-3-methylglutaryl; *OH*, hydroxy; *P*, phosphate; *PP*, pyrophosphate. Enzyme deficiencies are indicated by *solid bars*

# CHAPTER 29

# Disorders of Cholesterol Synthesis

Georg F. Hoffmann and Dorothea Haas

CONTENTS

Introduction ................................... 339
Mevalonic Aciduria (Mevalonate Kinase Deficiency)..... 339
Smith-Lemli-Opitz Syndrome ...................... 340
Sterol-$\Delta^8$-Isomerase Deficiency –
Chondrodysplasia Phenotypes .................... 341
Desmosterolosis................................ 342
References .................................... 342

Only four of over 30 possible defects of cholesterol biosynthesis have been described. Mevalonic aciduria due to mevalonate kinase deficiency is situated at the beginning of the pathway and compromises the production of nonsterol isoprenes in addition to cholesterol. Patients with this enzyme defect show a wide range of symptoms, particularly dysmorphic features, hypotonia, developmental delay, enteropathy, hepatosplenomegaly, and recurrent febrile crises. At the end of the pathway, a defect of 3β-hydroxysterol-$\Delta^7$-reductase could be delineated as the cause of Smith-Lemli-Opitz syndrome (SLOS), an important multiple malformation syndrome. Amongst these malformations are a characteristic facial appearance and syndactyly of toes 2–3. Deficiency of sterol-$\Delta^8$-isomerase can cause a chondrodysplasia punctata phenotype ranging from mild Conradi-Hünermann to a lethal chondrodysplasia. One child with a malformation syndrome different from SLOS and generalized accumulation of desmosterol can be assumed to have suffered from a deficiency of 3β-hydroxysterol-$\Delta^{24}$-reductase. Further defects are likely to be unraveled in due course and Table 29.1 summarizes indications for the two available tests: determination of mevalonic acid by urinary organic acid analysis and sterol analysis by gas chromatography–mass spectrometry (GCMS).

## Introduction

Disorders of cholesterol biosynthesis disrupt a complex, coordinated regulated pathway required for cellular structure, metabolism, and replication (Fig. 29.1). As cholesterol is inadequately transported through the placenta, cholesterol synthesis in the fetus starts very early, and a hallmark of defects of cholesterol biosynthesis are abnormalities of morphogenesis. Extent and pattern of malformations are likely to be determined by two mechanisms: first, a shortage of cholesterol and, second, a pathological accumulation of precursors, which is distinct for different defects.

## Mevalonic Aciduria (Mevalonate Kinase Deficiency)

### Clinical Presentation

Although patients with mevalonic aciduria have a recognizable phenotype of serious clinical manifestations [1], some are likely to remain undiagnosed as

Table 29.1. Routine chemical and clinical indications for cholesterol synthesis defects. Disorders caused by defects at the beginning of the biosynthetic pathway are searched for by determining mevalonic acid by urinary organic acid analysis (OA), whereas defects of the conversion of lanosterol to cholesterol are identified by gas chromatography–mass spectrometry sterol analysis (SA). The analyses should be considered particularly in cases of unexplained combinations of the named symptoms

Prenatal
   Low levels of maternal estriols (SA)
   Fetal malformations (SA)
   Growth retardation and microcephaly (OA, SA)
   Nuchal edema (SA)
Postnatal
   Variable reductions of serum cholesterol and/ or bile acids (OA, SA)
   Increased creatine kinase (OA)
   Failure to thrive (OA, SA)
   Enteropathy (OA, SA)
   Variable psychomotor retardation (OA, SA)
   Cataracts or retinitis pigmentosa (OA, SA)
   2,3 syndactyly or postaxial polydactyly (SA)
   Hypoplastic or ambiguous genitalia (SA)
   Major organ or skeletal malformations (SA)
   Cleft palate (SA)
   Recurrent exacerbations of a presumptive autoimmune disease (OA)
   Blood disorders suggestive of chronic leukemia or myelodysplastic syndrome (OA)

hypoglycemia, metabolic acidosis, and lactic acidosis, the usual concomitants of disorders of organic acid metabolism, are conspicuously absent. The most severely affected have died in infancy with severe failure to thrive, profound hypotonia, developmental delay, dysmorphic features, cataracts, hepatosplenomegaly, lymphadenopathy, and anemia, as well as diarrhea and malabsorption. The presenting picture may suggest congenital infections or chromosomal abnormalities. Less severely affected patients may be classified among patients with psychomotor retardation, failure to thrive, hypotonia, myopathy, and ataxia of unknown etiology. All patients have had recurrent febrile crises with vomiting and diarrhea, lymphadenopathy, increase in size of liver and spleen, arthralgia, edema, and morbilliform rashes, leukocytosis, elevated erythrocyte sedimentation rate and creatine kinase pointing to an infectious or autoimmune etiology. Sometimes hematological abnormalities prevail with normocytic hypoplastic anemia, leukocytosis, thrombocytopenia, and abnormal blood cell forms, leading to misdiagnoses of congenital infection, myelodysplastic syndromes, or chronic leukemia [2].

## Metabolic Derangement

Mevalonic aciduria is a consequence of the deficiency of mevalonate kinase, the first enzyme after 3-hydroxy-3-methylglutaryl (HMG)-CoA reductase in the biosynthesis of cholesterol and nonsterol isoprenes (Fig. 29.1). Clinical manifestations are thought to be due to an imbalance in the multilevel regulation of the biosynthetic pathway and a shortage of end products, especially nonsterol isoprenes [1, 3, 4].

## Diagnostic Tests

Serum levels of creatine kinase and transaminases are elevated in most patients. The only diagnostic biochemical abnormality is the gross elevation of mevalonic acid in all body fluids, detectable by organic acid analysis [5]. The diagnosis needs to be confirmed by assay of mevalonate kinase in white blood cells or cultured fibroblasts [1].

Carrier detection is possible by enzyme assay, though the results are sometimes equivocal [5] and by mutation analysis in investigated families. Heterozygotes were also found to show a significantly increased excretion of mevalonic acid in their urine [5]. Prenatal diagnosis is possible by assaying mevalonate kinase activity in amniocytes and chorionic villus cells, by direct assay of mevalonic acid in amniotic fluid using GCMS and molecular analysis [5].

## Treatment and Prognosis

Treatment is still experimental [1]. Dietary supplements of bile acids and cholesterol had to be discontinued because of worsening diarrhea and general malaise. Trials of corticosteroid therapy during clinical crises (2 mg prednisone/kg/day) resulted in positive responses. Additional long-term administration of ubiquinone-10 together with pharmaceutical doses of vitamins C and E appeared to further stabilize the clinical course and improve somatic and psychomotor development. The rationale is to correct the deficiency of ubiquinone-10 [4] and to increase small-molecule antioxidants.

The most severely affected patients have died in infancy. Generally, the clinical course appeared progressive during the first years of life with considerable phenotypic heterogeneity. From late childhood, patients displayed a stable clinical picture of borderline mental retardation, cerebellar ataxia, and muscular hypotonia.

## Genetics

Mevalonic aciduria is an autosomal recessive disorder. The gene for mevalonate kinase is localized on chromosome 12q24 [6]. Six inherited mutations could be identified in 10 of 20 known patients, all of which cluster in the C-terminal region of the protein (amino acid residues 243–334) [7]. All but two patients, whose homozygous mutation yields a protein with altered mevalonate binding [8], appear to be compound heterozygotes, suggesting the rarity of mutant alleles in the human population.

# Smith-Lemli-Opitz Syndrome (SLOS)

## Clinical Presentation

Patients afflicted with Smith-Lemli-Optiz Syndrome (SLOS) show a wide clinical variability ranging from intrauterine and neonatal death due to lethal malformations to mild mental retardation [9, 10]. Severe manifestations were originally classified as SLOS type II, milder SLOS type I [11]. However, a continuum exists between the two types.

Children afflicted with this disorder show a characteristic craniofacial appearance, limb abnormalities, organ and genital malformations and mental retardation [10, 12]. A sensitive clinical marker is syndactyly of toes 2–3 occurring in more than 98% of SLOS patients. More then 90% present with microcephaly and prenatal or postnatal growth retardation. The facial appearance of micro- or retrognathia, anteverted nares, low-set posteriorly rotated ears, blepharoptosis, cataracts, cleft palate and irregularity of the alveolar ridges is very characteristic (Fig. 29.2). Feet and hands often show postaxial polydactyly and equinovarus deformity. Genital abnormalities range from hypospadias and cryptorchism to ambiguous or female external genitalia in males. Major organs commonly involved are: the heart (atrial septal defects, endocardial cushion defects, anomalous pulmonary venous return and patent ductus arteriosus), the kidney (renal hypoplasia and microcystic kidney disease) and the gastrointestinal system (pyloric stenosis, Hirschsprung's disease or general dysmotility). Pathology of the central nervous system varies from frontal-lobe hypoplasia, white-matter hypoplasia, enlarged ventricles, agenesis of the corpus callosum, cerebellar hypoplasia to holoprosencephaly. Less frequently occurring abnormalities are sensorineural hearing loss and seizure disorders. The degree of mental retardation ranges from mostly severe to mild.

### Metabolic Derangement

SLOS is caused by a deficiency of $3\beta$-hydroxysterol-$\Delta^7$-reductase, which converts 7-dehydrocholesterol to cholesterol, the final step in cholesterol biosynthesis (Fig. 29.1). This leads to a decrease of cholesterol and an accumulation of the precursor 7-dehydrocholesterol (7-DHC) and its isomer 8-dehydrocholesterol in body fluids and tissues [13, 14].

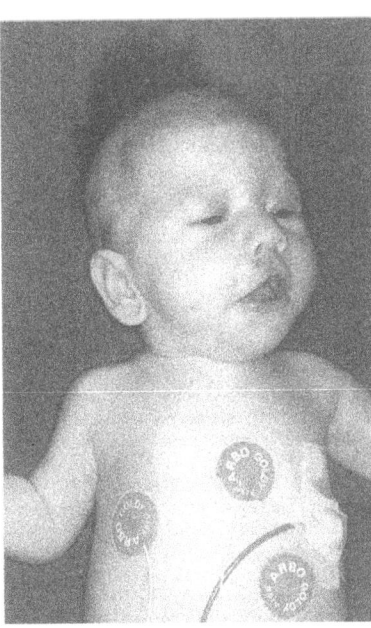

**Fig. 29.2.** Eight-month-old boy with Smith-Lemli-Opitz syndrome depicting the characteristic facial appearance, including microcephaly, ptosis, micrognathia, high palate, anteverted nares, and low-set posteriorly rotated ears

### Diagnostic Tests

The diagnosis is biochemically confirmed by an elevated 7-DHC in plasma measured by means of GCMS [14]. Routine laboratory measurement of cholesterol is not reliable, because mildly affected patients can have normal cholesterol levels and the abnormal sterols lead to overestimation of the cholesterol levels by colorimetric assays [14]. Prenatal diagnosis can be performed by quantifying the ratio of concentrations of 7-DHC to cholesterol in chorionic villus tissue or amniotic fluid [15, 16] and conceptually by molecular analysis. There are indications that SLOS fetuses cause low maternal serum estriol levels in pregnancy [12], which could be used as a screening test.

### Treatment and Prognosis

The discovery of a disorder of cholesterol biosynthesis as cause of SLOS has led to new treatment approaches, first of all cholesterol supplementation [17]. As cholesterol is a strong inhibitor of HMG-CoA reductase, the rationale is not only to compensate for the lack of cholesterol but also to inhibit the synthesis of the pathological sterols. Most treatment protocols provide cholesterol in the form of a purified powder in a dosage of 100 mg/kg/day. The benefit of cholesterol supplementation is a medium-term increase in weight and growth as well as an improvement of development and behavior. Bile acid supplementation has no positive effects [17].

Clinical severity and prognosis correlates inversely with the level of cholesterol or the ratio of cholesterol to total sterols at diagnosis, whereas the amount of abnormal sterols is less important. Plasma cholesterol levels less than 10 mg/dl are mostly seen in SLOS type II, and patients are stillborn or die shortly after birth. Patients with normal or low-normal cholesterol levels generally do not have severe malformations [10, 12], but invariably manifest mental retardation and behavioral abnormalities.

### Genetics

SLOS is an autosomal recessive disorder with an estimated incidence of 1:20,000 in Caucasians. The gene for $3\beta$-hydroxysterol-$\Delta^7$-reductase was located on chromosome 11q12-13 in 1997 [18] and, consequently, mutations described in patients with SLOS [19–21].

## Sterol-$\Delta 8$-Isomerase Deficiency – Chondrodysplasia Phenotypes

Chondrodysplasia punctata is a rare but well-recognized skeletal manifestation of different metabolic malformation syndromes. The rhizomelic type (short proximal limb) is classically caused by disorders of plasmalogen synthesis and can be clinically distinguished from the Conradi-Hünermann type following X-linked dominant, but also X-linked recessive or autosomal dominant modes of inheritance. In the Conradi-Hünermann syndrome, the limbs related to areas of punctuate mineralization in epiphyses are asymmetrically shortened. Pre- and postnatal growth and psychomotor development are usually impaired. Hydramnios and hydrops may develop prenatally, cataracts and ichtyosiform skin later on. Increased levels of 8(9)-cholestenol and 8-dehydrocholesterol were identified in probands affected with the X-linked form of Conradi-Hünermann syndrome or with lethal chondrodysplasia [22]. The assumed deficiency of 3β-hydroxysteroid-$\Delta^8$, $\Delta^7$-isomerase could be proven by identifying mutations in a sterol-$\Delta^8$-isomerase mapping to Xp22.3.

## Desmosterolosis

Recently, a generalized accumulation of desmosterol and a relative deficiency of cholesterol was documented in tissue samples of a patient with a malformation syndrome reminiscent of Raine syndrome and different from SLOS [23]. The baby girl had macrocephaly, a hypoplastic nasal bridge, thick alveolar ridges, gingival nodules, cleft palate, total anomalous pulmonary venous drainage, ambiguous genitalia, short limbs, and generalized osteosclerosis. She died shortly after birth and can be assumed to have suffered from a deficiency of 3β-hydroxysterol-$\Delta^{24}$-reductase.

## References

1. Hoffmann GF, Charpentier C, Mayatepek E et al. (1993) Clinical and biochemical phenotype in eleven patients with mevalonic aciduria. Pediatrics 91:915-921
2. Hinson DD, Rogers ZR, Hoffmann GF et al. (1998) Hematological abnormalities and cholestatic liver disease in two patients with mevalonate kinase deficiency. Am J Med Gen 78:408-412
3. Hoffmann GF, Wiesmann UN, Brendel S, Keller RK, Gibson KM (1997) Regulatory adaptation of isoprenoid biosynthesis and the LDL-receptor pathway in fibroblasts from patients with mevalonate kinase deficiency. Pediatr Res 41:541-546
4. Hübner C, Hoffmann GF, Charpentier C et al. (1993) Decreased plasma ubiquinon-10 concentration in patients with mevalonic aciduria. Pediatr Res 34:129-133
5. Hoffmann GF, Hunneman DH, Sweetman L et al. (1991) Facts and artefacts in mevalonic aciduria: development of a stable isotope dilution GCMS assay for mevalonic acid and its application to physiological fluids, tissue samples, prenatal diagnosis and carrier detection. Clin Chim Acta 198:209-228
6. Gibson KM, Hoffmann GF, Tanaka RD, Bishop RW, Chambliss KL (1997) Mevalonate kinase map position 12q24. Chromos Res 5:150
7. Hinson DD, Shaw JL, Kozich V et al. (1999) Identification of four new mutations in mevalonate kinase deficiency, including one in a patient of mennonite ancestry. Am J Hum Genet 65:327-335
8. Krisans S, Keller RK, Gibson KM (1997) Identification of an active site alanine in mevalonate kinase through characterization of a novel mutation in mevalonate kinase deficiency. J Biol Chem 42:26756-26760
9. Smith DW, Lemli L, Opitz JM (1964) A newly recognized syndrome of multiple congenital anomalies. J Pediatr 64:210-217
10. Cunniff C, Kratz LE, Moser A, Natowicz MR, Kelley RI (1997) Clinical and biochemical spectrum of patients with RSH/Smith-Lemli-Opitz syndrome and abnormal cholesterol metabolism. Am J Med Genet 68:263-269
11. Curry CJ, Carey JC, Holland JS et al. (1987) Smith-Lemli-Opitz syndrome type II: multiple congenital anomalies with male pseudohermaphroditism and frequent early lethality. Am J Med Genet 26:45-57
12. Kelley RI (1996) Smith-Lemli-Opitz syndrome. In: Moser HW (ed) Neurodystrophies and neurolipidoses. Elsvier Science, Amsterdam (Handbook of clinical neurology, vol 22/66)
13. Tint GS, Irons M, Elias ER et al. (1994) Defective cholesterol biosynthesis associated with the Smith-Lemli-Opitz syndrome. N Engl J Med 330:107-113
14. Kelley RI (1995) Diagnosis of Smith-Lemli-Opitz syndrome by gas chromatography/mass spectrometry of 7-dehydrocholesterol in plasma, amniotic fluid and coltured skin fibroblasts. Clin Chim Acta 236:45-58
15. Abuelo DN, Tint GS, Kelley R et al. (1995) Prenatal detection of the cholesterol biosynthetic defect in the Smith-Lemli-Opitz syndrome by the analysis of amniotic fluid sterols. Am J Med Genet 56:281-285
16. Mills K, Mandel H, Montemagno R et al. (1996) First trimester prenatal diagnosis of Smith-Lemli-Opitz syndrome (7-dehydrocholesterol reductase deficiency). Pediatr Res 39:816-819
17. Elias ER, Irons MB, Hurley AD, Tint GS, Salen G (1997) Clinical effects of cholesterol supplementation in six patients with the Smith-Lemli-Opitz syndrome (SLOS). Am J Med Genet 68:305-310
18. Moebius FF, Fitzky BU, Lee JN, Paik YK, Glossmann H (1998) Molecular cloning and expression of the human delta 7-sterol reductase. Proc Natl Acad Sci USA 95:1899-1902
19. Fitzky BU, Witsch-Baumgartner M, Erdel M et al. (1998) Mutations in the delta-7-sterol reductase gene in patients with the Smith-Lemli-Opitz syndrome. Proc Natl Acad Sci USA 95:8181-8186
20. Waterham HR, Wijburg FA, Hennekam RCM et al. (1998) Smith-Lemli-Opitz syndrome is caused by mutations in the 7-dehydrocholesterol reductase gene. Am J Hum Genet 63:329-338
21. Wassiff CA, Maslen C, Kachilele-Linjewile S et al. (1998) Mutations in the human sterol $\Delta7$-reductase gene at 11q12-13 cause Smith-Lemli-Opitz syndrome. Am J Hum Genet 63:55-62
22. Kelley RI, Wilcox WG, Smith M et al. (1999) Abnormal sterol metabolism in patients with Conradi-Hünermann-Haplle syndrome and sporadic lethal chondrodysplasia punctata. Am J Med Genet 83:213-219
23. FitzPatrick DR, Keeling JW, Evans MJ et al. (1998) Clinical phenotype of desmosterolosis. Am J Med Genet 75:145-152

# CHAPTER 30

## Bile-Acid Synthesis

Bile acids are biological detergents that are synthesised from cholesterol in the liver by modifications of the sterol nucleus and oxidation of the side chain; the major pathway is shown schematically in Fig. 30.1 [1, 2]. The enzyme defects discussed in this chapter are indicated by solid bars.

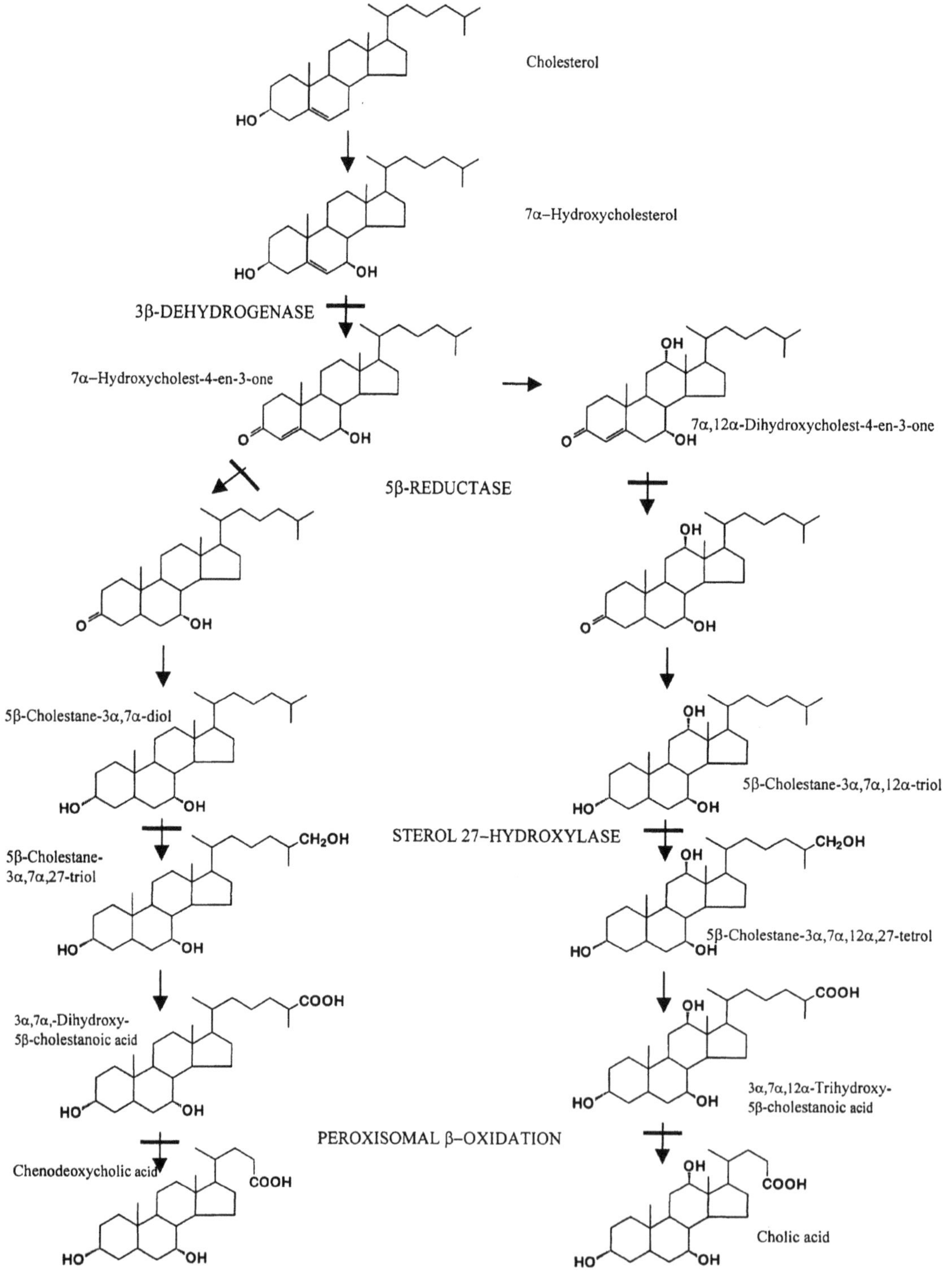

**Fig. 30.1.** Major reactions involved in the synthesis of bile acids from cholesterol. Enzyme defects are depicted by *solid bars* across the arrows

# CHAPTER 30

# Disorders of Bile Acid Synthesis

P.T. Clayton

CONTENTS

3β-Dehydrogenase Deficiency .................... 345
5β-Reductase Deficiency......................... 347
Cerebrotendinous Xanthomatosis ................. 348
Other Disorders................................. 349
References ..................................... 350

Two defined inborn errors of metabolism affect the modifications of the cholesterol nucleus in the major pathway for bile acid synthesis: 3β-hydroxy-$\Delta^5$-$C_{27}$-steroid-dehydrogenase (3β-dehydrogenase) deficiency and $\Delta^4$-3-oxo-steroid-5β-reductase (5β-reductase) deficiency. These disorders produce chronic cholestatic liver disease and malabsorption of fat and fat-soluble vitamins. Onset of symptoms is usually in the first year of life and, untreated, the liver disease can progress to cirrhosis. Treatment with chenodeoxycholic acid and cholic acid leads to dramatic improvement in the liver disease and the malabsorption.

Defective side-chain oxidation occurs in cerebrotendinous xanthomatosis (CTX; sterol 27-hydroxylase deficiency) and in peroxisomal disorders. In these disorders, neurological disease predominates. CTX typically presents with cataracts and mental retardation in childhood, followed by motor dysfunction and tendon xanthomata in the second or third decade. Death may be caused by progressive motor dysfunction and dementia or by premature atherosclerosis. Chenodeoxycholic acid has been shown to halt or even reverse neurological dysfunction. Peroxisomal disorders are described elsewhere (Chap. 37). In addition to the disorders mentioned above, disorders of amidation of bile acids and disorders of accessory pathways of bile-acid synthesis have been described in single patients.

## 3β-Dehydrogenase Deficiency

### Clinical Presentation

The first described patients with 3β-dehydrogenase deficiency presented with prolonged neonatal jaundice (conjugated bilirubin levels greater than 40 µM at the age of 2 months or older) associated with steatorrhoea. The stools were pale but not acholic. Rickets (due to malabsorption of vitamin D) was often apparent before the age of 6 months, and one patient developed a bleeding diathesis due to vitamin-K deficiency at the age of 9 months [3–5]. Routine investigations performed at age 2–6 months were not very helpful in distinguishing 3β-dehydrogenase deficiency from other causes of giant-cell hepatitis. The biochemical evidence of fat-soluble-vitamin malabsorption is perhaps more striking – e.g., plasma vitamin-E concentration consistently less than 4 µM (normal age range = 11.5–35 µM) – and the γ-glutamyl transpeptidase may be normal or only minimally elevated [despite a significantly elevated aspartate aminotransferase (AST)]. The liver biopsy shows a periportal inflammatory infiltrate (often including eosinophils), giant cells, some hepatocellular necrosis and bridging fibrosis or even early cirrhosis. In untreated patients, pruritus often becomes apparent from the age of 6 months, and the problems of steatorrhoea and malabsorption of fat-soluble vitamins continue. In 1994, Jacquemin et al. described a group of patients with 3β-dehydrogenase deficiency who presented with jaundice, hepatosplenomegaly and steatorrhoea (a clinical picture resembling progressive familial intrahepatic cholestasis) between the ages of 4 months and 46 months [6]. Pruritus was absent in these children, in contrast to other children with severe cholestasis. The authors noted normal γ-glutamyl-transpeptidase activities in plasma, low serum cholesterol concentrations and low vitamin-E concentrations. Presentation of 3β-dehydrogenase deficiency with chronic hepatitis in the second decade of life has also been described [7].

### Metabolic Derangement

3β-Dehydrogenase catalyses the second reaction in the major pathway of synthesis of bile acids: the conversion of 7α-hydroxycholesterol to 7α-hydroxycholest-4-en-3-one. When the enzyme is deficient, the accumulating 7α-hydroxycholesterol can undergo side-chain oxidation with or without 12α-hydroxylation to produce

3β,7α-dihydroxy-5-cholenoic acid and 3β,7α,12α-trihydroxy-5-cholenoic acid, respectively. These unsaturated $C_{24}$ bile acids are sulphated in the $C_3$ position; a proportion is conjugated to glycine, and they can be found in high concentrations in the urine. Concentrations of bile acids in the bile are low [8]. It is probable that the sulphated $\Delta^5$ bile acids cannot be secreted into the bile canaliculi and fuel bile flow in the same way as occurs with the normal bile acids. There are at least two possible ways in which this sequence of events might lead to damage to hepatocytes and, ultimately, to cirrhosis:

1. The abnormal metabolites produced from 7α-hydroxycholesterol may be hepatotoxic
2. Failure of bile acid-dependent bile flow may lead to hepatocyte damage, perhaps as a result of the accumulation of toxic compounds normally eliminated in the bile

## Diagnostic Tests

The diagnosis is established by demonstrating the presence of the characteristic $\Delta^5$ bile acids in plasma or urine. It is important to remember that bile acids with a $\Delta^5$ double bond and a 7-hydroxy group are acid labile. They may be destroyed by some of the methods that are used for solvolysis of sulphated bile acids prior to chromatographic analysis. Solvolysis is best performed using tetrahydrofuran–methanol–trifluoroacetic acid (900:100:1 volume ratio) [8]. Analysis by fast-atom-bombardment mass spectrometry (FAB-MS) overcomes the problem of lability [4, 9].

### Plasma

If plasma bile acids are analysed using a gas chromatography (GC)-MS method that does not include a solvolysis step, the profile of non-sulphated bile acids that is obtained shows concentrations of cholic and chenodeoxycholic acid, which are extremely low for an infant with *cholestasis*. The concentration of 3β,7α-dihydroxy-5-cholestenoic acid is increased. Inclusion of a solvolysis step reveals the presence of high concentrations of 3β,7α-dihydroxy-5-cholenoic acid (3-sulphate) and 3β,7α,12α-trihydroxy-5-cholenoic acid (3-sulphate). These can also be detected when plasma is analysed by FAB-MS or when a neonatal blood spot is analysed by electrospray-ionisation MS [10].

### Urine

Two rapid tests can be used to make a diagnosis from a urine sample; in both cases, the first step is extraction of the bile acids using a $C_{18}$ (octadecylsilane-bonded silica) cartridge. The dried urine extract will give an intense purple colour with Lifschütz reagent (glacial acetic acid/concentrated $H_2SO_4$ ratio = 10:1 v/v). Alternatively, reconstitution of the extract in glycerol/methanol and analysis by FAB-MS will show the characteristic ions of the diagnostic unsaturated bile acids: mass/charge ratios (m/z) = 469, 485, 526 and 542. Very occasionally, significantly sized ions with m/z values of 469 and 485 (steroid sulphates?) are seen in normal urine; if there is any uncertainty, the urine should be analysed by GC-MS following careful solvolysis.

### Fibroblasts

3β-Dehydrogenase can be assayed in cultured skin fibroblasts using tritiated 7α-hydroxycholesterol [11]. Patients show very low activity.

## Treatment and Prognosis

Untreated 3β-dehydrogenase deficiency has led to death from complications of cirrhosis before the age of 5 years; patients with milder forms of the disorder may survive with a chronic hepatitis into their second decade. The response to treatment depends upon the severity of the liver disease at the time of starting treatment. In patients with a bilirubin level less than 120 µM and an AST level less that 260 U/l, *chenodeoxycholic-acid therapy* has led to a dramatic improvement in symptoms and in liver-function tests within 4 weeks and to an improvement in the liver-biopsy appearances within 4 months. The dose of chenodeoxycholic acid that has been used is 12–18 mg/kg/day initially (for 2 months) followed by 9–12 mg/kg/day maintenance. In one infant with severe disease, chenodeoxycholic acid (15 mg/kg/day) led to a rise in bilirubin and AST. Her treatment regime was changed to 7 mg chenodeoxycholic acid/kg/day plus 7 mg cholic acid/kg/day. Over the course of 15 months, her bilirubin and transaminases returned to normal, and a repeat liver biopsy showed a more normal parenchyma and less inflammation. The combination of cholic acid and chenodeoxycholic acid is probably the treatment of choice for patients with severe liver damage. Bile-acid-replacement therapy may work in one of two ways:

1. By fuelling bile-acid-dependent flow (hence directly relieving cholestasis)
2. By suppressing the activity of cholesterol 7α-hydroxylase (thereby reducing the accumulation of potentially toxic metabolites of 7α-hydroxycholesterol)

## Genetics

3β-Dehydrogenase deficiency is probably inherited as an autosomal-recessive trait. Siblings have been affected even though the parents have no evidence of liver disease, and there is a high incidence of consanguinity among the parents. Levels of 3β-dehydrogenase in fibroblasts from the parents are at the lower end of or below the normal range [11]. The cDNA encoding 3β-dehydrogenase has not yet been characterised.

## 5β-Reductase Deficiency

### Clinical Presentation

In 1988, Setchell et al. [12] described male twins who presented with neonatal cholestatic jaundice associated with failure to thrive but no hepatosplenomegaly. Transaminases and alkaline phosphatase were mildly elevated, and the prothrombin time was slightly prolonged (and not responsive to vitamin K). Plasma tyrosine and methionine were slightly elevated. Liver biopsies showed lobular disarray as a result of giant cell and pseudoacinar transformation of hepatocytes, hepatocellular and canalicular bile stasis and a minimal lobular and portal cellular infiltrate. Electron microscopy showed small canaliculi with slit-like structure and few or absent microvilli. Setchell et al. showed that the twins were excreting large amounts of two 3-oxo-$\Delta^4$ bile acids (7α-hydroxy-3-oxo-4-cholenoic acid and 7α,12α-dihydroxy-3-oxo-4-cholenoic acid) in their urine; they postulated that these infants had a primary genetic deficiency of 5β-reductase. Others had previously urged caution in interpreting substantial excretion of 3-oxo-$\Delta^4$ bile acids in this way [13].

In 1994, Schneider et al. described two unrelated neonates with liver failure and histological features of neonatal haemochromatosis [14]. Because these infants excreted predominantly 3-oxo-$\Delta^4$ bile acids, the authors postulated a primary defect of 5β-reductase. However, this view was challenged [15] and, in 1997, Sumazaki et al. sequenced the 5β-reductase cDNA in an infant with neonatal haemochromatosis, liver failure and 3-oxo-$\Delta^4$ bile-acid excretion and found no mutation [16].

Meanwhile, in 1996, Clayton et al. described an infant who presented with prolonged neonatal jaundice and developed severe liver dysfunction at the age of 3 weeks [17]. Liver function improved spontaneously, but there was persistent cholestasis with steatorrhoea, failure to thrive and clinical rickets. The plasma vitamin-E concentration was low, and the γ-glutamyl-transpeptidase activity was normal at a time when the transaminases were markedly elevated. Cholic acid and chenodeoxycholic acid were undetectable in plasma and urine; concentrations of 3-oxo-$\Delta^4$ bile acids were considerably elevated. The patient's liver-function tests did not respond to ursodeoxycholic-acid therapy but did respond to treatment with chenodeoxycholic acid and cholic acid. Because of the similarity to patients with 3β-dehydrogenase deficiency (with regard to the clinical and laboratory findings and response to treatment), primary genetic deficiency of 5β-reductase was suggested and recently been proven.

### Metabolic Derangement

There is general agreement that excretion of 3-oxo-$\Delta^4$ bile acids is likely to result from reduced activity of the hepatic enzyme that brings about the 5β(H) saturation of the $\Delta^4$ double bonds of 7α-hydroxy-cholest-4-en-3-one and 7α,12α-dihydroxy-cholest-4-en-3-one. These intermediates may then undergo side-chain oxidation to produce the corresponding $C_{24}$ bile acids. Alternatively, they may be reduced by 3-oxo-$\Delta^4$-steroid-5α-reductase prior to side-chain oxidation, thus giving rise to 5α(H) bile acids (allochenodeoxycholic and allocholic acid). In some patients, defective 5β-reductase activity is clearly secondary to a known cause of hepatocyte damage (such as severe hepatitis-B infection), and the 3-oxo-$\Delta^4$ bile acids disappear when the hepatocytes recover from the primary insult [3, 13]. This observation creates considerable difficulty in the diagnosis of primary genetic deficiency of the 5β-reductase. Russell and Setchell [2] have demonstrated a reduced concentration of immunoreactive enzyme protein in the livers of the twins described above; however, this does not represent proof of a defective gene. The patient with neonatal hemochromatosis described by Sumazaki et al. had normal cDNA and a normal 5β-reductase protein on immunoblot [16]. Thus, primary genetic deficiency of normal 5β-reductase could be excluded. The mechanism of hepatocyte damage and cholestasis in 5β-reductase deficiency is unknown; as with 3β-dehydrogenase deficiency, toxicity of unsaturated intermediates and unsaturated bile acids and loss of bile-acid-dependent bile flow have been postulated.

### Diagnostic Tests

#### Plasma

GC-MS analysis of plasma bile acids from the twins described by Setchell et al. revealed a high concentration of chenodeoxycholic acid and a normal or low concentration of cholic acid, the two characteristic 3-oxo-$\Delta^4$ bile acids (10–20% of the total bile acid mixture) and allo bile acids (25–30% of the total) [12]. This pattern is qualitatively similar to that seen in patients with secondary 5β-reductase deficiency but, in these patients, the proportion of allo bile acids does not usually reach 25–30%. In the patient described by Clayton et al. [13], the plasma concentrations of 7α-hydroxy-3-oxo-4-cholenoic acid and 7α,12α-dihydroxy-3-oxo-4-cholenoic acid were 1.94 μM and 2.07 μM, respectively. The allocholic-acid concentration was 0.76 μM, and cholic acid and chenodeoxycholic acid were undetectable (<0.05 μM).

## Urine

Analysis of urine by FAB-MS shows the presence of major ions attributable to the glycine and taurine conjugates, of 7α-hydroxy-3-oxo-4-cholenoic acid (m/z = 444 and 494) and 7α,12α-dihydroxy-4-cholenoic acid (m/z = 460 and 510). These identities can be confirmed by GC-MS analysis following enzymatic deconjugation. In patients considered to have primary 5β-reductase deficiency, the 3-oxo-$\Delta^4$ bile acids have comprised more than 70% of the total urinary bile acids; a lower percentage is found in most children, whose excretion of 3-oxo-$\Delta^4$ bile acids is secondary to liver damage of other aetiology. An exception to this rule occurs in patients with liver failure secondary to neonatal hemochromatosis [16].

## Treatment and Prognosis

Untreated, the sibling of the twins described by Setchell et al. died of hepatic failure at 4 months of age [12]. The twins were treated with ursodeoxycholic acid (200 mg/day) followed by chenodeoxycholic acid and cholic acid (100 mg/day of each) and then by ursodeoxycholic acid and cholic acid (100 mg/day of each). The latter combination appeared to be the most successful; synthesis of 3-oxo-$\Delta^4$ bile acids and allo bile acids was suppressed, and liver-function tests and bile canalicular morphology normalised [7, 18]. The infant described by Clayton et al. showed an excellent response to treatment with chenodeoxycholic acid and cholic acid (8 mg/kg/day of each). She was asymptomatic, with normal liver function tests at the age of 1 year [17].

## Genetics

The putative cases of primary 5β-reductase deficiency have shown a pattern of inheritance compatible with an autosomal-recessive trait.

## Cerebrotendinous Xanthomatosis

### Clinical Presentation

The first symptom of CTX is often mental retardation detected during the first decade of life. *Cataracts* may also be present as early as 5 years of age. Wevers et al. [19] have documented four Dutch patients in whom persistent diarrhoea was present from early childhood. Motor dysfunction (spastic paresis, ataxia, expressive dysphasia) develops in approximately 60% of patients in the second or third decade of life. Tendon xanthomata may be detectable during the second decade of life but usually appear in the third or fourth decade. The Achilles tendon is the most common site; other sites include the tibial tuberosities and the extensor tendons of the fingers and the triceps. *Premature atherosclerosis* leading to death from myocardial infarction occurs in some patients. In others, death is caused by progression of the *neurological disease* with increasing spasticity, tremor and ataxia and pseudobulbar palsy. It is important to recognise that the neurological deterioration is very variable [20]. For example, some patients are normal intellectually but suffer from a neuropathy or mild spastic paresis; others have no neurological signs but present with psychiatric symptoms resembling schizophrenia. Magnetic-resonance imaging (MRI) of the brain may show diffuse cerebral atrophy and increased signal intensity in the cerebellar white matter on T2-weighted scans [21]. *Osteoporosis* is common in CTX and may produce pathological fractures; it is associated with low plasma concentrations of 25-hydroxy-vitamin D and 24,25-dihydroxy-vitamin D [22]. Patients with untreated CTX usually die from progressive neurological dysfunction or myocardial infarction between the ages of 30 years and 60 years.

### Metabolic Derangement

CTX is caused by a defect in the gene for *sterol 27-hydroxylase*, the mitochondrial enzyme that catalyses the first step in the process of side-chain oxidation, which is required to convert a $C_{27}$ sterol into a $C_{24}$ bile acid [23]. 5β-Cholestane-3α,7α,12α-triol cannot be hydroxylated in the $C_{27}$ position and accumulates in the liver. As a result, it is metabolised by an alternative pathway, starting with hydroxylation in the $C_{25}$ position (in the endoplasmic reticulum). Furthermore, hydroxylations – e.g. in the $C_{22}$ or $C_{23}$ position – result in the synthesis of the characteristic bile alcohols that are found (as glucuronides) in the urine. Bile-acid precursors other than 5β-cholestane-3α,7α,12α-triol also accumulate. Some of these (7α-hydroxy-cholest-4-en-3-one) are probably converted to cholestanol by a pathway involving 7α-dehydroxylation. Because patients with CTX have a reduced rate of bile-acid synthesis, the normal feedback inhibition of cholesterol 7α-hydroxylase by bile acids is disrupted. This further enhances the production of bile alcohols and cholestanol from bile-acid precursors. The major symptoms of CTX are produced by accumulation of cholestanol (and cholesterol) in almost every tissue of the body, particularly in the nervous system, atherosclerotic plaques and tendon xanthomata.

Recent research has indicated that sterol 27-hydoxylase is active in extrahepatic tissues, where it converts cholesterol into 27-hydroxycholesterol, which can be further metabolised and eliminated from cells. This pathway provides a route for the elimination of cholesterol; this route acts as an alternative to the high-density-lipoprotein-mediated reverse cholesterol transport [24]. Disruption of this pathway in CTX provides a further explanation for the accumulation of cholesterol in the tissues.

## Diagnostic Tests

### Plasma

The concentration of cholestanol in plasma can be determined by GC or high-performance liquid chromatography. Patients with CTX have plasma concentrations in the range of 30–400 µM (normal range = 2.6–16 µM). The plasma cholestanol-to-cholesterol ratio may be a better discriminant than the absolute cholestanol concentration. The following bile-acid precursors have been detected at increased concentrations in plasma:

- 7α-Hydroxycholesterol
- 7α-Hydroxy-cholest-4-en-3-one
- 7α,12α-Dihydroxy-cholest-4-en-3-one

Plasma concentrations of bile acids are low; plasma concentrations of bile-alcohol glucuronides are elevated.

### Urine

Rapid diagnosis of CTX can be achieved using FAB-MS. The major cholanoids in the urine are cholestane-pentol glucuronides, giving rise to an ion of m/z ratio 627 [3, 25]. The full bile-alcohol profile can be produced by analysing urine by GC-MS following treatment with *Helix pomatia* glucuronidase/sulphatase; the major alcohols are 3α,7α,12α,23,25-pentols and 3α,7α,12α,22,25-pentols. Increased bile-alcohol concentrations in the urine can be detected using a simple enzymatic assay based on 7α-hydroxysteroid dehydrogenase [26]. The urinary bile-alcohol excretion following cholestyramine administration has been used as a test for carriers of CTX [27].

### Fibroblasts

27-Hydroxylation of $C_{27}$ sterols can be measured in cultured skin fibroblasts, and the enzyme activity is virtually absent in fibroblasts from patients with CTX [28].

### DNA

In certain populations in which one or two common mutations predominate, DNA analysis may prove to be a rapid method for diagnosis of both homozygotes and carriers of CTX (see below).

## Treatment and Prognosis

The results of treatment with chenodeoxycholic acid were first reported in 1984 [29]. The rates of synthesis of cholestanol and cholesterol were reduced, and plasma cholestanol concentrations fell. A significant number of patients showed reversal of their neurological disability, with clearing of the dementia, improved orientation, a rise in intelligence quotient and enhanced strength and independence. The MRI appearances do not, however, show obvious improvement [30]. Urinary excretion of bile-alcohol glucuronides is markedly suppressed. Chenodeoxycholic acid almost certainly works by suppressing cholesterol 7α-hydroxylase activity; ursodeoxycholic acid, which does not inhibit the enzyme, is ineffective. Adults have usually been treated with a dose of 750 mg/day chenodeoxycholic acid. Other treatments that have been used in CTX include hydroxymethylglutaryl-coenzyme-A-reductase inhibitors, (statins such as lovastatin) [31] and low-density lipoprotein apheresis [32]. There is insufficient information available to assess these forms of treatment at the present time. The osteoporosis seen in patients with CTX appears to be resistant to chenodeoxycholic-acid therapy [33].

## Genetics

CTX is inherited as an autosomal-recessive trait. The cDNA encoding the 27-hydroxylase enzyme has been characterised, and the gene has been localised to chromosome 2q33-qter [23]. CTX can be caused by point mutations that lead to a production of an inactive enzyme (R362C and R446C) [34]. In Moroccan Jews, there appear to be two common mutations, both of which lead to failure of the production of sterol 27-hydroxylase mRNA. One is a frame-shift mutation, the other is a splice-junction mutation [35]. Many other mutations have now been described.

## Other Disorders

The use of FAB-MS to screen urine samples from infants with cholestatic liver disease and fat-soluble-vitamin malabsorption has revealed further defects in bile-acid synthesis. One neonate showed marked excretion of bile alcohols, particularly 5β-cholestane-3α,7α,12α,24S,25-pentol and low plasma bile-acid concentrations. His cholestatic liver disease improved with bile-acid-replacement therapy [36]. A 14-year-old boy was shown to excrete large amounts of unconjugated cholic acid in the urine, presumably due to a defect in one of the enzymes involved in amidation (formation of glycine and taurine conjugates). He originally presented with cholestatic liver disease and went on to show evidence of vitamin-K deficiency and rickets [37]. Recently, Setchell et al. have described a 10-week-old male infant with severe cholestasis, cirrhosis and liver synthetic failure who had a mutation in the gene encoding microsomal oxysterol 7α-hydroxylase, leading to inactivity of this enzyme and accumulation of 27-hydroxycholesterol, 3β-hydroxy-5-cholestenoic acid and 3β-hydroxy-5-cholenoic acid [38]. This finding confirms the importance of a second pathway of bile-acid synthesis (the acidic pathway, not shown in Fig. 30.1), which starts with side-chain oxidation of cholesterol and involves the oxysterol 7α-hydroxylase instead of cholesterol 7α-hydroxylase. The study of inborn errors of bile-acid synthesis has already taught us much about the biochemistry of the synthetic pathways in man; it is clear that this will continue in years to come.

## References

1. Björkhem I (1985) Mechanism of bile acid biosynthesis in mammalian liver. In: Danielsson H, Sjövall J (eds) Sterols and bile acids. New comprehensive biochemistry, vol 12. Elsevier, Amsterdam, pp 231–278
2. Russell DW, Setchell KDR (1992) Bile acid biosynthesis. Biochemistry 31:4737–4749
3. Clayton PT (1991) Inborn errors of bile acid metabolism. J Inherited Metab Dis 14:478–496
4. Clayton PT, Leonard JV, Lawson AM, Setchell KDR, Andersson S, Egestad B, Sjövall J (1987) Familial giant cell hepatitis associated with synthesis of 3β,7α-dihydroxy-and 3β,7α,12α-trihydroxy-5-cholenoic acids. J Clin Invest 79:1031–1038
5. Horslen SP, Lawson AM, Malone M, Clayton PT (1992) 3β-Hydroxy-Δ$^5$-C$_{27}$-steroid dehydrogenase deficiency; effect of chenodeoxycholic acid therapy on liver histology. J Inherited Metab Dis 15:38–46
6. Jacquemin E, Setchell KDR, O'Connell NC, Estrada A, Maggiore G, Schmitz J, Hadchouel M, Bernard O (1994) A new cause of progressive intrahepatic cholestasis: 3β-hydroxy-C$_{27}$-steroid dehydrogenase/isomerase deficiency. J Pediatr 125:379–384
7. Setchell KDR (1990) Disorders of bile acid synthesis. In: Walker WA, Durie PR, Hamilton JR, Walker-Smith JA, Watkins JB (eds) Pediatric gastrointestinal disease. Pathophysiology, diagnosis and management, vol 2. Dekker, Philadelphia, pp 922–1013
8. Ichimiya H, Egestad B, Nazer H, Baginski ES, Clayton PT, Sjövall J (1991) Bile acids and bile alcohols in a child with 3β-hydroxy-Δ$^5$-C$_{27}$-steroid dehydrogenase deficiency: effects of chenodeoxycholic acid treatment. J Lipid Res 32:829–841
9. Lawson AM, Madigan MJ, Shortland DB, Clayton PT (1986) Rapid diagnosis of Zellweger syndrome and infantile Refsum's disease by fast atom bombardment – mass spectrometry of urine bile salts. Clin Chim Acta 161:221–231
10. Mills K, Mushtaq I, Johnson A, Whitfield P, Clayton PT (1998) A method for the quantitation of conjugated bile acids in dried blood spots using electrospray ionization mass spectrometry. Pediatr Res 43:361–368
11. Buchmann MS, Kvittingen EA, Nazer H, Gunasekaran T, Clayton PT, Sjövall J (1990) Lack of 3β-hydroxy-Δ$^5$-C$_{27}$-steroid dehydrogenase/isomerase in fibroblasts from a child with urinary excretion of 3β-hydroxy-Δ$^5$-bile acids - a new inborn error of metabolism. J Clin Invest 86:2034–2037
12. Setchell KDR, Suchy FJ, Welsh MB, Zimmer-Nechemias L, Heubi J, Balistreri WF (1988) Δ$^4$-3-Oxosteroid 5β-reductase deficiency described in identical twins with neonatal hepatitis. A new inborn error in bile acid synthesis. J Clin Invest 82:2148–2157
13. Clayton PT, Patel E, Lawson AM, Carruthers RA, Tanner MS, Strandvik B, Egestad B, Sjövall J (1988) 3-Oxo-Δ$^4$ bile acids in liver disease. Lancet i:1283–1284
14. Schneider BL, Setchell KDR, Whittington PF, Neilson KA, Suchy FJ (1994) Δ$^4$-3-Oxosteroid 5β-reductase deficiency causing neonatal liver failure and neonatal hemochromatosis. J Pediatr 124:234–238
15. Clayton PT (1994) Δ$^4$-3-Oxosteroid 5β-reductase deficiency and neonatal hemochromatosis (letter). J Pediatr 125:845–846
16. Sumazaki R, Nakamura N, Shoda J, Kurosawa T, Tohma M (1997) Gene analysis in Δ$^4$-3-oxosteroid 5β-reductase deficiency. Lancet 349:329
17. Clayton PT, Mills KA, Johnson AW, Barabino A, Marazzi MG (1996) Δ$^4$-3-Oxosteroid 5β-reductase deficiency: failure of ursodeoxycholic acid treatment and response to chenodeoxycholic acid plus cholic acid. Gut 38:623–628
18. Balistreri WF (1991) Bile acid metabolism. J Inherit Metab Dis 14:459–477
19. Wevers RA, Cruysberg JRM, Van Heijst AFJ, Janssen-Zijlstra, Renier WO, Van Engelen BGM, Tolboom JJM (1992) Paediatric cerebrotendinous xanthomatosis. J Inherit Metab Dis 14:374–376
20. Kuriyama M, Fujiyama J, Yoshidome H, Takenaga S, Matsumoro M, Kasama T, Fukada K, Kuramoto T, Hoshita T, Seyama Y et al. (1991) Cerebrotendinous xanthomatosis: clinical features of eight patients and a review of the literature. J Neurol Sci 102:225–232
21. Bencze K, Polder DRV, Prockop LD (1990) Magnetic resonance imaging of the brain in CTX. J Neurol Neurosurg Psychiatry 53:166–167
22. Berginer VM, Shany S, Alkalay D, Berginer J, Dekel S, Salen G, Tint GS, Gazit D (1993) Osteoporosis and increased bone fractures in cerebrotendinous xanthomatosis. Metabolism 42:69–74
23. Cali JJ, Russell DW (1991) Characterisation of human sterol 27-hydroxylase: a mitochondrial cytochrome P-450 that catalyses multiple oxidations in bile acid biosynthesis. J Biol Chem 266:7774–7778
24. Babiker A, Andersson O, Lund E, Xin RJ, Deeb S, Reshef A, Leitersdorf E, Diczfalusy U, Björkhem I (1997) Elimination of cholesterol in macrophages and endothelial cells by the sterol 27-hydroxylase mechanism. Comparison with high density lipoprotein-mediated reverse cholesterol transport. J Biol Chem 272:26253–26261
25. Egestad B, Pettersson P, Skrede S, Sjövall J (1985) Fast atom bombardment mass spectrometry in the diagnosis of cerebrotendinous xanthomatosis. Scand J Clin Lab Invest 45:443–446
26. Koopman BJ, Molen JC, Wolthers BG, Waterreus RJ (1987) Screening for CTX by using an enzymatic method for 7α-hydroxylated steroids in urine. Clin Chem 33:142–143
27. Koopman BJ, Waterreus RJ, Brekel HWC, Wolthers BG (1986) Detection of carriers of CTX. Clin Chim Acta 158: 179–186
28. Skrede S, Björkhem I, Kvittingen EA, Buchmann MS, East C, Grundy S (1986) Demonstration of 26-hydroxylation of C$_{27}$-steroids in human skin fibroblasts, and a deficiency of this activity in CTX. J Clin Invest 78:729–735

29. Berginer VM, Salen G, Shefer S (1984) Long-term treatment of CTX with chenodeoxycholic acid therapy. N Engl J Med 311:1649–1652
30. Berginer VM, Berginer J, Korczyn AD, Tadmor R (1994) Magnetic resonance imaginng in cerebrotendinous xanthomatosis: a prospective clinical and neuroradiological study. J Neurol Sci 122:102–108
31. Lewis B, Mitchell WD, Marenah CB, Cortese C (1983) Cerebrotendinous xanthomatosis: biochemical response to inhibition of cholesterol synthesis. Br Med J 287:21–22
32. Mimura Y, Kuriyama M, Tokimura Y, Fujiyama J, Osame M, Takesako K, Tanaka N (1993) Treatment of cerebrotendinous xanthomatosis with low density lipoprotein (LDL)-apheresis. J Neurol Sci 114:227–230
33. Chang WN, Lui CC (1997) Failure in the treatment of long-standing osteoporosis in cerebrotendinous xanthomatosis. J Formos Med Assoc 96:225–227
34. Cali JJ, Hsieh C-L, Francke U, Russell DW (1991) Mutations in the bile acid biosynthetic enzyme sterol 27-hydroxylase underlie cerebrotendinous xanthomatosis. J Biol Chem 266:7779–7783
35. Leitersdorf E, Reshef A, Meiner V, Levitzki R, Schwartz SP, Dann EJ, Berkman N, Cali JJ, Klapholz L, Berginer VM (1993) Frameshift and splice-junction mutations in the sterol 27-hydroxylase gene cause cerebrotendinous xanthomatosis in Jews of Moroccan origin. J Clin Invest 91:2488–2496
36. Clayton PT, Casteels M, Mieli-Vergani G, Lawson AM (1995) Familial giant cell hepatitis with low bile acid concentrations and increased urinary excretion of specific bile alcohols: a new inborn error of bile acid synthesis? Pediatr Res 37: 424–431
37. Setchell KDR, Heubi JE, O'Connell NC, Hoffmann AF, Lavine JE (1997) Identification of a unique inborn error in bile and conjugation involving a deficiency in amidation. In: Paumgartner G, Stiehl A, Gerok W (eds) Bile acids in hepato biliary disease: basic and clinical applications. Kluwer Academic, Dordrecht, pp 43–47
38. Setchell KDR, Schwarz M, O'Connell NC, Lund EG, Davis DL, Lathe R, Thompson HR, Tyson RW, Sokol RJ, Russell DW (1998) Identification of a new inborn error in bile acid synthesis: mutation of the oxysterol 7α-hydroxylase gene causes severe neonatal liver disease. J Clin Invest 102:1690–1703

# PART VIII
# DISORDERS OF NUCLEIC ACID AND HEME METABOLISM

## Purine Metabolism

Purine nucleotides are essential cellular constituents; they intervene in energy transfer, metabolic regulation and the synthesis of DNA and RNA. Purine metabolism can be divided into three pathways:

1. The biosynthetic pathway, often termed de novo, starts with the formation of phosphoribosyl pyrophosphate and leads to the synthesis of inosine monophosphate (IMP). From IMP, adenosine monophosphate (AMP), guanosine monophosphate (GMP) and the other adenine and guanine nucleotides are formed. Deoxyribonucleotides are formed at the diphosphate level.
2. The catabolic pathway starts from GMP, IMP and AMP and produces uric acid, a poorly soluble compound which tends to crystallize once its plasma concentration surpasses 6.5–7 mg/dl (0.38–0.47 mmol/l).
3. The salvage pathway utilizes the purine bases, guanine, hypoxanthine and adenine (which are provided by food intake or the catabolic pathway) and reconverts them into GMP, IMP and AMP, respectively.

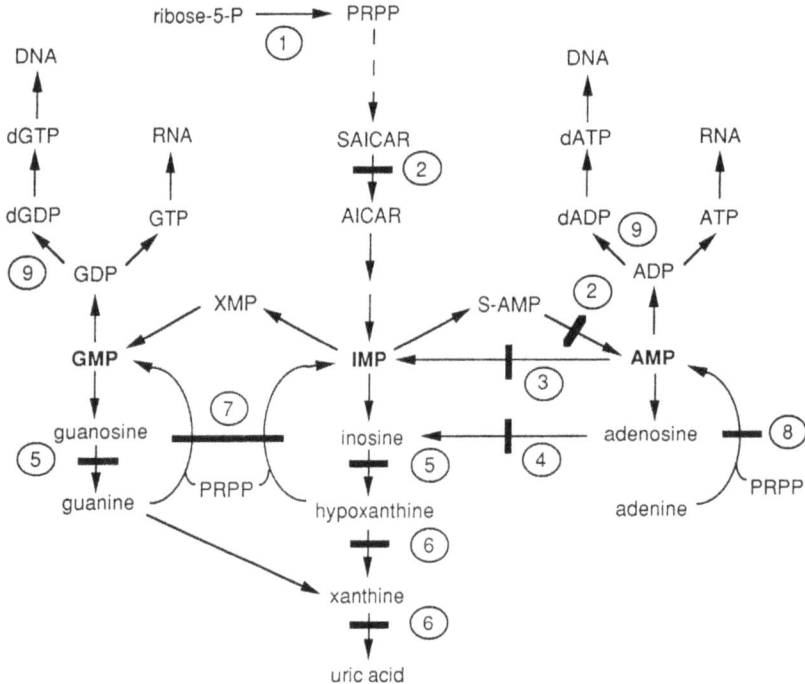

**Fig. 31.1.** Pathways of purine metabolism. *ADP*, adenosine diphosphate; *AICAR*, aminoimidazolecarboxamide ribotide; *d*, deoxy; *GDP*, guanosine diphosphate; *GMP*, guanosine monophosphate; *GTP*, guanosine triphosphate; *IMP*, inosine monophosphate; *PRPP*, phosphoribosyl pyrophosphate; *S-AMP*, adenylosuccinate; *SAICAR*, succinylaminoimidazolecarboxamide ribotide; *XMP*, xanthosine monophosphate; *1*, PRPP synthetase; *2*, adenylosuccinase; *3*, AMP deaminase; *4*, adenosine deaminase; *5*, purine nucleoside phosphorylase; *6*, xanthine oxidase; *7*, hypoxanthine-guanine phosphoribosyltransferase; *8*, adenine phosphoribosyltransferase; *9*, ribonucleotide reductase. Enzyme deficiencies are indicated by *solid bars*

# CHAPTER 31

# Disorders of Purine and Pyrimidine Metabolism

G. van den Berghe, M.-F. Vincent, and S. Marie

CONTENTS

Inborn Errors of Purine Metabolism ............... 355
  Phosphoribosyl Pyrophosphate-Synthetase Superactivity  355
  Phosphoribosyl Pyrophosphate-Synthetase Deficiency .. 357
  Adenylosuccinase Deficiency..................... 357
  Muscle Adenosine Monophosphate Deaminase Deficiency 358
  Adenosine Deaminase Deficiency ................. 359
  Adenosine Deaminase Superactivity ............... 360
  Purine Nucleoside Phosphorylase Deficiency ......... 360
  Xanthine Oxidase Deficiency..................... 361
  Hypoxanthine-Guanine Phosphoribosyl Transferase
  Deficiency ................................... 362
  Adenine Phosphoribosyl Transferase Deficiency....... 363

Inborn Errors of Pyrimidine Metabolism............. 365
  Uridine Monophosphate Synthase Deficiency
  (Hereditary Orotic Aciduria)...................... 365
  Dihydropyrimidine Dehydrogenase Deficiency........ 365
  Dihydropyrimidinase Deficiency................... 366
  Ureidopropionase Deficiency ..................... 366
  Pyrimidine-5'-Nucleotidase Deficiency .............. 367
  Cytosolic 5'-Nucleotidase Superactivity.............. 367

References....................................... 367

---

Inborn errors of the biosynthetic, catabolic and salvage pathways of purine and pyrimidine metabolism are depicted in Figs. 31.1 and 31.3, respectively. The major presenting signs and laboratory findings in these inborn errors are listed in Table 31.1.

## INBORN ERRORS OF PURINE METABOLISM

Inborn errors of purine metabolism comprise errors of:

- The synthesis of purine nucleotides: phosphoribosyl-pyrophosphate (PRPP)-synthetase superactivity and deficiency, adenylosuccinase (ADSL) deficiency
- Purine catabolism: the deficiencies of muscle AMP deaminase (AMP-DA, also termed myoadenylate deaminase), adenosine deaminase (ADA), purine nucleoside phosphorylase (PNP) and xanthine oxidase
- Purine salvage: the deficiencies of hypoxanthine-guanine-phosphoribosyl transferase (HGPRT) and adenine phosphoribosyl transferase (APRT)

With the exception of muscle AMP-DA deficiency, all these enzyme defects are very rare.

## Phosphoribosyl Pyrophosphate-Synthetase Superactivity

### Clinical Presentation

The disorder is mostly manifested by the appearance, in young adult males, of gouty arthritis and/or uric-acid lithiasis, potentially leading to renal insufficiency [1, 2]. Uricemia can be very high, reaching 10–15 mg/dl (0.60–0.90 mmol/l) [normal adult values: 2.9–5.5 mg/dl (0.17–0.32 mmol/l)]. The urinary excretion of uric acid is also increased, reaching up to 2400 mg (14 mmol)/24 h [normal adult values: 500–800 mg (3–4.7 mmol)/24 h]. Determination of the ratio of uric acid to creatinine (mg/mg) can also be informative, as described in the section discussing diagnostic tests of HGPRT deficiency. A few patients have been reported in which clinical signs of uric acid overproduction already appeared in infancy and were accompanied by neurologic abnormalities – mainly sensorineural deafness (particularly for high tones) but also hypotonia, locomotor delay, ataxia and autistic features [2].

### Metabolic Derangement

The enzyme forms PRPP from ribose-5-phosphate and ATP (Fig. 31.1). PRPP is the first intermediate of the de novo synthesis of purine nucleotides (not shown in detail in Fig. 31.1), which leads to the formation of inosine monophosphate (IMP), from which the other purine compounds are derived. PRPP synthetase is highly regulated. Various genetic regulatory and catalytic defects [1, 2] lead to superactivity, resulting in increased generation of PRPP. Because PRPP amidotransferase, the rate-limiting enzyme of the de novo pathway, is not physiologically saturated by PRPP, the synthesis of purine nucleotides (and hence the production of uric acid) increases. PRPP-synthetase superactivity is one of the few known examples of a hereditary anomaly that enhances the activity of the

**Table 31.1.** Main presenting clinical signs and laboratory data in inborn errors of purine and pyrimidine metabolism

| Clinical signs or laboratory data | Diagnostic possibilities |
| --- | --- |
| Arthritis | PRPP-synthetase superactivity |
|  | HGPRT deficiency (partial) |
| Ataxia | PNP deficiency |
|  | HGPRT deficiency (complete) |
|  | Cytosolic 5′-nucleotidase superactivity |
| Autistic features | PRPP-synthetase superactivity |
|  | Adenylosuccinase deficiency |
|  | Dihydropyrimidine-dehydrogenase deficiency |
|  | Cytosolic 5′-nucleotidase superactivity |
| Convulsions | Adenylosuccinase deficiency |
|  | PRPP-synthetase deficiency |
|  | Combined xanthine and sulfite-oxidase deficiency |
|  | Dihydropyrimidine-dehydrogenase deficiency |
|  | Dihydropyrimidinase deficiency |
|  | Cytosolic 5′-nucleotidase superactivity |
| Deafness | PRPP-synthetase superactivity |
| Growth retardation | Adenylosuccinase deficiency |
|  | ADA deficiency |
|  | UMP-synthase deficiency |
|  | Dihydropyrimidine-dehydrogenase deficiency |
|  | Cytosolic 5′-nucleotidase superactivity |
| Hypotonia | Adenylosuccinase deficiency |
|  | Muscle AMP-deaminase deficiency |
|  | Ureidopropionase deficiency |
| Kidney stones |  |
|   Uric acid | PRPP synthetase superactivity |
|  | HGPRT deficiency (complete or partial) |
|   Xanthine | Xanthine-oxidase deficiency (isolated or combined with sulfite oxidase deficiency) |
|   2,8-Dihydroxyadenine | APRT deficiency |
|   Orotic acid | UMP-synthase deficiency |
| Muscle cramps | Muscle-AMP deaminase deficiency |
| Muscle wasting | Adenylosuccinase deficiency |
| Psychomotor delay | PRPP-synthetase superactivity |
|  | Adenylosuccinase deficiency |
|  | Combined xanthine and sulfite-oxidase deficiency |
|  | HGPRT deficiency (complete) |
|  | UMP-synthase deficiency |
|  | Dihydropyrimidine-dehydrogenase deficiency |
|  | Dihydropyrimidinase deficiency |
|  | Ureidopropionase deficiency |
|  | Cytosolic 5′-nucleotidase superactivity |
| Recurrent infections | ADA deficiency |
|  | PNP deficiency |
|  | Cytosolic 5′-nucleotidase superactivity |
| Renal insufficiency | PRPP-synthetase superactivity |
|  | HGPRT deficiency (complete or partial) |
|  | APRT deficiency |
| Self-mutilation | HGPRT deficiency (complete) |
| Anemia |  |
|   Megaloblastic | PRPP-synthetase deficiency |
|  | UMP-synthase deficiency |
|   Hemolytic | ADA superactivity |
|  | Pyrimidine-5′-nucleotidase deficiency |
| Hyperuricemia | PRPP-synthetase superactivity |
|  | HGPRT deficiency (complete or partial) |
| Hypouricemia | PRPP-synthetase deficiency |
|  | PNP deficiency |
|  | Xanthine-oxidase deficiency (isolated or combined with sulfite-oxidase deficiency) |
| Lymphopenia |  |
|   B and T-cells | ADA deficiency |
|   T-cells | PNP deficiency |
| Orotic aciduria | PRPP-synthetase deficiency |
|  | UMP-synthase deficiency |

*ADA*, adenosine deaminase; *APRT*, adenine phosphoriboysltransferase; *HGPRT*, hypoxanthine–guanine phosphoribosyltransferase; *PNP*, purine nucleoside phosphorylase; *PRPP*, phosphoribosyl pyrophosphate; *UMP*, uridine monophosphate

enzyme it affects. The mechanism of the neurologic symptoms is unresolved.

## Diagnostic Tests

Diagnosis requires extensive kinetic studies of the enzyme, which are performed on erythrocytes and cultured fibroblasts in a few laboratories. The disorder should be differentiated from partial HGPRT deficiency, which gives similar clinical signs.

## Treatment and Prognosis

Patients should be treated with allopurinol, which inhibits xanthine oxidase, the last enzyme of purine catabolism (Fig. 31.1). This results in a decrease of the production of uric acid and in its replacement by hypoxanthine (which is approximately tenfold more soluble) and xanthine (which is slightly more soluble). The initial dosage of allopurinol is 10-20 mg/kg/day in children and 2-10 mg/kg/day in adults. It should be adjusted to the minimum required to maintain normal uric acid levels in plasma and should be reduced in subjects with renal insufficiency. In rare patients with a considerable increase in de novo synthesis, xanthine calculi can be formed during allopurinol therapy [3]. Additional measures to prevent crystallization are thus recommended. These include a low-purine diet (free of organ meats, sardines, dried beans and peas), high fluid intake and, since uric acid and xanthine are more soluble at alkaline pH than at acid pH, administration of sodium bicarbonate, potassium citrate or citrate mixtures to bring the urinary pH to 6.0-6.5. Adequate control of uricemia prevents gouty arthritis and urate nephropathy but does not correct the neurological symptoms.

## Genetics

The various forms of PRPP-synthetase superactivity are inherited as sex-linked traits. In families in which the anomaly is associated with sensorineural deafness, heterozygous females with gout and/or hearing impairment have also been found [2]. Studies of the gene in six families revealed a different single base change in each of them [4].

## Phosphoribosyl Pyrophosphate-Synthetase Deficiency

A defect of this enzyme in erythrocytes, associated with mental retardation, convulsions, hypouricemia, increased urinary excretion of orotic acid and megaloblastic bone-marrow changes, has been reported in a Japanese boy [5]. Convulsions responded to medication, particularly adrenocorticotropic hormone.

## Adenylosuccinase Deficiency

### Clinical Presentation

In the first reported presentation, often referred to as type I, patients display moderate to severe psychomotor retardation, frequently accompanied both by epilepsy after the first years and by autistic features (failure to make eye-to-eye contact, repetitive behavior, temper tantrums); more rarely, patients display severe growth retardation associated with muscular wasting [6, 7]. A single girl, referred to as type II, is only mildly retarded [7], whereas another girl displays profound muscle hypotonia accompanied by slightly delayed motor development [8]. More recently, patients with convulsions starting within the first days to weeks of life have been reported [9]. The marked clinical heterogeneity justifies systematic screening for the deficiency in unexplained, profound or mild psychomotor retardation and in neurological disease with convulsions and/or hypotonia.

### Metabolic Derangement

ADSL (also named adenylosuccinate lyase), catalyzes two steps in purine synthesis (Fig. 31.1): the conversion of succinylamino imidazolecarboxamide ribotide (SAI-CAR) into AICAR along the de novo pathway, and the conversion of adenylosuccinate (S-AMP) into AMP. Deficiency of ADSL results in accumulation in cerebrospinal fluid and urine of the succinylpurines, SAICA riboside and succinyladenosine (S-Ado), the products of the dephosphorylation of the two substrates of the enzyme by cytosolic 5'-nucleotidase. Present evidence indicates that the more severe presentations of ADSL deficiency tend to be associated with S-Ado/SAICA-riboside ratios around one whereas, in milder clinical pictures, these ratios are between two and four. This suggests that SAICA riboside is the offending compound and that S-Ado could protect against its toxic effects. The ADSL defect is marked in liver and kidney and is variably expressed in erythrocytes, muscle and fibroblasts [6, 7, 10]. The higher S-Ado/SAICA-riboside ratios might be explained by a more profound loss of activity of the enzyme toward S-AMP than toward SAICAR compared with a parallel deficiency of the activities in severely affected patients [10]. The symptoms of the deficiency remain unexplained, but positron-emission tomography reveals a marked decrease of the uptake of fluorodeoxyglucose in the cortical brain areas [11].

### Diagnostic Tests

Diagnosis is based on the presence (in cerebrospinal fluid and urine) of SAICA riboside and S-Ado, which are normally undetectable. These can be recognized by various techniques. For systematic screening, a modified Bratton-Marshall test [12], performed on urine, appears most practical. However, false positive results are recorded in patients who receive sulfonamides, for the measurement of which the test was initially devised. Several thin-layer-chromatographic methods are also available [13]. Final diagnosis requires HPLC with UV detection [6]. The activity of ADSL should preferably be measured on fresh biopsy specimens, owing to the sensitivity of the enzyme to freezing and thawing.

### Treatment and Prognosis

With the aim of replenishing hypothetically decreased concentrations of adenine nucleotides in ADSL-deficient tissues, some patients were treated with oral supplements of adenine (10 mg/kg/day) and allopurinol (5–10 mg/kg/day) for several months. Adenine can be incorporated into the adenine nucleotides via APRT (Fig. 31.1). Allopurinol is required to avoid conversion of adenine into minimally soluble 2,8-dihydroxyadenine (which forms kidney stones) by xanthine oxidase. No clinical or biochemical improvement was recorded with the exception of weight gain and some acceleration of growth [7]. Recently, oral administration of ribose (1.5 g/kg/day) was reported to reduce seizure frequency in an ADSL-deficient girl [14].

The prognosis for survival of ADSL-deficient patients is very variable. Mildly retarded patients have reached adult age, whereas several of those presenting with early epilepsy died within the first months of life.

### Genetics

The deficiency is transmitted as an autosomal-recessive trait [6, 7]. Studies of the ADSL gene, localized on chromosome 22, have led to the identification of about 20 mutations [15, 16]. Most were found in single families, with the exception of R426H, which was diagnosed in 12 families, six of them from the Netherlands.

## Muscle Adenosine Monophosphate Deaminase Deficiency

### Clinical Presentation

The deficiency of muscle AMP-DA (frequently referred to as *myoadenylate deaminase* in the clinical literature) can be a primary genetic defect or can be secondary to another neuromuscular disease. The primary defect typically presents with isolated muscular weakness, fatigue, cramps or myalgias following moderate to vigorous exercise, sometimes accompanied by an increase in serum creatine kinase and minor electromyographic abnormalities. Muscular wasting or histologic abnormalities are absent [17]. Primary AMP-DA deficiency was initially detected in young adults but, later, wide variability was observed with respect to age (1.5–70 years) at onset of the symptoms [18, 19]. Moreover, the enzyme defect has been detected in patients with hypotonia and/or cardiomyopathy and in asymptomatic family members of individuals with the disorder. Secondary muscle AMP-DA deficiency is found in several diseases, including amyotrophic lateral sclerosis, fascioscapulohumeral myopathy, Kugelberg-Welander syndrome, polyneuropathies and Werdnig-Hoffmann disease [18, 19].

### Metabolic Derangement

AMP-DA, Adenylosuccinate synthetase and ADSL form the purine-nucleotide cycle (Fig. 31.2). Numerous functions in muscle have been proposed for this cycle [20]: (1) removal of AMP formed during exercise in order to favor the formation of ATP from adenosine diphosphate by myokinase (adenylate kinase); (2) release of $NH_3$ and IMP, both stimulators of glycolysis and, hence, of energy production; and (3) production of fumarate, an intermediate of the citric-acid cycle, which also yields energy. Therefore, it has been proposed that the muscle dysfunction observed in primary AMP-DA deficiency is caused by impairment of energy production for muscle contraction. However, this tallies neither with the existence of asymptomatic AMP-DA-deficient individuals in families with the deficiency, nor with the frequent occurrence of the enzyme defect in other muscle disorders.

It should be noted that muscle, liver and erythrocytes contain different isoforms of AMP-DA. A regulatory mutation of liver AMP-DA has been proposed as a cause of primary gout with overproduction of uric acid [21]. Individuals with a complete (though totally asymptomatic) deficiency of erythrocyte AMP-DA have been detected in Japan, Korea and Taiwan [22].

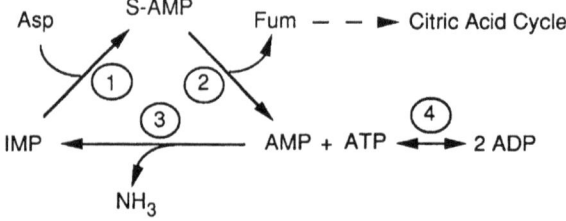

**Fig. 31.2.** The purine nucleotide cycle. *ADP*, adenosine diphosphate; *Asp*, aspartate; *Fum*, fumarate; *IMP*, inosine monophosphate; *S-AMP*, adenylosuccinate; 1, Adenylosuccinate synthetase; 2, adenylosuccinase; 3, AMP deaminase; 4, myokinase (adenylate kinase)

## Diagnostic Tests

Screening for the defect can be performed by an exercise test (Chap. 2). A several-fold elevation of venous plasma ammonia, seen in normal subjects, is absent in AMP-DA deficiency. Final diagnosis is established by histochemical or biochemical assay in a muscle biopsy. In the primary defect, the activity of AMP-DA is below 2% of normal, and little or no immunoprecipitable enzyme is found. In the secondary defect, the activity is 2–15% of normal, and appreciable immunoreactivity is usually present [23]. Remarkably, in several large series of muscle biopsies for diagnostic purposes, low enzyme activities were found in about 2% of all specimens [18, 19].

## Treatment and Prognosis

Patients may display a gradual progression of their symptoms, which may worsen until even dressing or walking a few steps leads to fatigue and myalgias. They should be advised to exercise with caution to prevent rhabdomyolysis and myoglobinuria. Administration of ribose (2–60 g/day orally, in divided doses) has been reported to improve muscular strength and endurance [24].

## Genetics

Primary AMP-DA deficiency is apparently transmitted as an autosomal-recessive trait. *AMPD1*, the gene encoding muscle AMP-DA, is located on chromosome 1. In nearly all patients with the primary deficiency investigated hitherto, the defect is caused by a nonsense c.34C → T mutation resulting in a stop codon [25]. Population studies suggest that this mutant allele is found with a high frequency in Caucasians. This accords with the finding that approximately 2% of diagnostic muscle biopsies are AMP-DA deficient and suggests that the mutation arose in a remote Western European ancestor.

# Adenosine Deaminase Deficiency

## Clinical Presentation

The majority of patients display, within the first weeks or months after birth, a profound impairment of both humoral and cellular immunity, known as *severe combined immunodeficiency disease* (SCID). Multiple, recurrent infections rapidly become life threatening [26–28]. Cases of delayed infantile onset, later childhood onset and even adult onset have been reported. Caused by a broad variety of organisms, infections are mainly localized in the skin and the respiratory and gastrointestinal tracts. In the latter, they often lead to intractable diarrhea, malnutrition and growth retardation. In affected children over 6 months of age, the hypoplasia or apparent absence of lymphoid tissue constitutes a suggestive sign. Bone abnormalities, clinically evident as prominence of the costochondral rib junctions and radiologically evident as cupping and flaring thereof, are found in about half of the patients. In a few affected children, neurologic abnormalities, including spasticity, head lag, movement disorders, nystagmus and inability to focus, are found. Recently, hepatic dysfunction has also been reported [28, 29].

SCID can be confirmed by relatively simple laboratory tests: (1) lymphopenia (usually less than 500 total lymphocytes per cubic millimeter) involving both B and T cells and (2) hypogammaglobulinemia are almost invariably present. Whereas the IgM deficiency may be detected early, the IgG deficiency becomes manifest only after the age of 3 months, when the maternal supply has been exhausted. More elaborate tests show a deficiency of antibody formation following specific immunization and an absence or severe diminution of the lymphocyte proliferation induced by mitogens. The disease is progressive, since residual B- and T-cell function, which may be found at birth, disappears later.

## Metabolic Derangement

The deficiency results in the accumulation of adenosine (Fig. 31.1) and deoxyadenosine (not shown in Fig. 31.1), a second substrate of ADA (derived from the catabolism of DNA) in body fluids. Both compounds are normally nearly undetectable. Among the mechanisms that have been proposed to explain the immunodeficiency, two are considered the most important: inhibition of ribonucleotide reductase and impairment of the transmethylation reactions [28]. Ribonucleotide reductase is an essential enzyme for the synthesis of DNA (Fig. 31.1), which has to proceed at a high rate during lymphocyte development and differentiation. Its inhibition in ADA-deficient lymphocytes is explained by accumulation of excess dATP formed from deoxyadenosine. Transmethylation reactions are vital in a number of physiological processes, such as lymphocyte function. Accumulation of deoxyadenosine has been shown to inactivate S-adenosylhomocysteine hydrolase, an essential enzyme of the transmethylation reaction sequence.

## Diagnostic Tests

The diagnosis is mostly performed on red blood cells. In general, severity of disease correlates with the loss of ADA activity. Children with neonatal onset of SCID display 0–1% residual activity; in individuals with later onset, 1–5% of normal ADA activity is found [26, 27]. It should be noted that only about 20–30% of the patients with the clinical and hematologic picture of inherited SCID are ADA-deficient. In the remaining patients, SCID is caused by other mechanisms. A few subjects with ADA deficiency in red blood cells but with normal immunocompetence have been described [27]. This is explained by the presence of residual ADA activity in their lymphocytes.

### Treatment and Prognosis

Bone-marrow transplantation, the first choice (provided a histocompatible donor is available), gives a good chance for complete cure, both clinically and immunologically [30]. The graft provides stem cells and, hence, T and B cells, which have sufficient ADA activity to prevent accumulation of adenosine and deoxyadenosine. Survival is, however, much lower with HLA-mismatched transplants. If no histocompatible bone-marrow donor is found, enzyme-replacement therapy can be accomplished with normal erythrocytes irradiated before use to prevent graft-versus-host disease. Repeated partial-exchange transfusions are performed at 2- to 4-week intervals. In some patients, marked clinical and immunological improvement is obtained but, in most, response is poor or is not sustained [30]. In addition, this therapy carries the risks inherent in repeated transfusions.

A much more effective enzyme-replacement therapy can be achieved with polyethylene-glycol-modified ADA (PEG-ADA). Covalent attachment of PEG to bovine ADA results in marked extension of its half-life and reduction of immunogenicity. Weekly to bi-weekly intramuscular injections of 15–30 units of PEG-ADA per kilogram usually result in marked clinical improvement. In vitro immune function also significantly improves [31].

The first approved clinical trial of gene therapy was performed in 1990 in a 4-year-old girl with ADA deficiency and, since then, preliminary reports on several patients have been published [32]. Peripheral-blood T cells of the patients were collected, cultured with interleukin-2, corrected by insertion of the ADA gene by means of a retroviral vector and reinfused. Patients improved, but the treatment has to be repeated at regular time intervals, because lymphocytes live only a few months. Moreover, all patients are still receiving PEG-ADA, so benefit cannot be attributed unequivocally to gene therapy. Future trials are aimed at the technically more demanding gene transfer into stem cells, which (in theory) have an unlimited life span.

Enzyme-replacement therapies have significantly improved the prognosis of ADA deficiency. Untreated, the defect invariably led to death, usually within the first year of life, unless drastic steps were taken, such as rearing in strictly sterile conditions beginning at birth.

### Genetics

Approximately one third of the cases of inherited SCID are X-linked, whereas two thirds are autosomal recessive. ADA deficiency is found only in the latter group, where it accounts for about 50% of the patients. The frequency of the deficiency is estimated at one per 100,000–500,000 births. Studies of the ADA gene, located on chromosome 20, have hitherto revealed over 40 mutations, the majority of which are single-nucleotide changes resulting in either an inactive or an unstable enzyme [28]. Most patients carry two different mutations on each chromosome 20, but others, mainly from inbred communities, are homozygous for the mutation. Recently, spontaneous in vivo reversion of a mutation on one allele to normal (as observed in tyrosinemia type I; Chap. 15) has been reported [33].

## Adenosine Deaminase Superactivity

A hereditary, approximately 50-fold elevation of red cell ADA has been shown to cause non-spherocytic hemolytic anemia [34]. The latter can be explained by an enhanced catabolism of the adenine nucleotides, including ATP, owing to the increased activity of ADA.

## Purine Nucleoside Phosphorylase Deficiency

### Clinical Presentation

Recurrent infections are usually of later onset, starting from the end of the first year up to 5–6 years of age, and are initially less severe than in ADA deficiency [35, 36]. A strikingly enhanced susceptibility to viral diseases, such as varicella, measles, cytomegalovirus and *Vaccinia*, has been reported, but severe *Candida* and pyogenic infections also occur. One-third of the patients have anemia, and two-thirds display neurologic symptoms, including spastic tetra- or diplegia, ataxia and tremor. Immunological studies reveal an increasing deficiency of cellular immunity, reflected by a marked reduction in the number of T cells. B-lymphocyte function is deficient in about one-third of the patients.

### Metabolic Derangement

In body fluids, the deficiency provokes an accumulation of the four normally nearly undetectable substrates of the enzyme: guanosine, inosine (Fig. 31.1), and their deoxy counterparts (not shown in Fig. 31.1), the latter derived from DNA breakdown. Thus, formation of uric acid is severely hampered. The profound impairment of cellular immunity, characterizing the disorder, has been explained by an accumulation of excess dGTP, particularly in T-cells. It is formed from deoxyguanosine and inhibits ribonucleotide reductase and, hence, cell division.

## Diagnostic Tests

Patients often display a striking decrease of the production of uric acid: plasma uric acid is usually below 1 mg/dl and may even be undetectable. However, in patients with residual PNP activity, uricemia may be at the borderline of normal. The urinary excretion of uric acid is usually also markedly diminished. Enzymatic diagnosis is mostly performed on red blood cells.

## Treatment and Prognosis

Most diagnosed patients have died (although at a later age than untreated ADA-deficient children) from overwhelming viral or bacterial infections. Treatments consisted of bone-marrow transplantation and repeated transfusions of normal, irradiated erythrocytes [30, 36]. More recently, successful matched bone-marrow transplantation has been reported [37]. Enzyme and gene therapy might become available in the near future.

## Genetics

The deficiency is inherited in an autosomal-recessive fashion. Studies of the PNP gene, located on chromosome 14, have revealed a number of molecular defects, among which a R234P mutation was most common [38].

# Xanthine Oxidase Deficiency

## Clinical Presentation

Two types of xanthine-oxidase (or -dehydrogenase) deficiency (also termed hereditary *xanthinuria*) are known: an isolated form [39] and a combined xanthine-oxidase and *sulfite-oxidase* deficiency [40]. Isolated xanthine-oxidase deficiency can be completely asymptomatic, although kidney stones are formed in about one-third of the cases. Most often not visible on X-ray, they may appear at any age. Myopathy may be present, associated with crystalline xanthine deposits. In combined xanthine-oxidase and sulfite-oxidase deficiency, the clinical picture of sulfite-oxidase deficiency (which is also found as an isolated defect [41]; Chap. 18), overrides that of xanthine-oxidase deficiency. The symptoms include neonatal feeding difficulties and intractable seizures, myoclonia, increased or decreased muscle tone, eye-lens dislocation and severe mental retardation.

## Metabolic Derangement

The deficiency results in the near total replacement of uric acid by hypoxanthine and xanthine as the end products of purine catabolism (Fig. 31.1). In combined xanthine-oxidase and sulfite-oxidase deficiency there is also accumulation of sulfite and of sulfur-containing metabolites and a diminution of the production of inorganic sulfate. The combined defect is caused by deficiency of a *molybdenum cofactor* required for the activity of both xanthine oxidase and sulfite oxidase.

## Diagnostic Tests

In both forms of the deficiency, plasma concentrations of uric acid below 1 mg/dl are measured; they may decrease to virtually undetectable values when the patient is on a low-purine diet. Urinary uric acid is reduced to a few percent of normal and is replaced by hypoxanthine and xanthine. In the combined defect, these urinary changes are accompanied by excessive excretion of sulfite and other sulfur-containing metabolites, such as S-sulfocysteine, thiosulfate and taurine, as detailed in Chap. 18. The enzymatic diagnosis requires liver or intestinal mucosa, the only human tissues that normally contain appreciable amounts of xanthine oxidase. Sulfite oxidase and the molybdenum cofactor can be assayed in liver and fibroblasts.

## Treatment and Prognosis

Isolated xanthine-oxidase deficiency is mostly benign. A low purine diet should be prescribed and fluid intake increased. In subjects with residual xanthine-oxidase activity, use of allopurinol to completely block the conversion of hypoxanthine into xanthine, which is about tenfold less soluble, has been advocated. The prognosis of combined xanthine-oxidase and sulfite-oxidase deficiency is very poor. So far, all therapeutic attempts, including low-sulfur diets, the administration of sulfate and molybdenum [40] and trials to bind sulfite with thiol-containing drugs, have been unsuccessful.

## Genetics

The inheritance of both isolated xanthine-oxidase deficiency and combined xanthine-oxidase and sulfite-oxidase deficiency is autosomal recessive. Studies of the xanthine-oxidase gene, localized on chromosome 2, have led to the identification of two mutations, one resulting in a nonsense substitution and the other in a termination codon [42].

## Hypoxanthine-Guanine Phosphoribosyltransferase Deficiency

### Clinical Presentation

HGPRT deficiency can present in two forms. Patients with complete or near-complete deficiency of HGPRT display the Lesch-Nyhan syndrome [43]. Affected children generally appear normal during the first months of life. At 3–4 months of age, a neurologic syndrome evolves, including delayed motor development, choreoathetoid movements, and spasticity with hyperreflexia and scissoring. In time, the patients develop a striking, compulsive, self-destructive behavior involving biting of their fingers and lips, which leads to mutilating loss of tissue. Speech is hampered by athetoid dysarthria. Whereas most patients have intelligence quotients around 50, some display normal intelligence. Approximately 50% of the patients have seizures. Sooner or later, they form uric-acid stones. Mothers of Lesch-Nyhan patients have reported the finding of orange crystals on diapers during the first few weeks after birth. Untreated, the uric acid nephrolithiasis progresses to obstructive uropathy and renal failure during the first decade of life. Exceptionally, the latter clinical picture may also be observed in early infancy.

Partial HGPRT deficiency is found in rare patients with gout. Most of them are normal on neurological examination, but (occasionally) spasticity, dysarthria and a spinocerebellar syndrome are found [44]. Whereas most patients with the Lesch-Nyhan syndrome do not develop gouty arthritis, this finding is common in partial HGPRT deficiency.

### Metabolic Derangement

The considerable increase of the production of uric acid is explained as follows: PRPP, which is not utilized at the level of HGPRT (Fig. 31.1), is available in increased quantities for the rate-limiting enzyme of de novo synthesis, PRPP amidotransferase (not shown in Fig. 31.1). Since the latter is normally not saturated with PRPP, its activity increases, and the ensuing acceleration of de novo synthesis results in the overproduction of uric acid.

The pathogenesis of the neurological symptoms is still not satisfactorily explained. A number of studies point to dopaminergic dysfunction involving decreases of the concentration of dopamine and of the activity of the enzymes required for its synthesis, although dopaminergic drugs are not useful. Recently, positron-emission tomography of the brain with [$^{18}$F]-fluorodopa, an analogue of the dopamine precursor levodopa, has revealed a generalized decrease of the activity of dopa decarboxylase [45]. How the HGPRT defect leads to the deficit of the dopaminergic system and how the latter results in the characteristic neuropsychiatric manifestations of Lesch-Nyhan syndrome remains to be clarified.

### Diagnostic Tests

Patients excrete excessive amounts of uric acid, ranging from 25 mg to 140 mg (0.15 to 0.85 mmol)/kg of body weight per 24 h, as compared with an upper limit of 18 mg (0.1 mmol)/kg per 24 h in normal children. Determination of the ratio of uric acid to creatinine (mg/mg) in morning samples of urine provides a screening test. This ratio is much higher in HGPRT deficiency than the normal upper limits of 2.5, 2.0, 1.0 and 0.6 for infants, children 2 years of age, children 10 years of age and adults, respectively [46]. Increased ratios are also found in other disorders with uric-acid overproduction, such as PRPP-synthetase superactivity, glycogenosis type I and lymphoproliferative diseases. The overproduction of uric acid is always accompanied by an increase of serum urate, which may reach concentrations as high as 18 mg/dl (1 mmol/l). However, occasionally (and particularly before puberty), uricemia may be in the normal or high-normal range.

Patients with the Lesch-Nyhan syndrome display nearly undetectable HGPRT activity in red blood cells [47]. In partial deficiencies, similar low values may be found [48]. Rates of incorporation of hypoxanthine into the adenine nucleotides of intact fibroblasts correlate better with the clinical symptomatology than HGPRT activities in erythrocytes do. Patients with complete Lesch-Nyhan syndrome incorporate less than 1.2% of the normal level, those with gout and neurological symptoms incorporate 1.2–10% of the normal level, and those with isolated gout incorporate 10–55% of the normal level [48].

### Treatment and Prognosis

Allopurinol, as detailed under "PRPP-Synthetase Superactivity", is thought to prevent urate nephropathy. However, allopurinol, even when given from birth, has no effect on the neurological symptoms, which have so far resisted all therapeutic attempts. Adenine has been administered, together with allopurinol, with the aim to correct a possible depletion of purine nucleotides. However, no or minimal changes in neurologic behavior were recorded [49]. Patients should be made more comfortable by appropriate restraints, including elbow splints, lip guards and even tooth extraction, to diminish self-mutilation. Diazepam, haloperidol and barbiturates may sometimes improve choreoathetosis. In a 22-year-old patient, bone-marrow transplantation restored erythrocyte HGPRT activity to normal but did not change neurologic symptoms [50].

### Genetics

Both Lesch-Nyhan syndrome and partial deficiencies of HGPRT are transmitted in a sex-linked, recessive manner. Studies of the HGPRT gene in large groups of unrelated patients have revealed a variety of defects ranging from point mutations provoking single amino acid substitutions (and therefore enzymes with altered stability and/or kinetic properties) to extensive deletions resulting in suppression of enzyme synthesis [51]. These studies have contributed a great deal to the understanding of the clinical variation observed in human inherited disease, and have provided support for the concept that, in X-linked disorders, new mutations constantly appear in the population. Presently, over 50 mutations of the HGPRT gene have been described, and molecular studies have led to precise prenatal diagnosis and efficient carrier testing of at-risk females [52].

## Adenine Phosphoribosyltransferase Deficiency

### Clinical Presentation

The deficiency may become clinically manifest in childhood [53], even from birth [54], but may also remain silent for several decades. Symptoms include urinary passage of gravel, small stones and crystals, frequently accompanied by abdominal colic, dysuria, hematuria and urinary-tract infection. Some patients may even present with acute anuric renal failure [55]. The urinary precipitates are composed of 2,8-dihydroxyadenine and are radiotranslucent and indistinguishable from uric-acid stones by routine chemical testing.

### Metabolic Derangement

The deficiency results in suppression of the salvage of adenine (Fig. 31.1) provided by food and by the polyamine pathway (not shown in Fig. 31.1). Consequently, adenine is oxidized into 2,8-dihydroxyadenine, a very poorly soluble compound (solubility in urine at pH 5 and 37°C is about 0.3 mg/dl compared with 15 mg/dl for uric acid), by xanthine oxidase.

The deficiency can be complete or partial. The partial deficiency is only found in the Japanese, among whom it is quite common [56]. Activities range from 10% to 30% of normal at supraphysiological concentrations of PRPP, but a 20- to 30-fold decrease in the affinity for PRPP results in near inactivity under physiological conditions.

### Diagnostic Tests

Identification of 2,8-dihydroxyadenine requires complex analyses, including UV and infrared spectrography, MS and X-ray crystallography [53, 54]. Therefore, it is usually easier to measure APRT activity in red blood cells.

### Treatment and Prognosis

In patients with symptoms, allopurinol should be given to inhibit the formation of 2,8-dihydroxyadenine, as detailed under "PRPP-Synthetase Superactivity". Both in patients with stones and in those without symptoms, dietary purine restriction and high fluid intake are recommended. Alkalinization of the urine is, however, not advised; unlike that of uric acid, the solubility of 2,8-dihydroxyadenine does not increase up to pH 9 [53].

Ultimate prognosis depends on renal function at the time of diagnosis: late recognition may result in irreversible renal insufficiency requiring chronic dialysis and early treatment in the prevention of stones. Unfortunately, kidney transplantation has been reported to be followed by recurrence of microcrystalline deposits and subsequent loss of graft function [57].

### Genetics

APRT deficiency is inherited as an autosomal-recessive trait. Approximately 70% of the Japanese patients carry the same c.2069T $\rightarrow$ C substitution in exon 5, resulting in a M136T change [56]. In Caucasians, about a dozen mutations have been identified, some of which seem more common, also suggesting founder effects [58].

# Pyrimidine Metabolism

Similar to that of the purine nucleotides, the metabolism of the pyrimidine nucleotides can be divided into three pathways:

1. The biosynthetic pathway starts with the formation of carbamoylphosphate by a cytosolic carbamoylphosphate synthetase (CPS-II), which is different from the mitochondrial CPS-I (which catalyzes the first step of ureogenesis; Fig. 17.1). This is followed by the synthesis of uridine monophosphate (UMP), and hence of cytidine monophosphate (CMP) and thymidine monophosphate (TMP).
2. The catabolic pathway starts from CMP, UMP and TMP and yields β-alanine and β-aminoisobutyrate, which are converted into intermediates of the citric-acid cycle.
3. The salvage pathway, composed of kinases, converts the pyrimidine nucleosides, cytidine, uridine and thymidine, into the corresponding nucleotides (Fig. 31.3).

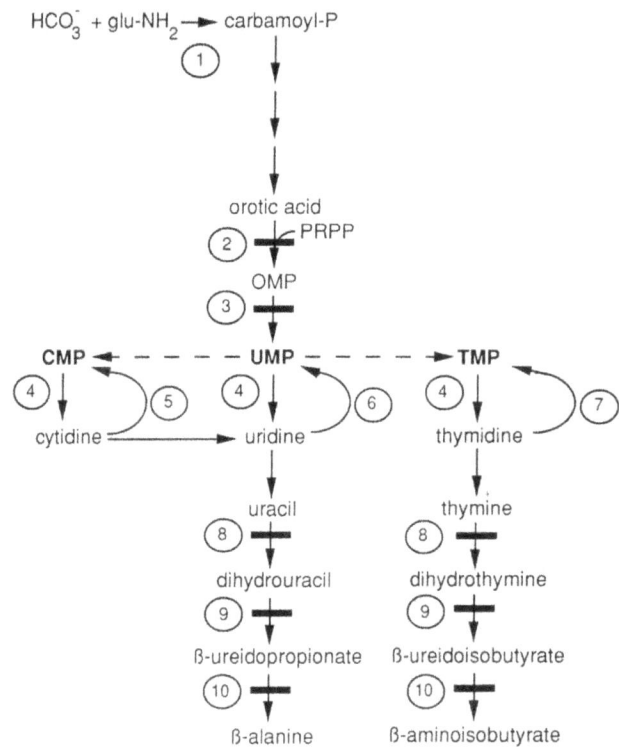

**Fig. 31.3.** Pathways of pyrimidine metabolism. *CMP*, cytidine monophosphate; *glu-NH₂*, glutamine; *OMP*, orotidine monophosphate; *PRPP*, phosphoribosylpyrophosphate; *TMP*, thymidine monophosphate; *UMP*, uridine monophosphate; 1, Carbamoylphosphate synthetase II; 2, orotate phosphoribosyltransferase; 3, orotidine decarboxylase (2 and 3 form UMP synthase); 4, pyrimidine (cytosolic) 5′-nucleotidase; 5, cytidine kinase; 6, uridine kinase; 7, thymidine kinase; 8, dihydropyrimidine dehydrogenase; 9, dihydropyrimidinase; 10, ureidopropionase. Enzyme deficiencies are indicated by *solid bars*

# INBORN ERRORS OF PYRIMIDINE METABOLISM

Inborn errors of pyrimidine metabolism include a defect of the synthesis of pyrimidine nucleotides (UMP-synthase deficiency) and three inborn errors of pyrimidine catabolism: the deficiencies of dihydropyrimidine dehydrogenase (DPD), dihydropyrimidinase (DHP), and pyrimidine 5'-nucleotidase. Recently, a fourth defect of pyrimidine catabolism, ureidopropionase deficiency, and superactivity of cytosolic 5'-nucleotidase have been reported.

## Uridine Monophosphate Synthase Deficiency (Hereditary Orotic Aciduria)

### Clinical Presentation

Megaloblastic anemia, which appears a few weeks or months after birth, is usually the first manifestation [59, 60]. Peripheral-blood smears often show anisocytosis, poikilocytosis, and moderate hypochromia. Bone-marrow examination reveals erythroid hyperplasia and numerous megaloblastic erythroid precursors. Characteristically, the anemia does not respond to iron, folic acid or vitamin $B_{12}$. Unrecognized, the disorder leads to failure to thrive and to retardation of growth and psychomotor development.

### Metabolic Derangement

UMP synthase is a bifunctional enzyme of the de novo synthesis of pyrimidines (Fig. 31.3). A first reaction, catalyzed by orotate phosphoribosyltransferase (OPRT), converts orotic acid into orotidine monophosphate (OMP), and a second, catalyzed by orotidine decarboxylase (ODC), decarboxylates OMP into UMP. The defect provokes a massive overproduction of orotic acid and a deficiency of pyrimidine nucleotides [60]. The overproduction is attributed to the ensuing decrease of the feedback inhibition exerted by the pyrimidine nucleotides on the first enzyme of their de novo synthesis, cytosolic CPS-II (Fig. 31.3). The deficiency of pyrimidine nucleotides leads to impairment of cell division, which results in megaloblastic anemia and retardation of growth and development.

### Diagnostic Tests

Urinary analysis reveals a massive overexcretion of orotic acid, reaching, in infants, 200- to 1000-fold the normal adult value of 1–1.5 mg per 24 h. Occasionally, orotic-acid crystalluria is noted, particularly upon dehydration. Enzymatic diagnosis can be performed on red blood cells. In all patients reported except one, both OPRT and ODC activities were deficient. This defect is termed type I. In a single patient, referred to as type II, only the activity of ODC was initially deficient, although that of OPRT also decreased later [60].

### Treatment and Prognosis

The enzyme defect can be bypassed by the administration of uridine, which is converted into UMP by uridine kinase (Fig. 31.3). An initial dose of 100–150 mg/kg, administered over a 1-day period, induces prompt hematologic response and acceleration of growth. Further dosage should be adapted to obtain the lowest possible output of orotic acid. In some cases, normal psychomotor development was achieved; this did not occur in other cases, possibly owing to delayed onset of therapy.

### Genetics

Hereditary orotic aciduria is inherited as an autosomal-recessive trait. The genetic lesion results in synthesis of an enzyme with reduced stability [61]. Three point mutations have been identified in two Japanese families [62].

## Dihydropyrimidine Dehydrogenase Deficiency

### Clinical Presentation

Two forms of DPD deficiency occur. The first is found in children. Approximately half of these display epilepsy and mental retardation, often accompanied by generalized hypertonia, hyperreflexia, growth retardation, dysmorphic features including microcephaly, motor retardation and autistic features [63, 64]. In these patients, the deficiency of DPD is complete or near complete. Nevertheless, the severity of the disorder is highly variable, because patients with epilepsy and normal intelligence (and even asymptomatic cases) have been identified. The second clinical picture is found in adults who receive the pyrimidine analog 5-fluorouracil for treatment of cancers of the breast, gastrointestinal tract or ovary [65, 66]. This causes severe toxicity, manifested by cytopenia, stomatitis, diarrhea and neurologic symptoms, including ataxia, paralysis and stupor. In these patients, DPD deficiency is partial and is only revealed by 5-fluorouracil therapy.

### Metabolic Derangement

The deficiency of DPD, which catalyzes the catabolism of uracil and thymine into, respectively, dihydrouracil and dihydrothymine (Fig. 31.3), leads to the accumulation of the former compounds [63]. Why a profound DPD deficiency becomes manifest in some pediatric patients but not in others is not known. How the defect leads to neurologic symptoms also remains elusive, but reduction of the concentration of β-alanine, a neurotransmitter, may play a role. The marked potentiation of the action (and therefore the toxicity) of the anticancer drug 5-fluorouracil is explained by a block of the catabolism of this pyrimidine analog via DPD.

### Diagnostic Tests

Patients excrete high amounts of uracil (56–683 mmol/mol creatinine compared with 3–33 mmol/mol in control urine) and thymine (7–439 mmol/mol creatinine compared with 0–4 mmol/mol in control urine). Elevations of uracil and thymine in plasma and cerebrospinal fluid are much less prominent [64]. Excretion of both compounds may also be less elevated in patients with high residual DPD activity. The pyrimidine catabolites can be detected by HPLC, gas chromatography–MS, and analysis of amino acids in urine before and after acid hydrolysis [67].

The enzyme defect can be demonstrated in the patients' fibroblasts, liver and blood cells, with the exception of erythrocytes [63, 64, 66]. In pediatric patients, DPD deficiency is complete or near-complete; in adult cancer patients experiencing acute 5-fluorouracil toxicity, it is only partial, with residual enzyme activities ranging from 3% to 30%.

### Treatment and Prognosis

No treatment is available for pediatric patients. Symptoms usually remain the same, but death of a more severely affected child in early infancy has been reported. In adult cancer patients, discontinuation of 5-fluorouracil results in slow resolution of the toxic symptoms [65, 66].

### Genetics

Both the infantile and the adult form are inherited as autosomal-recessive traits. The DPD gene is localized on chromosome 1. Several mutations have been identified, including a splice site mutation, found in five unrelated patients, which results in skipping of a complete exon [64]. Strikingly, patients who carry the same mutation may display widely variable clinical symptoms.

## Dihydropyrimidinase Deficiency

### Clinical Picture

DHP-deficiency disorder was first reported in a single male baby of consanguineous parents; the baby presented with convulsions and metabolic acidosis [68]. A few additional patients have been diagnosed since then [64]. As in DPD deficiency, the clinical picture varies from presentation with severe psychomotor retardation and epilepsy, dysmorphic features or microcephaly, to completely asymptomatic presentation.

### Metabolic Derangement

DHP catalyzes the cleavage of dihydrouracil and dihydrothymine into, respectively, β-ureidopropionate and β-ureidoisobutyrate (Fig. 31.3). Consequently, considerable quantities of dihydrouracil and dihydrothymine, which are normally found in small amounts, are excreted in urine [64]. There is also a moderate elevation of uracil and thymine excretion. As in DPD deficiency, the reasons for the appearance and the mechanisms of the symptoms remain unexplained, and reduced concentrations of the neurotransmitter β-alanine may play a role. Although not reported yet, increased sensitivity to 5-fluorouracil toxicity might also be present.

### Diagnostic Tests

Elevation of urinary dihydrouracil and dihydrothymine can be detected by the techniques used for measurement of uracil and thymine in DPD deficiency. Enzyme assay requires liver biopsy, because more accessible tissues do not possess DPH activity [68].

### Treatment and Prognosis

There is no therapy, and prognosis seems unpredictable. The first reported patient recovered completely and apparently displays normal physical and mental development [68]. In contrast, another patient had a progressive neurodegenerative clinical course [69].

### Genetics

The defect is inherited as an autosomal-recessive trait. Studies of the DPH gene, localized on chromosome 8, have led to the identification of one frame-shift and five missense mutations in one symptomatic and five asymptomatic individuals [70]. Enzyme expression showed no significant difference in residual activity between the mutations of the symptomatic and the asymptomatic individuals.

## Ureidopropionase Deficiency

Recently, in a single female infant of consanguineous parents, presenting with muscle hypotonia, dystonic movements and severe developmental delay, in vitro proton magnetic-resonance-imaging spectroscopy of urine revealed elevated ureidopropionic and ureidoisobutyric acid [71]. These findings have led to the presumptive diagnosis of ureidopropionase deficiency, a hitherto unreported enzyme defect.

## Pyrimidine-5′-Nucleotidase Deficiency

This defect, restricted to erythrocytes, provokes accumulation of pyrimidine nucleotides, resulting in basophilic stippling and chronic hemolytic anemia [72]. The mechanism by which the increased pyrimidine nucleotides cause hemolysis remains unknown.

## Cytosolic 5′-Nucleotidase Superactivity

Recently, four unrelated children with a syndrome including developmental delay, growth retardation, seizures, ataxia, recurrent infections, autistic features and hypouricosuria were described [73]. Studies in the patients' fibroblasts showed 6- to 20-fold elevations of the activity of cytosolic 5′-nucleotidase, measured either with a pyrimidine (UMP) or a purine (AMP) as the substrate. Based on the possibility that this increased catabolism might cause a deficiency of pyrimidine nucleotides, the patients were treated with uridine in a dose of 1 g/kg/day. Remarkable developmental improvement and a decrease in frequency of seizures and infections were recorded.

## References

1. Sperling O, Boer P, Persky-Brosh S, Kanarek E, De Vries A (1972) Altered kinetic property of erythrocyte phosphoribosylpyrophosphate synthetase in excessive purine production. Rev Eur Etud Clin Biol 17:703-706
2. Becker MA, Puig JG, Mateos FA, Jimenez ML, Kim M, Simmonds HA (1988) Inherited superactivity of phosphoribosylpyrophosphate synthetase: association of uric acid overproduction and sensorineural deafness. Am J Med 85:383-390
3. Kranen S, Keough D, Gordon RB, Emmerson BT (1985) Xanthine-containing calculi during allopurinol therapy. J Urol 133:658-659
4. Becker MA, Smith PR, Taylor W, Mustafi R, Switzer RL (1995) The genetic and functional basis of purine nucleotide feedback-resistant phosphoribosylpyrophosphate synthetase superactivity. J Clin Invest 96:2133-2141
5. Wada Y, Nishimura Y, Tanabu M et al. (1974) Hypouricemic, mentally retarded infant with a defect of 5-phosphoribosyl-1-pyrophosphate synthetase of erythrocytes. Tohoku J Exp Med 113:149-157
6. Jaeken J, Van den Berghe G (1984) An infantile autistic syndrome characterised by the presence of succinylpurines in body fluids. Lancet 2:1058-1061
7. Jaeken J, Wadman SK, Duran M et al. (1988) Adenylosuccinase deficiency: an inborn error of purine nucleotide synthesis. Eur J Pediatr 148:126-131
8. Valik D, Miner PT, Jones JD (1997) First U.S. case of adenylosuccinate lyase deficiency with severe hypotonia. Pediatr Neurol 16:252-255
9. Van den Bergh FAJTM, Bosschaart AN, Hageman G, Duran M, Poll-The BT (1998) Adenylosuccinase deficiency with neonatal onset severe epileptic seizures and sudden death. Neuropediatrics 29:51-53
10. Van den Berghe G, Vincent MF, Jaeken J (1997) Inborn errors of the purine nucleotide cycle: adenylosuccinase deficiency. J Inherit Metab Dis 20:193-202
11. De Volder AG, Jaeken J, Van den Berghe G et al. (1988) Regional brain glucose utilization in adenylosuccinase-deficient patients measured by positron emission tomography. Pediatr Res 24:238-242
12. Laikind PK, Seegmiller JE, Gruber HE (1986) Detection of 5′-phosphoribosyl-4-(N-succinylcarboxamide)-5-aminoimidazole in urine by use of the Bratton-Marshall reaction: identification of patients deficient in adenylosuccinate lyase activity. Anal Biochem 156:81-90
13. Sebesta I, Shobowale M, Krijt J, Simmonds HA (1995) Screening tests for adenylosuccinase deficiency. Screening 4:117-124
14. Salerno C, D'Eufemia P, Finocchiaro R et al. (1999) Effect of D-ribose on purine synthesis and neurological symptoms in a patient with adenylosuccinase deficiency. Biochim Biophys Acta 1453:135-140
15. Stone RL, Aimi J, Barshop BA et al. (1992) A mutation in adenylosuccinate lyase associated with mental retardation and autistic features. Nature Genet 1:59-63
16. Marie S, Cuppens H, Heutersprute M et al. (1999) Mutation analysis in adenylosuccinate lyase deficiency. Eight novel mutations in the re-evaluated full ADSL coding sequence. Hum Mutat 13:197-202
17. Fishbein WN, Armbrustmacher VW, Griffin JL (1978) Myoadenylate deaminase deficiency: a new disease of muscle. Science 200:545-548
18. Shumate JB, Katnik R, Ruiz M et al. (1979) Myoadenylate deaminase deficiency. Muscle Nerve 2:213-216
19. Mercelis R, Martin JJ, de Barsy T, Van den Berghe G (1987) Myoadenylate deaminase deficiency: absence of correlation with exercise intolerance in 452 muscle biopsies. J Neurol 234:385-389
20. Van den Berghe G, Bontemps F, Vincent MF, Van den Bergh F (1992) The purine nucleotide cycle and its molecular defects. Prog Neurobiol 39:547-561
21. Hers HG, Van den Berghe G (1979) Enzyme defect in primary gout. Lancet 1:585-586
22. Ogasawara N, Goto H, Yamada Y et al. (1987) Deficiency of AMP deaminase in erythrocytes. Hum Genet 75:15-18
23. Sabina RL, Fishbein WN, Pezeshkpour G, Clarke PR, Holmes EW (1992) Molecular analysis of the myoadenylate deaminase deficiencies. Neurology 42:170-179
24. Zöllner N, Reiter S, Gross M et al. (1986) Myoadenylate deaminase deficiency: successful symptomatic therapy by high dose oral administration of ribose. Klin Wochenschr 64:1281-1290
25. Morisaki T, Gross M, Morisaki H et al. (1992) Molecular basis of AMP deaminase deficiency in skeletal muscle. Proc Natl Acad Sci USA 89:6457-6461
26. Giblett ER, Anderson JE, Cohen F, Pollara B, Meuwissen HJ (1972) Adenosine-deaminase deficiency in two patients with severely impaired cellular immunity. Lancet 2:1067-1069
27. Hirschhorn R (1993) Overview of biochemical abnormalities and molecular genetics of adenosine deaminase deficiency. Pediatr Res 33 [Suppl 1]:S35-S41
28. Hershfield MS, Arredondo-Vega FX, Santisteban I (1997) Clinical expression, genetics and therapy of adenosine deaminase (ADA) deficiency. J Inherit Metab Dis 20:179-185
29. Bollinger ME, Arredondo-Vega FX, Santisteban I, Schwarz K, Hershfield MS, Lederman HM (1996) Brief report: hepatic dysfunction as a complication of adenosine deaminase deficiency. N Engl J Med 334:1367-1371
30. Markert ML, Hershfield MS, Schiff RI, Buckley RH (1987) Adenosine deaminase and purine nucleoside phosphorylase deficiencies: evaluation of therapeutic interventions in eight patients. J Clin Immunol 7:389-399
31. Hershfield MS (1995) PEG-ADA replacement therapy for adenosine deaminase deficiency: an update after 8.5 years. Clin Immunol Immunopathol 76:S228-S232

32. Blaese RM, Culver KW, Miller AD et al. (1995) T-lymphocyte-directed gene therapy for ADA-SCID: initial trial results after 4 years. Science 270:475-480
33. Hirschhorn R, Yang DR, Puck JM, Huie ML, Jinang CK, Kurlandsly LE (1996). Spontaneous in vivo reversion to normal of an inherited mutation in a patient with adenosine deaminase deficiency. Nature Genet 13:290-295
34. Valentine WN, Paglia DE, Tartaglia AP, Gilsanz F (1977) Hereditary hemolytic anemia with increased red cell adenosine deaminase (45- to 70-fold) and decreased adenosine triphosphate. Science 195:783-785
35. Giblett ER, Ammann AJ, Wara DW, Sandman R, Diamond LK (1975) Nucleoside phosphorylase deficiency in a child with severely defective T-cell immunity and normal B-cell immunity. Lancet 1:1010-1013
36. Markert ML (1991) Purine nucleoside phosphorylase deficiency. Immunodefic Rev 3:45-81
37. Carpenter PA, Ziegler JB, Vowels MR (1996) Late diagnosis and correction of purine nucleoside phosphorylase deficiency with allogeneic bone marrow transplantation. Bone Marrow Transplant 17:121-124
38. Markert ML, Finkel BD, McLaughlin TM et al. (1997) Mutations in purine nucleoside phosphorylase deficiency. Hum Mutat 9:118-121
39. Dent CE, Philpot GR (1954) Xanthinuria, an inborn error (or deviation) of metabolism. Lancet 1:182-185
40. Wadman SK, Duran M, Beemer FA et al. (1983) Absence of hepatic molybdenum cofactor: an inborn error of metabolism leading to a combined deficiency of sulphite oxidase and xanthine dehydrogenase. J Inherit Metab Dis 6 [Suppl 1]:78-83
41. Shih VE, Abroms IF, Johnson JL et al. (1977) Sulfite oxidase deficiency. Biochemical and clinical investigations of a hereditary metabolic disorder in sulfur metabolism. N Engl J Med 297:1022-1028
42. Ichida K, Amaya Y, Kamatani N, Nishino T, Hosoya T, Sakai O (1997) Identification of two mutations in human xanthine dehydrogenase gene responsible for classical type I xanthinuria. J Clin Invest 99:2391-2397
43. Lesch M, Nyhan WL (1964) A familial disorder of uric acid metabolism and central nervous system dysfuntion. Am J Med 36:561-570
44. Kelley WN, Greene ML, Rosenbloom FM, Henderson JF, Seegmiller JE (1969) Hypoxanthine-guanine phosphoribosyltransferase deficiency in gout. Ann Intern Med 70:155-206
45. Ernst M, Zametkin AJ, Matochik JA et al. (1996) Presynaptic dopaminergic deficits in Lesch-Nyhan disease. N Engl J Med 334:1568-1572
46. Kaufman JM, Greene ML, Seegmiller JE (1968) Urine uric acid to creatinine ratio - a screening test for inherited disorders of purine metabolism. Phosphoribosyltransferase (PRT) deficiency in X-linked cerebral palsy and in a variant of gout. J Pediatr 73:583-592
47. Seegmiller JE, Rosenbloom FM, Kelley WN (1967) Enzyme defect associated with a sex-linked human neurological disorder and excessive purine synthesis. Science 155:1682-1684
48. Page T, Bakay B, Nissinen E, Nyhan WL (1981) Hypoxanthine-guanine phosphoribosyltransferase variants: correlation of clinical phenotype with enzyme activity. J Inherit Metab Dis 4:203-206
49. Watts RWE, McKeran RO, Brown E, Andrews TM, Griffiths MI (1974) Clinical and biochemical studies on treatment of Lesch-Nyhan syndrome. Arch Dis Child 49:693-702
50. Nyhan WL, Parkman R, Page T et al. (1986) Bone marrow transplantation in Lesch-Nyhan disease. Adv Exp Med Biol 195A:167-170
51. Davidson BL, Tarlé SA, Van Antwerp M et al. (1991) Identification of 17 independent mutations responsible for human hypoxanthine-guanine phosphoribosyltransferase (HPRT) deficiency. Am J Hum Genet 48:951-958
52. Alford RL, Redman JB, O'Brien WE, Caskey CT (1995) Lesch-Nyhan syndrome: carrier and prenatal diagnosis. Prenat Diagn 15:329-338
53. Cartier P, Hamet M (1974) Une nouvelle maladie métabolique: le déficit complet en adénine-phosphoribosyltransférase avec lithiase de 2,8-dihydroxyadénine. C R Acad Sci Paris 279D:883-886
54. Van Acker KJ, Simmonds HA, Potter C, Cameron JS (1977) Complete deficiency of adenine phosphoribosyltransferase. Report of a family. N Engl J Med 297:127-132
55. Greenwood MC, Dillon MJ, Simmonds HA, Barratt TM, Pincott JR, Metreweli C (1982) Renal failure due to 2,8-dihydroxyadenine urolithiasis. Eur J Pediatr 138:346-349
56. Hidaka Y, Tarlé SA, Fujimori S, Kamatani N, Kelley WN, Palella TD (1988) Human adenine phosphoribosyltransferase deficiency. Demonstration of a single mutant allele common to the Japanese. J Clin Invest 81:945-950
57. Gagne ER, Deland E, Daudon M, Noel LH, Nawar T (1994). Chronic renal failure secondary to 2,8-dihydroxyadenine deposition: the first report of recurrence in a kidney transplant. Am J Kidney Dis 24:104-107
58. Sahota A, Chen J, Stambrook PJ, Tischfield JA (1991) Mutational basis of adenine phosphoribosyltransferase deficiency. Adv Exp Med Biol 309B:73-76
59. Huguley CM, Bain JA, Rivers SL, Scoggins RB (1959) Refractory megaloblastic anemia associated with excretion of orotic acid. Blood 14:615-634
60. Smith LH (1973) Pyrimidine metabolism in man. N Engl J Med 288:764-771
61. Perry ME, Jones ME (1989) Orotic aciduria fibroblasts express a labile form of UMP synthase. J Biol Chem 264:15522-15528
62. Suchi M, Mizuno H, Kawai Y et al. (1997) Molecular cloning of the human UMP synthase gene and characterization of point mutations in two hereditary orotic aciduria families. Am J Hum Genet 60:525-539
63. Berger R, Stoker-de Vries SA, Wadman SK et al. (1984) Dihydropyrimidine dehydrogenase deficiency leading to thymine-uraciluria. An inborn error of pyrimidine metabolism. Clin Chim Acta 141:227-234
64. Van Gennip AH, Abeling NGGM, Vreken P, van Kuilenburg ABP (1997) Inborn errors of pyrimidine degradation: clinical, biochemical and molecular aspects. J Inherit Metab Dis 20:203-213
65. Tuchman M, Stoeckeler JS, Kiang DT, O'Dea RF, Ramnaraine ML, Mirkin BL (1985) Familial pyrimidinemia and pyrimidinuria associated with severe fluorouracil toxicity. N Engl J Med 313:245-249
66. Diasio RB, Beavers TL, Carpenter JT (1988) Familial deficiency of dihydropyrimidine dehydrogenase. Biochemical basis for familial pyrimidinemia and severe 5-fluorouracil-induced toxicity. J Clin Invest 81:47-51
67. Van Gennip AH, Driedijk PC, Elzinga A, Abeling NGGM (1992) Screening for defects of dihydropyrimidine degradation by analysis of amino acids in urine before and after acid hydrolysis. J Inherit Metab Dis 15:413-415
68. Duran M, Rovers P, de Bree PK et al. (1991) Dihydropyrimidinuria: a new inborn error of pyrimidine metabolism. J Inherit Metab Dis 14:367-370
69. Putman CW, Rotteveel JJ, Wevers RA, van Gennip AH, Bakkeren JA, De Abreu RA (1997) Dihydropyrimidinase deficiency: a progressive neurological disorder? Neuropediatrics 28:106-110
70. Hamajima N, Kouwaki M, Vreken P et al. (1998) Dihydropyrimidinase deficiency: structural organization, chromosomal localization, and mutation analysis of the human dihydropyrimidinase gene. Am J Hum Genet 63:717-726
71. Assmann B, Göhlich-Ratmann G, Bräutigam C et al. (1998) Presumptive ureidopropionase deficiency as a new defect in pyrimidine catabolism found with in vitro H-NMR spectroscopy. J Inherit Metab Dis 21 [Suppl 2]:1
72. Valentine WN, Fink K, Paglia DE, Harris SR, Adams WS (1974) Hereditary hemolytic anemia with human erythrocyte pyrimidine 5'-nucleotidase deficiency. J Clin Invest 54:866-879
73. Page T, Yu A, Fontanesi J, Nyhan WL (1997) Developmental disorder associated with increased cellular nucleotidase activity. Proc Natl Acad Sci USA 94:11601-11606

# CHAPTER 32

## The Heme Biosynthetic Pathway

Heme (iron protoporphyrin) is a metalloporphyrin with iron as the central metal atom. The heme biosynthetic pathway (Fig. 32.1) consists of eight enzymes and their substrates and products. The first and last three enzymes are located in the mitochondria and the other four in the cytosol. The pathway is regulated differently in bone marrow and liver, which are the tissues that make the largest amounts of heme. Most of the heme synthesized in bone marrow is used for hemoglobin formation, whereas most heme synthesized in liver is used for cytochrome P450 enzymes. Factors that influence liver cytochrome P450 enzyme synthesis (drugs, diet, and certain hormones) have little effect on hemoglobin synthesis in the bone marrow. This explains why such factors are important determinants of the clinical expression in hepatic porphyrias but not in erythropoietic porphyrias.

The first enzyme of the pathway, δ-aminolevulinic acid synthase, is rate limiting in the liver, and is induced by a variety of drugs, steroids and other chemicals that also induce cytochrome P450 enzymes (1, 2). The liver enzyme is subject to negative feedback by heme, which represses its synthesis and its import into mitochondria. By contrast, erythroid δ-aminolevulinic acid synthase, which is encoded by a separate gene located on the X chromosome, is induced by heme and erythropoietin. The gene for this erythroid enzyme and those for at least three other enzymes of the pathway, contain DNA sequences that provide for erythroid-specific regulation of heme synthesis. Mutations of the erythroid-specific form of δ-aminolevulinic acid synthase, the first enzyme in the pathway, are found in X-linked sideroblastic anemia. Mutations in genes for the other seven enzymes are found in the porphyrias.

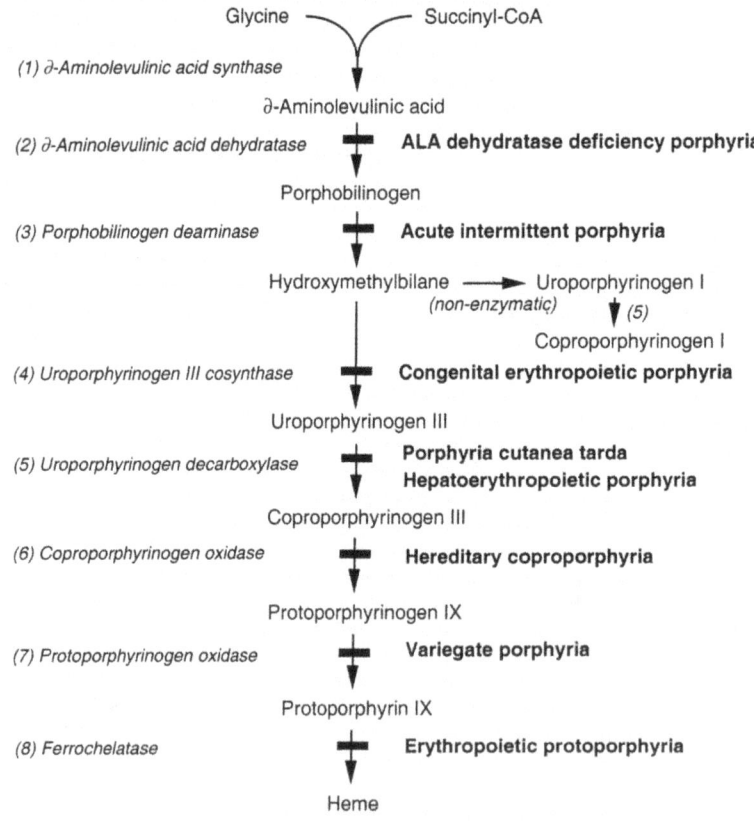

**Fig. 32.1.** Pathway of heme biosynthesis. Intermediates and enzymes of the heme biosynthetic pathway are listed. The porphyrias caused by the various enzyme deficiencies (indicated by *solid bars*) are given in bold. ALA, δ-aminolevulinic acid

# CHAPTER 32

# The Porphyrias

N.G. Egger, D.E. Goeger, and K.E. Anderson

CONTENTS

The Porphyrias .................................. 371
   Classification and Diagnosis ..................... 371
      Diagnostic Tests ............................. 372
   δ-Aminolevulinic acid dehydratase
   deficiency Porphyria ........................... 373
Acute Intermittent Porphyria...................... 374
Congenital Erythropoietic Porphyria
(Gunther Disease) ............................... 376
Porphyria Cutanea Tarda ......................... 377
Hepatoerythropoietic Porphyria ................... 378
Hereditary Coproporphyria and Variegate Porphyria .... 379
Erythropoietic Protoporphyria .................... 380
References ...................................... 381

---

Porphyrias are metabolic disorders due to deficiencies of specific enzymes of the heme biosynthetic pathway. They are associated with striking accumulations and excess excretion of heme pathway intermediates and their oxidized products. Symptoms and signs are almost all due to effects on the nervous system or skin. The three most common porphyrias, *acute intermittent porphyria, porphyria cutanea tarda and erythropoietic protoporphyria*, differ considerably from each other. The first presents with acute neurovisceral symptoms and can be aggravated by some drugs. Its treatment comprises intravenous heme administration and liberal amounts of glucose. In the latter two porphyrias the skin is affected, though with different lesions. Treatment differs also, from reducing excess iron for the first, and administration of β-carotene for the last porphyria. These disorders are more often manifest in adults than are most metabolic diseases. All porphyrias are inherited, with the exception of porphyria cutanea tarda, which is due to an acquired enzyme deficiency in liver, although an inherited deficiency is a predisposing factor in some cases.

## THE PORPHYRIAS

A diagnosis of porphyria should be considered in patients with unexplained neurovisceral symptoms, such as abdominal pain, or cutaneous photosensitivity. All porphyrias are inherited except for porphyria cutanea tarda which is primarily an acquired disorder. Because porphyrias may be clinically latent with no excess accumulation of pathway intermediates even throughout life, the family history is often negative. In clinically active stages, porphyrin precursors (δ-aminolevulinic acid and porphobilinogen), porphyrinogens (reduced forms of porphyrins), and porphyrins accumulate and are excreted in excess amounts. In contrast, only small amounts of heme pathway intermediates are excreted in normal individuals, and they differ strikingly in their chemical properties and routes of excretion. δ-Aminolevulinic acid, porphobilinogen and porphyrins with a large number of carboxyl side chains (e.g. uroporphyrin, an octacarboxyl porphyrin) are water soluble and are excreted almost entirely in urine. Protoporphyrin (a dicarboxyl porphyrin) is not soluble in water and is excreted entirely in bile and feces. Coproporphyrin (a tetracarboxyl porphyrin) is excreted partly in urine and bile, and its urinary excretion increases when hepatobiliary function is impaired. Porphyrinogens undergo autooxidation when they leave the intracellular environment and are excreted primarily as the corresponding porphyrins. δ-Aminolevulinic acid, porphobilinogen and porphyrinogens are colorless and non-fluorescent, whereas oxidized porphyrins are reddish and fluoresce when exposed to ultraviolet light (3).

### Classification and Diagnosis

The porphyrias are classified with regard to the tissue where the metabolic defect is primarily expressed (hepatic and erythropoietic porphyrias), the specific enzyme deficiency, or the clinical presentation (acute neurovisceral or cutaneous porphyrias) (Table 32.1).

**Table 32.1.** Classification of porphyrias and their associated enzyme deficiencies. The most precise classification is according to the specific enzyme deficiencies. Other classifications based on the major tissue site of overproduction of heme pathway intermediates (hepatic vs. erythropoietic) or the type of major symptoms (acute neurovisceral vs. cutaneous) are useful but not precise or mutually exclusive. (Modified from (4), with permission)

| Disease | Enzyme | Porphyria classifications | | | |
|---|---|---|---|---|---|
| | | Hepatic | Erythropoietic | Acute | Cutane |
| δ-Aminolevulinic acid dehydratase-deficiency porphyria | ALA dehydratase | ?X | | X | |
| Acute intermittent porphyria | Porphobilinogen deaminase[1] | X | | X | |
| Congenital erythropoietic porphyria | Uroporphyrinogen III cosynthase | | X | | X |
| Porphyria cutanea tarda[2] | Uroporphyrinogen decarboxylase | X | | | X |
| Hepatoerythropoietic porphyria | Uroporphyrinogen decarboxylase | X | X | | X |
| Hereditary coproporphyria | Coproporphyrinogen oxidase | X | | X | X |
| Variegate porphyria | Protoporphyrinogen oxidase | X | | X | X |
| Erythropoietic proto porphyria | Ferrochelatase | | X | | X |

[1] This enzyme is also known as hydroxymethylbilane synthase, and formerly as uroporphyrinogen I synthase
[2] Inherited deficiency of uroporphyrinogen decarboxylase is partially responsible for familial (type II) porphyria cutanea tarda
ALA, δ-aminolevulinic acid

Some porphyrias may cause both neurovisceral and cutaneous manifestations. Now that each porphyria is known to be due to a specific enzyme deficiency a precise classification and diagnosis of every case is desirable. In general, diagnosis of active cases is still based primarily on measurement of porphyrin precursors and porphyrins in urine, blood and feces, rather than on measurement of deficient enzymes or DNA methods. The diverse symptoms of the porphyrias are nonspecific and can be mimicked by many other more common disorders. Therefore, it is important to maintain a high index of suspicion and to avoid delay in diagnosis. The major clinical manifestations of porphyrias are either neurovisceral (e.g. abdominal pain, neuropathy, and mental disturbances) or cutaneous. The neurologic symptoms can be life-threatening, are poorly understood in terms of mechanism, and occur only in types of porphyria in which porphyrin precursors (especially δ-aminolevulinic acid) accumulate (acute intermittent porphyria, variegate porphyria, hereditary coproporphyria and δ-aminolevulinic acid dehydratase-deficiency porphyria). These disorders, which are characterized by acute attacks, are collectively referred to as *acute porphyrias*. Porphyrias accompanied by skin manifestations are termed *cutaneous porphyrias*. Cutaneous features are elicited by long-wave ultraviolet light (UV-B) found in sunlight. Excitation of excess porphyrins in the skin by light leads to generation of singlet oxygen and cell damage.

### Diagnostic Tests

Appropriate laboratory testing is essential for confirming or excluding a diagnosis of porphyria in patients with suggestive but nonspecific symptoms. In contrast to the nonspecific nature of symptoms, laboratory tests, if properly chosen and interpreted, can be both sensitive and specific (4). It is not difficult to confirm or rule out porphyria by these tests when a patient has symptoms. Diagnosis may be more difficult when symptoms have been absent for a long period of time. Some tests, particularly urinary porphyrin measurements, may be abnormal in other diseases and are subject to overuse and misinterpretation, which can lead to unnecessary expense and delay in achieving a correct diagnosis. Measurements of the deficient enzyme or DNA testing may be useful when asymptomatic relatives are being screened for porphyria.

The initial presentation determines the type of initial laboratory testing (Table 32.2). Measurement of urinary porphyrin precursors (δ-aminolevulinic acid and porphobilinogen) and total porphyrins is recommended when neurovisceral symptoms are suggestive of acute porphyria. In a severely ill patient it is very important to confirm a diagnosis of acute porphyria promptly, because the success of treatment is highly dependent on starting specific treatment soon after the onset of symptoms. Urinary porphobilinogen (and δ-aminolevulinic acid) is always markedly increased

**Table 32.2.** First-line laboratory tests for screening for porphyrias and second-line tests for further evaluation when initial testing is positive. (Modified from (4), with permission)

| Testing | Symptoms suggesting porphyria | |
|---|---|---|
| | Acute neurovisceral symptoms | Cutaneous photosensitivity |
| First-line | Urinary δ-aminolevulinic acid, porphobilinogen & total porphyrins (quantitative; random or 24 hour urine) | Total plasma porphyrins[1] |
| Second-line | Urinary δ-aminolevulinic acid, porphobilinogen and total porphyrins[2] (quantitative; 24 hour urine) | Erythrocyte porphyrins Urinary δ-aminolevulinic acid, Porphobilinogen and total porphyrins[2] (quantitative, 24 hour urine) |
| | Total fecal porphyrins[2] Erythrocyte porphobilinogen deaminase Total plasma porphyrins[1] | Total fecal porphyrins[2] |

[1] The preferred method is by direct fluorescent spectrophotometry
[2] Fractionation of urinary and fecal porphyrins is usually not helpful unless the total is increased

during attacks of acute intermittent porphyria but may be less increased in hereditary coproporphyria and variegate porphyria. Normal levels of δ-aminolevulinic acid, porphobilinogen and total porphyrins effectively excludes all acute porphyrias as potential causes of current symptoms.

Total plasma porphyrins are virtually always increased in patients with active skin lesions due to porphyrias, and should be measured when a cutaneous porphyria is suspected. A direct fluorometric method is most suitable for measuring total plasma porphyrins (5). Recording the fluorescence spectrum of plasma porphyrins at neutral pH is very useful in differentiating several porphyrias rapidly, and especially for distinguishing porphyria cutanea tarda from variegate porphyria (6). Normal plasma porphyrin levels (in unhemolyzed blood samples) exclude porphyria as a cause of cutaneous symptoms.

Further laboratory evaluation is necessary if the initial tests are positive because it is essential for management and genetic counseling to distinguish between the different types of porphyria.

## δ-Aminolevulinic Acid Dehydratase Deficiency Porphyria

### Clinical Presentation

This is an extremely rare disorder with symptoms that resemble acute intermittent porphyria, including abdominal pain and neuropathy. The disease may begin in childhood and be accompanied by failure to thrive and anemia. Other causes of δ-aminolevulinic acid dehydratase deficiency and increased urinary δ-aminolevulinic acid need to be excluded, such as lead poisoning and hereditary tyrosinemia; these conditions can also present with symptoms resembling those in acute porphyrias.

### Metabolic Derangement

This disorder is due to a homozygous or compound heterozygous deficiency of δ-aminolevulinic acid dehydratase, the second enzyme in the heme biosynthetic pathway (Figure 32.1). The enzyme is markedly reduced (1–2% of normal) in affected individuals, and approximately half-normal in both parents, which is consistent with autosomal recessive inheritance (Table 32.1). This is the most recently described porphyria, and only 4 well documented cases have been reported. Lead poisoning can be distinguished by showing reversal of the inhibition of δ-aminolevulinic acid dehydratase in erythrocytes by the in-vitro addition of dithiothreitol. Hereditary tyrosinemia type 1, resulting from a deficiency of fumarylacetoacetase, leads to accumulation of succinylacetone (2,3-dioxoheptanoic acid). This intermediate is a structural analog of δ-aminolevulinic acid and a potent inhibitor of the dehydratase. Other heavy metals and styrene can also inhibit δ-aminolevulinic acid dehydratase.

### Diagnostic Tests

The finding of increased urinary δ-aminolevulinic acid and coproporphyrin, normal urinary porphobilinogen, and a marked decrease in erythrocyte δ-aminolevulinic acid dehydratase is consistent with this disorder, if other causes of δ-aminolevulinic acid dehydratase deficiency are excluded. The increase in urinary coproporphyrin (mostly isomer III) is probably due to metabolism of δ-aminolevulinic acid via the heme biosynthetic pathway in tissues other than the liver. Coproporphyrin III is known to increase in normal subjects after loading with exogenous δ-aminolevulinic acid (7). Erythrocyte zinc protoporphyrin content is also increased, as in other homozygous cases of porphyria. If this type of porphyria is indicated by the testing described above, confirmation by immunological studies of the mutant enzyme and DNA studies should be done for further confirmation (8).

### Treatment and Prognosis

There is little experience in treating this porphyria. In general, the approach is the same as in acute intermittent porphyria. Heme therapy has not been uniformly effective, however. It is prudent to avoid drugs that are harmful in other acute porphyrias.

### Genetics

As discussed above, this is an autosomal recessive disorder. Affected individuals are homozygous or compound heterozygous for mutations at the δ-aminolevulinic acid dehydratase locus. All well-documented cases were unrelated, and most had different mutations. Immunological studies to date have indicated that most mutant alleles produce a defective enzyme protein (8).

## Acute Intermittent Porphyria

### Clinical Presentation

This autosomal dominant condition is the most common of the acute porphyrias, with a prevalence estimated to be approximately 5 per 100,000 in northern European populations. Most heterozygotes remain clinically asymptomatic for all or most of their lives. Factors that can contribute to clinical expression include certain drugs, steroid hormones and nutrition. Symptoms are rare in children. Acute attacks of neurovisceral symptoms and signs is the most common presentation, although subacute and chronic manifestations can also occur. Symptoms are more common in women during the reproductive period than in men. Attacks usually last for several days, often require hospitalization, and are followed by complete recovery. Severe attacks may be much more prolonged and are sometimes fatal, especially if the diagnosis is delayed. Abdominal pain, the most common symptom, is usually steady and poorly localized, but is sometimes crampy. Tachycardia, hypertension, restlessness, fine tremors, and excess sweating suggest sympathetic overactivity. Other common manifestations may include nausea, vomiting, constipation, pain in the limbs, head, neck or chest, muscle weakness and sensory loss. Dysuria and bladder dysfunction may occur as well as ileus, with abdominal distention and decreased bowel sounds. However, increased bowel sounds and diarrhea may occur. Because the abdominal symptoms are neurological rather than inflammatory, tenderness, fever and leukocytosis are characteristically mild or absent. A peripheral neuropathy that is primarily motor can develop. This is manifested by muscle weakness that most often begins proximally in the upper extremities, sometimes progressing to involve all extremities, respiratory muscles and even leading to bulbar paralysis. Muscle weakness is sometimes focal and asymmetric. Tendon reflexes may be little affected or hyperactive in early stages, but are usually decreased or absent with advanced neuropathy. Cranial and sensory nerves can be affected. Advanced motor neuropathy and death are rare unless porphyria is not recognized and appropriate treatment not instituted. Seizures may occur as an acute neurological manifestation of acute porphyrias, as a result of hyponatremia, or due to other causes unrelated to porphyria. Hyponatremia can be due to electrolyte depletion from vomiting or diarrhea, poor intake, sodium loss, or inappropriate antidiuretic hormone secretion. Persistent hypertension and impaired renal function may occur over the long term. Chronic abnormalities in liver function tests, particularly transaminases, are common, although few patients develop significant hepatic impairment. The risk of hepatocellular carcinoma is increased in this and other acute porphyrias, as well as in porphyria cutanea tarda (3, 9, 10).

### Metabolic Derangement

This disorder results from an approximately 50% deficiency of porphobilinogen deaminase (also known as hydroxymethylbilane synthase and formerly as uroporphyrinogen I synthase), the third enzyme in the heme biosynthetic pathway (Figure 32.1, Table 32.1). Most individuals with the enzyme deficiency remain asymptomatic with normal levels of urinary porphyrin precursors. Clinical expression of the disease is accompanied by accumulation of heme pathway intermediates in liver and excretion primarily in urine. Therefore, acute intermittent porphyria is classified as one of the hepatic porphyrias. Apparently, the partial deficiency of porphobilinogen deaminase does not of itself greatly impair hepatic heme synthesis or lead to induction of δ-aminolevulinic acid synthase, the rate limiting enzyme for heme synthesis in the liver. However, when the demand for hepatic heme is increased by drugs, hormones, or nutritional factors, the deficient enzyme can become limiting for heme synthesis. Induction of hepatic δ-aminolevulinic acid synthase is then accentuated and δ-aminolevulinic acid and porphobilinogen accumulate. Excess porphyrins originate nonenzymatically from porphobilinogen, and perhaps enzymatically from δ-aminolevulinc acid transported to tissues other than the liver. Most drugs that are harmful in this and other acute hepatic porphyrias are known to have the capacity to induce the synthesis of cytochrome P450 enzymes and δ-aminolevulinic acid synthase in the liver. An exception are the sulfonamide antibiotics, which may be harmful by inhibiting porphobilinogen deaminase (3).

## Diagnostic Tests

The finding of a substantial increase in urinary porphobilinogen is a sensitive and specific indication that a patient has either acute intermittent porphyria, hereditary coproporphyria or variegate porphyria (Table 32.2). Porphobilinogen remains increased between attacks of acute intermittent porphyria unless there have been no symptoms for a prolonged period. Fecal total porphyrins are generally normal or minimally increased in acute intermittent porphyria, and markedly increased in the other two conditions. Variegate porphyria is also characterized by increased total plasma porphyrins, as discussed later, whereas this is not characteristic of acute intermittent porphyria. Urinary porphyrins, and particularly coproporphyrin is generally more increased in hereditary coproporphyria and variegate porphyria, although some patients with acute intermittent porphyria have substantial increases in coproporphyrin. Urinary uroporphyrin can be increased in all of these disorders, especially when porphobilinogen is increased.

Decreased erythrocyte porphobilinogen deaminase confirms a diagnosis of acute intermittent porphyria and excludes the possibility of other acute porphyrias. However, falsely low activity may occur if there is a problem with processing or storing the sample. The erythrocyte enzyme is not deficient in all patients because some mutations of the porphobilinogen deaminase gene only reduce the nonerythroid (housekeeping) form of the enzyme and the erythroid enzyme is not affected (11). Furthermore, erythrocyte porphobilinogen deaminase has a wide normal range (up to 3-fold) that overlaps the range of patients with acute intermittent porphyria.

Measuring erythrocyte porphobilinogen deaminase is very useful for detecting asymptomatic carriers, if it is known that the propositus has a deficiency of the erythrocyte enzyme. Urinary porphobilinogen should also be measured when relatives are screened for this porphyria.

## Treatment and Prognosis

Most acute attacks are severe enough to require hospitalization for administration of intravenous glucose, and heme. Patients should also be observed for the development of neurological complications and electrolyte imbalances. Narcotic analgesics are commonly required for abdominal, back or extremity pain, and small doses of a phenothiazine are useful for nausea, vomiting, anxiety, and restlessness. High doses or prolonged administration of a phenothiazine after recovery from the attack is seldom useful. Chloral hydrate can be administered for insomnia. Diazepam in low doses is safe if a minor tranquilizer is required, although it needs to be kept in mind that benzodiazepines have some inducing effect on hepatic heme synthesis and may act in an additive fashion to other inducing influences. Bladder distention may require catheterization. Intravenous heme therapy (hematin, heme albumin or heme arginate) and carbohydrate loading are considered specific therapies for acute attacks of porphyria because they repress hepatic δ-aminolevulinic acid synthase. Heme therapy, 3–4 mg heme arginate per kg body weight infused intravenously once daily for 4 days, is most effective and should be initiated early, but only after the finding of a marked increase in urinary porphobilinogen. A longer course of treatment is seldom necessary if treatment is started early. Efficacy is reduced and recovery less rapid when treatment is delayed and neuronal damage is more advanced. Heme therapy is not effective for chronic symptoms of acute porphyrias. Heme arginate is the preferred form of heme for intravenous administration (12). In countries where heme arginate is not available, hematin can be reconstituted with human albumin, which stabilizes the heme as heme albumin and confers some of the advantages of heme arginate (13). Hematin administration commonly causes phlebitis at the site of infusion and a transient anticoagulant effect.

Carbohydrate loading (at least 300 g daily) may suffice for mild attacks and can be given orally or, if necessary, intravenously. Complete parenteral nutrition should be considered for patients when oral intake is not possible.

When an attack resolves, abdominal pain may disappear within hours, and paresis begins to improve within days. After a prolonged attack with severe motor neuropathy, muscle weakness may resolve completely, but sometimes there is some residual weakness. Identification and correction of precipitating factors such as certain drugs, inadequate nutrition, cyclic or exogenous hormones (particularly progesterone and progestins), and intercurrent infections can hasten recovery from an attack and prevent future attacks. Frequent clear-cut cyclical attacks occur in some women during the luteal phase of the cycle when progesterone levels are highest, and can be prevented by administration of a gonadotropin-releasing hormone analogue to prevent ovulation (14).

Treatment of seizures is problematic, because almost all antiseizure drugs can exacerbate acute porphyrias. Bromides, gabapentin and probably vigabatrin can be given safely (15). β-adrenergic blocking agents may control tachycardia and hypertension in acute attacks of porphyria, but do not have a specific effect on the underlying pathophysiology of the disease (3).

If acute attacks are treated promptly and precautions are taken to prevent further attacks, the outlook for patients with acute porphyrias is usually excellent. Fatal attacks have become much less common than in the past (9). In some such patients attacks occur in the absence of identifiable precipitating factors.

Occasional patients develop chronic pain and other symptoms, and some may become addicted to narcotic analgesics. Such patients need to be followed closely because there is often coexistent depression and an increased risk of suicide.

### Genetics

Acute intermittent porphyria is an autosomal dominant inherited disorder that is due to an approximately 50% deficiency of porphobilinogen deaminase. More than 100 different mutations of the porphobilinogen deaminase gene have been identified in unrelated families (11). Two forms of this enzyme, an erythroid-specific and a housekeeping form are derived from the same gene. The gene has two promoters, one of which is erythroid-specific, and two mRNAs are derived by alternative splicing of two primary transcripts (11). Mutations at many locations in the gene can lead to a deficiency of both isozymes, although a deficiency of the housekeeping form in nonerythroid tissues and especially the liver is essential for causing acute intermittent porphyria. Mutations located in or near the first of the 15 exons in this gene can impair the synthesis of the housekeeping form but not the erythroid-specific form of porphobilinogen deaminase. Characterization of the mutation in a propositus is useful to identify additional carriers of the gene in the family. Homozygous cases of acute intermittent porphyria are extremely rare, but should be suspected particularly if the disease is active early in childhood.

## Congenital Erythropoietic Porphyria (Gunther Disease)

### Clinical Presentation

Although it is usually a severe disease with manifestations noted soon after birth, or even in utero, clinical expression is variable, determined in part by the degree of enzyme deficiency. In milder cases symptoms have sometimes been first noted in adult life. Cutaneous features resemble those in porphyria cutanea tarda but are much more severe in most cases. Lesions include bullae and vesicles on sun-exposed skin, hypo- or hyperpigmented areas, hypertrichosis, and scarring. The teeth are reddish brown (erythrodontia) because of porphyrin deposition, and may fluoresce when exposed to long-wave ultraviolet light. Porphyrins are also deposited in bone. Hemolysis is almost invariably present and results from the markedly increased erythrocyte porphyrin levels, and is accompanied by splenomegaly. Life expectancy is often shortened by infections or hematological complications. There are no neurological manifestations.

Congenital erythropoietic porphyria can present in utero as nonimmune hydrops (16). When this is recognized, intrauterine transfusion is possible, and after birth severe photosensitivity can be prevented by avoiding phototherapy for hyperbilirubinemia.

### Metabolic Derangement

This rare disorder is due to a severe deficiency of uroporphyrinogen III cosynthase, the fourth enzyme of the heme synthesis pathway (Fig. 32.1, Table 32.1). There is considerable accumulation of hydroxymethylbilane (the substrate of the deficient enzyme), which is converted nonenzymatically to uroporphyrinogen I, a nonphysiological intermediate, which cannot be metabolized to heme. Therefore, uroporphyrin, coproporphyrin and other porphyrins accumulate in bone marrow, plasma, urine, and feces. In erythroid cells this accumulation results in intramedullary and intravascular hemolysis which leads to increased erythropoiesis and heme synthesis in spite of the inherited enzyme deficiency. Although the porphyrins that accumulate in this disease are primarily type I, type III porphyrins are also increased.

### Diagnostic Tests

Erythrocyte and plasma porphyrins are markedly increased and usually consist mostly of uroporphyrin I. Porphyrins in urine are primarily uroporphyrin I and coproporphyrin I, and in feces mostly coproporphyrin I. Porphyrin precursors are not increased. The diagnosis should be confirmed by a markedly deficient uroporphyrinogen III cosynthase activity and by mutation analysis.

### Treatment and Prognosis

Protection of the skin from sunlight is essential. Minor trauma, which can lead to denudation of fragile skin, should be avoided and secondary bacterial infections treated promptly to prevent scarring and mutilation. Improvement in hemolysis has been reported after splenectomy. Oral charcoal may be helpful by increasing fecal excretion of porphyrins. Blood transfusions sufficient to suppress erythropoiesis and bone marrow transplantation may be the most effective current therapies. Gene therapy may eventually be possible (17).

### Genetics

Congenital erythropoietic porphyria is an autosomal recessive disorder. Patients have either homozygous or compound heterozygous mutations of the gene that encodes uroporphyrinogen III cosynthase. Like other porphyrias, this disease is genetically heterogeneous, and many different mutations have been identified (17). Parents and other heterozygotes display intermediate deficiencies of the cosynthase. The disease can be diagnosed in utero by porphyrin measurements and DNA methods.

## Porphyria Cutanea Tarda

### Clinical Presentation

This is the most common and readily treated form of porphyria and is manifested primarily by chronic, blistering skin lesions, especially on the backs of the hands, forearms, face and (in women) the dorsa of the feet. Sun-exposed skin is also friable, and minor trauma may precede the formation of bullae or cause denudation of the skin. Small white plaques ("milia") may precede or follow vesicle formation. Hypertrichosis and hyperpigmentation are also noted. Thickening, scarring and calcification of affected skin may be striking, and is referred to as "pseudoscleroderma". Neurological effects are not observed. A number of acquired and inherited factors contribute to the development of porphyria cutanea tarda, and multiple factors are commonly identified in an individual patient. A normal or increased amount of hepatic iron is a requirement for the disease. Other factors include moderate or heavy alcohol intake, hepatitis C infection, estrogen use and smoking. Infection with HIV is a less common association. There are geographic differences in the association with hepatitis C; in some locations more than 80% of patients are infected with this virus. The contribution of genetic factors is discussed below (18, 19).

### Metabolic Derangement

This porphyria is caused by a profound deficiency of hepatic uroporphyrinogen decarboxylase, the fifth enzyme of the heme biosynthetic pathway (Figure 32.1, Table 32.1). "Sporadic" (type 1) and "familial" (types II and III) forms of the disease have been described, which do not differ substantially in terms of clinical features or treatment. In all cases, the major metabolic derangement appears to be inactivation or inhibition of uroporphyrinogen decarboxylase specifically in the liver. This appears to occur by an iron-dependent oxidative mechanism. Alcohol, hepatitis C, estrogens, smoking, induction of certain cytochrome P450 enzymes, HIV and low levels of ascorbic acid and carotenoids presumably contribute to this oxidative damage within hepatocytes. The mechanisms by which these factors work are not clearly understood (20). An additional mechanism for porphyrin accumulation may be oxidation of uroporphyrinogen – the substrate for the enzyme – to uroporphyrin. Individuals with type II disease have since birth a half normal amount of the normal enzyme activity and are therefore more susceptible to developing a more profound enzyme deficiency in the liver.

The pattern of porphyrins that accumulate in this disease is complex and characteristic. Uroporphyrinogen, (an octacarboxyl porphyrinogen) undergoes a sequential, four-step decarboxylation to coproporphyrinogen (a tetracarboxyl porphyrinogen). When uroporphyrinogen decarboxylase is markedly deficient, uroporphyrinogen and hepta-, hexa-, and pentacarboxyl porphyrinogens accumulate. To complicate the porphyrin pattern further, pentacarboxyl porphyrinogen can be metabolized by coproporphyrinogen oxidase to a tetracarboxyl porphyrinogen termed isocoproporphyrinogen. All of these porphyrinogens are mostly oxidized to the corresponding porphyrins, which first accumulate in the liver, then appear in plasma and are excreted in urine, bile and feces. Successful treatment may require some time before the massive porphyrin accumulations in liver are cleared.

A large outbreak of porphyria, with blisters, scarring, hyperpigmentation and hypertrichosis of sun exposed skin occurred in eastern Turkey in the 1950s from ingestion of wheat that was intended for planting, and had been previously treated with hexachlorobenzene as a fungicide. Porphyria cutanea tarda has been reported after exposure to other chemicals including di- and trichlorophenols and 2, 3, 7, 8-tetrachlorodibenzo-p-dioxin (TCDD, dioxin). These halogenated polycyclic aromatic hydrocarbons have since been shown to induce experimental porphyria in laboratory animals that biochemically closely resembles human porphyria cutanea tarda. Uroporphyria in animals and cultured liver cells treated with such chemicals is iron-dependent and is accompanied by induction of cytochrome P450 enzymes. Although these chemicals provide useful animal models for the study of porphyria cutanea tarda, in most human cases such toxic exposures are not evident.

### Diagnostic Tests

Genetic factors that can contribute include inheritance of a partial deficiency of uroporphyrinogen decarboxylase in patients classified as having familial (type II) porphyria cutanea tarda and inheritance of the mutations associated with familial hemochromatosis. Skin lesions are indistinguishable clinically from all other cutaneous porphyrias, except for erythropoietic protoporphyria (see later discussion). Cases of so-called "pseudoporphyria" have skin lesions resembling porphyria cutanea tarda but no significant increases in porphyrins; presumably other photosensitizers are responsible. Skin histopathology in this and other cutaneous porphyrias is not specific. Therefore, a skin biopsy does not establish a definitive diagnosis of porphyria cutanea tarda. It is important to differentiate these conditions by laboratory testing before starting therapy.

Plasma porphyrins are virtually always increased in patients with skin lesions due to any type of porphyria; the fluorescence spectrum of plasma porphyrins can distinguish variegate porphyria and erythropoietic protoporphyria from porphyria cutanea tarda (Table 32.2). Total fecal porphyrins are usually less increased than in hereditary coproporphyria and variegate porphyria.

### Treatment and Prognosis

Repeated phlebotomy or low-dose chloroquine is highly effective. Patients are also advised to discontinue alcohol, estrogens, iron supplements, and other contributing factors. The purpose of phlebotomies is to reduce hepatic iron content, which can be achieved by reducing the serum ferritin to near the lower limit of normal (approximately 20 ng/ml). This can usually be achieved by removal of only 5-6 units (450 mL each) of blood at 1-2 week intervals. Further iron depletion is of no additional benefit and may cause anemia and associated symptoms. Many more phlebotomies may be needed in patients who also have familial hemochromatosis. Plasma or serum porphyrin levels as well as ferritin should be followed; the porphyrin level falls somewhat more slowly than ferritin. With treatment and remission of the disease the activity of hepatic uroporphyrinogen decarboxylase gradually increases to normal. After remission, ferritin can return to pretreatment values without recurrence, in most cases. Postmenopausal women who have been treated for porphyria cutanea tarda can usually resume estrogen replacement without recurrence. Relapses seem to be more common in patients who resume alcohol intake, but will respond to another course of phlebotomies.

Low-dose chloroquine is a suitable alternative when phlebotomy is contraindicated or difficult. This is the preferred treatment in some centers. Standard doses exacerbate photosensitivity and hepatocellular damage, and should not be used. A low dose of chloroquine (125 mg twice weekly) or hydroxychloroquine (100 mg twice weekly) for several months gradually removes excess porphyrins from the liver and avoids the adverse effects of standard doses. The mechanism by which these drugs remove porphyrins from the liver in this condition is not known (21). This treatment is not effective in other porphyrias.

### Genetics

Porphyria cutanea tarda results from a liver-specific, apparently acquired deficiency of uroporphyrinogen decarboxylase. Genetic factors including an inherited partial deficiency of this enzyme may contribute. Human uroporphyrinogen decarboxylase is encoded by a single gene. No mutations in this gene have been found in sporadic (type I) porphyria cutanea tarda. The amount of hepatic uroporphyrinogen decarboxylase protein in type I disease, as measured immunochemically, is normal, suggesting that an acquired process has inactivated or inhibited the enzyme activity. In approximately 20% of patients with porphyria cutanea tarda, erythrocyte uroporphyrinogen decarboxylase is approximately 50% of normal in erythrocytes, and this feature is inherited as an autosomal dominant trait affecting all tissues. These cases are classified as familial or type II; the individual cases are clinical identical to type I, and the disease is not manifest unless the hepatic enzyme activity falls to a level much lower than half-normal. Type II disease is responsive to treatment by phlebotomy or low-dose chloroquine. A number of mutations of the gene encoding uroporphyrinogen decarboxylase have been identified in type II disease. Cases classified as type III disease, which are rare, have normal erythrocyte uroporphyrinogen decarboxylase activity but one or more relatives also have the disease. A genetic defect has not been clearly identified in type III (18).

## Hepatoerythropoietic Porphyria

### Clinical Presentation

This rare disease is clinically similar to congenital erythropoietic porphyria and usually presents with red urine and blistering skin lesions shortly after birth. Mild cases may present later in life and more closely resemble porphyria cutanea tarda. Concurrent conditions, such as viral hepatitis, may accentuate porphyrin accumulation.

### Metabolic Derangement

Hepatoerythropoietic porphyria results from a substantial deficiency of uroporphyrinogen decarboxylase, with intermediate deficiencies of the enzyme in parents, as expected for an autosomal recessive disorder (Figure 32.1, Table 32.1). The disease has features of both hepatic and erythropoietic porphyrias. As discussed above, a liver-specific, primarily acquired deficiency of the same enzyme is found in porphyria cutanea tarda.

### Diagnostic Tests

The excess porphyrins found in urine, plasma and feces in this condition are similar to those in porphyria cutanea tarda. In addition, erythrocyte zinc protoporphyrin is increased, as in a number of other autosomal recessive porphyrias. This finding probably reflects an earlier accumulation of uroporphyrinogen in erythroblasts, which after completion of hemoglobin synthesis is metabolized to protoporphyrin. Hepatoerythropoietic porphyria is differentiated from congenital erythropoietic porphyria also by excess isocoproporphyrins in feces and urine, and by decreased uroporphyrinogen decarboxylase activity, as most conveniently measured in erythrocytes.

### Treatment and Prognosis

Therapeutic options are essentially the same as in congenital erythropoietic porphyria.

### Genetics

This porphyria results from a homozygous or compound heterozygous state for mutations of the gene encoding uroporphyrinogen decarboxylase. The disease is genetically heterogeneous. Although the enzyme deficiency is usually severe, there is some residual activity, so heme formation can occur. With a few exceptions, uroporphyrinogen decarboxylase mutations found in this disease are not found in type II porphyria cutanea tarda (18). The mutations in hepatoerythropoietic porphyria may be less severe in the heterozygous state and are associated with severe deficiencies only in the homozygous or compound heterozygous state.

## Hereditary Coproporphyria and Variegate Porphyria

### Clinical Presentation

These disorders are classified as acute hepatic porphyrias because they can present with acute attacks that are identical to those in acute intermittent porphyria. However, unlike the latter disease, they are also cutaneous porphyrias, because they may cause blistering skin lesions that are indistinguishable from those of porphyria cutanea tarda. Factors that exacerbate acute intermittent porphyria are important in both of these porphyrias. Homozygous cases of hereditary coproporphyria and variegate porphyria have been described, and in such cases clinical manifestations may begin in childhood. Symptoms in heterozygotes are most common after puberty. Variegate porphyria is particularly common in South Africa where most cases are descendants of a couple who emigrated from Holland and arrived in Capetown in 1688 (22, 23).

### Metabolic Derangement

Hereditary coproporphyria and variegate porphyria result from a deficiency of coproporphyrinogen oxidase and of protoporphyrinogen oxidase, respectively, which are the sixth and seventh enzyme of the heme biosynthetic pathway (Figure 32.1, Table 32.1). Heterozygotes have approximately 50% deficiencies of these enzymes. In hereditary coproporphyria there is marked accumulation of coproporphyrin III (derived from autooxidation of coproporphyrinogen III), and urinary porphyrin precursors and uroporphyrin are increased particularly in association with acute attacks. Similar abnormalities are seen in variegate porphyria, but in addition protoporphyrin (derived from autooxidation of protoporphyrinogen) is increased in feces (and bile), and plasma porphyrins are increased. A close association of coproporphyrinogen oxidase and protoporphyrinogen oxidase in mitochondria may explain the accumulation of both coproporphyrinogen and protoporphyrinogen (mostly as the corresponding porphyrins) in variegate porphyria. Protoporphyrinogen has been shown to inhibit porphobilinogen deaminase, which along with induction of hepatic δ-aminolevulinic acid synthase, may account for the increase in porphyrin precursors during acute attacks, at least in variegate porphyria.

### Diagnostic Tests

Urinary δ-aminolevulinic acid and porphobilinogen are increased during acute attacks of these porphyrias, although the increases may be less and more transient than in acute intermittent porphyria. Urinary coproporphyrin increases may be more prominent and prolonged. Plasma porphyrins are commonly increased in variegate porphyria, and the fluorescence spectrum of plasma porphyrins (at neutral pH) is characteristic and very useful for rapidly distinguishing this disease from the other porphyrias. This is at least as sensitive as fecal porphyrin measurement for detecting variegate porphyria, although not as sensitive as a reliable assay for lymphocyte protoporphyrinogen oxidase (24). A marked, isolated increase in fecal coproporphyrin (especially isomer III) is distinctive for hereditary coproporphyria. Fecal coproporphyrin and protoporphyrin are about equally increased in variegate porphyria.

The finding of a small or even moderate increase in urinary coproporphyrin is not a strong indication that a patient has hereditary coproporphyria or variegate porphyria, because coproporphyrinuria is a highly nonspecific finding. It can be seen in many medical conditions, especially when hepatic or bone marrow function is affected. Reliable assays for protoporphyrinogen oxidase and coproporphyrinogen oxidase in cultured fibroblasts or lymphocytes are available only in a few research laboratories. Erythrocytes cannot be used to measure these mitochondrial enzymes, because mature erythrocytes do not contain mitochondria. Misdiagnosis of these porphyrias, and especially hereditary coproporphyria, has resulted from unreliable assays offered by some commercial laboratories using erythrocytes and inappropriate substrates.

### Treatment and Prognosis

Attacks of neurologic symptoms are treated as in acute intermittent porphyria (see above). Cutaneous symptoms are more difficult to treat, and therapies that are effective for porphyria cutanea tarda (phlebotomy and low-dose chloroquine) are not effective in these conditions. Protection from sunlight is important.

### Genetics

Both of these porphyrias are autosomal dominant conditions in which affected individuals and latent carriers have approximately 50% activity of the affected enzyme. Homozygous cases are rare. Genetic heterogeneity is a feature of these porphyrias, and many different mutations have been identified. As expected, a single mutation accounts for the many descendants with variegate porphyria in South Africa, which is an example of the founder effect (23).

## Erythropoietic Protoporphyria

### Clinical Presentation

Erythropoietic protoporphyria is the third most common porphyria. Cutaneous manifestations usually begin in childhood. However, in contrast to other cutaneous porphyrias, blistering, milia, friability, and chronic skin changes such as scarring and hypertrichosis are not prominent. Burning, itching, erythema, and swelling can occur within minutes of sun exposure. Diffuse edema of sun-exposed areas may resemble angioneurotic edema. Other more chronic skin changes may include lichenification, leathery pseudovesicles, labial grooving, and nail changes. There is no fluorescence of the teeth and no neuropathic manifestations. Mild anemia with hypochromia and microcytosis is noted in some cases.

The severity of the symptoms is remarkably stable over time. Drugs that exacerbate hepatic porphyrias are not known to worsen this disease, although they are generally avoided as a precaution. Some patients develop liver disease, which can progress rapidly to death from hepatic failure. This complication is accompanied by marked deposition of protoporphyrin in liver and increased levels in plasma and erythrocytes. Gallstones containing protoporphyrin may also develop (25).

### Metabolic Derangement

This disease is due to an inherited deficiency of ferrochelatase, the eighth and last enzyme in the heme biosynthetic pathway (Figure 32.1, Table 32.1). Ferrochelatase is deficient in all tissues, but becomes rate-limiting for protoporphyrin metabolism primarily in bone marrow reticulocytes, which are the primary source of the excess protoporphyrin. Circulating erythrocytes and perhaps the liver contribute smaller amounts. Protoporphyrin is increased in plasma and is excreted in bile, and feces. Excess erythrocyte protoporphyrin in this disease is not complexed with zinc and diffuses more readily into plasma than does zinc protoporphyrin. Excess zinc protoporphyrin, as found in lead poisoning, iron deficiency, and homozygous forms of porphyria, remains in the erythrocyte for its full life span. Normal erythrocytes also contain some zinc protoporphyrin, but not free protoporphyrin. Protoporphyrin is excreted in bile and may undergo enterohepatic circulation. In the minority of patients who develop liver failure, the excess protoporphyrin deposited in the liver appears to be derived primarily from the bone marrow. Excess protoporphyrin may itself have cholestatic effects and damage hepatocytes.

### Diagnostic Tests

Protoporphyrin is increased in bone marrow, plasma, bile, and feces. The most useful screening tests for this disorder are erythrocyte protoporphyrin and total plasma porphyrin determinations. Measurement of erythrocyte protoporphyrin is most sensitive, but may give false positive results especially in patients with iron deficiency, lead poisoning and other erythrocyte disorders. An increase in total plasma porphyrins is a highly specific indication that a patient has a cutaneous form of porphyria. However, because excess protoporphyrin found in plasma is particularly sensitive to light exposure, samples must be shielded from light. The fluorescence spectrum of plasma porphyrins at neutral pH can distinguish erythropoietic protoporphyria from other porphyrias, if the plasma sample is not hemolyzed. The finding of an increased erythrocyte protoporphyrin value should be followed by a determination whether the protoporphyrin is free or complexed with zinc, because an increase in free protoporphyrin is specific for erythropoietic protoporphyria. This can be done by a simple ethanol extraction method, which is not widely available. Many other assays said to measure "free erythrocyte protoporphyrin" actually measure total erythrocyte porphyrins or protoporphyrin, including zinc protoporphyrin. The finding of increased fecal porphyrins (mostly protoporphyrin) helps to confirm a diagnosis of protoporphyria. Urinary porphyrins and porphyrin precursors are normal, unless the patient has liver impairment, in which case urinary porphyrins may increase. Hepatic complications of the disease are often preceded by increasing levels of erythrocyte and plasma protoporphyrin, abnormal liver function tests, marked deposition of protoporphyrin in liver cells and bile canaliculi, and increased photosensitivity.

### Treatment and Prognosis

Photosensitivity is managed by avoidance of sunlight. Oral β-carotene improves tolerance to sunlight in some patients. It may quench singlet oxygen or free radicals, and seems to be more effective in erythropoietic protoporphyria than in other cutaneous porphyrias. Cholestyramine may reduce protoporphyrin levels by interrupting its enterohepatic circulation. Iron deficiency, caloric restriction, and drugs or hormone preparations that impair hepatic excretory function should be avoided.

Treatment of liver complications is difficult. Transfusions or heme therapy may suppress erythroid and hepatic protoporphyrin production. Liver transplantation is sometimes required, but there is some risk that the new liver will also accumulate excess protoporphyrin and develop impaired function. Operating room lights have produced severe skin and peritoneal burns in some patients with protoporphyria, liver failure, and marked increases in erythrocyte and plasma protoporphyrin concentrations. A patient with erythropoietic protoporphyria who underwent bone marrow transplantation for leukemia experienced complete remission of the porphyria (26). Therefore, there is potential benefit from bone marrow replacement and gene therapy in this and other erythropoietic porphyrias.

### Genetics

Many mutations in the ferrochelatase gene have been associated with protoporphyria. Most of these mutant alleles express little or no ferrochelatase. The pattern of inheritance has been considered to be autosomal dominant. However, some obligate carriers have little or no increase in red cell protoporphyrin, and affected individuals often have ferrochelatase activities that are less than half-normal. Such findings are not explained by an autosomal dominant inherited trait. As proposed in 1984, and supported by recent molecular evidence, patients with clinically manifest disease have coinherited a weak ferrochelatase gene mutation on the "normal" allele (27-29). This weak mutation is commonly found in normal individuals, but by itself has no clinical consequences. Therefore, patients with manifest erythropoietic protoporphyria may be compound heterozygotes rather than heterozygotes, and the disease should not always be considered a truly autosomal dominant condition.

## References

1. Granick S (1966) The induction in vitro of the synthesis of δ-aminolevulinic acid synthetase in chemical porphyria: a response to certain drugs, sex hormones, and foreign chemicals. J Biol Chem 241:1359-1375
2. Anderson KE, Freddara U, Kappas A (1982) Induction of hepatic cytochrome P-450 by natural steroids: relationships to the induction of δ-aminolevulinate synthase and porphyrin accumulation in the avian embryo. Arch Biochem Biophys 17:597-608
3. Anderson KE (1996) The porphyrias (Chapter 14). In: Zakim D, Boyer T, (ed) Hepatology. Philadelphia, W.B. Saunders Co 417-463
4. Anderson KE (2000) The porphyrias. In: Goldman L, Bennett CJ (ed) Cecil Textbook of Medicine. Philadelphia, W.B. Saunders Co. 1123-1132
5. Poh-Fitzpatrick MB (1976) Lamola AA. Direct spectrophotometry of diluted erythrocytes and plasma: a rapid diagnostic method in primary and secondary porphyrincmias. J Lab Clin Med 87:362-370
6. Poh-Fitzpatrick MB (1980) A plasma porphyrin fluorescence marker for variegate porphyria. Arch Dermatol 116:543-547
7. Shimizu Y, Ida S, Naruto H, Urata G (1978) Excretion of porphyrins in urine and bile after the administration of delta-aminolevulinic acid. J Lab Clin Med 92:795-802
8. Sassa S (1998) ALAD porphyria. Semin Liver Dis 18:95-101
9. Kauppinen R, Mustajoki P (1992) Prognosis of acute porphyria: occurrence of acute attacks, precipitating factors, and associated diseases. Medicine 71:1-13
10. Andant C, Puy H, Faivre J, Deybach JC (1998) Acute hepatic porphyrias and primary liver cancer [letter]. N Engl J Med 338:1853-4
11. Grandchamp B (1998) Acute intermittent porphyria. Semin Liver Dis 18:17-24
12. Tenhunen R, Mustajoki P (1998) Acute porphyria: treatment with heme. Semin Liver Dis 18:53-5
13. Bonkovsky HL, Healey BS, Lourie AN, Gerron GG (1991) Intravenous heme-albumin in acute intermittent porphyria: evidence for repletion of hepatic hemoproteins and regulatory heme pools. Am J Gastroenterol 86:1050-1056
14. Anderson KE, Spitz IM, Bardin CW, Kappas A (1990) A GnRH analogue prevents cyclical attacks of porphyria. Arch Int Med 150:1469-1474
15. Hahn M, Gildemeister OS, Krauss GL et al. (1997) Effects of new anticonvulsant medications on porphyrin synthesis in cultured liver cells: potential implications for patients with acute porphyria. Neurology 49:97-106
16. Verstraeten L, Van Regemorter N, Pardou A et al. (1993) Biochemical diagnosis of a fatal case of Gunther's disease in a newborn with hydrops-fetalis. Eur J Clin Chem Clin Biochem 31:121-128

17. Desnick RJ, Glass IA, Xu W, Solis C, Astrin KH (1998) Molecular genetics of congenital erythropoietic porphyria. Semin Liver Dis 18:77–84
18. Elder GH (1998) Porphyria cutanea tarda. Semin Liver Dis 18:67–75
19. Bonkovsky HL, Poh-Fitzpatrick M, Pimstone N et al. (1998) Porphyria cutanea tarda, hepatitis C, and HFE gene mutations in North America. Hepatology 27:1661–1669
20. Sinclair PR, Gorman G, Shedlofsky SI et al. (1997) Ascorbic acid deficiency in porphyria cutanea tarda. J Lab Clin Med 130:197–201
21. Egger NG, Goeger DE, Anderson KE (1996) Effects of chloroquine in hematoporphyrin-treated animals. Chemico-Biological Interactions 102:69–78
22. Martasek P (1998) Hereditary coproporphyria. Semin Liver Dis 18:25–32
23. Kirsch RE, Meissner PN, Hift RJ (1998) Variegate porphyria. Semin Liver Dis 18:33–41
24. Da Silva V, Simonin S, Deybach JC, Puy H, Nordmann Y (1995) Variegate porphyria: diagnostic value of fluorometric scanning of plasma porphyrins. Clin Chim Acta 238:163–8
25. Cox TM, Alexander GJ, Sarkany RP (1998) Protoporphyria. Semin Liver Dis 18:85–93
26. Lichtin A, Anderson K, Bloomer J et al. (1998) Correction of erythropoietic protoporphyria (EPP) phenotype by allogenic bone marrow transplant. American Society of Hematology Annual Meeting, December 7, 1998
27. Went LN, Klasen EC (1984) Genetic aspects of erythropoietic protoporphyria. Ann Hum Genet 48:105–17
28. Wang X, Kurtz L, Bloomer J, Christiano A, Poh-Fitzpatrick M (1998). Haplotype analysis in four families with an exon skipping mutation in the ferrochelatase gene. J Invest Dermatol 110:618
29. Poh-Fitzpatrick MB, Piomelli S, Deybach J-C, Gouya W, Wang X (1997) Erythropoietic protoporphyria: a triallelic inheritance model. J Invest Dermatol 108:598

# PART IX
# DISORDERS OF METAL TRANSPORT

## Copper Transport

Copper, in addition to iron, is one of the few trace metals for which the transport pathways are beginning to be identified. Copper is ingested by organisms in a large excess, and a homeostatic mechanism ensures that the trace amounts needed are maintained, and the remainder is eliminated from the body. Copper is an essential component of a number of enzymes.

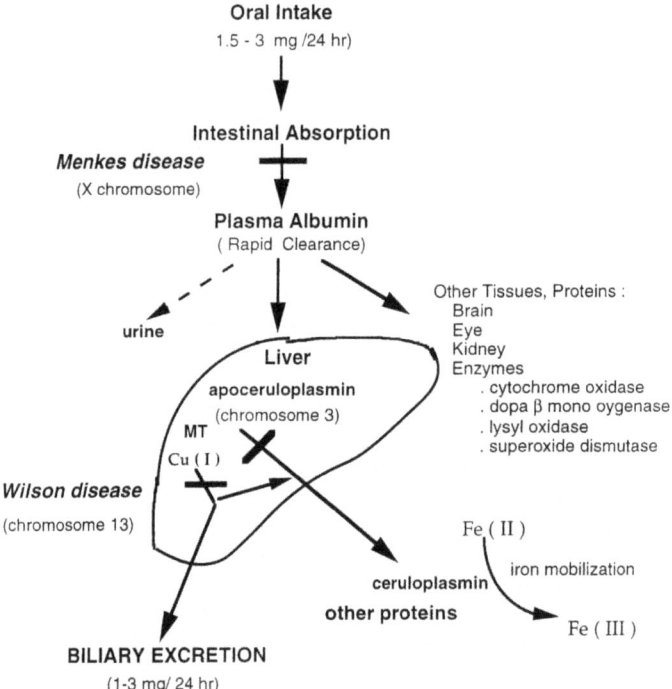

**Fig. 33.1.** Copper is absorbed in the intestinal cell, from which it is exported with the aid of a P-type adenosine triphosphatase, ATP7A. Copper in the plasma is bound to albumin. Proteins CTR1 and CTR2 are involved in the uptake of copper [in the reduced or Cu(I) form] into the cell. Low-molecular-weight chaperones transfer copper to specific proteins: CCS transfers copper to superoxide dismutase, COX17 transfers it to cytochrome C-oxidase in the mitochondria, and ATOX1 transfers it to the P-type transporters ATP7A and ATP7B. ATP7B, the protein defective in Wilson disease, is localized in the trans-Golgi network and is responsible for excretion of copper out of the hepatocyte and the incorporation of copper into ceruloplasmin. Tissue copper accumulation can induce metallothioneins, which bind excess copper. Ceruloplasmin may be involved in uptake of copper into the cell but mainly serves as a ferroxidase to mobilize iron (modified from Cox DW 1995, 56:828–834). Enzyme defects are depicted by *solid bars*

# Copper Transport Disorders: Wilson Disease and Menkes Disease

Diane W. Cox, Zeynep Tümer, and Eve A. Roberts

CONTENTS

Wilson Disease ................................ 385
   Clinical Presentation ........................... 385
   Metabolic Derangement ......................... 386
   Diagnostic Tests ............................... 386
   Treatment and Prognosis ....................... 387
   Genetics ...................................... 388
Menkes Disease ................................ 388
   Clinical Presentation ........................... 388
      Classical Severe Form ....................... 388
      Milder Forms ............................... 388
   Metabolic Derangement ......................... 389
   Diagnostic Tests ............................... 389
      Prenatal Diagnosis .......................... 390
      Carrier Identification ....................... 390
   Treatment and Prognosis ....................... 390
   Genetics ...................................... 390
      Mutation Spectrum .......................... 390
References ...................................... 391

Copper is essential for a number of important enzymes. An efficient system for copper transport has developed to provide an appropriate body balance. Two human diseases, in which copper transport is disrupted, are known: Wilson disease and Menkes disease. The clinical features are very different, due to copper accumulation in Wilson disease and widespread tissue copper deficiency in Menkes disease.

*Wilson disease*, also called hepatolenticular degeneration, results from the inability to transport copper out of the liver via the bile. As a result, copper accumulates, first in the liver and gradually in other organs, particularly the brain but also the kidney and the cornea. The main clinical manifestations are hepatic and neurologic. Treatment with decoppering drugs or oral zinc is effective if started early.

*Menkes disease* is a multisystem X-linked disorder, usually lethal in infancy or childhood. It is characterized by progressive neurodegeneration and marked connective-tissue disturbances due to a functional copper deficiency. Early treatment with copper histidine is partially effective.

## Wilson Disease

### Clinical Presentation

The clinical manifestations of Wilson disease, an autosomal recessive disorder, are highly variable. Age of onset ranges from 3 years to more than 50 years of age. The initial onset of symptoms can be hepatic, neurological, psychiatric or as an acute hemolytic crisis. Hepatic and neurological presentations occur with approximately equal frequency, and other initial manifestations are less frequent.

*Hepatic symptoms* vary widely. Mutation studies indicate that the spectrum of clinical symptoms may be wider than previously suspected. Wilson disease is a common cause of chronic liver disease in childhood and is one of the few causes with an effective treatment. Hepatic symptoms manifest most frequently between 8 years and 20 years of age, but can occur as early as 3 years of age or as late as the fifth and sixth decades of life. Manifestation may be as chronic or fulminant liver disease. In some cases, the presentation may suggest autoimmune hepatitis. Several episodes of jaundice may occur before a critical fulminant episode. Subtle neurological involvement, particularly mood changes, changes in school performance, changes in handwriting and clumsiness, may occur in patients with predominant hepatic manifestations.

*Neurological presentation* tends to occur in the second and third decades but has been reported in children 6–10 years old. The neurological involvement manifests either as a movement disorder or rigid dystonia. Movement disorders include tremors, loss of fine motor control and poor coordination. The tremor may be fine or may occur as marked flapping movements. Those with the spastic form generally develop a mask-like face, rigidity and an abnormal gait. Dysarthria and swallowing difficulties are particularly frequent. Wilson disease does not lead to developmental delay or deterioration of intellect.

The high copper concentration of plasma may lead to an acute bout of hemolysis. Recurrent bouts of hemolysis may lead to the development of gallstones in children. The kidney is involved to some degree in almost half of patients. Microscopic hematuria, amino aciduria, phosphaturia and inadequate acidification have been reported. Osteoporosis, arthritis and/or cardiomyopathy are only occasionally consequences of Wilson disease.

As many as 20% of patients may have *psychiatric symptoms* only. Depression is a common symptom, although other abnormal behaviors, such as compulsive behaviors, phobias, aggression or anti-social behavior have been reported.

A classic sign of Wilson disease is the Kayser-Fleischer ring, a golden brown granular pigmentation at the limbus of the cornea, due to deposition of copper in Desemet's membrane. Identification may be possible only with a slip-lamp examination. These rings disappear with chelation therapy. They are often absent in patients with exclusively hepatic involvement and are unlikely in pre-symptomatic patients. Although most patients with a neurological or psychiatric manifestation of Wilson disease have Kayser-Fleischer rings, at least 5% do not. While generally an indication of Wilson disease, Kayser-Fleischer rings may also be found in other types of chronic cholestatic liver disease, including primary biliary cirrhosis and familial cholestatic syndromes. Further information on Wilson disease can be found in reviews [1, 2].

The diagnosis should be suspected on clinical grounds in the presence of chronic liver disease, tremor or dystonia and Kayser-Fleischer rings, although patients with this classic combination of symptoms (originally described by S.A. Kinnear Wilson in 1912) are uncommon. Wilson disease should be considered in young adults with movement disorders and in patients of all ages with progressive liver disease. Abnormalities of liver function, frequently mild, are usually found in patients with predominately neurological disease. Kayser-Fleischer rings are usually (but not always) present in patients with neurological symptoms. At least two thirds of patients with liver disease proven by mutation analysis have been shown to lack Kayser-Fleischer rings [3].

## Metabolic Derangement

The characteristic defects observed in Wilson disease are reduced excretion of copper into the bile (resulting in a toxic accumulation of copper in the liver and an apparently secondary increase of copper excretion in the urine) and a reduced incorporation of copper into ceruloplasmin (Fig. 33.1). Although the plasma ceruloplasmin concentration may fall within the normal range, the incorporation of radioactive or stable isotopes of copper always indicates a defect in the incorporation of copper into ceruloplasmin.

The cloning of the gene, a copper-transporting adenosine triphosphatase (ATPase) [4, 5], has helped explain the biochemical and clinical changes observed. The copper transporter appears to traffic from the trans-Golgi membranes to cytoplasmic vesicles (and possibly directly to the bile canalicular membrane) in the presence of high cellular copper in experimental systems. As a result of mutations in the copper-transporting ATPase, copper accumulates in the liver, first inducing the production of metallothionein, which can apparently maintain copper in a relatively non-toxic state. The accumulation of copper causes damage to mitochondria. Copper is also deposited in renal tubules, and kidney damage occurs to varying degrees. Copper also accumulates in the basal ganglia of the brain, causing neurological disease. It has not been determined whether the neurological damage is due to expression of the gene in the basal ganglia, high copper levels in circulating plasma or both. The Long-Evans Cinnamon rat and toxic-milk mouse are animal models with a deletion or a missense mutation, respectively, in the homologous *ATP7B* gene.

## Diagnostic Tests

As the clinical spectrum of Wilson disease unfolds (with proof of diagnosis by mutation analysis), interpretation of biochemical results is changing. The serum ceruloplasmin concentration, considered to be below the normal range in 95% of patients, has been found, in at least 50% of patients with liver disease, to be in the normal and even upper normal range [3]. As copper accumulates in the liver, there may be some mechanism to bypass the necessity for incorporation of copper by the copper-transporting ATPase. Non-ceruloplasmin-bound copper is elevated, as estimated by subtracting the amount of copper associated with ceruloplasmin from the total serum concentration of copper (Table 33.1). Copper and ceruloplasmin measurements must be accurate for this calculation to be meaningful. Low serum ceruloplasmin can, in rare cases, be due to inherited aceruloplasminemia, which results in tissue iron storage, leading to diabetes and neurological abnormalities [6, 7].

With stringent precautions against contamination, measurement of the urinary copper excretion (preferably via three 24-h collections) is useful. Basal 24-h urinary copper excretion is elevated (Table 33.1). Heterozygotes usually have normal 24-h urinary copper excretion, although this may be elevated to borderline levels. Pre-symptomatic individuals often have normal copper excretion. Urinary copper excretion after penicillamine administration is generally elevated and may aid in diagnosis [8]. However, this test tends to yield borderline excretion levels when other biochemical assays also yield borderline values, and there is overlap with results of tests for other types of liver disease.

Hepatic tissue copper concentration (measured by neutron activation analysis or atomic absorption spectrometry) is usually markedly elevated (Table 33.1). A borderline normal elevation can also be observed in heterozygotes. This measurement is more reliable in the early stages of the disease, when copper is distributed evenly throughout the hepatocyte. In later stages of the disease, copper is distributed unequally in the liver. In some cases, coagulation defects may make liver biopsy inappropriate. Diagnosis through mutation analysis is already possible and will improve as the whole spectrum of mutations for this disease becomes known.

## Treatment and Prognosis

The prognosis is excellent when treatment is initiated sufficiently early to avoid severe tissue damage. The substances generally used for treatment are penicillamine, trien (trientine) and zinc. The use of tetrathiomolybdate as a chelator is relatively recent.

Penicillamine has been in use since proposed by J.M. Walshe in 1956. The initial dose is 1–1.5 g/day in four divided doses, given with pyridoxine (25 mg/day). Penicillamine can cause initial worsening of neurological symptoms. Urinary excretion of copper is greatly increased, but copper already accumulated in the liver is not removed well. Penicillamine inhibits collagen cross-linking and may act as an immunosuppressant. Some patients develop an allergic type of reaction and can have penicillamine therapy instituted at lower doses along with corticosteroid. Up to 30% of patients with Wilson disease develop some type of side effect. When treated for long periods of time, chronic skin changes may result, with elastosis serpiginosa and loss of skin elastin. These changes could be due in part to excessively severe decoppering. The treatment of children in the pre-symptomatic phase may have fewer side effects with zinc than with penicillamine.

Trientine (triethylene tetraminehydrochloride)-2,2,2-teteramine), called trien, chelates copper by forming a stable complex, which is subsequently excreted in the urine. Side effects appear to be less common with this type of therapy [9].

Oral zinc, used since 1979 [10], has been more widely used in Europe than in North America. High doses of zinc induce metallothionein in the intestinal cells. Metallothionein, a zinc- and copper-binding protein, has a greater affinity for copper than for zinc. Thus, the former metal is preferentially bound in the intestine and then is shed with normal cellular turnover. There are few serious side effects with zinc therapy, although intestinal upsets sometime occur [11]. These seem to be reduced with the use of zinc acetate rather than sulfate. Zinc may not be effective sufficiently quickly in patients with severe clinical symptoms, and many researchers recommend initial decoppering with another agent. There is no rationale for the combined use of both zinc and penicillamine, which could cause side effects from both and may result in either severe copper depletion or inadequate decoppering.

Ammonium tetrathiomolybdate, originally used by S.M. Walshe (based on its extensive use in the decoppering of sheep), binds to plasma copper with high affinity. It is not fully known whether copper might be displaced to other tissues. However, this therapy seems effective for rapid removal of copper in severe cases of neurological disease or severe hepatic dysfunction [12]. Initial deterioration of neurological symptoms, often seen with penicillamine therapy, may be avoided [13]. Risks with this therapy are not yet evaluated.

With any of these decoppering agents, copper depletion must be avoided. Depletion of copper can be recognized when the enzymatic (not immunological) serum concentration of ceruloplasmin is consistently zero. Iron storage (typical of inherited aceruloplasminemia) with increased plasma ferritin would be a predicted consequence.

The importance of the role of antioxidants needs greater consideration, since copper can induce free-radical production [14]. Copper-loaded animals and patients with Wilson disease have enhanced free-

Table 33.1. Results of copper assays in patients with Wilson disease and in normal adults

|  | Wilson disease[a] | Normal adults |
|---|---|---|
| Serum ceruloplasmin (mg/l)[b] | 0–200 | 200–350 |
| Serum copper (µg/l) | 190–640 | 700–1520 |
| (µmol/l) | 3–10 | 11–24 |
| Urinary copper (µg/day) | 100–1000 | <40 |
| (µmol/day) | >1.6 | <0.6 |
| Urinary copper (after penicillamine; µmol/day) | >20 | <15 |
| Liver copper (µg/g dry weight) | >250 | 20–50 |

[a] There may be overlap with some heterozygotes for any of these assays
[b] Ceruloplasmin-bound copper in milligrams/liter = ceruloplasmin in milligrams × 3.15 (amount of copper in micrograms/liter in ceruloplasmin). Convert to milligrams/liter by multiplying by 63.5, the formula weight of copper (modified from [2])

radical production in tissues, with consequent low levels of α-tocopherol [15]. Modest use of antioxidants (such as α-tocopherol in doses several times the daily requirement) to restore the normal plasma level may be helpful in the prevention and even reversal of liver damage. Initially, higher doses might be needed to restore plasma α-tocopherol to normal levels.

For fulminant hepatic failure or end-stage liver disease, liver transplantation is the final option [16]. Transplantation should be reserved for patients with severe decompensated liver disease unresponsive to combinations of chelating and antioxidant therapies. If medication is stopped when the disease seems to be under control, new neurological abnormalities may arise. Furthermore, liver failure may occur rapidly after stopping chelation therapy.

## Genetics

Wilson disease is an autosomal recessive trait due to mutations in the *ATP7B* gene located at chromosome 13q14. The predicted product of the gene is a membrane P-type ATPase with a molecular mass of 160 kDa. There are six N-terminal copper-binding domains. Other domains common to membrane transporting P-type ATPases are a phosphorylation domain, a transduction domain for the transfer of energy, an ATP-binding region and eight transmembrane domains. All functionally important regions are conserved in bacteria and yeast [17]. The human gene is 57% identical (in amino acids) compared with the Menkes gene, but the expression pattern is very different. The Wilson-disease gene is expressed predominately in the liver and kidney, with minor expression in brain, lungs and placenta [4, 5]. The protein product has a trans-Golgi localization [18]. Mutations have now been identified in hundreds of patients of different racial groups [19–21]. More than 150 mutations have been identified [22] and are currently listed in the Human Genome Organization (HUGO) Wilson Disease Mutation Database (www.medgen.med.ualberta.ca). One of the original mutations to be described, His1069Gln [5], is the only mutation found relatively frequently in populations of European origin. The mutation Arg778Leu [19] is found frequently in Asian populations. Most patients are compound heterozygotes, i.e., they carry a different mutation on each of their two chromosomes. This has made the study of correlations of genotypes and clinical features (phenotypes) difficult. In general, an early onset of disease results from mutations that completely destroy the function of the gene. The common mutation His1069Gln has an age of onset of about 20 years of age, and many of these are of neurological onset. Mutation detection should become an effective diagnostic tool as more rapid and efficient approaches are developed.

Within families, several close markers can be used for pre-symptomatic diagnosis once a definite diagnosis has been made in a patient. Markers D13S314, D13S301 and D13S316 are among those used [23]. These markers provide the most reliable method for the identification of pre-symptomatic patients.

## Menkes Disease

### Clinical Presentation

Menkes disease [24] shows clinical variability, although about 90–95% of the patients suffer from the severe, lethal form. *Occipital-horn syndrome* represents a subgroup distinct from those including mildly affected patients with different degrees of nervous-system or connective-tissue involvement [25, 26].

### Classical Severe Form

Premature delivery is frequent, and the birth weight is low. The developmental milestones are usually normal for the first few months, although hypothermia and subtle hair changes may already be present. A genetic disease is rarely suspected at this stage. From about 2–3 months of age, convulsions begin and are resistant to therapy. Hypotonia, vomiting, diarrhea and feeding problems are other initial symptoms. Developmental regression becomes obvious around 5–6 months of age. Hypotonia is later replaced by spasticity, which finally develops into paresis. Spontaneous movements become limited, and drowsiness and lethargy emerge. At this age, Menkes disease is often suspected due to the unusual kinky hair (shafts tangled around one another). The facial appearance (with pudgy cheeks and frontal or occipital bossing) also becomes prominent (Fig. 33.2). Significantly decreased serum copper and ceruloplasmin levels support the clinical diagnosis.

Connective-tissue symptoms include vascular abnormalities, multiple bladder diverticulae and inguinal hernia. The joints are hyperextensive, and the skin is loose and dry. Patients have skeletal changes, such as pectus excavatum or pectus carrinatum, widening of the flared metaphyses and osteoporosis. Late manifestations include blindness, subdural hematomas and respiratory failure. Most of the patients die before the third year of life, due to infections or vascular complications, such as vessel rupture.

### Milder Forms

The most distinctive and mildest form of Menkes disease is *occipital-horn syndrome*, described in 1975

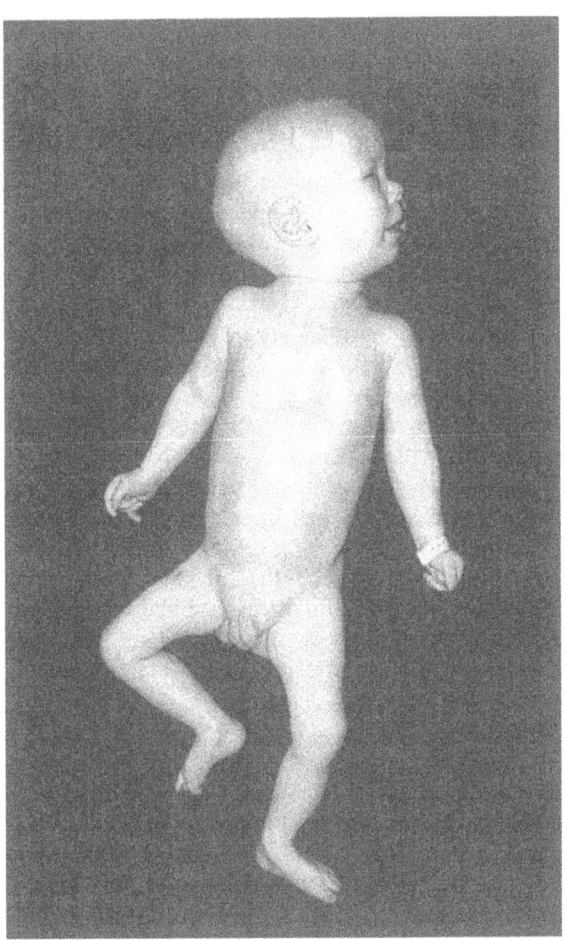

**Fig. 33.2.** A 9-month-old boy affected with the severe form of Menkes disease, showing typical hair changes, hypopigmentation and hypotonia. Courtesy of Dr. Nina Horn, John F. Kennedy Institute, Denmark

[27]. Neurological symptoms are relatively few. The intellectual capacity is described as low to borderline normal. The patients suffer orthostatic hypotension and chronic diarrhea. Skeletal and connective-tissue abnormalities, including occipital horns, bladder diverticulae, joint laxity and loose skin, are numerous. The facial appearance is distinctive, with a long, thin face, high forehead, down-slanting eyes, prominent nose, long philtrum and large ears. The hair is usually not unusual. The life span of these patients is almost normal. Menkes disease and occipital-horn syndrome have been shown to be allelic by the finding of mutations within the same gene, namely *ATP7A* [28, 29].

## Metabolic Derangement

In Menkes disease, cellular copper uptake is normal, but copper cannot be exported from the cell (due to a defect in the membrane copper transporter, ATP7A), and copper-requiring enzymes cannot receive the copper necessary for their normal function (Fig. 33.1). The gene defective in Menkes disease, *ATP7A*, is expressed in all tissues analyzed (though in trace amounts, if any, in liver). This result is consistent with the copper distribution in the disease and with the multisystemic involvement. Copper accumulates in extra-hepatic tissues (except for the brain, due to the blood–brain barrier). However, intracellular accumulation of copper does not reach the toxic state, as copper entering the body is trapped in the intestinal epithelium. Therefore, the amount of copper reaching the liver is low, as are serum copper and ceruloplasmin levels. Thus, in Menkes disease, a true copper deficiency occurs in the brain while, in other extra-hepatic tissues, the deficiency is secondary to lack of copper export into critical copper enzymes.

Multisystemic manifestations of Menkes disease are attributable to dysfunction of one or more copper-requiring enzymes that are involved in metabolic processes, including neurotransmitter biosynthesis (dopamine β-monooxygenase), connective-tissue cross-linking (lysyl oxidase), cellular respiration (cytochrome-C oxidase), neuropeptide maturation (peptidyl α-amidating enzyme), antioxidant defense (Cu–Zn superoxide dismutase) and pigmentation (tyrosinase). The mottled-mouse variants are models for Menkes disease and have defects in the homologous gene, *Atp7a*.

## Diagnostic Tests

An initial diagnosis of Menkes disease is suggested by the clinical features (especially the typical hair changes) and is supported by the reduced levels of serum copper and ceruloplasmin [25]. However, interpretation of these signs may be difficult in the first months of life, as serum copper and ceruloplasmin levels may also be low in normal infants in this period. Biochemical diagnosis is based on the intracellular accumulation of copper due to impaired efflux. Postnatal diagnosis is evaluated in cultured fibroblasts by measuring radioactive copper ($^{64}$Cu) retention after a 20-h pulse, and impaired efflux is directly determined after a 24-h pulse–chase. However, these techniques require expertise and are carried out only in a few specialized centers in the world. Demonstration of a defect in the *ATP7A* gene is the ultimate diagnostic proof. However, it should be kept in mind that mutation detection in Menkes disease is challenging; the gene is very large, and the genetic defect shows great variety. As an X-linked, recessive trait lethal for males, only a few common mutations have been detected among approximately 200 unrelated families analyzed.

## Prenatal Diagnosis

Prenatal diagnosis by biochemical means is carried out by determining the total copper content in chorionic villi by neutron activation analysis in the first trimester and by $^{64}$Cu accumulation in cultured amniotic-fluid cells in the second trimester [25]. Prenatal diagnosis by mutation analysis is possible only when the mutation in the family has been identified prior to pregnancy; otherwise, there would be insufficient time for mutation analysis. In certain families for which the mutation is not yet known, DNA studies may still be informative by linkage analysis utilizing two intragenic, polymorphic markers (CA repeats) within *ATP7A* [30].

## Carrier Identification

Carrier determination by measuring radioactive-copper accumulation in cultured fibroblasts is possible but, due to random inactivation of one of the X-chromosomes, negative results are not reliable and, therefore, mutation analyses will provide the ultimate proof of heterozygosity [25]. In informative families, the intragenic, polymorphic markers may be used for carrier diagnosis.

## Treatment and Prognosis

Menkes disease is a progressive disorder leading to death in early childhood in the severe forms, though some patients survive above 5 years of age. The treatment is mainly symptomatic, but copper treatment has been applied to a few patients who were diagnosed very early.

The objective of treatment for Menkes disease is to provide extra copper to the tissues and the enzymes requiring copper for their normal function. Copper uptake is normal, but a defective ATP7A transporter disturbs the intracellular copper balance. Copper cannot be extruded from the cell, and the copper pool shifts to metallothionein; meanwhile, copper enzymes are deficient in copper. Oral administration of copper is ineffective, as copper is trapped in the intestines, the first step defective in the overall copper-transport pathway. In blood, the main copper-carrying molecules are ceruloplasmin, albumin and copper amino acid complexes, including histidine. The copper-histidine complex has, therefore, been preferred in the treatment of Menkes-disease patients [31]. Daily intravenous administration of copper as copper-histidine [31] restores the serum copper and ceruloplasmin levels and is the most effective form of treatment when compared with various other copper preparations. However, such a treatment should start very early in life. The course of the disease does not seem to be improved in patients receiving copper after the first few months of age, though survival may be prolonged. Studies with the mottled mouse imply that there is a critical stage at which copper is essential in brain development. This suggests that the therapy should be initiated prior to the occurrence of neurodegeneration, which is irreversible. However, it is difficult to diagnose sporadic cases in the neonatal period, and discontinuation of the pregnancy may, for some parents, be the preferred option.

Copper-histidine treatment has been started with favorable results in four unrelated patients from 3 days to 7 weeks of age [32]. In these patients, the devastating neurological symptoms were corrected at ages between 10 months and 20 months and all patients showed borderline normal intellectual development. However, connective-tissue abnormalities, orthostatic hypotension, and diarrhea were persistent. In one of these patients, hypotension and diarrhea could be corrected by administration of L-threo-3,4-dihydroxy phenylserine. Copper treatment, although helpful, cannot be regarded as a cure for Menkes disease at present.

## Genetics

The gene defective in Menkes disease, designated *ATP7A*, is localized to the proximal long arm of the X chromosome (Xq13.3) [25]. The *ATP7A* gene covers an approximately 150-kb genomic region and is organized into 23 exons [33]. The protein product (ATP7A) is a 1500-amino-acid, 165-kDa, copper-binding membrane protein localized to the trans-Golgi membranes [34, 35].

### Mutation Spectrum

Mutations of *ATP7A* show great variety, and every type of genetic mutation has been observed. To date, approximately 200 mutations have been identified in unrelated patients with the classical severe form or with one of the atypical phenotypes [36].

Chromosome abnormalities affecting *ATP7A* have been described in one male and six female patients. One of the female patients was mosaic for the Turner phenotype; the remainder had X-autosome translocations. Cytogenetic analysis should, therefore, be carried out when a female is diagnosed with Menkes disease. Approximately 15% of patients have partial gene deletions of *ATP7A*, while a large proportion of patients have small base-pair changes (point mutations). Point mutations are detected throughout the gene, and an accumulation of mutations can be recognized in the middle of the gene. More than 60 of these mutations have now been published [36]. As our knowledge of the mutations in the various forms increases, the different clinical forms may be distinguished genotypically. For example, in occipital-horn patients, mainly point mutations have been described, and these mutations affect the normal splicing of the gene. However, a normal transcript, though in reduced amount, is always present.

## References

1. Brewer GJ, Yuzbasiyan-Gurkan V (1992) Wilson disease. Medicine 71:139-164
2. Cox DW, Roberts EA (1998) Wilson disease. In: Feldman M, Scharschmidt BF, Sleisenger MF (eds) Sleisenger and Fordtran's gastrointestinal and liver disease. 6th edn. W.B. Saunders, 1104-1111
3. Steindl P, Ferenci P, Dienes HP et al. (1998) Wilson's disease in patients with liver disease: a diagnostic challenge. Gastroenterology 113:212-218
4. Bull PC, Thomas GR, Rommens JM, Forbes JR, Cox DW (1993) The Wilson disease gene is a putative copper transporting P-type ATPase similar to the Menkes gene [erratum appears in Nat Genet 1994, 6:214]. Nature Genet 5:327-337
5. Tanzi RE, Petrukhin KE, Chernov I et al. (1993) The Wilson disease gene is a copper transporting ATPase with homology to the Menkes disease gene. Nature Genet 5:344-350
6. Yoshida K, Furihata K, Takeda S et al. (1995) A mutation in the ceruloplasmin gene is associated with systemic hemosiderosis in humans. Nature Genet 9:267-272
7. Harris ZL, Takahashi Y, Miyajima H et al. (1995) Aceruloplasminemia: molecular characterization of this disorder of iron metabolism. Proc Natl Acad Sci USA 92:2539-2543
8. Martins da Costa C, Baldwin D, Portmann B et al. (1992) Value of urinary copper excretion after penicillamine challenge in the diagnosis of Wilson's disease. Hepatology 15:609-615
9. Walshe JM (1973) Copper chelation in patients with Wilson's disease. A comparison of penicillamine and triethylene tetramine dihydrochloride. Q J Med 42:441-452
10. Hoogenraad TU, Koevoet R, DeRuyter-Korver EGWM (1979) Oral zinc sulphate as long term treatment in Wilson's disease (hepatolenticular degeneration). Eur Neurol 18:205-211
11. Brewer GJ, Yuzbasiyan-Gurkan V (1989) Wilson's disease: an update, with emphasis on new approaches to treatment. Dig Dis 7:178-193
12. Danks DM (1990) Copper-induced dystonia secondary to cholestatic liver disease. Lancet 42:21466
13. Brewer GJ, Johnson V, Dick RD et al. (1996) Treatment of Wilson disease with ammonium tetrathiomolybdate. II. Initial therapy in 33 neurologically affected patients and follow-up with zinc therapy. Arch Neurol 53:1017-1025
14. Britton RS (1996) Metal-induced hepatotoxicity. Semin Liver Dis 16:3-12
15. Sokol R, Twedt D, McKim JM et al. (1994) Oxidant injury to hepatic mitochondria in patients with Wilson's disease and Bedlington terriers with copper toxicosis. Gastroenterology 107:1788-1798
16. Schilsky ML, Scheinberg IH, Sternlieb I (1994) Liver transplantation for Wilson's disease: indications and outcome. Hepatology 19:583-587
17. Vulpe C, Levinson B, Whitney S, Packman S, Gitschier J (1993) Isolation of a candidate gene for Menkes disease and evidence that it encodes a copper transporting ATPase. (erratum) Nature Genet 3:7-13
18. Hung IH, Suzuki M, Yamaguchi Y et al. (1997) Biochemical characterization of the wilson disease protein and functional expression in the yeast saccharomyces cerevisiae. J Biol Chem 272:21461-21466
19. Thomas GR, Forbes JR, Roberts EA, Walshe JM, Cox DW (1995) The Wilson disease gene: spectrum of mutations and their consequences [published erratum appears in Nat Genet 1995, 9:451. Nature Genet 9:210-217
20. Shah AB, Chernov I, Zhang HT et al. (1997) Identification and analysis of mutations in the wilson disease gene (atp7b): population frequencies, genotype-phenotype correlation, and functional analyses. Am J Hum Genet 61:317-328
21. Loudianos G, Dessi V, Angius A et al. (1996) Wilson disease mutations associated with uncommon haplotypes in Mediterranean patients. Hum Genet 98:640-642
22. Cox DW (1996) Molecular advances in Wilson disease. In: Boyer JL, Ockner RK (eds) Progress in liver diseases. Saunders, Philadelphia, pp 245-264
23. Thomas GR, Roberts EA, Walshe JM, Cox DW (1995) Haplotypes and mutations in Wilson disease. Am J Hum Genet 56:1315-1319
24. Menkes JH, Alter M, Steigleder G, Weakley DR, Sung JH (1962) A sex-linked recessive disorder with retardation of growth, peculiar hair, and focal cerebral and cerebellar degeneration. Pediatrics 29:764-779
25. Tumer Z, Horn N (1997) Menkes disease: recent advances and new aspects. J Med Genet 34:265-274
26. Horn N, Tønnesen T, Tumer Z (1995) Variability in clinical in clinical expression of an X-linked copper disturbance, Menkes disease. In: Sarkar B (ed) Genetic response to metals. Marcel and Dekker, New York, pp 285-303
27. Lazoff SG, Rybak JJ, Parker BR, Luzzatti L (1975) Skeletal dysplasia, occipital horns, diarrhea and obstructive uropathy - a new hereditary syndrome. Birth Defects Orig Artic Ser 11:71-74
28. Kaler SG, Gallo LK, Proud VK et al. (1994) Occipital horn syndrome and a mild Menkes phenotype associated with splice site mutations at the MNK locus. Nat Genet 8:195-202
29. Das S, Levinson B, Vulpe C et al. (1995) Similar splicing mutations of the Menkes/mottled copper transporting ATPase gene in X-linked cutis laxa and the blotchy mouse. Am J Hum Genet 56:570-576
30. Begy CR, Dierick HA, Innis JW, Glover TW (1995) Two highly polymorphic CA repeats in the Menkes gene (ATP7A). Hum Genet 96:355-356
31. Sarkar B, Lingertat-Walsh K, Clarke JTR (1993) Copperhistidine therapy for Menkes disease. J Pediatr 123:828-830
32. Christodoulou J, Danks DM, Sarkar B et al. (1998) Early treatment of Menkes disease with parenteral copper-histidine: long-term follow-up of four treated patients. Am J Med Genet 76:154-164
33. Tumer Z, Vural B, Tønnesen T et al. (1995) Characterization of the exon structure of the Menkes disease gene using vectorette PCR. Genomics 26:437-442
34. Dierick HA, Adam AN, Escara-Wilke JF, Glover TW (1997) Immunocytochemical localization of the Menkes copper transport protein (ATP7A) to the trans-Golgi network. Hum Mol Genet 6:409-416
35. Petris MJ, Mercer JF, Culvenor JG et al. (1996) Ligand-regulated transport of the Menkes copper P-type ATPase efflux pump from the Golgi apparatus to the plasma membrane: a novel mechanism of regulated trafficking. EMBO J 15:6084-6095
36. Tumer Z, Moller LB, Horn N (1999) Mutation spectrum of ATP7A, the gene defective in Menkes disease. Adv Exp Med Biol 448:83-95

# CHAPTER 34

# Genetic Defects Related to Metals Other Than Copper

F. Jochum and I. Lombeck

CONTENTS

Magnesium .................................... 393
    Primary Hypomagnesemia ...................... 393
    Magnesium-Losing Kidney ..................... 394
Selenium ..................................... 395
Zinc.......................................... 395
    Acrodermatitis Enteropathica................... 396
    Hyperzincemia with Functional Zinc Depletion....... 397
    Hyperzincemia (Without Symptoms)............... 397
Other Metals.................................. 397
References.................................... 397

---

MAGNESIUM, SELENIUM AND ZINC. The clinical symptoms of *primary hypomagnesemia* develop within the first months of life. Different neurological signs, such as irritability, jitteriness evolving into tetany and carpopedal spasm, were reported. Opisthotonus, generalized convulsions or raised intracranial pressure may develop. The disease is caused by impaired absorption of magnesium. Treatment with high-dose magnesium supplementation is very effective.

The symptoms of the *magnesium-losing kidney* appear usually later in childhood. The patients present with a large clinical heterogeneity ranging from an asymptomatic status to tetany, nephrocalcinosis, dermatitis and deafness. This disease is caused by a renal-tubular reabsorption defect. Treatment with magnesium shows little or no effect on the clinical symptoms and never affects progressive renal failure. Kidney transplantation was shown to normalize the electrolyte excretion, but there has been no long-term experience.

*Selenocysteine* is a constituent of the three iodothyronine deiodinases that convert T4 into T3. It is also an integral part of the glutathione peroxidases and other enzymes and proteins. The functions of some selenoproteins and enzymes are still not known sufficiently. Erythrocyte glutathione-peroxidase deficiency was reported to cause mild hemolytic disease, drug-induced hemolysis and neonatal jaundice. *Keshan disease and Kashin-Beck disease* are preventable by selenium supplementation. It is not known whether the two endemic diseases are also genetically determined.

*Acrodermatitis enteropathica* starts, typically at infancy after weaning (mother's milk), with a characteristic skin rash, diarrhea, infections, alopecia and failure to thrive. Impaired zinc absorption due to an intestinal transport defect leads to hypozincemia, low activity of the numerous zinc metalloenzymes and immunodeficiency. Zinc therapy is very effective.

The rare *hereditary hyperzincemia with functional zinc depletion* shows some clinical similarity to acrodermatitis enteropathica. Treatment is not yet feasible. *Hyperzincemia without clinical symptoms*, dominantly inherited, appears to be a rare non-disease.

---

## MAGNESIUM

Two inherited disorders are known: primary hypomagnesemia and magnesium-losing kidney.

### Primary Hypomagnesemia

*Clinical Presentation*

Since the first description of primary hypomagnesemia in 1965 [1], more than 30 infants with the condition from different parts of the world have been observed. Most affected infants are apparently healthy at birth. The initial symptoms usually start between the third week and the fourth month of life. The infants become irritable, develop sleeping and feeding difficulties, jitteriness, hyperreactivity, tetany with facial twitching and carpopedal spasm. The signs of Chvostek and Trousseau are positive. Sometimes, generalized convulsions develop. Opisthotonus, hypotonicity and areflexia may be present. Occasional peripheral edema, raised intracranial pressure or bulging fontanels occur.

An increased occipitofrontal circumference was observed once.

## Metabolic Derangement

Primary, chronic hypomagnesemia is caused by impaired intestinal absorption or inadequate Mg handling by the kidney. Tracer doses of $^{28}$Mg (half-life = 21.3 h) reveal Mg malabsorption, whereas secretion into the gut and the renal clearance are normal [2, 3]. A reduced Mg retention is also proven by Mg-balance studies. The biochemical mechanism responsible for the malabsorption has not been found, but a defect or absence of a specific protein facilitating active Mg transport in the gut is probable. Hypocalcemia is considered to be secondary to impaired synthesis, secretion or end-organ response to parathormone (PTH). Moderate Mg deficiency is believed to stimulate PTH release, and severe deficiency is believed to inhibit PTH release and cause end-organ resistance to PTH.

## Diagnostic Tests

The main laboratory findings in these infants are hypomagnesemia [with Mg values from 0.15 mmol/l to 0.30 mmol/l (normal range = 0.7–1.00 mmol/l)] and hypocalcemia [with total calcium values from 1.2 mmol/l to 1.6 mmol/l (normal range = 2.2–2.7 mmol/l)]. Urinary Mg excretion is markedly reduced during hypomagnesemia. Estimates of reduced Mg levels in erythrocytes and/or leukocytes are not well established and, in most centers, normal values for cells of different ages are lacking. Serum inorganic phosphorus levels are elevated in patients with primary hypomagnesemia, and alkaline phosphatase levels are normal. Values of circulating PTH are variable. All other biochemical parameters, including glomerular filtration and tubular reabsorption tests, do not reveal any dysfunction.

## Treatment and Prognosis

Without treatment, primary hypomagnesemia usually leads to death within the first year of life. Adequate treatment consists of high-dose Mg supplementation. During the acute phase, Mg must be administered parenterally as Mg sulfate, gluconate or chloride. Usually, MgSO$_4$ (10%) is used at a dose of 0.4–1.0 mmol/kg body weight/day. This treatment produces a rapid clinical remission. After Mg supplementation alone, a spontaneous return of plasma calcium to normal levels occurs. Initially, Mg should be given intravenously by slow infusion (6–24 h). Intramuscular injections are also effective. After clinical stabilization, subsequent oral therapy [in a dose that must be adjusted to the clinical response and side effects (diarrhea)] is adequate. Supplements of 1.5–2.0 (or up to 5.0) mmol/kg body weight/day are needed. In addition to MgSO$_4$ 7H$_2$O (1 mol = 246 g), MgCl$_2$ 6H$_2$O (1 mol = 203 g), Mg(C$_2$H$_3$O$_2$)$_2$ 4H$_2$O (1 mol = 214 g), trimagnesium dicitrate [Mg$_3$(C$_6$H$_5$O$_7$)$_2$ 14H$_2$O (1 mol = 703 g)] or (more rarely) Mg hydroxide, lactate, gluconate, aspartate or glycerophosphate are used. The oral supplementation should be given in three to five divided doses to prevent diarrhea and establish stable blood Mg levels. The treatment must be life long. Interruption results in a recurrence of hypomagnesemia and hypocalcemia, with tetanic convulsions within 1–4 weeks depending on body Mg stores. Usually, it is difficult to achieve normal plasma Mg values. Neither hypocalcemia nor clinical symptoms respond to calcium, vitamin D or PTH.

The prognosis of primary hypomagnesemia, if promptly diagnosed and treated, is good. Under continuous Mg supplementation, normal growth and psychomotor development is achieved. In some patients, caries (and, in a few, epilepsy) was observed. Because of the relatively small number of published patients, it is not clear whether epilepsy occurs more frequently than in the normal population.

## Genetics

This rare disease is most frequently observed in boys (ratio of boys to girls = 3:1). Nevertheless, it appears to be an autosomal-recessive disorder. Parenteral consanguinity was observed in five families. The condition was diagnosed four times among siblings. In at least three other families, older siblings had died from similar clinical disorders. One female was found to have multiple congenital abnormalities and a balanced translocation [46-XX t(9; x)(q12; p22)] in addition to the typical clinical signs.

## Magnesium-Losing Kidney

### Clinical Presentation

The clinical symptoms differ markedly from those of primary hypomagnesemia [4–7]. Many patients suffer from tetany beginning in childhood, but the diagnosis is often delayed until adulthood. About 50% of adult patients present with nephrocalcinosis. Chondrocalcinosis and osteochondrosis are also observed. Patients with additional hypokalemia complain of muscular weakness. This type of hypomagnesemia due to renal wasting is difficult to distinguish from Bartter's syndrome (see "Diagnostic Tests"). Occasionally, dermatitis, sensorineural deafness, schizoid behavior, and oligospermia were reported. Four asymptomatic patients were discovered during family surveys. The clinical heterogeneity seems to correlate with different genetic forms (see below) [8].

## Metabolic Derangement

The primary defect is a tubular Mg-reabsorption deficiency. This can be proven by isotope studies or intravenous Mg loading [6]. Hypomagnesemia of hereditary renal origin comprises different congenital disorders. Some patients also suffer from primary or secondary defects of other tubular transport systems.

## Diagnostic Tests

Mg levels in hypomagnesemia vary from 0.29 mmol/l to 0.60 mmol/l. Moderate hypocalcemia is present in about 40% of cases and hypokalemia in about 50% of cases. The urinary Mg excretion is inappropriately high. During Mg depletion, Mg excretion above 0.46 mmol/24 h is considered to be high in adults. Renal Mg wasting may be associated with other tubular dysfunctions, incomplete tubular acidosis, intermittent glucosuria, increased levels of amino acids in blood and urine, hypercalciuria and hyperkaluria. Three patients had elevated PTH levels. Creatinine clearance was moderately reduced in 50% of patients. Renal biopsies reveal various pictures: some show patchy interstitial fibrosis, but most show a normal histology. The clinical signs of Bartter's syndrome and those of magnesium-losing kidney can be difficult to differentiate. Both diseases may present with hypokalemia, hypokaluria, elevated plasma renin activity, high aldosterone levels and prostaglandinuria. They differ in hypomagnesemia and hypercalciuria, which is characteristic for magnesium-losing kidney [9].

## Treatment and Prognosis

Mg supplementation was tried in doses similar to those used in patients with intestinal Mg malabsorption. Amelioration of the tetanic symptoms was achieved in less than 50% of the patients. Hypercalciuria was either reduced or increased by Mg therapy. In spite of Mg supplementation, the renal function of eight patients with renal Mg wasting, nephrocalcinosis and incomplete tubular acidosis deteriorated to end-stage renal failure. Five patients have received a kidney graft. The transplanted patients showed normal serum electrolyte levels and a regular renal electrolyte excretion after transplantation.

## Genetics

Inherited Mg-losing kidney probably comprises different tubular absorption defects with a high phenotype variability [8, 10]. Both sexes are afflicted. The mode of inheritance, whether autosomal recessive or autosomal dominant, is not known.

# SELENIUM

Selenium (in the form of selenocysteine) is an integral part of at least four different glutathione peroxidases (cellular, plasma, phospholipid hydroperoxide and gastrointestinal). Its function is not completely known, but a protective intra- and extramembranous effect against oxidative damage to tissues and membranes is suggested. It is also a constituent of other enzymes and proteins, such as the three iodothyronine deiodinases that convert T4 into T3 in different tissues. All three deiodinase isoenzymes have been characterized, cloned and shown to have frame TGA codons specifying for selenocysteine [11–13]. However, the influence of Se on the activities of these three enzymes is different. The type-II 5'-deiodinase was not affected, even by severe selenium deficiency, whereas the activity of the type-I 5'-deiodinase was markedly reduced. Mutations in the deiodinase genes raise speculations about involvement of thyroid-hormone metabolism in various disorders. Mutations in the DIo1 gene that cause euthyroid hyperthyroxinemia [14] are well documented.

Two different genetic defects concerning glutathione peroxidase were described approximately 30 years ago. Necheles et al. [15] reported moderate or severe erythrocyte glutathione-peroxidase deficiency resulting in compensated hemolytic disease, drug-induced hemolysis and neonatal jaundice. The glutathioneperoxidase-I gene was then localized to chromosome locus 3p21. Genetic studies of patients with hemolytic diseases from different ethnic groups are lacking. Karpatkin and Weiss [16] reported a reduced activity of platelet glutathione peroxidase with high levels of reduced glutathione in three patients with Glanzman thrombasthenia. The inheritance is assumed to be autosomal recessive. The heterozygous state is usually asymptomatic. In China, two endemic diseases, Keshan disease (a cardiomyopathy [17]) and Kashin-Beck disease (an osteochondroarthropathy [18]) are observed in remote areas where the food has a low selenium content. They can be prevented by Se supplementation. It is still speculative which other factors – genetics, nutrition, environment, infection – may play a role. People from other countries with low dietary selenium during parenteral or semisynthetic feeding [19] seldom reveal signs of Se-responsive myopathy or skeletal or cardiac abnormalities.

# ZINC

Three inherited disorders are known: AE [20], a hereditary hyperzincemia with functional zinc depletion [21] and a hereditary hyperzincemia that presents as a non-disease without symptoms [22].

## Acrodermatitis Enteropathica

### Clinical Presentation

The most obvious clinical feature is skin rash, which has a characteristic symmetrical, circumorificial, retroauricular and acral distribution. The skin lesions are erythematous in acute stages. Later, vesiculobullous, pustular or hyperkeratotic changes may become prominent. Secondary infection is common, usually with *Candida* or *Staphylococci*, which may lead to a wrong diagnosis. Mucosal lesions include gingivitis, stomatitis and glossitis. Symptoms usually present in infancy. Onset is delayed in breast-fed infants until after weaning, whereas babies fed on infant formula develop the syndrome as early as the first 2-4 weeks of life. During early infancy, frequent passage of watery stool, anorexia and failure to thrive often precede the skin lesions without being recognized as typical symptoms of this genetic disorder. Total alopecia, i.e. loss of scalp and superciliary hair, occurs frequently. Nail deformities and ophthalmologic problems, including blepharitis, conjunctivitis, photophobia and impaired dark adaptation, may also occur. Mood changes, irritability, lethargy or depression are also early features of zinc deficiency, as are recurrent infections. All clinical features are aggravated during infections and physiological stress, during growth spurts in early childhood, and during puberty. After puberty, men are less vulnerable to zinc deficiency than women are. One third of pregnancies in untreated patients ended in spontaneous abortion or congenital defects of the skeletal or central nervous system [23]. Although fluctuation in the clinical course occurred, it usually became progressively worse before the advent of zinc therapy. A few patients who suffered from a variant of AE with severe diarrhea, occasional cheilosis, growth failure and normal plasma zinc concentration have been described [24]. Their symptoms exacerbated after withdrawal of zinc therapy.

### Metabolic Derangement

The disturbance of zinc homeostasis results from a partial block in the intestinal absorption [25]. This was first demonstrated in vivo after oral application of tracer doses of $^{65}$Zn [26] or $^{69m}$Zn [27]. Atherton et al. [28] also showed (with biopsies) that in vitro $^{65}$Zn accumulation in jejunal mucosa is markedly reduced. Recent investigations suggest that the binding of zinc to the cell surface and its translocation across the plasma membrane into the cell (possible mediated through a defective anion-exchange mechanism) are both impaired [29]. Ultrastructural studies of duodenal biopsies revealed characteristic inclusion bodies in Paneth cells. Reduced zinc absorption due to a transport defect results in severe zinc deficiency, with an impairment of the function of many zinc metalloenzymes that regulate metabolic pathways. The clinical picture of severe zinc deficiency results from a disrupted metabolism in many tissues.

### Diagnostic Tests

In most patients, plasma zinc (serum zinc is about 15% higher) is reduced to 20-40% of the values for age-matched controls (3-6 µmol/l; normal range = 9-20 µmol/l). Blood assays for zinc should be taken at fasting, and the patient should preferably be without infection. The sample should be centrifuged within 2 h, avoiding hemolysis and contamination. The diagnosis of AE can never be proven beyond doubt by analyzing plasma zinc. In some samples, plasma zinc may be normal because of zinc released from catabolized tissues. However, plasma zinc may be low because of acquired zinc deficiency or the redistribution of zinc in other body pools during stress and infection. For practical reasons, neutrophils, lymphocytes and red cells are unsuitable for zinc determination. Usually, urinary zinc excretion is decreased if plasma zinc is lowered, and sometimes plasma copper levels are in the upper normal range. In general, plasma alkaline phosphatase parallels plasma zinc during severe zinc deficiency. Hair zinc is unreliable in AE, because hair growth is often impaired. Raised blood ammonia, hypo-β-lipoproteinemia and an altered fatty-acid pattern also occur. In many patients, impaired immune responses are associated with both depressed humoral and cell-mediated immunities. In the future, this characteristic might make it possible to find an easily measurable immunological parameter for the evaluation of the zinc status [30].

AE can be suspected if patients with the characteristic clinical picture have markedly reduced plasma zinc values and low renal Zn excretion or, preferably, if absorption tests reveal defective intestinal absorption. The diagnosis is established if, after successful zinc therapy and clinical remission, a withdrawal of zinc leads to a relapse. This is the best way to differentiate the recessively inherited AE (with a low plasma zinc level) and its variant form (with a normal zinc plasma level) from acquired zinc deficiency.

### Treatment and Prognosis

Before zinc deficiency was known to cause the clinical symptoms of AE, treatment mainly consisted of feeding the patient human milk and administering hydroxylated quinolines. This resulted in partial or total remission of the clinical symptoms in several patients. We now know that the bioavailability of zinc in human milk is higher than in other dietary sources.

Since 1973, zinc supplementation has been used. Usually, 1 week after zinc therapy has started, skin lesions disappear, and plasma zinc, urinary zinc excretion and alkaline phosphatase activity increase to normal values [31]. The usual therapeutic dose is 30–50 (or up to 100) mg zinc/day (10–30 µmol Zn/kg body weight/day). Zinc is not very toxic; thus, higher amounts (50–200 mg) do not cause the typical signs of Zn intoxication (nausea, vomiting, abdominal pain, pancreatitis, microhematuria, lethargy and confusion). Plasma copper should be monitored to avoid hypocupremia. The zinc salts used are: sulfate, gluconate, aspartate, orotate and acetate. Usually, zinc sulfate is used ($ZnSO_4$ $7H_2O$; 200 mg = ~45 mg Zn). It is administered in individually prepared capsules or tablets or as a sweet solution. If gastric problems occur, it should be given in at last three divided doses per day. In many patients, the total dose per day remains constant throughout childhood. Sometimes, a higher dose is needed during growth spurts; a higher dose is always needed during pregnancy and lactation (100 mg Zn/day). With this treatment, women treated for AE had uncomplicated pregnancies and deliveries and gave birth to healthy babies. Because of the supplementation regime, the zinc content of their milk was kept within the normal range. Since zinc supplementation was established, the prognosis of AE has been good. In patients with variant forms of AE, similar zinc doses were used and led to an amelioration of the clinical symptoms.

### Genetics

AE is a recessively inherited defect. The gene defect has not yet been identified. For that reason, heterozygous carriers cannot be detected at present. The HLA pattern was investigated in a few patients. A characteristic pattern has not emerged.

### Hyperzincemia with Functional Zinc Depletion

Sampson et al. [21] reported clinical symptoms resembling AE in a boy and his mother. Further investigations showed a gross hyperzincemia and a previously unknown plasma Zn-binding protein (110,000–300,000 kDa) in both patients. Studies with stable isotopes demonstrated an increased Zn exchange and a rapid flux from the plasma to a stable pool. The Zn and Cu concentrations in the liver and skeletal muscle were found to be raised, but the histology of the tissues was normal.

### Hyperzincemia (Without Symptoms)

In 1976, Smith et al. [22] reported a dominantly inherited defect with elevated plasma zinc levels (2500–4350 ng/ml). There were no clinical or biochemical abnormalities. The zinc contents of the erythrocytes and hair were normal.

## OTHER METALS

Manganese-related disease (prolidase deficiency) is discussed in Chap. 22; molybdenum-related disease (combined deficiency of sulfite oxidase and xanthine oxidase) is discussed in Chap. 31.

## References

1. Paunier L, Radde IC, Kooh SW, Fraser D (1965) Primary hypomagnesemia with secondary hypocalcemia. J Pediatr 67:945
2. Lombeck I, Ritzl F, Schnippering HG, Michael H, Bremer HJ, Feinendegen LE, Kosenow W (1975) Primary hypomagnesemia. I. Absorption Studies. Z Kinderheilkd 118:249–258
3. Stromme JH, Nesbakken R, Normann T, Skjorten F, Skyberg D, Johannessen B (1969) Familial hypomagnesemia. Biochemical, histological and hereditary aspects. Acta Paediatr Scand 58:433–444
4. Evans RA, Carter JN, George CR, Walls RS, Newland RC, McDonnell GD, Lawrence JR (1981) The congenital "magnesium-losing kidney". Report of two patients. Q J Med New Ser L 50:39–52
5. Hedemann L, Strunge P, Munck V (1986) The familial magnesium-losing kidney. Acta Med Scand 219:133–136
6. Manz F, Scharer K, Janka P, Lombeck I (1978) Renal magnesium wasting, incomplete tubular acidosis, hypercalciuria and nephrocalcinosis in siblings. Eur J Pediatr 128:67–79
7. Praga M, Vara J, Gonzalez Parra F, Andres A, Alamo C, Araque A, Ortiz A, Rodicio JL (1995) Familial hypomagnesemia with hypercalciuria and nephrocalcinosis. Kidney Int 47:1419–1425
8. Bettinelli A, Bianchetti MG, Borella P, Volpini E, Metta MG, Basilico E, Selicorni A, Bargellini A, Grassi MR (1995) Genetic heterogeneity in tubular hypomagnesemia-hypokalemia with hypocalcuria (Gitelman's syndrome). Kidney Int 47:547–551
9. Mehrotra R, Nolph KD, Kathuria P, Dotson L (1997) Hypokalemic metabolic alkalosis with hypomagnesuric hyperrmagnesemia and severe hypocalciuria: a new syndrome? Am J Kidney Dis 29:106–114
10. Geven WB, Monnens LA, Willems HL, Buijs WC, ter Haar BG (1987) Renal magnesium wasting in two families with autosomal dominant inheritance. Kidney Int 31:1140–1144
11. Behne D, Kyriakopoulos A, Meinhold H, Kohrle J (1990) Identification of type I iodothyronine 5'-deiodinase as a selenoenzyme. Biochem Biophys Res Commun 173:1143–1149
12. Croteau W, Davey JC, Galton VA, St Germain DL (1996) Cloning of the mammalian type II iodothyronine deiodinase. A selenoprotein differentially expressed and regulated in human and rat brain and other tissues. J Clin Invest 98:405–417
13. Salvatore D, Low SC, Berry MJ, Maia AL, Harney JW, Croteau W, St. Germain DL, Larsen PR (1995) Type III 5'-idothyronine deiodinase. Cloning, in vitro expression, and functional analysis of the placental selenoenzyme. J Clin Invest 96:2421–2430
14. Kleinhaus N, Faber J, Kahana L, Schneer J, Scheinfeld M (1988) Euthyroid hyperthyroxinemia due to a generalized 5'-deiodinase defect. J Clin Endocrinol Metab 66:684–688
15. Necheles TF, Steinberg MH, Cameron D (1970) Erythrocyte glutathione-peroxidase deficiency. Br J Haematol 19:605–612

16. Karpatkin S, Weiss HJ (1972) Deficiency of glutathione peroxidase associated with high levels of reduced glutathione in Glanzmann's thrombasthenia. N Engl J Med 287:1062-1066
17. Chen X, Guangqui Y, Chen J, Chen Y, Wen Z, Ge K (1980) Studies on the relations of selenium and Keshan disease. Biol Trace Elem Res 2:91-107
18. Mo D (1984) Pathology and selenium deficiency in Kashin-Beck disease. 3rd international symposium on selenium and biology in medicine, 28 May 1984, Beijing
19. Lombeck I, Kasperek K, Harbisch HD, Becker K, Schumann E, Schroter W (1978) The selenium state of children. II. Selenium content of serum, whole blood. Eur J Pediatr 128:213-223
20. Danboldt M, Closs K (1942) Acrodermatits enteropathica. Acta Derm Venerol (Stockh) 23:127-169
21. Sampson B, Kovar IZ, Rauscher A, Fairweather Tait S, Beattie J, McArdle HJ, Ahmed R, Green C (1997) A case of hyperzincemia with functional zinc depletion: a new disorder? Pediatr Res 42:219-225
22. Smith JC, Zeller JA, Brown ED, Ong SC (1976) Elevated plasma zinc: a heritable anomaly. Science 193:496-498
23. Hambidge KM, Neldner KH, Walravens PA (1975) Letter: zinc, acrodermatitis enteropathica, and congenital malformations. Lancet 1:577-578
24. Krieger I, Evans GW, Zelkowitz PS (1982) Zinc dependency as a cause of chronic diarrhea in variant acrodermatitis. Pediatrics 69:773-777
25. Moynahan EJ, Barnes PM (1973) Zinc deficiency and a synthetic diet for lactose intolerance. Lancet 1:676-677
26. Lombeck I, Schnippering HG, Ritzl F, Feinendegen LE, Bremer HJ (1975) Absorption of zinc in acrodermatitis enteropathica (letter). Lancet 1(7911):855
27. van den Hamer DJH, Cornelisse C, Hoogenraad TU, van Wouwe JP (1985) Use of 69mZn loading test for monitoring of zinc malabsorption. In: Mills CF, Bremner I, Chester JK (eds) Trace elements in man and animals-TEMA 5. Commonwealth Agricultural Bureaux, Slough UK, pp 689-691
28. Atherton DJ, Muller DP, Aggett PJ, Harries JT (1979) A defect in zinc uptake by jejunal biopsies in acrodermatitis. Clin Sci 56:505-507
29. Vazquez F, Grider A (1995) The effect of the acrodermatitis enteropathica mutation on zinc uptake in. Biol Trace Elem Res 50:109-117
30. Jochum F (1998) Spurenelemente in der Immunologie. Labormedizin 2/98:14
31. Neldner KH, Hambidge KM (1975) Zinc therapy of acrodermatitis enteropathica. N Engl J Med 292:879-882

# PART X
# ORGANELLE-RELATED DISORDERS:
# LYSOSOMES, PEROXISOMES, and GOLGI AND PRE-GOLGI SYSTEMS

## Sphingolipid Metabolism

Sphingolipids are complex membrane lipids composed of one molecule of the $C_{18}$ amino alcohol sphingosine, one molecule of a long-chain fatty acid attached to the C-2 amino group of sphingosine, and various polar head groups attached to the OH group at C-1 by a β-glycosidic linkage. There are three classes of sphingolipids, all derivatives of ceramide: cerebrosides, sphingomyelins, and gangliosides.

Cerebrosides have a single sugar, either glucose or galactose, and an additional sulfate group on the galactose in the sulfatides. Sphingomyelins contain phosphorylcholine or phosphorylethanolamine. Gangliosides, the most complex sphingolipids, contain several sugar units and one (monosialogangliosides) or more (disialogangliosides, etc.) sialic-acid residue (N-acetylneuraminic acid). Two monosialogangliosides, GM1 and GM2, are primarily involved in lysosomal storage.

The catabolism of the sphingolipids requires lysosomal hydrolases. Their deficiency results in sphingolipidoses, which are classified according the compound stored (Fig. 35.1).

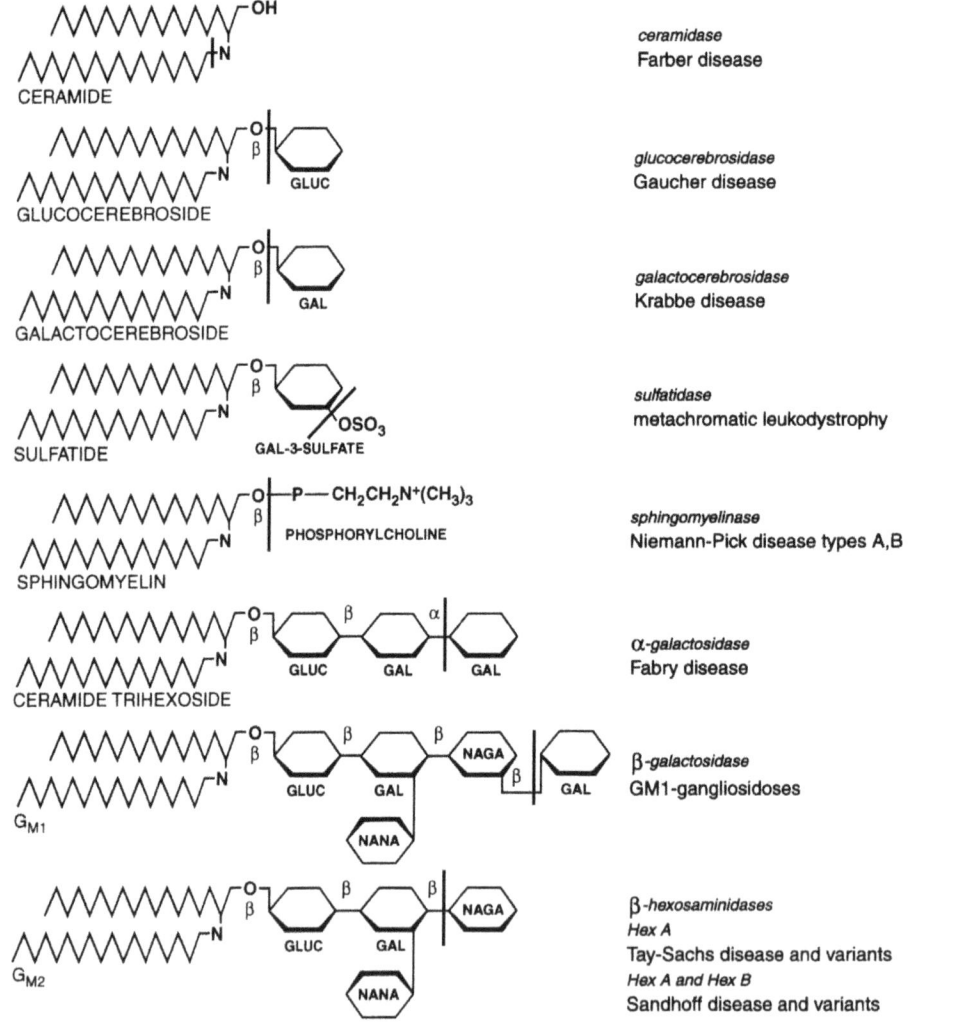

**Fig. 35.1.** Structure of the sphingolipids that accumulate in the sphingolipidoses. The uppermost formula belongs to ceramide, which is composed of the $C_{18}$ amino alcohol sphingosine (at the top of the ceramide structure). A long-chain fatty acid is attached to the amino group at C-2 (at the bottom of the ceramide structure). All the sphingolipids are derived from ceramide and are distinguished by their different polar groups at C-1. $G_M$, monosialoganglioside; *NAGA*, N-acetyl galactosamine; *NANA*, N-acetyl neuraminic acid. Enzyme defects are depicted by *bars*

… CHAPTER 35

# Disorders of Sphingolipid Metabolism

Peter G. Barth

CONTENTS

Niemann-Pick Disease Types A and B. . . . . . . . . . . . . . . . 401
Niemann-Pick Disease Types C and D . . . . . . . . . . . . . . 402
Gaucher Disease. . . . . . . . . . . . . . . . . . . . . . . . . . . . . . . . . 403
Fabry Disease. . . . . . . . . . . . . . . . . . . . . . . . . . . . . . . . . . . 404
Farber Disease . . . . . . . . . . . . . . . . . . . . . . . . . . . . . . . . . . 405
GM1 Gangliosidosis . . . . . . . . . . . . . . . . . . . . . . . . . . . . . 405
GM2 Gangliosidosis . . . . . . . . . . . . . . . . . . . . . . . . . . . . . 406
Metachromatic Leukodystrophy. . . . . . . . . . . . . . . . . . . . 407
Krabbe Disease . . . . . . . . . . . . . . . . . . . . . . . . . . . . . . . . . 409
References . . . . . . . . . . . . . . . . . . . . . . . . . . . . . . . . . . . . . 410

Sphingolipidoses are a subgroup of lysosomal storage disorders. They are characterized by relentless progressive storage in affected organs and concomitant functional impairments. No overall screening procedure for these disorders is available. Their course and appearance, however, are usually characteristic and, together with relevant technical procedures, such as magnetic resonance imaging (MRI), clinical neurophysiology, ophthalmoscopic examination, etc., a provisional diagnosis ("educated guess") can be made, after which enzymatic diagnosis can close the gap in the diagnostic process. Subgroups of sphingolipidoses are grouped together, such as: disorders with prominent hepatosplenomegaly (Niemann-Pick A, B and Gaucher disease), disorders with central and peripheral demyelination [metachromatic leukodystrophy (MLD) and Krabbe disease], and disorders with prominent neuronal storage (the gangliosidoses). Farber disease and Fabry disease are unique in themselves. Fundamentally different etiologies separate Niemann-Pick types C and D from types A and B. For traditional reasons, they are still grouped together in this edition. The last decade has seen hopeful progress in therapeutic strategies, especially for Gaucher disease. Therefore, emphasis has been placed on these new developments.

## Niemann-Pick Disease Types A and B

### Clinical Presentation

Disorders with storage of sphingomyelin were originally grouped together under the eponym Niemann-Pick disease. Originally, types A, B, C, and D were defined [1] in the so-called Crocker classification. However, only types A and B are deficient in sphingomyelinase (enzyme activity 10% or less); types C and D are biochemically and genetically entirely different. A revised nomenclature groups A and B together as type I, and C and D as type II [2], but this nomenclature has not replaced the original one, which is followed here.

Type A is the infantile type. Onset is insidious, with feeding difficulties and dystrophy in the first months of life. Respiratory infections are common, and X-ray examination of the thorax often shows a mottled pattern resembling miliary tuberculosis. Hepatomegaly is constant and predominates over splenomegaly, in contrast to the presentation in Gaucher disease. Neurological deterioration takes place, usually in the second half of the first year of life, with loss of visual contact and muscle hypotonia followed by spastic pareses. Tendon reflexes are diminished. In about half of the cases, a cherry-red spot is seen in the macula region. This red spot represents a normal macula that contrasts with the surrounding retina, which changes to a grayish color due to storage compounds. Foam cells (Niemann-Pick cells) in the bone marrow are typically present. In the brain and spinal cord, neuronal storage is widespread, leading to cytoplasmic swelling and atrophy of the cerebellum. The progressive course leads to death in early childhood.

Type B is a chronic disease presenting with hepatosplenomegaly or splenomegaly after infancy. Disease results from visceral involvement. Liver involvement may be complicated by cirrhosis. Pancytopenia may result from hypersplenism. Chronic pulmonary involvement due to macrophage infiltration in alveolar septa, as demonstrable by thoracic X-ray, leads to impairment of gas diffusion and may cause cor pulmonale. Statural growth is usually diminished in late childhood. Clinical involvement of the nervous system usually does not occur. However, ataxia and retained intelligence may sometimes occur, and a small minority may show macular changes. Hypercholesterolemia is often present.

### Metabolic Derangement

Niemann-Pick types A and B are due to a deficiency of lysosomal sphingomyelinase, which catalyzes the breakdown of sphingomyelin to ceramide and phosphorylcholine [3]. Residual activity tends to be higher in the chronic type than in the acute type, but overlap exists.

### Diagnostic Tests

Sphingomyelinase assay in leukocytes or in fibroblasts provides a reliable diagnosis. Differential diagnosis between types A and B is not possible on this basis. Mutation analysis of the lysosomal sphingomyelinase gene on chromosome 11 (11p15.4-p15.1) [4] may help to assign the phenotype and may help in an individuals prognosis in case of a frequent mutation.

### Treatment and Prognosis

No specific therapy is available for types A and B. Symptomatic therapy for type B should include monitoring for hypersplenism and pulmonary involvement. Patients with type A usually die before the age of 3 years. Survival of patients with type B is generally normal, but complications, such as liver cirrhosis and hypersplenism, may sometimes occur.

### Genetics

The mode of inheritance in Niemann-Pick disease types A and B is autosomal recessive. The gene for lysosomal sphingomyelinase is localized on chromosome 11p15.1-p15.4 [4] and has been sequenced. So far, three mutations have been found to be prevalent in Ashkenazi Jews carrying type A. Other mutations prevail in the Mediterranean region and southern Europe, causing the type-B phenotype [2, 5]. Most mutations found are heteroallelic and, because of their rarity, do not allow prediction of the associated phenotype.

## Niemann-Pick Disease Types C and D

### Clinical Presentation

Niemann-Pick disease types C and D are genetically identical. Type C is found worldwide; type D, the Nova Scotia variant, represents a single mutation prevalent in Nova Scotia as the result of a founder effect. The clinical, biochemical and genetic profiles have been charted by comprehensive studies [5-7] but the pathomechanism has not been clarified completely. The earliest symptom of Niemann-Pick type C is neonatal conjugated hyperbilirubinemia (in about half the cases). This symptom usually resolves spontaneously, to be followed by neurological symptoms later in childhood. In some cases, however, early liver involvement has a rapidly fatal course and is often misdiagnosed as fetal hepatitis. Onset of neurological deterioration in Niemann-Pick type C may occur at any time throughout childhood [8] and occasionally in adulthood [9]. In the severe infantile form of type C, hepatosplenomegaly is accompanied by hypotonia and mental regression by the age of 1-1.5 years. This is often followed by spasticity. Epileptic seizures and supranuclear palsy are rare. In the juvenile form of type C, early-onset supranuclear vertical (upward and downward) gaze paralysis is a characteristic symptom. Another important sign may be cataplexy (sudden loss of tone, associated with laughter). Hepatosplenomegaly may be mild or absent. Dementia, cerebellar ataxia, epileptic seizures, and (in some cases) dystonia are found. Dystonia with onset in childhood should always alert the clinician to the possibility of Niemann-Pick disease type C. Death usually occurs in the second decade. Clinical staging has led to the delineation of two subgroups, characterized as preschool-onset and school-age-onset types, with a higher mortality in the former [8]. Patients with Niemann-Pick type D may have early-onset hyperbilirubinemia. Neurological involvement usually starts in the second half of the first decade [10].

### Metabolic Derangement

The basic defect in Niemann-Pick type C has not been completely determined. There is impaired intracellular transport of exogenous low-density-lipoprotein (LDL)-derived cholesterol, impaired esterification, trapping of unesterified cholesterol in lysosomes and abnormal enrichment of cholesterol in trans-Golgi cisternae [7]. Niemann-Pick disease type C is not a sphingolipidosis in the strict sense [7, 8]. Sphingomyelinase activity is normal or elevated in most tissues but is partially deficient (60-70%) in fibroblasts from the majority of patients, as determined by conventional methods. Indeed, storage of sphingomyelin in tissues in Niemann-Pick types C and D is much less than that seen in types A and B and is accompanied by additional storage of unesterified cholesterol, bis(monacylglycero)phosphate and glucosylceramide. In the brain, neither sphingomyelin nor cholesterol accumulates, but increased levels of lactosylceramide, glucosylceramide, GM2-ganglioside, and asialo-GM2 have been reported [7]. The abnormality in the brain leads to swelling of the proximal neurite segment in the cerebral cortex (similar to the pathology of gangliosidoses and Batten disease [11]) and neurofibrillary tangles containing paired helical filaments (similar to the pathology of Alzheimer disease [7]).

## Diagnostic Tests

Diagnosis of Niemann-Pick disease types C and D in cultured fibroblasts is a staged procedure making use of the dye filipin (which stains the accumulated cholesterol in 90% of the cases) as a screening method, followed by in vitro measurement of LDL-derived intracellular cholesterol esterification [5, 7].

## Treatment and Prognosis

There is no specific treatment available. Prognosis is variable and is best correlated with the age of onset [8].

## Genetics

Somatic cell hybridization and complementation in fibroblast lines from patients have shown that most patients with Niemann-Pick type C harbor the same gene defect, which maps to 18q11-12. The same gene is mutated in Niemann Pick type D (Nova Scotia variant). A minority that exhibits a new phenotype with predominant pulmonary involvement but the same storage profile in fibroblasts does not link to this region, but its chromosomal localization has yet to be determined [6]. The gene involved in Niemann-Pick disease types C and D, the NPC1 gene, has been cloned, and mutations have been identified [12].

## Gaucher Disease

### Clinical Presentation

Gaucher disease (all types) results from storage of glucocerebroside in visceral organs; the brain is affected in two of the three types [13]. Type 1 is by far the most prevalent and mainly occurs in the Ashkenazi Jewish population (genotype frequency 1:10). It causes hepatosplenomegaly, anemia, thrombocytopenia, and (occasionally) leukopenia and skeletal changes. The bleeding tendency causes epistaxis and bleeding gums. Subcapsular splenic infarctions may cause attacks of acute abdominal pain. Bleeding tendency may be enhanced by impaired hepatic synthesis of coagulation factors. Frank cirrhosis is unusual. The lungs may be involved by diffuse infiltration, requiring treatment with oxygen. Patients may have abnormal diffuse yellow-brown skin pigmentation on their face and legs and may experience delays of growth, menarche, and dentition. Skeletal involvement causes the typical Erlenmeyer deformity of the femur. This is seen only in a minority of patients. Aseptic necrosis of the femoral head is more common. Medullary infarction of long bones may cause severely painful crises that have to be treated by powerful analgesics. Osteopenia involves long bones and the vertebral column and may cause spontaneous fractures. Hematologic and bone disease apparently progress independently of each other [14]. The CNS is not clinically affected by storage.

Type 2 causes hepatosplenomegaly and severe CNS involvement in infancy, with death in early childhood. Hepatosplenomegaly appears in the course of the first 6 months. Dystrophy is usual. CNS involvement causes spastic quadriplegia. Characteristic for Gaucher disease type 2 is involvement of the bulbar motor centers. It causes regurgitation and aspiration of food. Convergent squinting and horizontal gaze palsy are also characteristic.

Type 3, also known as Norbottnian type (from the Swedish region of Norbotten) [15], has a clinical expression intermediate in severity between types 1 and 2. Convergent squinting and horizontal gaze palsy are early findings. Splenomegaly may be extreme. Neurological deterioration is slow and may be enhanced by splenectomy.

A neonatal form of Gaucher disease with rapidly lethal course has been described. Typical cases have collodion or ichthyotic skin, hepatosplenomegaly, and (sometimes) fetal hydrops [16].

### Metabolic Derangement

All three major types of Gaucher disease are caused by a deficiency of the lysosomal enzyme glucocerebrosidase, which splits glucose from cerebroside, yielding ceramide and glucose. The disease leads to storage in the liver, spleen, lungs, and (in two of three types) in the brain [13–18]. A minority of the patients with Gaucher disease have a deficiency of saposin C (SAP-2, sphingolipid activator protein-2), a small glycoprotein that binds to and activates glucocerebrosidase [19]. This will result in a type of Gaucher disease similar to type 3. Severe neonatal-onset disease due to deficiency of prosaposin, the common precursor of four saposins (A, B, C, and D) has features in common with Gaucher type-2 disease.

### Diagnostic Tests

Bone-marrow examination may reveal Gaucher cells, reticuloendothelial cells with displaced nuclei and a "crumpled paper"-like cytoplasm that carries the storage product [17]. In Gaucher disease, diagnosis may be suspected by an increase of one of the following enzymes in plasma: angiotensin-converting enzyme or non-tartrate-inhibitable acid phosphatase. Diagnosis rests on the finding of reduced glucocerebrosidase activity in leukocytes or in cultured fibroblasts making use of artificial substrate. More recently, the enzyme chitotriosidase has been found to be highly elevated in Gaucher disease [20]. It may also be used to monitor the effect of enzyme replacement therapy or allogenic bone-marrow transplantation [21]. Gaucher-like disease with normal glucocerebrosidase may result from SAP-2 or from prosaposin deficiency. SAP-2 deficiency can be diagnosed by immunochemical means (enzyme-linked immunosorbent assay) in tissue samples.

**Treatment and Prognosis**

Enzyme replacement has become an effective treatment. Use is made of modified placental glucocerebrosidase (alglucerase) in Gaucher disease types 2 and 3 [13, 14]. The enzyme is modified to expose a mannose terminus for lysosomal targeting after uptake by macrophages. Allogenic bone-marrow transplantation (if tolerated) offers an alternative, which obviates the necessity for frequent and expensive enzyme treatment [14]. Gene therapy consisting of the transfer of the glucocerebrosidase cDNA to autologous bone-marrow-precursor cells by a retroviral vector is under intensive study but has yet to pass the experimental stage [18]. Splenectomy in Gaucher disease may help to correct thrombocytopenia and anemia but enhances the risk of serious infection and may accelerate the progression of the disease at other sites.

**Genetics**

Gaucher disease is transmitted as an autosomal-recessive disorder caused by a deficiency of lysosomal glucocerebrosidase. With the rare exception of the SAP-2-deficient type, all Gaucher cases result from mutations in the glucocerebrosidase gene on 1q21 [17]. Several mutations are frequently seen and are associated with a particular phenotype, e.g., patients who are homozygous for the N370S (asparagine-to-serine substitution) usually express mild type-1 disease [22]. This mutation accounts for the majority of Ashkenazi Jews who have type-1 disease. Results with mutations causing the other phenotypes (types 2 and 3) are less straightforward.

## Fabry Disease

### Clinical Presentation

Fabry disease results from storage of glycolipids, mainly ceramidetrihexoside (globotriaosylceramide, Gb3), in endothelial, perithelial, and smooth-muscle cells of blood vessels, cardiac myocytes, the autonomic spinal ganglia, glomerular endothelium and epithelial cells of glomeruli and tubuli of the kidney. The disorder is X-linked. Its typical presentation is in a young male during later childhood or adolescence, with crises of severe pain in the extremities (acroparesthesia) provoked by exertion or temperature changes. Unexplained bouts of fever and hypohidrosis occur. Characteristic skin lesions called angiokeratoma consist of clusters of small, dark red angiectases on the lower part of the abdomen, buttocks and scrotum. Although angiokeratoma may be found in other lysosomal storage disorders, such as fucosidosis and galactosialidosis, its finding in a mentally normal young male or female who complains of acroparesthesia or has renal or cardiovascular disease is most helpful in the diagnostic work-up. Cerebrovascular disease frequently complicates Fabry syndrome [23, 24]. This is caused by progressive deposition of glycosphingolipids in heart valves and myocardial myocytes. Other symptoms that may be encountered are arthralgia, priapism, hypertrophic cardiomyopathy, mitral valve insufficiency, diarrhea, weight loss, and disturbed temperature sensation (especially for cold) in the extremities. Progressive renal disease due to glycosphingolipid deposition in the kidney results in proteinuria, progressive tubular dysfunction and, eventually, renal failure. Eye symptoms include corneal opacities, visible by slit-lamp microscopy, which can be found both in male hemizygotes and female heterozygotes. Lenticular opacities have also been described [25]. In the absence of a positive family history, the diagnosis may be easily missed. Female heterozygotes may have clinical expression of the disease, especially acroparesthesiae and angiokeratoma. A late-onset cardiac variant has been described in hemizygous males. In this category, adult-onset hypertrophic cardiomyopathy is the presenting symptom; the typical angiokeratoma, pain syndrome, and autonomic neuropathy are absent [26].

### Metabolic Derangement

Fabry disease is caused by the deficiency of the lysosomal enzyme α-galactosidase A, which splits the terminal α-galactoside, releasing galactose from ceramide trihexoside (globotriaosylceramide). Five point mutations in one single gene have each been associated with Fabry disease [24].

### Diagnostic Tests

Birefringent lipid deposits in the form of Maltese crosses can be observed in the urinary sediment by polarization microscopy and may be helpful in the initial steps of diagnosis. A definite diagnosis of Fabry disease can be made by determination of α-galactosidase A in leukocytes or cultured fibroblasts.

## Treatment and Prognosis

Involvement of the cardiovascular system, including cerebral and renal vessels, is the cause of limited life expectancy, which used to be 40 years or less. Cerebrovascular disease is an important cause of illness and limited life expectancy [24]. Dialysis and renal transplantation are the main therapeutic instruments that have improved the outlook. The alleviation of pain is important in Fabry disease (crises). Prophylactic treatment with diphenylhydantoin [27] or carbamazepine [28, 29] has been advocated. Carbamazepine should be preferred because of the well-known side effects of prolonged administration of diphenylhydantoin. Opioids may be needed occasionally in acute treatment of severe crises [30].

## Genetics

Inheritance is X-linked. The gene is localized on the X-chromosome at Xq22.1. The full sequence of the gene is known. Mutations have been found in all exons. The excess of "private" mutations generally makes phenotype predictions on the basis of an observed mutation difficult [31, 32].

## Farber Disease

### Clinical Presentation

Farber lipogranulomatosis results mainly from storage of ceramide in various organs. It presents as an early-onset disease with joint swelling, swelling over bone prominences, and hoarseness due to laryngeal involvement; it causes early death in most cases. Due to storage of ceramide and gangliosides, there is variable involvement of the brain. Lymph nodes, heart, and lungs may be affected. Most patients die before the end of their first year of life. However, a later-onset type with progressive neurological involvement has been recognized. Seven types have been delineated.

### Metabolic Derangement

Farber lipogranulamotosis is caused by deficiency of the lysosomal enzyme ceramidase.

### Diagnostic Tests

It is important to consider Farber disease in the differential diagnosis of atypical juvenile rheumatoid arthritis. Histopathological examination of excised granulomas may reveal a typical lysosomal storage of curvilinear profiles called Farber bodies. Ceramidase activity can be measured using artificial substrates.

### Treatment and Prognosis

No treatment is available. Life expectancy generally depends on the subtype. It varies between death in early childhood and survival into adolescence. Treatment is mainly supportive and should be directed at analgesic and anti-inflammatory treatment.

### Genetics

Farber disease has an autosomal-recessive mode of inheritance. A cDNA containing the full length of the ceramidase gene has been cloned, and a homoallelic point mutation has been demonstrated [33].

## GM1 Gangliosidosis

### Clinical Presentation

Ganglioside storage diseases comprise two main groups: GM1 and GM2 gangliosidoses. In all gangliosidoses, storage in neurons causes displacement of the nucleus towards the periphery and a characteristic swelling of the proximal part of the axon, the so-called meganeurite. The storage material is contained in lysosomes that take the form of concentric lamellated bodies or membranous cytoplasmic bodies. GM1 gangliosidosis comprises three general phenotypes, depending on age of onset: an infantile type, a juvenile type and an adult or chronic type.

Infantile GM1 gangliosidosis or Landing disease [34–36] causes delayed head control from birth. Babies learn to make eye pursuit movements and usually learn to grasp, although insufficiently. Most develop hepatosplenomegaly, a puffy face, a wide upper lip, maxillary hyperplasia, hypertrophied gums, large low-set ears, and moderate macroglossia. The wrists are thickened. At the age of half a year, striking Hurler-like bone changes are seen with vertebral beaking in the thoracolumbar zone, broadening of the shafts of the long bones with distal tapering, and widening of the metacarpal shafts with proximal pinching of the four lateral metacarpals. About half have a cherry-red spot in the macular region, similar to the presentation in Tay-Sachs disease. Rapid neurological regression is usual after the first year of life, with generalized seizures, swallowing disorder, decerebrate posturing, and death, usually before the second birthday. In addition to the infantile type, a severe neonatal-onset type with cardiomyopathy has been described [37]. GM1 gangliosidosis can present (rarely) as non-immune fetal hydrops prenatally and in the newborn [38].

Juvenile GM1 gangliosidosis has no external distinguishing features. Patients are normal until 1 year of age, then lose manipulative skills and become dull, with autistiform behavior. Ataxia, epilepsy, and spastic pareses develop progressively after the first year. Death usually occurs before the tenth birthday. Mild vertebral changes (vertebral beaking) may occur.

Adult GM1 gangliosidosis has its onset during childhood, mainly as an extrapyramidal disorder with dystonic or Parkinsonian features or atypical spinocerebellar ataxia. Severe dementia is usual. There are no external distinguishing symptoms, and vertebral changes or cherry-red macula are not distinguishing features of this subtype.

**Metabolic Derangement**

The lysosomal hydrolase acid β-galactosidase catalyzes the removal of the terminal galactose from GM1 ganglioside to generate GM2 ganglioside. All GM1 gangliosidoses are caused by the deficiency of the lysosomal enzyme β-galactosidase, resulting in the storage of the substrates in brain and visceral organs. The impaired enzyme activity also affects degradation of oligosaccharides, resulting in the excretion of various galactose-rich oligosaccharides and keratan sulfate, which can be used for diagnostic purposes. Several mutations may affect the intracellular activity of acid β-galactosidase. Incomplete processing of the precursor protein or intralysosomal degradation may affect the overall activity of the enzyme [39]. Co-existent deficiencies of acid β-galactosidase and neuraminidase characterize a separate genetic disorder, galactosialidosis, which is due to deficiency of a protecting protein necessary for the function of both enzymes.

**Diagnostic Tests**

Routine morphological examination of peripheral blood may reveal vacuolated lymphocytes in GM1 gangliosidosis. In the infantile type, bone abnormalities of a Hurler-like character may suggest the diagnosis. Brain imaging usually gives non-specific results, suggesting cerebral atrophy. In the adult type, MRI may reveal lesions in the basal ganglia [40]. In the infantile type, thalamic hyperdense lesions have been observed on computed-tomography (CT) scanning [41]. Examination of the urine may reveal galactose-rich oligosaccharides and keratan sulfate. Keratan sulfate is another substrate for β-galactosidase, which is also excreted in large amounts in Morquio-B disease; its presence in GM1 gangliosidosis illustrates the biochemical relationship between the two disorders, although the mutation in Morquio-B disease is different from those in GM1 gangliosidoses. An enzymatic test making use of chromogenic or fluorogenic artificial substrate is available for the diagnosis of β-galactosidase deficiency. The test should preferably be performed in cellular material (leukocytes, fibroblasts), because the activity may sometimes be increased in serum or plasma [42]. A cell line deficient for β-galactosidase should also be screened for neuraminidase deficiency to disprove galactosialidosis.

**Treatment and Prognosis**

No treatment is available.

**Genetics**

The gene affected in GM1 gangliosidosis is localized on chromosome 3p21.33 and has been fully identified and characterized. Mutations have been found in each clinical subtype [43–47].

## GM2 Gangliosidosis

### Clinical Presentation

There are four biochemical variants, known as variant B or Tay-Sachs disease, variant O or Sandhoff disease, variant AB or activator-protein deficiency, and variant B1, with altered substrate specificity to artificial substrate. Tay-Sachs disease and Sandhoff disease are largely similar in neurological expression. Tay-Sachs disease is mostly, although not exclusively, seen in Ashkenazi Jews. Sandhoff disease appears to be ubiquitous. Hepatosplenomegaly or storage macrophages are sometimes found in Sandhoff disease but not in Tay-Sachs disease. The infantile type of both disorders becomes manifest during the first 6 months, with exaggerated response to loud noises [48]. Some functions, such as manual reaching and even sitting, may be achieved but are subsequently lost before the first birthday. Loss of visual attentiveness, with roving eye movements, is also seen early. A typical cherry-red spot in the macula region is seen on fundoscopy in all patients with Tay-Sachs or Sandhoff disease. After the first year, spasticity, disordered swallowing, and seizures develop, the latter usually amenable to standard treatment. Typically, progressive macrocephaly appears during the second year and is due to swelling of cerebral white matter. Death, often due to aspiration pneumonia, occurs between 2 years and 4 years of age. Later-onset GM2 gangliosidoses are mainly due to a deficiency of hexosaminidase A (Hex A). In juvenile

GM2 gangliosidosis due to Hex-A deficiency, onset of ataxia and dementia occurs during the second year or later. Cherry-red spots are only seen in a minority of patients. An adult type of the disease also exists. Presentations of the adult type include psychosis [49, 50], dystonia [51], and a syndrome of supranuclear ophthalmoplegia, cerebellar ataxia, and spinal anterior horn disease [52]. Adult onset hex AB deficiency may present as spinocerebellar degeneration [53]. The B1 variant may have a presentation similar to that of classic Tay-Sachs disease or may present as a juvenile progressive neurodegenerative disorder. The cherry-red spot, characteristic of the infantile-onset forms, may appear late in the late-onset forms [54, 55]. The GM2 activator-protein deficiency may also take the form of classical Tay-Sachs disease or may present at a later age [56].

## Metabolic Derangement

Three gene deficiencies cause GM2 gangliosidoses. Two of these genes encode polypeptides that can only act catalytically on GM2 by forming dimers. The gene HEX-A encodes polypeptide (subunit) $\alpha$; the gene HEX-B encodes polypeptide (subunit) $\beta$. The enzyme $\beta$-hexosaminidase A (Hex-A) consists of the $\alpha\beta$ heterodimer, and the enzyme $\beta$-hexosaminidase B (Hex-B) consists of the $\beta\beta$ homodimer. Both enzymes act on GM2 (the substrate) by $\beta$-hydrolytic cleavage of a terminal $N$-acetylglucosamine. In Tay-Sachs disease, Hex-A is deficient; in Sandhoff disease, Hex-A and Hex-B are both deficient. The third gene encodes the GM2-A (or -activator). The GM2-activator is a small, soluble glycoprotein that complexes with GM2 as an initial step in the catalytic breakdown of GM2. In a nomenclature still in use, Tay-Sachs disease is known as variant B, Sandhoff disease is variant 0, and activator-protein deficiency is variant AB. Variant B1 is a mutant Hex-A with impaired affinity for the natural substrate, normal affinity for the artificial substrate 4-methyl-umbelliferyl-$\beta$-$N$-acetylglucosaminide, and diminished activity for the sulfated artificial substrate.

## Diagnostic Tests

The cherry-red macula is useful in diagnosis but is not specific, since it may also be present in Niemann-Pick disease A and in GM1 gangliosidosis. Oligosaccharide excretion patterns in the urine are abnormal in Sandhoff disease but not in Tay-Sachs disease. Skin biopsies may reveal concentric lamellated bodies in nerve endings, representing secondary lysosomes with the storage product. Reliable diagnostic tests are based on the determination of residual enzyme activities in leukocytes or fibroblasts by the use of artificial fluorogenic or colorimetric glucosaminide substrates.

## Treatment and Prognosis

No treatment is available.

## Genetics

HEX-A is localized at 15q23-4 and HEX-B at 5q11.2-13.3. Their gene structures are largely homologous. Close to 90 mutations have been identified in the Hex-A gene [57]. Three mutations are frequently seen in Ashkenazi Jews [55]. More variability is seen in non-Ashkenazi groups [58]. The B1 HEX-A variant is caused by a unique mutation that may have its founder origin in northern Portugal [59]. HEX-B mutations are more rare than HEX-A mutations. A recent publication lists 20 different mutations in the HEX-B gene [60]. The GM2 activator protein is linked to chromosome 5.

# Metachromatic Leukodystrophy

## Clinical Presentation

The typical presentation of MLD is in an infant about to walk without support who starts with weak, valgus feet. At this stage, tendon reflexes may be absent due to concomitant peripheral neuropathy. Walking then rapidly becomes more difficult and, in the course of the second year, becomes impossible. The next stage is dementia and spastic quadriplegia with retained ability to perform some voluntary acts, followed by a vegetative stage and death, usually between 3 years and 6 years of age. A detailed description by Hagberg [61] is based on this author's large personal experience. Storage products may lead to radiculopathy with severe pains, which requires analgesic treatment. Peripheral nerve conduction is strongly delayed due to demyelination, and CSF protein content is elevated. CT, or preferably MRI (Fig. 35.2), shows central demyelination and, often, cerebellar atrophy.

### Juvenile Metachromatic Leukodystrophy During Childhood or Adolescence

Patients may be "dull" before deterioration sets in. Initial symptoms may be ataxia, spasticity, and absent or low tendon reflexes together with positive Babinski signs. Spastic quadriplegia and dementia will ensue within a few years after onset. Speech disturbances, bulbar symptoms, ataxia, epilepsy, and tremor may occur. Decrease of motor-nerve conduction velocities of peripheral nerves is seen, similar to the presentation in late-infantile cases [62]. In adult MLD, psychiatric symptoms are often the presenting symptom [63, 64], followed by dementia and spastic pareses. Optic

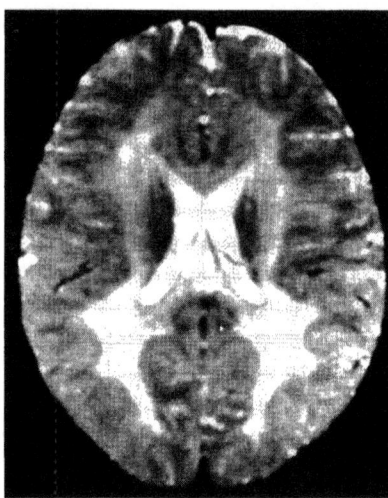

**Fig. 35.2.** Magnetic-resonance imaging axial T2-weighted section from a patient with infantile metachromatic leukodystrophy. There is increased signal in the central white matter, with subcortical sparing. Abnormalities predominate in the posterior halves of the hemispheres

atrophy and variable degrees of dystonia may occur. In adults, decreased nerve conduction velocity may not always be present.

## Metabolic Derangement

Sulfatide (galactocerebroside sulfate) is a constituent of the normal myelin sheath, both in peripheral nerves and in the CNS. Inability to hydrolyze the sulfate bond results in storage of sulfatide and induces a demyelinating disease called MLD or sulfatidosis. The name derives from early neuropathologists' microscopic observation that the storage product appears brown rather than the usual red on staining with a thiazine stain ("metachromasia"). Storage of sulfatide results in MLD, or sulfatidosis. The storage entails both central and peripheral demyelination. The lysosomal enzyme cerebroside sulfatase has two portions, of which arylsulfatase A is the heat-labile portion. Adequate catalytic function requires the presence of a heat-stable activator protein, sphingolipid activator protein, SAP-1 (also known as saposin B). Therefore, MLD can be caused by deficiencies of either of two proteins: arylsulfatase A or SAP-1. The common cause is deficiency of arylsulfatase A. SAP-1 deficiency [65] appears to be very rare, but this may be due (in part) to the difficult diagnosis. Arylsulfatase A catalyzes the degradation of sulfated glycolipids, especially cerebroside-3-sulphate, which is mainly present in myelin. A large number of allelic mutations may cause arylsulfatase A-deficient MLD, with age of onset related to residual activity. Infantile, juvenile, and adult types exist. A double homozygous mutation in the arylsulfatase A gene is associated with low arylsulfatase-A activity against the artificial substrate $p$-nitrocatechol sulfate, but the activity is sufficient to prevent clinical MLD [66]. The gene is known as pseudodeficiency (PD) gene. It occurs in MLD heterozygotes (MLD/PD) or in pseudodeficiency homozygotes (PD/PD) [67]. In both conditions, the $p$-nitrocatechol-sulfate-based arylsulfatase-A assay will demonstrate decreased activity that may be in the homozygote-MLD range. Therefore, a cautious approach is advocated in atypical cases showing decreased arylsulftase-A activity (see below).

## Diagnostic Tests

Motor-nerve conduction velocities of peripheral nerves are strongly delayed, indicating demyelination. MRI of the brain shows demyelination, especially in the central white matter of the hemispheres. Cerebellar atrophy is often seen. CSF protein is usually elevated.

Biopsies containing myelinated peripheral-nerve fibers usually show segmental demyelination with metachromatic material. Electron microscopy will show typical prismatic inclusions. Final diagnosis rests on the determination of arylsulfatase-A deficiency, making use of nitrocatechol sulfate as an artificial substrate. In cases where the patient displays a Hurler-like aspect or ichtyosis, multiple sulfatase deficiency must be excluded by appropriate means (Chap. 36). It is particularly important to exclude pseudodeficiency. Differentiation of pseudodeficiency from MLD may require additional examinations, such as tests for signs of storage (sulfatide excretion tests, in vitro sulfatide loading tests, cutaneous nerve biopsies), especially when clinical symptoms are equivocal or absent. Excretion of sulfatide in the urine has been found to be in the control range in cases of pseudodeficiency [68]. Mutation analysis of the MLD gene is similarly helpful in distinguishing MLD from the pseudodeficiency state (see "Genetics" below). The impaired turnover of radiolabeled sulfatide in fibroblasts distinguishes true MLD from the pseudodeficiency state [69]. This test will also detect SAP-1 deficiency [69]. This auxiliary test is also applied in prenatal diagnosis [70].

## Treatment and Prognosis

MLD may cause a secondary deficiency of γ-aminobutyric acid. Administration of the glutamic-acid-transaminase inhibitor vigabatrine, better known as an antiepileptic drug, has been of particular use in a small number of patients with these leukodystrophies [71]. This drug may alleviate painful spasms. Radiculopathy may cause severe pains that may need to be treated by analgesics. A number of reports indicate beneficial effects of allogenic bone-marrow transplantation [72-77].

## Genetics

The arylsulfatase-A gene is localized to chromosome 22q13.31-qter. The gene has been fully characterized [78], and mutations have been identified [79-85]. Use can be made of mutation analysis in predicting or explaining the phenotype. Mutations that prevent synthesis of mRNA, such as the splice donor-site mutation in intron 2, cause the severe infantile-onset disease when present on both alleles. According to one study [81], non-identical alleles were present in adult (90%), juvenile (50%), and late-infantile (36%) variants. Pseudodeficiency is found in 1% of the population who have arylsulfatase-A activities in the MLD range. Gene analysis can be helpful in the diagnosis of pseudodeficiency. A pseudodeficiency allele is caused by one of two independent mutations [85]. Homozygous pseudodeficiency state (PD/PD) leads to reduction of arylsulfatase-A activity to 8%.

## Krabbe Disease

### Clinical Presentation

Krabbe disease or globoid cell leukodystrophy is caused by storage of galactocerebroside and, like MLD, causes both central and peripheral demyelination. The denomination "globoid cell leukodystrophy" is derived from the presence of large numbers of multinuclear macrophages in cerebral white matter.

The disease usually starts within the first 3-6 months, with irritable behavior, crying, tonic spasms on light or noise stimulation, symptoms of early-onset cerebral palsy, blindness, optic atrophy, and deafness. Periods of unexplained fever may occur. This stage is followed by permanent opisthotonic posturing, with flexed upper extremities and extended lower extremities. Hyperpyrexia, hypersalivation, frequent seizures, and loss of all social contact are seen at this stage. This is followed by loss of bulbar functions, hypotonia, and death from hyperpyrexia, respiratory complications, or aspiration at a median age of 13 months [86]. CSF protein is elevated, and nerve conduction velocities are delayed because of coincident segmental demyelination of peripheral nerves [87]. Later onset is described in patients whose disease starts between 6 months and 3 years of age or even later, during childhood. In the late-infantile form, presentation may include failing vision, irritability, motor deterioration, ataxia, and stiffness. In the late-childhood form, failing vision appears to be the most common symptom but, surprisingly, an asymmetric presentation with hemiparesis may be the first symptom [88]. While peripheral neuropathy is always seen in the early-infantile form, normal peripheral nerve findings in the late-childhood and adult forms, together with a hemiplegic presentation, may cause confusion and diagnostic delay. MRI will display cerebral white-matter degeneration.

### Metabolic Derangement

The basic defect is the deficiency of a lysosomal galactocerebroside β-galactosidase. The substrate galactocerebroside, however, does not accumulate. Instead, there is an accumulation of the toxic intermediate galactosylsphingosine (psychosine), which is held responsible for damage to oligodendrocytes [89].

### Diagnostic Tests

Demyelination in peripheral nerves leads to severe slowing of nerve conduction velocities. This laboratory finding in an infant with a rapidly progressive cerebral disorder is useful in making a presumptive diagnosis of Krabbe disease. Protein in CSF obtained on lumbar puncture is usually elevated. Delayed nerve conduction velocities can also be found in the juvenile form but may be absent in the adult type. Demyelination observed on MRI of the brain is also an important clue in diagnosis. Specific abnormalities on brain imaging include calcifications that may sometimes be observed in the thalamus, basal ganglia, and periventricular white matter [90, 91]. Increase of the T1 signal is caused by increased water content due to destructive changes. The ultimate diagnosis is made by diagnosis of the enzyme defect in leukocytes or cultured fibroblasts.

### Treatment and Prognosis

There is no treatment in the rapidly progressive infantile disease. Recent data in a series of five patients (one infantile and four childhood-onset cases) indicate that allogenic bone-marrow transplantation may be effective in preventing onset or halting progression of the disease [92]. Supportive treatment in advanced disease should be aimed at analgesic treatment of the often severe pain due to peripheral demyelination that may result from radiculopathy.

### Genetics

The gene has been cloned and is localized to chromosome 14q.31. The gene has been fully characterized, and many mutations have been identified [93, 94].

## References

1. Crocker AC (1961) The cerebral defect in Tay-Sachs disease and Niemann-Pick disease. J Neurochem 7:69-80
2. Vanier MT, Suzuki K (1996) Niemann-Pick diseases. In: Moser HW (ed) Neurodystrophies and Neurolipidoses. In: Vinken PJ, Bruyn GW (eds) Handbook of clinical neurology, vol 66. Revised series, vol 22. Elsevier Science, Amsterdam, pp 133-162
3. Brady RO, Kanfer JN, Mock MB, Fredrickson DS (1966) The metabolism of sphingomyelin II. Evidence of an enzymatic deficiency in Niemann-Pick disease. Proc Natl Acad Sci USA 55:366-369
4. Schuchman EH, Levran O, Pereira LV, Desnick RJ (1992) Structural organization and complete nucleotide sequence of the gene encoding human acid sphingomyelinase (SMPD1). Genomics 12:197-205
5. Vanier MT, Pentchev P, Rodriguez-Lafrasse C, Rousson R (1991) Niemann-Pick disease type C: an update. J Inherit Metab Dis 14:580-595
6. Vanier MT, Duthel S, Rodriguez-Lafrasse C, Pentchev P, Carstea ED (1996) Genetic heterogeneity in Niemann-Pick C disease: a study using somatic cell hybridization and linkage analysis. Am J Hum Genet 58:118-125
7. Vanier MT, Suzuki K (1998) Recent advances in elucidating Niemann-Pick C disease. Brain Pathol 8:163-174
8. Higgins JJ, Patterson MC, Dambrosia JM et al. (1992) A clinical staging classification for type C Niemann-Pick disease. Neurology 42:2286-2290
9. Lossos A, Schlesinger I, Okon E et al. (1997) Adult-onset Niemann-Pick type C disease. Clinical, biochemical, and genetic study. Arch Neurol 54:1536-1541
10. Jan MM, Camfield PR (1998) Nova Scotia Niemann-Pick disease (type D): clinical study of 20 cases. J Child Neurol 13:75-78
11. Braak H, Braak E, Goebel HH (1983) Isocortical pathology in type C Niemann-Pick disease. A combined Golgi-pigmentarchitectonic study. J Neuropathol Exp Neurol 42:671-687
12. Carstea ED, Morris JA, Coleman KG et al. (1997) Niemann-Pick C1 disease gene: homology to mediators of cholesterol homeostasis. Science 277:228-231
13. Beutler E (1991) Gaucher's disease. N Engl J Med 325:1354-1360
14. Beutler E (1993) Modern diagnosis and treatment of Gaucher's disease. Am J Dis Child 147:1175-1183
15. Dreberg S, Erikson A, Hagberg B (1980) Gaucher disease - Norbottnian type. I. General clinical description. Eur J Pediatr 133:107-118
16. Sidransky E, Sherer DM, Ginns EI (1992) Gaucher disease in the neonate: a distinct Gaucher phenotype is analogous to a mouse model created by targeted disruption of the glucocerebrosidase gene. Pediatr Res 32:494-498
17. Brady RO, Barton NW, Grabowski GA (1993) The role of neurogenetics in Gaucher's disease. Arch Neurol 50:1212-1224
18. Brady RO (1996) Gaucher Disease. In: Moser HW (ed) Neurodystrophies and neurolipidoses. In: Vinken PJ, Bruyn GW (eds) Handbook of clinical neurology, vol 66. Revised series, vol 22. Elsevier Science, Amsterdam, pp 123-132
19. Kishimoto Y, Hiraiwa M, O'Brien JS (1992) Saposins: structure, function, distribution, and molecular genetics. J Lipid Res 33:1255-1267
20. Hollak CE, van Weely S, van Oers MHJ, Aerts JFMG (1994) Marked elevation of plasma chitotriosidase activity. A novel hallmark of Gaucher disease. J Clin Invest 93:1288-1292
21. Young E, Chatterton C, Vellodi A, Winchester B (1997) Plasma chitotriosidase activity in Gaucher disease patients who have been treated either by bone marrow transplantation or by enzyme replacement therapy with alglucerase. J Inherit Metab Dis 20:595-602
22. Mistry PK, Cox TM (1993) The glucocerebrosidase locus in Gaucher's disease: molecular analysis of a lysosomal enzyme. J Med Genet 30:889-894
23. Grewal RP, Barton NW (1992) Fabry's disease presenting with stroke. Clin Neurol Neurosurg 94:177-179
24. Mitsias P, Levine SR (1996) Cerebrovascular complications of Fabry's disease. Ann Neurol 40:8-17
25. Font RL, Fine BS (1972) Ocular pathology in Fabry's disease. Histochemical and electron microscopic observations. Am J Ophthalmol 73:419-430
26. Nakao S, Takenaka T, Maeda M et al. (1995) An atypical variant of Fabry's disease in men with left ventricular hypertrophy. N Engl J Med 333:288-293
27. Lockman LA, Hunninghake DB, Krivit W, Desnick RJ (1973) Relief of pain of Fabry's disease by diphenylhydantoin. Neurology 23:871-875
28. Shibasaki H, Tabira T, Inoue N, Goto I, Kuroiwa Y (1973) Carbamazepine for painful crises in Fabry's disease. J Neurol Sci 18:47-51
29. Filling-Katz MR, Merrick HF, Fink JK et al. (1989) Carbamazepine in Fabry's disease: effective analgesia with dose-dependent exacerbation of autonomic dysfunction. Neurology 39:598-600
30. Gordon KE, Ludman MD, Finley GA (1995) Successful treatment of painful crises of Fabry disease with low dose morphine. Pediatr Neurol 3:250-251
31. Eng CM, Resnick-Silverman LA, Niehaus DJ, Astrin KH, Desnick RJ (1993) Nature and frequency of mutations in the alpha-galactosidase A gene that cause Fabry disease. Am J Hum Genet 53:1186-1197
32. Eng CM, Ashley GA, Burgert TS et al. (1997) Fabry disease: thirty-five mutations in the alpha-galactosidase A gene in patients with classic and variant phenotypes. Mol Med 3: 174-182
33. Koch J, Gartner S, Li CM et al. (1996) Molecular cloning and characterization of a full-length complementary DNA encoding human acid ceramidase. Identification of the first molecular lesion causing Farber disease. J Biol Chem 271:33110-33115
34. Landing BH, Silverman FN, Craig MM et al. (1964) Familial neurovisceral lipidosis. Am J Dis Child 108: 503-522
35. O'Brien JS (1970) Generalized gangliosidosis. In: Vinken PJ, Bruyn GW (eds) Handbook of clinical neurology, vol 10. North Holland, Amsterdam, pp 462-483
36. Suzuki K, Suzuki K (1996) The gangliosidoses. In: Moser HW (ed) Neurodystrophies and Neurolipidoses. In: Vinken PJ, Bruyn GW (eds) Handbook of clinical neurology, vol 66. Revised series, vol 22. Elsevier Science, Amsterdam, pp 247-280
37. Kohlschütter A, Sieg K, Schulte FJ, Hayek HW, Goebel HH (1982) Infantile cardiomyopathy and neuromyopathy with - galactosidase deficiency. Eur J Pediatr 139:75-81
38. Bonduelle M, Lissens W, Goossens A et al. (1991) Lysosomal storage diseases presenting as transient or persistent hydrops fetalis. Genet Couns 2:227-232
39. Oshima A, Yoshida K, Itoh K et al. (1994) Intracellular processing and maturation of mutant gene products in hereditary beta-galactosidase deficiency (beta-galactosidosis). Hum Genet 93:109-114
40. Uyama E, Terasaki T, Watanabe S et al. (1992) Type 3 GM1 gangliosidosis: characteristic MRI findings correlated with dystonia. Acta Neurol Scand 86:609-615
41. Kobayashi O, Takashima S (1994) Thalamic hyperdensity on CT in infantile GM1 gangliosidosis. Brain Dev 16:472-474
42. Ishii N, Oshima A, Sakuraba H, Fukuyama Y, Suzuki Y (1994) Normal serum beta-galactosidase in juvenile GM1 gangliosidosis. Pediatr Neurol 10:317-319
43. Boustany RM, Qian WH, Suzuki K (1993) Mutations in acid beta-galactosidase cause GM1 gangliosidosis in American patients. Am J Hum Genet 53:881-888
44. Chakraborty S, Rafi MA, Wenger DA (1994) Mutations in the lysosomal beta-galactosidase gene that causes the adult form of GM1 gangliosidosis. Am J Hum Genet 54:1004-1013
45. Morrone A, Morreau H, Zhou XY et al. (1994) Insertion of a T next to the donor splice of intron 1 causes aberrantly spliced mRNA in a case of infantile GM1-gangliosidosis. Hum Mutat 3:112-120
46. Mosna G, Fattore S, Tubiello G et al. (1992) A homozygous missense arginine to histidine substitution at position 482 of the beta-galactosidase in an Italian infantile GM1 gangliosidosis patient. Hum Genet 90:247-250

47. Nishimoto J, Nanba E, Inui K, Okada S, Suzuki K (1991) GM1 gangliosidosis (genetic beta-galactosidase deficiency): identification of four mutations in different clinical phenotypes among Japanese patients. Am J Hum Genet 49:566–574
48. Volk BW, Schneck L, Adachi M (1970) Clinic, pathology and biochemistry of Tay-Sachs disease. In: Vinken PJ, Bruyn GW (eds) Handbook of clinical neurology, vol 10, North Holland, Amsterdam, pp 385–426
49. Argov Z, Navon R (1984) Clinical and genetic variations in the syndrome of adult GM2 gangliosidosis resulting from hexosaminidase A deficiency. Ann Neurol 16:14–20
50. Navon R, Argov Z, Frisch A (1986) Hexosaminidase A deficiency in adults. Am J Med Genet 24:179–196
51. Meek D, Wolfe LS, Andermann E, Andermann F (1984) Juvenile progressive dystonia: a new phenotype of GM2-gangliosidosis. Ann Neurol 15:348–352
52. Harding AE, Young EP, Schon F (1987) Adult onset supranuclear ophthalmoplegia, cerebellar ataxia and neurogenic proximal muscular weakness in a brother and sister: another hexosaminidase A deficiency syndrome. J Neurol Neurosurg Psychiatr 50:687
53. Bolhuis PA, Oonk JGW, Kamp PE et al. (1987) Ganglioside storage, hexosaminidase lability, and urinary oligosaccharides in adult Sandhoff's disease. Neurology 37:75–81
54. Goebel HH, Stolte G, Kustermann-Kuhn B, Harzer K (1989) B1 variant of GM2 gangliosidosis in a 12-year old patient. Pediatr Res 25:1–1893
55. Maia M, Alves D, Ribeiro G, Pinto R, Sa Miranda MC (1990) Juvenile GM2 gangliosidosis variant B1: clinical and biochemical study in seven patients. Neuropediatrics 21:18–23
56. Goldman JE, Yamanaka T, Rapin I et al. (1980) The AB-variant of GM2-gangliosidosis. Acta Neuropathol (Berl) 52:189–202
57. Myerowitz R (1997) Tay-Sach disease-causing mutations and neutral polymorphisms in the Hex A gene. Hum Mutat 9:195–208
58. Paw BH, Tieu PT, Kaback MM, Lim J, Neufeld EF (1990) Frequency of three Hex A mutant alleles among Jewish and non-Jewish carriers identified in a Tay-Sachs screening program. Am J Hum Genet 47:698–705
59. Dos Santos MR, Tamala A. S Miranda MC et al. (1991) GM2-gangliosidosis B1 variant: analysis of β-hexosaminidase α gene mutations in eleven patients from a defined regio in Portugal. Am J Hum Genet 49:886–890
60. Kolodny EH (1997) GM2 gangliosidoses. In: Rosenberg RN, Prusiner SB, DiMauro S, Barchi RL (eds) The molecular and genetic basis of neurological disease. Butterworth-Heinemann, Boston, pp 473–490
61. Hagberg B (1963) Clinical symptoms, signs and tests in metachromatic leucodystrophy. In: Folch-Pi J, Bauer H (eds) Brain lipids and lipoproteins and the leukodystrophies. Elsevier, Amsterdam, pp 134–146
62. Haltia T, Palo J, Haltia M, Icén A (1980) Juvenile metachromatic leukodystrophy. Clinical, biochemical, and neuropathologic studies in nine cases. Arch Neurol 37:42–46
63. Shapiro EG, Lockman LA, Knopman D, Krivit W (1994) Characteristics of the dementia in late-onset metachromatic leukodystrophy. Neurology 44:662–665
64. Hyde TM, Ziegler JC, Weinberger DR (1992) Psychiatric disturbances in metachromatic leukodystrophy. Arch Neurol 49:401–406
65. Hahn AF, Gordon BA, Feleki V, Hinton GG, Gilbert JJ (1982) A variant form of metachromatic leukodystrophy without arylsulfatase A deficiency. Ann Neurol 12:33–36
66. Gieselmann V, Polten A, Kreysing J et al. (1991) Molecular genetics of metachromatic leukodystrophy (review). Dev Neurosci 13:222–227
67. Francis GS, Bonni A, Shen N et al. (1993) Metachromatic leukodystrophy: multiple nonfunctional and pseudodeficiency alleles in a pedigree: problems with diagnosis and counseling. Ann Neurol 34:212–218
68. Lugowska A, Tylki-Szymanska A, Berger J, Molzer B (1997) Elevated sulfatide excretion in compound heterozygotes of metachromatic leukodystrophy and ASA-pseudodeficiency allele. Clin Biochem 30:325–331
69. Wenger DA, DeGala G, Williams C et al. (1989) Clinical pathological, and biochemical studies on an infantile case of sulfatide/GM1 activator protein deficiency. Am J Med Genet 33:255–265
70. Kihara H, Ho CK, Fluharty AL, Tsay KK, Hartlage PL (1980) Prenatal diagnosis of metachromatic leukodystrophy in a family with pseudoarylsulfatase A deficiency by the cerebroside sulfate loading test. Pediatr Res 14:224–227
71. Jaeken J, De Cock P, Casaer P (1991) Vigabatrin as spasmolytic drug. Lancet 338:1603 (letter)
72. Bayever E, Ladisch S, Philippart M et al. (1985) Bone marrow transplantation for metachromatic leucodystrophy. Lancet 2:471–473
73. Krivit W, Shapiro E, Kennedy W et al. (1990) Treatment of late infantile metachromatic leukodystrophy by bone marrow transplantation. N Engl J Med 322:28–32
74. Dhuna A, Toro C, Torres F, Kennedy WR, Krivit W (1992) Longitudinal neurophysiological studies in a patient with metachromatic leukodystrophy following bone marrow transplantation. Arch Neurol 49:1088–1092
75. Malm G, Ringden O, Winiarski J et al. (1996) Clinical outcome in four children with metachromatic leukodystrophy treated by bone marrow transplantation. Bone Marrow Transplant 17:1003–1008
76. Navarro C, Fernandez JM, Dominguez C, Fachal C, Alvarez M (1996) Late juvenile metachromatic leukodystrophy treated with bone marrow transplantation: a 4-year follow-up study. Neurology 46:254–256
77. Kidd D, Nelson J, Jones F et al. (1998) Long-term stabilization after bone marrow transplantation in juvenile metachromatic leukodystrophy. Arch Neurol 55:98–99
78. Kreysing HJ, von Figura K, Gieselmann V (1990) The structure of the arylsulfatase A gene. Eur J Biochem 191:627–631
79. Gieselmann V, Zlotogora J, Harris A et al. (1994) Molecular genetics of metachromatic leukodystrophy. Hum Mut 4:233–242
80. Polten A, Fluharty AL, Fluharty CB, Kappler J, von Figura K, Gieselmann V (1991) Molecular basis of different forms of metachromatic leukodystrophy. N Engl J Med 324:18–22
81. Berger J, Loschl B, Bernheimer H, Lugowska A, Tylki-Szymanska A, Gieselmann V, Molzer B (1997) Occurrence, distribution, and phenotype of arylsulfatase A mutations in patients with metachromatic leukodystrophy. Am J Med Genet 69:335–340
82. Draghia R, Letourneur F, Drugan C et al. (1997) Metachromatic leukodystrophy: identification of the first deletion in exon 1 and of nine novel point mutations in the arylsulfatase A gene. Hum Mutat 9:234–242
83. Zlotogora J, Bach G, Bosenberg C et al. (1995) Molecular basis of late infantile metachromatic leukodystrophy in the Habbanite Jews. Hum Mutat 5:137–143
84. Regis S, Filocamo M, Stroppiano M, Corsolini F, Gatti R (1996) Molecular analysis of the arylsulphatase A gene in late infantile metachromatic leukodystrophy patients and healthy subjects from Italy. J Med Genet 33:251–252
85. Harvey JS, Carey WF, Morris CP (1998) Importance of the glycosylation and polyadenylation variants in metachromatic leukodystrophy pseudodeficiency phenotype. Hum Mol Genet 7:1215–1219
86. Hagberg B, Kollberg H, Sourander P, kesson HO (1969) Infantile globoid cell leucodystrophy. A clinical and genetic study of 32 Swedish cases 1953–1967. Neuropadiatrie 1:74–88
87. Lake BD (1968) Segmental demyelination of peripheral nerves in Krabbe's disease. Nature 217:171–12l
88. Loonen MCB, van Diggelen OP, Janse H, Kleijer WJ, Arts WFM (1985) Late-onset globoid cell leucodystrophy (Krabbe's disease) Clinical and genetic delineation of two forms and their relation to the early-infantile form. Neuropediatrics 16:137–142
89. Suzuki K (1998) Twenty five years of the "psychosine hypothesis": a personal perspective of its history and present status. Neurochem Res 23:251–259
90. Baram TZ, Goldman AM, Percy AK (1986) Krabbe disease: specific MRI and CT findings. Neurology 36:111–115

91. Barone R, Brühl K, Stoeter P et al. (1996) Clinical and neuroradiological findings in classic infantile and late-onset globoid-cell leukodystrophy (Krabbe disease). Am J Med Genet 63:209–217
92. Krivit W, Shapiro EG, Peters C et al. (1998) Hematopoietic stem-cell transplantation in globoid-cell leukodystrophy. N Engl J Med 338:1119–1126
93. Wenger DA, Rafi MA, Luzi P (1997) Molecular genetics of Krabbe disease (globoid cell leukodystrophy): diagnostic and clinical implications. Hum Mutat 10:268–279
94. Kleijer WJ, Keulemans JL, van der Kraan M et al. (1997) Prevalent mutations in the GALC gene of patients with Krabbe disease of Dutch and other European origin. J Inherit Metab Dis 20:587–594

# CHAPTER 36

## Mucopolysaccharides

Mucopolysaccharides (now preferentially called glycosaminoglycans) are essential constituents of connective tissue, including cartilage and vessel walls. They are composed of long sugar chains containing highly sulfated, alternating uronic acid and hexosamine residues assembled into repeating units. The polysaccharide chains are bound to specific core proteins within complex macromolecules called proteoglycans. Depending on the composition of the repeating units, several mucopolysaccharides are known (Fig. 36.1). Their degradation takes place inside the lysosomes and requires several acid hydrolases. Deficiencies of specific degradative enzymes have been found to be the cause of a variety of eponymous disorders collectively termed mucopolysaccharidoses.

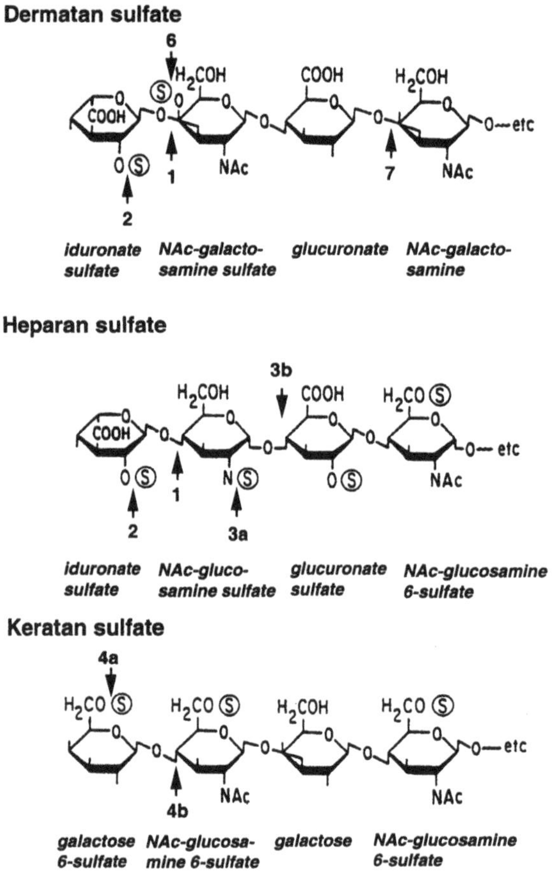

**Fig. 36.1.** Main repeating units in mucopolysaccharides and locations of the enzyme defects in the mucopolysaccharidoses. *MPS*, mucopolysaccharidosis; *NAc*, *N*-acetyl; *S*, sulfate; 1, α-Iduronidase (MPS-I; Hurler and Scheie disease). 2, Iduronate sulfatase (MPS-II; Hunter disease). 3a, Heparan *N*-sulfatase (MPS-IIIa; Sanfilippo-A disease). 3b, α-*N*-acetylglucosaminidase (MPS-IIIb; Sanfilippo-B disease). 4a, Galactose-6-sulfatase (MPS-IVa; Morquio-A disease). 4b, β-Galactosidase (MPS-IVb; Morquio-B disease). 6, NAc-galactosamine 4-sulfatase (MPS-VI; Maroteaux-Lamy disease). 7, β-Glucuronidase (MPS-VII; Sly disease)

# CHAPTER 36

# Mucopolysaccharidoses and Oligosaccharidoses

M. Beck

## CONTENTS

| | |
|---|---|
| The Mucopolysaccharidoses. . . . . . . . . . . . . . . . . . . . . . . . | 415 |
| Clinical Presentation . . . . . . . . . . . . . . . . . . . . . . . . . . . | 415 |
| The Oligosaccharidoses . . . . . . . . . . . . . . . . . . . . . . . . . . | 417 |
| Clinical Presentation . . . . . . . . . . . . . . . . . . . . . . . . . . . | 417 |
| Mucopolysaccharidoses and Oligosaccharidoses . . . . . . . . | 417 |
| Metabolic Derangement . . . . . . . . . . . . . . . . . . . . . . . . | 417 |
| Sialic Acid Storage Disorders . . . . . . . . . . . . . . . . . . . | 419 |
| Diagnostic Tests . . . . . . . . . . . . . . . . . . . . . . . . . . . . . | 419 |
| Treatment and Prognosis . . . . . . . . . . . . . . . . . . . . . . . | 420 |
| Genetics. . . . . . . . . . . . . . . . . . . . . . . . . . . . . . . . . . . . . | 420 |
| References . . . . . . . . . . . . . . . . . . . . . . . . . . . . . . . . . . . . | 420 |

Genetic defects of enzymes that are involved in the intralysosomal degradation of mucopolysaccharides (Fig. 36.1) and oligosaccharides lead to chronic and progressive storage disorders that share many clinical features varying from facial dysmorphism, bone and joint dysplasias, corneal clouding, hepatosplenomegaly, dwarfism, neurological abnormalities, mental regression and a reduced life span to an almost normal phenotype and life expectancy. Mucopolysaccharidoses (MPS) and oligosaccharidoses are transmitted in an autosomal-recessive manner, except for the X-linked MPS-II. Diagnosis of these lysosomal storage disorders can be made by enzyme assays in serum, leukocytes or fibroblasts. Partially degraded mucopolysaccharides and oligosaccharide fragments are excreted in the urine. Prenatal diagnosis is possible. As yet, treatment is mostly supportive. In some cases (especially in MPS-I), bone-marrow transplantation may improve the clinical course, but the value of this procedure is limited by the high risk and uncertain long-term neurological outcome. In the future, recombinant enzymes produced by genetically engineered cell lines may be available for MPS patients, at least for those in whom the central nervous system is not involved. Other therapeutic strategies, such as gene transfer into hematopoietic cells, are under consideration.

## The Mucopolysaccharidoses

### Clinical Presentation

Deficiency of the lysosomal enzyme α-L-iduronidase is the cause of MPS-I. The spectrum of clinical phenotypes ranges from the severe form (Hurler disease) to the adult variant (Scheie disease). *Hurler disease* represents the prototype of all MPS: in the first year of life, a delay in motor and mental development and frequent ear, nose and throat infections are observed. Thereafter, the patient develops the full clinical picture, with coarse facial features, macrocephaly, thick skin and corneal clouding (Fig. 36.4). The liver and spleen are enlarged. The bone dysplasia leads to growth retardation and severe deformities of the trunk and extremities. Radiological examination of the skeleton reveals a characteristic pattern called dysostosis multiplex. Its major features are a large skull with a deep, elongated sella, deformed, hook-shaped lower thoracic and lumbar vertebrae, pelvic dysplasia and shortened tubular bones (Figs. 36.5, 36.6). Patients with the severe form of MPS-I die within the first two decades of life.

Patients with *Scheie disease* are of almost normal height and do not show mental retardation. Typical symptoms are stiff joints, corneal opacities, carpal-tunnel syndrome and mild skeletal changes. In most cases, the heart and vascular system are involved in the storage process. Mitral and aortic regurgitation (caused by thickening of the valves) together with chronic respiratory infections results in acute or chronic cardiac failure. Some patients with α-iduronidase deficiency exhibit symptoms that are intermediate between those of Hurler and Scheie diseases. The different phenotypes are caused by allelic mutations of the α-iduronidase gene (*IDUA*) and various states of double heterozygosity.

*Hunter disease* (MPS-II) differs from all other known MPS by an X-linked mode of inheritance. It is caused by a deficiency of iduronate sulfatase. The clinical picture resembles that of Hurler disease, but corneal clouding is absent. There are Hunter patients who have only moderate or no mental retardation

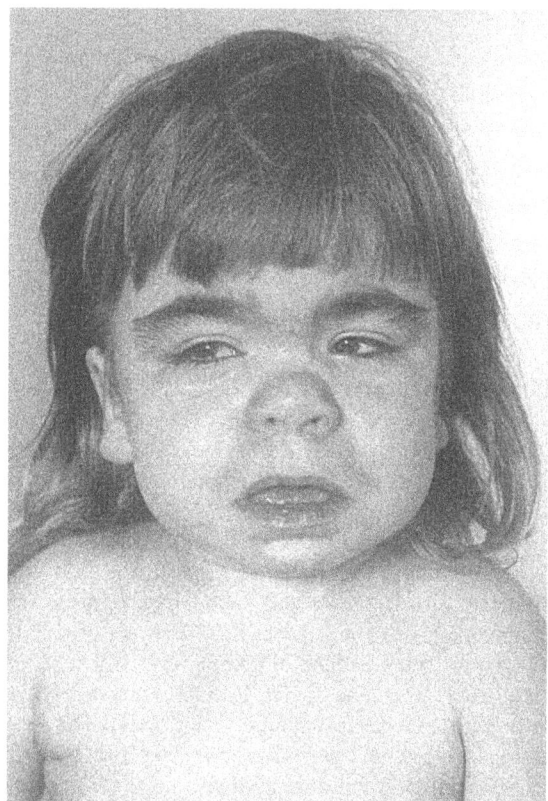

**Fig. 36.4.** A 4-year-old girl with MPS-I (Hurler disease). Coarse facial features, depressed nasal bridge, macroglossia

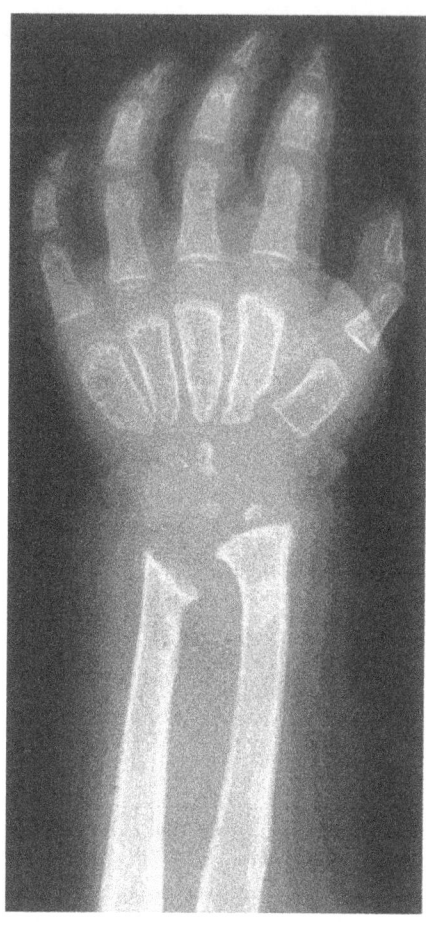

**Fig. 36.5.** Dysostosis multiplex (Hurler disease, 4-year-old patient). Short, deformed metacarpals and phalanges. The ulna and radius are slanted towards each other

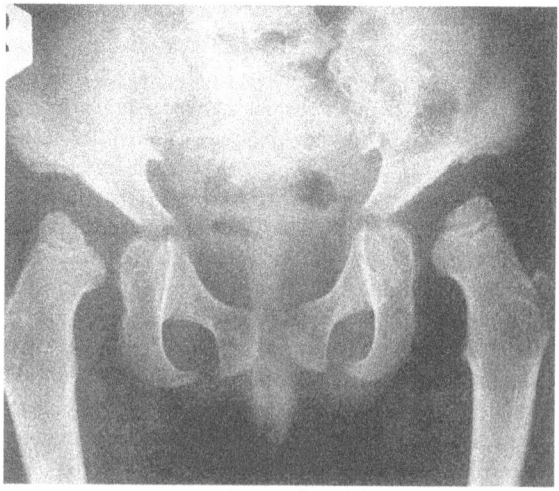

**Fig. 36.6.** Dysostosis multiplex. Hypoplastic basilar portions of the iliac bones, small capital femoral epiphyses

and a higher life expectancy (Hunter disease type B). The severe type A and the mild type B represent extremes of a broad phenotypic spectrum expressing numerous allelic mutations of the iduronate sulfatase gene (*IDS*) [1].

The cause of all four subtypes of MPS-III (*Sanfilippo disease*) is a disturbance in the intralysosomal degradation of heparan sulfate (Fig. 36.1). The clinical appearance is dominated by central-nervous-system symptoms, whereas the somatic features are moderate. After normal development in the first three or four years of life, patients with Sanfilippo disease lose acquired skills, such as speech, normal gait and toilet training. Later, aggressive behavior and hyperactivity occur. The hair is coarse, often blond, and hirsutism may be present. Bone changes are mild. MPS-III patients are of normal height, sometimes have seizures and become tetraspastic in the first or second decade of life; they usually die from aspiration pneumonia.

*Morquio disease* (MPS-IV) is caused by a genetic defect in the degradation of keratan sulfate. Patients with *Morquio disease type A* (*N*-acetyl-galactosamine-6-sulfatase deficiency) are characterized by disproportionate dwarfism, joint contractures, kyphoscoliosis and corneal clouding. In *Morquio disease type B*, β-galactosidase activity is absent, and the course is milder. An identical enzyme defect is responsible for

GM$_1$-gangliosidosis, a neurodegenerative disorder with predominant involvement of the central nervous system.

Patients with MPS-VI (*Maroteaux-Lamy disease*) have somatic features resembling those of Hurler patients but without neurological impairment. *Sly disease* (MPS-VII) shows the broadest phenotypic spectrum of MPS, ranging from lethal hydrops fetalis to almost normal individuals. Most MPS-VII patients exhibit symptoms almost identical to those of Hurler disease, such as coarse facial features, dysostosis multiplex, and hepatosplenomegaly.

## The Oligosaccharidoses

### Clinical Presentation

Oligosaccharidoses or glycoprotein-storage disorders clinically resemble MPS, but urinary mucopolysaccharide excretion is normal. In α-*mannosidosis*, mild skeletal deformities, coarse facial features and moderate to marked mental retardation are present. Patients are often deaf and have an increased susceptibility to infections. Only a few patients with β-*mannosidosis* have been reported. All of them showed a relatively mild course without major skeletal lesions or facial dysmorphism but with prominent hearing loss. Swallowing difficulties seem to be a characteristic symptom, too [2]. Clinical symptoms of *fucosidosis* include progressive neurological degeneration and mental impairment, seizures, angiokeratoma and mild skeletal dysplasia. A wide, continuous spectrum of phenotypes reflects different mutations of the α-fucosidosis gene [3].

The clinical spectrum of *sialidosis* (neuraminidase deficiency) ranges from a congenital form with hydrops fetalis to the comparatively mild cherry-red spot myoclonus syndrome, which is characterized by a life expectancy of over 30 years of age. In *galactosialidosis*, two lysosomal enzymes (β-galactosidase and neuraminidase) are deficient due to a malfunction of a common protective protein with cathepsin-A like activity. Patients with the early infantile form have symptoms similar to those with GM$_1$-gangliosidosis and sialidosis; they are hypotonic from birth on, with facial edema, coarse features and cherry-red retinal spots. In juvenile galactosialidosis, skeletal dysplasia, corneal opacities and mental retardation are the leading symptoms.

*Aspartylglucosaminuria* has a high prevalence in Finland and is rare in other countries. Clinical findings include progressive psychomotor retardation, coarse facial features and mild skeletal dysplasia.

*Sialic storage diseases* are characterized by abnormal accumulation of free (unbound) sialic acid in lysosomes of different tissues. The two main phenotypes are the "Finnish" Salla disease, presenting with mental retardation, ataxia and near-normal life span, and an infantile form that presents with severe visceral involvement, skeletal dysplasia, psychomotor retardation and early death. Both clinical forms have been observed both in Finland and in other ethnic groups [4].

The two original patients with α-*N*-acetyl-galactosaminidase (α-NAGA) deficiency (*Schindler disease*) had progressive psychomotor deterioration, blindness and seizures [5]. Since the first report, only a few patients with α-NAGA deficiency have been described. There is a broad clinical heterogeneity among these patients; at one end of the clinical spectrum are patients who exhibit symptoms of neuroaxonal dystrophy, whereas mildly affected patients have angiokeratoma without any overt neurological abnormality [6].

Patients with *mucolipidosis II* ("I-cell disease") resemble those with severe Hurler disease; they rarely survive the first decade of life [7]. Symptoms of *mucolipidosis III* are less severe, with variable dysostosis multiplex, growth retardation and joint stiffness. Mental retardation is moderate or may be absent.

## Mucopolysaccharidoses and Oligosaccharidoses

### Metabolic Derangement

The deficient activity of one or more lysosomal enzymes (Tables 36.1, 36.2) leads to the accumulation of the substrate mucopolysaccharides or oligosaccharides. The non-degraded or partially degraded substances are stored in lysosomes and excreted in urine. The storage process becomes visible in blood smears that show abnormal cytoplasmic inclusions in lymphocytes; storage cells are also detectable in bone marrow. The clinical phenotype partially depends on the type and amount of storage substance. For example, the accumulation of heparan sulfate, an essential component of nerve cell membranes, results in progressive mental deterioration (MPS-III, Sanfilippo disease) whereas, in Morquio disease (MPS-IV), a defect in keratan sulfate degradation causes severe skeletal deformities. Other symptoms, such as organomegaly or coarse face, are non-specific storage phenomena.

The primary defects in most of the lysosomal storage disorders are mutations of genes that encode single lysosomal enzymes. Multiple lysosomal enzyme deficiencies in mucolipidosis II and mucolipidosis III result from post-translational defects; after their synthesis in the rough endoplasmic reticulum, lysosomal

## Oligosaccharides/Glycoproteins

Almost all the secreted and membrane-associated proteins of the body are glycosylated, as are numerous intracellular proteins, including the lysosomal acid-hydrolases. A great variety of oligosaccharide chains is attached to the protein backbone via the hydroxyl group of serine or threonine (O-linked) or via the amide group of asparagine (N-linked) to form tree-like structures (Fig. 36.2). The chains usually have a core composed of N-acetylglucosamine and mannose, often contain galactose, fucose and N-acetylgalactosamine and frequently possess terminal sialic acids (N-acetylneuraminic acid). Oligosaccharide chains with a terminal mannose-6-phosphate are involved in the targeting of lysosomal enzymes to lysosomes. This recognition marker is synthesized in two steps from uridine diphosphate-N-acetylglucosamine (Fig. 36.3). Deficiencies of the enzymes required for the degradation of the oligosaccharide chains cause glycoprotein-storage diseases. Defects of the synthesis of the mannose-6-phosphate recognition marker result in the mislocalization of lysosomal enzymes. Defects of the synthesis of the oligosaccharide chains are discussed in Chap. 38 (congenital defects of glycosylation).

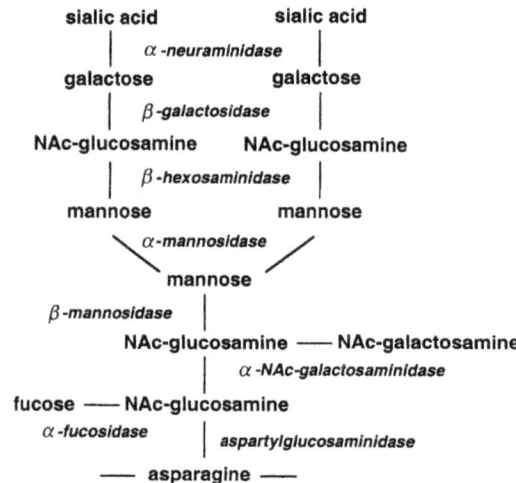

**Fig. 36.2.** General composite example of a glycoprotein oligosaccharide chain. *NAc*, N-acetyl. Degradative enzymes are listed in *italics*

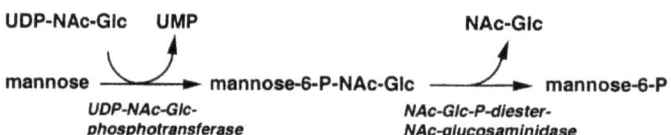

**Fig. 36.3.** Synthesis of the mannose-6-phosphate recognition marker. *NAc-Glc*, N-acetylglucosamine; *UDP*, uridine diphosphate; *UMP*, uridine monophosphate. Enzymes are listed in *italics*

**Table 36.1.** Mucopolysaccharidoses: some diagnostic data [23–35]

| Disorder | Eponym | Defective enzyme | Assay |
|---|---|---|---|
| MPS-I-H | Hurler | α-Iduronidase | L, F, AC, CV |
| MPS-I-S | Scheie | α-Iduronidase | L, F, AC, CV |
| MPS-I-H/S | Variants | α-Iduronidase | L, F, AC, CV |
| MPS-II | Hunter | Iduronate-2-sulfatase | S, F, AF, AC |
| MPS-III-A | Sanfilippo A | Sulfamidase | L, F, AC, CV |
| MPS-III-B | Sanfilippo B | α-N-acetylglucosaminidase | S, F, AC, CV |
| MPS-III-C | Sanfilippo C | AcCoA:α-glucosaminide-N-Ac-transferase | F, AC |
| MPS-III-D | Sanfilippo D | N-Acetylglucosamine-6-sulfatase | F, AC |
| MPS-IV-A | Morquio A | Galactosamine-6-sulphatase | L, F, AC |
| MPS-IV-B | Morquio B | β-Galactosidase | L, F, AC, CV |
| MPS-VI | Maroteaux-Lamy | N-Acetylgalactosamine-4-sulfatase | L, F, AC |
| MPS-VII | Sly | β-Glucuronidase | S, F, AC |

*Ac*, acetyl; *AC*, cultured amniotic cells; *AcCoA*, acetyl coenzyme A; *AF*, amniotic fluid; *CV*, chorionic villi; *F*, cultured fibroblasts; *L*, leucocytes; *S*, serum

**Table 36.2.** Oligosaccharidoses: some diagnostic data [23–35]

| Disorder | Defective enzyme | Assay |
|---|---|---|
| α-Mannosidosis | α-Mannosidase | S, L, F, AC |
| β-Mannosidosis | β-Mannosidase | S, L, F, AC |
| α-Fucosidosis | α-Fucosidase | S, L, F, AC |
| Sialidosis, type I | α-Neuraminidase | F, AC |
| Sialidosis, type II | α-Neuraminidase | F, AC |
| Sialidosis, congenital | α-Neuraminidase | F, AC |
| Galactosialidosis | "Protective protein" | F, AC |
| Aspartylglucosaminuria | Aspartylglycosaminidase | F, AC |
| Schindler disease | α-N-Ac-galactosaminidase | F |
| Free neuraminic storage disorder | "Proton-driven carrier" | F |
| Mucolipidosis-II (I-cell disease) | N-acetylglucosamine-1-phosphotransferase | S, F, AF, AC |
| Mucolipidosis-III (pseudo-Hurler) | N-acetylglucosamine-1-phosphotransferase | S, F, AF, AC |

*Ac*, acetyl; *AC*, cultured amniotic cells; *AF*, amniotic fluid; *CV*, chorionic villi; *F*, cultured fibroblasts; *L*, leucocytes; *S*, serum

enzymes are subjected to a series of post-translational modifications. Thus, N-acetylglucosamine-1-phosphotransferase transfers phosphate groups onto oligosaccharide units of lysosomal enzyme precursors. In its absence, the common phosphomannosyl recognition marker of acid hydrolases is not generated, and multiple enzymes are not targeted to the lysosomes. As a consequence, the enzymes are secreted in the extracellular space, where high activities are found; inside the cells, the enzyme levels are considerably reduced.

The combined neur⁻/β-gal⁻ defect is caused by lack of a 32-kDa glycoprotein ("protective protein") that is responsible for the stabilization of an α-neuraminidase/β-galactosidase complex. The "protective protein" is identical to a multifunctional protein that exhibits cathepsin-A activity [8].

### Sialic Acid Storage Disorders

Sialic acid storage disorders are produced by a defect of a proton-driven carrier that is responsible for the efflux of sialic acid (and other acidic monosaccharides) from the lysosomal compartment. Because the carrier has a wide substrate affinity, storage of different compounds may be involved in the pathogenesis of these disorders.

### Diagnostic Tests

Because most patients excrete increased amounts of mucopolysaccharides (glycosaminoglycans) or oligosaccharides, analysis of urine for the presence of these substances is recommended as the first diagnostic step [9]. For the qualitative and quantitative measurement of glycosaminoglycans, many methods have been described; spot tests are quick and inexpensive but may produce false negative and false positive results. In particular, MPS-III and MPS-IV are often missed. For quantitative measurement of urinary glycosaminoglycans, several tests are available (determination of uronic acid based on the carbazol method or the spectrophotometric assay using the dye dimethylmethylene blue). One- or two-dimensional electrophoresis allows discrimination between classes of glycosaminoglycans (chondroitin, heparan, keratan and dermatan sulfate). To detect abnormal urinary excretion of oligosaccharides, thin-layer chromatography is available. Free sialic acid is excreted in increased amounts in free-sialic-acid-storage disorders and can be determined in urine by thin-layer chromatography or spectrophotometric assay [10].

All urinary tests can give false negative results, especially in older patients with mild clinical manifestation. Patients with mucolipidosis II and mucolipidosis III are usually missed. Therefore, the definitive diagnosis has to be confirmed by enzymatic assay in serum, leukocytes and/or fibroblasts [11].

## Treatment and Prognosis

Presently, only palliative treatment is available. However, supportive management of behavioral disturbances can improve the patient's quality of life and the patient–parent relationships (Chap. 4). Examples are the management of sleep disturbances in MPS-III by sedative drugs (chlorpromazine, haloperidol, melatonin) or behavioral intervention (Chap. 4) [12]. To treat hyperactivity, methylphenidate may be helpful. Corneal transplantation may become necessary in MPS-I (Scheie disease) and MPS-VI. In patients with MPS-IV (Morquio disease), a cervical fusion must be performed to prevent atlantoaxial subluxation. A release operation is recommended in carpal tunnel syndrome. Hydrocephalus, observed in some cases of MPS-I (Hurler disease) has to be relieved by a shunt operation.

Bone marrow has been transplanted into a great number of patients with various lysosomal storage disorders [13]. After successful engraftment, leukocyte enzyme activity normalized, organomegaly decreased and joint mobility increased. Skeletal abnormalities remained unchanged. Whether brain function can be improved in patients with mental retardation remains questionable. Some patients maintained their learning ability or intelligence quotient, while others continued to deteriorate. Broad application of this therapy is limited by the high morbidity and mortality of the procedure and the low availability of compatible donors and transplantation facilities.

Correction of the metabolic defect by the administration of the missing enzyme has been successful in animal models [14]. Clinical trials with enzyme replacement have been initiated in MPS-I [15], but long-term results have not been obtained. Gene transfer using retroviral vectors corrected the metabolic defect in animal and human MPS-VI cells [16]. Identical results have been obtained in bone-marrow cells of MPS-I [17] and fibroblasts of MPS-III patients [18]. Ohashi et al. studied adenovirus-mediated transfer of the human β-glucuronidase gene in MPS-VII mice; the abnormalities of the liver and spleen improved, whereas β-glucuronidase activity in the kidney and brain was not significantly increased [19].

## Genetics

With the exception of MPS-II (Hunter disease) and Fabry disease (α-galactosidase deficiency), the lysosomal storage disorders are inherited in an autosomal-recessive manner. The genes coding for most of the lysosomal enzymes have been localized. Numerous allelic mutations have been detected, explaining the clinical variability of these disorders. Gene analysis enables the reliable identification of carriers in Hunter disease [20]. Prenatal diagnosis is possible in all lysosomal storage disorders by measurement of enzyme activity and/or gene analysis in chorionic villi, cultured trophoblasts or cultured amniotic cells [21].

For MPS, the total incidence of all types has been estimated to be one in 25,000. A summary of published incidence data is given by Nelson [22].

## References

1. Froissart R, Maire I, Millat G et al. (1998) Identification of iduronate sulfatase gene alterations in 70 unrelated Hunter patients. Clin Genet 53:362–368
2. Gourrier E, Thomas MP, Munnich A et al. (1997) β-Mannosidose: une nouvelle observation. Arch Pediatr 4:147–151
3. Cragg H, Williamson M, Young E et al. (1997) Fucosidosis: genetic and biochemical analysis of eight cases. J Med Genet 34:105–110
4. Sewell AC, Poets CF, Degen I, Stoss H, Pontz BF (1996) The spectrum of free neuraminic acid storage disease in childhood: clinical, morphological and biochemical observations in three non-Finnish patients. Am J Med Genet 63:203–208
5. Van Diggelen OP, Schindler D, Kleijer WJ et al. (1987) Lysosomal α-N-acetylgalactosaminidase deficiency: a new inherited metabolic disorder. Lancet 2:804
6. Keulemans JL, Reuser AJ, Kroos MA et al. (1996) Human α-N-acetylgalactosaminidase (α-NAGA) deficiency: new mutations and the paradox between genotype and phenotype. J Med Genet 33:458–464
7. Beck M, Barone R, Hoffmann R et al. (1995) Inter- and intrafamilial variability in mucolipidosis II (I-cell disease). Clin Genet 47:191–199
8. van der Spoel A, Bonten E, Azzo A (1998) Transport of human lysosomal neuraminidase to mature lysosomes requires protective protein/cathepsin A. EMBO J 17:1588–1597
9. Sewell AC, Gehler J, Spranger J (1979) Comprehensive urinary screening for inborn errors of complex carbohydrate metabolism. Klin Wochenschr 57:581–585
10. Renlund M (1984) Clinical and laboratory diagnosis of Salla disease in infancy and childhood. J Pediatr 104:232–236
11. Thompson JN, Nowakowski RW (1991) Enzymatic diagnosis of selected mucopolysacharidoses. In: Hommes FA (ed) Techniques in diagnostic human biochemical genetics. Wiley-Liss, New York, pp 567–586
12. Colville GA, Watters JP, Yule W, Bax M (1996) Sleep problems in children with Sanfilippo syndrome. Dev Med Child Neurol 38:538–544
13. Hoogerbrugge PM, Brouwer OF, Bordigoni P et al. (1995) Allogeneic bone marrow transplantation for lysosomal storage diseases. Lancet 345:1398–1402
14. Crawley AC, Niedzielski KH, Isaac EL et al. (1997) Enzyme replacement therapy from birth in a feline model of mucopolysaccharidosis type VI. J Clin Invest 99:651–662
15. Kakkis E, Muenzer J, Tiller G et al. (1998) Recombinant α-L-iduronidase replacement therapy in mucopolysaccharidosis I:

results of a human clinical trial. Annual meeting of the Society for Human Genetics, Denver, Oct 1998, abstract 128
16. Fillat C, Simonaro CM, Yeyati PL et al. (1996) Arylsulfatase B activities and glycosaminoglycan levels in retrovirally transduced mucopolysaccharidosis type VI cells. Prospects for gene therapy. J Clin Invest 98:497-502
17. Fairbairn LJ, Lashford LS, Spooncer E et al. (1996) Long-term in vitro correction of alpha-L-iduronidase deficiency (Hurler syndrome) in human bone marrow. Proc Natl Acad Sci USA 93:2025-2030
18. Bielicki J, Hopwood JJ, Anson DS (1996) Correction of Sanfilippo A skin fibroblasts by retroviral vector-mediated gene transfer. Hum Gene Ther 7:1965-1970
19. Ohashi T, Watabe K, Uehara K et al. (1997) Adenovirus-mediated gene transfer and expression of human β-glucuronidase gene in the liver, spleen, and central nervous system in mucopolysaccharidosis type VII mice. Proc Natl Acad Sci USA 94:1287-1292
20. Gal A, Beck M, Sewell AC et al. (1992) Gene diagnosis and carrier detection in Hunter syndrome by the iduronate-2-sulphatase cDNA probe. J Inherit Metab Dis 15:342-346
21. Wenger DA, Williams C (1991) Screening for lysosomal disorders. In: Hommes FA (ed) Techniques in diagnostic human biochemical genetics. Wiley-Liss, New York, pp 587-617
22. Nelson J (1997) Incidence of the mucopolysaccharidoses in Northern Ireland. Hum Genet 101:355-358
23. Stirling JL, Robinson D, Fensom AH, Benson PF, Baker JE (1978) Fluorimetric assay for prenatal detection of Hurler and Scheie homozygotes or heterozygotes (letter). Lancet 1:147
24. Hopwood JJ (1979) alpha-L-iduronidase, beta-D-glucuronidase, and 2-sulfo-L-iduronate-2-sulfatase: preparation and characterization of radioactive substrates from heparin. Carbohydr Res 69:203-216
25. Karpova EA, Voznyi Ya V, Keulemans JL et al. (1996) A fluorimetric enzyme assay for the diagnosis of Sanfilippo disease type A (MPS-IIIA). J Inherit Metab Dis 19:278-285
26. Marsh J, Fensom AH (1985) 4-Methylumbelliferyl α-N-acetylglucosaminidase activity for diagnosis of Sanfilippo B disease. Clin Genet 27:258-262
27. He W, Voznyi Ya V, Huijmans JG et al. (1994) Prenatal diagnosis of Sanfilippo disease type C using a simple fluorometric enzyme assay. Prenat Diagn 14:17-22
28. He W, Voznyi Ya V, Boer AM, Kleijer WJ, van Diggelen OP (1993) A fluorimetric enzyme assay for the diagnosis of Sanfilippo disease type D (MPS-IIID). J Inherit Metab Dis 16:935-941
29. Yuen M, Fensom AH (1985) Diagnosis of classical Morquio's disease: N-acetylgalactosamine 6-sulphate sulphatase activity in cultured fibroblasts, leukocytes, amniotic cells and chorionic villi. J Inherit Metab Dis 8:80-86
30. Wenger DA, Sattler M, Clark C, Wharton C (1976) I-cell disease: activities of lysosomal enzymes toward natural and synthetic substrates. Life Sci 19:413-420
31. Baum H, Dodgson KS, Spencer B (1959) The assay of arylsulfatase A and B in human urine. Clin Chim Acta 4:453-455
32. Glaser JH, Sly WS (1973) Beta-glucuronidase deficiency mucopolysaccharidosis: methods for enzymatic diagnosis. J Lab Clin Med 82:969-977
33. Panday RS, van Diggelen OP, Kleijer WJ, Niermeijer MF (1984) β-Mannosidase in human leukocytes and fibroblasts. J Inherit Metab Dis 7:155-156
34. Beck M, Scheuring E, Voelter HU, Brandt J, Harzer K (1996) Neuraminidase assay in cultured human fibroblasts: in situ versus in vitro procedures. Clin Chim Acta 251:163-171
35. Voznyi Ya V, Keulemans JL, Kleijer WJ et al. (1993) Applications of a new fluorimetric enzyme assay for the diagnosis of aspartylglucosaminuria. J Inherit Metab Dis 16:929-934

## Peroxisomal Functions

Peroxisomes are cell organelles that derive their name from the presence of catalase, which converts hydrogen peroxide into oxygen and water. Like lysosomes, they are found in all human cells except erythrocytes; however, unlike lysosomes, they possess both anabolic and catabolic functions. Peroxisomes are mainly involved in lipid metabolism: they synthesize ether phospholipids called plasmalogens, which are important constituents of cell membranes and myelin; they β-oxidize very long-chain fatty acids; and, through the mevalonate pathway, they are involved in the formation of cholesterol, bile acids and their derivatives dolichol and ubiquinone. Peroxisomes are also involved in the oxidation of phytanic acid, a chlorophyl derivative. They also intervene (not shown in Fig. 37.1) in the catabolism of lysine via pipecolic acid and glutaric acid and of glyoxylate.

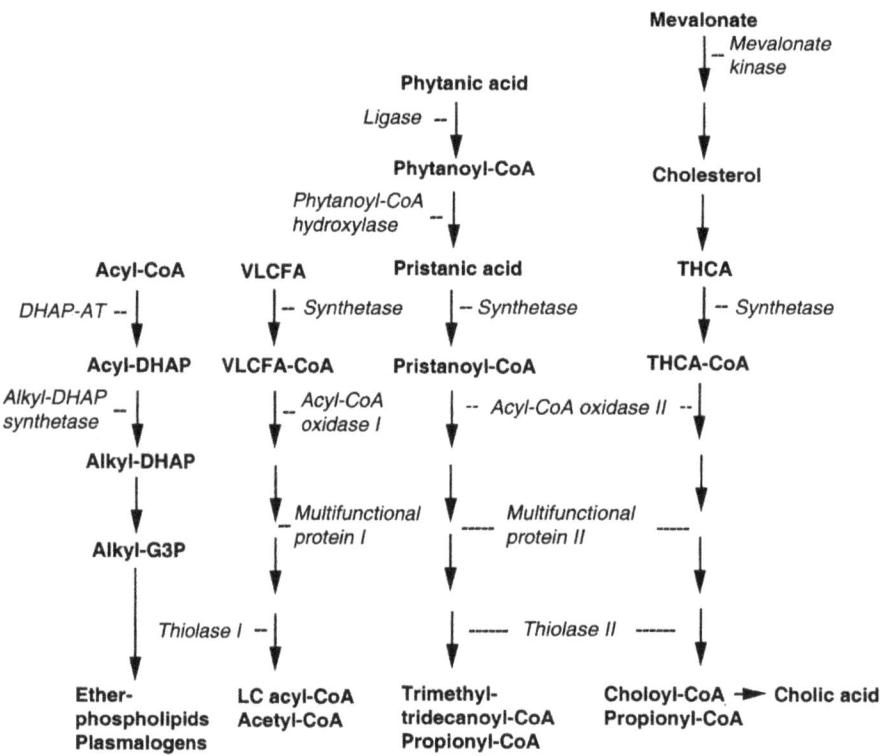

Fig. 37.1. Schematic representation of the main peroxisomal functions. *alkyl-G3P*, alkyl glycerol-3-phosphate; *DHAP*, dihydroxyacetone phosphate; *DHAPAT*, dihydroxyacetone phosphate acyltransferase; *LC*, long chain; *THCA*, trihydroxycholestanoic acid; *VLCFA-CoA*, very-long-chain fatty acid-coenzyme A

# CHAPTER 37

# Peroxisomal Disorders

B.T. Poll-The and J.-M. Saudubray

CONTENTS

Clinical Presentation ........................... 423
    The Neonatal Period........................ 424
    First Six Months of Life..................... 425
    Between Six Months and Four Years ............ 426
    Beyond Four Years of Age .................... 426
Metabolic Derangements ....................... 426
Diagnostic Tests............................. 428
Treatment and Prognosis....................... 429
Genetics.................................... 430
References .................................. 430

---

Peroxisomal disorders can be recognized by the presence of dysmorphias, neurologic abnormalities and hepato-intestinal dysfunction. Widely different features are the following:

- Craniofacial abnormalities, skeletal dysmorphias, shortened proximal limbs, calcific stippling of the epiphyses
- Encephalopathy, fits, peripheral neuropathy, abnormal gait, hypotonia
- Ocular abnormalities such as retinopathy, optic-nerve dysplasia, cataracts
- Hepato-intestinal dysfunction, hepatomegaly and cholestasis

Possibilities for (dietary) treatment are limited.

---

## Clinical Presentation

At least 21 clinically and biochemically markedly heterogeneous disorders linked to peroxisomal dysfunction have been identified (Fig. 37.1) [1, 2]. In general, the onset of the symptoms is not accompanied by an acute event or by abnormal routine laboratory tests indicating metabolic derangement. Most often, the presentation is associated either with chronic encephalopathy beginning in infancy or early childhood or with progressive neurologic manifestations starting in the school-age period. Given the diversity of the clinical and biochemical abnormalities of peroxisomal disorders, it is easier to regard the clinical diagnosis as a function of both the age of the patient (Table 37.1) and the following predominant general features of recognition: dysmorphia, neurologic dysfunction and hepatodigestive manifestations [3].

POLYMALFORMATIVE SYNDROME AND DYSMORPHIA. Craniofacial abnormalities, including large fontanelles, high forehead, epicanthus and abnormal ears, may be mistaken for chromosomal aberrations, such as Down syndrome. They are frequently associated with other abnormalities: rhizomelic shortening of limbs in rhizomelic chondrodysplasia punctata (RCDP); stippled calcifications of epiphyses in RCDP and classical Zellweger syndrome (CZ); renal cysts in CZ; or abnormalities in neuronal migration (gyral abnormalities, neuronal heterotopias) and cerebral myelination. These congenital manifestations point to dysmorphogenesis during the prenatal period, as observed in some inborn errors affecting energy-producing pathways of the fetus and in metabolic dysfunctions of the mother during pregnancy [4].

NEUROLOGIC DYSFUNCTION. At birth, the predominant symptoms are often a severe hypotonia with areactivity (which can be mistaken for a neuromuscular disorder), a disorder of the CNS and autonomic nervous system, and malformation syndromes. An increasing number of inborn errors of metabolism without evident biochemical abnormalities on routine laboratory screening should also be considered in the diagnosis [4]. Severe axial hypotonia may be associated with seizures and neurologic distress with hypertonia of the limbs. It may be difficult to differentiate it from a mitochondrial respiratory chain disorder. An important difference is that peroxisomal disorders are not associated with an acute metabolic derangement or abnormal routine laboratory tests, such as metabolic acidosis or lacticaciduria.

HEPATODIGESTIVE MANIFESTATIONS. The predominant manifestations may be hepatomegaly, cholestasis, hyperbilirubinemia and prolonged jaundice, especially in isolated di- and trihydroxycholestanoic acidurias [5].

**Table 37.1.** Clinical symptoms of peroxisomal disorders related to age

| Symptoms | Disorder |
|---|---|
| **Neonatal period** | |
| Hypotonia, areactivity, seizures | ZS, ZS variants[a] |
| Craniofacial dysmorphia | Neonatal ALD |
| Skeletal abnormalities | Pseudo-neonatal ALD |
| Conjugated hyperbilirubinemia | Acyl-CoA oxidase deficiency, multifunctional enzyme deficiency, RCDP (typical/atypical), THC aciduria, pipecolic aciduria, mevalonic aciduria |
| **First 6 months of life** | |
| Failure to thrive | IRD, pseudo-IRD |
| Hepatomegaly, prolonged jaundice | Pipecolic aciduria, neonatal ALD, milder |
| Digestive problems, hypocholesterolemia | forms of ZS |
| Vitamin-E deficiency | Atypical chondrodysplasia |
| Visual abnormalities | Mevalonic aciduria |
| **Six months to 4 years** | |
| Failure to thrive | IRD, pseudo-IRD |
| Neurologic presentation | Pipecolic aciduria, neonatal |
| Psychomotor retardation | ALD, milder forms of ZS |
| Visual and hearing impairment (ERG, VEP, BAEP), | Atypical chondrodysplasia, THC aciduria |
| Osteoporosis | |
| **Beyond 4 years of age** | |
| Behavior changes, deterioration of intellectual functions, white-matter demyelination | X-linked ALD |
| Visual and hearing impairment | Classical Refsum disease |
| Peripheral neuropathy, gait abnormality | Atypical biogenesis defects |

*ALD*, adrenoleukodystrophy; *BAEP*, brain auditory-evoked potentials; *ERG*, electroretinogram; *IRD*, infantile Refsum disease; *RCDP*, rhizomelic chondrodysplasia punctata; *THC*, trihydroxycholestanoic; *VEP*, visual-evoked potentials; *ZS*, Zellweger syndrome
[a] ZS variants include ZS-like disease and pseudo-ZS (peroxisomal thiolase)

## The Neonatal Period

In order to facilitate the recognition of peroxisomal disorders, their clinical presentation is categorized according to sequential periods of life. Two prototypes of neonatal presentation are CZ, which is the most severe condition, and RCDP. Their phenotypes are distinct from the other disorders and should not cause difficulties in the differential diagnosis. Other disorders with less typical neonatal presentation are neonatal adrenoleukodystrophy (NALD), hyperpipecolic aciduria, di- and trihydroxycholestanoic acidurias and mevalonic aciduria.

CZ SYNDROME. *CZ syndrome* is characterized by:

- Errors of morphogenesis
- Severe neurological dysfunction, including generalized hypotonia and seizures
- Sensorineural hearing loss
- Ocular abnormalities
- Degenerative changes
- Hepatodigestive involvement with failure to thrive
- Absence of recognizable hepatic peroxisomes (presence of peroxisomal "ghosts")
- Death, usually in the first year

The patients show typical facial dysmorphia (Fig. 37.2a), which may become less characteristic if the patient survives beyond the first year of life. Although certain milestones develop, only some "older" CZ patients attain the ability to sit without support, and subsequently develop peripheral hypertonia.

*Classical Rhizomelic Chondrodysplasia Punctata.* Classical RCDP is characterized by the presence of shortened proximal limbs, facial dysmorphia, cataracts, psychomotor retardation, coronal clefts of vertebral bodies, and stippled foci of calcification of the epiphyses during infancy, which may disappear after the age of 2 years. The chondrodysplasia punctata is more widespread than in CZ and may involve extraskeletal tissues. Some patients have ichthyosis. Peroxisomal structures appear to be intact in fibroblasts whereas, in the liver, these organelles may be fewer or absent in some hepatocytes and may be enlarged in size in others [6]. A variant of chondrodysplasia punctata is associated with the characteristic peroxisomal defects observed in classical RCDP; however, it lacks the rhizomelic shortening of the limbs [7–9]. Conversely, patients were identified with the typical clinical phenotype of classical RCDP but with a single enzyme deficiency [10, 11]. Classical RCDP and its variants must be distinguished from other forms of chondrodysplasia punctata, such as Conradi-Hünermann syndrome (Chap. 29) and the X-linked dominant and recessive forms of chondrodysplasia punctata.

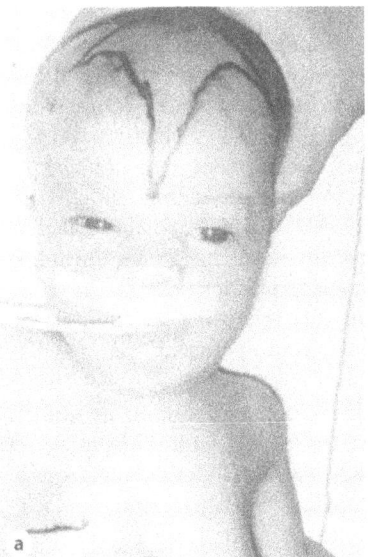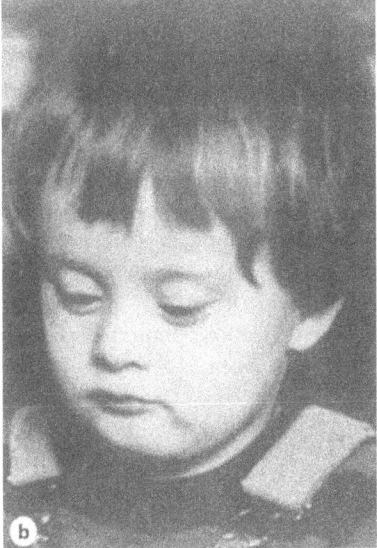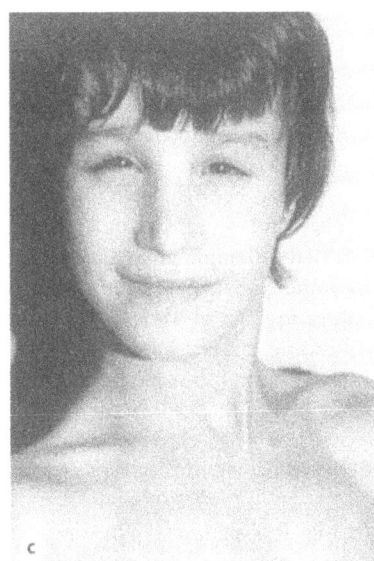

Fig. 37.2. Three patients with multiple enzyme defects and defective peroxisome assembly. a Classical Zellweger syndrome at 2 weeks of age. b Infantile Refsum disease at 2 years of age; note facial dysmorphism resembling Down syndrome. c Infantile Refsum disease at 11 years of age

*Neonatal Adrenoleukodystrophy.* NALD patients are somewhat less severely affected than CZ patients [12]. Facial dysmorphia is not always present, and patients may show some development before their progressive deterioration begins; this is usually followed by death before the age of 6 years. Some patients can mimic Werdnig-Hoffmann syndrome [13]. Cerebral demyelination is more prominent than dysmyelination and gray-matter heterotopia. Computed tomography (CT) scan of the brain may show abnormal contrast enhancement around demyelination areas. Chondrodysplasia punctata and renal cysts are absent. Patients with a single enzyme defect but with clinical manifestations resembling those of NALD [14], pseudo-NALD, acyl-coenzyme A (CoA)-oxidase deficiency [15], bi(tri)functional (multifunctional)-enzyme deficiency [16], deficiencies of the pseudo-Zellweger syndrome and peroxisomal-thiolase deficiency [17] have been described. Liver peroxisomes were normal (multifunctional-enzyme deficiency) or appeared to be enlarged in size (acyl-CoA oxidase and peroxisomal-thiolase deficiency) whereas, in CZ syndrome and NALD, they are morphologically absent or severely decreased in number.

*Hyperpipecolic Aciduria.* The term hyperpipecolic aciduria was assigned to patients exhibiting accumulation of pipecolic acid prior to the discovery of the generalized peroxisomal defects. However, hyperpipecolic aciduria should only be assigned to patients with elevated pipecolic acid values in body fluids but no other symptoms. Hyperpipecolic aciduria associated with a Joubert syndrome has been observed in three siblings.

*Di- and Trihydroxycholestanoic Acidurias.* Di- and trihydroxycholestanoic acidurias have been reported in patients with predominantly hepatic manifestations associated with neurologic involvement [5].

*Mevalonic Aciduria.* Mevalonic aciduria is a disorder characterized by dysmorphic features and cataracts and should probably be considered a peroxisomal disorder, because mevalonate kinase is predominantly localized in peroxisomes (Chap. 29) [18, 19].

### First Six Months of Life

During this period of life, the predominant symptoms may be hepatomegaly associated (or not associated) with prolonged jaundice, liver failure and nonspecific digestive problems (anorexia, vomiting, diarrhea) leading to failure to thrive and osteoporosis. Hypocholesterolemia, hypolipoproteinemia and decreased levels of fat-soluble vitamins, symptoms that resemble a malabsorption syndrome, are frequently present (Table 37.1). These patients can be erroneously diagnosed as having a congenital defect of glycosylation (CDG) (Chap. 38). Most CZ patients develop hepatomegaly and seizures and do not survive beyond this period.

*Infantile Refsum Disease.* Infantile Refsum disease (IRD) is similar to CZ biochemically and in its absence of (or its decreased number of) liver peroxisomes [12]. However, IRD patients differ clearly from CZ patients; IRD patients have a later onset of initial symptoms, less CNS involvement and longer survival. Little or no facial dysmorphia is noted in early childhood (Fig. 37.2b). Early developmental milestones are usually normal before slowing begins between the ages of 1 years and 3 years. This is followed by completely arrested development associated with autistic behavior in some patients. Most patients walk independently before the age of 3 years. A patient with pseudo-IRD exhibiting clinical similarity to IRD, but with somewhat different biochemical abnormalities has been described [20].

### Between Six Months and Four Years

During this period of life, severe psychomotor retardation becomes evident (Table 37.1). Sensorineural hearing loss is associated with abnormal brain-stem auditorily evoked responses. Various ocular abnormalities can be observed, including cataracts, retinitis pigmentosa, optic nerve atrophy, glaucoma and brushfield spots. The electroretinogram and visually evoked responses are frequently disturbed, and this may precede the fundoscopic abnormalities. Retinitis pigmentosa associated with hearing loss, developmental delay and dysmorphia may be mistaken for other diseases, including malformative syndromes [21]. In this respect, it has to be realized that the boundaries between malformative syndromes and inborn errors are not well delineated. This fact is confirmed by the recent finding of a defective cholesterol biosynthesis in Smith-Lemli-Opitz syndrome [22] and Conradi-Hünermann syndrome (Chap. 29). Most NALD patients do not survive beyond this period.

### Beyond Four Years of Age

*X-Linked Adrenoleukodystrophy/Adrenomyeloneuropathy* is the most common peroxisomal disorder. Considerable clinical variability exists even within the same family [23]. The childhood form is the most severe phenotype, with onset of neurologic involvement usually between 5–10 years of age, leading to a vegetative state and death in a few years. The affected males may present with school failure, attention-deficit disorder or behavior changes as first manifestations, followed by visual impairment and quadriplegia; seizures are usually a late symptom. Hypoglycemic episodes and a dark discoloration of the skin may reflect adrenal insufficiency, which may precede, coincide with or follow the onset of neurologic involvement. Most childhood patients show characteristic symmetric cerebral lesions on CT or magnetic-resonance imaging involving the white matter in the posterior and occipital lobes. Following intravenous injection of contrast, a garland-like contrast enhancement adjacent to hypodense lesions is shown by CT. The CNS demyelination has a mostly caudorostral progression. Liver peroxisomes are normal. Behavior changes associated with visual impairment may initially be mistaken for psychiatric manifestations. Intellectual deterioration in this period of life may be related to various other regressive encephalopathies, including Sanfilippo disease, Niemann-Pick disease type C, Wilson's disease, subacute sclerosing panencephalitis, multiple sclerosis and ceroid lipofuscinosis.

*Classical Refsum Disease. Classical Refsum disease* is another peroxisomal disorder with clinical onset during the school-age period. Retinitis pigmentosa, peripheral polyneuropathy, cerebellar ataxia and elevated cerebrospinal-fluid protein level are the main features. Less constant are nerve deafness, anosmia, ichthyosis and skeletal and cardiac abnormalities. Mental retardation, liver dysfunction and dysmorphia are absent. The onset of clinical manifestations varies from childhood to the fifth decade of life.

Recently, patients with atypical Refsum disease have been reported. In addition to increased plasma levels of phytanic acid, those of pipecolic acid [24] and pristanic acid [25] are also increased.

## Metabolic Derangements

From a biochemical viewpoint, peroxisomal disorders can be divided into two groups: those in which multiple or several peroxisomal enzymes are lost (also termed peroxisome assembly deficiencies) and those with a single deficient peroxisomal enzyme (Table 37.2) [26]. Elucidation of the pathogenesis of peroxisomal disorders has proven arduous, because a marked clinical heterogeneity exists in the expression of diseases with similar biochemical defects; however, at the same time, similar clinical phenotypes may be associated with different biochemical lesions. For example, classical CZ syndrome and IRD have similar generalized enzyme deficiencies, and both peroxisome assembly and single-enzyme defects underlie RCDP. Recently, new insights into the pathogenesis of peroxisomal disorders have been gained by the finding that, to be imported into the peroxisome, newly synthesized enzymes must contain short amino acid sequences termed peroxisome-targeting signals (PTS). Recognition of these signals by PTS receptors localized on the peroxisomal membrane is essential for entry of the enzymes into the peroxisomes. The PTS receptors are encoded by genes called PEX genes.

Subsequent work has shown that the peroxisome *assembly deficiencies* are caused by defects of PEX genes encoding PTS receptors common to several peroxisomal enzymes [27]. Most of the enzymes that fail to enter the peroxisomes are degraded in the cytosol. Catalase and thiolase are exceptions, and the latter protein is present in its 44-kDa, unprocessed form in the liver and fibroblasts of patients with variant RCDP or classical RCDP. When the enzyme defects are generalized (Table 37.2), the peroxisomes, which are normally clearly visible under the microscope owing to their average diameter of 0.2–1 µM, are reduced to virtually absent, barely detectable ghosts. This is best observed in liver and kidney, where peroxisomes are more abundant and bigger than in, e.g., brain or cultured skin fibroblasts. The clinical phenotype of RDCP is mostly associated with a defect in the PTS2 receptor (resulting in a deficient import of several enzymes) but can also be found in patients with an isolated enzyme deficiency.

*Single-enzyme deficiencies* are caused by defects of PEX genes encoding PTS receptors that are required by a single enzyme for entry of the enzyme into the peroxisome. Single-enzyme deficiencies may also be caused by defects of enzyme-encoding genes. Peroxisomes can be normal or variably modified (Table 37.2). In X-linked ALD, a defect has been suggested in a protein involved in the transport of very-long-chain fatty acid (VLCFA)-CoA synthase into the peroxisomal membrane or in a protein associated with VLCFA-CoA synthase in the membrane [28]. Mistargeting of a peroxisomal protein to the mitochondria has been demonstrated in some patients with hyperoxaluria type 1 [29]. As outlined in part in Fig. 37.1, deficiencies of peroxisomal enzymes affect several important physiological functions.

PLASMALOGEN BIOSYNTHES. The peroxisomal enzymes dihydroxyacetone phosphate acyltransferase (DHAPAT) and alkyl-DHAP synthase are essential for the synthesis of plasmalogens. Although the physiological function of these ether phospholipids (which are particularly abundant in nervous tissue) has not been completely clarified, they are known to be involved in platelet activation and free-radical scavenging.

β-OXIDATION OF FATTY ACIDS. A number of compounds, including straight-chain VLCFA (22 carbons or more) and branched-chain fatty acids [such as phytanic acid and trihydroxycholestanoic acid (THCA)], require a distinct peroxisomal β-oxidation system for chain shortening [25]. VLCFA are changed into the corresponding acyl-CoA esters by acyl-CoA synthases and are degraded in the peroxisomal matrix via a β-oxidation cycle, which consists of acyl-CoA oxidase 1, multifunctional protein 1 (containing enoyl-CoA hydratase, 3-OH-acyl-CoA dehydrogenase and isomerase activities) and peroxisomal 3-ketoacyl-CoA thiolase 1.

**Table 37.2.** Classification of perox-isomal disorders

| | Peroxisomes | Enzyme defect |
|---|---|---|
| **Peroxisome assembly deficiencies** | | |
| Classical Zellweger syndrome | Absent | Generalized |
| Neonatal adrenoleukodystrophy | | |
| Infantile Refsum disese | | |
| Pseudo-infantile Refsum disease | Absent | Generalized |
| Zellweger-like syndrome | Present | VLCFA oxidation, THCA oxidation, DHAPAT, phytanic acid oxidation |
| Rhizomelic chondrodysplasia punctata (classical/atypical phenotype), | Abnormal | DHAPAT, alkyl DHAP synthase, phytanic acid oxidase, unprocessed peroxisomal thiolase |
| Atypical Refsum disease | | Phytanic acid oxidase, pipecolic acid oxidase |
| **Single peroxisomal-enzyme deficiencies** | | |
| Rhizomelic chondrodysplasia punctata | | Isolated DHAPAT or alkyl DHAP synthase, phytanic acid oxidase |
| X-linked adrenoleukodystrophy | Normal | VLCFA-CoA synthase transport |
| Pseudo-neonatal adrenoleukodystrophy | Enlarged | Acyl-CoA oxidase |
| Bifunctional enzyme deficiency | Normal | Bi(tri)functional enzyme |
| Pseudo-Zellweger syndrome | Enlarged | Peroxisomal thiolase |
| Trihydroxycholestanoic acidemia | | THCA-CoA oxidase |
| Isolated pipecolic acidemia | Abnormal | Pipecolic acid oxidase |
| Mevalonic aciduria | | Mevalonate kinase |
| Classical Refsum disease | | Phytanoyl-CoA hydroxylase |
| Glutaric aciduria type III | Normal | Peroxisomal glutaryl-CoA oxidase |
| Hyperoxaluria type I | Reduced | Alanine:glyoxylate amino-transferase |
| Hyperoxaluria type I | | Mistargeting |
| Acatalasemia | Normal | Catalase |

*CoA*, coenzyme A; *DHAP*, dihydroxyacetone phosphate; *DHAPAT*, dihydroxyacetone phosphate acyltransferase; *THCA*, trihydroxycholestanoic acid; *VLCFA*, very long chain fatty acids

PHYTANIC-ACID OXIDATION. The degradation of the branched-chain fatty acid phytanic acid, which is exclusively of dietary origin, involves first an activation to phytanoyl-CoA. This is followed by a three-step cycle of α-oxidation, which produces pristanic acid [30]. Subsequently, pristanic acid is activated to pristanoyl-CoA and undergoes β-oxidation by acyl-CoA oxidase II, multifunctional protein II and thiolase II.

BILE-ACID AND CHOLESTEROL SYNTHESIS. The β-oxidation of THCA, an intermediate in the synthesis of bile acids from cholesterol, is carried out by the same enzymes that carry out the β-oxidation of pristanoyl-CoA. Owing to the presence of phosphomevalonate kinase [19] and the sterol carrier protein X [31], peroxisomes are also involved in the synthesis of cholesterol (Chap. 29).

LYSINE CATABOLISM. Lysine catabolism involves a major route via saccharopine (Chap. 20), and a minor route via pipecolic acid (Fig. 20.1). The first step in the further degradation of pipecolic acid is catalyzed by L-pipecolic acid oxidase, a peroxisomal enzyme in humans. The catabolism of both saccharopine and L-pipecolic acid proceeds via glutaric acid and a mitochondrial glutaryl-CoA dehydrogenase, but a peroxisomal glutaryl-CoA oxidase probably also exists [32].

GLYOXYLATE METABOLISM. Glyoxylate, the most important precursor of oxalate (Chap. 40), can be transaminated to glycine in a reaction catalyzed by alanine:glyoxylate aminotransferase, a vitamin-$B_6$-requiring enzyme in liver peroxisomes [29].

## Diagnostic Tests

A variety of assays (Table 37.3) exist for the diagnosis of peroxisomal disorders. *Urinary pipecolic-acid excretion, medium- and long-chain dicarboxylic aciduria, hyperoxaluria and mevalonic aciduria can be detected by an overall metabolic screening.* Twelve of the 17 peroxisomal disorders with neurologic involvement are associated with an accumulation of VLCFA and/or impaired de novo synthesis of ether phospholipids (plasmalogens), which suggests that assays of plasma VLCFA and plasmalogens in erythrocytes should be used as primary tests. If both results are abnormal, a disorder of peroxisome biogenesis should be substantiated by further appropriate analysis in plasma, liver and/or fibroblasts. If only the VLCFA

**Table 37.3.** Diagnostic assays in peroxisomal disorders

| Disease | Material | Type of assay |
|---|---|---|
| Classical ZS  Neonatal ALD  Infantile Refsum disease  Zellweger-like syndrome  Pseudo-infantile Refsum disease | Plasma  RBCs  Fibroblasts | VLCFA, bile acids, phytanic acid,  pristanic acid, pipecolic acid,  polyunsaturated fatty acids  Plasmalogens  DHAPAT, alkyl DHAP synthase  Particle-bound catalase  VLCFA β-oxidation  Immunoblotting β-oxidation proteins,  phytanic-acid oxidation |
| Rhizomelic chondrodysplasia  punctata (classical/atypical  phenotypes) | Plasma  RBCs  Fibroblasts | Phytanic acid  Plasmalogens  DHAPAT, alkyl DHAP synthase,  phytanic-acid oxidation |
| Isolated peroxisomal β-oxidation  defects | Plasma  Fibroblasts | VLCFA, bile acids  VLCFA β-oxidation, immunoblotting  β-oxidation proteins |
| Isolated defect of bile-acid synthesis | Plasma  Liver | Bile acids  THCA–CoA oxidase |
| Isolated pipecolic aciduria | Plasma  Liver | Pipecolic acid  Pipecolic-acid oxidase |
| Mevalonic aciduria | Plasma, urine  Fibroblasts, lymphocytes | Organic acids  Mevalonate kinase |
| Classical Refsum | Plasma  Fibroblasts | Phytanic acid  Phytanic-acid oxidation |
| Glutaric aciduria type III | Urine  Liver | Organic acids  Glutaryl-CoA oxidase |
| Hyperoxaluria type I | Urine  Liver | Organic acids  AGT |
| Actalasemia | RBCs | Catalase |

*AGT*, alanine:glyoxylate aminotransferase; *DHAP*, dihydroxyacetone phosphate; *DHAPAT*, dihydroxyacetone phosphate acyltransferase; *RBCs*, red blood cells; *THCA*, trihydroxycholestanoic acid; *VLCFA*, very-long-chain fatty acids

assay is abnormal, this indicates an isolated deficiency of peroxisomal β-oxidation at the level of acyl-CoA oxidase, multifunctional protein or thiolase; therefore, additional analysis in plasma (bile acids, phytanic acid and pristanic acid) and studies in liver and/or fibroblasts should be performed. If the VLCFA concentration is normal and the plasmalogen assay is abnormal, a disorder in the group of (rhizomelic) chondrodysplasias is suspected; plasma phytanic acid and the enzymes DHAPAT and alkyl-DHAP synthase in fibroblasts should be assayed.

The clinical presentation of the typical phenotypes of RCDP (abnormal levels of phytanic acid and plasmalogens) and classical Refsum (abnormal levels of phytanic acid) are distinct from those of the other disorders and should not cause difficulties in their diagnosis. In order to elucidate whether the accumulation of VLCFA in a patient's plasma results from a defect in peroxisome biogenesis or is caused by a defect in one of the peroxisomal β-oxidation enzyme activities, additional assay procedures must be carried out, particularly assays of plasmalogen levels and immunoblotting of peroxisomal β-oxidation proteins. Therefore, it should be stressed that it is no longer possible to screen all peroxisomal disorders only by measuring plasma VLCFA. It would be advisable to carry out assays of plasma bile-acid intermediates, phytanic, pristanic and pipecolic acid, plasmalogens in red blood cells and DHAPAT and alkyl-DHAP synthase in cultured skin fibroblasts [33]. In some patients with variant forms, the enzymatic deficit(s) are only expressed in the liver (not in cultured fibroblasts). Extensive peroxisomal investigations are necessary (even when the clinical phenotype is very typical), since some disorders may be associated with very atypical biochemical phenotypes.

Although, in some cases, levels of metabolites in CSF from patients exceed the control range, measurements of VLCFA, bile acids, pristanic acid and phytanic acid do not seem to provide a diagnostic advantage, as all measurements can be performed more conveniently in plasma [34]. For some disorders, a retrospective diagnosis can be obtained by analyzing stored blood spots collected during neonatal screening [35, 36].

HISTOLOGICAL DETECTION. Using the diaminobenzidine staining procedure (which reacts with the peroxisomal marker enzyme catalase) and immunochemical techniques with antibodies against matrix and membrane peroxisomal proteins facilitates the histological detection of peroxisomes [37]. The abundance, size and structure of liver peroxisomes should be studied. When peroxisomes are lacking, virtually all of the catalase is present in the cytosolic fraction instead of the particulate fraction. In some patients (mild variants), a peroxisome mosaicism in the liver and in fibroblasts can be observed [38].

PRENATAL DIAGNOSIS. A variety of techniques are available. Almost all peroxisomal disorders can be identified prenatally either by using (cultured) chorionvillous samples or amniocytes or by direct analysis of levels of VLCFA and bile-acid intermediates in amniotic fluid. Measurement of VLCFA and/or assays of plasmalogen synthesis are the most useful methods except in cases of THCA-CoA oxidase deficiency, isolated pipecolic aciduria, glutaric aciduria type III and hyperoxaluria type I (as detected in fetal liver biopsies). Other approaches include the cytochemical staining of peroxisomes in chorion-villus samples and mutational analysis (when the mutation is known in the index case).

HETEROZYGOTE IDENTIFICATION. Heterozygote identification is available for X-linked ALD using VLCFA analysis, DNA-linkage analysis with a specific marker (when studies of the necessary relatives are informative) or mutational analysis [39]. Mutational analysis is also a possibility in affected families with other disorders when the mutation has been demonstrated in the proband.

## Treatment and Prognosis

In classical Refsum disease, reduction of plasma phytanic acid levels by a low-phytanate diet (especially prohibition of ruminant meats and ruminant fats), with or without plasmapheresis, has been successful in arresting the progress of the peripheral neuropathy. The stored phytanic acid is exclusively of exogenous origin. However, when the diet is too strict, it may lead to a reduction of the energy intake, weight loss and a paradoxical rise in plasma phytanic acid levels followed by clinical deterioration. This is due to the mobilization of phytanate from lipids stored in adipose tissue.

In X-linked ALD patients, it was demonstrated that it is possible to normalize the plasma VLCFA levels by using a regimen that combines fat restriction and oral supplementation of monounsaturated fatty acids in triglyceride form (oleic- and erucic-acid from olive oil and rapeseed oil, respectively; glyceroltrioleate/glyceroltrierucate). However, the clinical benefit of the dietary therapy has been disappointing, and final conclusions will hopefully be available in the near future [23]. The usefulness of bone-marrow transplantation has been encouraging in clinically mildly affected patients with the childhood form of X-linked ALD [40]. Adrenocortical insufficiency should be treated with steroid-hormone substitution.

For patients with abnormal peroxisomal assembly and defects that originate in fetal life, the possibilities for treatment are very poor. Supplementation of docosahexaenoic acid [41] or other regimens [42] is now being tested in patients with the milder forms of multiple peroxisomal dysfunction or atypical chondrodysplasia punctata. Major reliance is focused on supportive therapy.

## Genetics

With the exception of X-linked ALD, the pattern of inheritance of peroxisomal disorders is autosomal recessive. There may exist a subset of neonatal ALD with an X-linked inheritance.

Most of the hitherto available information about the genetics of peroxisomal disorders has been obtained by complementation analysis. The underlying principle of complementation analysis is that cell lines that complement each other represent distinct genotypes [43–46]. Somatic cell fusion experiments with fibroblasts from patients have led to the identification of at least ten complementation groups, some of which are already known to be related to specific PEX-gene mutations. The lack of correlation between genotype and clinical phenotype in these complementation groups suggests that the currently used clinical denomination is not sufficiently distinctive. Identification of mutations in the various PEX genes is currently in progress and will help to elucidate the phenotypic variation in the patients [47].

## References

1. Fournier B, Smeitink JAM, Dorland L et al. (1994) Peroxisomal disorders: a review. J Inherit Metab Dis 17:470–486
2. Wanders RJA, Heymans HSA, Schutgens RBH et al. (1988) Peroxisomal disorders in neurology. J Neurol Sci 88:1–39
3. Baumgartner MR, Poll-The BT, Jakobs C et al. (1998) Clinical approach to inherited peroxisomal disorders. Consecutive series of 27 patients. Ann Neurol 44:720–730
4. Poll-The BT, Saudubray JM, Ogier H et al. (1987) Clinical approach to inherited peroxisomal disorders. In: Vogel F, Sperling (eds) Human genetics. Springer, Berlin Heidelberg New York, pp 345–351
5. Ten Brink HJ, Wanders RJA, ChristensenE, Brandt NJ, Jakobs C (1994) Heterogeneity in di/trihydroxycholestanoic acidemia. Ann Clin Biochem 31:195–197
6. De Craemer D, Zweens MJ, Lyonnet S et al. (1991) Very large peroxisomes in distinct peroxisomal disorders (rhizomelic chondrodysplasia punctata and acyl-CoA oxidase deficiency): novel data. Virchows Arch [A] Pathol Anat 419:523–525
7. Poll-The BT, Maroteaux P, Narcy C et al. (1991) A new type of chondrodysplasia punctata associated with peroxisomal dysfunction. J Inherit Metab Dis 14:361–363
8. Smeitink JAM, Beemer FA, Espeel M et al. (1992) Bone dysplasia associated with phytanic acid accumulation and deficient plasmalogen synthesis: a peroxisomal entity amenable to plasmapheresis. J Inherit Metab Dis 15: 377–380
9. Barth PG, Wanders RJA, Schutgens RBH, Staalman CR (1996) Variant rhizomelic chondrodysplasia punctata (RCDP) with normal plasma phytanic acid: clinico-biochemical delineation of a subtype and complementation studies. Am J Med Genet 62:164–168
10. Wanders RJA, Schumacher H, Heikoop J, Schutgens RBH, Tager JM (1992) Human dihydroxyacetonephosphate acyltransferase deficiency: a new peroxisomal disorder. J Inherit Metab Dis 15:389–391
11. Wanders RJA, Dekker C, Hovarth VAP et al. (1994) Human alkyldihydroxyacetonephosphate synthase deficiency: a new peroxisomal disorder. J Inherit Metab Dis 17:315–318
12. Poll-The BT, Saudubray JM, Ogier H et al. (1987) Infantile Refsum disease: an inherited peroxisomal disorder. Comparison with Zellweger syndrome and neonatal adrenoleukodystrophy. Eur J Pediatr 146:477–483
13. Baumgartner R, Verhoeven NM, Jakobs C, Roels F, Espeel M, Martinez M, Rabier D, Wanders RJ, Saudubray JM (1998) Defective peroxisome biogenesis with a neuromuscular disorder resembling Werdnig-Hoffmann disease. Neurology 51:1427–1432
14. Poll-The BT, Roels F, Ogier H et al. (1988) A new peroxisomal disorder with enlarged peroxisomes and a specific deficiency of acyl-CoA oxidase (pseudo-neonatal adrenoleukodystrophy). Am J Hum Genet 42:422–434
15. Suzuki Y, Shimozawa N, Yajima S et al. (1994) Novel subtype of peroxisomal acyl-CoA oxidase deficiency and bifunctional enzyme deficiency with detectable enzyme protein: identification by means of complementation analysis. Am J Hum Genet 54:36–43
16. Watkins PA, Chen WN, Harris CJ et al. (1989) Peroxisomal bifunctional enzyme deficiency. J Clin Invest 83:771–777
17. Schram AW, Goldfischer S, van Roermund CWT et al. (1987) Human peroxisomal 3-oxoacyl-coenzyme A thiolase deficiency. Proc Natl Acad Sci USA 84:2494–2496
18. Hoffmann G, Gibson KM, Brandt IK et al. (1986) Mevalonic aciduria – an inborn error of cholesterol and nonsterol isoprene biosynthesis. N Engl J Med 314:1610–1614
19. Biardi L, Sreedhar A, Zokaei A et al. (1994) Mevalonate kinase is predominantly localized in peroxisomes and is defective in patients with peroxisome deficiency disorders. J Biol Chem 269:1197–1205
20. Aubourg P, Kremser K, Roland MO, Rocchiccioli F, Singh I (1993) Pseudo infantile Refsum's disease: catalase-deficient peroxisomal particles with partial deficiency of plasmalogen synthesis and oxidation of fatty acids. Pediatr Res 34:270–276
21. Poll-The BT, Billette de Villemeur T, Abitbol M, Dufier JL, Saudubray JM (1992) Metabolic pigmentary retinopathies: diagnosis and therapeutic attempts. Eur J Pediatr 151:2–11
22. Tint GS, Irons M, Elias E et al. (1994) Defective cholesterol biosynthesis associated with the Smith-Lemli-Opitz syndrome. N Engl J Med 330:107–113
23. Moser HW, Moser AB, Smith KD et al. (1992) Adrenoleukodystrophy: phenotype variability and implications for therapy. J Inherit Metab Dis 15:645–664
24. Tranchant C, Aubourg P, Mohr M et al. (1993) A new peroxisomal disease with impaired phytanic and pipecolic acid oxidation. Neurology 43:2044–2048
25. Reddy JK, Mannaerts GP (1994) Peroxisomal lipid metabolism. Annu Rev Nutr 14:343–370
26. Van den Bosch H, Schutgens RBH, Wanders RJA, Tager JG (1992) Biochemistry of peroxisomes. Annu Rev Biochem 61:157–197
27. Subramani S (1997) PEX genes on the rise. Nature Genet 15:331–333
28. Mosser J, Douar AM, Sarde CO et al. (1993) Putative X-linked adrenoleukodystrophy gene shares unexpected homology with ABC transporters. Nature 361:726–730
29. Danpure CJ, Copper PJ, Wise PJ, Jennings PR (1989) An enzyme trafficking defect in two patients with primary hyperoxaluria type 1: peroxisomal alanine: glyoxylate aminotransferase rerouted to mitochondria. J Cell Biol 108: 1345–1352
30. Verhoeven NM, Wanders RJA, Poll-The BT, Saudubray JM, Jakobs C (1998) The metabolism of phytanic and pristanic acid in man: a review. J Inherit Metab Dis 21:697–728
31. Wanders RJA, Denis S, Wouters F, Wirtz KWA, Seedorf U (1997) Sterol carrier protein x (SCPx) is a peroxisomal branched-chain β-ketothiolase specifically reacting with

3-oxo-pristanoyl-CoA: a new, unique role for SCPx in branched-chain fatty acid metabolism in peroxisomes. Biochem Biophys Res Commun 236:565-569
32. Bennett MJ, Pollitt RJ, Goodman SI, Hale DE, Vamecq J (1991) Atypical riboflavin-responsive glutaric aciduria, and deficient peroxisomal glutaryl-CoA oxidase activity: a new peroxisomal disorder. J Inherit Metab Dis 14:165-173
33. Mandel H, Espeel M, Roels F et al. (1994) A new type of peroxisomal disorder with variable expression in liver and fibroblasts. J Pediatr 125:549-555
34. Ten Brink HJ, van den Heuvel CMM, Poll-The BT, Wanders RJA, Jakobs C (1993) Peroxisomal disorders: concentrations of metabolites in cerebrospinal fluid compared with plasma. J Inherit Metab Dis 16:587-590
35. Jakobs C, van den Heuvel CMM, Stellaard F et al. (1993) Diagnosis of Zellweger syndrome by analysis of very long-chain fatty acids in stored blood spots collected at neonatal screening. J Inherit Metab Dis 16:63-66
36. Ten Brink HJ, van den Heuvel CMM, Christensen E, Largillière C, Jakobs C (1993) Diagnosis of peroxisomal disorders by analysis of phytanic and pristanic acids in stored blood spots collected at neonatal screening. Clin Chem 39:1904-1906
37. Roels F, Espeel M, De Craemer D (1991) Liver pathology and immunocytochemistry in congenital peroxisomal diseases: a review. J Inherit Metab Dis 14:853-875
38. Espeel, M, Mandel H, Poggi F, et al. (1995) Peroxisome mosaicism in the livers of peroxisomal deficient patients. Hepatology 22:497-504
39. Feigenbaum B, Lombard-Platet G, Guidoux S et al. (1996) Mutational and protein analysis of patients and heterozygous woman with X-linked adrenoleukodystrophy. Am J Hum Genet 58:1135-1144

40. Aubourg P, Blanche S, Jambaqué I et al. (1990) Reversal of early neurologic and neuroradiologic manifestations of X-linked adrenoleukodystrophy by bone marrow transplantation. N Engl J Med 322:1860-1866
41. Martinez M (1992) Treatment with docosahexaenoic acid favorably modifies the fatty acid composition of erythrocytes in peroxisomal patients. In: Coats PM, Tanaka K (eds) New developments in fatty acid oxidation. Wiley-Liss, New York, pp 389-397
42. Robertson EF, Poulos A, Sharp P et al. (1988) Treatment of infantile phytanic acid storage disease: clinical, biochemical and ultra-structural findings in two children treated for 2 years. Eur J Pediatr 147:133-142
43. Brul S, Westerveld A, Strijland A et al. (1988) Genetic heterogeneity in the cerebrohepatorenal (Zellweger) syndrome and other inherited disorders with generalized impairment of peroxisomal functions. A study using complementation analysis. J Clin Invest 81:1710-1715
44. Poll-The BT, Skjeldal OH, Stokke O et al. (1989) Phytanic acid alpha-oxidation and complementation analysis of classical Refsum and peroxisomal disorders. Hum Genet 81:175-181
45. Roscher AA, Hoefler S, Hoefler G et al. (1989) Genetic and phenotypic heterogeneity in disorders of peroxisome biogenesis - a complementation study involving cell lines from 19 patients. Pediatr Res 26:67-72
46. Heikoop JC, Wanders RJA, Strijland A et al. (1992) Genetic and biochemical heterogeneity in patients with the rhizomelic form of chondrodysplasia punctata - a complementation study. Hum Genet 89:439-444
47. Braverman N, Steel G, Obie C et al. (1997) Human PEX7 encodes the peroxisomal PTS$_2$ receptor and is responsible for rhizomelic chondrodysplasia punctata. Nature Genet 15:369-376

## Synthesis of N-Glycans

N-glycans are tree- or antenna-like oligosaccharide structures (Fig. 36.2) which, when attached to the amide group of asparagine on a protein backbone, make up N-linked glycoproteins. Their synthesis proceeds in three stages (Fig. 38.1):

1. Synthesis of nucleotide-linked sugars (mainly guanosine diphosphate-mannose) in the cytosol
2. Assembly of the dolichol (Dol)–pyrophosphate (PP) oligosaccharide precursor Dol–PP–N-acetyl-glucosamine$_2$–mannose$_9$–glucose$_3$ in the endoplasmic reticulum
3. Transfer of this precursor onto the nascent protein, followed by final processing of the antenna in the Golgi complex by trimming and attachment of various sugar units.

In recent years, several disorders of the synthesis of the oligosaccharide chains of N-linked glycoproteins, which are now named congenital defects of glycosylation, (CDG), have been identified. The defect of oligosaccharide-chain processing that leads to deficiency of the mannose-6-phosphate recognition marker of the lysosomal enzymes and causes mucolipidosis II or III is discussed in Chap. 36.

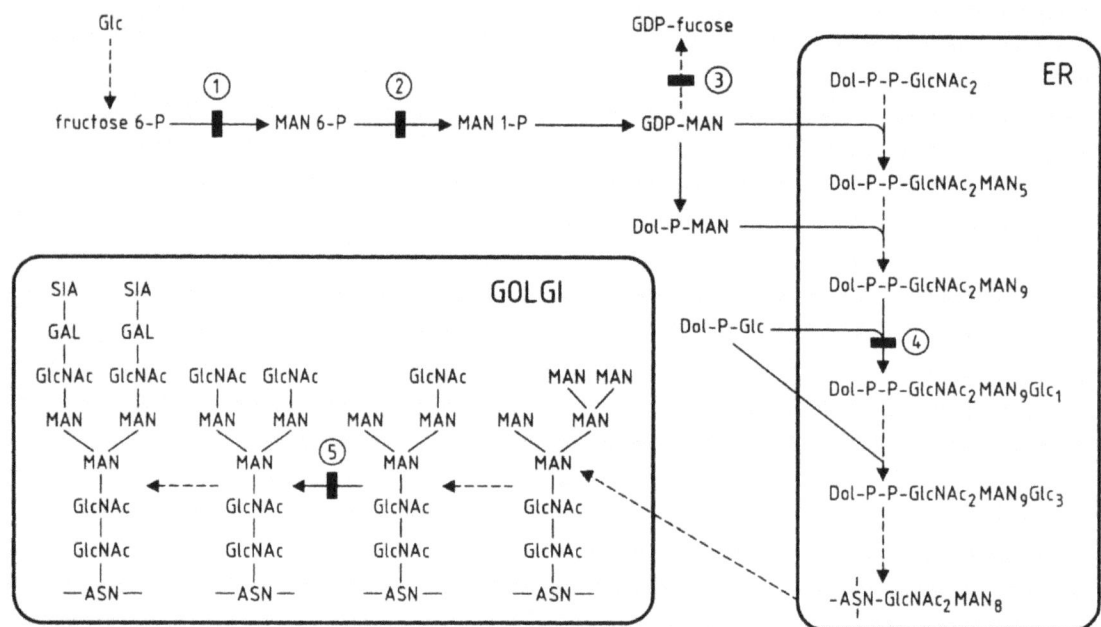

**Fig. 38.1.** Schematic representation of the synthesis of N-glycans. Known defects are indicated by *bars* and *circled figures*. *ASN*, asparagine; *Dol*, dolichol; *ER*, endoplasmic reticulum; *GAL*, galactose; *Glc*, glucose; *GlcNAc*, N-acetylglucosamine; *GDP*, guanosine diphosphate; *MAN*, mannose; *P*, phosphate; *SIA*, sialic acid. 1, Phosphomannose-isomerase deficiency; 2, phosphomannomutase deficiency; 3, leukocyte-adhesion-deficiency syndrome type II; 4, endoplasmic-reticulum-localized glucosyltransferase deficiency; 5, Golgi-localized N-acetylglucosaminyltransferase-II deficiency

# CHAPTER 38

# Congenital Defects of Glycosylation: Disorders of *N*-Glycan Synthesis

Jaak Jaeken

## CONTENTS

Phosphomannomutase Deficiency (CDG-Ia) . . . . . . . . . . . 433
Phosphomannose Isomerase Deficiency (CDG-Ib) . . . . . . . 434
Leukocyte Adhesion Deficiency Syndrome Type II . . . . . . 435
Glucosyltransferase I Deficiency (CDG-Ic) . . . . . . . . . . . . 435
*N*-acetylglucosaminyltransferase II Deficiency
(CDG-IIa) . . . . . . . . . . . . . . . . . . . . . . . . . . . . . . . . . . . . . 436
Unexplained Glycosylation Defects . . . . . . . . . . . . . . . . . . 436
References . . . . . . . . . . . . . . . . . . . . . . . . . . . . . . . . . . . . . 437

---

Five inborn errors in the pathway of *N*-glycan synthesis [also called congenital defects of glycosylation, CDG syndromes] are known and are depicted in Fig. 38.1. *Phosphomannomutase (PMM) deficiency*, reported in the large majority of these patients, is a multisystem disease with, as a rule, severe brain involvement.

*Phosphomannose isomerase deficiency* has a hepatic intestinal presentation and responds to mannose treatment. The main characteristic of *leukocyte adhesion deficiency syndrome type II (LAD-II)* are recurrent bacterial infections (without pus formation) due to a deficiency of neutrophil recruiting and other fucose containing glycoproteins. The two reported families with an endoplasmic reticulum (ER)-localized *glucosyltransferase I deficiency* mainly show a mild neurological picture. The Golgi defect *N-acetylglucosaminyl transferase-II deficiency* is associated with craniofacial dysmorphy and severe brain disease. A rapidly growing number of patients with CDG have unknown basic defects.

---

## Phosphomannomutase Deficiency (CDG-Ia)

### Clinical Presentation

At least 130 patients with phosphomannomutase (PMM) deficiency are known worldwide. The symptomatology can be recognized shortly after birth. The nervous system is affected in all patients, and most other organs are involved in a variable way. The neurologic picture comprises alternating internal strabism and other abnormal eye movements, axial hypotonia, psychomotor retardation (intelligence quotient mostly between 40 and 60), ataxia and hyporeflexia. After infancy, symptoms include retinitis pigmentosa; often, stroke-like episodes are seen, as is epilepsy (sometimes). There is no regression. During the first year(s) of life, there are feeding problems (anorexia, vomiting, diarrhea) that can result in severe failure to thrive. Other features are a variable dysmorphy, [large, hypoplastic/dysplastic ears, abnormal subcutaneous adipose-tissue distribution (fat pads, "orange peel" skin), inverted nipples], mild to moderate hepatomegaly, skeletal abnormalities and hypogonadism. Some infants develop pericardial effusion and/or cardiomyopathy. Patients often have an extroverted and happy appearance. Some 20% have died due to severe infection or liver, cardiac or renal insufficiency during the first years of life. Neurotechnical investigation reveals (olivopontoǃ)cerebellar hypoplasia, variable cerebral hypoplasia and peripheral neuropathy. Liver pathology is characterized by fibrosis and steatosis, and electron microscopy shows myelin-like lysosomal inclusions in hepatocytes but not in Kupffer cells [1–4].

### Metabolic Derangement

PMM deficiency is a (cytosolic) defect in the second step of the mannose pathway (transforming mannose-6-phosphate into mannose-1-phosphate), which normally leads to the synthesis of guanosine diphosphate (GDP)–mannose [5, 6]. This nucleotide sugar is the donor of mannose used in the ER to assemble the dolichol-pyrophosphate oligosaccharide precursor (Fig. 38.1). Deficiency of GDP–mannose causes hypoglycosylation (and hence deficiency, increase and/or dysfunction) of numerous glycoproteins, including serum proteins, lysosomal enzymes and membranous glycoproteins [7, 8].

## Diagnostic Tests

The diagnosis of CDG in general (and of PMM deficiency in particular) is usually made by isoelectrofocusing and immunofixation of serum transferrin [9]. Normal serum transferrin is mainly composed of tetrasialotransferrin and small amounts of mono-, di-, tri-, penta- and hexasialotransferrins. The partial deficiency of sialic acid (a negatively charged and end-standing sugar) in CDG causes a cathodal shift. Two main types of cathodal shift can be recognized: type I is characterized by an increase of both asialo- and disialotransferrin and a decrease of tetra-, penta- and hexasialotransferrins (type-I pattern); in type II, there is also an increase of the tri- and/or monosialotransferrin bands (type-II pattern). In PMM deficiency, a type-I pattern is found. Similar changes are noted in chronic alcoholism, untreated galactosemia and hereditary fructose intolerance. These changes can be quantified by densitometry of transferrin. In order to exclude isoelectrofocusing changes due to a transferrin-protein variant, this test should be supplemented by isoelectrofocusing of another serum glycoprotein, such as hexosaminidase A or thyroxin-binding globulin. The carbohydrate-deficient transferrin (CDT) assay enables the quantification of the total sialic-acid-deficient serum transferrin.

In addition to the above-mentioned serum glycoprotein abnormalities, laboratory findings include elevation of serum transaminase levels, hypoalbuminemia and tubular proteinuria. Prenatal diagnosis is possible with amniocytes and chorionic villus cells; this should be combined with mutation analysis of the $PMM_2$ gene (see "Genetics"). To confirm the diagnosis, the activity of PMM should be measured in leukocytes or fibroblasts.

## Treatment and Prognosis

No efficient treatment is available. The promising finding that mannose is able to correct glycosylation in fibroblasts with PMM deficiency could not be substantiated in patients [10, 11].

## Genetics

PMM deficiency is inherited as an autosomal-recessive trait due to mutations of the $PMM_2$ gene on chromosome 16p13. At least 36 mutations have been identified (mainly missense mutations). The most frequent mutation causes the R141H substitution (which, remarkably, has not yet been found in the homozygous state). The frequency of this mutation in the normal Belgian population is as high as 1/50 [12]. The incidence of PMM deficiency is not known; it has been estimated at 1:40,000 in Sweden [3].

# Phosphomannose-Isomerase Deficiency (CDG-Ib)

## Clinical Presentation

Eight patients with phosphomannose isomerase (PMI) deficiency belonging to six families and five nationalities have been reported [13–15]. They had a mainly hepatic-intestinal presentation without notable dysmorphy and without or with only minor neurological involvement. Symptoms started at ages between 1 month and 11 months. One patient had recurrent vomiting and liver disease that disappeared after the introduction of solid food at age 3 months. In the others, symptoms persisted and consisted of various combinations of recurrent vomiting, abdominal pain, protein-losing enteropathy, recurrent thromboses, gastrointestinal bleeding, liver disease and symptoms of (hyperinsulinemic) hypoglycemia. In 1985, four infants from Quebec were reported with a similar syndrome who probably had the same disease [16]. Several explanations for the differences in tissue involvement between PMI and PMM deficiency and for the fact that the clinical manifestations of PMI deficiency are mainly hepatic and intestinal have been provided [15].

## Metabolic Derangement

PMI deficiency is a defect of the first committed step in the synthesis of the nucleotide sugar GDP–mannose. Hence, the blood biochemical abnormalities are indistinguishable from those found in PMM deficiency (Table 38.1).

## Diagnostic Tests

As for other CDG, the diagnostic test "par excellence" is isoelectrofocusing and immunofixation of serum transferrin. A type-I pattern is found, as in PMM deficiency and in ER-localized glucosyltransferase deficiency. To confirm the diagnosis, the activity of PMI should be measured in leukocytes or fibroblasts.

## Treatment and Prognosis

PMI deficiency is the only known CDG that can be efficiently treated; mannose is the therapeutic agent. Indeed, hexokinases phosphorylate mannose to mannose-6-phosphate, thus bypassing the defect. This was successfully demonstrated in a patient who normalized under oral mannose therapy (up to 150 mg/kg five times per day) [13].

Two of the eight patients with proven PMI deficiency (one undergoing mannose treatment and the other not receiving mannose treatment) died at 2 years and 4 years. The four reported patients from Quebec with suspected PMI deficiency all died between 4 months and 21 months of age [16].

### Genetics

Inheritance of PMI deficiency is autosomal recessive. The gene has been localized to chromosome 15q22. Five (missense) mutations have been found in four patients studied [13, 15]. In one of these (with transient liver disease), no mutation could be identified in the structural gene.

## Leukocyte Adhesion Deficiency Syndrome Type II (LAD-II)

### Clinical Presentation

LAD-II has been reported in two unrelated boys with a syndrome featuring recurrent bacterial infections without pus formation and an unusual facial appearance, short stature and encephalopathy [17].

### Metabolic Derangement

The biochemical basis of this disease is a failure to convert GDP–mannose into GDP–fucose, resulting in a generalized deficiency of fucose. The precise biochemical defect is not yet known. The neutrophils of these patients lack sialyl-Lewis X, a fucose-containing carbohydrate ligand of the selectin family of cell-adhesion molecules required for normal recruitment of neutrophils to sites of inflammation. Other fucose-containing carbohydrate sequences (such as A, B, O and Lewis-A blood groups) are also absent.

### Diagnostic Tests

This disease has to be considered in infections accompanied by very high leukocyte counts (up to 150,000 cells/µl) without pus formation. Contrary to the other CDG, LAD-II shows a normal isoelectrofocusing pattern of serum transferrin because there is no sialic acid deficiency. To confirm the diagnosis, GDP–fucose deficiency or generalized fucose deficiency should be demonstrated.

### Treatment and Prognosis

Apart from the symptomatic treatment of the recurrent infections, no efficient treatment is available for this syndrome. Both reported children showed severe psychomotor and physical retardation but were still alive at ages 4.5 years and 2.5 years.

### Genetics

Both children were born to related parents, suggesting autosomal-recessive inheritance.

## Glucosyltransferase-I Deficiency (CDG-Ic)

### Clinical Presentation

Five patients with an ER-localized glucosyltransferase-I deficiency (CDG-Ic) have been reported [18, 19]. Four of them belonged to a highly inbred Dutch family and showed mild to moderate psychomotor retardation and epilepsy. They had some impression of the temporal head regions but no marked dysmorphy. There was no cerebellar hypoplasia or polyneuropathy nor was there clinical involvement of other organs.

### Metabolic Derangement

This glucosyltransferase-I deficiency is a defect in the attachment of the first glucose (of three) to the dolichol-linked mannose$_9$-$N$-acetylglucosamine$_2$ ER intermediate. It causes hypoglycosylation of serum glycoproteins, because non-glucosylated oligosaccharides are a sub-optimal substrate for the oligosaccharyltransferase and are, therefore, transferred to proteins with a reduced efficiency. For an unknown reason, some of the glycoproteins have unusually low blood levels (particularly of factor XI and coagulation inhibitors, such as antithrombin and protein C). The reason the clinical picture in these patients is much milder than that of PMM deficient patients may be because a deficiency in glucosylation of the dolichol-linked oligosaccharides does not affect the biosynthesis of GDP–mannose and, hence, does not affect the biosynthesis of GDP–fucose or the biosynthesis of glycosylphosphatidylinositol-anchored glycoproteins.

### Diagnostic Tests

This disease illustrates that, in cases of mild psychomotor retardation without any specific dysmorphy, a CDG has to be considered and isoelectrofocusing of serum sialotransferrins has to be performed. When a type-I pattern is found and PMM and PMI deficiency has been excluded, electron microscopy of liver or another tissue should be performed. The presence of lysosomal inclusions strongly indicates that the defect is in the ER or in the pre-ER compartment. The next step is analysis of the dolichol-linked oligosaccharides in fibroblasts. If the major fraction of these oligosaccharides consists of nine mannose and two N-acetylglucosamine residues without the normally present three glucose residues, this specific glucosyltransferase activity has to be measured in fibroblasts (done in very few laboratories) and, in case of deficiency, this must be completed by mutation analysis.

### Treatment and Prognosis

No efficient treatment is available. Limited experience with this disease does not permit a valuable prognosis.

### Genetics

Inheritance of this glucosyltransferase deficiency is autosomal recessive. The chromosomal localization of the gene is unknown. Two mutations have been identified among the five reported patients.

## N-Acetylglucosaminyltransferase-II Deficiency (CDG-IIa)

### Clinical Presentation

CDG-IIa is the most severe of the known defects of N-glycan synthesis. Four unrelated patients are known, two of which have been reported [20-22]. Both showed dysmorphy, particularly of the face, with high forehead, beaked nose with upturned alae nasi, receding chin, protruding upper incisivae and dysplastic, large, low-set ears. They had a prominent stereotypic behavior, including hand-washing movements like those seen in Rett syndrome. They also showed growth retardation, severe psychomotor retardation and poorly manageable epilepsy. There was no polyneuropathy. Evoked potentials were normal, as was electron microscopy of a liver biopsy.

### Metabolic Derangement

CDG-IIa is caused by a deficiency of N-acetylglucosaminyltransferase II, a Golgi transmembrane enzyme that attaches an N-acetylglucosamine residue specifically to one of the mannoses of the glycan core. This defect prevents the oligosaccharide antenna on this mannose from being completed. The antenna on the other mannose is normally formed. The lysosomal enzymes possess recognition markers that are also mainly made from high-mannose glycans but are synthesized by a different Golgi pathway. Therefore, they are normally targeted to (and are active in) lysosomes, which explains the absence of lysosomal inclusions in this disease. This is in contrast to findings in ER and most pre-ER glycosylation disorders.

### Diagnostic Tests

Isoelectrofocusing of serum transferrin shows a type-II pattern (see "PMM Deficiency"), with the special features that tetrasialotransferrin is nearly absent and that disialotransferrin is the predominant sialotransferrin fraction. Decreased blood or serum levels of a large number of glycoproteins are found. Unlike the case in PMM deficiency, there is no decrease of serum albumin and cholesterol levels and no increase of arylsulfatase A. Interestingly, serum glutamic oxaloacetic transaminase is increased but glutamic pyruvate transaminase is normal. The diagnosis is confirmed by measuring the activity of N-acetylglucosaminyltransferase II in fibroblasts, lymphocytes or lymphoblasts (done in very few laboratories).

### Treatment and Prognosis

There is no efficient treatment available. Several affected family members of these patients died at an early-infantile stage. In this context, it is interesting to note that the knock-out mouse shows a high mortality rate at an early stage.

### Genetics

CDGS-IIa is an autosomal-recessive disease. The gene (*MGAT2*) maps to chromosome 14q21. Among three patients studied, four point mutations have been identified [22].

## Unexplained Glycosylation Defects

The number of patients with an unidentified glycosylation defect is rapidly growing. They have various clinical phenotypes ranging from relatively mild to extremely severe, including presentations not previously reported in CDG (for example, hydrops fetalis) [23-29]. Therefore, a broad screening using the CDT assay (or, preferably, isoelectrofocusing of serum sialotransferrins) is recommended in any unexplained clinical disorder. The majority of patients with an unexplained glycosylation disorder show a type-I sialotransferrin pattern. These type-I patients can have a defect in any part of the glycosylation apparatus (pre-ER, ER or Golgi). In order to make a further differentiation, it is suggested that electron microscopy of liver or other tissues be performed. The presence of lysosomal inclusions indicates that the defect is probably in a pre-Golgi compartment. Patients with a type-II pattern probably have Golgi-localized defects. Further biochemical work should reveal the basic (enzymatic or transport) defect.

**Note added in proof:** Recently two new defects have been reported: CDG-Id due to an ER-localised mannosyltransferase deficiency [30] and CDG-Ie due to a defect in the ER-localised synthesis of dolichol-phosphate mannose.

## References

1. Jaeken J, Vanderschueren-Lodeweyckx M, Casaer P et al. (1980) Familial psychomotor retardation with markedly fluctuating serum proteins, FSH and GH levels, partial TBG-deficiency, increased serum arylsulphatase A and increased CSF protein: a new syndrome? Pediatr Res 14:179
2. Jaeken J, Carchon H, Stibler H (1993) The carbohydrate-deficient glycoprotein syndromes: pre-Golgi and Golgi disorders? Glycobiology 3:423-428
3. Jaeken J, Stibler H, Hagberg B (1991) The carbohydrate-deficient glycoprotein syndrome: a new inherited multisystemic disease with severe nervous system involvement. Acta Paediatr Scand [Suppl] 375:1-71
4. Jaeken J, Matthijs G, Barone R, Carchon H (1997) Carbohydrate deficient glycoprotein (CDG) syndrome type I. J Med Genet 34:73-76
5. Van Schaftingen E, Jaeken J (1995) Phosphomannomutase deficiency is a cause of carbohydrate-deficient glycoprotein syndrome type I. FEBS Lett 377:318-320
6. Jaeken J, Besley G, Buist N et al. (1996) Phosphomannomutase deficiency is the major cause of carbohydrate-deficient glycoprotein syndrome type I. J Inherit Metab Dis 19 [Suppl 1]:6
7. de Zegher F, Jaeken J (1995) Endocrinology of the carbohydrate-deficient glycoprotein syndrome type 1 from birth through adolescence. Pediatr Res 37:395-401
8. Van Geet C, Jaeken J (1993) A unique pattern of coagulation abnormalities in carbohydrate-deficient glycoprotein syndrome. Pediatr Res 33:540-541
9. Jaeken J, van Eijk HG, van der Heul C et al. (1984) Sialic acid-deficient serum and cerebrospinal fluid transferrin in a newly recognized genetic syndrome. Clin Chim Acta 144:245-247
10. Panneerselvam K, Freeze HH (1996) Mannose corrects altered N-glycosylation in carbohydrate-deficient glycoprotein syndrome fibroblasts. J Clin Invest 97:1478-1487
11. Kjaergaard S, Kristiansson B, Stibler H et al. (1998) Failure of short-term mannose therapy of patients with carbohydrate-deficient glycoprotein syndrome type I A. Acta Paediatr 87:884-888
12. Matthijs G, Schollen E, Van Schaftingen E, Cassiman J-J, Jaeken J (1998) Lack of homozygotes for the most frequent disease allele in carbohydrate-deficient glycoprotein syndrome type 1 A. Am J Hum Genet 62:542-550
13. Niehues R, Hasilik M, Alton G et al. (1998) Carbohydrate-deficient glycoprotein syndrome type Ib: phosphomannose isomerase deficiency and mannose therapy. J Clin Invest 101:1414-1420
14. de Koning TJ, Dorland L, van Diggelen OP et al. (1998) A novel disorder of N-glycosylation due to phosphomannose isomerase deficiency. Biochem Biophys Res Commun 245:38-42
15. Jaeken J, Matthijs G, Saudubray J-M et al. (1985) Phosphomannose isomerase deficiency: a carbohydrate-deficient glycoprotein syndrome with hepatic-intestinal presentation. Am J Hum Genet 62:1535-1539
16. Pelletier VA, Galeano N, Brochu P et al. (1985) Secretory diarrhea with protein-losing enteropathy, enterocolitis cystica superficialis, intestinal lymphangiectasia and congenital hepatic fibrosis: a new syndrome. J Pediatr 108:61-65
17. Sturla L, Etzioni A, Bisso A et al. (1998) Defective intracellular activity of GDP-D-mannose-4, 6-dehydratase in leukocyte adhesion deficiency type II syndrome. FEBS Lett 429:274-278
18. Burda P, Borsig L, de Rijk-van Andel J et al. (1998) A novel carbohydrate-deficient glycoprotein syndrome characterized by a deficiency in glucosylation of the dolichol-linked oligosaccharide. J Clin Invest 102:647-652
19. Körner C, Knauer R, Holzbach U et al. (1998) Carbohydrate-deficient glycoprotein syndrome type V: deficiency of dolichyl-P-Glc:Man$_9$ GlcNAc$_2$-PP-dolichyl glucosyltransferase. Proc Natl Acad Sci USA 95:13200-13205
20. Ramaekers VT, Stibler H, Kint J, Jaeken J (1991) A new variant of the carbohydrate deficient glycoproteins syndrome. J Inherit Metab Dis 14:385-388
21. Jaeken J, Schachter H, Carchon H et al. (1994) Carbohydrate-deficient glycoprotein syndrome type II: a deficiency in Golgi localised N-acetyl-glucosaminyltransferase II. Arch Dis Child 71:123-127
22. Tan J, Dunn J, Jaeken J, Schachter H (1996) Mutations in the MGAT2 gene controlling complex N-glycan synthesis cause carbohydrate-deficient glycoprotein syndrome type II, an autosomal recessive disease with defective brain development. Am J Hum Genet 59:810-817
23. Stibler H, Westerberg B, Hanefeld F, Hagberg B (1993) Carbohydrate-deficient glycoprotein (CDG) syndrome - a new variant, type III. Neuropediatrics 24:51-52
24. Eyskens F, Ceuterick C, Martin J-J, Janssens G, Jaeken J (1994) Carbohydrate-deficient glycoprotein syndrome with previously unreported features. Acta Paediatr 83:892-896
25. Stibler H, Stephani U, Kutsch U (1995) Carbohydrate-deficient glycoprotein syndrome - a fourth subtype. Neuropediatrics 26:235-237
26. Skladal D, Sperl W, Henry H, Bachmann C (1996) Congenital cataract and familial brachydactyly in carbohydrate-deficient glycoprotein syndrome. J Inherit Metab Dis 19:251-252
27. Assmann B, Hoffmann GF, Köhler M et al. (1997) CDG-syndrome type I with normal phosphomannomutase activity and unusually mild clinical manifestations. Amino Acids 12:387
28. Dorland L, de Koning TJ, Toet M et al. (1997) Recurrent non-immune hydrops fetalis associated with carbohydrate-deficient glycoprotein syndrome. J Inherit Metab Dis 20 [Suppl 1]:88
29. Gasparini P, Miraglia del Giudice E, Delaunay J et al. (1997) Localization of the congenital dyserythropoietic anemia II locus to chromosome 20q11.2 by genomewide search. Am J Hum Genet 61:1112-1116
30. Körner C, Knauer R, Stephani U et al. (1999) Carbohydrate deficient glycoprotein syndrome type IV: deficiency of dolichyl-P-Man:Man$_5$GlcNAc$_2$-PP-dolichylmannosyltransferase. EMBO J 18:6816-6822

## Lysosomal Porters for Cystine and Related Compounds

Intralysosomal cystine is formed by protein catabolism in the organelle and is normally exported by a cystine porter containing a membrane protein, *cystinosine*. A defect in its production causes lysosomal accumulation of cystine. Cysteamine can pass through the lysosomal membrane and combine with cystine. This results in the formation of cysteine (which can be exported by the cysteine porter) and of the mixed disulfide cysteine–cysteamine (which, due to its structural analogy, can be exported by the lysine porter) (Fig. 39.1).

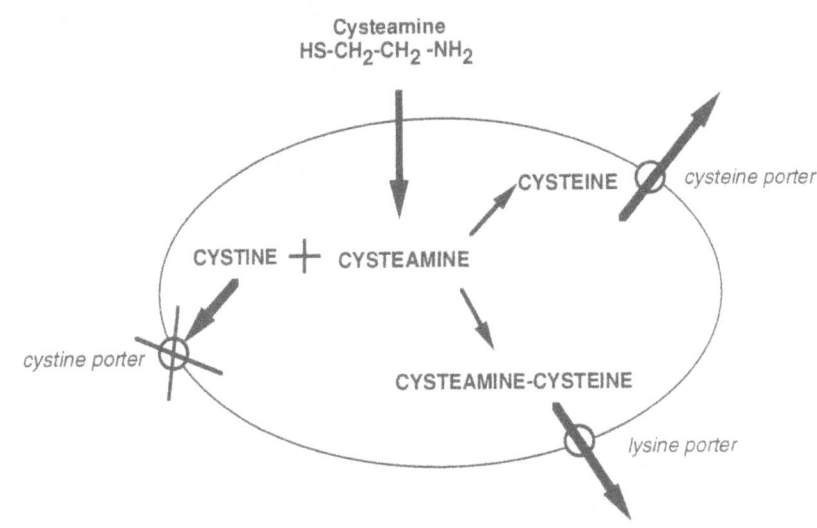

**Fig. 39.1.** Mechanisms of lysosomal export of cystine and related compounds

# CHAPTER 39

# Cystinosis

M. Broyer

CONTENTS

| | |
|---|---|
| Infantile Cystinosis | 439 |
|    Clinical Presentation | 439 |
|       First Stage | 439 |
|       End-Stage Renal Failure | 439 |
|       Late Symptoms | 440 |
|    Metabolic Derangement | 441 |
|    Diagnostic Tests | 441 |
|    Treatment | 441 |
|       Symptomatic Treatment of the Tubular Losses due to Fanconi's Syndrome | 441 |
|       Renal Replacement Therapy | 442 |
|       Symptomatic Treatment of Extrarenal Complications | 442 |
|       Specific Therapy | 442 |
|    Genetics | 442 |
| Adolescent Cystinosis | 443 |
| Adult Benign Cystinosis | 443 |
| References | 443 |

> Cystinosis is a generalized lysosomal storage disease caused by an incompletely understood defect of the export of cystine from lysosomes. The cystine accumulation leads to cellular dysfunction of many organs, the most serious of which involves the kidneys. Three phenotypic forms are discerned: the nephropathic or infantile form, an intermediate or juvenile-onset form, and a benign or adult form. The infantile form is the most frequent and is the main subject of this chapter.

## Infantile Cystinosis

### Clinical Presentation

#### First Stage

Usually, the first 3-6 months of life are symptom free. The first symptoms mostly develop before the age of 1 year [1]. They include anorexia, vomiting, polyuria and failure to thrive. If the diagnosis is delayed, severe rickets develops after 10-18 months despite correct vitamin-D supplementation. The diagnosis can immediately be suspected if both glucose and protein are found in the urine. When the disease has become symptomatic, the full expression of Fanconi syndrome is generally present at the first examination. It includes normoglycemic glycosuria, generalized aminoaciduria, tubular proteinuria (with massive excretion of $\beta_2$-microglobulin and lysozyme), phosphaturia with hypophosphatemia, hyponatremia, acidosis and excessive loss of potassium and sodium bicarbonate leading to hypokalemia. Hypercalciuria is also massive, and hypouricemia is constantly found. Tubular loss of carnitine may cause carnitine depletion. A severe concentrating defect develops simultaneously, leading rapidly to a polyuria of 2-5 l/day. The urine of cystinotic patients is characteristic, being pale and cloudy with a peculiar odor, probably due to aminoaciduria. Early onset of renal failure without marked Fanconi syndrome has also been reported [2].

The general reabsorptive defect of the proximal tubule may explain severe hydroelectrolyte imbalance, which may be life threatening. Episodes of fever, probably related to dehydration, are also commonly noted. Lithiasis related to the high urinary excretion of urate, calcium and organic acids has been reported in rare cases, and nephrocalcinosis may be observed [3]. Blond hair and a fair complexion with difficulty tanning after exposure to the sun are often noted in white cystinotic children.

Involvement of the eye, starting with photophobia, which usually appears at 2 years or 3 years of age and is more or less marked, is a primary symptom of cystinosis. Ophthalmologic examination with a slit lamp and a biomicroscope reveals cystine crystal deposits. There are also fundus abnormalities with typical retinopathy and subsequent alteration of the retinogram.

#### End-Stage Renal Failure

The natural history of the disease includes severe stunting of growth and a progressive decrease of the glomerular filtration rate, leading to end-stage renal failure (ESRF) between 6 years and 12 years of age. This evolution may be delayed by cysteamine treatment, especially when started in the first months of life. This treatment also improves growth velocity. During the course of renal deterioration, the decrease in glomerular filtration is reflected by a spurious improvement of urinary losses and a regression of Fanconi's syndrome. At ESRF, severe renal hypertension may have developed. Repeated nasal bleeding is sometimes observed in cystinotic patients on dialysis [4]. After kidney transplantation, there is no recurrence of Fanconi's syndrome, even if cystine crystals are seen in the graft, where they are carried inside macrophages or leukocytes. Tubular symptoms in grafted patients are, in fact, due to a rejection reaction.

## Late Symptoms

The advent of renal replacement therapy and transplantation has uncovered the continued cystine accumulation in extrarenal organs and has emphasized the multisystemic nature of cystinosis, which may also involve the eyes, thyroid, liver, spleen, pancreas, muscle and CNS [4–6].

## Ocular Complications

The severity of eye involvement differs from one patient to another [7, 8]. Corneal deposits accumulate progressively in the stroma of the cornea and iris in all patients and on the surface of the anterior lens and retina in some. Photophobia, watering and blepharospasm may become disabling; these symptoms are often related to erosion of corneal epithelium, leading eventually to keratopathy. Photophobia may be prevented and even cured by cysteamine eyedrops [9]. Sight may be progressively reduced, leading to blindness in a few patients who already had major ocular symptoms at an early age.

## Endocrine Disturbances

HYPOTHYROIDISM. Thyroid dysfunction usually appears, between 8 years and 12 years of age, but it may be earlier or later. It is rarely overt with clinical symptoms but rather discovered by systematic assessment of thyroid function [4, 10], and it may be partly responsible for the growth impairment. Cysteamine was reported to delay or prevent thyroid dysfunction [11].

GONADAL FUNCTION. Abnormalities in the pituitary testicular axis, with a low plasma testosterone level and a high follicle-stimulating-hormone/lutenizing-hormone level [12], seem common in male patients with cystinosis. These abnormalities may preclude attainment of full pubertal development. Female patients exhibit pubertal delay but seem to have more normal gonadal functions.

ENDOCRINE PANCREAS. Postoperative hyperglycemia and permanent insulin-dependent diabetes have been reported in several series of cystinotic patients after kidney transplantation [13]. In patients not treated by cysteamine, 50% had diabetes according to the World Health Organisation definition [14]. Exocrine pancreas was not affected except in one reported case with steatorrea [15].

## Liver and Spleen Involvement

Hepatomegaly and splenomegaly occur after 15 years of age in one-third to one-half of the cases who did not receive cysteamine [4]. Hepatomegaly is related to enlarged Kupffer's cells that transform into large foam cells containing cystine crystals. This enlargement may be the cause of portal hypertension with bleeding from gastroesophageal varices. Splenomegaly is also related to the development of foam cells in the red pulp. Hematological symptoms of hypersplenism may be noted. A recent study showed that cysteamine prevented this type of complication (personal observation).

## Muscle

A distinctive myopathy with generalized muscle atrophy and weakness (mainly of distal muscles) of all limbs, with more severe involvement of the interossei muscles and the muscles of the thenar eminence, was reported in some patients [16, 17]. Pharyngeal and oral dysfunction observed in some patients was imputed to muscle dysfunction [18].

## Central Nervous System

Several kinds of neurologic complication have been reported in cystinosis. Convulsions may occur at any age, but it is difficult to evaluate whether cystinosis is the direct cause or whether it is related to uremia, electrolyte dysequilibrium, drug toxicity, etc. A subtle and specific visuoperceptual defect and lower cognitive performances, sometimes with subtle impairment of visual memory and tactile recognition, were reported recently [19–21]. More severe CNS abnormalities with various defects have also been reported [4–6, 22]. The clinical symptoms include hypotonia, swallowing and speech difficulties, development of bilateral pyramidal signs and walking difficulties, cerebellar symptoms and a progressive intellectual deterioration. In other cases, acute ischemic episodes with hemiplegia or aphasia may occur. This cystinotic encephalopathy was only observed in patients above 19 years of age and, at present, it is difficult to know its actual incidence. The effectiveness of cysteamine treatment for the prevention of CNS involvement is also unknown. Cysteamine treatment was associated (in some cases) with an improvement of neurologic symptoms [22]. Brain imaging in cystinosis may show several types of abnormalities. Brain atrophy, calcifications and abnormal features of white matter are commonly observed after 15–20 years of age on magnetic-resonance-imaging examination [4, 21–23].

## Metabolic Derangement

The primary defect causing lysosomal accumulation of cystine in many tissues (including kidney, bone marrow, conjunctiva, thyroid, muscle, choroid plexus, brain parenchyma and lymph nodes) is not completely understood. Extensive searches for an enzyme defect of cystine breakdown have remained negative. However, it has been shown that the movement of cystine out of cystinotic lysosomes was significantly decreased in comparison to that of normal lysosomes [24, 25]. This abnormality is supposed to be related to a molecular defect of the protein involved as a transport carrier for cystine across the lysosomal membrane. The gene involved in cystinosis, CTNS, was recently cloned [26]. This gene encodes an integral membrane protein (cystinosine) with features of a lysosomal-membrane protein. Its defect is associated with lysosomal accumulation of cystine. The reason lysosomal cystine accumulation leads to cellular dysfunction is not known. It was shown that cystine loading of proximal tubular cells in vitro was associated with ATP depletion [27]. This partial knowledge has led to the to use of cysteamine, which increases the transport of cystine out of the lysosome.

## Diagnostic Tests

Cystinosis is ascertained by the assay of the free-cystine content (usually in leukocytes) which, in patients with nephropathic cystinosis, is approximately 10–50 times the normal value [28]. The assay, using a protein-binding technique on polymorphonuclear cells, is very sensitive and may be carried out on small blood samples. In cystinosis, the level is usually 5–15 nmol half cystine/mg protein. This technique enables even heterozygous carriers to be detected (0.5–1.4 nmol half cystine/mg protein) [29]. The cystine content of control subjects is usually undetectable or less than 0.4 nmol half cystine/mg protein. This assay may also be carried out on fibroblasts, conjunctiva and muscle. S-labeled cystine incorporation in fibroblasts cultured from the skin, amniotic cells or chorionic villi enables a prenatal diagnosis to be made during the first trimester [30]. The diagnosis may also be made by DNA assay if a deletion is found in the locus D17-S829 or another mutation in that gene. The use of markers close to this locus when the first child is affected could also allow prenatal diagnosis.

## Treatment

The therapy of nephropathic cystinosis is both symptomatic and specific.

### Symptomatic Treatment of the Tubular Losses due to Fanconi's Syndrome

Several abnormalities have to be corrected.

### Water

The water intake must be adjusted to diuresis, short-term weight variation and, if necessary, plasma protein concentration. Fluid requirement increases with external temperature and with fever. It is also dependent on the required mineral supplements.

### Acid–Base Equilibrium

Sodium and potassium bicarbonate, which have a better gastric tolerance than citrate, have to be given to obtain a plasma bicarbonate level between 21 mmol/l and 24 mmol/l. This is sometimes difficult and may require large amounts of buffer (up to 10–15 mmol/kg/day).

### Sodium

Sodium losses sometimes remain uncompensated after achieving acid–base equilibrium. This point is documented by a persistent hyponatremia with failure to thrive.

### Potassium

Hypokalemia requires potassium supplements in order to maintain serum potassium above 3 mmol/l. To achieve this goal, 4–10 mmol/kg/day are usually necessary. Prescription of amiloride (2–5 mg/day) may help in some cases.

### Phosphorus

Hypophosphatemia must be corrected with a supplement of sodium/potassium phosphate (0.3–1 g/day). The aim is to obtain a plasma phosphate level just above 1.0–1.2 mmol/l. This supplement, if poorly tolerated, may be gradually withdrawn after some months or years. Excessive phosphorus prescription may favor nephrocalcinosis.

### Vitamin-D Supplementation

Since tubular 1α hydroxylation is diminished in this disease, administration of 1α- or 1α-25-hydroxy vitamin $D_3$ (0.10–0.50 μg/day) is justified, especially in cases of symptomatic rickets. These prescriptions must be carefully adjusted by regular follow-up of serum calcium.

### Carnitine Supplementation

Carnitine supplementation (100 mg/kg/day in four divided doses) has been proposed in order to correct muscle carnitine depletion [31].

All supplements used to treat tubular losses due to Fanconi's syndrome have to be given regularly in order to replace the losses, which are permanent. A good way to achieve this goal is to prepare in advance all the supplements except vitamin D in a bottle containing the usual amount of water for the day. Losses of water, potassium and sodium may be drastically reduced by the prescription of Indomethacin (1.5–3 mg/kg/day in two separate doses) [32]. When renal deterioration progresses and the glomerular filtration rate decreases, the drug must be stopped; at this time, tubular losses also decrease, and the mineral supplements must be adjusted and progressively reduced in order to avoid overload, especially with sodium and potassium. At the dialysis stage, mineral supplements are no longer necessary. Feeding problems may require tube feeding and, in some cases, continuous or intermittent total parenteral nutrition [33].

### *Renal Replacement Therapy*

There is no specific requirement for cystinotic children at this stage. Hemodialysis or continuous ambulatory/cyclic peritoneal dialysis are both effective and are applied according to the circumstances. As for any child with ESRF, kidney transplantation is considered the best approach. Results of kidney transplantation in the European Dialysis and Transplant Association pediatric registry were better for infantile cystinosis than for any other primary renal disease in children [4, 34].

### *Symptomatic Treatment of Extrarenal Complications*

Hypothyroidy has to be compensated by L-thyroxine supplementation, even when asymptomatic. Growth stunting, one of the most striking complications of nephropathic cystinosis, was reported to be improved by administration of recombinant growth hormone at a dose of 1 U/kg/week [35]. Portal hypertension may lead to ascites and bleeding esophageal varices, rendering a portal bypass necessary. Hypersplenism with permanent leukopenia and/or thrombopenia may be an indication for splenectomy. As for ophthalmologic treatment, photophobia and watering may be improved by local symptomatic therapy, such as vitamin-A eyedrops, artificial tears, topical lubricants, and thin-bandage soft contact lenses. It has been shown that eyedrops containing 0.5% cysteamine were able to prevent corneal deposits [9] and may decrease the deposits already present. Corneal graft has been performed (rarely), with variable results.

### *Specific Therapy*

Several attempts have been made to suppress lysosomal cystine storage, which is the basic lesion of cystinosis. Dietary restriction of sulfur amino acids has no effect; ascorbic acid, which is able to reduce cystine accumulation in vitro, is of no clinical value and even worsens the renal prognosis. Dithiothreitol was also ineffective. Only one drug, cysteamine ($HS-CH_2-NH_2$), has been employed in cystinosis with apparent benefit, as shown in a prospective study [36]. Nevertheless, the prescription of cysteamine raises some problems, since its odor and taste make its administration unattractive. It is also causes the breath to have an unpleasant smell. Phosphocysteamine would have the same efficacy, with a better odor and taste [37]. Cysteamine is now commercially available in North America and Europe as cysteamine bitartrate (Cystagon). The dose is progressively increased from 10 mg/kg/day to 50 mg/kg/day. Cysteamine is rapidly absorbed, and its maximum effect (assessed by cystine assay in leukocytes) occurs after 1–2 h and lasts generally no longer than 6 h. Therefore, it has to be given in four separate doses (one every 6 h) in order to obtain the best prevention of cystine accumulation. Careful monitoring of polymorphonuclear cystine content is essential since the response to cysteamine is variable. Polymorphonuclear cystine content should be determined just prior to the next dose; the aim is to keep this content at less than 2 nmol half cystine/mg protein or, preferably, less than 1 nmol half cystine/mg protein. The drug should be started as soon as the diagnosis is confirmed [38, 39]. The good results obtained for patients with nephropathy have encouraged us to also give cysteamine to patients who are at risk of developing extrarenal complications.

### Genetics

Nephropathic cystinosis is an autosomal-recessive disorder. The gene was first mapped to chromosome 17 [43] and was cloned 2 years later [26]. This gene, named *CTNS*, consists of 12 exons and encodes a protein of 367 amino acids; it has the structure of an integral membrane protein with six or seven membrane-spanning domains; 30% of patients of European origin were homozygously deleted. In the other cases, a mutation or deletion was found in both alleles.

Adolescent and adult forms have the same mode of inheritance. In somatic cell hybrids, lack of complementation between fibroblasts from patients with nephropathic and benign cystinosis supports the hypothesis that the genes of these three diseases are alleles on the same locus [44]. However, adult and infantile forms have never been observed in the same family.

## Adolescent Cystinosis

This corresponds to a very rare, milder form of the disease, with a later clinical onset and delayed evolution to ESRF [40, 41]. The first symptom usually appears after 6-8 years of age. Proteinuria may be misleading, because its severity is sometimes in the nephrotic range. Fanconi's syndrome may be absent [41] or is moderate, and tubular losses are less important than in infantile cystinosis. The same is true for extrarenal symptoms. ESRF usually develops at approximately 15 years of age in most patients. The diagnosis is ascertained by the assessment of the cystine content of leukocytes, which has been found to be similar to that seen in infantile cases.

## Adult Benign Cystinosis

Adult or benign cystinosis was first reported by Cogan et al. in 1957 [42]. This exceptional autosomal-recessive disorder is characterized by the presence of cystine crystals in the eye and bone marrow. Crystals in the cornea are usually found by chance examination. The level of cystine in leukocytes is between those of heterozygotes and homozygotes of nephropathic cystinosis. The affected patients are asymptomatic.

## References

1. Broyer M, Guillot M, Gubler MC, Habib R (1981) Infantile cystinosis: a reappraisal of early and late symptoms. Adv Nephrol 10:137-166
2. Van't Hoff WG, Ledermann SE, Waldron M, Trompeter RS (1995) Early-onset chronic renal failure as a presentation of infantile nephropathic cystinosis. Pediatr Nephrol 9:483-484
3. Theodoropoulos DS, Shawker TH, Heinrichs C, Gahl WA (1995) Medullary nephrocalcinosis in nephropathic cystinosis. Pediatr Nephrol 9:412-418
4. Broyer M, Tete MJ, Gubler MC (1987) Late symptoms in infantile cystinosis. Pediatr Nephrol 1:519-524
5. Gahl WA, Kaiser-Kupfer MI (1987) Complications of nephropathic cystinosis after renal failure. Pediatr Nephrol 1:260-268
6. Theodoropoulos DS, Krasnewich D, Kaiser-Kupfer MI, Gahl WA (1993) Classic nephropathic cystinosis as an adult disease. JAMA 270:2200-2204
7. Kaiser-Kupfer MI, Caruso RC, Monkler DS, Gahl WA (1986) Long-term ocular manifestations in nephropathic cystinosis. Arch Ophthalmol 104:706-711
8. Dufier JL, Dhermy P, Gubler MC, Gagnadoux MF, Broyer M (1987) Ocular change in long term evolution of infantile cystinosis. Ophthalmic Paediatr Genet 8:131-137
9. Jones NP, Postlethwaite RJ, Noble JL (1991) Clearance of corneal crystals in nephropathic cystinosis by topical cysteamine 0.5%. Br J Ophthalmol 75:311-312
10. Lucky AW, Howley PM, Megyesi K, Spielberg SP, Schulman JD (1977) Endocrine studies in cystinosis: compensated primary hypothyroidism. J Pediatr 91:204-210
11. Kimonis VE, Troendle J, Rose SR et al. (1995) Effects of early cysteamine therapy on thyroid function and growth in nephropathic cystinosis. J Clin Endocrinol Metab 80:3257-3261
12. Chik CL, Friedman A, Merriam GR, Gahl WA (1993) Pituitary-testicular function in nephropathic cystinosis. Ann Intern Med 119:568-575
13. Bakchine H, Niaudet P, Gagnadoux MF, Broyer M, Czernichow P (1984) Diabète induit par les corticoides chez 6 enfants après transplantation rénale. Arch Fr Pediatr 41:261-224
14. Robert JJ, Tête MJ, Guest G et al. (1999) Diabetes mellitus in patients with infantile cystinosis after renal transplantation. Pediatr Nephrol 13:524-529
15. Fivusch B, Flick J, Gahl WA (1988) Pancreatic exocrine insufficiency in a patient with nephropathic cystinosis. J Pediatr 112:49-51
16. Gahl WA, Dalakas M, Charnas L, Chen K (1988) Myopathy and cystine storage in muscles in a patient with nephropathic cystinosis. N Engl J Med 319:1461-1464
17. Charnas LR, Luciano CA, Dalakas M et al. (1994) Distal vacuolar myopathy in nephropathic cystinosis. Ann Neurol 35:181-188
18. Sonies BC, Ekman EF, Andersson HC et al. (1990) Swallowing dysfunction in nephropathic cystinosis. N Engl J Med 323:565-570
19. Trauner D, Chase C, Scheller J, Katz B, Schnsider J (1988) Neurologic and cognitive deficits in children with cystinosis. J Pediatr 112:912-914
20. Colah S, Trauner DA (1997) Tactile recognition in infantile nephropathic cystinosis. Dev Med Child Neurol 39:409-413
21. Nichols S, Press G, Schneider J, Trauner D (1990) Cortical atrophy and cognitive performance in infantile nephropathic cystinosis. Pediatr Neurol 6:379-381
22. Broyer M, Tete MJ, Guest G et al. (1996) Clinical polymorphism of cystinosis encephalopathy. Results of treatment with cysteamine. J Inherit Metab Dis 19:65-75
23. Cochat P, Drachman R, Gagnadoux MF, Pariente D, Broyer M (1986) Cerebral atrophy and nephropathic cystinosis. Arch Dis Child 61:401-403
24. Gahl WA, Bashan N, Tietze F (1982) Cystine transport is defective in isolated leukocyte lysosomes from patients with cystinosis. Science 217:1263-1265
25. Gahl WA, Tietze F, Bashan N (1983) Characteristics of cystine countertransport is normal and cystinotic lysosome-rich granular fractions. Biochem J 216:393-400
26. Town M, Jean G, Cherqui S et al. (1998) A novel gene encoding an integral membrane protein is mutated in nephropathic cystinosis. Nature Genet 18:319-324
27. Foreman JW, Bowring MA, Lee J, States B, Segal S (1987) Effect of cystine dimethyl ester on renal solute handling and isolated renal tubule transport in the rat: a new model of the Fanconi syndrome. Metabolism 36:1185-1191
28. Schneider JA, Wong V, Bradley K, Seegmiler JE (1968) Biochemical comparisons of the adult and childhood forms of cystinosis. N Engl J Med 279:1253-1257
29. Smolin LA, Clark KF, Schneider JA (1987) An improved method for heterozygote detection of cystinosis, using polymorphonuclear leukocyte. Am J Hum Genet 41:266-275

30. Patrick AD, Young EP, Mossman J et al. (1987) First trimester diagnosis of cystinosis using intact chorionic villi. Prenat Diagn 7:71-74
31. Gahl WA, Bernardini IM, Dalakas MC et al. (1993) Muscle carnitine repletion by long-term carnitine supplementation in nephropathic cystinosis. Pediatr Res 34:115-119
32. Haycock GB, Al-dahhan J, Mak RHK, Chantler C (1982) Effect of indomethacin on clinical progress and renal function in cystinosis. Arch Dis Child 57:934-939
33. Elenberg E, Norling LL, Kleinman RE, Ingelfinger JR (1998) Feeding problems in cystinosis. Pediatr Nephrol 12:365-370
34. Broyer M, on behalf of the EDTA registry committee (1989) Kidney transplantation in children, data from the EDTA registry. Transplant Proc 21:1985-1988
35. Wuhl E, Haffner D, Gretz N et al. (1998) Treatment with recombinant human growth hormone in short children with nephropathic cystinosis: no evidence for increased deterioration rate of renal function. The European Study Group on growth hormone treatment in short children with nephropathic cystinosis. Pediatr Res 43:484-488
36. Gahl WA, Reed G, Thoene JG et al. (1987) Cysteamine therapy for children with nephropathic cystinosis. N Engl J Med 316:971-977
37. Smolin LA, Clark KF, Thoene JG, Gahl WA, Schneider JA (1988) A comparison of the effectiveness of cysteamine and phosphocysteamine in elevating plasma cysteamine concentration and decreasing leukocyte free cystine in nephropathic cystinosis. Pediatr Res 23:616-620
38. Da Silva VA, Zurbrug RP, Lavanchu P et al. (1985) Long term treatment of infantile nephropathic cystinosis with cysteamine. N Engl J Med 313:1460-1463
39. Markello TC, Bernardini IM, Gahl WA (1993) Improved renal function in children with cystinosis treated with cysteamine. N Engl J Med 328:1157-1162
40. Hauglustaine D, Corbeel L, Vandamme B, Serrus M, Michelsen P (1976) Glomerulonephritis in late onset cystinosis. Report of 2 cases and review of literature. Clin Nephrol 6:529-535
41. Hory B, Billerey C, Royer J, Saint Hillier Y (1994) Glomerular lesions in juvenile cystinosis: report of 2 cases. Clin Nephrol 42:327-330
42. Cogan DG, Kuwabara T, Kinoshita J, Sheehan L, Merola L (1957) Cystinosis in a adult. JAMA 164:394
43. The cystinosis Collaborative Research Group (1995) Linkage of the gene for cystinosis to markers on the short arm of chromosome 17. Nature Genet 10:246-248
44. Pellet OL, Smith ML, Greene AA, Schneidet KA (1988) Lack of complementation in somatic cell hydrids between fibroblasts from patients with different forms of cystinosis. Proc Natl Acad Sci USA 85:3531-3534

# CHAPTER 40

## Oxalate Metabolism

Oxalate is a very poorly soluble end-product of the metabolism of a number of amino acids (particularly glycine) and of other compounds, such as sugars and ascorbic acid. The immediate precursors of oxalate are glyoxylate and glycolate. The main site of synthesis of glyoxylate and oxalate is within the hepatic peroxisomes, which can also detoxify glyoxylate by reconversion into glycine, catalyzed by alanine:glyoxylate aminotransferase. In the cytosol, glyoxylate can be converted into oxalate by lactic acid dehydrogenase (LDH). It can also be converted into glycolate by glyoxylate reductase (GR) and into glycine by glutamate:glyoxylate aminotransferase. Glycolate can also be formed from hydroxypyruvate, a catabolite of glucose and fructose. Hydroxypyruvate can be converted into L-glycerate by LDH and into D-glycerate by D-glycerate dehydrogenase, which also has a GR activity.

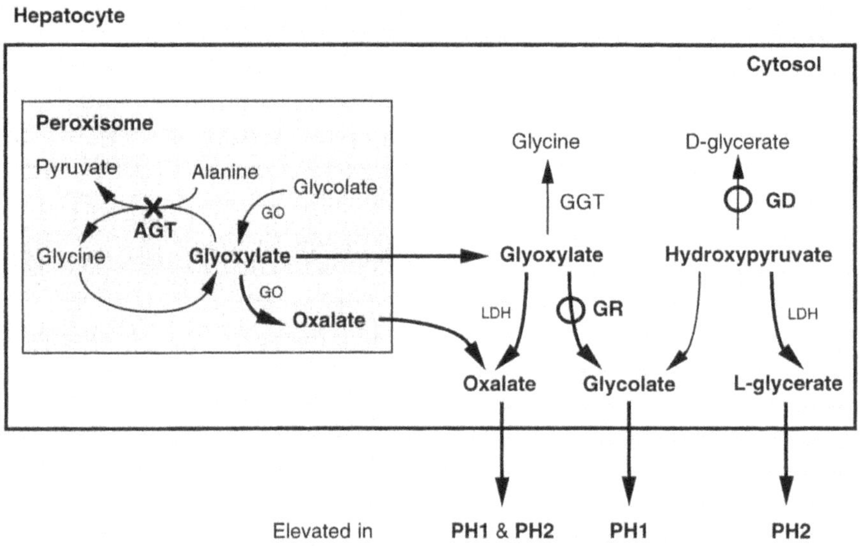

**Fig. 40.1.** Major reactions involved in oxalate, glyoxylate and glycolate metabolism in the human hepatocyte. *AGT*, alanine:glyoxylate aminotransferase; *GD*, D-glycerate dehydrogenase; *GGT*, glutamate:glyoxylate aminotransferase; *GO*, glycolate oxidase; *GR*, glyoxylate reductase; *LDH*, lactate dehydrogenase; ×, metabolic block in primary hyperoxaluria type 1 (PH1); O, metabolic block in primary hyperoxaluria type 2

# Primary Hyperoxalurias

Pierre Cochat and Marie-Odile Rolland

CONTENTS

Primary Hyperoxaluria Type 1 .................... 447
   Clinical Presentation ......................... 447
      Renal Involvement ........................ 447
      Extrarenal Involvement ..................... 447
   Metabolic Derangement ....................... 448
   Diagnostic Tests ............................ 448
   Treatment and Prognosis ..................... 449
      Conservative Treatment .................... 449
      Renal Replacement Therapy ................ 449
      Enzyme Replacement Therapy .............. 449
      Combined Liver-Kidney Transplantation ......... 450
      Prospect for Gene Therapy ................. 450
   Genetics .................................. 450
Primary Hyperoxaluria Type 2 .................... 451
   Clinical Presentation ......................... 451
   Metabolic Derangement ....................... 451
   Diagnostic Test ............................. 451
   Treatment and Prognosis ..................... 451
   Genetics .................................. 452
Primary Hyperoxaluria Type 3 .................... 452
References .................................. 452

---

Primary hyperoxalurias (PH) are rare diseases characterized by overproduction and accumulation of oxalate in tissues. Three types have been identified.

PH1, caused by deficiency or mistargeting of alanine:glyoxylate aminotransferase (AGT) in liver peroxisomes, is the most frequent and most severe form. Deposits of calcium-oxalate crystals in the kidney lead to stones, nephrocalcinosis and deteriorating kidney function, while bone disease is the most severe extrarenal involvement. Careful conservative treatment (high fluid intake, calcium-oxalate crystallization inhibitors and pyridoxine) should be started early, as it significantly prolongs kidney survival. Liver and kidney transplantation are the final current options. Hyperoxaluria and hyperglycoluria are indicative of PH1.

PH2, caused by D-glycerate dehydrogenase (GD) deficiency in the liver, is less frequent and less severe, and treatment is less demanding. Hyperoxaluria without hyperglycoluria and increased urinary excretion of L-glycerate differentiate PH2 from PH1. PH3 is likely to exist, but its pathophysiology has not been identified.

## Primary Hyperoxaluria Type 1

### Clinical Presentation

#### Renal Involvement

PH1 presents with symptoms referable to the urinary tract (loin pain, hematuria, urinary tract infection or passage of a stone) in more than 80% of cases. Calculi – multiple, bilateral and radio-opaque – are composed of calcium oxalate. Nephrocalcinosis, best demonstrated by ultrasound, is present on plain abdomen X-ray at an advanced stage (Fig. 40.2). The median age at initial symptoms is 5 years, ranging from birth to the sixth decade; end-stage renal failure (ESRF) is reached by the age of 15 years in half of the patients [1]. The infantile form of PH1 often presents as a life-threatening disease because of both early oxalate load and immature glomerular filtration rate (GFR) [2]. However, some patients are asymptomatic, and PH1 may be discovered by family history.

#### Extrarenal Involvement

When the GFR of PH1 patients falls to below 40 ml/min/1.73 m$^2$, continued overproduction of oxalate by the liver concurrent with reduced oxalate excretion by the kidneys leads to increasing oxalate deposition in many organs [3]. Bone is the major compartment of the insoluble oxalate pool. Calcium-oxalate crystals accumulate first in the metaphyseal area and form dense suprametaphyseal bands on X-rays, which are almost pathognomonic [4]. Later, the whole skeleton becomes dense, and spontaneous fractures occur, ultimately leading to severe disability. Along with the skeleton, systemic involvement includes many organs: heart (cardiomyopathy, conduction defect), nerves (peripheral neuropathy, mononeuritis multiplex), joints (synovitis), arteries (disseminated occlusive lesions, gangrene), skin (ulcerating subcutaneous calcinosis, livedo reticularis), soft tissues and retina (flecked retinopathy; Fig. 40.3).

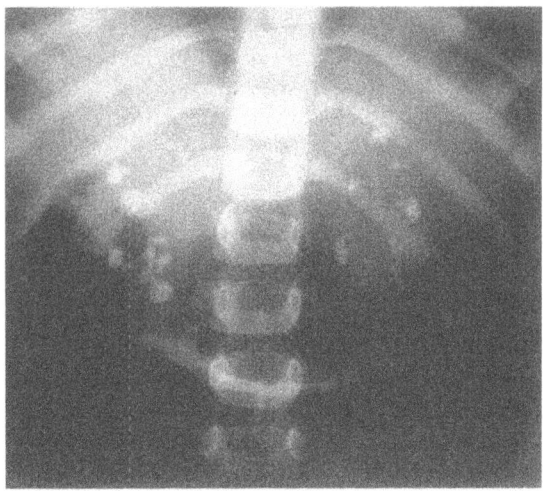

**Fig. 40.2.** Plain abdomen X-ray in a 5-year-old boy with primary hyperoxaluria type 1, showing bilateral nephrocalcinosis

## Metabolic Derangement

PH1 is caused by a deficiency of the liver-specific AGT. The resulting decreased transamination of glyoxylate into glycine leads to a subsequent increase in its oxidation to oxalate (Fig. 40.1), a poorly soluble end product. In patients with a presumptive diagnosis of PH, 10–15% are identified as non-PH1 because AGT activity and immunoreactivity are normal [5]. Among PH1 patients, two thirds have undetectable enzyme activity ($enz^-$), and the majority of these also have no immunoreactive protein (cross-reacting material, $crm^-$). In the rare, $enz^-/crm^+$ PH1 patients, a catalytically inactive but immunoreactive AGT is found within the peroxisomes. The remaining PH1 patients have AGT activities in the range of 5–50% of the mean normal activity ($enz^+$), and the level of immunoreactive protein parallels the level of enzyme activity [5]. In $enz^+/crm^+$ patients, the disease is caused by a mistargeting of AGT; about 90% of the immunoreactive AGT is localized in the mitochondria instead of in the peroxisomes, where only 10% of the activity is found [5, 7]. Surprisingly, human hepatocyte AGT, which is normally exclusively localized within the peroxisomes, is unable to function when diverted to the mitochondria, its normal localization in several other species (cat, dog, frog) [6].

## Diagnostic Tests

Concomitant hyperoxaluria and hyperglycoluria are indicative of PH1 (Table 40.1), but 20–30% of patients with PH1 do not have hyperglycoluria [11]. In some cases, the presence of monohydrated calcium-oxalate crystals (whewhellite) in urine or in tissues (kidney, bone marrow) can be assessed by polarized-light microscopy or infrared spectrophotometry [12]. Only AGT activity and its subcellular localization from a percutaneous liver biopsy will confirm the diagnosis. Prenatal diagnosis can be performed from DNA obtained from chorionic villi (10–12 weeks gestation) or amniocytes (16 weeks gestation) using mutation detection (if the index case has been investigated) or analysis of several microsatellites flanking the *AGXT* gene (if the analysis has been performed for both the index case and the parents) [5].

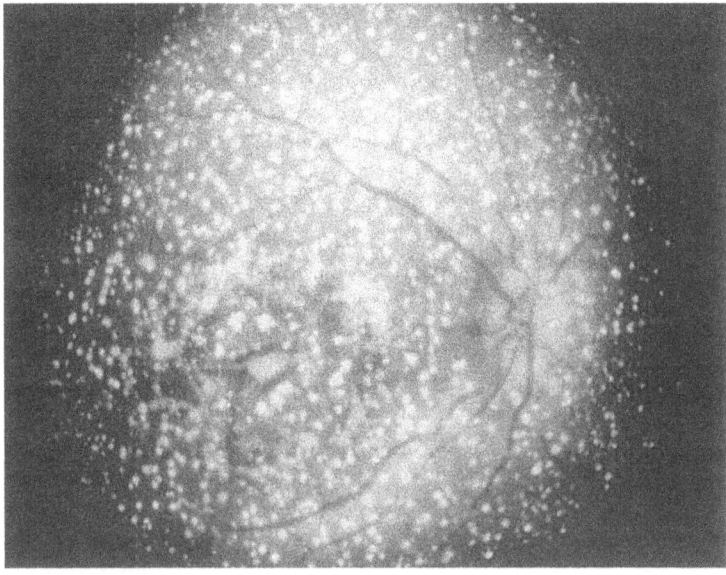

**Fig. 40.3.** Fundoscopy in a 14-year-old girl with primary hyperoxaluria type 1

## Treatment and Prognosis

### Conservative Treatment

When urine oxalate exceeds 0.4 mmol/l, the risk of stone formation is increased, especially if urine calcium exceeds 4 mmol/l; therefore, supportive therapy should be adapted to keep the concentrations of oxalate and calcium below these limits [13]. Low concentrations can be achieved by a high fluid intake (>2 l/m²/day) supported by calcium-oxalate-crystallization inhibitors, i.e., sodium/potassium citrate 150 mg/kg/day [14, 15]. Diuretics require careful management, since frusemide is able to maintain a high urine output, but with the risk of an increased calciuria, whereas the diuretic effect of hydrochlorothiazide is less marked but is associated with an appreciable decrease of calcium excretion. A combination of both diuretics would contribute to reduce the urinary concentration of calcium. Restriction of dietary oxalate intake (strawberries, rhubarb, spinach, coffee, tea, nuts) has only minor influence on the disease. The effects of conservative measures can be assessed by serial determinations of crystalluria score and calcium-oxalate supersaturation software [15, 16].

In some PH1 patients, pyridoxine – a cofactor of AGT – may produce a significant reduction in oxalate excretion when given at a pharmacological dose (5–15 mg/kg/day), so its efficacy must be tested for each patient at an early stage of the disease [5, 6, 15]. It should be kept in mind that megadoses of pyridoxine may induce sensory neuropathy. Recently, an attempt to inhibit hepatic synthesis of oxalate by using 2-oxothiazolidine-4-carboxylate has been reported and is under evaluation [17].

The surgical treatment of stones should clearly be limited because of the known risk of further decline in GFR. The use of extracorporeal shock-wave lithotripsy may be an available option in selected patients, but the presence of nephrocalcinosis may be responsible for parenchymal damage.

### Renal Replacement Therapy

#### Dialysis

Conventional dialysis is unsuitable for PH1 patients who have reached ESRF, because it cannot clear sufficient amounts of oxalate [6]. Theoretically, daily hemodialysis (6–8 h/session) would be required, but such a strategy cannot be routinely used. However, it may be helpful before and after isolated kidney or combined liver–kidney transplantation. Conventional long-term hemodialysis is generally regarded as contraindicated, because it serves only to prolong a miserable existence: instead of dying quickly from uremia, patients promptly experience a deterioration in their quality of life and die miserably because of the progression of extrarenal oxalate deposition.

### Kidney Transplantation

Kidney transplantation allows significant removal of soluble oxalate. However, because the biochemical defect is in the liver, overproduction of oxalate and subsequent deposition in tissues continues unabated. The high rate of urinary oxalate excretion originates from both ongoing oxalate production from the native liver and oxalate deposits in tissues. Due to oxalate accumulation in the transplanted kidney, overall graft survival is often short. In 1990, data from the European Dialysis and Transplantation Association Registry showed that, 3 years after renal transplantation, only 23% of living-donor and 17% of cadaver-donor grafts were functioning; 26% of the recipients had died after 3 years [18]. The long-term outcome of PH1 after renal transplantation remains uncertain, with a 5- to 10-year patient survival rate ranging from 10% to 50% [19]. In addition, renal transplantation does not prevent the progression of skeletal and vascular complications. The chances of a successful transplantation are unrelated to residual AGT activity [20], but success is improved only if the transplant is performed when there is substantial renal function (i.e., a GFR ranging from 20 ml/min/1.73 m² to 30 ml/min/1.73 m²) and in the absence of important extrarenal involvement. In selected patients, good results have been reported after early renal transplantation and vigorous perioperative dialysis [21]; however, living related donors should be avoided because the overall results are poor [19]. In spite of the supposed former pyridoxine resistance, pyridoxine should be re-tested in patients after isolated renal transplantation.

### Enzyme Replacement Therapy

Ideally, any kind of transplantation should precede advanced systemic oxalate storage. Therefore, further assessment of the oxalate burden needs to be predicted by sequentially monitoring GFR, plasma oxalate and, possibly, systemic involvement (bone histology, dual-energy X-ray absorptiometry) [22].

### Rationale for Liver Transplantation

Since the liver is the only organ responsible for glyoxylate detoxification by AGT, the excessive production of oxalate will continue as long as the native liver is left in place. Therefore, any form of enzyme replacement is successful only when the deficient host liver is removed concomitantly [23]. Liver transplantation can supply the missing enzyme in the correct organ (liver), cell (hepatocyte) and intracellular compartment (peroxisome) [5, 23]. The ultimate goal of organ replacement is to change a positive whole-body accretion rate into a negative one by reducing endogenous oxalate synthesis and providing good oxalate clearance via either the native or the transplanted kidney [24].

### Combined Liver-Kidney Transplantation

In Europe, eight to ten combined liver–kidney transplantations per year have been reported in the PH1 Transplant Registry Report [19]. The results are encouraging, as patient survival approximates 80% at 5 years and 70% at 10 years. In addition, despite the potential risks to the grafted kidney due to oxalate release from the body stores, kidney survival is about 95% 3 years after transplantation, and the GFR ranges between 40 ml/min/1.73 m$^2$ and 60 ml/min/1.73 m$^2$ after 5-10 years [19, 22, 23].

### Isolated Liver Transplantation

Isolated liver transplantation might be the first-choice treatment in selected patients before an advanced stage of chronic renal failure occurs, i.e., while the GFR is between 60 ml/min/1.73 m$^2$ and 40 ml/min/1.73 m$^2$ [24, 25]. Such a strategy has a strong theoretical basis but raises ethical controversies. Approximately 20 patients have received an isolated liver transplant without uniformly accepted recommendations, because the course of the disease is unpredictable and a sustained improvement can follow a phase of rapid decrease in GFR.

### Outcome of Renal and Extrarenal Involvement after Transplantation

Deposits of calcium oxalate in tissues can be remobilized by decreasing the synthesis and increasing the clearance of oxalate; these processes depend on the accessibility of the oxalate burden to the bloodstream [23]. After combined transplantation, the plasma oxalate level returns to normal before the urine oxalate level does; indeed oxaluria can remain elevated for several weeks or months after transplantation [5, 19]. Combined liver-kidney transplantation is, therefore, able to normalize urine oxalate without the associated risk of recurrent nephrocalcinosis or renal calculi, which might compromise graft function. Glycolate, which is more soluble than oxalate and does not accumulate, is excreted in normal amounts immediately after liver transplantation.

Whatever the transplantation strategy, the kidney must be protected against the damage that can be induced by heavy oxalate load suddenly released from tissues. Forced fluid intake (5 l/1.73 m$^2$/day), supported by diuretics and the use of crystallization inhibitors, is the most important strategy (see "Conservative Treatment"); crystalluria is a helpful tool in renal graft management after combined liver–kidney transplantation [16]. The benefit of daily high-efficiency (pre- and) post-transplant hemodialysis/filtration is still debated; it should be proposed on the basis of a rapid drop in plasma oxalate. However, it can increase the risk of urine calcium-oxalate supersaturation and, therefore, should be limited to patients with severe systemic involvement.

Combined transplantation should be planned when the GFR ranges between 20 ml/min/1.73 m$^2$ and 40 ml/min/1.73 m$^2$ because, at this level, oxalate retention increases rapidly [4]. Of course, in patients with ESRF, vigorous hemodialysis should be started, and urgent liver-kidney transplantation should be performed. Even at these late stages, damaged organs (skeleton, heart) might benefit from enzyme replacement [24].

### Donors for Combined Liver–Kidney Transplantation

The type of donor – cadaver or living, related – depends mainly on the physician and the country where the patient is treated. However, due to the timing of the transplant procedure relative to the course of the disease, a living related donor should be considered because of the restricted number of potential bi-organ cadaver donors for synchronous liver-kidney transplantation.

### Prospect for Gene Therapy

Although gene therapy has the potential to overcome most of the problems associated with organ transplantation, it is likely to give rise to a whole new set of problems, some associated with vector technology in general and some specifically associated with PH1 [5].

## Genetics

PH1 is the most common form of PH (1:120,000 live births in France). Due to autosomal-recessive inheritance, it is much more frequent when parental consanguinity is present, i.e., in developing countries.

It is responsible for less than 0.5% of ESRF in children in Europe versus 13% in Tunisia [25].

Human liver AGT cDNA and genomic DNA have been cloned and sequenced; the normal AGT gene (*AGXT*) maps to chromosome 2q37.3 [5]. Polymorphic variations have been identified in *AGXT* (74-bp duplication within intron 1 and $Pro_{11}Leu$ substitution) [5]. Over 20 mutations have been identified to date and might play a role in enzyme trafficking; e.g., G170R substitution is found in one third of European and North American patients, appears to act with a very common, normally occurring P11L polymorphism and leads to peroxisome-to-mitochondrion AGT mistargeting [26]. However, there is no clear relationship between genotype, residual AGT activity and disease severity. The study of DNA among different ethnic groups has shown relevant features in Japanese, Turkish and Pakistani populations [5, 25, 27].

Prenatal diagnosis using a combination of linked polymorphism and detection of the two most common mutations has an accuracy of more than 99% and can be performed during the first trimester of pregnancy with chorionic villus biopsy [28].

## Primary Hyperoxaluria Type 2

### Clinical Presentation

PH2 has been documented in less than 30 published cases, but there are probably some unreported cases [8]. Median age at onset of first symptoms is 15 years [29]. In most patients, the classical presentation is urolithiasis, including infection and obstruction, but stone-forming activity is lower than in PH1, nephrocalcinosis is less frequent, and systemic involvement is exceptional.

### Metabolic Derangement

Deficiency of the enzyme GD, which also has glyoxylate reductase (GR) activity, is believed to be the underlying defect (Fig. 40.1) [30]. Analysis of liver and lymphocyte samples from patients with PH2 showed that GR activity was either very low or undetectable, while GD activity was reduced in liver but within the normal range in lymphocytes [31].

### Diagnostic Test

In the presence of hyperoxaluria without hyperglycoluria, a diagnosis of PH2 should be considered, especially when AGT activity is normal. However, hyperoxaluria in PH2 tends to be less pronounced than in PH1. The biochemical hallmark is the increased urinary excretion of L-glycerate (Table 40.1) and the definitive diagnosis requires measurement of GR and GD activities in a liver biopsy [31].

### Treatment and Prognosis

The overall long-term prognosis is better than for PH1. ESRF occurs in 12% of patients between 23 years and 50 years of age [8, 29]. As in PH1, supportive treatment includes high fluid intake, crystallization inhibitors and prevention of complications; there is no reason to use pyridoxine. Kidney transplantation has been performed in some ESRF patients, often leading to recurrence (nephrocalcinosis) including hyperoxaluria and L-glycerate excretion [8]. The concept of liver transplantation has, therefore, been suggested, but more data are needed concerning the tissue distribution of the deficient enzyme and the biochemical impact of hepatic GD/GR deficiency before such a strategy can be recommended.

**Table 40.1.** Plasma and urine concentrations of oxalate, glycolate and L-glycerate: normal values [6, 8–10]

| | | | |
|---|---|---|---|
| Urine | Oxalate[a] (per day) | Child | <0.46 mmol/1.73 m$^2$ |
| | | Adult | <0.40 mmol/1.73 m$^2$ |
| | Oxalate:creatinine | <1 year | <0.25 mmol/mmol |
| | | 1–4 years | <0.13 mmol/mmol |
| | | 5–12 years | <0.07 mmol/mmol |
| | | Adult | <0.08 mmol/mmol |
| | Glycolate[b] (per day) | Child | <0.55 mmol/1.73 m$^2$ |
| | | Adult | <0.26 mmol/1.73 m$^2$ |
| | Glycolate:creatinine | <1 year | <0.07 mmol/mmol |
| | | 1–4 years | <0.09 mmol/mmol |
| | | 5–12 years | <0.05 mmol/mmol |
| | | Adult | <0.04 mmol/mmol |
| | L-Glycerate:creatinine | | <0.03 mmol/mmol |
| Plasma | Oxalate[a] | Child | <7.4 µmol/l |
| | | Adult | <5.4 µmol/l |
| | Oxalate:creatinine | Child | <0.19 µmol/µmol |
| | | Adult | <0.06 µmol/µmol |

[a] Oxalate (COOH–COOH): 1 mmol = 90 mg
[b] Glycolate (COOH–CH$_2$OH): 1 mmol = 76 mg

## Genetics

There is evidence for autosomal-recessive transmission of the disease, and the human GD gene has recently been located on chromosome 9q11 [32].

## Primary Hyperoxaluria Type 3

To date, there is no clearly identified PH3, but some patients with PH have been shown to have neither AGT nor GD/GR deficiency [25]. Primary oxalate hyperabsorption unrelated to any identified intestinal disease and without any other organic aciduria has been advocated, but this seems unlikely. A novel type of PH has been suggested on the basis of rare cases of hyperoxaluria with hyperglycoluria in the absence of AGT deficiency [33]. Therefore, it is likely that there is at least one other unexplained form of PH.

## References

1. Cochat P, Deloraine A, Rotily M et al. (1995) Epidemiology of primary hyperoxaluria type 1. Nephrol Dial Transplant 10 [Suppl 8]:3–7
2. Cochat P, Koch Nogueira PC, Mahmoud AM et al. (1999) Primary hyperoxaluria in infants: medical, ethical and economical issues. J Pediatr 135:746–750
3. Morgan SH, Purkiss P, Watts RWE, Mansell MA (1987) Oxalate dynamics in chronic renal failure. Comparison with normal subjects and patients with primary hyperoxaluria. Nephron 46:253–257
4. Schnitzler C, Kok JA, Jacobs DWC et al. (1991) Skeletal manifestations of primary oxalosis. Pediatr Nephrol 5:193–199
5. Danpure CJ, Rumsby G (1995) Enzymological and molecular genetics of primary hyperoxaluria type 1. Consequences for clinical management. In: Khan SR (ed) Calcium oxalate in biological systems. CRC Press, Boca Raton, pp 189–205
6. Barratt TM, Danpure CJ (1994) Hyperoxaluria. In: Holliday MA, Barratt TM, Avner ED (eds) Pediatric nephrology. Williams and Wilkins, Baltimore, pp 557–572
7. Danpure CJ, Cooper PJ, Wise PJ, Jennings PR (1989) An enzyme trafficking defect in two patients with primary hyperoxaluria type 1: peroxisomal alanine: glyoxylate aminotransferase rerouted to mitochondria. J Cell Biol 108:1345–1352
8. Kemper MJ, Conrad S, Müller-Wiefel DE (1997) Primary hyperoxaluria type 2. Eur J Pediatr 156:509–512
9. Gaulier JM, Cochat P, Lardet G, Vallon JJ (1997) Serum oxalate microassay using chemiluminescence detection. Kidney Int 52:1700–1703
10. Matos V, Van Melle G, Werners D, Bardy D, Guignard JP (1999) Urinary oxalate and urate to creatinine ratios in a healthy pediatric population. Am J Kidney Dis 34:e1
11. Latta K, Brodehl J (1991) Primary hyperoxaluria type 1. Eur J Pediatr 149:518–522
12. Daudon M, Estepa L, Lacour B, Jungers P (1998) Unusual morphology of calcium oxalate calculi in primary hyperoxaluria. J Nephrol 11 [Suppl 1]:51–55
13. Hallson PC (1988) Oxalate crystalluria. In: Rose GA (ed) Oxalate metabolism in relation to urinary stone. Springer, Berlin Heidelberg New York, pp 131–166
14. Leumann E, Hoppe B, Neuhaus T (1993) Management of primary hyperoxaluria: efficacy of oral citrate administration. Pediatr Nephrol 7:207–211
15. Milliner DS, Eickholt JT, Bergstrahl EJ, Wilson DM, Smith LH (1994) Results of long-term treatment with orthophosphate and pyridoxine in patients with primary hyperoxaluria. N Engl J Med 331:1553–1558
16. Jouvet P, Priquelier L, Gagnadoux MF et al. (1998) Crystalluria: a clinically useful investigation in children with primary hyperoxaluria post-transplantation. Kidney Int 53:1412–1416
17. Holmes RP (1998) Pharmacological approaches in the treatment of primary hyperoxaluria. J Nephrol 11 [Suppl 1]:32–35
18. Broyer M, Brunner FP, Brynger H et al. (1990) Kidney transplantation in primary oxalosis: data from the EDTA registry. Nephrol Dial Transplant 5:332–336
19. Jamieson NV (1998) The results of combined liver/kidney transplantation for primary hyperoxaluria (PH1) 1984–1997. The European PH1 transplant registry report. European PH1 Transplantation Study Group. J Nephrol 11 [Suppl 1]:36–41
20. Katz A, Freese D, Danpure CJ, Scheinman JI, Mauer SM (1992) Success of kidney transplantation in oxalosis is unrelated to residual hepatic enzyme activity. Kidney Int 42:1408–1411
21. Scheinman JI, Najarian JS, Mauer SM (1984) Successful strategies for renal transplantation in primary oxalosis. Kidney Int 25:804–811
22. Watts RWE, Morgan SH, Danpure CJ et al. (1991) Combined hepatic and renal transplantation in primary hyperoxaluria type I: clinical report of 9 cases. Am J Med 90:179–188
23. Danpure CJ (1991) Scientific rationale for hepato-renal transplantation in primary hyperoxaluria type 1. In: Touraine JL, Traeger J, Betuel H, et al. (eds) Transplantation and clinical immunology, vol XXII. Elsevier, Amsterdam, pp 91–95
24. Cochat P, Shärer K (1993) Should liver transplantation be performed before advanced renal insufficiency in primary hyperoxaluria type 1? Pediatr Nephrol 7:212–2181
25. Latta A, Müller-Wiefel DE, Sturm E et al. (1996) Transplantation procedures in primary hyperoxaluria type 1. Clin Nephrol 46:21–23
26. Cochat P (1999) Primary hyperoxaluria type 1. Kidney Int 55:2533–2547
27. von Schnakenburg C, Hulton SA, Milford DV, Roper HP, Rumsby G (1998) Variable presentation of primary hyperoxaluria type 1 in 2 patients homozygous for a novel combined deletion and insertion mutation in exon 8 of the AGXT gene. Nephron 78:485–488
28. Rumsby G (1998) Experience in prenatal diagnosis of primary hyperoxaluria type 1. J Nephrol 11 [Suppl 1]:13–14
29. Milliner DS, Wilson DM, Smith LH (1998) Clinical expression and long term outcomes of primary hyperoxaluria types 1 and 2. J Nephrol 11 [Suppl 1]:56–59
30. Giafi CF, Rumsby G (1998) Primary hyperoxaluria type 2: enzymology. J Nephrol 11 [Suppl 1]:29–31
31. Giafi CF, Rumsby G (1998) Kinetic analysis and tissue distribution of human D-glycerate dehydrogenase/glyoxylate reductase and its relevance to the diagnosis of primary hyperoxaluria type 2. Ann Clin Biochem 35:104–109
32. Cramer SD, Ferree PM, Lin K, Milliner DS, Holmes RP (1999) The gene encoding hydroxypyruvate reductase (GRHPR) is mutated in patients with primary hyperoxaluria type II. Hum Mol Genet 8:2063–2069
33. van Acker KJ, Eyskens FJ, Espeel MF et al. (1996) Hyperoxaluria with hyperglycoluria not due to alanine: glyoxylate aminotransferase defect: a novel type of primary hyperoxaluria. Kidney Int 50:1747–1752

# CHAPTER 41

## Leukotriene Metabolism

Leukotrienes (LTs) are a group of biologically highly active compounds derived from arachidonic acid in the 5-lipoxygenase pathway [1, 2]. Biosynthesis of LTs is limited to a small number of human cells, including brain tissue [3]. Peroxisomes have been identified as the main cell organelle performing degradation of LTs [4, 5]. In addition to their well-known function in the mediation of inflammation and host defense, cysteinyl LTs have neuromodulatory and neuroendocrine functions in the brain [6–9]. The first two committed steps in LT synthesis are catalyzed by the enzyme 5-lipoxygenase, which requires the presence of the 5-lipoxygenase-activating protein. The resulting unstable epoxide intermediate, $LTA_4$, can be further metabolized to either $LTB_4$ or $LTC_4$. The latter step is specifically catalyzed by the enzyme $LTC_4$ synthase and requires the presence of glutathione. $LTC_4$ and its metabolites $LTD_4$ and $LTE_4$ are termed cysteinyl LTs. The rate-limiting step in the synthesis of cysteinyl LTs is the conversion of $LTA_4$ to $LTC_4$, which is catalyzed by $LTC_4$ synthase (Fig. 41.1).

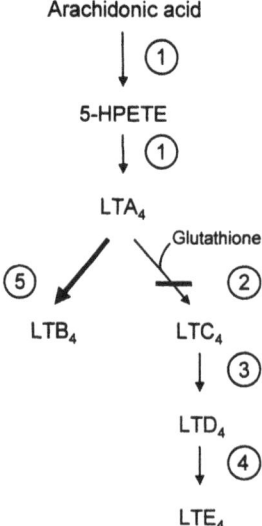

**Fig. 41.1.** Pathway of leukotriene metabolism. *FLAP*, 5-lipoxygenase activating protein; *5-HPETE*, hydroperoxy-eicosatetraenoic acid; *LT*, leukotriene; 1, 5-lipoxygenase/FLAP. 2, $LTC_4$ synthase. 3, γ-glutamyl transpeptidase. 4, dipeptidase. 5, $LTA_4$ hydrolase. Enzyme defect is depicted by *solid bar*

# Leukotriene $C_4$-Synthesis Deficiency

E. Mayatepek

CONTENTS

Clinical Presentation .......................... 455
Metabolic Derangement ....................... 455
Diagnostic Tests .............................. 456
Treatment and Prognosis ..................... 456
Genetics ..................................... 456
References ................................... 457

---

Leukotriene $C_4$-synthesis deficiency is a newly recognized neurometabolic disorder that is characterized by a fatal developmental syndrome with rapidly progressing neurological symptoms, such as muscular hypotonia, severe psychomotor retardation, failure to thrive, and microcephaly. To date, two patients have been identified. Symptoms occurred at birth or in early infancy. Both patients died at the age of 6 months. Absence of $LTC_4$ and its metabolites in cerebrospinal fluid and other body fluids as well as defective synthesis of $LTC_4$ in monocytes and platelets are biochemical features of the disorder. The primary lesion is thought to be a defect in the enzyme $LTC_4$ synthase, which catalyzes the conversion of $LTA_4$ to $LTC_4$, the rate-limiting step in the synthesis of cysteinyl leukotrienes. No specific treatment is available.

---

## Clinical Presentation

To date, $LTC_4$-synthesis deficiency has been identified in two infants in association with a fatal developmental syndrome [10, 11]. The first patient was a girl born to consanguineous Turkish parents [10]. At birth, weight, length and head circumference were below the third percentile. At 2 months of age, muscular hypotonia and psychomotor retardation became apparent. During the next months, muscular hypotonia progressed rapidly. There were no pyramidal-tract signs, and there were reduced deep-tendon reflexes. Spontaneous movements of extremities were minimal, with a slight symmetrical extension in the legs. There was poor visual contact and no head control.

Electroencephalogram (EEG) patterns revealed widespread slow waves. Motor-nerve conduction velocity was decreased. Echocardiography and magnetic resonance imaging of the brain were unremarkable.

In the course of the disease, failure to thrive and microcephaly became more prominent. Developmental delay and the general clinical condition rapidly worsened. The infant died at 6 months of age.

The second patient was a girl born at 33 weeks of gestation to consanguineous Italian parents [11]. Delivery was done by caesarian section because of prolaps of the umbilical cord. She had to be intubated and ventilated. From the first day of life, the patient exhibited a generalized muscular hypotonia with lack of facial expression. After the first week of life, she developed symmetrical extension in the lower extremities, including stretching of the trunk musculature. During the course of the disease, severe muscular hypotonia, failure to thrive and tendency to microcephaly became rapidly progressive. There were no pyramidal-tract signs, and deep-tendon reflexes were not present.

Repeated EEG patterns revealed widespread slow waves. Motor-nerve conduction velocity was slightly decreased. Echocardiography and magnetic resonance imaging scans of the brain at the age of 2 months were normal.

There was no development or visual or social contact at any time. She was never able to swallow and had no head control. At no time it was possible to extubate the patient. The infant died at 6 months of age.

## Metabolic Derangement

The absence of cysteinyl LTs in body fluids and the cellular defect of $LTC_4$-synthesis deficiency are presumably due to defective $LTC_4$ synthase, which catalyzes the transformation of $LTA_4$ and glutathione into $LTC_4$ (Fig. 41.1). $LTC_4$ synthase is predominantly localized in human platelet homogenates, which were shown to lack $LTC_4$-synthesis capacity in an affected patient [10]. Human $LTC_4$ synthase is a glutathione-S-transferase specific for $LTA_4$ [12]. It has been purified and characterized as a protein with a mass of 18 kDa [13]. This enzyme represents a member of a novel superfamily designated "membrane-associated proteins in eicosanoid and glutathione metabolism" [12].

The metabolic block also leads to an increased production of $LTB_4$. However, its concentration is only moderately raised and, remarkably, there is no urinary excretion of $LTB_4$. Therefore, it is unlikely that these moderately increased $LTB_4$ concentrations cause pathophysiologically relevant biological effects. The absence of $LTB_4$ in the urine is notable, because increased urinary excretion of $LTB_4$ has been detected, e.g., in patients with peroxisome deficiency disorders, reflecting the fact that the degradation of LTs by β-oxidation proceeds in peroxisomes [4, 5].

Consistent with the absence of cysteinyl LTs in CSF and other body fluids, formation of $LTC_4$ in peripheral blood cells is not detectable in $LTC_4$-synthesis deficiency. Glutathione-synthetase deficiency is the only previously reported human inborn error affecting LT synthesis [14]. Patients with this disorder show an impaired production of cysteinyl LTs secondary to generalized intracellular glutathione deficiency. However, in this condition, synthesis of $LTC_4$ and other cysteinyl LTs is still present in adequate amounts. In $LTC_4$-synthesis deficiency, intracellular glutathione concentrations were shown to be within normal ranges [10].

The pathophysiology of $LTC_4$-synthesis deficiency is only poorly understood. It is postulated that absence of $LTC_4$, especially in the brain, is responsible for neurological symptoms. $LTC_4$ is concentrated in the choroid plexus by an active-transport system [15]. $LTC_4$ has been found in high concentrations in the hippothalamus and median eminence, whereas $LTB_4$ production was distributed more uniformly throughout the brain [7, 16]. Neuromodulatory effects of $LTC_4$ include prolonged excitation of cerebellar Purkinje cells. $LTC_4$ has also been shown to mediate an inhibitory effect of the neuropeptide somatostatin on pyramidal neurons in the hippocampus [6, 9]. The cysteinyl LTs have at least two receptors, which are present in airway smooth muscle and pulmonary blood vessels. They have been studied most extensively in lung but, to date, nothing is known about the receptor's status in human brain.

The observation of $LTC_4$-synthesis deficiency in association with severe clinical symptoms provides new insights into the pathobiological role of cysteinyl LTs in humans. Whereas overproduction LTs have been considered to act as mediators of inflammation and host defense, low (but constant) concentrations of $LTC_4$ and other LTs seem to be important, especially in the central nervous system. Further studies have to clarify the role of LTs in the brain and the extent to which LTs are essential for normal development.

## Diagnostic Tests

When neonates and infants develop severe muscular hypotonia, psychomotor retardation, failure to thrive and microcephaly, and when these symptoms cannot readily be explained by other commonly encountered pediatric problems or neurometabolic diseases, $LTC_4$-synthesis deficiency should be considered, and CSF LTs should be analyzed as a first step. Other crucial features include: normal glutathione concentrations, absence of any other biochemical abnormalities and uninformative detailed investigations of other inborn errors of metabolism. In $LTC_4$-synthesis deficiency, the $LTC_4$, $LTD_4$ and $LTE_4$ levels in CSF are below the detection limit of about 5 pg/ml, whereas $LTB_4$ is within normal ranges or is slightly elevated. This profile of LTs in CSF is pathognomonic for $LTC_4$-synthesis deficiency.

In addition, concentrations of $LTC_4$ and its metabolites are below the detection limit in plasma and urine. $LTC_4$ can not be generated in calcium-ionophore-A23187-stimulated monocytes, whereas $LTB_4$ synthesis is increased in these cells. Finally, tritium-labeled $LTC_4$ cannot be made from labeled $LTA_4$ in the patient's monocytes and platelets. In these cells, the $LTC_4$ synthesis rate is 0%, whereas it is approximately 50% in controls [10].

## Treatment and Prognosis

At present there exists no specific treatment. Before any theoretical approach can be suggested, it is necessary to reach a more profound understanding of the role of $LTC_4$ in the brain and in the pathophysiology of this newly recognized inborn error of metabolism. Clinical symptoms of the two known patients were rapidly progressive. Both infants died at the age of 6 months, indicating a severe prognosis and an unfavorable outcome.

## Genetics

Since the parents of both patients were related cousins, an autosomal-recessive trait is suggested. The prevalence is not known. Recently, the $LTC_4$-synthase gene was cloned and is located in the distal region of the long arm of chromosome 5 in 5q35 [17]. It was shown that the gene does not contain elements of a typical regulated gene and may, therefore, contain novel regulatory elements [17, 18]. Due to the lack of patient DNA, molecular studies have not been performed.

# References

1. Samuelsson B, Dahlén SE, Lindgren JA, Rouzer CA, Serhan CN (1987) Leukotrienes and lipoxins: structures, biosynthesis, and biological effects. Science 237:1171–1176
2. Mayatepek E, Hoffmann GF (1995) Leukotrienes: biosynthesis, metabolism and pathophysiological significance. Pediatr Res 37:1–9
3. Simmet T, Luck W, Delank WK, Peskar BA (1988) Formation of cysteinyl leukotrienes by human brain tissue. Brain Res 456:344–349
4. Jedlitschky G, Huber M, Völkl A et al. (1991) Peroxisomal degradation of leukotrienes by $\beta$-oxidation from the $\omega$-end. J Biol Chem 266:24763–24772
5. Mayatepek E, Lehmann W-D, Fauler J et al. (1993) Impaired degradation of leukotrienes in patients with peroxisome deficiency disorders. J Clin Invest 91:881–888
6. Lammers CH, Schweitzer P, Facchinetti P et al. (1996) Arachidonate 5-lipoxygenase and its activating protein: prominent hippocampal expression and its role in somatostatin signaling. J Neurochem 66:147–152
7. Lindgren JA, Hökfelt T, Dahlén SE, Patrono C, Samuelsson B (1984) Leukotrienes in the rat central nervous system. Proc Natl Acad Sci U S A 81:6212–6216
8. Kiesel L, Przylipiak AF, Habenicht AJR, Przylipiak MS, Runnebaum B (1991) Production of leukotrienes in gonadotropin-releasing hormone stimulated pituitary cells: potential role in luteinizing hormone release. Proc Natl Acad Sci U S A 88:8801–8805
9. Schweitzer P, Madamba S, Champagnat J, Siggins GR (1993) Somatostatin inhibition of hippocampal CA1 pyramidal neurons: mediation by arachidonic acid and its metabolites. J Neurosci 13:2033–2049
10. Mayatepek E, Flock B (1998) Leukotriene $C_4$-synthesis deficiency: a new inborn error of metabolism linked to a fatal developmental syndrome. Lancet 352:1514–1517
11. Mayatepek E, Lindner M, Zelezny R et al. (1999) A severely affected infant with absence of cysteinyl leukotrienes in cerebrospinal fluid: further evidence that leukotriene $C_4$-synthesis deficiency is a new neurometabolic disorder. Neuropediatrics 30:5–7
12. Jakobsson P-J, Morgenstern R, Mancini J, Ford-Hutchinson A, Persson B (1999) Common structural features of MAPEG. A widespread superfamily of membrane associated proteins with highly divergent functions in eicosanoid and glutathione metabolism. Protein Sci 8:689–692
13. Penrose JF, Gragnon L, Goppelt-Streube M et al. (1992) Purification of human leukotriene $C_4$ synthase. Proc Natl Acad Sci U S A 89:11603–11606
14. Mayatepek E, Hoffmann GF, Carlsson B, Larsson A, Becker K (1994) Impaired synthesis of lipoxygenase products in glutathione synthetase deficiency. Pediatr Res 35:307–310
15. Spector R, Goetzl EL (1986) Role of concentrative leukotriene transport systems in the central nervous system. Biochem Pharmacol 35:2849–2853
16. Miyamato T, Lindgren JA, Hökfelt T, Samuelsson B (1987) Regional distribution of leukotriene and monohydroxyeicosanoic acid production in the rat brain. FEBS Lett 216:123–127
17. Lam BK, Penrose JF, Freeman GJ, Austen KJ (1994) Expression cloning of a cDNA for human leukotriene $C_4$ synthase, an integral membrane protein conjugating reduced glutathione to leukotriene $A_4$. Proc Natl Acad Sci U S A 91: 7663–7667
18. Bigby TD, Hodulik CR, Arden KC, Fu L (1996) Molecular cloning of the human leukotriene $C_4$ synthase gene and assignment to chromosome 5q35. Mol Med 2:637–646

# Subject Index

## A

abdominal pain 14, 373, 374
abetalipoproteinemia 329–330
- acanthocytosis 330
- diet 330
- exceedingly low total cholesterol levels 329
- fat malabsorption 330
- neurological problems 330
acanthocytosis 36, 330, 331
acrocyanosis 25, 34
acrodermatitis enteropathica 396–397
- gingivitis 396
- glossitis 396
- skin
- - lesions 396
- - rash 396
- stomatitis 396
- zinc
- - reduced in plasma 396
- - supplementation 397
acute
- fatty-liver-of-pregnancy syndrome 144
- intermittent porphyria 374–376
- - abdominal pain 374
- - carbohydrate loading 375
- - heme therapy 375
- - hypertension 374
- - neurovisceral symptoms 374
- - periphal neuropathy 374
- - porphobilinogen deaminase deficiency 375
- - tachycardia 374
- - urinary porphobilinogen increase 375
- pancreatitis 36
- tyrosinaemia 187
acylcarnitines 145
- profile 145
adenine phosphoribosyltransferase deficiency 363–364
- allopurinol 363
- dietary purine restriction 363
- 2,8-dihydroxyadenine 363
- uric-acid lithiasis 363
adenosylcobalamin (see combined deficiency of adenosylcobalamin and methylcobalamin) 288
adenylosuccinase deficiency 357–358
- convulsions 358
- hypotonia 358
- neurological disease 358
adult GM1 gangliosidosis 406
alkaptonuria 192–193
- arthritis 192
- ochronosis 192
allopurinol 92, 357, 363
alopecia 34

alternative pathways for nitrogen excretion 219
amino acid transport at the cell membrane 266
2-amino-/2-oxo-adipic aciduria 244–245
aminoaciduria 269, 313, 439
- generalized 313
δ-aminolevulinic acid dehydratase deficiency porphyria 373–374
- abdominal pain 373
- neuropathy 374
- urinary (increased)
- - δ-aminolevulinic acid 373
- - coprophorphyrin 374
amylopectinosis (see branching-enzyme deficiency) 96–97
anemias 36, 37, 90
- megaloblastic 36
- non macrocytic, hemolytic 37
angiokeratoma 404
angiokeratosis 34
anion gap 7
anxiety 16
apneas 6
apoA-I mutations 331
- corneal clouding 331
- premature CAD 331
- very low HDL cholesterol levels 331
apoC-II deficiency 325
- hypertriglyceridemia 325
- pancreatitis 325
- xanthomas 325
aromatic L-aminoacid decarboxylase deficiency 305, 307
- L-dopa 307
- extrapyramidal movement disorder 305
- 5-HTP 307
- 3-methoxytyrosine 307
- oculogyric crisis 305
- vanillactic acid 307
arrhythmia 7, 15, 34
artherosclerosis 331
arthritis 40, 192
aspartylglucosaminuria 417
ataxia 14, 29, 30
atherosclerosis 90, 331
athetoid dysarthria 362
auto-aggressiveness 33

## B

babies born to mothers with HELLP-syndrome 36
*Bartter's* syndrome 395
behavior disturbances 30, 31
benign hyperphenylalaninaemia 172
betaine 228
beta-oxidation defects 142–144

- acute fatty-liver-of-pregnancy syndrome 144
bile-acid
- sequestrants 333
- synthesis 344
- - disorders 345 pp.
biopterin metabolism defects 180–183
- defects of
- - recycling 181
- - synthesis 181
biotin 280, 281
- cycle 276
biotinidase deficiency 278–281
- neurologic symptoms 278
- oral biotin 280
- skin lesions 278
bleeding tendency 15, 37
bone abnormalities 359
adenosine deaminase superactivity 361
- hemolytic anemia 361
bone
- crisis 15, 40
- dysplasia 415
- marrow transplantation 70, 420
- necrosis 40
boxing movements 6
bradycardia 6
branched-chain amino acids, catabolism of 196
branching-enzyme deficiency (GSD-IV, amylopectinosis) 96–97
- hepatic form 96
- myopathic form 96
brittle hair 34

## C

calcification 31
carbohydrate-deficient transferrin (CDT) assay 434
cardiac failure 15
cardiomyopathy 7, 22, 34
carnitine deficiency (secondary) 248
carnitine therapy 147, 205, 208, 249, 442
carnitine-cycle defects 141–142
carnosinuria 317
cataracts 38, 233, 233, 340
CDG (see congenital defects of glycosylation 437 pp.
CDG-Ia (see phosphomannomutase (PMM) deficiency) 433–434
CDG-Ib (see phosphomannose-isomerase deficiency) 433, 434–435
CDG-Ic (see glucosyltransferase-I deficiency) 435–436
CDG-IIa (see N-acetylglucosaminyltransferase-II deficiency) 436
ceramidase deficiency 405
ceramidetrihexoside 404

cerebellar
- ataxia  30
- hypoplasia  31
cerebral edema  60
- continuous hemofiltration  58
- exchange transfusion  57
- hemodialysis  58
- peritoneal dialysis  57
- toxin-removal procedures  54 p., 57, 61
cerebrotendinous xanthomatosis  348–350
- chenodeoxycholic-acid therapy  349
- cholestanol, accumulation of  349
- neurological disease  348
- osteoporosis  348
- premature atherosclerosis  348
- sterol 27-hydroxylase defect  349
- xanthomata  348
cerebrovascular disease  404
ceruloplasmin  388
cherry-red spot  31, 39, 402, 406, 407, 417
cholestatic jaundice  7, 37
cholesteryl ester transfer protein deficiency  331–332
- elevated HDL-C levels  332
cholesteryl-ester storage disease  327
- hepatosplenomegaly  327
- statin  327
- steatorrhea  327
cholestyramine  381
choreoathetosis  363
chorioretinal atrophy  233
chronic and progressive general symptoms  20–32
- abnormal eye movement  33
- ataxia  30
- cerebellar hypoplasia  31
- cherry-red spot  31
- corpus callosus agenesis  31
- deafness  31
- deterioration related to age  22
- digestive symptoms  22
- extrapyramidal signs  25, 26, 31
- hypotonia in the neonatal period  32
- intracranial calcifications  31
- *Leigh* syndrome  32
- macrocephaly  26, 32
- mental regression  32
- microcephaly  32
- muscular symptoms  33
- myoclonic epilepsy  32
- neurological symptoms  22
- nystagmus  33
- periphal neuropathy  33
- retinitis pigmentosa  33
- self mutilation
- spastic paraplegia  33
chronic diarrhea (*see* diarrhea)
cirrhosis  37
coarse facies  35
cobalamin  284 pp.
- disorders  285 pp.
- defective transport by enterocytes (*Imerslund-Gräsbeck* syndrome)  286
- - megaloblastic anemia  286
- - neurologic abnormalities  286
- - proteinuria  286
- metabolism  284
- transport  284
cofactor responsive disorders, treatment and dosage  81
collapse  15
coma  6, 11

combined deficiency of adenosylcobalamin and methylcobalamin  288–290
- cblC  288
- cblF  288
- early-onset group  288
- late-onset group  289
- multisystem pathology  289
- pancytopenia  289
combined xanthine-oxidase and sulfite-oxidase deficiency  361
compulsive and self-destructive behavior  362
conduction defects  7, 15, 34
congenital erythrodermia  34
congenital erythropoietic porphyria (*Gunther* disease)  376–377
- cutaneous lesions  376
- porphyrins in urine  377
- protection of the skin  377
- reddish brown teeth  376
- uroporphyrinogen III cosynthase deficiency  376
congenital malformations  35
connective-tissue symptoms  388
*Conradi-Hünermann* syndrome  342
continuous nocturnal intragastric feedings  147
convulsions  6, 358
copper transport  384
coproporphyrinogen oxidase deficiency  379
corneal
- clouding  415
- opacities  39, 331, 405
corpus callosus agenesis  31
cramps  358
creatine deficiency  238
creatine metabolism (*see* ornithine and creatine metabolism)  232
- disorders  237–239
- - accumulation of guanidinoacetate  238
- - creatine deficiency  238
- - encephalopathy  237
- - oral supplementation of creatine monohydrate  238
crisis intervention  64
- coping at times of predictable crisis  64
γ-cystathionase deficiency  229–230
- cystathioninuria  229
cysteinyl-glycinase deficiency  316
cystinosis  439 pp.
- adolescent  443
- adult benign  443
- infantile (*see there*)  439–442
cystinuria  267–268
- alkalinization of urine  268
- cystine stones  267
- excessive hydration  268
- subtypes  268
- urolithiasis  267
cytosolic 5′-nucleotidase superactivity  367
- hypouricosuria  367
cytosolic acetoacetyl-CoA-thiolase deficiency  155

D

deafness  31, 409
debranching-enzyme deficiency (GSD-III)  95–96

- combined hepatic-myogenic form  95
- purely hepatic form  95
dehydration  15
3β-dehydrogenase deficiency  345–347
- chenodeoxycholic-acid therapy  346
- prolonged neonatal jaundice  345
- rickets  345
- steatorrhoea  345
dementia  31, 408
demyelination  408
desmosterolosis  342
deterioration related to age  22
development delay/regression  25, 388
di- and trihydroxycholestanoic acidurias  425
diabetes  36
diarrhea chronic  23, 36
dicarboxylicaciduria  146
digestive symptoms  22
dihydrolipoamide-dehydrogenase deficiency  134–135
dihydropyrimidinase deficiency  366–367
- dysmorphic features  366
- urinary
- - dihydrouracil  366
- - dihydrothymine  366
dihydropyrimidine dehydrogenase deficiency  364, 366
- dysmorphic features  364
- epilepsy  364
- neurological symptoms  366
- uracil excretion  366
dihydroxycholestanoic acidurias  425
DNPH test  7
dysarthria  29
dysbetalipoproteinemia (hyperlipoproteinemia type III)  329
- elevation in both cholesterol and triglycerides  329
- fibrate  329
- low-fat diet  329
- premature artherosclerosis  329
- xanthomas  329
dyslipidemias  321 pp.
dysmorphia  35
dysostosis multiplex  415

E

ectopia lentis (*see also* lens dislocation)  39, 225
EEG
- bursts  6
- periodic pattern  6
electron-transfer defects  144
elevated lipoprotein(a)  332
- thrombosis  332
emergency diet  206
emergency protocol for investigations  8
emergency treatments  53 pp.
encephalopathic crisis  248
encephalopathy comprising congenital microcephaly  263
enzyme-replacement therapy  81
epilepsy  29, 364
erythrocyte glutathione-synthetase deficiency  315
- hemolytic anemia  315
- splenomegaly  315
erythropoitietic protoporphyria  380–381
- cholestyramine  381
- ferrochelatase deficiency  380
- oral β-carotene  381
- plasma protoporphyrin increase  380
essential fructosuria  111

- fructokinase 111
exercise intolerance 15, 37, 90
- cramps 15
- muscle pain 15
- myoglobinuria 15
external ophtalmoplegia 39
extrapyramidal
- disorder 406
- signs/symptoms 25–26, 29, 31
eye 225
- abnormal movements 33, 39
- symptoms 190

## F

Fabry disease 404–405
- angiokeratoma 404
- birefringent lipid deposits in the urinary sediment 405
- ceramidetrihexoside 404
- cerebrovascular disease 404
- corneal opacities 405
- α-galactosidase A 405
- glycolipids 404
- pain in the extremities 404
- progressive renal disease 405
- skin lesions 404
failure to thrive 23, 36
familial combined hyperlipidemia and the small dense-LDL syndromes 326–327
- combination therapy of statin with fibrate or nicotinic acid 327
- diet (reduction in total fat, saturated fat and cholesterol) 327
- hyperapoB 326
- premature CAD 326
familial hypercholesterolemia 327–328
- diet (low cholesterol and low saturated fat) 328
- HMG-CoA-reductase inhibitor 328
- increased LDL cholesterol levels 328
- premature CAD 327
- xanthomas 327
familial hypertriglyceridemia 325–326
- hypercholesterolemia 325
- obesity 326
- periphal vascular disease 326
familial hypoalphalipoproteinemia 330–331
- CAD, increased prevalence of 330
- low level of HDL cholesterol 330
familial ligand-defective apoB 328
- hypercholesterolemia 328
Fanconi-Bickel syndrome 99, 108
Farber disease 405
- brain involvement 405
- ceramidase deficiency 405
- ceramide storage 405
- joint swelling 405
- lipogranulomatosis 405
fasting test 45, 146
fat loading 146
- test 47
fat malabsorption 330
fatty-acid oxidation 140
fatty-alcohol metabolism defects 149
fatty-aldehyde-dehydrogenase deficiency 149
ferrochelatase deficiency 380
fetal growth retardation 7
fibrates 334
fish-eye disease (see also lecithin:cholesterol acyl transferase deficiency) 331

foam cells in the bone marrow 402
focal
- glomerulosclerosis 89
- signs 11
folate metabolism 292
folinic acid responsive seizures 308
- generalized seizures 308
fructose
- metabolism 110
- test 47
fructose-1,6-bisphosphatase deficiency 113–115
fructosuria 7
fucosidosis 417
fumarase deficiency 135–136
- cerebral dysgenesis 135
- complex II of succinate-ubiquinone oxidoreductase 136
- increased
- - 2-ketoglutaric acid 136
- - succinic acid 136
- - urinary fumaric acid 136
- iron-sulfur clusters 136
- isoforms
- - cytosolic 135
- - mitochondrial 135
function tests 43 pp.

## G

galactokinase deficiency 103–104
- cataracts 103
- galactitol 103
galactose test 46
galactose-1-phosphate uridyltransferase deficiency 104–107
- antenatal diagnosis 105
- classical galactosemia 104
- diet, exclusion of all galactose 105
- Duarte variant 105
- galactitol 104
- kidney failure 104
- liver failure 104
- mass screening 104
- ovarian dysfunctions 106
- partial transferase deficiency 104, 105
- - treatment 106–107
- reducing substance in urine 104
- self-intoxication 105
galactosialidosis 417
α-galactosidase A 405
β-galactosidase deficiency 409
galactosuria 7
gamma amino butyric acid deficiency 302–303
- convulsions 303
- hyperreflexia 303
- hypotonia 303
gastrointestinal symptoms 21
Gaucher cells in bone marrow 404
Gaucher disease 403–404
- bleeding tendency 403
- bone marrow transplantation 404
- CNS involvement 403
- enzyme replacement 404
- lung infiltration 403
- glucocerebrosidase deficiency 404
- glucocerebroside, storage in visceral organs 403
- hepatosplenomegaly 403
- osteopenia 403
- type I 403
- type II 403
gene therapy 83

generalized glutathione-synthetase deficiency 314–315
- acidosis 314
- hemolytic anemia 314
- L-5-oxoproline 314
- progressive CNS damage 314
gingivitis 396
globoid cell leukodystrophy 409
glossitis 396
glucocerebrosidase deficiency 404
glucocerebroside, storage in visceral organs 403
glucose test 46
glucose-6-phosphatase deficiency (GSD-Ia, Von Gierke disease) 87–93
- anemia 90
- artherosclerosis 90
- focal glomerulosclerosis 89
- gout 90
- liver adenoma 89
- osteopenia 90
- pancreatitis 90
- polycystic ovaries 90
- starch-tolerance test 91
- treatment of complications 92
- uncooked starch 91
- vascular abnormalities 90
- xanthomas 90
glucose-6-phosphate-translocase deficiency (GSD-Ib) 93–94
glucose-induced insulin secretion and its modulation 118
glucosuria 7
glucosyltransferase-I deficiency 433, 435–436
- craniofacial dysmorphy 433
- severe brain disease 433
glutamate-formiminotransferase deficiency 293–294
- megaloblastic anemia 293
- mental retardation 293
γ-glutamyl-cysteine synthetase deficiency 313–314
- cerebellar involvement 313
- generalized aminoaciduria 313
- hemolytic anemia 313
γ-glutamyl-transpeptidase deficiency 315–316
- glutathionuria 315
glutaric aciduria type 1 245–250
- emergency treatment 249
- encephalopathic crisis 248
- 3-hydroxyglutaric acid 248
- low-protein diet 249
- macrocephalus 245
- metabolic crisis 248
- neuropharmaceutical agents 250
- oral supplementation with carnitine 249
- retinal hemorrhages 247
- secondary carnitine deficiency 248
- severe neurological disease 245
- subdural hemorrhages 247
glutaric aciduria type 2 deficiency 144
glutaryl-CoA-dehydrogenase deficiency (see glutaric aciduria type I) 245–250
glutathione metabolism 312
glycine metabolism 251
glycine-cleavage system (CGS) 255
glycogen metabolism 86, 102
glycogenolytic defects located in muscles 99

glycogen-storage diseases (*see* GSD) 87 pp.
glycogen-synthetase deficiency (GSD-0) 99
glycolytic defects located in muscles 99
glycoproteins 418
glycoprotein-storage disorders 417
GM1 gangliosidosis 406
- adult GM1 gangliosidosis 406
- bone changes 406
- cherry-red spot 406
- extrapyramidal disorder 406
- hepatosplenomegaly 406
- infantile GM1 gangliosidosis (Landing disease) 406
- juvenile GM1 gangliosidosis 406
- maxillary hyperplasia 406
- neurological regression 406
- vacuolated lymphocytes 406
GM2 gangliosidosis 407
- cherry-red spot 407
- hepatosplenomegaly 407
- hexosaminidase A deficiency 407
- progressive macrocephaly 407
- variant B (*Tay-Sachs* disease) 407
- variant O (*Sandhoff* disease) 407
gout 90, 362
gouty arthritis 357
growth-hormone deficiency 36
GSD (*see* glycogen-storage diseases) 87 pp.
- classification of 88
GSD-0 (glycogen-synthetase deficiency) 99
GSD-Ia (*see* glucose-6-phosphatase deficiency) 87-93
GSD-Ib (glucose-6-phosphate-translocase deficiency) 93-94
GSD-II (*see* lysosomal α-1,4-glucosidase deficiency) 94-95
GSD-III (*see* debranching-enzyme deficiency) 95-96
GSD-IV (*see* branching-enzyme deficiency) 96-97
GSD-VI (*see* phosphorylase deficiency of the liver) 97-98
GSD-IX (*see* phosphorylase-B-kinase deficiency of the liver) 97-98
guanosine triphosphate cyclohydroxylase I deficiency 308
- defective biosynthesis of
- - catecholamines 308
- - serotonin 308
- dopa-responsive dystonia 308
*Gunther* disease (*see* congenital erythropoietic porphyria) 376-377
gyrate atrophy of the choroid and retina (*see* hyperornithinemia due to aminotransferase deficiency) 233-236

# H

hallucinations 16
*Hartnup*-disease 268-270
- hyperaminoaciduria 269
- neurological symptoms 268
- neutral amino acids 269
- oral nicotinamide 269
- pellagra-like dermatitis 268
hawkinsinuria 192
hemangiomas 34
hematological
- abnormalities 15, 340
- symptoms 15

heme biosynthetic pathway 370
hemolysis 385
hemolytic
- anemia 313, 361
- uremic syndrome 38
hemorrhagic syndromes 15
hepatic coma 12, 13-14
- cirrhosis 13, 189
- exsudative enteropathy 13
- hemolytic jaundice 13
- hepatomegaly 19
- *Reye* syndrome 13
hepatic uroporphyrinogen decarboxylase deficiency 377
hepatic lipase deficiency 329
- hypercholesterolemia 329
- hypertriglyceridemia 329
- low-fat diet 329
- pancreatitis 329
- premature cardiovascular disease 329
- xanthomas 329
hepatoerythropoietic porphyria 379
- blistering skin lesions 379
- red urine 379
- porphyrins in urine and plasma 379
- uroporphyrinogen decarboxylase deficiency 377
hepatomegaly 10, 11, 87, 270, 402
hepatorenal GSD with the *Fanconi-Bickel* syndrome 99
- GLUT$_2$ 99
hepatosplenomegaly 324, 327, 340, 403, 406, 407
hereditary coproporphyria and variegate porphyria 379-380
- coproporphyrinogen oxidase deficiency 379
- cutaneous symptoms 380
- neurological symptoms 380
- porphobilinogen increase 380
- protoporphyrinogen oxidase deficiency 379
- urinary
- - δ-aminolevulinic acid 380
- - coproporphyrin increase 380
hereditary folate malabsorption 291, 293
- megaloblastic anemia 291
- neurological deterioration 291
hereditary fructose intolerance 112-113
- aldolase B 112
hereditary intrinsic factor deficiency 285-286
- megaloblastic anemia 285
- neurologic abnormalities 286
hereditary orotic aciduria (*see* uridine monophosphate synthase deficiency) 365
hereditary tyrosinaemia type I 187-190
- acute 187
- chronic 188
- deficiency of fumarylacetoacetase 188
- liver transplantation 189
- NTBC treatment 188
- succinylacetone 188
hereditary tyrosinaemia type II 190-191
- eye symptoms 190
- skin lesions 190
hereditary tyrosinaemia type III 191-192
hexosaminidase A deficiency 407
hiccups 6
HMG-CoA-reductase inhibitor 328
holocarboxylase-synthetase deficiency 277-281
- biotin therapy 280

- metabolic acidosis 277
- psychomotor retardation 277
- skin lesions 277
homocarnosinosis 317
homocystinuria 291, 294
homocystinuria due to cystathione-β-synthase deficiency 225-229
- betaine 228
- central nervous system 226
- ectopia lentis 225
- eye 225
- folic acid 228
- low-methionine/high-cystine diet 228
- pyridoxine 226, 228
- - nonresponder 226
- - responder 226
- skeleton 225
- vascular system 226
homozygous hypobetalipoproteinemia 330
*Hunter* disease (MPS II)/syndrome 68, 415
*Hurler* disease 415
3-hydroxy-3-methylglutaryl-CoA-lyase deficiency 153, 154
3-hydroxy-3-methylglutaryl-CoA-synthase deficiency 154
γ-hydroxybutyric aciduria 303
L-2-hydroxyglutaric aciduria 250-251
- progressive
- - ataxia 250
- - mental retardation 250
D-2-hydroxyglutaric aciduria 251-252
3-hydroxyisobutyric aciduria 210
hydroxylysine catabolism 242
hyperaminoaciduria (*see* aminoaciduria)
hyperammonaemia (*see* urea cycle disorders) 7, 8, 20, 215, 270
hypercholesterolemia 328, 329
hyperekplexia 304-305
- clonazepam 304
- startle disease 304
- stiffness 304
hyperglycemia 9
hyperglycoluria 448
hyperimidodipeptiduria 318
hyperkeratosis 34
hyperlactacidemia 7, 18, 19
diagnostic approach
- L/P ratio 19
- 3OHB/AA 19
- redox potential states 19
hyperlipoproteinemia type III (*see* dysbetalipoproteinemia) 329
hyperlysinemia/saccharopinuria 244
hyperornithinemia due to aminotransferase deficiency 233-236
- arginine-restricted diet 235
- cataracts 233
- chorioretinal atrophy 233
- low-protein diet 235
- muscle pathology 234
- pyridoxine
- - nonresponder 234
- - responder 234
- - treatment 235
hyperornithinemia-homocitrullinuria-hyperammonemia-syndrome 236-237
- developmental delay 236
- intolerance to protein feeding 236
- low-protein diet 237
hyperoxaluria 448
- primary (*see there*) 447 pp.

hyperphenylalaninaemias 171 pp.
hyperpipecolic aciduria 425
hyperprolinemia type I 261–262
hyperprolinemia type II 262
- recurrent seizures 262
hypertension 374
hyperthyroidism 36
hypertriglyceridemia 324, 325, 329
hypertyrosinemia 7, 187 pp.
hyperventilation 15
- attacks 15–16
hyperzincemia 397
- with functional zinc depletion 397
hypobetalipoproteinemia 330
- reduced risk for premature
    atherosclerosis 330
hypocalcemia 7
hypoglycemia 20, 21
- general approach 21
hypogonadism 36
hypoparathyroidsm 36
hypothyroidism 440
hypotonia 6, 7, 358
- in the neonatal period 32
hypouricosuria 367
hypoxanthine-guanine phosphoribosyl-
    transferase deficiency 362–363
- allopurinol 363
- athetoid dysarthria 362
- choreoathetosis 363
- compulsive and self-destructive
    behavior 362
- gout 362
- neurological syndrome 362
- uric-acid lithiasis 362

## I
I-cell disease 417
ichthyosis 34, 263
iduronate sulfatase deficiency 415
α-L-iduronidase deficiency 415
*Imerslund-Gräsbeck* syndrome (*see*
    cobalamin, defective transport
    by enterocytes) 286
imidazole-dipeptide metabolism 312
infantile cystinosis 439–443
- adolescent 443
- adult benign 443
- aminociduria 439
- carnitine supplementation 442
- cysteamine
-- bitartate 442
-- eyedrops 442
- hypothyroidism 440
- multisystemic nature 440
- ophtalmologic symptoms 439
- photophobia 439
- rickets 439
- urinary
-- glucose 439
-- protein 439
- vitamin-D supplementation 442
infantile GM1 gangliosidosis (*Landing*
    disease) 406
intracranial
- calcifications 31
- hypertension 11
isolated sulfite-oxidase deficiency 230
- lens dislocation 230
- refractory convulsions 230
isovaleric aciduria (IVA) 197 p.,
    200, 204
- acute intermittent late-onset form 198
- chronic progressive forms 199

- complications 199
- dietary therapy 204
- glycine and carnitine 204
- severe neonatal-onset form 197
- toxin-removal procedures 204

## J
jaundice
- cholestatic 37
- hemolytic 13
- liver failure syndrome 7
- prolonged neonatal 347
joint contractures 40
juvenile GM1 gangliosidosis 406

## K
*Kayser-Fleischer* ring 386
keratitis 39
ketoacidosis 7
ketogenesis 152
2-ketoglutarate-dehydrogenase
    deficiency 134
ketolysis 152
ketone-synthesis defects 144
ketosis 9, 18
ketotic hyperglycinemia 256
kidney
- damage 386
- failure 104
kinky hair 388
*Krabbe* disease 409–410
- deafness 409
- demyelination, central and
    periphal 409
- galactocerebroside storage 409
- β-galactosidase deficiency 409
- globoid cell leukodystrophy 409
- optic atrophy 409
- protein in CSF 409

## L
lactic acidosis 9
*Landing* disease (infantile GM1 gang-
    liosidosis) 406
large-amplitude tremors 6
later-onset acute and recurrent attacks
    (childhood and beyond) 11 pp.
- abdominal pain 14
- anxiety 16
- arrhythmias 15
- ataxia 14
- bone crisis 15
- cardiac failure 15
- cerebellar hemorrhage 13
- collapse 15
- coma 11
- cyclic vomiting syndrome 13
- dehydration 15
- exercise intolerance (*see there*) 15
- extrapyramidal disease 13
- hallucinations 16
- hematological symptoms 15
- hemorrhagic symptoms 15
- hyperventilation 15
- lethargy 11
- liver failure (*see there*) 16
- MELAS 13
- metabolic derangements 16
- peripheral neuropathy 14
- petechiae 13
- psychiatric symptoms 14, 16
- *Reye* syndrome 16
- skin rashes 16
- strokes 11, 13

- sudden infant death 16
- vomiting 11, 14, 16
laxity 35
LDL (*see* low density lipoprotein)
lecithin:cholesterol acyl transferase
    deficiency 331
- artherosclerosis 331
- corneal opacifications 331
- plasma free-cholesterol-to-total-
    cholesterol 331
*Leigh* disease/syndrome 32, 132
lens dislocation (*see also* ectopia
    lentis) 230
*Lesch-Nyhan* syndrome 69
lethargy 6, 11, 12
leukocyte-adhesion-deficiency syndrome
    type II 433, 435
- recurrent bacterial infections 433
- very high leukocyte counts 435
leukodystrophies 69
leukopenia 37
leukotriene
- $C_4$-synthesis deficiency 455 pp.
-- defective $LTC_4$ synthase 455
-- microcephaly 455
-- neurometabolic disorder 455
- metabolism 454
lipogranulomatosis 405
lipoprotein lipase deficiency
    324–325
- colic 324
- hepatosplenomegaly 324
- hypertriglyceridemia 324
- pancreatitis 324
lipoprotein metabolism 320
- endogenous 322–323
- exogenous 321
liver failure 16, 37, 104
- adenoma 89
- ascites 16
- carcinoma 189
- edema 16
- hydrops fetalis 16
liver-kidney transplantation 450
low density lipoprotein lowering
    drugs 333–334
- bile-acid sequestrants 333
- fibrates 334
- niacin (nicotinic acid) 333
- statins 333, 334
lysine catabolism 242
- disorders 243 pp.
lysinuric protein intolerance 270–272
- aversion to high-protein foods 270
- citrulline, daily supplement 271
- decreased availability of lysine, arginine
    and ornithine 271
- growth failure 270
- hepatomegaly 270
- hyperammonemia 270
- osteoporosis 270
- splenomegaly 270
lysosomal α-1,4-glucosidase deficiency
    (GSD-II, *Pompe's* disease) 94–95
- adult form 94
- infantile form 94
- juvenile form 94
lysosomal-acid-lipase deficiency 327
- hepatosplenomegaly 327
- statin 327
- steatorrhea 327
lysosomal porters for cystine and related
    compounds 438

# M

macrocephaly 25-26, 32, 245, 415
magnesium-losing kidney 395-396
- *Bartter* syndrome 395
- high urinary excretion 395
- nephrocalcinosis 395
- tetany 395
malonic aciduria 208
- secondary inhibition of fatty-acid β-oxidation 209
management of intercurrent decompensation 206
α-mannosidosis 417
β-mannosidosis 417
maple syrup urine disease (MSUD) 197 p., 200, 204
- acute intermittent late-onset form 198
- chronic progressive forms 199
- complications 199
- dietary therapy 204
- severe neonatal-onset form 197
- toxin-removal procedures 204
- treatment 202
- vitamin therapy 204
*Maroteaux-Lamy* disease 417
maternal phenylketonuria 178-180
- counselling 179
maxillary hyperplasia 406
medium-chain acyl-coenzyme A dehydrogenase (MCAD) deficiency 143
megaloblastic anemia 285-287, 290, 291, 293, 364
*Menkes* disease 388-391
- connective-tissue symptoms 388
- copper-histidine treatment 390
- decreased
- - ceruloplasmin 388
- - serum copper 388
- developmental regression 388
- kinky hair 388
- multisystemic manifestations 389
- occipital-horn syndrome 388
mental
- deterioration 29
- regression 32
- retardation 293, 417
α-mercaptopropionylglycine 268
metabolic acidosis 7, 8, 16, 17
metabolic coma 11, 12
- acidosis (metabolic) 12
- hyperammonemia 12
- hyperlactacidemia 12
- hypoglycemia 12
- ketosis 12
metabolic crisis 248
metabolic derangements 7-11, 16
metabolic distress, five major types 8
metabolic profile over the course of a day 43 pp.
metachromatic leukodystrophy 407-409
- dementia 408
- demyelination 408
- spastic quadriplegia 408
- vigabatrine 409
S-methionine-adenosyltransferase deficiency 229
- high methionine in plasma and urine 229
methionine 229
methylacetoacetyl-CoA-thiolase deficiency 154, 155
methylcobalamin (*see also* combined deficiency of adenosylcobalamin and methylcobalamin) 288
- deficiency (cblE; cblG) 290-291
- - homocystinuria 291
- - megaloblastic anemia 290
- - neurological disease 290
3-methylcrotonyl glycinuria 207
- glycine and carnitine therapies 208
methylenetetrahydrofolate deficiency 294, 295
- betaine treatment 295
- homocystinuria 294
- progressive encephalopathy 294
3-methylglutaconic aciduria type I 209-210
methylmalonic aciduria (MMA) 197 p., 201, 205
- acute intermittent late-onset form 198
- carnitine therapy 205
- chronic progressive forms 199
- complications 199
- metronidazole therapy 205
- severe neonatal-onset form 197
- toxin-removal procedures 205
- treatment 202
- vitamin therapy 205
mevalonic aciduria (mevalonate kinase deficiency) 339-340, 425
- cataracts 340
- dysmorphic features 340
- hematological abnormalities 340
- hepatosplenomegaly 340
- psychomotor retardation 340
- recurrent febrile crisis 340
- ubiquinone-10 340
microcephaly 32, 340, 455
microcornea 39
monoamines 306
monoamine oxidase A deficiency 307-308
- 3-methoxytyramine 307
- normetanephrine 307
- serotonin 307
- tyramine 307
- violent behavior 307
*Morquio* disease 416
mucopolysaccharides 414
mucopolysaccharidoses (MPS) 415-417
- bone dysplasia 415
- central nervous system 416
- coarse
- - facial features 415
- - hair 416
- corneal
- - clouding and opacities 415, 416
- - dysostosis multiplex 415
- dwarfism 416
- heart and vascular system involvement 415
- *Hunter* disease (MPS II) 415
- *Hurler* disease 415
- iduronate sulfatase deficiency 415
- α-L-iduronidase deficiency 415
- macrocephaly 415
- *Maroteaux-Lamy* disease 417
- *Morquio* disease 416
- *Sanfilippo* disease 416
- *Scheie* disease 415
- skeletal changes 415
- *Sly* disease 417
- stiff joints 415
- thick skin 415
- Type-I 415
- Type-II 415
- Type-III 416
- Type-IV 68, 416
- Type-VI 417
- Type-VII 417
multiple acyl-CoA-dehydrogenase deficiency 144
muscle adenosine monophosphate deaminase deficiency 358-359
- cramps 358
- myalgias 358
- myoadenylate deaminase deficiency 358
muscle GSDs 98-99
- exercise
- - intolerance 98
- - test 98
muscle pain 37
muscular symptoms 33
myalgias 358
adenosine deaminase deficiency 359-361
- bone
- - abnormalities 359
- - marrow transplantation 360
- gene therapy 360
- severe combined immunodeficiency disease 359
myoadenylate deaminase deficiency 358
myoclonic
- epilepsy 32
- jerks 6
myoclonus 29
myoglobinuria 37
myopathy 29, 38

# N

N-acetylglucosaminyltransferase-II deficiency (CDG-IIa) 436
neonatal adrenoleukodystrophy 424, 425
neonatal hyperphenylalaninaemia 173
neonatal period and early infancy
- acute symptoms 5 pp.
- - cardiac presentation 7
- - coma 6
- - hepatic presentation 7
- - hypotonia 6
- - lethargy 6
- - metabolic derangements 7-11
- - neurologic deterioration 6
- classification of inborn errors 10, 11
nephrocalcinosis 38, 395, 447
nephrolithiasis 38
nephropathy 38
nephrotic syndrome 38
neurologic coma 11
- cerebral edema 12
- extrapyramidal signs 12
- focal signs 11, 12
- hemiplegia 12
- intracranial hypertension 11, 12
- seizures 11, 12
- thromboembolic accidents 12
neurologic deterioration 6, 9, 291
- cholestatic jaundice 10
- chronic diarrhea 10
- energy-deficiency type 10
- hepatocellular disturbances 10
- hepatomegaly 10
- hepatosplenomegaly 11
- hypoglycemia 10
- intoxication type 10
- special odor 10
- storage signs 11
neurological

- distress 9
- symptoms 22, 366
neuropathy 374
neurotransmitters 300
- disorders 301 pp.
neurovisceral symptoms 374
N-glycans 432
niacin (nicotinic acid) 333
*Niemann-Pick* disease
- types A and B 401–402
-- cherry-red spot in the macula 402
-- foam cells in the bone marrow 402
-- hepatomegaly 402
-- neurological deterioration 402
-- sphingomyelin storage
-- sphingomyelinase deficiency 401
- types C and D 402–403
-- unesterified cholesterol storage 403
nitrogen excretion, alternative
    pathways 219
nodules 35
nonketotic hyperglycinemia 255 pp.
- characteristic electroencephalogram
    (EEG) pattern 255
- glycine-cleavage system (CGS) 255
- late-onset type 255
- ketotic hyperglycinemia 256
- neonatal type 255
- transient neonatal 255
nystagmus 33

## O

occipital-horn syndrome 388
ochronosis 192
odor
- abnormal body odor 6
- abnormal urine odor 6
oligosaccharides 418
oligosaccharidoses 417–420
- aspartylglucosaminuria 417
- bone marrow transplantation 420
- cherry-red retinal spots 417
- coarse facial features 417
- fucosidosis 417
- galactosialidosis 417
- glycoprotein-storage disorders 417
- α-mannosidosis 417
- β-mannosidosis 417
- mental retardation 417
- mucolipidosis II (I-cell disease) 417
- *Schindler* disease 417
- sialic storage diseases 417
- sialidosis 417
- skeletal deformities 417
ophisthotonus 6
optic atrophy 409
organ symptoms 33 pp.
- cardiology 34
- dermatology 34
- dysmorphology 35
- endocrinology 36
- gastroenterology 36
- hematology 36
- hepatology 37
- myology 37
- nephrology 38
- neurology 38
- ophtalmology 38
- osteology 40
- osteoporosis 36
- pneumology 40
- psychiatry 40
- rheumatology 40
- vascular symptoms 40

organ transplantation 72
ornithine and creatine metabolism 232
orotic acid 217, 364
osteopenia 40, 90, 403
osteoporosis 36, 270, 348
ovarian cysts 90
ovarian dysfunctions 106
oxalate
- deposition in organs 447
- metabolism 446
5-oxoprolinase deficiency 316
5-oxoprolinuria 316

## P

pancreatitis 90, 324, 325, 329
pancytopenia 37, 287
paraplegia 26
pedalling movements 6
pellagra-like dermatitis 268
peripheral neuropathy 33, 331, 374
peroxisomal
- disorders 423 pp.
-- abnormalities
--- craniofacial 423
--- ocular 423
-- clinical symptoms 424
-- di- and trihydroxycholestanoic
    acidurias 425
-- hepato-intestinal dysfunctions 423
-- hyperpipecolic aciduria 425
-- Refsum disease
--- classical 426
--- infantile 425
-- mevalonic aciduria 425
-- neonatal adrenoleukodystrophy
    424, 425
-- periphal neuropathy 423
-- peroxisome-targeting signals
    (PTS) 426
--- receptors 426
--- PEX genes 426
-- rhizomelic chondrodysplasia
    punctata 424
-- skeletal dysmorphies 423
-- x-linked adrenoleukodystrophy 426
- functions 422
peroxisome-targeting signals (PTS) 426
- receptors 426
persistent hyperinsulinemic hypoglycemia
    in infancy (PHHI) 119 pp.
- diffuse 119
- focal 119
- hyperammonemia/hyperinsulinism
    syndrome 120
- macrosomy 119
- pancreatic
-- arteriography 121
-- venous catheterization 121
petechia 25, 35
PEX genes 426
phenylalanine metabolism 170
phenylalanine-hydroxylase
    deficiency 172–178
- benign hyperphenylalaninaemia 172
- causes of neonatal
    hyperphenylalaninaemia 173
- neonatal screening 173
phenylketonuria 65 p., 172 pp.
phenylpropionate loading test 146
PHHI (*see* persistent hyperinsulinemic
    hypoglycemia in infancy)
    119 pp.
phosphoenolpyruvate-carboxykinase
    deficiency 130–134

- congenital malformation of the
    brain 132
- decreased activity second to respiratory-
    chain dysfunction 130
- dichloroacetate 133
- $E_1$, α- or 2-ketoacid dehydrogenase 131
- $E_2$, dihydrolipoamide
    acyltransferase 131
- $E_3$, dihydrolipoamide
    dehydrogenase 131
- isoforms
-- cytosolic 130
-- mitochondrial 130
- *Leigh* disease 132
- thiamin 133
3-phosphoglycerate-dehydrogenase
    deficiency 263
- encephalopathy comprising congenital
    microcephaly 263
- treatment with L-serine 263
phosphomannomutase (PMM) deficiency
    (CDG-Ia) 433–434
- carbohydrate-deficient transferrin
    (CDT) assay 434
- multisystem syndrome 433
- severe brain involvement 433
phosphomannose-isomerase deficiency
    (CDG-Ib) 433, 434–435
- hepatic-intestinal presentation 433
- mannose treatment 433
phosphoribosyl pyrophosphate synthetase
- deficiency 357
- superactivity 357
-- allopurinol 357
-- gouty arthritis 357
-- low purine diet 357
-- uric acid lithiasis 357
phosphorylase deficiency of the liver
    (GSD-VI) 97–98
phosphorylase-B-kinase deficiency of the
    liver (GSD-IX) 97–98
- variants 97
phosphoserine-phosphatase
    deficiency 263–264
- ichthyosis 263
- polyneuropathy 263
- oral serine 263
photophobia 439
photosensitivity 35
pili torti 35
PKU (*see* phenylketonuria) 172 pp.
pneumopathy 40
polycystic ovaries 90
polymyoclonia 29, 32
polyneuropathy 30, 263
*Pompe's* disease (*see* lysosomal α-1,4-
    glucosidase deficiency) 94–95
porphobilinogen deaminase
    deficiency 375
porphyria cutanea tarda 377–379
- chronic blistering skin lesions 377
- hepatic uroporphyrinogen decarboxy-
    lase deficiency 377
- low-dose chloroquine 378
- plasma porphyrin increase 378
- repeated phlebotomy 378
porphyrias 371 pp.
postmortem protocol 49
premature cardiovascular disease 329
primary hyperoxaluria 447
- type I 447–451
-- hyperglycoluria 448
-- hyperoxaluria 448
-- liver-kidney transplantation 450

- – nephrocalcinosis 447
- – oxalate deposition in organs 447
- – renal failure 447
- – suprametaphyseal bands on X-rays 447
- type 2 451–452
- – L-glycerate, increased urinary excretion of 451
- – supportive treatment 451
- – urolithiasis 451
- type 3 452
primary hypomagnesemia 394–395
- carpopedal spasm 394
- magnesium
- – reduced urinary excretion 394
- – supplementation 394
- tetany with facial twitching 394
progressive encephalopathy 294
progressive macrocephaly 407
progressive neurologic and mental deterioration related to age 22 pp.
- ataxia 27
- autistic behavior/feature 26, 27
- cerebral palsy 26
- choreoathetosis 28
- convulsions 27
- coarse facies 27
- developmental arrest 26
- dystonia 28
- early infancy 22
- extrapyramidal signs 26, 28
- hypsarrhythmia 26
- infantile spasms 26
- late childhood to adolescence 29
- late infancy to early childhood 25
- macrocephaly 26
- myoclonus 27
- neurological crisis 26
- paraplegia 27
- *Parkinson* syndrome 28
- peripheral neuropathy 27
- polymyoclonia 28
- polyneuropathy 28
- onset in adult 30
- schizophrenia 28
prolidase deficiency 317–318
- characteristic face 317
- hyperimidodipeptiduria 318
- skin lesions
proline metabolism 260
- inborn errors of 261 pp.
proline-oxidase deficiency 261–262
prolonged neonatal jaundice 345, 347
rickets 345, 347
steatorrhoea 345, 347
propionic aciduria (PA) 197 p., 200, 201, 205
- acute intermittent late-onset form 198
- carnitine therapy 205
- chronic progressive forms 199
- complications 199
- dietary therapy 205
- metronidazole therapy 205
- severe neonatal-onset form 197
- toxin-removal procedures 205
- treatment 202
- vitamin therapy 205
protoporphyrinogen oxidase deficiency 379
pseudodiabetes 36
psychiatric symptoms 14, 16
psychic regression 29
psychological care of the child and family 63 pp.

ptosis 39
PTS (peroxisome-targeting signals) 426
- receptors 426
punctate epiphyseal calcifications 40
purine metabolism 354
- disorders 355 pp.
purine nucleoside phosphorylase deficiency 361
- bone-marrow transplantation 361
- impairment of cellular immunity 361
- neurological symptoms 361
- recurrent infections 361
- viral diseases, enhanced susceptibility to 361
purpuras 35
pyridoxine-responsive and –unresponsive putative glutamic-acid-decarboxylase deficiency 302
- abnormal eye movements 302
- brain GABA deficiency 302
pyridoxine-responsive convulsions 302
pyrimidine metabolism 365
pyrimidine-5'-nucleotidase deficiency 367
D'-pyrroline-5-carboxylate-dehydrogenase deficiency 262
- recurrent seizures 262
D'-pyrroline-5-carboxylate-synthase deficiency 262
pyruvate metabolism and tricarboxylic-acid cycle 126
- clinical features of defects of enzymes of 128
- metabolic abnormalities 129
pyruvate-carboxylase deficiency 128–130

## R
*Raynaud* syndrome 40
5β-reductase deficiency 347–348
- chenodeoxycholic-acid therapy 348
- prolonged neonatal jaundice 347
- rickets 347
- steatorrhoea 347
refractory convulsions 230
Refsum disease
- classical 426
- infantile 425
renal
- failure 447
- polycystosis 38
resins 328
respiratory alkalosis 7
respiratory chain 158
- defects 159 pp.
- – screening 163
- – symptoms 160
retinal hemorrhages 247
retinitis pigmentosa 33, 39
*Reye* syndrome 13, 16, 37
rhizomelic chondrodysplasia punctata 424
riboflavin 148
rickets 439

## S
salt-losing syndrome 36
*Sandhoff* disease 407
*Sanfilippo* disease/syndrome 68, 416
scarring 35
*Scheie* disease 415
*Schindler* disease 417
seizures 6, 11
self mutilation 33
semi-emergency diet 206

serine metabolism 260
serum carnosinase deficiency 317
severe combined immunodeficiency disease 359
sexual ambiguity 36
short stature 36
sialic storage diseases 417
sialidosis 417
sitosterolemia 328–329
- premature CAD 328
- resins 328
- xanthomas 328
skin lesions 190, 278, 396
skin rash 16, 35, 396
- photosensitivity 16
*Smith-Lemli-Opitz* syndrome 340–341
- cholesterol
- – decrease of 341
- – supplementation 341
- 7-dehydrocholesterol, accumulation of 341
- malformations 340
- microcephaly 340
- syndactyly 340
spastic
- paraplegia 33
- quadriplegia 408
spasticity 26
sphingolipid metabolism 400
sphingomyelin storage 403
sphingomyelinase deficiency 401
splenomegaly 270
startle disease 304
statins 333
steatorrhea 327
sterility 36
sterol 27-hydroxylase defect 349
sterol-δ8-isomerase deficiency-chondrodysplasia phenotypes 341–342
- *Conradi-Hünermann* syndrome 342
stiff joints 415
stiffness 304
stridor 40
strokes 11
subdural hemorrhages 247
succinic semialdehyde dehydrogenase deficiency 303–304
- γ-hydroxybutyric aciduria 303
- neurological course 304
- vigabatrin 304
succinyl-CoA oxoacid-transferase deficiency 153, 154
sudden infant death syndrome 16, 34
sulfur-containing aminoacids, metabolism of 224
syndactyly 340
syndromes/diseases (names only)
- *Bartter* syndrome 395
- *Conradi-Hünermann* syndrome 342
- *Fabry* disease (*see there*) 404–405
- *Fanconi-Bickel* syndrome 99, 108
- *Farber* disease (*see there*) 405
- *Gaucher* disease (*see there*) 403–404
- *Gunther* disease (*see* congenital erythropoietic porphyria) 376–377
- *Hartnup*-disease (*see there*) 268–270
- *Hunter* disease (MPS II)/syndrome 68, 415
- *Hurler* disease 415
- *Imerslund-Gräsbeck* syndrome (*see* cobalamin, defective transport by enterocytes) 286
- *Krabbe* disease (*see there*) 409–410
- *Landing* disease 406

- *Leigh* disease/syndrome  32, 132
- *Lesch-Nyhan* syndrome  69
- *Maroteaux-Lamy* disease  417
- *Menkes* disease *(see there)*  388-391
- *Morquio* disease  416
- *Niemann-Pick* disease *(see there)*  401-402
- *Parkinson* syndrome  28
- *Pompe* disease *(see* lysosomal α-1,4-glucosidase deficiency*)*  94-95
- *Raynaud* syndrome  40
- *Sandhoff* disease  407
- *Sanfilippo* disease/syndrome  68, 416
- *Scheie* disease  415
- *Schindler* disease  417
- *Smith-Lemli-Opitz* syndrome *(see there)*  340-341
- *Tangier* disease *(see there)*  331
- *Tay-Sachs* disease  407
- *Von Gierke* disease *(see* glucose-6-phosphatase deficiency*)*  87-93
- *Wilson* disease *(see there)*  385-388
- *Wolman* disease *(see there)*  327

## T

tachycardia  374
*Tangier* disease  331
- atherosclerosis  331
- periphal neuropathy  331
- tonsils, enlarged orange-yellow  331
*Tay-Sachs* disease  407
teleangiectasias  35
tetrahydrobiopterin test  47
thrive failure to  7
thromboembolic accidents  40
thrombopenia  37
thrombosis  332
transcobalamin I CR-binder deficiency  287
transcobalamin II deficiency  287
- megaloblastic anemia  287
- pancytopenia  287
transient hyperammonemia of the newborn  218
transient neonatal nonketotic hyperglycinemia  255
transplantation  82
transport defect of fatty acids  148

treatment of inborn errors  75 pp.
- medication used in  76-79
tremors, large-amplitude  6
tricarboxylic-acid cycle  126
trichorrhexis nodosa  35
triglyceride lowering drugs  334-335
- other lipid lowering drugs  335
cholesterol synthesis  338
tryptophan catabolism  242
tubulopathy  38
tyrosine hydroxylase deficiency  305
- extrapyramidal symptoms  305
- CSF  305
- L-Dopa treatment  305
- 5-HIAA  305
- HVA  305
- MHPG  305
tyrosine metabolism  186

## U

UDP-galactose 4'-epimerase deficiency  107-108
ulceration  35
- skin ulcers  35
unexplained glycosylation defects  436-437
urea cycle  214
- disorders  215 pp.
- - adults  216
- - alternative pathways for nitrogen excretion  219
- - assessment for treatment  220
- - children  216
- - diet  218
- - emergency treatment  221
- - hyperammonaemia  215
- - - differential diagnosis  218
- - infantile presentation  216
- - neonatal presentation  215
- - orotic acid  217
- - orotidine  217
- - transient hyperammonia of the newborn  218
ureidopropionase deficiency  367
uric acid lithiasis  357, 362
uridine monophosphate synthase deficiency (hereditary orotic synthase aciduria)  365

- megaloblastic anemia  364
- orotic acid, massive overproduction of  364
- uridine, administration of  364
urine
- color  38
- odors  7, 38
urolithiasis  267
uroporphyrinogen III cosynthase deficiency  376

## V

vacuolated lymphocytes  37, 406
variegate porphyria *(see* hereditary coproporphyria*)*  379-380
vascular system  226
vesiculo bullous skin lesions  35
vigabatrine  409
vomiting  11, 12, 16, 23
*Von Gierke* disease *(see* glucose-6-phosphatase deficiency*)*  87-93

## W

*Wilson* disease  385-388
- hemolysis  385
- *Kayser-Fleischer* ring  386
- kidney damage  386
- oral zinc  387
- penicillamine  387
- symptoms
- - hepatic  385
- - neurological symptoms  385
- - psychiatric symptoms  385, 386
- trientine  387
- urinary copper excretion  386
*Wolman* disease  327
- hepatosplenomegaly  327
- statin  327
- steatorrhea  327

## X

xanthine-oxidase deficiency  361-362
- combined xanthine-oxidase and sulfite-oxidase deficiency  361
- low purine diet  362
- xanthinuria  361
xanthomas  90, 325, 327-329, 348
x-linked adrenoleukodystrophy  426

GPSR Compliance
The European Union's (EU) General Product Safety Regulation (GPSR) is a set of rules that requires consumer products to be safe and our obligations to ensure this.

If you have any concerns about our products, you can contact us on

ProductSafety@springernature.com

In case Publisher is established outside the EU, the EU authorized representative is:

Springer Nature Customer Service Center GmbH
Europaplatz 3
69115 Heidelberg, Germany